The Universe of Quadrics

Boris Odehnal · Hellmuth Stachel · Georg Glaeser

The Universe of Quadrics

Boris Odehnal
Department of Geometry
University of Applied Arts Vienna
Vienna, Austria

Hellmuth Stachel
Institute of Discrete Mathematics and Geometry
Vienna University of Technology
Vienna, Austria

Georg Glaeser
Department of Geometry
University of Applied Arts Vienna
Vienna, Austria

ISBN 978-3-662-61055-8   ISBN 978-3-662-61053-4 (eBook)
https://doi.org/10.1007/978-3-662-61053-4

This Springer imprint is published by the registered company Springer-Verlag GmbH, DE part of Springer Nature.
The registered company address is: Heidelberger Platz 3, 14197 Berlin, Germany

Preface

In 2016, the authors published the first of the two intended books about curves and surfaces of degree two, "The Universe of Conics". The plan was to come up with the second volume about *surfaces* of degree two "as soon as possible".

The ulterior motive was to write two compendia containing important geometric knowledge that seems in danger of getting lost. The curves of degree two have been of interest to mathematicians ever since the advent of geometry, and the ancient Greeks already had a better understanding of them than many aspiring mathematics students nowadays. In the course of centuries, many famous mathematicians contributed to this body of knowledge. Just to name one of them, the ingenious BLAISE PASCAL published his "Essay pour les Coniques" at the age of sixteen. The heydays of these simplest algebraic curves might have been in the 17th century, probably inspired by the fact that conics appear regularly in the universe as orbits of planets and other celestial bodies. ISAAC NEWTON used his profound understanding of those curves when he proved Kepler's laws, which had only been conjectures until that point.

In the following centuries, mathematics became more advanced – mainly through the introduction of infinitesimal calculus. Famous mathematicians shifted their focus onto logical generalization. CARL FRIEDRICH GAUSS, who had previously achieved fame by predicting the re-appearance of the dwarf planet Ceres, transferred the remarkable properties of conics into space and found wonderful examples for his famous theorems about Differential Geometry. The 19th century (CHASLES, DUPIN, MONGE, CAYLEY, KLEIN, and many others) was probably the time that saw the most advances in knowledge about the simplest algebraic surfaces – the quadrics.

To come back to our attempt to write compendia about conics and quadrics: During the work on the first book, we soon figured out that there is so much knowledge about conics that even a thousand pages would not have been enough to cover the entire topic (the book has "only" some 500 pages). And – alas – an additional 500 pages cannot cover the entire body of knowledge about quadrics. Still, three years after the release of the first book, we think that we have collected a lot of theorems that might "stay alive" by means of this compendium. Some of the theorems may even be new, although they are just the consequence of what has been accumulated by others in the course of the time. Anyway, the saying "Good things come to those who wait" can be applied to the result. As probably most book writers can confirm, "it was much more work than expected". Eventually, however, it is very satisfying to see how essential the surfaces of degree two are for the entire framework of mathematics (see Figure 0.1), which is based on the profound knowledge about the most elementary surfaces in space.

FIGURE 0.1. Hyperboloid in space – textured with space, symbolizing that the quadrics are essential when one deals with non-Euclidean geometry, Projective Geometry, Algebraic Geometry, Differential Geometry, higher dimensional geometry, and many other part disciplines of mathematics.

Table of Contents

1 Introduction

From the conics in 2-space ...

As already mentioned in the preface, this book builds up on its predecessor, "The Universe of Conics – From the ancient Greeks to 21st century developments" (Springer Spektrum, Heidelberg, 2016). There, the authors showed the substantial role of conics (conic sections) in many sciences as well as in the whole universe. Since conics have a perfect shape, designers and architects use "true conics" instead of "just some freeform curve" when they deal with certain shapes. We will soon see that this also holds for the generalization of conics in 3-space, the so-called quadrics.

Points in the Euclidean plane can be given by Cartesian (x, y)-coordinates. Algebraically speaking, conics can be given by the equation

$$Ax^2 + Bxy + Cy^2 + Dx + Ey + F = 0.$$

The trivial case $A = B = C = 0$ has to be excluded, since then, the equation is linear and describes a straight line. In the general case, the equation describes the classic conics (ellipse, parabola, or hyperbola), which can be given by five points (if $F \neq 0$, one can divide by F and then solve a system of linear equations with five unknowns). The curves can also degenerate to pairs of straight lines.

B. Odehnal et al., *The Universe of Quadrics*,
https://doi.org/10.1007/978-3-662-61053-4_1

In the algebraic sense, a conic c – a curve of degree two – always has two points of intersection S_1 and S_2 with a given straight line s. They can both be real or complex conjugate. The limiting case $S_1 = S_2$ arises when s is a tangent of c.

Conics are also curves of *class* two, which means that, from each point in the conic's plane, we have two tangents to the conic (in the algebraic sense). When the point lies on the conic, the tangents coincide. The set of all points where there are no real tangents to the conic can be called the interior of a conic. The points where the tangents are complex conjugate and self-perpendicular are called focal points.

Depending on the number of real intersection points with the line at infinity, we can distinguish three types of conics: Ellipses (no real intersection points), parabolas (the line at infinity touches the curve) and hyperbolas (two real intersection points).

Conics are perfectly suitable to introduce the concept of duality, where points become straight lines and vice versa. Each theorem about conics that only contains the terms point, straight line, to touch, and to intersect remains true if we interchange the terms point and straight line (and vice versa) but keep the relation of being incident. A conic considered as its set of points and tangents is a self-dual term. The dual of a conic is still a conic, the set of its tangent lines.

... to the quadrics in 3-space

In Euclidean 3-space, points can be determined by three Cartesian (x, y, z)-coordinates. Then, a quadric can be given by the equation

$$Ax^2 + By^2 + Cz^2 + Dxy + Exz + Fyz + Gx + Hy + Iz + J = 0.$$

Clearly, if the first six coefficients vanish, the above equation is linear and describes just a plane in space. In the general case, the equation describes the classic quadrics (ellipsoids, paraboloids, or hyperboloids). If $J \neq 0$, the above equation can be divided by J, and if we insert the coordinates of nine points, we can solve a system of linear equations with nine unknowns. However, quadrics can also degenerate to quadratic cones, quadratic cylinders, and pairs of planes.

In the algebraic sense, a quadric Q is a surface of degree two that always has two points S_1 and S_2 of intersection with a given straight line s

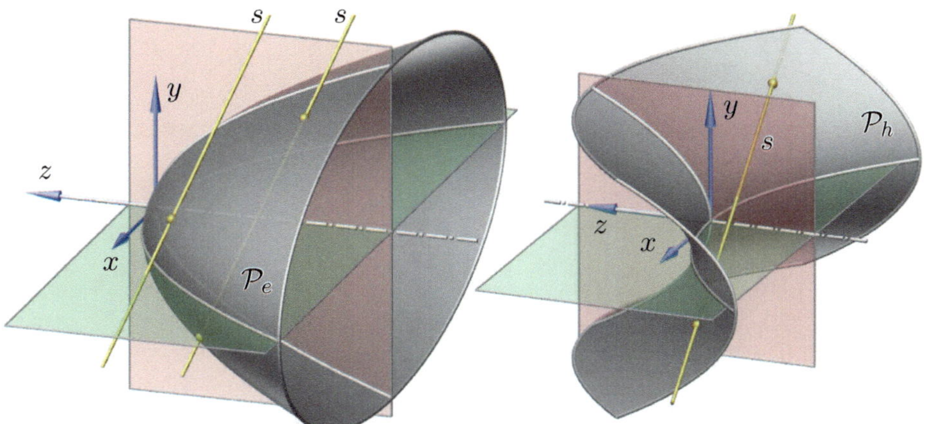

FIGURE 1.1. An elliptic paraboloid $\mathcal{P}_e$ and a hyperbolic paraboloid $\mathcal{P}_h$ with the implicit equations $\frac{1}{9}\,y^2 \pm \frac{4}{9}\,z^2 + 2x = 0$ and the intersection with straight "test lines" s (all with two real intersection points in the algebraic sense). Paraboloids (not of revolution) have two planes of symmetry. Quadratic cylinders have two planes of symmetry parallel to their generators (and infinitely many orthogonal to the generators), while triaxial quadrics have three.

(Figure 1.1). They can both be real or complex conjugate. The limiting case $S_1 = S_2$ arises when s is a tangent of Q.

Quadrics are also surfaces of *class* two, which means that, from each point in space, we have infinitely many tangent planes to the quadric that envelop a quadratic cone in the algebraic sense (Figure 1.2).

When the point lies on the quadric, there is a unique tangent plane. The set of all points where there are no real tangent planes to the quadric, can be called the interior of a quadric. The points where the tangential cones are cones of revolution are called *focal points.* These cones can also be imaginary. The set of focal points of a quadric consists of a pair of focal conics. Such *focal curves* play a comparably dominant role as the focal points of the conics, and they will appear in many theorems about quadrics.

Depending on the intersection with the plane at infinity, we can distinguish three types of quadrics: ellipsoids (no real intersection conic), paraboloids (the plane of infinity touches the quadric), and hyperboloids (a real intersection conic).

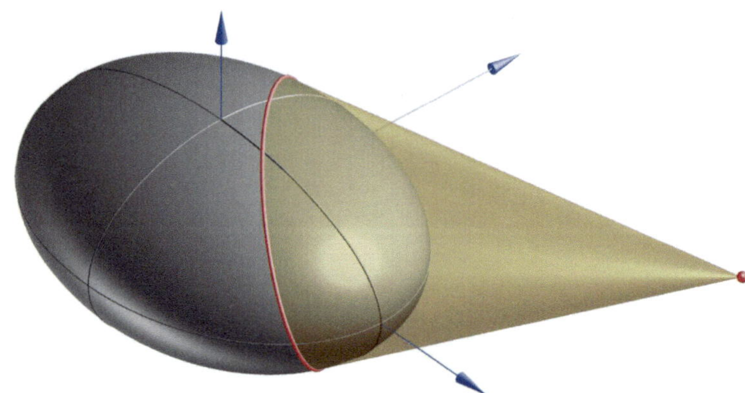

FIGURE 1.2. Since quadrics are surfaces of class two, the tangential cone from any point is quadratic. If the point is in the interior, the cone is imaginary, and it degenerates to the tangential plane when the point lies on the quadric.

Quadrics are perfectly suitable to introduce the concept of duality in 3-space, where points become planes (and vice versa), and straight lines are self-dual. Each theorem about quadrics that only contains the terms point, straight line, plane, to touch, and to intersect remains, therefore, true if we interchange the terms point and plane (and vice versa) but keep the relation of being incident. A quadric considered as its set of points together with its tangential planes is a self-dual term. The dual of a quadric is still a quadric, the set of its tangent planes.

From a mathematician's point of view, surfaces of degree two have, of course, absolutely perfect shapes. Even a deviation of a permille from it can only be called an approximation of such a surface. Strictly speaking, quadrics are simply theoretical shapes that only appear in more or less good approximations in arts (Figure 1.3), architecture and nature. When we look at Figure 1.4, Barcelona might be called the "city of quadrics". Especially Antoni Gaudí, the most famous architect of the area, put a great deal of effort into explaining his constructions.

In nature, ellipsoids (or, again, better approximations of them) seem to be quite common, *e.g.*, when we talk about egg shapes. Interestingly, the shapes are usually formed by at least two different ellipsoids (Figure 1.5) that fit rather well together. The more elongated the egg's shape is, the less likely the egg is to roll off or away – instead, it will move in circles. This is important for birds that hatch on rocks.

FIGURE 1.3. Aesthetic lamps "Queen Titania", designed by Alberto Meda and Paolo Rizzatto. The lamps are perfect ellipsoids with three axes. The aesthetics of ellipsoids seems inspire artists.

FIGURE 1.4. Left to right: Torre Glòries (formerly called Torre Agbar) by Jean Nouvel (upper part half of an ellipsoid); airport tower of Barcelona designed by Bruce Fairbanks (framework of a hyperboloid); detail of the roof of the Sagrada Familia by Antoni Gaudí (parts of a hyperboloid).

The example in Figure 1.6 shows approximations of a hyperbolic paraboloid: It can be proven that a hyperbolic paraboloid is never a

FIGURE 1.5. Egg shapes are often formed by at least two different ellipsoids. The more elongated the egg's shape is, the less likely the egg is to roll off or away.

minimal surface, but very close to being one. Gossamers of caterpillars, *e.g.*, tend to converge to minimal surfaces, since they are under tension, and therefore come close to hyperbolic paraboloids as well.

FIGURE 1.6. Left: A minimal surface (gray) can be approximated quite well by a hyperbolic paraboloid (yellow). Right: Gossamers of caterpillars come close to minimal surfaces, and therefore, also to hyperbolic paraboloids.

2 Quadrics in Euclidean 3-space

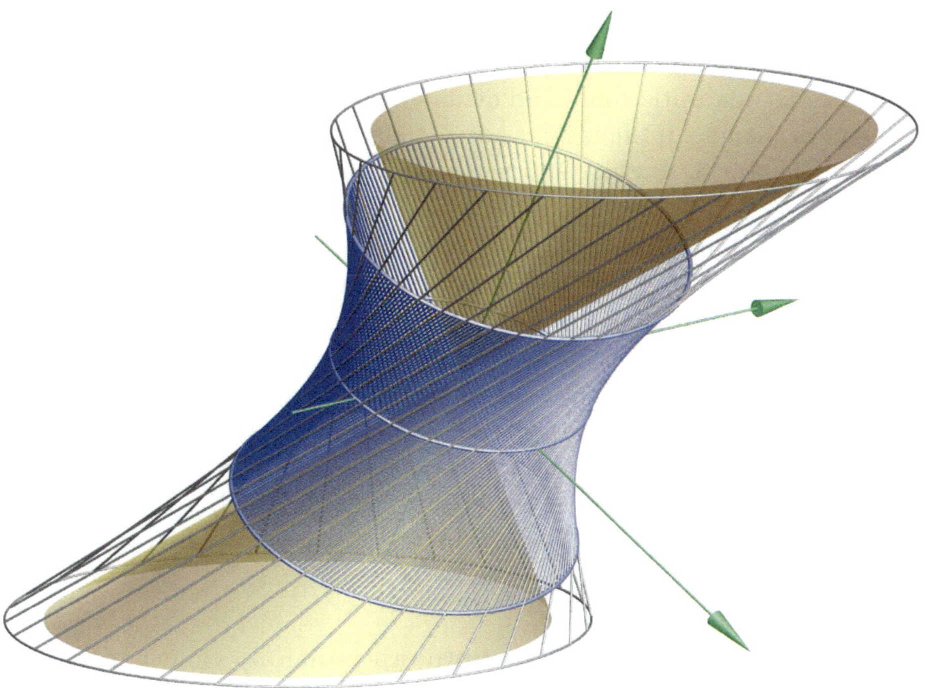

Among the quadrics in $\mathbb{E}^3$, the one-sheeted hyperboloids are characterized as ruled surfaces with a center. They even carry two families of rulings and two families of circles, provided that they are no surfaces of revolution. The picture shows a one-sheeted hyperboloid with one regulus as well as its asymptotic cone and the axes of symmetry.

B. Odehnal et al., *The Universe of Quadrics*,
https://doi.org/10.1007/978-3-662-61053-4_2

To begin with, we present the different types of quadrics in Euclidean space $\mathbb{E}^3$ in an intuitive way, based on their standard equations. Contrary to conics, there is no way to define all quadrics simultaneously by a metric relation such that the different properties of quadrics are easy to achieve.[1] In this section, the term *quadric* stands for ellipsoids (including spheres), hyperboloids, and paraboloids. Sometimes, we use the term *regular quadric* in order to emphasize the difference to *singular* surfaces of degree 2 like quadratic cones and cylinders or *degenerate cases* like planes or pairs of planes. It should be noted that in some literature as well as later in Chapter 3, singular and reducible cases are also called quadrics.

A common definition for all types of quadrics will be given in Chapter 3 based on Linear Algebra. As with conics, studying quadrics in the framework of Projective Geometry leads to a much deeper understanding of their properties. This follows in Chapter 4, which also contains a uniform definition of quadrics via polarities. However, in this approach, *empty* quadrics, *i.e.*, quadrics without real points, are also included. After the study of conics, the reader will also expect focal properties of quadrics in Euclidean space. This will be the goal of Section 7.1. Here we will meet the pairs of focal conics again, but this time in quite a different setting.

Types of quadrics in Euclidean space $\mathbb{E}^3$

Each irreducible surface of degree two other than a cylinder or cone is called *quadric*. Due to principal axis transformation, which will be discussed in Chapter 3, a Cartesian coordinate frame can be found for each quadric such that the quadric satisfies one of the following standard equations with real constants $a, b, c > 0$:

From the algebraic point of view, there exists an additional standard equation, namely

$$\frac{x^2}{a^2} + \frac{y^2}{b^2} + \frac{z^2}{c^2} + 1 = 0.$$

However, it cannot be satisfied by any single real point. Only in the complex extension of $\mathbb{E}^3$ this defines a quadric. Hence, it is called an *empty* or an *imaginary quadric*.

[1] Only JACOBI's focal property, presented in Theorem 2.4.3, provides a universal metric definition. A metric definition for ruled quadrics is given in Exercise 2.3.10.

TABLE 2.1. The quadrics in Euclidean 3-space

1	$\frac{x^2}{a^2}+\frac{y^2}{b^2}+\frac{z^2}{c^2}-1=0$	*ellipsoid*
2	$\frac{x^2}{a^2}+\frac{y^2}{b^2}-\frac{z^2}{c^2}-1=0$	*one-sheeted hyperboloid*
3	$\frac{x^2}{a^2}-\frac{y^2}{b^2}-\frac{z^2}{c^2}-1=0$	*two-sheeted hyperboloid*
4	$\frac{x^2}{a^2}+\frac{y^2}{b^2}-2z=0$	*elliptic paraboloid*
5	$\frac{x^2}{a^2}-\frac{y^2}{b^2}-2z=0$	*hyperbolic paraboloid*

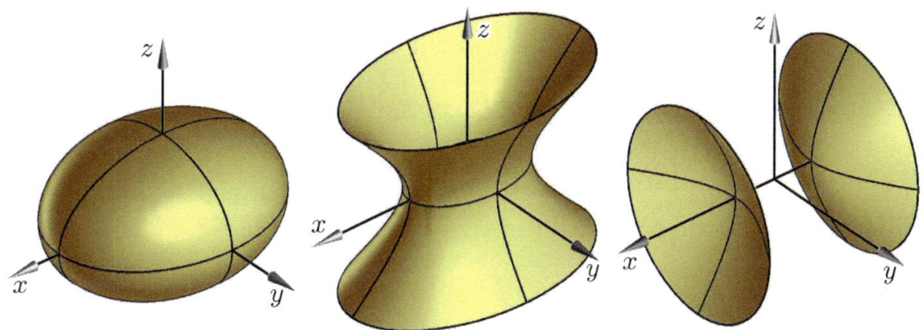

ellipsoid - one-sheeted hyperboloid - two-sheeted hyperboloid

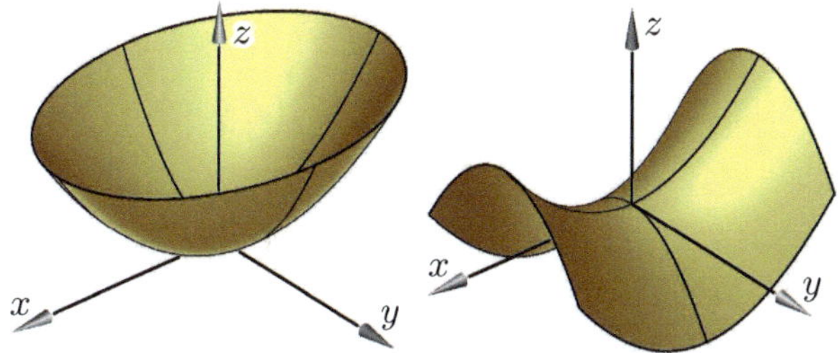

elliptic paraboloid - hyperbolic paraboloid

FIGURE 2.1. The five types of quadrics in $\mathbb{E}^3$.

All ellipsoids and hyperboloids (types 1 to 3) are symmetric with respect to (henceforth abbreviated as w.r.t.) each coordinate plane (see Figure 2.1, first row). Therefore, rotations through 180° about any coordinate axis and the reflection in the origin $(x,y,z) \mapsto (-x,-y,-z)$ map these quadrics onto themselves. We also call them *central quadrics* for short. The points of intersection with any coordinate axis are the *vertices*; the curves of intersection with the planes of symmetry are the *principal sections* of the quadric. The constants a,b,c showing up in the standard equations are called *semiaxes*.

The two paraboloids (types 4 and 5) are only symmetric w.r.t. the $[x,z]$- and $[y,z]$-plane. Hence, only the z-axis is an axis of symmetry. Its point of intersection with the paraboloid, the origin, is the unique *vertex* of the paraboloid (Figure 2.1, second row).

The standard equations, as given above, show, on the left-hand side, a polynomial of degree two in the unknowns x, y, and z. The second order terms in this polynomial define a quadratic form. In the following, we denote the corresponding polar form with $\sigma(\mathbf{x}_1,\mathbf{x}_2)$. It is a symmetric bilinear form on $\mathbb{R}^3$. For example, in the case of an ellipsoid, we have

$$\sigma(\mathbf{x}_1,\mathbf{x}_2) := \frac{x_1x_2}{a^2} + \frac{y_1y_2}{b^2} + \frac{z_1z_2}{c^2}, \tag{2.1}$$

where $\mathbf{x}_i = (x_i, y_i, z_i)$ for $i = 1,2$. Thus, the standard equation of an ellipsoid can also be written briefly as $\sigma(\mathbf{x},\mathbf{x}) - 1 = 0$.

2.1 Ellipsoids

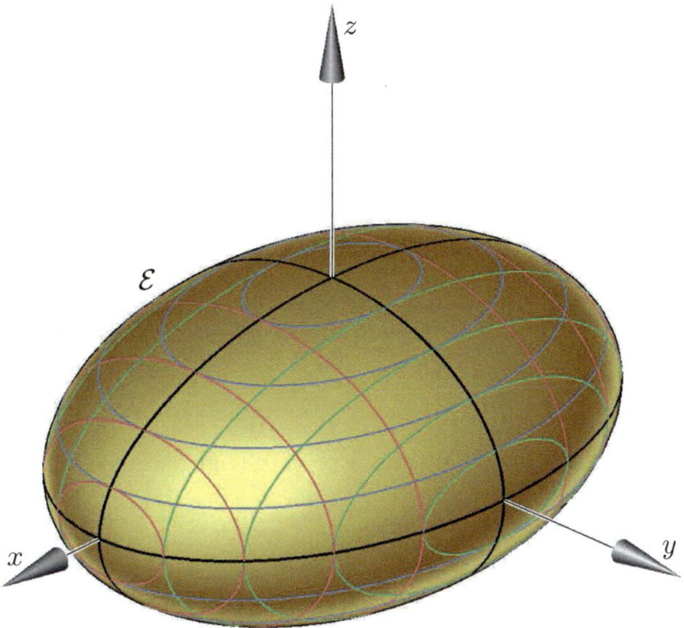

FIGURE 2.2. Triaxial ellipsoid $\mathcal{E}$ with planar sections orthogonal to the axes.

The ellipsoid $\mathcal{E}$ with the standard equation

$$\mathcal{E}\colon\ \frac{x^2}{a^2} + \frac{y^2}{b^2} + \frac{z^2}{c^2} - 1 = 0 \tag{2.2}$$

intersects the coordinate axes at the vertices $A_i = (\pm a, 0, 0)$, $B_i = (0, \pm b, 0)$, and $C_i = (0, 0, \pm c)$, $i = 1, 2$. The constants a, b, and c are the semiaxes of $\mathcal{E}$. If they are mutually different, the ellipsoid is called *triaxial.* If two of them are equal, say $a = b$, then $\mathcal{E}$ is an *ellipsoid of revolution* with the z-axis as axis of rotation. There are two types of ellipsoids of revolution (Figure 2.3): In the case $a > c$, the ellipsoid is called *oblate*, and otherwise *prolate*. Only the second type has two real focal points $(0, 0, \pm\sqrt{c^2 - a^2})$ on the axis of revolution. The ellipsoid with $a = b = c$ is a sphere with radius a.

Planar sections of ellipsoids

The ellipsoid $\mathcal{E}$ with semiaxes a, b, and c can be obtained from the unit sphere S^2 satisfying $x'^2 + y'^2 + z'^2 = 1$ by scaling the coordinates according

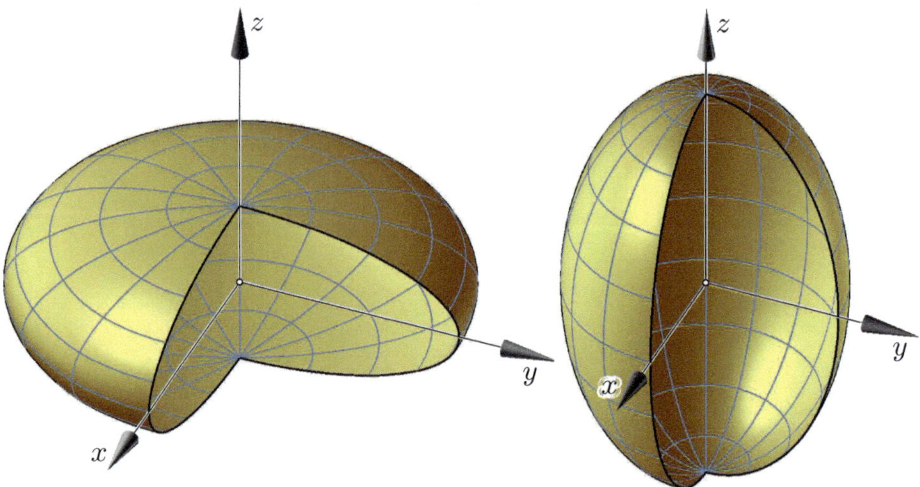

FIGURE 2.3. An oblate (left) and a prolate (right) ellipsoid of revolution.

to

$$\alpha\colon \begin{pmatrix} x' \\ y' \\ z' \end{pmatrix} \mapsto \begin{pmatrix} x \\ y \\ z \end{pmatrix} = \begin{pmatrix} ax' \\ by' \\ cz' \end{pmatrix}. \tag{2.3}$$

This scaling defines a bijective affine transformation α, an *affinity* in brief. We recall that affinities preserve parallelities between lines or planes as well as the affine ratios of triplets of collinear points.

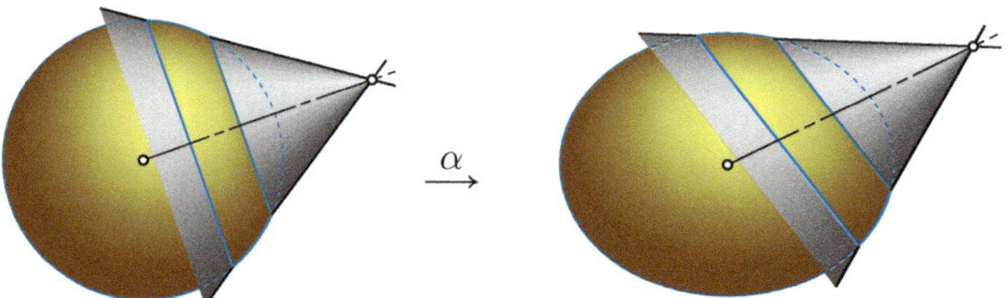

FIGURE 2.4. The sections of an ellipsoid with parallel planes are homothetic.

The affinity α transforms the circles c' on S^2 into ellipses e on $\mathcal{E}$. Two circles of S^2 in parallel planes can be connected by a cone of revolution with an axis through the origin and an apex A'. The circles are corresponding under a dilatation with center A'. The affinity α maps the circles onto two ellipses on $\mathcal{E}$ in parallel planes and corresponding under a dilation with center A, the α-image of A'. Therefore, the two ellipses have parallel major axes and parallel minor axes, the same ratio of semiaxes, *i.e.*, they are *homothetic*. Furthermore, their centers are aligned with the origin. If one of the two circles on S^2 tends to a zero-circle, because its plane contacts S^2, the axes of the corresponding 'zero-ellipse' are so-called *principal curvature tangents* of $\mathcal{E}$. Their pre-images are parallel to conjugate diameters of circles. Thus, these are orthogonal lines which remain orthogonal under α. We will meet principal curvature tangents again in the Chapters 7, 8, and 9).

Is it possible that a planar section of a triaxial $\mathcal{E}$ with semiaxes $a > b > c$ is a circle, *i.e.*, a *circular section* of the ellipsoid?

Theorem 2.1.1 *There are only two sets of parallel planes which intersect the triaxial ellipsoid $\mathcal{E}$ with semiaxes $a > b > c$ along circles. For the ellipsoid with the standard equation (2.2), these planes are parallel to the y-axis and the tangent plane at one of the four umbilical points of $\mathcal{E}$.*

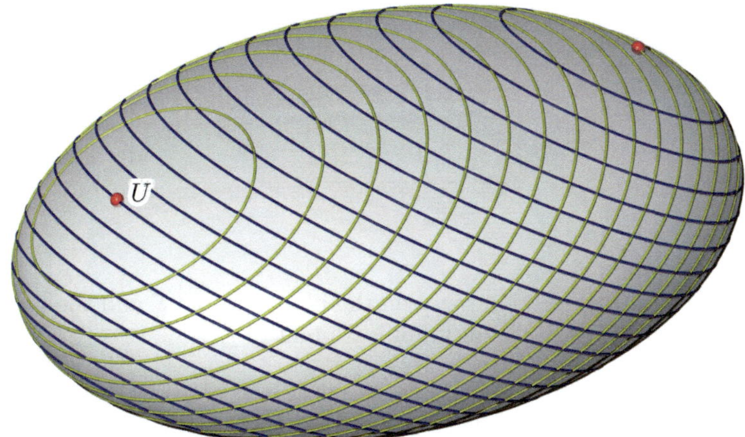

FIGURE 2.5. Triaxial ellipsoids carry two families of circular sections, including zero-circles at the four umbilical points.

Proof: We obtain the circular sections by looking for the intersection between $\mathcal{E}$ and the concentric sphere $\mathcal{S}$ with radius b. Points of this curve of intersection must also satisfy the difference of the respective equations, *i.e.*,

$$\left(\frac{x^2}{b^2} + \frac{y^2}{b^2} + \frac{z^2}{b^2} - 1\right) - \left(\frac{x^2}{a^2} + \frac{y^2}{b^2} + \frac{z^2}{c^2} - 1\right) = 0,$$

which can be decomposed into a product of two linear forms

$$\left(\frac{\sqrt{a^2 - b^2}}{ab}\,x + \frac{\sqrt{b^2 - c^2}}{bc}\,z\right)\left(\frac{\sqrt{a^2 - b^2}}{ab}\,x - \frac{\sqrt{b^2 - c^2}}{bc}\,z\right) = 0. \tag{2.4}$$

Hence, common points of $\mathcal{E}$ and the sphere $\mathcal{S}$ belong to at least one of these two planes through the y-axis (Figure 2.6).

Conversely, each point which is located in one of these planes and also on the sphere $\mathcal{S}$ satisfies the equation of $\mathcal{E}$. Thus, we have found two circles centered at the origin and completely placed on $\mathcal{E}$. In addition, all planes parallel to one of these diameter planes and sufficiently close to the origin intersect $\mathcal{E}$ along circles, too. The centers of zero-circles included in these sets are the so-called *umbilical points* of $\mathcal{E}$ (note point U in Figures 2.5 and 2.6) with coordinates

$$\left(\pm a\frac{\sqrt{a^2-b^2}}{\sqrt{a^2-c^2}},\ 0,\ \pm c\frac{\sqrt{b^2-c^2}}{\sqrt{a^2-c^2}}\right). \tag{2.5}$$

At these points, each tangent line is a principal curvature tangent.

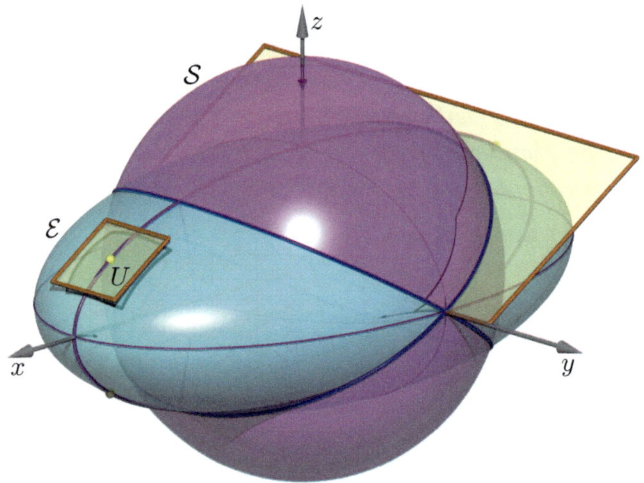

FIGURE 2.6. The intersection between the blue ellipsoid $\mathcal{E}$ and the pink sphere $\mathcal{S}$ splits into two circles. U is an umbilical point of $\mathcal{E}$.

If we intersect $\mathcal{E}$ with a concentric sphere with radius a or c, the difference of the related equations is reducible only in the complex extension of $\mathbb{E}^3$.
Suppose there is another circle on the triaxial ellipsoid $\mathcal{E}$. Then, after translating its plane through the origin, we obtain a circle with radius r centered at the origin. This circle is the equator of a sphere with radius $r \neq b$. However, this sphere can never share a circle with the ellipsoid $\mathcal{E}$, since, after the elimination of the y-coordinate from the related equations, we obtain the equation

$$\frac{a^2-b^2}{a^2}x^2 - \frac{b^2-c^2}{c^2}z^2 + b^2 - r^2 = 0$$

of a hyperbola in the $[x,z]$-plane which can never be an orthogonal view of a circle. ■

Obviously, on an ellipsoid of revolution, the only circular sections are those in planes orthogonal to the axis of revolution, the so-called *parallel circles* of the surface.

An oblate ellipsoid $\mathcal{E}$ of revolution serves as a mathematical standard model of the Earth.[2] Figure 2.7 illustrates (to an exaggerated degree) a remarkable difference to a sphere model: The shortest connection of two points A, B on a sphere follows the connecting great circle. It lies in the plane which is uniquely defined by A and B, provided that they are no antipodes. This plane is orthogonal to the sphere at A and at B.

Quite contrary, for the ellipsoid $\mathcal{E}$, the surface normals n_A and n_B at points A, B on different parallel circles and meridians are skew. Consequently, a plane through A and B cannot be simultaneously orthogonal to $\mathcal{E}$ at A and B. The *geodesic line* as the shortest connection on $\mathcal{E}$ of A and B is no longer a planar section. We recall that geodesics on a surface are either straight line segments or at each point the osculating plane is orthogonal to the surface. A particular characterization of geodesics on quadrics will be presented in Theorem 7.1.4; note also Section 9.6.

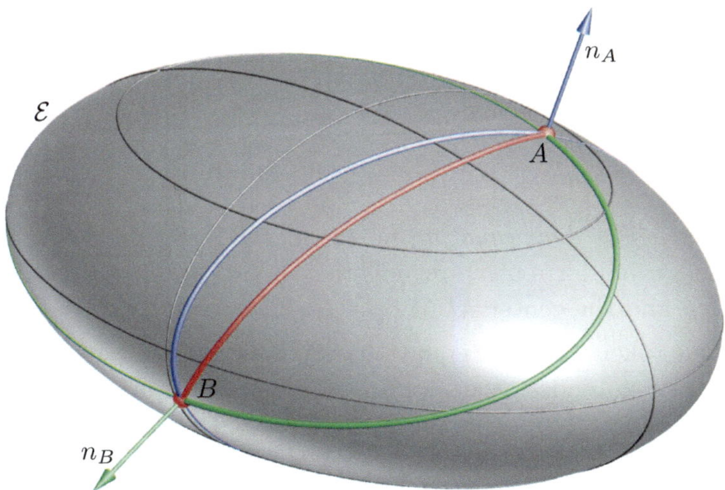

FIGURE 2.7. The normals n_A, n_B to the ellipsoid $\mathcal{E}$ at the points A, B are skew. The red geodesic connecting A and B lies between the blue and green planar sections through A and B, orthogonal to $\mathcal{E}$ either at A or at B.

---

[2]In WGS 84, the *World Geodetic System*, the meridians of the reference ellipsoid have the semiaxes $a = 6378.137$ km and $b = 6356.752$ km. However, the *geoid*, which is everywhere orthogonal to the force of gravity, deviates up to ± 100 m from this ellipsoid.

Remark 2.1.1 Suppose the points A and B are mountain tops, and one places a well-calibrated theodolite at A with the sight axis directed towards B. When the inclination of the sight axis is modified, it hits the ground along a curve in a plane which is vertical at A. This path seems to be 'straight', and therefore the shortest from A to B. However, as illustrated in Figure 2.7, we obtain a different path when looking from B towards A, and it deviates from the geodesic connection.

Tangent cones of ellipsoids

Let us return to the affinity α in (2.3) between the unit sphere S^2 and the ellipsoid $\mathcal{E}$. This affinity maps tangent planes of S^2 to tangent planes of $\mathcal{E}$. Therefore, the tangent plane τ_P to $\mathcal{E}$ at the point $P \in \mathcal{E}$ with position vector $\mathbf{p} = (x_P, y_P, z_P)$ satisfies, by virtue of (2.1), the linear equation

$$\sigma(\mathbf{p}, \mathbf{x}) - 1 = \frac{x_P}{a^2}\,x + \frac{y_P}{b^2}\,y + \frac{z_P}{c^2}\,z - 1 = 0, \tag{2.6}$$

where $\mathbf{x} = (x, y, z)$. In other words, if we rewrite this equation as

$$ux + vy + wz - 1 = 0, \tag{2.7}$$

the coefficients (u, v, w) must satisfy

$$a^2u^2 + b^2v^2 + c^2w^2 - 1 = 0. \tag{2.8}$$

This is called the *dual equation* or *tangential equation* of the ellipsoid. The Hesse normal form of τ_P reveals that $1/\sqrt{u^2 + v^2 + w^2}$ is the distance between τ_P and the center M of $\mathcal{E}$.

By the same token, the dual equation even makes sense when one semiaxis of the ellipsoid $\mathcal{E}$ tends to zero and $\mathcal{E}$ becomes an elliptic disk. In the case $c = 0$, we obtain an ellipse in the plane $z = 0$ with semiaxes a and b, and the remaining equation $a^2u^2 + b^2v^2 - 1 = 0$ means that the plane with coefficients (u, v, w) passes through a tangent line of this ellipse.

The affine transformation α maps pairs of orthogonal diameters of the sphere S^2 onto a pair of *conjugate diameters* of the ellipsoid $\mathcal{E}$. Their direction vectors $\mathbf{d}_1$ and $\mathbf{d}_2$ satisfy

$$\sigma(\mathbf{d}_1, \mathbf{d}_2) = 0. \tag{2.9}$$

Three mutually orthogonal diameters of S^2 are transferred onto a *triplet of conjugate diameters* of $\mathcal{E}$. They have the property that the tangent planes at the endpoints of each diameter are parallel to the plane spanned by the other two diameters.

We follow the usual notation and already call the segment MP, bounded by the center M and any point $P \in \mathcal{E}$, a *diameter* or *diameter segment* of $\mathcal{E}$, though it is obviously only half of a diameter. In case of ambiguities, the full line $[M, P]$ is called a *diameter line* of $\mathcal{E}$.

Theorem 2.1.2 *If P_1, P_2, and P_3 are the endpoints of three mutually conjugate diameters of the ellipsoid $\mathcal{E}$ with center M and semiaxes a, b, and c, then the lengths of the three diameters satisfy*

$$\overline{MP_1}^2 + \overline{MP_2}^2 + \overline{MP_3}^2 = a^2 + b^2 + c^2.$$

The parallelepiped spanned by the vectors $\overrightarrow{MP_1}$, $\overrightarrow{MP_2}$, $\overrightarrow{MP_3}$ has the constant volume abc.

Proof: We still use the coordinate frame of (2.2) and the *axial scaling* α of (2.3). We write the coordinates (x_i, y_i, z_i) of the endpoints P_i, $i = 1, 2, 3$, as column vectors in a 3×3-matrix $\mathbf{P}$ and obtain

$$\mathbf{P} = \begin{pmatrix} x_1 & x_2 & x_3 \\ y_1 & y_2 & y_3 \\ z_1 & z_2 & z_3 \end{pmatrix} = \begin{pmatrix} ax_1' & ax_2' & ax_3' \\ by_1' & by_2' & by_3' \\ cz_1' & cz_2' & cz_3' \end{pmatrix} = \mathbf{DP}',$$

where the matrix $\mathbf{D}$ is the diagonal matrix diag(a, b, c) and matrix $\mathbf{P}'$ is orthogonal. Since the row-vectors in $\mathbf{P}'$ also have unit length, we obtain

$$\overline{MP_1}^2 + \overline{MP_2}^2 + \overline{MP_3}^2 = (x_1^2 + y_1^2 + z_1^2) + \cdots + (x_3^2 + y_3^2 + z_3^2) = a^2 + b^2 + c^2$$

as the squared Frobenius norm of the matrices $\mathbf{P} = \mathbf{DP}'$, *i.e.*, as the sum of the squared entries. The proof of the final statement in Theorem 2.1.2 is left as an exercise (Exercise 2.1.1). ∎

Remark 2.1.2 An ellipsoid $\mathcal{E}$ is uniquely defined by its center and the endpoints of a triplet of conjugate diameters. This is a consequence of principal axis transformation, which is discussed in Chapter 3. However, there is no ruler-and-compass construction which could serve as an analogue to Rytz construction [46, p. 358]. Principal axis transformation needs the eigenvalues of a 3×3-matrix, *i.e.*, the roots of a cubic polynomial, which cannot be determined graphically with ruler and compass, in general. Note the construction of Chasles and further discussions in [61].

The affinity α maps the right cone or cylinder which contacts the unit sphere S^2 along a circle c' onto a quadratic cone or elliptic cylinder, which is tangent to $\mathcal{E}$ along the ellipse c. Conversely, we learn from α: When looking for planes tangent to $\mathcal{E}$ and passing through a given point Q in the exterior of $\mathcal{E}$ or being parallel to a given direction, we obtain a quadratic cone or an elliptic cylinder, and the contact points form an ellipse $c \subset \mathcal{E}$.

The plane spanned by c is the so-called *polar plane* π_Q of the *pole* Q and satisfies an equation analogue to (2.6), provided that $\mathbf{p}$ is replaced by the position vector $\mathbf{q} := (x_Q, y_Q, z_Q)$ of Q.[3]

This can easily be verified: If the point $\mathbf{q}$ satisfies (2.6),

$$\sigma(\mathbf{p},\mathbf{q}) - 1 = \frac{x_P}{a^2}\,x_Q + \frac{y_P}{b^2}\,y_Q + \frac{z_P}{c^2}\,z_Q - 1 = 0, \tag{2.10}$$

i.e., lies in the tangent plane τ_P to $\mathcal{E}$ at the point $\mathbf{p}$, then $\mathbf{p}$ lies in the polar plane of $\mathbf{q}$. This follows directly from the symmetry $\sigma(\mathbf{p},\mathbf{q}) = \sigma(\mathbf{q},\mathbf{p})$ of the bilinear form σ.

By analogy with conics (note [46, p. 263]), we call two points $\mathbf{p}$ and $\mathbf{q}$ *conjugate* w.r.t. the ellipsoid $\mathcal{E}$ if one of these points lies in the polar plane of the other, *i.e.*, if (2.10) holds. Only the points $\mathbf{q} \in \mathcal{E}$ are *self-conjugate*, *i.e.*, they lie in their polar plane and satisfy $\sigma(\mathbf{q},\mathbf{q}) = 1$.

As a consequence, in each orthogonal or oblique view, the visual contour of an ellipsoid is an ellipse. In each central projection with its center outside $\mathcal{E}$, the visual contour of $\mathcal{E}$ is an ellipse, a parabola, or a hyperbola (note [46, p. 153ff]).

Lemma 2.1.1 (i) *If the pole Q moves along a line g, the corresponding polar plane π_Q either rotates about a line g^*, which is called the polar line of g, or it undergoes a translation. In the second case, the polar line g^* is defined as a line at infinity (Figure 2.8).*

(ii) *If a plane ε intersects the ellipsoid $\mathcal{E}$ along an ellipse e, then the relation of conjugate points w.r.t. e within ε is exactly the restriction to ε of the relation of conjugate points w.r.t. $\mathcal{E}$.*

Proof: (i) Let the line g connect the points Q_1 and Q_2. Then, the corresponding position vectors of Q, Q_1, and Q_2 satisfy

$$\mathbf{q} = (1-\lambda)\mathbf{q}_1 + \lambda\mathbf{q}_2, \quad \lambda \in \mathbb{R},$$

which implies, for the polar plane π_Q, the equation

$$\begin{aligned}\sigma(\mathbf{q},\mathbf{x}) - 1 &= [(1-\lambda)\sigma(\mathbf{q}_1,\mathbf{x}) + \lambda\sigma(\mathbf{q}_2,\mathbf{x})] - 1\\ &= (1-\lambda)\,[\sigma(\mathbf{q}_1,\mathbf{x}) - 1] + \lambda\,[\sigma(\mathbf{q}_2,\mathbf{x}) - 1] = 0.\end{aligned}$$

Hence, π_Q either passes through the line of intersection $g^* = \pi_{Q_1} \cap \pi_{Q_2}$, or, in the case of parallelity, π_Q is parallel to π_{Q_1} and π_{Q_2}. The latter case shows up when the position vectors

[3] For the sake of simplicity, we identify, in the following, points Q with their position vectors $\mathbf{q}$. Therefore, we speak of the point $\mathbf{q}$ for short.

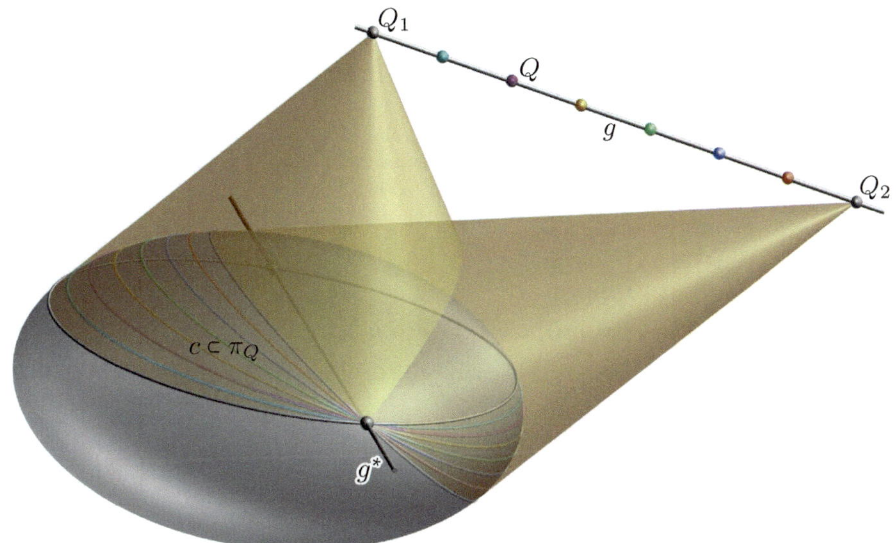

FIGURE 2.8. While the pole Q moves along a line g, the corresponding polar planes rotate about the polar line g^*.

$\mathbf{q}_1$ and $\mathbf{q}_2$ are linearly dependent, *i.e.*, the line $g = [Q_1, Q_2]$ passes through the center of the ellipsoid. Of course, the relation between g and g^* is also symmetric.

(ii) For points $Q \in \varepsilon$ in the exterior of $\mathcal{E}$, the tangent cone from Q to $\mathcal{E}$ contacts $\mathcal{E}$ along a conic in the polar plane π_Q. This tangent cone intersects the plane ε through the apex Q in the two tangents drawn from Q to the conic e. Therefore, the polar line of Q w.r.t. e lies in π_Q, and consequently, points conjugate to Q w.r.t. e are also conjugate to Q w.r.t. $\mathcal{E}$. On the other hand, points of e are self-conjugate w.r.t. e and $\mathcal{E}$. According to [46, Sec. 7.1], this is sufficient for proving that conjugacy w.r.t. $e \subset \varepsilon$ equals the restriction of conjugacy w.r.t. $\mathcal{E}$ to the plane ε.[4] ■

As mentioned above, for all points Q on a diameter line g of $\mathcal{E}$, the polar planes π_Q are mutually parallel. The polar line g^* of g is the line at infinity of π_Q. If π_Q intersects $\mathcal{E}$ along an ellipse e, then the center of e is conjugate to all points of g^* w.r.t. e and, by virtue of Lemma 2.1.1, also w.r.t. $\mathcal{E}$. Thus, each ellipse $e = \pi_Q \cap \mathcal{E}$ has its midpoint on the line connecting the ellipsoid's center with the pole Q of π_Q.

Lemma 2.1.2 *The quadratic cone formed by the tangents from the point* $\mathbf{q}$ *to the ellipsoid* $\mathcal{E}$ *satisfies, in terms of the bilinear form* (2.1)*, the*

[4]As already announced, a deeper insight into the polarity w.r.t. quadrics follows in Chapter 4.

equation

$$[\sigma(\mathbf{q},\mathbf{q})-1]\,[\sigma(\mathbf{x},\mathbf{x})-1]-[\sigma(\mathbf{q},\mathbf{x})-1]^2=0. \tag{2.11}$$

Proof: We verify: All points $\mathbf{p}\in\mathcal{E}$, which lie in the polar plane of $\mathbf{q}$ w.r.t. $\mathcal{E}$, *i.e.*, with $\sigma(\mathbf{p},\mathbf{p})=1$ and $\sigma(\mathbf{q},\mathbf{p})=1$, satisfy (2.11). However, the same holds for all points $\mathbf{x}=(1-\lambda)\mathbf{p}+\lambda\mathbf{q}$, $\lambda\in\mathbb{R}$, on the line connecting $\mathbf{p}$ with $\mathbf{q}$, as shown below:

$$\begin{aligned}
&[\sigma(\mathbf{q},\mathbf{q})-1]\,[\sigma((1-\lambda)\mathbf{p}+\lambda\mathbf{q},\ (1-\lambda)\mathbf{p}+\lambda\mathbf{q})-1]-[\sigma(\mathbf{q},\ (1-\lambda)\mathbf{p}+\lambda\mathbf{q})-1]^2\\
=\ &[\sigma(\mathbf{q},\mathbf{q})-1]\left[(1-\lambda)^2\sigma(\mathbf{p},\mathbf{p})+2(1-\lambda)\lambda\sigma(\mathbf{p},\mathbf{q})+\lambda^2\sigma(\mathbf{q},\mathbf{q})-1\right]\\
&-[(1-\lambda)\sigma(\mathbf{q},\mathbf{p})+\lambda\sigma(\mathbf{q},\mathbf{q})-1]^2\\
=\ &[\sigma(\mathbf{q},\mathbf{q})-1]\,\lambda^2\,[-1+\sigma(\mathbf{q},\mathbf{q})]-\lambda^2\,[-1+\sigma(\mathbf{q},\mathbf{q})]^2=0.
\end{aligned}$$

For points $\mathbf{q}\in\mathcal{E}$, (2.11) is reduced to the square of (2.6) of the tangent plane to $\mathcal{E}$ at $\mathbf{q}$. Hence, this cone, if seen as a set of points, degenerates into the tangent plane, but counted with multiplicity two.

The dual representation of this cone, *i.e.*, the condition for its tangent planes with equations $ux+vy+wz-1=0$, consists of two equations, (2.8) and $x_Qu+y_Qv+z_Qw=1$ (note the dual equation as used in Exercise 2.1.4). ∎

Let us substitute $\mathbf{q}=\lambda\mathbf{r}$ in (2.11). Then, we obtain

$$\left[\lambda^2\sigma(\mathbf{r},\mathbf{r})-1\right]\,[\sigma(\mathbf{x},\mathbf{x})-1]-[\lambda\sigma(\mathbf{r},\mathbf{x})-1]^2=0.$$

For fixed $\mathbf{r}$ we obtain as the limit for $\lambda\to\infty$

$$\sigma(\mathbf{r},\mathbf{r})\,[\sigma(\mathbf{x},\mathbf{x})-1]-\sigma^2(\mathbf{r},\mathbf{x})=0. \tag{2.12}$$

This is the equation of a quadratic cylinder tangent to $\mathcal{E}$ and with generators parallel to the vector $\mathbf{r}$. The points of contact belong to the plane satisfying $\sigma(\mathbf{r},\mathbf{x})=0$.

Below, we present a result attributed to GASPARD MONGE (cf. [127, p. 186]). This is a three-dimensional analogue to the director circle of an ellipse, which was discussed in [46, p. 47].

Theorem 2.1.3 *The point S is the intersection of three mutually orthogonal tangent planes τ_1, τ_2, and τ_3 of the ellipsoid $\mathcal{E}$ with center M and semiaxes a,b,c if, and only if,*

$$\overline{MS}^2=a^2+b^2+c^2=\overline{M\tau_1}^2+\overline{M\tau_2}^2+\overline{M\tau_3}^2. \tag{2.13}$$

Proof: We refer to the tangential equation (2.8) of the ellipsoid $\mathcal{E}$. Let three mutually orthogonal tangent planes τ_i, $i=1,2,3$, be given with coordinate vectors $\mathbf{u}_i:=(u_i,v_i,w_i)$ satisfying

(2.7). We combine these vectors as columns in a 3×3-matrix $\mathbf{T}$. Then, after normalization of the column vectors, we obtain the orthogonal matrix

$$\mathbf{M} = \begin{pmatrix} u_1 & u_2 & u_3 \\ v_1 & v_2 & v_3 \\ w_1 & w_2 & w_3 \end{pmatrix} \begin{pmatrix} \frac{1}{\|\mathbf{u}_1\|} & 0 & 0 \\ 0 & \frac{1}{\|\mathbf{u}_2\|} & 0 \\ 0 & 0 & \frac{1}{\|\mathbf{u}_3\|} \end{pmatrix} = \mathbf{TN}$$

with $\mathbf{N} = \operatorname{diag}(\frac{1}{\|\mathbf{u}_1\|}, \frac{1}{\|\mathbf{u}_2\|}, \frac{1}{\|\mathbf{u}_3\|})$.

The three tangent planes intersect at the point S with position vector $\mathbf{s}$. Hence, $\mathbf{s}$ satisfies the equations $\langle \mathbf{u}_i, \mathbf{s} \rangle = 1$ of τ_i for $i = 1, 2, 3$, where $\langle \ , \ \rangle$ denotes the standard scalar product in $\mathbb{R}^3$. We set $\mathbf{s} = \lambda_1 \mathbf{u}_1 + \lambda_2 \mathbf{u}_2 + \lambda_3 \mathbf{u}_3$ and, by virtue of $\langle \mathbf{u}_i, \mathbf{u}_j \rangle = 0$ for $i \neq j$, we obtain $\lambda_i = 1/\langle \mathbf{u}_i, \mathbf{u}_i \rangle$, *i.e.*,

$$\mathbf{s} = \frac{1}{\|\mathbf{u}_1\|^2}\mathbf{u}_1 + \frac{1}{\|\mathbf{u}_2\|^2}\mathbf{u}_2 + \frac{1}{\|\mathbf{u}_3\|^2}\mathbf{u}_3. \tag{2.14}$$

Therefore,

$$\overline{MS}^2 = \|\mathbf{s}\|^2 = \frac{1}{\|\mathbf{u}_1\|^2} + \frac{1}{\|\mathbf{u}_2\|^2} + \frac{1}{\|\mathbf{u}_3\|^2}. \tag{2.15}$$

On the other hand, all three vectors $\mathbf{u}_i$ satisfy the dual equation (2.8). This means, in matrix form,

$$\mathbf{T}^{\mathrm{T}}\mathbf{A}\,\mathbf{T} = \begin{pmatrix} \mathbf{u}_1^{\mathrm{T}} \\ \mathbf{u}_2^{\mathrm{T}} \\ \mathbf{u}_3^{\mathrm{T}} \end{pmatrix} \begin{pmatrix} a^2 & 0 & 0 \\ 0 & b^2 & 0 \\ 0 & 0 & c^2 \end{pmatrix} (\mathbf{u}_1\ \mathbf{u}_2\ \mathbf{u}_3) = \begin{pmatrix} 1 & * & * \\ * & 1 & * \\ * & * & 1 \end{pmatrix}.$$

We substitute $\mathbf{T} = \mathbf{MN}^{-1}$ and obtain, after multiplication with $\mathbf{N}$ from both sides,

$$\mathbf{M}^{\mathrm{T}}\mathbf{A}\,\mathbf{M} = \mathbf{N}\begin{pmatrix} 1 & * & * \\ * & 1 & * \\ * & * & 1 \end{pmatrix}\mathbf{N} = \begin{pmatrix} \frac{1}{\|\mathbf{u}_1\|^2} & * & * \\ * & \frac{1}{\|\mathbf{u}_2\|^2} & * \\ * & * & \frac{1}{\|\mathbf{u}_3\|^2} \end{pmatrix}. \tag{2.16}$$

Since $\mathbf{M}^{\mathrm{T}} = \mathbf{M}^{-1}$, the trace of the left-most matrix in (2.16) equals the trace of $\mathbf{A}$, *i.e.*, $a^2 + b^2 + c^2$, and also the trace of the right-hand matrix. However, by virtue of (2.15), the latter is equal to $\overline{MS}^2$. Thus, together with $\overline{M\tau_i}^2 = \frac{1}{\|\mathbf{u}_i\|^2}$, the equation (2.13) is confirmed.

Conversely, the point S satisfies (2.13). Then, S is an exterior point of $\mathcal{E}$, since its distance to the center M is larger than the maximal semiaxis a of $\mathcal{E}$. Tangent planes parallel to $[M, S]$ envelop a tangent cylinder of $\mathcal{E}$. We use one of these planes, τ_1, as image plane of an orthogonal projection. Then, the image S^n of S lies outside the visual contour of $\mathcal{E}$. We select one tangent line through S^n to this contour. This is the image of a tangent plane τ_2 orthogonal to τ_1. We denote the unique third plane through S orthogonal to τ_1 and τ_2 with ε. By virtue of (2.13), its distance to M satisfies

$$\overline{M\varepsilon}^2 = a^2 + b^2 + c^2 - \overline{M\tau_1}^2 - \overline{M\tau_2}^2.$$

On the other hand, among the tangent planes of $\mathcal{E}$, exactly two are parallel to ε. According to the first part of this proof, their distance to M equals that of ε. Consequently, ε must be one of these tangent planes, which finally confirms the statement of Theorem 2.1.3. ∎

Remark 2.1.3 The proofs above reveal that the Theorems 2.1.2 and 2.1.3 are similarly valid in higher-dimensional spaces for *hyper-ellipsoids*, the higher-dimensional counterparts of ellipsoids.

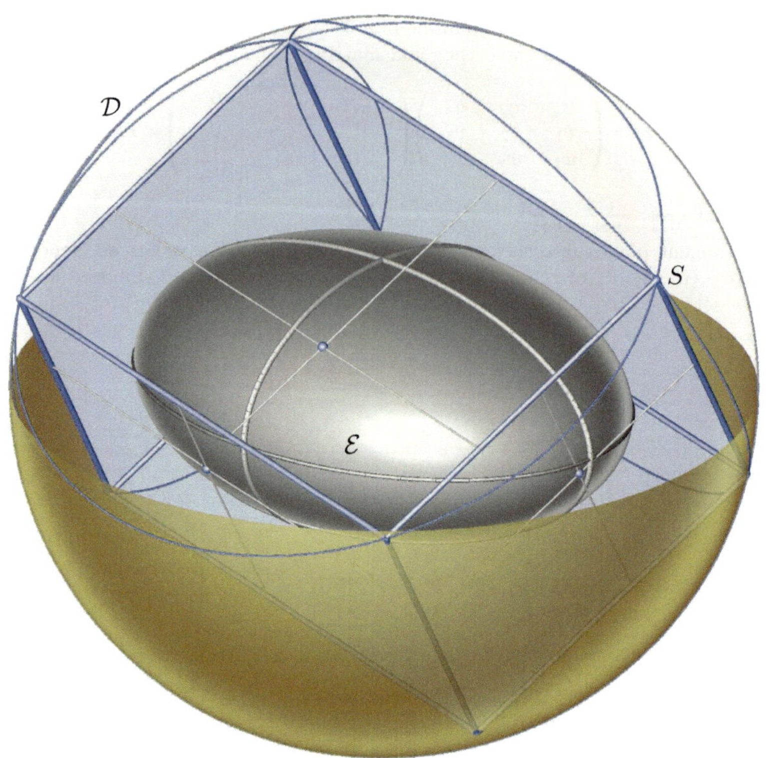

FIGURE 2.9. A box circumscribed to a triaxial ellipsoid $\mathcal{E}$ is inscribed in the director sphere $\mathcal{D}$.

As claimed in Theorem 2.1.3, the common points S of the orthogonal triplets of tangent planes of $\mathcal{E}$ are located on a concentric sphere with radius $\sqrt{a^2 + b^2 + c^2}$, called *director sphere* $\mathcal{D}$. After reflection of the triplet (τ_1, τ_2, τ_3) in the center M of $\mathcal{E}$, we obtain the six faces of a box circumscribed to $\mathcal{E}$ and inscribed in the sphere $\mathcal{D}$, since each vertex of this box is the meet of three mutually orthogonal tangent planes. The three edge lengths of this box are $2\,\overline{M\tau_i}$. All diagonals are diameters of the director sphere.

In order to determine a single box, we specify one tangent plane τ_1 and project $\mathcal{E}$ orthogonally into τ_1. We obtain an ellipse, the visual contour of $\mathcal{E}$, and a one-parameter set of circumscribed rectangles (see [46, p. 47]). Any two orthogonal sides of such a rectangle can be chosen as orthogonal views of tangent planes τ_2 and τ_3 which form with τ_1 an orthogonal triplet.

This proves that there is a three-parameter family of circumscribed boxes.

Each point $S \in \mathcal{D}$ is a vertex of a circumscribed box. There is even a one-parameter family of triplets (τ_1, τ_2, τ_3) of orthogonal tangent planes passing through S. They perform a constrained motion which keeps S and M fixed. During this motion the distances $1/\|\mathbf{u}_i\|$ of these planes to the center M of $\mathcal{E}$ vary, according to (2.14), while they envelop the quadratic cone drawn from S tangent to $\mathcal{E}$. This is a special quadratic cone: It is dual (in the sense of planar Projective Geometry) to an equilateral cone (see [46, p. 459–463]), which carries triplets of orthogonal generators and whose quadratic form has a vanishing trace.

A parameter count reveals: We have a three-parameter family of boxes circumscribed to $\mathcal{E}$, but, due to Theorem 2.1.3, only a two-parameter set of admissible dimensions. Hence, we cannot fix an ellipse in such a box, except in the case when one edge length equals the smallest or the largest diameter of $\mathcal{E}$. In other words:

Theorem 2.1.4 *Let $\mathcal{E}$ be a triaxial ellipsoid with semiaxes $a > b > c$. Then, each rigid box circumscribed to $\mathcal{E}$ with edge lengths smaller than $2a$ and bigger than $2c$ can move, while all six faces remain tangent to $\mathcal{E}$.*

In Corollary 7.1.3, we will learn that only for apices on the focal conics of $\mathcal{E}$, the tangent cone is a cone of revolution (note Figure 7.6).

Each ellipsoid bounds a convex body. In the literature, there are many theorems which characterize ellipsoids among all strictly convex bodies by various properties, *e.g.*, by planar contours. For more details, see [48].

Parametrizations of ellipsoids

In order to obtain a parametrization of the ellipsoid $\mathcal{E}$ with standard equation (2.2), we recall the well-known parametrization of a sphere by its geographic longitude λ and latitude τ. We can transfer it to the parametrization of $\mathcal{E}$ by applying the affinity α in (2.3). Thus, we obtain

$$\mathcal{E}: \ \mathbf{x}(\lambda, \tau) = (a\cos\lambda\cos\tau, \ b\sin\lambda\cos\tau, \ c\sin\tau) \tag{2.17}$$

for $0 \le \lambda < 2\pi$ and $-\frac{\pi}{2} \le \tau \le \frac{\pi}{2}$. This parametrization is one-to-one apart from the two singularities, the vertices $\tau = \pm\frac{\pi}{2}$ on the z-axis.

The half-angle substitutions

$$\sin\lambda = \frac{1-u^2}{1+u^2}\,,\ \cos\lambda = \frac{2u}{1+u^2}\ \text{ with } u = \tan\frac{\lambda}{2}\ \text{ and } v = \tan\frac{\tau}{2}$$

give rise to a rational parametrization of the ellipsoid:

$$\mathbf{x}(u,v) = \left(\frac{2au}{1+u^2}\frac{2v}{1+v^2}, \frac{1-u^2}{1+u^2}\frac{2bv}{1+v^2}, \frac{c(1-v^2)}{1+v^2}\right) \tag{2.18}$$

for $u \in \mathbb{R}$ and $-1 \le v \le 1$, which is one-to-one for $|v| \ne 1$. A stereographic projection of $\mathcal{E}$ into a plane (see page 75) can be used for deriving a rational parametrization of $\mathcal{E}$ of even lower degree, namely

$$\mathbf{x}(u,v) = \left(\frac{2au}{1+u^2+v^2}, \frac{2bv}{1+v^2+v^2}, \frac{c(1-u^2-v^2)}{1+u^2+v^2}\right). \tag{2.19}$$

Due to a standard result of Differential Geometry, each parametrization of an ellipsoid has a singularity.

• **Exercise 2.1.1** Conjugate diameters span a constant volume.
Prove the final statement of Theorem 2.1.2: For all triplets of an ellipse, the parallelepiped spanned by conjugate diameter segments MP_1, MP_2, MP_3 has the constant volume abc.

• **Exercise 2.1.2** Affinely invariant orthogonality.
Prove the following theorem: *Pairs of orthogonal lines which remain orthogonal when transformed by the affine mapping $(x, y, z) \mapsto (px, qy, rz)$ with mutually different $p, q, r \in \mathbb{R}\setminus\{0\}$, are parallel to a pair of principal curvature tangents at a point of an ellipsoid (see page 2.1).*
Hint: Orthogonal lines through the origin are conjugate diameters of the unit sphere; compare with [111].

• **Exercise 2.1.3** Three mutually orthogonal tangents.
Prove the following theorem: *Given an ellipsoid $\mathcal{E}$ with semiaxes a, b, and c, the locus of points P, from which three mutually orthogonal tangents of $\mathcal{E}$ can be drawn, is a coaxial ellipsoid $\mathcal{E}^*$ with semiaxes*

$$a^* = \sqrt{a^2 + \frac{b^2c^2}{b^2+c^2}},\quad b^* = \sqrt{b^2 + \frac{a^2c^2}{a^2+c^2}},\quad c^* = \sqrt{c^2 + \frac{a^2b^2}{a^2+b^2}}.$$

Hint: By virtue of [46, p. 459ff], the cone with apex P and tangent to $\mathcal{E}$ must be equilateral. In this case, the trace of the quadratic form, which shows up in the Cartesian equation of the tangent cone (2.11), must vanish.

• **Exercise 2.1.4** Another proof of Theorem 2.1.3.

Three mutually orthogonal tangent planes of the ellipsoid $\mathcal{E}$ are passing through S if, and only if, the tangent cone with apex S is dual (in the sense of two-dimensional Projective Geometry)

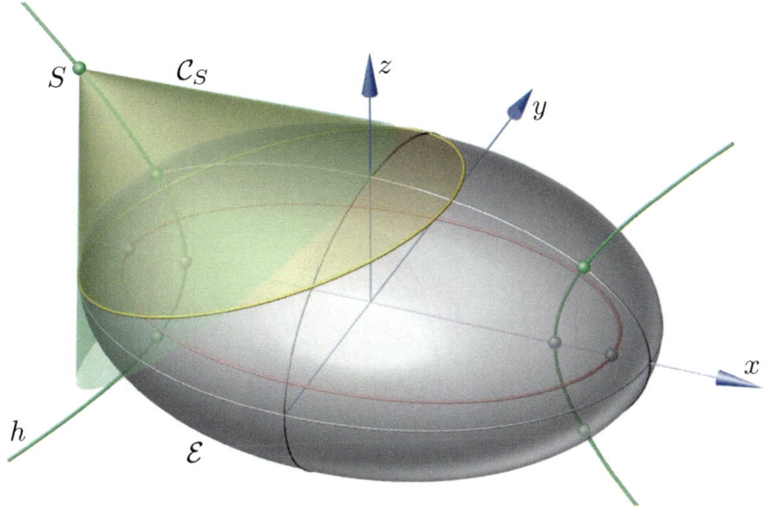

FIGURE 2.10. Tangent cone $\mathcal{C}_S$ as a cone of revolution (Exercise 2.1.5).

to an equilateral cone, a so-called *dual-equilateral* cone. Therefore, by virtue of [46, p. 459ff], the tangential equation of this cone — dual to [46, (10.13)] — must have zero trace. Show that this trace equals $(a^2 - x_S^2) + (b^2 - y_S^2) + (z^2 - z_S^2)$.

Hint: Translate the coordinate frame such that S is the new origin. This means that the homogeneous coordinates $(u_0 : u_1 : u_2 : u_3) = (-1 : u : v : w)$ of the tangent plane, as used in (2.8), are replaced by $(u_0' : u_1' : u_2' : u_3')$ where $u_0' = -u_0 + x_S u_1 + y_S u_2 + z_S u_3$ and $u_i' = u_i$ for $i = 1, 2, 3$. Plug this into (2.8). Then, after setting $u_0' = 0$, one obtains the requested dual equation of the tangent cone with apex S.

• **Exercise 2.1.5** Tangent cones of revolution.

Prove the following theorem (Figure 2.10): *Let the tangent cone of the triaxial standard ellipsoid $\mathcal{E}$ in (2.2) with apex S in the plane $y = 0$ be a cone $\mathcal{C}_S$ of revolution. Then, S lies on the hyperbola h satisfying*

$$h\colon\ \frac{x^2}{a^2 - b^2} - \frac{z^2}{b^2 - c^2} = 1.$$

This is the focal hyperbola of $\mathcal{E}$ (see (7.3)).

Hint: A cone of revolution with apex $S = (\xi, 0, \zeta)$ intersects the unit sphere centered at S along a pair of circles. Their planes are symmetric w.r.t. S and orthogonal to the plane $y = 0$. In the translated coordinate frame with origin S, the tangent cone $\mathcal{C}_S$ satisfies, by virtue of (2.11), the homogeneous equation

$$\left(\frac{\zeta^2}{c^2} - 1\right)\frac{x^2}{a^2} + \left(\frac{\xi^2}{a^2} + \frac{\zeta^2}{c^2} - 1\right)\frac{y^2}{b^2} + \left(\frac{\xi^2}{a^2} - 1\right)\frac{z^2}{c^2} - \frac{2\xi\zeta}{a^2c^2} = 0.$$

The claim for a cone of revolution yields the necessary condition

$$\left[(b^2 - c^2)\xi^2 - (a^2 - b^2)\zeta^2 - (a^2 - b^2)(b^2 - c^2)\right]\left[c^2\xi^2 + a^2\zeta^2 - a^2c^2\right] = 0.$$

2.2 Hyperboloids

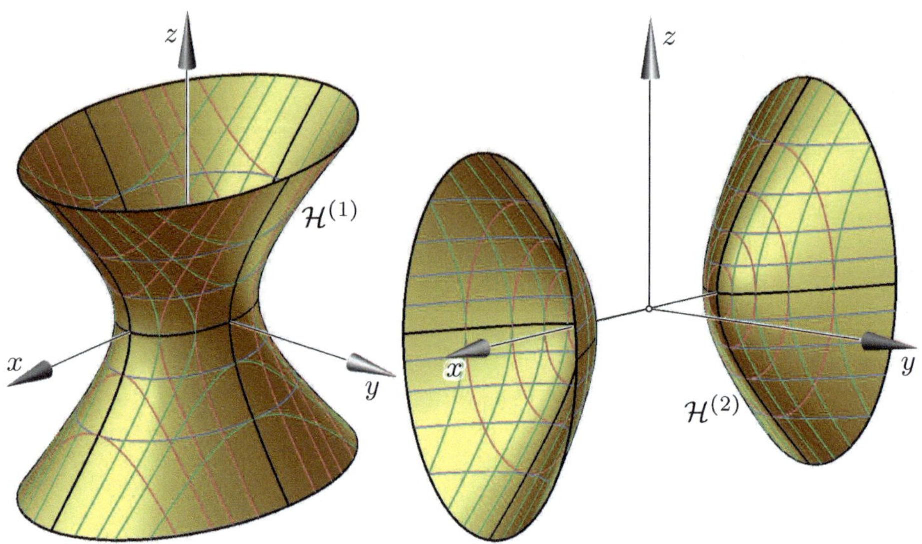

FIGURE 2.11. One-sheeted hyperboloid $\mathcal{H}^{(1)}$, two-sheeted hyperboloid $\mathcal{H}^{(2)}$.

The equations of the one-sheeted hyperboloid $\mathcal{H}^{(1)}$ and the two-sheeted hyperboloid $\mathcal{H}^{(2)}$,

$$\mathcal{H}^{(1)}\colon \ \frac{x^2}{a^2}+\frac{y^2}{b^2}-\frac{z^2}{c^2}-1=0 \ \text{ and } \ \mathcal{H}^{(2)}\colon \ \frac{x^2}{a^2}-\frac{y^2}{b^2}-\frac{z^2}{c^2}-1=0, \qquad (2.20)$$

differ only in one sign. Nevertheless, their appearances are quite different and already expressed in their names (Figure 2.11).

In the case $b=a$, the one-sheeted hyperboloid $\mathcal{H}^{(1)}$ is a surface of revolution w.r.t. the z-axis. On the other hand, under $b=c$ the two-sheeted hyperboloid $\mathcal{H}^{(2)}$ is a surface of revolution w.r.t. the x-axis. One obtains these surfaces from a hyperbola with semiaxes a and c in the $[x,z]$-plane by rotation, either about its secondary axis or about its first axis. The rotations of the asymptotes generate cones of revolution, the corresponding *asymptotic cones.* By rotation about the secondary axis, the vertices of the hyperbola in the $[x,z]$-plane trace the *gorge circle* of the one-sheeted hyperboloid of revolution.

The special cases of hyperboloids with two equal semiaxes $a=b$ or $b=c$ can be transformed in all other hyperboloids given in (2.20) by respective axial scalings

$$\begin{aligned}\alpha_1\colon (x, y, z) &\mapsto (x', y', z') = \left(x, \frac{b}{\mathrm{a}}\, y, z\right) \quad \text{for } \mathcal{H}^{(1)},\\ \alpha_2\colon (x, y, z) &\mapsto (x', y', z') = \left(x, \frac{b}{\mathrm{c}}\, y, z\right) \quad \text{for } \mathcal{H}^{(2)}.\end{aligned} \tag{2.21}$$

The asymptotic cones of revolution are sent hereby onto the respective *asymptotic cones* satisfying

$$\mathcal{C}^{(1)}\colon \frac{x^2}{a^2} + \frac{y^2}{b^2} - \frac{z^2}{c^2} = 0 \ \text{ and } \ \mathcal{C}^{(2)}\colon \frac{x^2}{a^2} - \frac{y^2}{b^2} - \frac{z^2}{c^2} = 0. \tag{2.22}$$

These quadratic cones have their apices at the center M of the hyperboloids. In the one-sheeted case, the affine transformation α_1 transforms the gorge circle into the *gorge ellipse* of $\mathcal{H}^{(1)}$.

Besides the different appearances of the two hyperboloids, there is another fundamental difference: Only one-sheeted hyperboloids $\mathcal{H}^{(1)}$ are *ruled surfaces*, *i.e.*, swept by a line moving in space. They even carry two families of lines. Each family is called a *regulus* of $\mathcal{H}^{(1)}$, and each line on $\mathcal{H}^{(1)}$ is called a *generator* or *ruling* of $\mathcal{H}^{(1)}$.

For each regulus, we get a single generator by intersecting $\mathcal{H}^{(1)}$ from (2.20) with the plane $y = b$: The (x, z)-coordinates of the common points satisfy

$$0 = \frac{x^2}{a^2} - \frac{z^2}{c^2} = \left(\frac{x}{a} + \frac{z}{c}\right)\left(\frac{x}{a} - \frac{z}{c}\right).$$

Thus, the curve of intersection splits into two lines with the slopes $\pm c/a$.[5] A one-sheeted hyperboloid of revolution can be generated by revolving one of these lines about the axis of revolution (Figure 2.12), a result which is attributed to Sir Christopher Wren (1669), who designed St. Paul's Cathedral in London. We study this case first and the apply the affine transformation α_1 for deriving properties of the two rulings in the general case.

One-sheeted hyperboloid of revolution

Let two skew lines be given: An axis s and a generator e. First we prove that during the rotation about s the line e sweeps out a one-sheeted hyperboloid $\mathcal{H}^{(1)}$. The converse was already proved above: Each one-sheeted hyperboloid $\mathcal{H}^{(1)}$ of revolution can be obtained in this way.

[5] In the case of a two-sheeted hyperboloid, these two lines are complex conjugate.

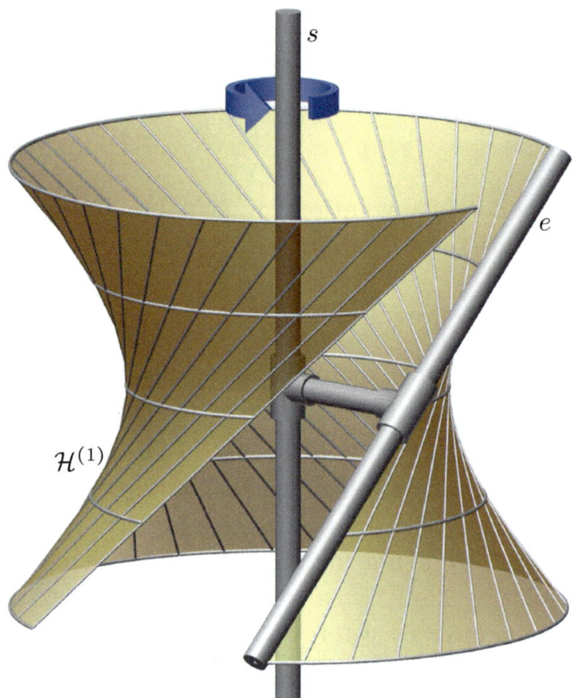

FIGURE 2.12. 1-sheeted hyperboloid $\mathcal{H}^{(1)}$, generated by rotation of e about s.

Lemma 2.2.1 *Let a line e rotate about an axis s which is neither parallel to e nor orthogonal or intersecting. Then, the poses of e form one regulus of a one-sheeted hyperboloid* $\mathcal{H}^{(1)}$ *of revolution. The second regulus can be obtained by reflection in any plane through s.*

The shortest distance between e and s equals the radius of the gorge circle of $\mathcal{H}^{(1)}$*; the angle between the two given lines equals half of the angle of aperture of the asymptotic cone.*

In Figure 2.13, we visualize the rotation of e about s in top and front view. For the sake of simplicity, we specify the axis s as the vertical z-axis and choose an initial pose e_1 of e parallel to the image plane of the front view, the $[y, z]$-plane (note [46, Fig. 4.1]). The x-axis is placed along the common normal of s and e_1. The following proof of Lemma 2.2.1 is of constructive nature and based on both views, top view and front view.

Proof: The poses of the line e generate the *e-regulus* on the swept surface. We recognize the shape of this surface when we look for its *meridians*, *i.e.*, its mutually congruent intersection

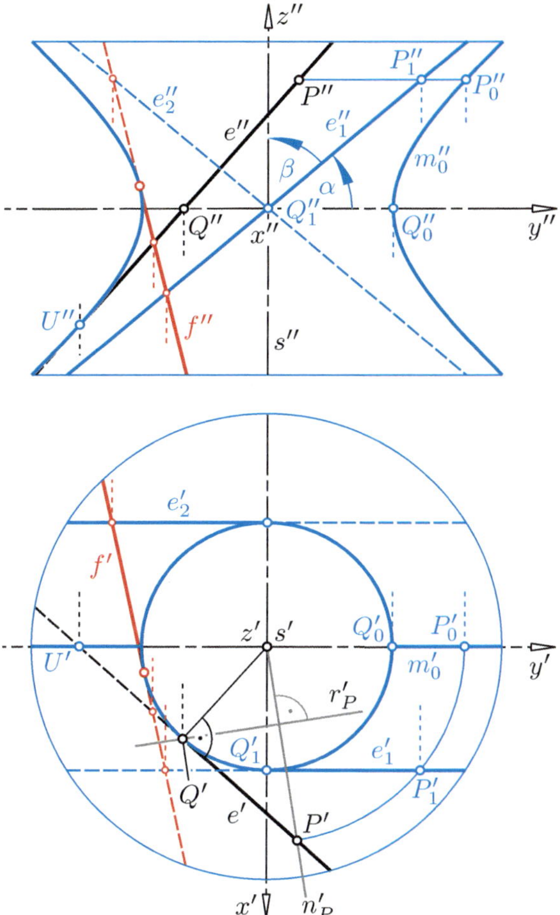

FIGURE 2.13. Top and front view visualizing the rotation of the given line e about the vertical axis s.

curves with planes passing through the axis s. In particular, we focus on the meridian m_0 in the $[y, z]$-plane.

Each point $P \in e$ traces a horizontal circle, a *parallel circle* on $\mathcal{H}^{(1)}$ which intersects the $[y, z]$-plane in two points symmetric w.r.t. s. The smallest parallel circle, the gorge circle of $\mathcal{H}^{(1)}$, is the trajectory of the point $Q \in e$ on the common perpendicular of s and e. We denote the distance between s and e with d.

Let the z-coordinate of any point $P_1 \in e_1$ be given. Then, in terms of the slope angle α of e_1 and e w.r.t. the horizontal $[x, y]$-plane, the coordinates of $P_1 \in e_1$ are $(d,\ z\cot\alpha,\ z)$ (see Figure 2.13), where $0 < \alpha < \frac{\pi}{2}$. We rotate P_1 about s to the point P_0 on the right half of the meridian m_0 in the $[y, z]$-plane and obtain the coordinates $P_0 = (0, y, z)$, where

$y^2 = d^2 + z^2\cot^2\alpha$. This shows that P_0 satisfies the equation

$$\frac{y^2}{d^2} - \frac{z^2}{(d\tan\alpha)^2} = 1$$

of a hyperbola with the center M at the origin.

Conversely, for each hyperbola in the $[y,z]$-plane that is centered at the origin and has a and c as its semiaxes, we immediately find the corresponding pair (d,α) where $a = d$ and $b = d\tan\alpha$ so that each one-sheeted hyperboloid of revolution can be generated by rotating a line e about a skew axis. The parallels to the poses of e through the center M of $\mathcal{H}^{(1)}$ form the asymptotic cone of $\mathcal{H}^{(1)}$.

The meridian m_0 of $\mathcal{H}^{(1)}$ appears in the front view of Figure 2.13 as the *contour* of the hyperboloid. At each point of m_0, the tangent plane to $\mathcal{H}^{(1)}$ is orthogonal to the $[y,z]$-plane. Therefore, the front view e'' of any pose e is tangent to m_0'', and the point of contact is the front view of the point U of intersection between e and the $[y,z]$-plane. For e_1 this point lies at infinity which reveals that e_1'' is an *asymptote* of m_0'' and tangent to m_0'' at a point at infinity.

Since, under the rotations about s through angles between 0° and 360°, no point outside the axis remains fixed, any two different poses of the generator e must be skew.

The reflection in any plane through the axis s maps $\mathcal{H}^{(1)}$ onto itself while each pose of e is mapped onto another line $f \subset \mathcal{H}^{(1)}$ which meets e at the finite or infinite point of intersection between e and the mirror plane. Thus, we obtain a second regulus on our hyperboloid, which we call the *f-regulus*. It is the *complementary regulus* of the e-regulus.

The reflection of e in the $[y,z]$-plane yields a generator f with the front view f'' coinciding with e'' and tangent to m_0'' at the contour point U''. This reveals that the tangent plane τ_U is spanned by the two generators e and f passing through U, and this is valid for each point $U \in \mathcal{H}^{(1)}$. ■

The two reguli can be distinguished from the viewpoint of chirality, according to their relative position to the axis. We can move each generator into the axis s by a helical transformation, composed of a translation of length a along the common perpendicular and by a rotation about this perpendicular through an angle β between $-\pi/2$ and $\pi/2$, where $|\cot\beta| = a/c$. With reference to the notation in Figure 2.13, this yields a left-handed screw for e-generators, since $\beta < 0$, and a right-handed screw for f-generators. Therefore, e-generators are called *left-twisted* to the axis s and f-generators *right-twisted.*[6]

Any two generators e, f taken out from different reguli of $\mathcal{H}^{(1)}$ must be intersecting or parallel. We find the point of intersection in the bisector plane of the trace points of e and f in the plane of the gorge circle. This is a meridian plane and a plane of symmetry between e and f.

[6] Note that being left-twisted is a symmetric relation: When e_1 is left-twisted to s, then s is left-twisted to e_1, too.

We can control the complanarity of each e- and each f-generator in Figure 2.13. Let the top view f' be given, tangent to the top view of the gorge circle. Then, the corresponding front view f'' must pass through the front views of all points of intersection between our particular f-generator and all e-generators, in particular those parallel to the $[y,z]$-plane, which appear as asymptotes of m_0''.

Another property of our hyperboloid $\mathcal{H}^{(1)}$ becomes apparent when inspecting Figures 2.13 and 2.14: We know that all surface normals of a surface of revolution meet the axis s. Hence, while a point P runs along the generator e, the surface normal n_P as well as the tangent plane τ_P to $\mathcal{H}^{(1)}$ must rotate about e (Figure 2.13). In the top view, we see the trace r_P' of this tangent plane τ_P passing through the trace point Q of e and perpendicular to $n_P' = [P', s']$. When P tends along e to infinity, the tangent plane τ_U becomes the *asymptotic plane* of e. Its trace equals $[Q, s']$, and this proves that the asymptotic plane passes through the center M of $\mathcal{H}^{(1)}$ and is tangent to the asymptotic cone. The asymptotic plane is orthogonal to the tangent plane τ_Q at the point Q on the gorge circle.

Conversely, for each plane τ_X through e, there exists one, and only one, point $X \in e$ where the given plane τ_X contacts the hyperboloid (Figure 2.13). We find the top view of this point on the line n_X' perpendicular to the trace of τ_X. If, for example, the front view of τ_X coincides with e'', its point of contact coincides with the unique contour point U'' of e'' on m_0''.

Let us look in more detail at the relation between points $P \in e$ and their tangent planes $\tau_P \supset e$. In Figure 2.15, the front view (left) shows the projection into the meridian plane which is parallel to the generator e, while, in the side view (right), the line e is depicted as a point. This side view shows the angle ψ between different surface normals in true size, *i.e.*, undistorted. We compare the normals n_Q at the point Q on the gorge circle and n_P at the point P at distance u to Q. In terms of the angle α with $\tan\alpha = c/a$, we derive

$$\tan\psi = \frac{u\cot\alpha}{a} = \frac{u}{a}\frac{a}{c} = \frac{u}{c}, \text{ hence } u = c\tan\psi. \tag{2.23}$$

As indicated in Figure 2.15, the length u of the translation along e from Q to P and the angle ψ of the rotation about e from τ_Q to τ_P are equipped with a sign. In the case of the left-twisted generator e, this motion is

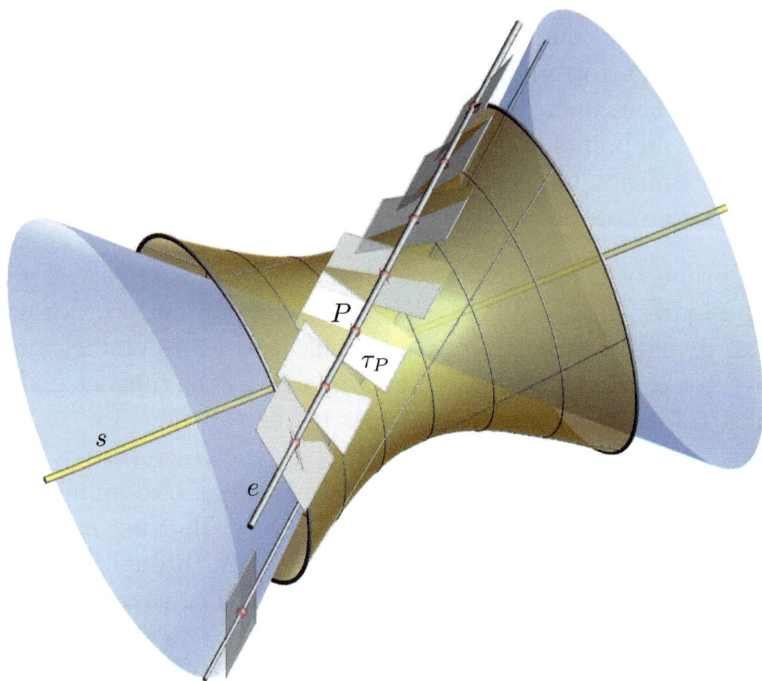

FIGURE 2.14. The distribution of tangent planes τ_P along the generator e.

a right-handed screw motion, while, for right-twisted generators f, this screw motion is left-handed.

We will learn in Chapter 9 that an equation similar to $u = c \tan \psi$ is even valid for general *ruled surfaces* $\mathcal{R}$, which are defined as surfaces swept out by any line moving in space (note Figure 9.7 on page 398). The poses of this line e are again called *generators* or *rulings*. Through each point $P \in \mathcal{R}$ passes a generator e, which also belongs to the tangent plane τ_P at P. When P moves along e, then τ_P is either constant or varies from point to point. In the first case, the generator e is called *torsal*, otherwise *skew*.

Ruled surfaces which include only torsal generators are called *developables* or *developable surfaces*[7]. Cones or cylinders are examples of developables.

[7]According to a theorem from Differential Geometry (Chapter 9, page 399), developables are the only surfaces which can be flattened, *i.e.*, for which there exists an isometric mapping into the plane. Smooth developables are either cylindrical and conical surfaces or tangent surfaces, *i.e.*, generated by the tangent lines of a space curve.

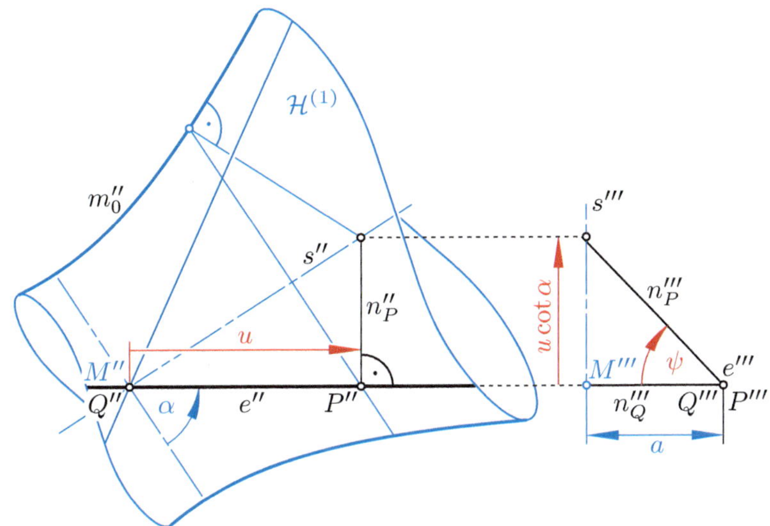

FIGURE 2.15. Surface normals n_P at points P of the generator e.

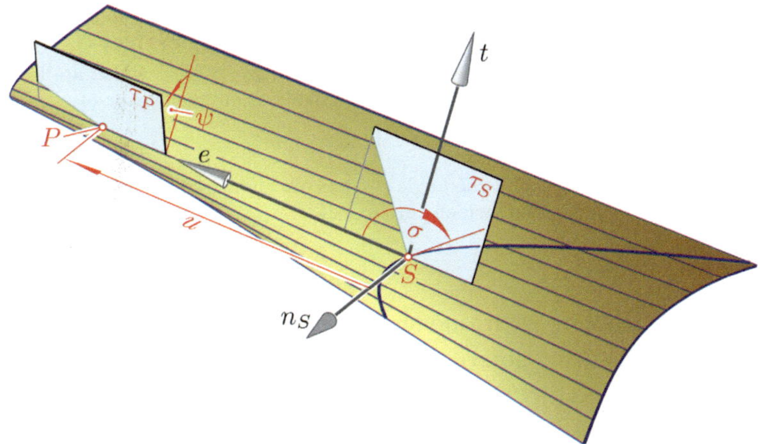

FIGURE 2.16. The tangent plane τ_P at the point $P \in e$ is defined by the distribution parameter δ via $u = -\delta \tan \psi$.

Ruled surfaces with only non-torsal generators are called *skew ruled surfaces.* The reguli on one-sheeted hyperboloids are examples of skew ruled surfaces.

On each skew generator e, there exists a particular point S, the *central* or *striction point* and a constant $\delta \in \mathbb{R}$ such that for the tangent plane τ_P

at any point $P \in e$ holds

$$\overline{PS} = \delta \tan\psi \text{ with } \psi = \sphericalangle \tau_P \tau_S$$

(note Figure 2.16). The rate δ is called *distribution parameter*[8], since it controls the distribution of tangent planes along a generator. We obtain $\delta = \pm\overline{PS}$ for $\psi = 45°$. The limit $\overline{PS} \to \infty$ shows that τ_S is orthogonal to the asymptotic plane, *i.e.*, the tangent plane at the ideal point of e.

Corollary 2.2.2 *For all generators of a one-sheeted hyperboloid of revolution $\mathcal{H}^{(1)}$, the absolute value of the distribution parameter $|\delta|$ equals the secondary semiaxis c of the meridians. The distribution parameter δ is positive for generators that are right-twisted to the axis and otherwise negative.*

Proof: All generators of $\mathcal{H}^{(1)}$ have their striction point on the gorge circle, since the asymptotic planes are tangent to the asymptotic cone. The rest follows from (2.23).

As noted above, δ can be equipped with a sign by the definition that δ is positive if, and only if, the translation along the generator and the rotation of the respective tangent planes corresponds to a left-handed screw (note [106, p. 266, Remark 5.3.4]). By virtue of Figure 2.15, we obtain a negative distribution parameter for generators $e \subset \mathcal{H}^{(1)}$ which are left-twisted to the axis s. Therefore, the signed distribution parameter of any generator of $\mathcal{H}^{(1)}$ satisfies

$$\delta = a \cot\beta, \tag{2.24}$$

when both the distance $a = \overline{es}$ and the angle $\beta = \sphericalangle es$ are signed according to the screw motion along the common perpendicular, which carries e to s. ■

Finally, we recall that each one-sheeted hyperboloid can be transformed by the affine transformation α_1 from (2.21) into a one-sheeted hyperboloid of revolution. Therefore, we can summarize some properties that were originally derived only for the revolute-case (see Figure 2.17).

Theorem 2.2.1 *Each one-sheeted hyperboloid $\mathcal{H}^{(1)}$ carries two one-parameter families of lines, the e-regulus and the f-regulus. Any two different generators of the same regulus are skew. Any two generators taken from different reguli are intersecting or parallel. They span either*

[8] The distribution parameter of e equals the limit of the ratio of the signed distance and angle (in radians) between e and a neighboring generator e_1 which converges towards e, *i.e.*, $\delta = \lim_{e_1 \to e} \dfrac{\overline{e_1 e}}{\sphericalangle e_1 e}$, and this is valid for all skew generators of any ruled surface.

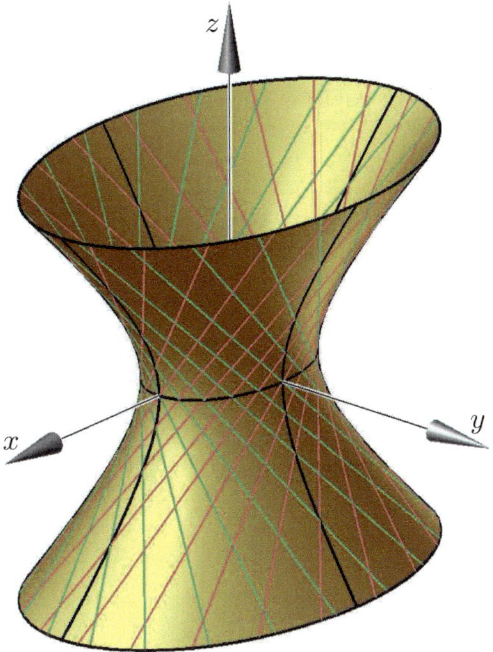

FIGURE 2.17. One-sheeted hyperboloid with its two reguli and the axes of symmetry.

the tangent plane τ_X at their point of intersection X or the asymptotic plane at their common point at infinity, and this plane passes through the center M of $\mathcal{H}^{(1)}$. The asymptotic cone of $\mathcal{H}^{(1)}$, formed by lines passing through M parallel to any generator, is the envelope of the asymptotic planes of the generators.

In Chapter 4, we will learn more generally, that the set of lines meeting simultaneously three given and mutually skew lines is always a regulus.

Three theorems on one-sheeted hyperboloids of revolution

First, we focus on a quadrilateral formed by two e- and two f-generators on a one-sheeted hyperboloid $\mathcal{H}^{(1)}$. By virtue of Theorem 2.2.1, this quadrilateral must be skew, *i.e.*, non-planar.

Theorem 2.2.2 *The sides of a non-planar quadrilateral are located on a one-sheeted hyperboloid of revolution if, and only if, an alternating sum of its side lengths vanishes.*

Proof: From left to right:
Let $X_1, \ldots, X_4$ be the vertices of a quadrilateral with sides $[X_1, X_2]$, ..., $[X_4, X_1]$ being generators of a one-sheeted hyperboloid $\mathcal{H}^{(1)}$, and let $z_1, \ldots, z_4$ be the respective z-coordinates w.r.t. the coordinate frame used in Figure 2.13. Since all generators enclose the same angle $90° - \alpha$ with the z-axis, we can express the length s_i of the side X_iX_{i+1}, i modulo 4, as

$$s_i = |z_{i+1} - z_i| / \sin\alpha.$$

The identity

$$(z_2 - z_1) + (z_3 - z_2) + (z_4 - z_3) + (z_1 - z_4) = 0$$

gives rise to a vanishing sum of the side lengths $s_1, \ldots, s_4$, each with an adjusted sign. No single side length can be equal to the sum of the other three, since in this case the four vertices would be aligned, which contradicts our assumption of non-planarity. Therefore, in this vanishing sum, two side lengths must have a plus and two a minus.
The same can be concluded from the fact that the top view in the $[x, y]$-plane shows a quadrilateral with side lengths $s_1 \cos\alpha$, ..., $s_4 \cos\alpha$ circumscribed to the gorge circle.
We distinguish two cases, depending on whether the quadrilateral surrounds the axis of $\mathcal{H}^{(1)}$ or not. According to this, by choosing an appropriate initial point X_1, we can set up our alternating sum, without restricting generality, as either

$$\text{(a)} \quad s_1 - s_2 + s_3 - s_4 = 0 \quad \text{or} \quad \text{(b)} \quad s_1 - s_2 - s_3 + s_4 = 0.^9 \tag{2.25}$$

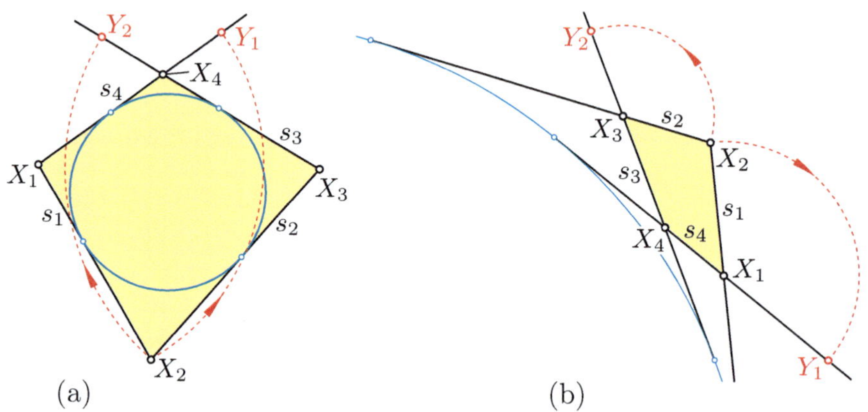

FIGURE 2.18. Proof of the sufficiency of the conditions (a) or (b), as stated in Theorem 2.2.2.

From right to left: Let a skew quadrilateral with consecutive vertices $X_1, \ldots, X_4$ and a vanishing alternating sum of type (a) or (b) be given. Now, we protract from X_1 on the line $[X_1, X_4]$ the side length s_1, namely, in case (a) in the direction X_1X_4, in case (b) in the opposite direction. This yields the point Y_1. Similarly, we protract from X_3 along $[X_3, X_4]$ the length s_2, and obtain Y_2 (see Figure 2.18). Then, in both cases, we have equal distances

$$\overline{X_4Y_1} = \overline{X_4Y_2},$$

[9] According to [46, Exercise 2.2.9], in these cases X_2 and X_4 are focal points of a quadric of revolution, a two-sheeted hyperboloid, or an ellipsoid which passes simultaneously through X_1 and X_3, and vice versa. Note also [46, Fig. 4.17].

since under (a) $\overline{X_4Y_1} = |s_1 - s_4| = |s_2 - s_3| = \overline{X_4Y_2}$, and under (b), $\overline{X_4Y_1} = s_1 + s_4 = s_2 + s_3 = \overline{X_4Y_2}$.
There are two cases to be distinguished:

(1) If $Y_1 \neq Y_2$, these two points, together with X_2, span a plane ε, since, in the case of collinearity, the quadrilateral would be planar. Let us assume that the plane ε is horizontal. Then, we find three isosceles triangles with a horizontal basis: $\triangle X_2X_1Y_1$, $\triangle Y_2X_4Y_1$, and $\triangle X_2X_3Y_2$. Hence, through reflections in vertical planes, we can transform $[X_2, X_1] \mapsto [Y_1, X_1] = [Y_1, X_4] \mapsto [Y_2, X_4] = [Y_2, X_3] \mapsto [X_2, X_3]$. Consequently, the four sides of the quadrilateral have the same inclination angle $\alpha < 90°$ w.r.t. ε.

In the top view in ε, as shown in Figure 2.18, the four side lengths, each multiplied with $\cos\alpha$, satisfy an equation analogue to (2.25). Therefore, the top views of the four sides are tangent to a circle. Consequently, the four vertical planes of symmetry between any two neighboring sides have a vertical axis s in common. The side line $[X_1, X_2]$ can be transformed into the other side lines either by a rotation about s or by a reflection in a plane through s. This confirms that the four sides are located on a one-sheeted hyperboloid of revolution with axis s.

(2) An identity $Y_1 = Y_2$ is only possible in case (a), when $s_1 = s_4$, and therefore, also $s_2 = s_3$. Then, the quadrilateral admits a planar symmetry: The perpendicular bisector σ_{24} of X_2 and X_4 passes through X_1 and X_3. The interior bisector plane of the angle $\sphericalangle X_1X_2X_3$ intersects σ_{24} in a line s which has the same properties as the axis s in the previous case (1). ■

Theorem 2.2.3 *For any two skew generators e_1, e_2 on a one-sheeted hyperboloid $\mathcal{H}^{(1)}$ of revolution with the secondary semiaxis c holds a relation between the shortest distance $d := \overline{e_1e_2}$ and the enclosed angle $\varphi := \sphericalangle e_1e_2$, as given below:*

$$\frac{d}{2}\cot\frac{\varphi}{2} = c.$$

Proof: We refer to the particular position of $\mathcal{H}^{(1)}$ as used in Figure 2.13. Let e_1 and e_2 be two poses with trace points Q_1, Q_2 symmetric w.r.t. the $[x, z]$-plane. Then, the rotation about the x-axis through 180° exchanges the two generators e_1 and e_2, and their common normal is orthogonal to the x-axis. If a denotes the radius of the gorge circle, we can set

$$e_1: \begin{pmatrix} a\cos\psi \\ -a\sin\psi \\ 0 \end{pmatrix} + \mathbb{R}\begin{pmatrix} a\sin\psi \\ a\cos\psi \\ c \end{pmatrix}, \quad e_2: \begin{pmatrix} a\cos\psi \\ a\sin\psi \\ 0 \end{pmatrix} + \mathbb{R}\begin{pmatrix} -a\sin\psi \\ a\cos\psi \\ c \end{pmatrix}.$$

The common normal passes through $(a\cos\psi, 0, 0)$ and has the direction of $(0, c, -a\cos\psi)$, which gives for the distance $d = \overline{e_1e_2} = 2ac\sin\psi/\sqrt{a^2\cos^2\psi + c^2}$.

On the other hand, we obtain for the angle $\varphi = \sphericalangle e_1e_2$ between the two generators

$$\cos\varphi = \frac{a^2(\cos^2\psi - \sin^2\psi) + c^2}{a^2 + c^2} \quad \text{and} \quad \tan\frac{\varphi}{2} = \sqrt{\frac{1-\cos\varphi}{1+\cos\varphi}} = \frac{a\sin\psi}{\sqrt{a^2\cos^2\psi + c^2}},$$

which immediately yields the statement of Theorem 2.2.3. ■

Theorem 2.2.3 offers another possibility to prove Corollary 2.2.2 by computing the limit $e_2 \to e_1$.

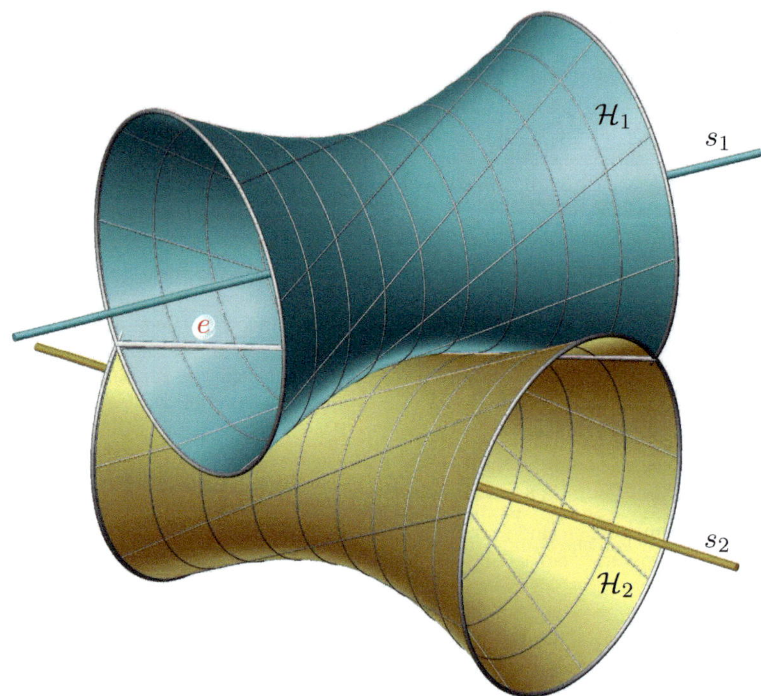

FIGURE 2.19. Two hyperboloids of revolution in contact along the line e.

Theorem 2.2.4 *Any two skew lines s_1 and s_2 are respective axes of two one-sheeted hyperboloids of revolution which contact each other along a common generator e. All possible lines e of contact form a Plücker conoid (= cylindroid) passing through s_1 and s_2 (Figure 2.20).*

Proof: We use the common normal of the given lines (s_1, s_2) as the z-axis of a Cartesian coordinate frame and set $2\beta := \sphericalangle s_1 s_2$ with $0 < \beta < \frac{\pi}{2}$ and, for the shortest distance, $2d := \overline{s_1 s_2}$ (Figure 2.20). Then, we can define the two axes by the equations

$$z = \pm d, \qquad x : y = \pm\cos\beta : \sin\beta.$$

The contact along the generator e implies that the two hyperboloids share the distribution parameters (note Figure 2.16) and the striction points of e, *i.e.*, the points of intersection Q_1, Q_2 of e with the respective gorge circles. Consequently, the common surface normal n_Q at $Q_1 = Q_2$ coincides with the common perpendicular of s_1 and s_2, which was chosen as the z-axis. Thus, we can specify the line e of contact by the equations

$$z = z_0 \quad \text{and} \quad x : y = \cos\varphi_0 : \sin\varphi_0.$$

Then, the gorge circles of the two hyperboloids have the respective radii $|d \mp z_0|$, and the angles between e and the given axes s_1, s_2 amount to $|\beta \mp \varphi_0|$. By virtue of Corollary 2.2.2 and (2.24), equally signed distribution parameters of the two one-sheeted hyperboloids yield the condition

$$(d - z_0)\cot(\beta - \varphi_0) = (d + z_0)\cot(\beta + \varphi_0),$$

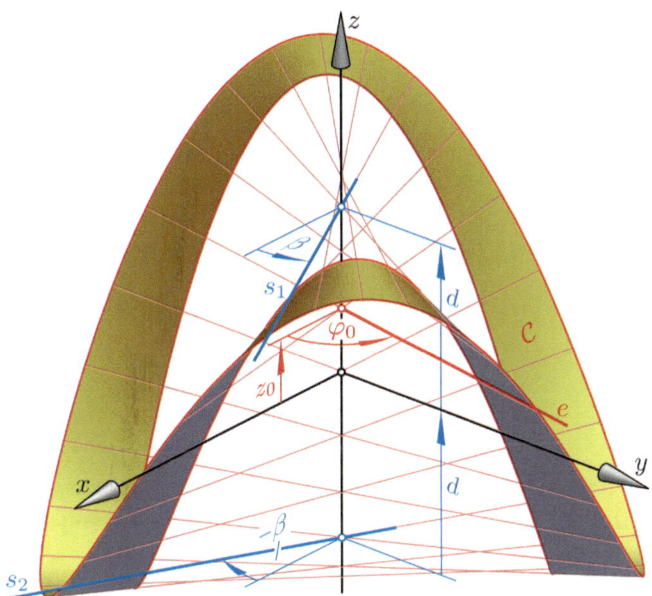

FIGURE 2.20. The axes s_1, s_2, the line e of contact, and a portion of the Plücker conoid $\mathcal{C}$.

and, furthermore,

$$(d - z_0)\cos(\beta - \varphi_0)\sin(\beta + \varphi_0) = (d + z_0)\cos(\beta + \varphi_0)\sin(\beta - \varphi_0),$$

hence

$$\frac{1}{2}(d - z_0)(\sin 2\beta + \sin 2\varphi_0) = \frac{1}{2}(d + z_0)(\sin 2\beta - \sin 2\varphi_0).$$

Thus, we end up with the equation[10]

$$d \sin 2\varphi_0 = z_0 \sin 2\beta, \quad z_0 = \frac{d}{\sin 2\beta} \sin 2\varphi_0. \tag{2.26}$$

The line of contact e can be parametrized as $(x, y, z) = (t\cos\varphi_0,\ t\sin\varphi_0,\ z_0)$, where $t \in \mathbb{R}$, $-\pi < \varphi_0 \leq \pi$ and $z_0 = z_0(\varphi_0)$ satisfies (2.26). All these lines form a cubic ruled surface called *Plücker conoid* or *cylindroid* $\mathcal{C}$ fulfilling the equation

$$\mathcal{C}\colon\ z = \frac{2d}{\sin 2\beta}\,\frac{xy}{x^2 + y^2}\,. \tag{2.27}$$

Obviously, $\mathcal{C}$ passes through the given axes s_1 and s_2 and through the x- and y-axis. The double line is part of the z-axis. The conoid $\mathcal{C}$ is bounded by the tangent planes $z = \pm d/\sin 2\beta$ along two torsal generators. By the same token, in orthogonal views, as that in Figure 2.20, the visual contour of $\mathcal{C}$, *i.e.*, the envelope of the depicted generators of $\mathcal{C}$, is affine to a hypocycloid of three cusps, called *Steiner curve* (cf. [90, p. 220]). More about the contour of Plücker's conoid can be found in Exercise 6.9.6. ■

---

[10] For an alternative proof of Theorem 2.2.4, note Exercise 2.2.2.

Remark 2.2.1 Such pairs of contacting one-sheeted hyperboloids play an important role in spatial gearing as the axodes of a uniform transmission between the rotations about skew axes.

By the same token, the complete intersection between the two contacting hyperboloids consists of the line of contact e with multiplicity 2 together with two complex conjugate generators of the f-regulus (cf. [105, p. 119–122]).

Planar sections and tangent cones of hyperboloids

Based on the standard equations of hyperboloids $\mathcal{H}$ in (2.20), we can define a symmetric bilinear form σ as

$$\sigma(\mathbf{x}_1, \mathbf{x}_2) := \frac{x_1 x_2}{a^2} \pm \frac{y_1 y_2}{b^2} - \frac{z_1 z_2}{c^2} \tag{2.28}$$

such that the equations of the hyperboloids can be expressed briefly as $\sigma(\mathbf{x}, \mathbf{x}) - 1 = 0$ with the associated asymptotic cones given by $\sigma(\mathbf{x}, \mathbf{x}) = 0$.

Similar to ellipsoids, we assign to each point P with position vector $\mathbf{p} = (x_P, y_P, z_P)$ the polar plane given by

$$\pi_P\colon\ \sigma(\mathbf{p}, \mathbf{x}) - 1 = \frac{x_P}{a^2}\, x \pm \frac{y_P}{b^2}\, y - \frac{z_P}{c^2}\, z - 1 = 0. \tag{2.29}$$

All points $Q \in \pi_P$, *i.e.*, with position vector $\mathbf{q}$ satisfying $\sigma(\mathbf{p}, \mathbf{q}) - 1 = 0$, are called *conjugate* to P w.r.t. $\mathcal{H}$, and this is a symmetric relation.

We can extend the relation of conjugacy even to points at infinity. For the point $\mathbf{q} = \lambda\mathbf{p}$, the limit $\lambda \to \infty$ yields the equation $\sigma(\mathbf{p}, \mathbf{x}) = 0$ for the polar plane. This is a plane through the center M of $\mathcal{H}$, a *diameter plane* of $\mathcal{H}$.

By analogy with (2.9), we call two lines through M with respective direction vectors $\mathbf{d}_1$ and $\mathbf{d}_2$ *conjugate diameters* of $\mathcal{H}$ when $\sigma(\mathbf{d}_1, \mathbf{d}_2) = 0$. This is equivalent to the statement that all points of one diameter are conjugate to the ideal point of the conjugate diameter. For example, the three axes of our coordinate frame are mutually conjugate w.r.t. $\mathcal{H}$. By virtue of (2.22), the self-conjugate diameters form the asymptotic cone of $\mathcal{H}$. Note that if two diameters are conjugate w.r.t. $\mathcal{H}$, they are also conjugate w.r.t. all quadrics with equations $\sigma(\mathbf{x}, \mathbf{x}) - k = 0$, $k \in \mathbb{R} \setminus \{0\}$. Moreover, they are also conjugate w.r.t. the asymptotic cone $\sigma(\mathbf{x}, \mathbf{x}) = 0$, when the bundle of lines with carrier M is seen as a projective plane (see [46, p. 180]) with the asymptotic cone being a conic.

For $P \in \mathcal{H}$, the polar plane equals the tangent plane τ_P at P to $\mathcal{H}$. Let (u, v, w) be the coefficients of x, y, z in the equation of τ_P. Then, $P \in \mathcal{H}$, *i.e.*, $\sigma(\mathbf{p}, \mathbf{p}) - 1 = 0$, is equivalent to the *dual equation* or *tangential equation*

$$a^2u^2 \pm b^2v^2 - c^2w^2 - 1 = 0 \tag{2.30}$$

of the hyperboloids given in (2.20).

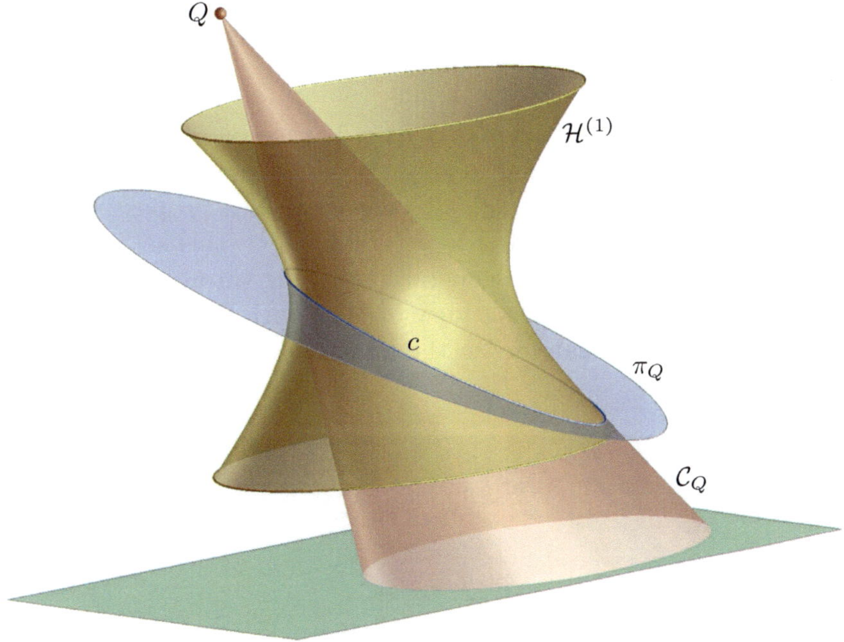

FIGURE 2.21. Pole Q, polar plane π_Q w.r.t. the hyperboloid $\mathcal{H}^{(1)}$ and the tangent cone $\mathcal{C}_Q$.

For any point $Q \in \tau_P \backslash \mathcal{H}$, the polar plane π_Q intersects $\mathcal{H}$ along the contact curve c with the tangent cone $\mathcal{C}_Q$ with apex Q (Figure 2.21). This follows from the symmetry of conjugacy, since $X \in c \subset \mathcal{H}$ is conjugate to Q if, and only if, $Q \in \tau_X$. The equation of the tangential cone, as a point set, is the same as the one given in (2.11) for ellipsoids, namely

$$\mathcal{C}_Q\colon\ [\sigma(\mathbf{q}, \mathbf{q}) - 1]\,[\sigma(\mathbf{x}, \mathbf{x}) - 1] - [\sigma(\mathbf{q}, \mathbf{x}) - 1]^2 = 0.$$

For Q at the center of $\mathcal{H}$, we obtain $\mathbf{q} = \mathbf{0}$, hence $\mathcal{C}_Q\colon \sigma(\mathbf{x}, \mathbf{x}) = 0$. This shows that the asymptotic cone equals the tangent cone $\mathcal{C}_M$ with its apex at the hyperboloid's center M. This is the spatial analogue to the fact that the asymptotes of a hyperbola are tangents at the point at infinity [46, p. 31].

Remark 2.2.2 A one-sheeted hyperboloid $\mathcal{H}^{(1)}$ separates the space intuitively into a tube-like interior and the ring-like exterior. However, one should note that, in the sense of the definition, as given above, the *'exterior'* is the complete space except $\mathcal{H}^{(1)}$. This makes sense, since, for all points $Q \notin \mathcal{H}^{(1)}$, the tangent cone $\mathcal{C}_Q$ is non-empty, because each plane connecting Q with a generator of $\mathcal{H}^{(1)}$ contacts the hyperboloid. The dual statement is also valid: Each plane ε which is not tangent to $\mathcal{H}^{(1)}$ intersects the hyperboloid along a non-empty conic. It is the set of points, where ε meets the generators of $\mathcal{H}^{(1)}$.

For an apex $Q \in \mathcal{H}^{(1)}$, *i.e.*, for $\sigma(\mathbf{q}, \mathbf{q}) = 1$, the cone $\mathcal{C}_Q$ degenerates to the tangent plane τ_Q with multiplicity two. From the dual point of view, the tangent cone $\mathcal{C}_Q$ consists of all tangent planes τ_P passing through Q. In the case of a one-sheeted hyperboloid $\mathcal{H}^{(1)}$, this set splits to two pencils of planes with the two generators passing through Q as the axes. Figure 2.22 shows a perspective view of $\mathcal{H}^{(1)}$ with the center fairly close to $\mathcal{H}^{(1)}$, in order to reveal why in the limit the depicted generators form two pencils of lines.

FIGURE 2.22. For points $Q \in \mathcal{H}^{(1)}$, the tangent cone $\mathcal{C}_Q$ splits, as a set of tangent planes, into two pencils. The depicted perspective image shows how $\mathcal{H}^{(1)}$ appears when seen from a center close to the surface. In the limit, the views of the two reguli appear as two pencils of lines.

Theorem 2.2.5 *Each non-empty intersection between the hyperboloid $\mathcal{H}$ and a non-tangential plane ε is a conic c. If the plane $\overline{\varepsilon} \parallel \varepsilon$ through the hyperboloid's center M intersects the asymptotic cone $\mathcal{C}_M$ along two lines, one line, or only at the single point M, the conic $c = \varepsilon \cap \mathcal{H}$ is a hyperbola, a parabola, or an ellipse.*

The conic c is homothetic to the intersection $\overline{c} = \overline{\varepsilon} \cap \mathcal{C}_M$. This means for $M \notin \varepsilon$ that the two conics c and $\overline{c}$ share the axes of symmetry and, in the case of an ellipse or hyperbola, the ratio of semiaxes.

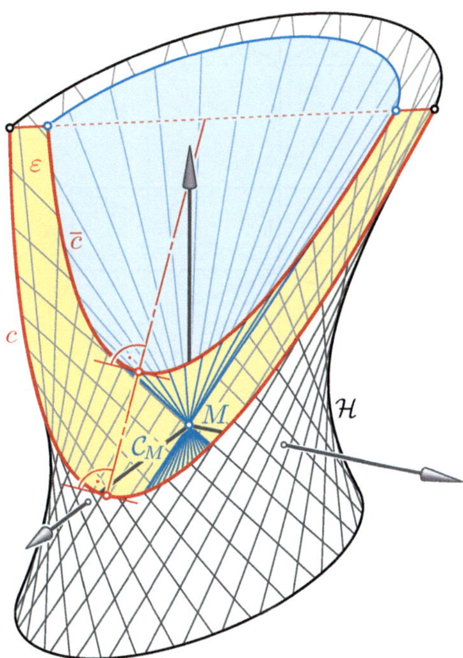

FIGURE 2.23. The intersections c and $\overline{c}$ of the hyperboloid $\mathcal{H}$ and its asymptotic cone $\mathcal{C}_M$ with the plane ε are congruent parabolas with a common axis.

Proof: The curve $\varepsilon \cap \mathcal{H}$ is of degree two and, due to the restrictions, non-degenerate — hence, a conic c (note Remark 2.2.2). By virtue of Lemma 2.1.1, which is also valid for hyperboloids, two conjugate diameters of c are also conjugate w.r.t. the asymptotic cone, since all points of one diameter are conjugate to the ideal point of the other diameter. Therefore, in the cases of an ellipse or hyperbola, the conics c and $\overline{c}$ share the involution of conjugate diameters ([46, p. 269]). In other words, the conics are homothetic. In the case of parabolas, they share the axis, since this is the unique diameter with the pole in orthogonal direction (Figure 2.23). The two parabolas c and $\overline{c}$ are even congruent. The reason is that, in the projective setting, $\mathcal{H}$ and the asymptotic cone $\mathcal{C}_M$ span a contact-pencil of quadrics which intersects ε in a pencil of hyperosculating conics (see [46, Sec. 7.3]), and two parabolas with a four-point contact at their point at infinity are congruent. ■

We learned in [46, p. 155] that each quadratic cone has circular sections. Consequently, the same must hold for hyperboloids $\mathcal{H}$ (see Figure 2.26). We find these sections also directly, similar to the proof of Theorem 2.1.1

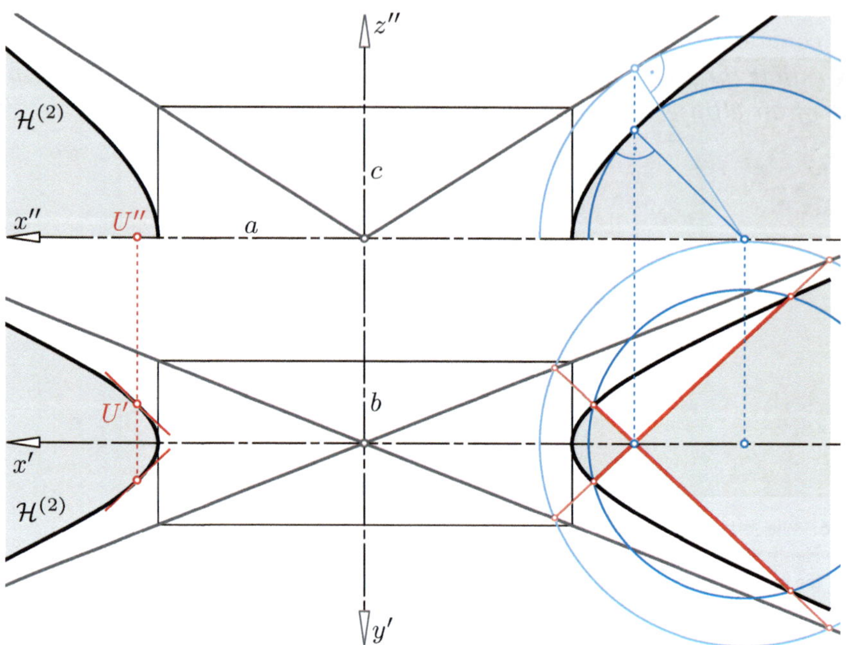

FIGURE 2.24. Finding circular sections and umbilical points U on a two-sheeted hyperboloid $\mathcal{H}^{(2)}$ in top and front view.

(note Figure 2.6), by intersecting $\mathcal{H}$ with a two-fold contacting sphere (Figures 2.24 and 2.25). Based on the standard equations (2.20) of hyperboloids, the planes of the circular sections are parallel to the diameter planes satisfying

$$\begin{aligned} &\text{for } \mathcal{H}^{(1)} \text{ with } a > b: \quad cy\sqrt{a^2 - b^2} \pm bz\sqrt{a^2 + c^2} = 0, \\ &\text{for } \mathcal{H}^{(2)} \text{ with } c > b: \quad bx\sqrt{a^2 + c^2} \pm ay\sqrt{c^2 - b^2} = 0. \end{aligned} \tag{2.31}$$

On two-sheeted hyperboloids, there exist umbilical points. They are characterized by tangent planes parallel to the planes of the circular sections. Therefore, they lie on the respective polar diameters. We obtain as coordinates of the umbilical points on $\mathcal{H}^{(2)}$ with $c > b$ (Figure 2.24)

$$\left(\pm a\frac{\sqrt{a^2 + c^2}}{\sqrt{a^2 + b^2}},\ \pm b\frac{\sqrt{c^2 - b^2}}{\sqrt{a^2 + b^2}},\ 0 \right). \tag{2.32}$$

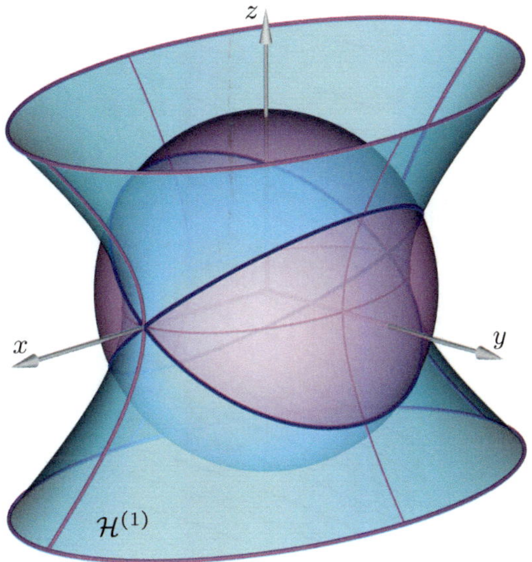

FIGURE 2.25. The intersection between the blue hyperboloid and the pink sphere splits into two circles.

Theorem 2.2.6 *There are two sets of parallel planes which intersect hyperboloids $\mathcal{H}$ along circles. On two-sheeted hyperboloids, there exist four umbilical points. Their coordinates are given in* (2.32).

For hyperboloids, there also exists a director sphere. The related theorem, as given below, is analogue to Theorem 2.1.3.

Theorem 2.2.7 *A point S is the intersection of three mutually orthogonal tangent planes of a hyperboloid $\mathcal{H}$ satisfying* (2.20) *if, and only if, the distance of S to the center M of $\mathcal{H}$ satisfies*

$$\overline{MS}^2 = a^2 \pm b^2 - c^2. \tag{2.33}$$

For the sake of brevity, a proof like the one in Exercise 2.1.4, is given in Exercise 2.2.6.

By virtue of the necessary condition given in Theorem 2.2.7, three mutually orthogonal tangent planes exist for a two-sheeted hyperboloid only if $a^2 > b^2 + c^2$.

For a one-sheeted hyperboloid $\mathcal{H}^{(1)}$, the condition $a^2 + b^2 > c^2$ is necessary. In the case $a^2 > c^2$, the director sphere given in (2.33) intersects $\mathcal{H}^{(1)}$ even

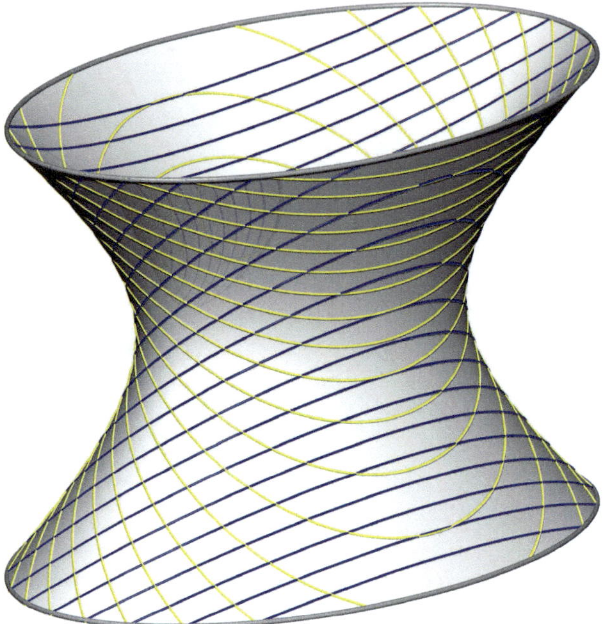

FIGURE 2.26. Hyperboloids carry two families of circular sections. Umbilical points exist only on two-sheeted hyperboloids.

along a curve $c_\perp$. For points S of this curve $c_\perp$, each of the three mutually orthogonal tangent planes must contain a generator of $\mathcal{H}^{(1)}$. Hence, three mutually orthogonal tangent planes through $S \in \mathcal{H}^{(1)}$ exist if, and only if, the two generators through S are orthogonal (see also [105, p. 116, Fig. 4.02]).

Theorem 2.2.8 *On a one-sheeted hyperboloid $\mathcal{H}^{(1)}$ with semiaxes a, b, and c, the locus of intersection points between two orthogonal generators is the curve of intersection $c_\perp$ between $\mathcal{H}^{(1)}$ and the director sphere, which is concentric with $\mathcal{H}^{(1)}$ and has the radius $\sqrt{a^2+b^2-c^2}$.*

The orthogonal projection into the $[x,y]$-plane maps the curve $c_\perp$ onto the conic with equation

$$c'_\perp: \frac{(a^2+c^2)x^2}{(a^2+b^2)a^2} + \frac{(b^2+c^2)y^2}{(a^2+b^2)b^2} = 1.$$

If a spatial polygon with sides along generators of $\mathcal{H}^{(1)}$ and right angles at each corner closes after $2n$ vertices, then it closes for each initial point on $c_\perp$. The reason behind this porism is that, in the top view, we obtain a polygon inscribed in $c'_\perp$ and circumscribed to the gorge ellipse of $\mathcal{H}^{(1)}$ (cf. [46, p. 422ff]. Figure 2.27 shows a case with closing hexagons. Here the asymptotic cone of $\mathcal{H}^{(1)}$ is equilateral (see [46, Theorem 10.1.11]). Other examples can be found in [149].

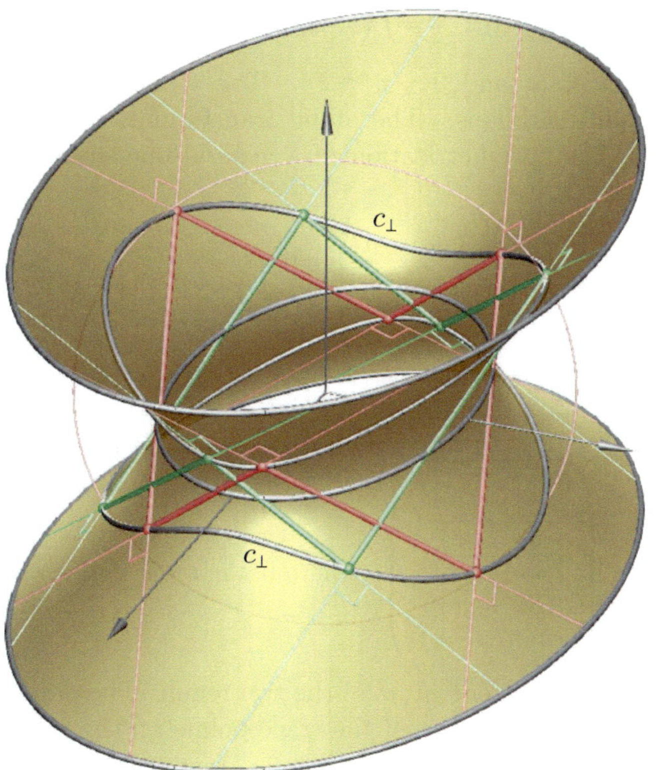

FIGURE 2.27. On this hyperboloid, each skew hexagon with sides along generators and right angles closes.

Parametrizations of hyperboloids

We refer to the standard equation of a one-sheeted hyperboloid of revolution with a as the radius of the gorge circle and with c as the secondary semiaxis,

$$\frac{x^2+y^2}{a^2}-\frac{z^2}{c^2}=1.$$

The parametrization of one branch of the meridian in the $[x,z]$-plane by $(a\cosh u,\ 0, c\sinh u)$ and the rotation about the z-axis through the angle v leads to the parametrization

$$\mathbf{x}(u,v)=\begin{pmatrix}\cos v & -\sin v & 0\\ \sin v & \cos v & 0\\ 0 & 0 & 1\end{pmatrix}\begin{pmatrix}a\cosh u\\ 0\\ c\sinh u\end{pmatrix}=\begin{pmatrix}a\cosh u\cos v\\ a\cosh u\sin v\\ c\sinh u\end{pmatrix}, \quad (2.34)$$

where $-\infty < u < \infty$ and $0 \le v < 2\pi$. The v-lines are parallel circles, the u-lines half-meridians. The affine transformation α_1, according to (2.21), yields the parametrization of a general one-sheeted hyperboloid with semiaxes a, b, c as

$$\mathcal{H}^{(1)}\colon\ \mathbf{x}(u,v)=\begin{pmatrix}a\cosh u\cos v\\ b\cosh u\sin v\\ c\sinh u\end{pmatrix}\ \text{for}\ -\infty<u<\infty,\ 0\le v<2\pi. \quad (2.35)$$

The v-lines are ellipses in planes z = const.. The u-lines are branches of hyperbolas in planes through the z-axis.

In the following parametrization of the one-sheeted hyperboloid $\mathcal{H}^{(1)}$,

$$\mathbf{x}(u,v)=\begin{pmatrix}\dfrac{a\cos(u+v)}{\cos(u-v)}\\ \dfrac{b\sin(u+v)}{\cos(u-v)}\\ c\tan(u-v)\end{pmatrix},\ \text{where}\ 0\le u,v\le\pi\ \text{and}\ u\ne v, \quad (2.36)$$

the u- and v-lines are generators of the two reguli. The parameters u and v are half of the polar angles of the trace points on the gorge ellipse.

We find the same net of parameter lines in the following parametrization, which is even rational,

$$(u,v)\in(\mathbb{R}\cup\{\infty\})^2\ \mapsto\ \mathbf{x}(u,v)=\begin{pmatrix}\dfrac{a(1-uv)}{1+uv}\\ \dfrac{b(u+v)}{1+uv}\\ \dfrac{c(u-v)}{1+uv}\end{pmatrix} \quad (2.37)$$

with $\mathbf{x}(\infty,v)=\lim_{u\to\infty}\mathbf{x}(u,v)$, $\quad \mathbf{x}(u,\infty)=\lim_{v\to\infty}\mathbf{x}(u,v)$.

It is left as an exercise to prove that in this case the parameter curves are lines (Exercise 2.2.1). For readers who are familiar with the projective extension of $\mathbb{E}^3$ and its properties this is obvious, since the homogeneous coordinates are linear in u and v.

Similar to (2.34), we obtain a parametrization of one sheet of the two-sheeted hyperboloid $\mathcal{H}^{(2)}$ given in (2.20) as

$$\mathbf{x}(u,v) = \begin{pmatrix} a\cosh u \\ b\sinh u\cos v \\ c\sinh u\sin v \end{pmatrix} \text{ for } 0 \le v < 2\pi,\ -\infty < v < \infty. \tag{2.38}$$

The v-lines are ellipses in planes $x = \text{const.}$. The u-lines are branches of hyperbolas in planes through the x-axis.

The half-angle substitutions $\widetilde{u} := \tanh\frac{u}{2}$, $\widetilde{v} := \tan\frac{v}{2}$ (compare with (2.18)) together with

$$\sinh u = \frac{2\widetilde{u}}{1-\widetilde{u}^2},\quad \cosh u = \frac{1+\widetilde{u}^2}{1-\widetilde{u}^2},\quad \sin v = \frac{2\widetilde{v}}{1+\widetilde{v}^2},\quad \cos v = \frac{1-\widetilde{v}^2}{1+\widetilde{v}^2}$$

lead to rational parametrizations of one shell of $\mathcal{H}^{(2)}$,

$$\mathbf{x}(\widetilde{u},\widetilde{v}) = \left(\frac{a(1+\widetilde{u}^2)}{1-\widetilde{u}^2}, \frac{2b\widetilde{u}(1-\widetilde{v}^2)}{(1-\widetilde{u}^2)(1+\widetilde{v}^2)}, \frac{4c\widetilde{u}\widetilde{v}}{(1-\widetilde{u}^2)(1+\widetilde{v}^2)}\right), \ (\widetilde{u},\widetilde{v}) \in \mathbb{R}^2,$$

as well as for the one-sheeted hyperboloid $\mathcal{H}^{(1)}$ as

$$\mathbf{x}(\widetilde{u},\widetilde{v}) = \left(\frac{a(1+\widetilde{u}^2)(1-\widetilde{v}^2)}{(1-\widetilde{u}^2)(1+\widetilde{v}^2)}, \frac{2b(1+\widetilde{u}^2)\widetilde{v}}{(1-\widetilde{u}^2)(1+\widetilde{v}^2)}, \frac{2c\widetilde{u}}{1-\widetilde{u}^2}\right), \ (\widetilde{u},\widetilde{v}) \in \mathbb{R}^2.$$

The parameterizations

$$\mathbf{x}(u,v) = \begin{pmatrix} a\cos v \\ b\sin v \\ 0 \end{pmatrix} \pm \begin{pmatrix} -a\sin v \\ b\cos v \\ c \end{pmatrix} u$$

underline that $\mathcal{H}^{(1)}$ is a ruled surface, even with two reguli.

- **Exercise 2.2.1** Parametrization of the reguli on $\mathcal{H}^{(1)}$.

Why are the parameter curves of the parametrization (2.37) straight lines?

Hint: Verify that the parameter line $v = v_0 = \text{const.}$ can be rewritten as

$$\mathbf{x}(u,v_0) = \begin{pmatrix} \dfrac{a(1-v_0^2)}{1+v_0^2} \\ \dfrac{2bv_0}{1+v_0^2} \\ 0 \end{pmatrix} + \lambda(u) \begin{pmatrix} \dfrac{-2av_0}{1+v_0^2} \\ \dfrac{b(1-v_0^2)}{1+v_0^2} \\ c \end{pmatrix} \text{ with } \lambda(u) = \frac{u-v_0}{1+uv_0}.$$

A similar representation for rulings of the second regulus follows after exchanging u with v and changing the sign of the z-coordinate.

- **Exercise 2.2.2** Alternative proof of Theorem 2.2.4.

Prove Theorem 2.2.4 by applying the condition that, at points $P \in e$, the surface normals n_P to the two hyperboloids must be aligned. This means that each plane orthogonal to e must intersect the lines s_1, s_2, and e at three collinear points.

- **Exercise 2.2.3** Altitudes of a tetrahedron and equilateral hyperboloid.

Prove the following statement: *The four altitudes of a tetrahedron are either concurrent or located on a one-sheeted hyperboloid whose asymptotic cone is equilateral* ([46, p. 459]).

Hint: For each triangular face, check the perpendicular through the triangle's orthocenter. For further details, see [55].

Tetrahedra with concurrent altitudes are called *orthocentric* and characterized by (at least) two pairs of orthogonal opposite edges. A hyperboloid with an equilateral asymptotic cone is called *equilateral hyperboloid.* In this case, each plane orthogonal to any generator intersects the hyperboloid along an equilateral hyperbola or two orthogonal lines (see [115, p. 293, Ex. 19]).

- **Exercise 2.2.4** A spatial version of Thales's theorem.

Prove the following statement (Figure 2.28): *Let two skew and non-orthogonal lines e_1 and e_2 be given. Then, the set of lines f which intersect e_1 and e_2 such that the planes connecting f with e_1 and e_2 are orthogonal is a regulus on a one-sheeted hyperboloid $\mathcal{H}^{(1)}$ with a normal asymptotic cone* (see [46, p. 459]).

What is the set of lines in the case of orthogonality of e_1 and e_2?

Hint: Choose the common perpendicular of e_1 and e_2 as x-axis and their axes of symmetry as y- and z-axis such that

$$e_1\colon \begin{pmatrix} a \\ 0 \\ 0 \end{pmatrix} + \mathbb{R}\begin{pmatrix} 0 \\ \cos\alpha \\ \sin\alpha \end{pmatrix}, \quad e_2\colon \begin{pmatrix} -a \\ 0 \\ 0 \end{pmatrix} + \mathbb{R}\begin{pmatrix} 0 \\ \cos\alpha \\ -\sin\alpha \end{pmatrix}, \quad 0 < \alpha < \frac{\pi}{2}.$$

Then, the hyperboloid $\mathcal{H}^{(1)}$ satisfies the equation

$$\frac{x^2}{a^2} + \frac{y^2}{a^2(1-\cot^2\alpha)} - \frac{z^2}{a^2(\tan^2\alpha - 1)} = 1.$$

- **Exercise 2.2.5** Circular sections of a hyperboloid.

Referring to Figure 2.29, explain the depicted construction of the second family of circular sections of a hyperboloid, provided that a circular section of the first family is given.

- **Exercise 2.2.6** Proof of Theorem 2.2.7.

Three mutually orthogonal tangent planes of the hyperboloid $\mathcal{H}$ pass through S if, and only if, the tangent cone $\mathcal{C}_S$ is dual-equilateral (see Exercise 2.1.4). Therefore, by virtue of [46, p. 459ff], the tangential equation of this cone — dual to [46, (10.13)] — must have a zero trace. Show that this trace equals $(a^2 - x_S^2) \pm (b^2 - y_S^2) - (z^2 - z_S^2)$. Note the hint given in Exercise 2.1.4.

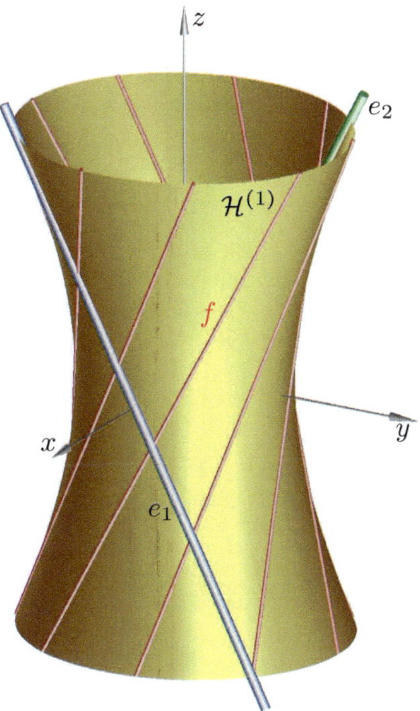

FIGURE 2.28. A spatial version of Thales's Theorem: The two planes connecting $f \subset \mathcal{H}^{(1)}$ with e_1 and e_2 are orthogonal.

• **Exercise 2.2.7** Pascal's Theorem by spatial interpretation.

Let six given points $T_1, \ldots, T_6$ of a conic be interpreted as the trace points of the sides $e_1, f_2, e_3, f_1, e_2, f_3$ formed by three e- and three f-generators of a one-sheeted hyperboloid $\mathcal{H}^{(1)}$. Prove Pascal's Theorem [46, p. 220] by studying the traces of the tangent planes to $\mathcal{H}^{(1)}$ at the vertices of the skew hexagon with the given sides ([144, II, p. 25]).

• **Exercise 2.2.8** Brianchon's Theorem by spatial interpretation.

Any six given tangent lines of a conic can be interpreted as a projection of three e- and three f-generators $e_1, f_2, e_3, f_1, e_2, f_3$ of a one-sheeted hyperboloid $\mathcal{H}^{(1)}$. Prove Brianchon's theorem [46, p. 222] by studying the tangent planes at the intersection points $e_i \cap f_i$ between opposite sides of the skew hexagon on $\mathcal{H}^{(1)}$ ([144, II, pp. 25–26]).

• **Exercise 2.2.9** Hyperboloids of revolution through a skew quadrangle.

Let $X_1 \ldots X_4$ be a skew quadrangle with side lengths $s_1, \ldots, s_4$ satisfying the conditions (a) or (b), as listed in (2.25). How many hyperboloids of revolution are passing through the sides of this quadrangle?

Hint: Note that the results differ between the cases (a) and (b). See also Section 8.3.

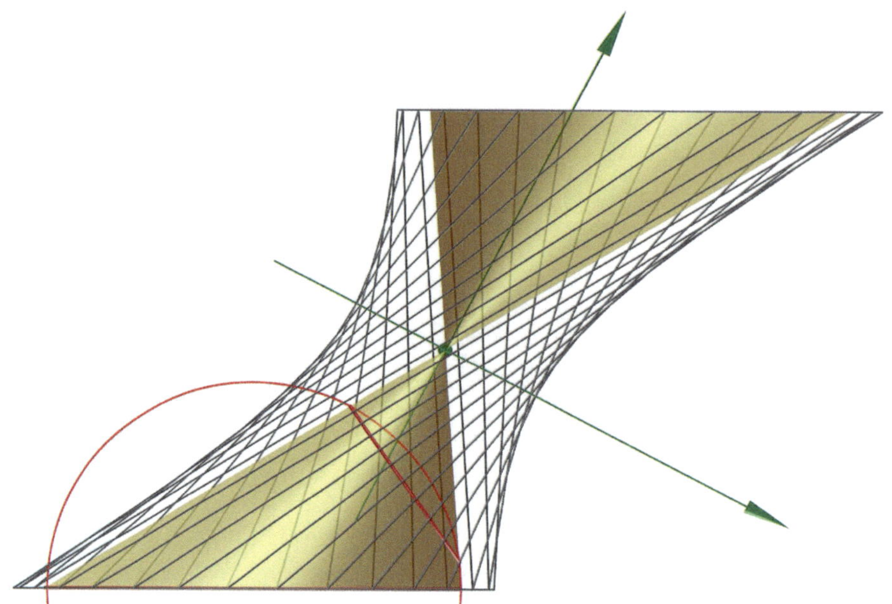

FIGURE 2.29. One-sheeted hyperboloid and circular sections.

2.3 Paraboloids

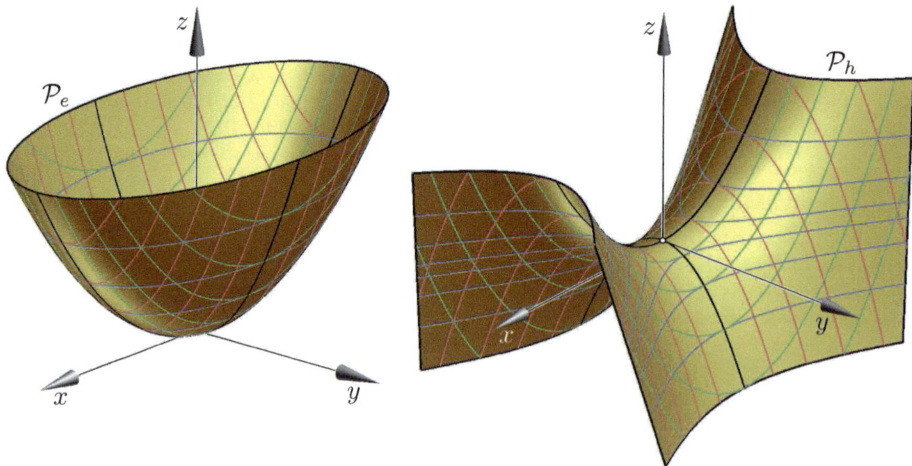

FIGURE 2.30. Elliptic and hyperbolic paraboloid.

Among the quadrics in $\mathbb{E}^3$ (Table 2.1 on page 9), there are two types of paraboloids, the elliptic paraboloid $\mathcal{P}_e$ and the hyperbolic paraboloid $\mathcal{P}_h$ with the equations

$$\mathcal{P}_e\colon\ \frac{x^2}{a^2}+\frac{y^2}{b^2}-2z=0 \quad\text{and}\quad \mathcal{P}_h\colon\ \frac{x^2}{a^2}-\frac{y^2}{b^2}-2z=0, \tag{2.39}$$

respectively (Figure 2.30). The different names correspond to the types of the intersection curves with the level planes $z=k=$ const., $k\neq 0$, as well as to the types of surface points [46, p. 120ff]. The elliptic paraboloid with $a=b$ is a paraboloid of revolution; the hyperbolic paraboloid with $a=b$ is called *equilateral* or *orthogonal.*

Referring to (2.39), the origin is the vertex and the z-axis is the axis of the paraboloids $\mathcal{P}_e$ and $\mathcal{P}_h$. The sections with the $[x,z]$-plane and the $[y,z]$-plane, the principal sections of the paraboloids, are parabolas (Figure 2.30). The paraboloids in (2.39) remain congruent when the term $-2z$ is replaced by $+2z$.

The ruled paraboloid

Though the equations of the two types $\mathcal{P}_e$ and $\mathcal{P}_h$ of paraboloids in (2.39) are similar, there is one fundamental difference: The surface $\mathcal{P}_h$ is a ruled surface.

Theorem 2.3.1 *The hyperbolic paraboloid $\mathcal{P}_h$ carries two families of generators, called reguli. At each point $X\in\mathcal{P}_h$, two generators meet, and they span the tangent plane τ_X to $\mathcal{P}_h$ at X.*

Generators of the same regulus are mutually skew and located in parallel planes. The points of intersection between the generators $e_1, e_2, \dots$ of one regulus and two generators f and f' of the other regulus define an affine transformation $f\to f'$ between the point ranges. Conversely, for any affine transformation between two skew point ranges, the lines connecting corresponding points cover a hyperbolic paraboloid.

Proof: (a) The common points of the hyperbolic paraboloid

$$\mathcal{P}_h\colon\ 2z=\left(\frac{x^2}{a^2}-\frac{y^2}{b^2}\right)=\left(\frac{x}{a}+\frac{y}{b}\right)\left(\frac{x}{a}-\frac{y}{b}\right)$$

and the plane $\varepsilon(k)\colon\ \frac{x}{a}+\frac{y}{b}=k$, $k=$ const. are also located in a second plane satisfying

$$2z=k\left(\frac{x}{a}-\frac{y}{b}\right),$$

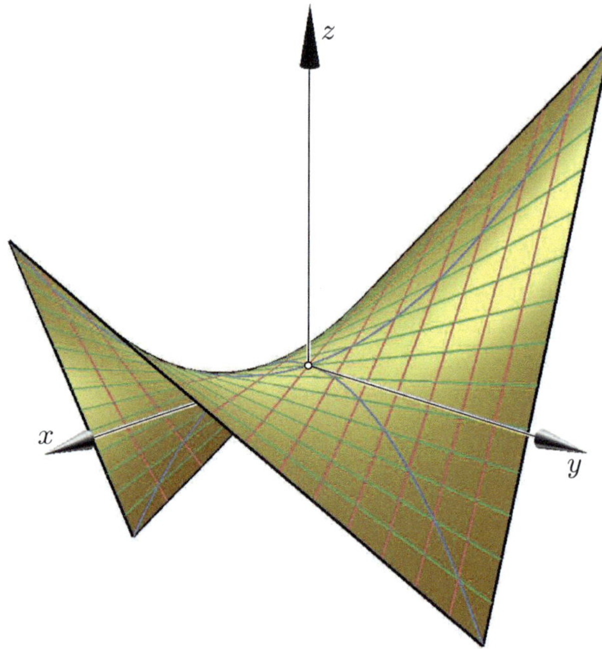

FIGURE 2.31. The hyperbolic paraboloid carries two reguli.

which is not parallel to $\varepsilon(k)$. Therefore, the two planes intersect along a straight line.[11] By variation of k, we obtain the first family of generators located in parallel planes $\varepsilon(k)$.

Analogously, the planes $\varphi(k): \frac{x}{a} - \frac{y}{b} = k$ intersect the surface $\mathcal{P}_h$ along generators of the second family.

Similar to the one-sheeted hyperboloid, generators of the same regulus are mutually skew. Two generators taken from different families are coplanar and, therefore, intersecting. Because of the parallelity of the planes $\varepsilon(k)$ as well as that of the set $\varphi(k)$, any three generators e_1, e_2, e_3 of one regulus intersect any two generators f, f' of the other regulus at points $E_1, \ldots$ and $E_1', \ldots$, respectively, such that the affine ratios are equal, *i.e.*, $\mathrm{ar}(E_1, E_2, E_3) = \mathrm{ar}(E_1', E_2', E_3')$, where $|\mathrm{ar}(E_1, E_2, E_3)| = \overline{E_1E_3} : \overline{E_2E_3}$ (note [46, p. 194]). We speak of *affine ranges* f and f' of points.

(b) Conversely, let the affinity between two ranges f and f' be given by two pairs of corresponding points, $A \mapsto A'$ and $B \mapsto B'$. We assign to the skew quadrangle $ABB'A'$ a Cartesian coordinate frame in such a way that the four sides satisfy one of the two equations

$$\frac{x^2}{a^2} - \frac{y^2}{b^2} \mp 2z = 0. \tag{2.40}$$

with appropriate parameters a and b. Then, according to the first part of this proof, all lines connecting corresponding points of the affine ranges satisfy (2.40).

[11] In the projective extension of the 3-space, the complete intersection $\mathcal{P}_h \cap \varepsilon(k)$ also includes a line at infinity.

Firstly, the connection of the midpoints of AB' and $A'B$ defines the direction of the z-axis. As a consequence, the orthogonal projection parallel to the z-axis into the $[x,y]$-plane maps $ABB'A'$ onto a parallelogram. Secondly, the angle bisector at A of this parallelogram defines the direction of the x-axis. Then, we can suppose that f and f' are parallel to the planes $\varepsilon(k)$ and the sides AA' and BB' parallel to the planes $\varphi(k)$, as used above. This defines the coordinate frame up to a translation. This means that the coordinates (x_A, y_A, z_A) of A are still unknown, while the coordinates of B, A' and B' relative to A are given. Hence, we can set

$$\begin{pmatrix} x_B \\ y_B \\ z_B \end{pmatrix} = \begin{pmatrix} x_A - a\lambda \\ y_A + b\lambda \\ z_B \end{pmatrix}, \quad \begin{pmatrix} x_{A'} \\ y_{A'} \\ z_{A'} \end{pmatrix} = \begin{pmatrix} x_A + a\mu \\ y_A + b\mu \\ z_{A'} \end{pmatrix}, \quad \begin{pmatrix} x_{B'} \\ y_{B'} \\ z_{B'} \end{pmatrix} = \begin{pmatrix} x_A + a(\mu - \lambda) \\ y_A + b(\mu + \lambda) \\ z_{B'} \end{pmatrix}$$

in terms of further unknowns λ, μ. This yields

$$-a\lambda = x_B - x_A, \quad b\lambda = y_B - y_A, \quad a\mu = x_{A'} - x_A.$$

The request that B, A' and B' satisfy (2.40) yields

$$z_B - z_A = \mp\left(\frac{x_A}{a} + \frac{y_A}{b}\right), \; z_{A'} - z_A = \pm\left(\frac{x_A}{a} - \frac{y_A}{b}\right), \; z_{B'} - z_A = \mp\left(\frac{x_A}{a} + \frac{y_A}{b}\right) \mp \left(\frac{x_A}{a} + \frac{y_A}{b}\right) \mp 2\lambda\mu.$$

After some computations, we end up with

$$\mp a^2\left[(z_B - z_A) + (z_{B'} - z_A) - (z_{B'} - z_A)\right] = 2(x_B - x_A)(x_{A'} - x_A).$$

The term in brackets on the left-hand side cannot vanish, since the quadrangle $ABB'A'$ is skew. Thus, we find $a > 0$ and the correct sign in (2.40). Consequently, λ, μ, b, x_A, and y_A are uniquely defined as well as the z-coordinates of B, A' and B' in terms of z_A, which finally follows from (2.40). This confirms that the four sides of the quadrangle can be placed on an appropriate hyperbolic paraboloid.

In Chapter 4, we will learn that, in the projective setting, for each projectivity between two skew ranges f and f', the lines connecting corresponding points constitute a regulus. ■

By the same token, the one-sheeted hyperboloid and the hyperbolic paraboloid are the only irreducible algebraic surfaces of degree > 1 which carry two one-parameter families of straight lines.

The generators passing through the vertex A of $\mathcal{P}_h$ and spanning the tangent plane $\tau_A \colon z = 0$ are called *vertex generators*. In the case $a = b$ of an orthogonal paraboloid, all generators of one regulus are orthogonal to the vertex generator of the other regulus.

Corollary 2.3.1 *Given a skew (= non-planar) quadrilateral in $\mathbb{E}^3$, there exists a unique hyperbolic paraboloid passing through the four sides.*

Proof: Let $\mathbf{a}\,\mathbf{b}\,\mathbf{c}\,\mathbf{d}$ be the sequence of vertices of the given quadrilateral. By virtue of Theorem 2.3.1, we select opposite sides, *e.g.*, $\mathbf{ab}$ and $\mathbf{cd}$. Then, we subdivide them uniformly by $\mathbf{p}(u) = u\,\mathbf{a} + (1-u)\mathbf{b}$ and $\mathbf{q}(u) = u\,\mathbf{d} + (1-u)\mathbf{c}$ and connect, for each $u \in [0,1]$, the corresponding points $\mathbf{p}(u)$ and $\mathbf{q}(u)$ by a straight line segment.

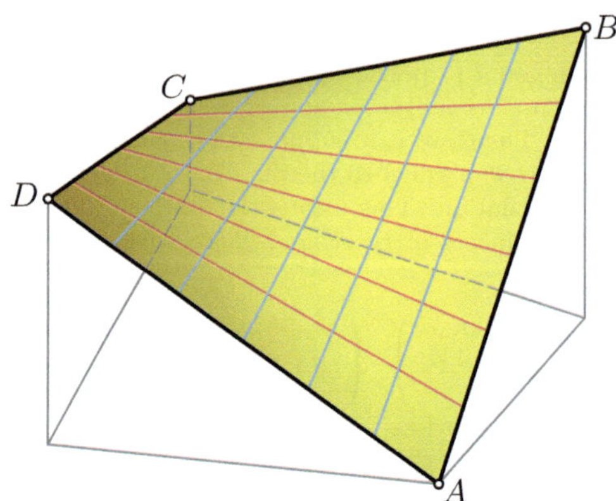

FIGURE 2.32. Hyperbolic paraboloid through a skew quadrilateral $ABCD$.

The following parametrization confirms again that both pairs of opposite sides define the same surface:

$$\begin{aligned}\mathbf{x}(u,v) &= v\,[u\,\mathbf{a}+(1-u)\,\mathbf{b}]+(1-v)\,[u\,\mathbf{d}+(1-u)\,\mathbf{c}]\\ &= uv\,\mathbf{a}+(1-u)v\,\mathbf{b}+(1-u)(1-v)\,\mathbf{c}+u(1-v)\,\mathbf{d}\\ &= u\,[v\,\mathbf{a}+(1-v)\,\mathbf{d}]+(1-u)\,[v\,\mathbf{b}+(1-v)\,\mathbf{c}]\end{aligned} \tag{2.41}$$

for $u,v \in [0,1]$. Therefore, the hyperbolic paraboloid is uniquely defined. ■

We continue with two comments on this filling hyperbolic paraboloid.

(1) The second line in (2.41) reveals that the filling paraboloid is a particular *Bézier tensor-product surface* (cf. [63]).

(2) The generators u = const. are parallel to the plane spanned by $\mathbf{a}-\mathbf{d}$ and $\mathbf{b}-\mathbf{c}$. Those of the second regulus v = const. are parallel to the plane spanned by $\mathbf{a}-\mathbf{b}$ and $\mathbf{c}-\mathbf{d}$. The axis is parallel to both planes, and hence orthogonal to $(\mathbf{a}-\mathbf{d})\times(\mathbf{b}-\mathbf{c})$ and $(\mathbf{a}-\mathbf{b})\times(\mathbf{c}-\mathbf{d})$. Thus, we can confirm that the vector

$$\mathbf{s} := \mathbf{a}-\mathbf{b}+\mathbf{c}-\mathbf{d} \tag{2.42}$$

is parallel to the axis of the hyperbolic paraboloid. This vector cannot vanish for a given skew quadrilateral, and it has the direction of the line which connects the midpoints of the diagonals AC and BD. Furthermore, based on the parametrization (2.41), the point $\mathbf{x}(u,v)$ with

$$u = \frac{\langle \mathbf{c}-\mathbf{b},\, \mathbf{s}\rangle}{\|\mathbf{s}\|^2}, \quad v = \frac{\langle \mathbf{c}-\mathbf{d},\, \mathbf{s}\rangle}{\|\mathbf{s}\|^2} \tag{2.43}$$

is the vertex of the paraboloid.

In the example below, we show how to find the filling hyperbolic paraboloid by means of a graphical construction.

■ **Example 2.3.1** Hyperbolic paraboloid filling a skew quadrangle

Given: Skew quadrangle $ABCD$ in front and top view (Figure 2.33). The top view $A'B'C'D'$ is a parallelogram (note Exercise 2.3.1).

Wanted: Generators and contour of the paraboloid $\mathcal{P}_h$ with generators $[A,B]$, $[C,D]$ of the e-regulus and $[A,C]$, $[B,D]$ of the f-regulus.

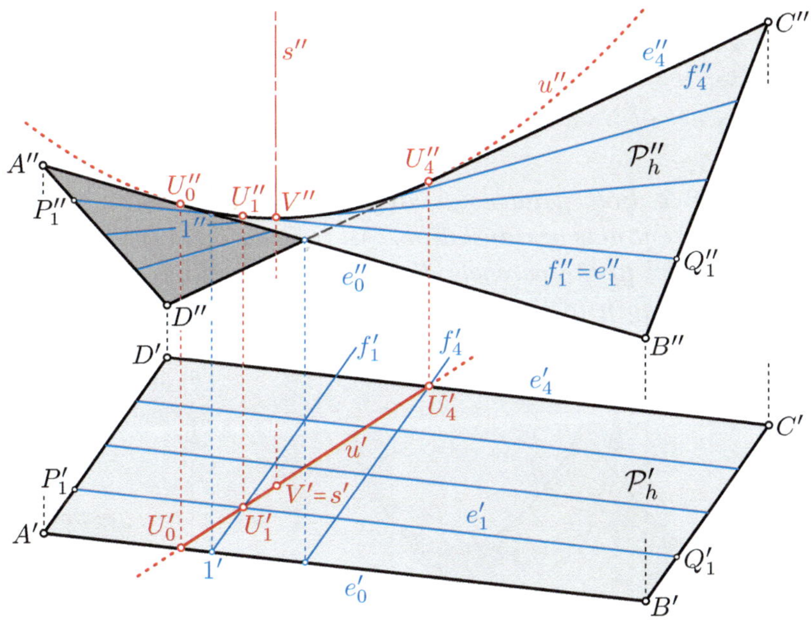

FIGURE 2.33. Hyperbolic paraboloid filling the given skew quadrilateral $ABCD$; the top view of this quadrilateral is a parallelogram (note Example 2.3.1.)

We uniformly subdivide the opposite sides AD and BC of the given quadrilateral into the same number n of points $P_0 = A$, P_1, ..., $P_n = D$ and $Q_0 = B$, Q_1, ..., $Q_n = C$. Then, the connecting lines $e_i = [P_i, Q_i]$ for $i = 1, \dots, n-1$ are generators of the first regulus of $\mathcal{P}_h$. Their top views are parallel.

In the front view, the paraboloid's contour point of e_1 is the front view of the point of intersection between e_1 and the generator f_1 taken from the second regulus, but with coinciding front views $e_1'' = f_1''$. The generator f_1 intersects the lines $[A,B]$ at point 1. All contour points belong to a parabola u, which appears as the line u' in the top view. The vertex V of the parabola u is the vertex of $\mathcal{P}_h$ with the front view V'' being the vertex of u''. The axis s of $\mathcal{P}_h$ is vertical.

Parametrizations of paraboloids

Both paraboloids share a property related to the following:

Definition 2.3.1 A surface $\mathcal{T}$ is called *surface of translation* if it can be parametrized as

$$\mathbf{x}(u,v) = \mathbf{c}(u) + \mathbf{d}(v), \quad (u,v) \in I \times J, \; I, J \subset \mathbb{R}.$$

This surface is swept by the curve $\mathbf{c}(u)$, $u \in I$, under the translations $\mathbf{x} \mapsto \mathbf{x}' = \mathbf{x} + \mathbf{d}(v)$, $v \in J$. However, $\mathcal{T}$ is also swept by the curve $\mathbf{d}(v)$, $v \in J$, under $\mathbf{x} \mapsto \mathbf{x}'' = \mathbf{x} + \mathbf{c}(u)$, $u \in I$. The u- and v-curves together constitute a *net of translation* on the surface $\mathcal{T}$.

Theorem 2.3.2 *Both paraboloids are surfaces of translation with a continuum of different nets of translation. All translation curves are parabolas with axes parallel to the paraboloid's axis. At the elliptic paraboloid, the u- and v-curves open to the same side; at the hyperbolic paraboloid these parabolas open in opposite directions.*

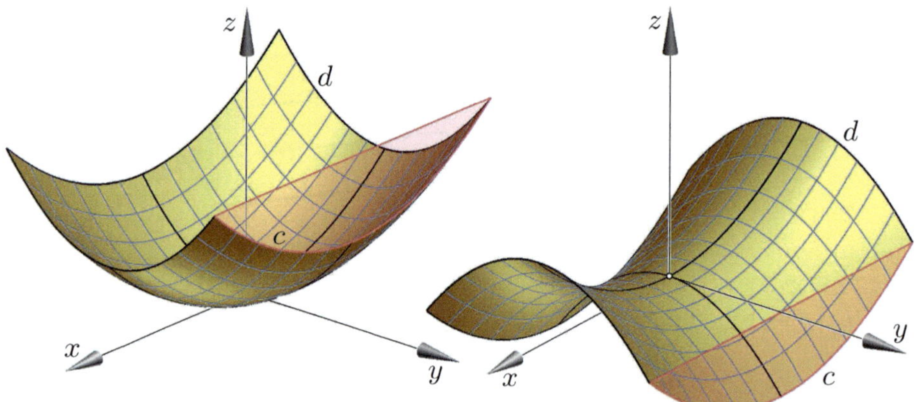

FIGURE 2.34. Both paraboloids are surfaces of translation.

Proof: The intersection of the paraboloid (2.39) with the plane $x = u =$ const. satisfies

$$\frac{y^2}{b^2} = \pm\left(2z - \frac{u^2}{a^2}\right).$$

We obtain this curve when the principal section $u = 0$ in the plane $x = 0$ is transformed by the translation

$$\begin{pmatrix} 0 \\ y \\ z \end{pmatrix} \mapsto \begin{pmatrix} 0 \\ y \\ z \end{pmatrix} + \begin{pmatrix} u \\ 0 \\ -\frac{u^2}{2a^2} \end{pmatrix}.$$

The vertices $(u,\, 0,\, u^2/2a^2)$ of all translated parabolas trace the principal section in $y = 0$ satisfying $z = x^2/2a^2$. The axes of all these parabolas are parallel to the z-axis.

The same holds true for the sections with the planes $y = v = \text{const}$. When replacing (x, y) in (2.39) with (u, v), we obtain for both paraboloids the parametrization

$$\mathbf{x}(u,v) = \begin{pmatrix} u \\ v \\ \frac{u^2}{2a^2} \pm \frac{v^2}{2b^2} \end{pmatrix} = \begin{pmatrix} u \\ 0 \\ \frac{u^2}{2a^2} \end{pmatrix} + \begin{pmatrix} 0 \\ v \\ \pm\frac{v^2}{2b^2} \end{pmatrix} = \mathbf{c}(u) + \mathbf{d}(v), \ (u,v) \in \mathbb{R}^2. \tag{2.44}$$

In order to find other translation curves on the paraboloids, we use the linear mapping

$$l(t)\colon\ \mathbb{R}^3 \to \mathbb{R}^3,\ \begin{pmatrix} x \\ y \\ z \end{pmatrix} \mapsto \begin{pmatrix} x' \\ y' \\ z' \end{pmatrix} = \begin{pmatrix} \cos t & -\frac{a}{b}\sin t & 0 \\ \frac{b}{a}\sin t & \cos t & 0 \\ 0 & 0 & 1 \end{pmatrix} \begin{pmatrix} x \\ y \\ z \end{pmatrix} \ \text{for } 0 \le t < \pi,$$

with the inverse

$$\begin{pmatrix} x' \\ y' \\ z' \end{pmatrix} \mapsto \begin{pmatrix} x \\ y \\ z \end{pmatrix} = \begin{pmatrix} \cos t & \frac{a}{b}\sin t & 0 \\ -\frac{b}{a}\sin t & \cos t & 0 \\ 0 & 0 & 1 \end{pmatrix} \begin{pmatrix} x' \\ y' \\ z' \end{pmatrix}.$$

The mappings $l(t)$ represent a one-parameter set of affine transformations. Each of them maps the elliptic paraboloid onto itself because

$$\frac{x^2}{a^2} + \frac{y^2}{b^2} - 2z = \frac{(x'\cos t + y'\frac{a}{b}\sin t)^2}{a^2} + \frac{(-x'\frac{b}{a}\sin t + y'\cos t)^2}{b^2} - 2z' = \frac{x'^2}{a^2} + \frac{y'^2}{b^2} - 2z'.$$

By virtue of the linearity of $l(t)$, we have $l(t)(\mathbf{c} + \mathbf{d}) = l(t)(\mathbf{c}) + l(t)(\mathbf{d})$. Hence, for each $t \in \mathbb{R}$, the elliptic paraboloid $\mathcal{P}_e$ can also be parametrized as

$$\mathbf{x}(u,v) = l(t)\,(\mathbf{c}) + l(t)\,(\mathbf{d}) = \begin{pmatrix} u\cos t \\ \frac{b}{a}\, u\sin t \\ \frac{u^2}{2a^2} \end{pmatrix} + \begin{pmatrix} -\frac{a}{b}\, v\sin t \\ v\cos t \\ \frac{v^2}{2b^2} \end{pmatrix}. \tag{2.45}$$

Analogously, the hyperbolic paraboloid is mapped onto itself by

$$l(t)\colon\ \mathbb{R}^3 \to \mathbb{R}^3,\ \begin{pmatrix} x \\ y \\ z \end{pmatrix} \mapsto \begin{pmatrix} x' \\ y' \\ z' \end{pmatrix} = \begin{pmatrix} \cosh t & -\frac{a}{b}\sinh t & 0 \\ -\frac{b}{a}\sinh t & \cosh t & 0 \\ 0 & 0 & 1 \end{pmatrix} \begin{pmatrix} x \\ y \\ z \end{pmatrix} \ \text{for } t \in \mathbb{R}.$$

Thus, we obtain for each $t \in \mathbb{R}$ another net of translation curves

$$\mathbf{x}(u,v) = \begin{pmatrix} u\cosh t \\ -\frac{b}{a}\, u\sinh t \\ \frac{u^2}{2a^2} \end{pmatrix} + \begin{pmatrix} -\frac{a}{b}\, v\sinh t \\ v\cosh t \\ -\frac{v^2}{2b^2} \end{pmatrix} \tag{2.46}$$

on the hyperbolic paraboloid $\mathcal{P}_h$. ■

We recall that (2.41) with $(u, v) \in \mathbb{R}^2$ provides a parametrization of the hyperbolic paraboloid passing through the sides of given quadrangle **abcd** with the parameter net consisting of the two reguli. The same net of parameter lines shows up at the parametrization

$$\mathbf{x}(u,v) = \begin{pmatrix} a(u+v) \\ b(u-v) \\ 2uv \end{pmatrix}, \quad (u,v) \in \mathbb{R}^2 \tag{2.47}$$

of the hyperbolic paraboloid $\mathcal{P}_h$ in (2.39).

Tangent planes and tangent cones of paraboloids

The implicit equations of the two paraboloids in (2.39) can be rewritten as

$$F(\mathbf{x}) := \sigma(\mathbf{x},\mathbf{x}) - 2z = 0 \quad \text{with} \quad \sigma(\mathbf{x},\mathbf{x}) = \frac{x^2}{a^2} \pm \frac{y^2}{b^2}\,.$$

Since the gradient of F,

$$\mathbf{grad}\,F(\mathbf{x}) = 2\left(\frac{x}{a^2},\ \pm\frac{y}{b^2},\ -1\right),$$

defines the direction of the surface normal at any point $\mathbf{x}$ on the paraboloid, we obtain the equation of the tangent plane τ_P at the point P with position vector $\mathbf{p} = (x_P,\, y_P,\, z_P)$ on the paraboloid, *i.e.*, with $\sigma(\mathbf{p},\mathbf{p}) = 2z_P$, as

$$\sigma(\mathbf{p},\mathbf{x}) - (z + z_P) = \frac{x_P}{a^2}\,x \pm \frac{y_P}{b^2}\,y - z - z_P = 0. \tag{2.48}$$

In the case $z_P \neq 0$, we can rewrite this equation as

$$ux + vy + wz - 1 = 0.$$

Then, analogously to (2.8), the coefficients (u, v, w) satisfy the paraboloid's *dual* or *tangential equation*

$$a^2u^2 \pm b^2v^2 + 2w = 0. \tag{2.49}$$

Since the coefficient of z in (2.48) is -1, tangent planes of the paraboloid $\mathcal{P}$ are never parallel to the z-axis.

Similar to ellipsoids, we use (2.48) to define the *polar plane* of the *pole* $\mathbf{p}$ w.r.t. the paraboloids $\mathcal{P}_e$ and $\mathcal{P}_h$. Of course, for $\mathbf{p}$ on the paraboloid, this is the tangent plane τ_P. Otherwise, we notice that if any point $\mathbf{q}$ lies on the paraboloid and in the polar plane of $\mathbf{p}$, *i.e.*, $\sigma(\mathbf{q},\mathbf{q}) = 2z_Q$ and $\sigma(\mathbf{p},\mathbf{q}) - (z_P + z_Q) = 0$, then conversely, $\mathbf{p}$ lies in the tangent plane of $\mathbf{q}$. Also w.r.t. paraboloids, we call two points $\mathbf{p}$, $\mathbf{q}$ *conjugate* w.r.t. $\mathcal{P}$ if one lies in the polar plane of the other, *i.e.*, if

$$\sigma(\mathbf{p},\mathbf{q}) - z_{\mathbf{p}} - z_{\mathbf{q}} = 0. \tag{2.50}$$

When the pole P runs along a *diameter line*, *i.e.*, parallel to the z-axis, the polar plane π_P remains parallel. In the sense of Lemma 2.1.1, diameter lines of a paraboloid are polar to lines at infinity. Since z is linear in the paraboloid's equation, each diameter line has a single point of intersection with the paraboloid. Hence, for each plane, not parallel to the z-axis, there exists a single parallel plane which contacts the paraboloid $\mathcal{P}$.

Lemma 2.3.2 *The tangent cone $\mathcal{C}_Q$ of a point Q (with position vector* $\mathbf{q}$*) to the paraboloid $\mathcal{P}$ satisfies, in terms of the bilinear form* (2.1)*, the equation*

$$[\sigma(\mathbf{q},\mathbf{q}) - 2z_Q]\,[\sigma(\mathbf{x},\mathbf{x}) - 2z] - [\sigma(\mathbf{q},\mathbf{x}) - z - z_Q]^2 = 0. \qquad (2.51)$$

Readers are invited to derive a proof similar to the one of Lemma 2.11 in Exercise 2.3.5.

In (2.51), the tangent cone $\mathcal{C}_Q$ is represented as a point set. Then, in the case $Q \in \mathcal{P}$, the quadratic cone degenerates to the tangent plane τ_Q, counted twice. The tangent cone of Q could also be defined in a dual way, as the set of tangent planes of $\mathcal{P}$ passing through Q. Then, in the case of a hyperbolic paraboloid, the tangent cone of $Q \in \mathcal{P}_h$ consists of two pencils of planes with the generators through Q as respective axes.

Theorem 2.3.3 *The point S is the intersection of three mutually orthogonal tangent planes τ_1, τ_2, and τ_3 of the paraboloid $\mathcal{P}$ in* (2.39) *if, and only if, the point S is located in the plane*

$$z = -\frac{1}{2}\left(a^2 \pm b^2\right). \qquad (2.52)$$

In the case of an equilateral hyperbolic paraboloid, this is the tangent plane at the vertex.

Readers will supply the proof, which is similar to those in Exercises 2.1.4 and 2.2.6, in Exercise 2.3.6.

According to Theorem 2.3.2, all sections of the paraboloids in (2.39) with planes parallel to the z-axis are parabolas or straight lines. Other planes intersect the hyperbolic paraboloid $\mathcal{P}_h$ along hyperbolas or pairs of lines. This reason lies in the fact that, for each plane ε that is not parallel to the z-axis, there exists a parallel tangent plane of $\mathcal{P}_h$, and this plane contains

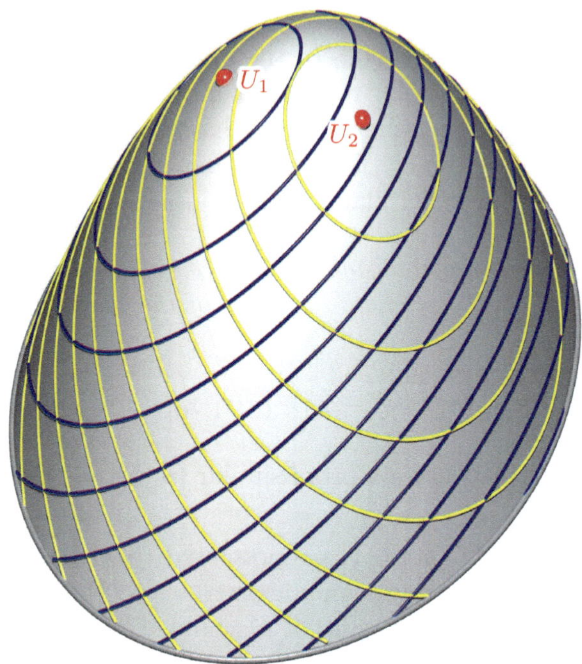

FIGURE 2.35. Elliptic paraboloids carry two families of circular sections, including null circles at the two umbilical points U_1 and U_2.

two generators parallel to ε. The intersection $\mathcal{P}_h \cap \varepsilon$ is the set of points where the generators of $\mathcal{P}_h$ meet the plane ε. Therefore, the intersection curve can never be empty. From the dual point of view, tangent cones $\mathcal{C}_Q$ of $\mathcal{P}_h$ can never be empty.

Besides parabolas, non-empty intersection curves of an elliptic paraboloid $\mathcal{P}_e$ with non-contacting planes are ellipses or even circles.

Theorem 2.3.4 *There are two one-parameter families of parallel planes which intersect the elliptic paraboloid $\mathcal{P}_e$ — thus satisfying the first equation in* (2.39) *with $a > b$ — along circles. These planes are parallel to the x-axis. The coordinates of the two umbilic points $\mathcal{P}_e$ are listed in* (2.53).

Proof: The sphere $x^2 + y^2 + (z - z_1 - a^2)^2 = x_1^2 + a^4$, centered on the z-axis, contacts the paraboloid $\mathcal{P}_e$ at the two points $(x_1, 0, z_1)$ where $2a^2 z_1 = x_1^2$. The intersection between $\mathcal{P}_e$ and the sphere splits into two circles in planes parallel to one of the planes

$$bz \pm y\sqrt{a^2 - b^2} = 0.$$

The two umbilic points

$$U_{1,2} = \left(0,\ \pm b\sqrt{a^2 - b^2},\ \frac{1}{2}(a^2 - b^2)\right)^{\mathrm{T}} \tag{2.53}$$

of $\mathcal{P}_e$ are characterized as points of contact with tangent planes included in the two pencils of parallel planes (see Figure 2.35). ■

The orthogonal hyperbolic paraboloid

As already mentioned, in the case of an orthogonal hyperbolic paraboloid, all generators of one regulus are orthogonal to the vertex generator of the complementary regulus. Therefore, any plane parallel to all generators of one regulus (note Theorem 2.3.1) must be orthogonal to any plane parallel to all generators of the complementary regulus. Consequently, the particular top view, as shown in Figure 2.33, must be a rectangle. On the other hand, this is sufficient for a hyperbolic paraboloid to be orthogonal.

Theorem 2.3.5 *The hyperbolic paraboloid passing through a skew quadrilateral with vertices $ABCD$ is orthogonal if, and only if, the four vertices are located on a cylinder of revolution with an axis connecting the midpoints of the diagonals AC and BD.*

Each orthogonal hyperbolic paraboloid is mapped onto itself by a rotation about a vertex generator through 180°.

Figure 2.16 illustrates the following statement: Along each skew generator g of a ruled surface $\mathcal{R}$, the correspondence between tangent planes τ_P and points of contact P obeys a common formula, depending only on the distribution parameter δ of g. The vertex generators of the hyperbolic paraboloid $\mathcal{P}_h$ in (2.39) with $b = a$ have the distribution parameter $\delta = \pm a^2$ and the vertex as the striction point (Exercise 2.3.8). Therefore, along each skew generator $g \in \mathcal{R}$, there exists a contacting orthogonal hyperbolic paraboloid with vertex generator g. A rotation about g through 90° reveals that *the surface normals along g form one regulus of an orthogonal hyperbolic paraboloid*[12]. Generators of the complementary regulus are axes of one-sheeted hyperboloids of revolution which contact the given ruled surface $\mathcal{R}$ along g.

[12]We will encounter this result again in Chapter 9 (note Figure 9.7).

Definition 2.3.2 For two given point sets S_1, S_2 in $\mathbb{E}^2$ or $\mathbb{E}^3$, the set of points being equidistant to S_1 and S_2 is called *bisector* of S_1 and S_2.

In the case of two given points $P, Q \in \mathbb{E}^3$, the bisector is the orthogonal bisector plane σ_{PQ} of P and Q. The standard definition of a parabola in $\mathbb{E}^2$ as bisector of its focal point and directrix reveals that each paraboloid of revolution in $\mathbb{E}^3$ is the bisector of a point F and a plane θ not passing through F. However, the equilateral hyperbolic paraboloid is also a bisector, as reported, *e.g.*, in [109, p. 154] and stated in the theorem below.

Theorem 2.3.6 *Let g and h be two skew lines in $\mathbb{E}^3$ with $2\varphi := \sphericalangle gh$ and the shortest distance $2d := \overline{gh}$.*

1. *The bisector of g and h is an orthogonal hyperbolic paraboloid $\mathcal{P}_h$. If g and h are given by $z = \pm d$ and $x \sin\varphi = \pm y \cos\varphi$, then*

$$\mathcal{P}_h\colon\ z + \frac{\sin 2\varphi}{2d}\,xy = 0\,. \tag{2.54}$$

2. *The axes of symmetry e_0 and f_0 of the two skew lines (g, h) are the vertex generators of $\mathcal{P}_h$; the common perpendicular of g and h is the paraboloid's axis s. The lines g and h are polar w.r.t. $\mathcal{P}_h$, i.e., each point $G \in g$ is conjugate w.r.t. $\mathcal{P}_h$ to all points $H \in h$, and vice versa.*
3. *At any point $X \in \mathcal{P}_h$, the tangent plane τ_X to $\mathcal{P}_h$ is the plane of symmetry of the pedal points G, H of X w.r.t. g and h, respectively. Hence, $\mathcal{P}_h$ is the envelope of all bisecting planes of points $G \in g$ and $H \in h$.*
4. *The generators of $\mathcal{P}_h$ are exactly the axes of rotations in $\mathbb{E}^3$ which send the line g to the line h. Therefore, they are also the axes of one-sheeted hyperboloids of revolution passing through the given skew pair or lines (g, h). All hyperboloids have the same secondary semiaxis $b = d \cot\varphi$.*

Proof: 1: Let the line g be given in vector form as $\mathbf{p} + \mathbb{R}\mathbf{v}$ with $\|\mathbf{v}\| = 1$. Then, its distance to any point X with position vector $\mathbf{x}$ satisfies

$$\overline{Xg}^{\,2} = \|\mathbf{x} - \mathbf{p}\|^2 - \langle \mathbf{x} - \mathbf{p},\, \mathbf{v}\rangle^2. \tag{2.55}$$

We specify

$$\mathbf{p} = (0, 0, \pm d) \ \text{ and } \ \mathbf{v} = (\cos\varphi,\ \pm\sin\varphi,\ 0)\,.$$

Then, $\overline{Xg} = \overline{Xh}$ is equivalent to

$$x^2 + y^2 + (z - d)^2 - (x\cos\varphi + y\sin\varphi)^2 = x^2 + y^2 + (z + d)^2 - (x\cos\varphi - y\sin\varphi)^2$$

and, furthermore, to

$$dz + xy\cos\varphi\sin\varphi = 0.$$

The rotation $(x, y, z) \mapsto (x', y', z')$ about the z-axis through $\pi/4$ with

$$x = \frac{1}{\sqrt{2}}(x' - y'), \quad y = \frac{1}{\sqrt{2}}(x' + y'), \quad z = z',$$

yields the standard equation

$$2z' + \frac{\sin 2\varphi}{2d}(x'^2 - y'^2) = 0$$

of an orthogonal hyperbolic paraboloid (Figure 2.37).

2. The corresponding polar form (2.50),

$$\frac{\sin 2\varphi}{2d}(x'_G x'_H - y'_G y'_H) + (z'_G + z'_H),$$

can be transformed into the polar form

$$\frac{\sin 2\varphi}{2d}(x_G y_H + x_H y_G) + (z_G + z_H)$$

of (2.54). Obviously, we obtain zero when we plug in the coordinates $(r_G\cos\varphi,\ r_G\sin\varphi,\ d)$ of $G \in g$ and $(r_H\cos\varphi,\ -r_H\sin\varphi,\ -d)$ of $H \in h$.

The common points of $\mathcal{P}_h$ and the vertex plane $z = 0$ satisfy $xy = 0$.

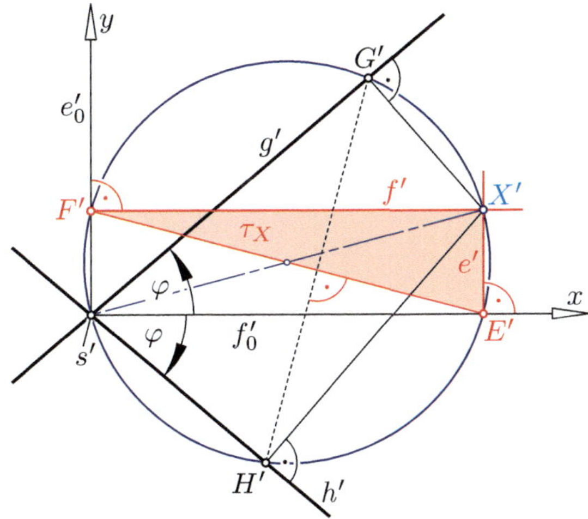

FIGURE 2.36. At the point X, the orthogonal bisector plane of G and H is tangent to the bisecting paraboloid of the lines g and h.

3. Let $G \in g$ and $H \in h$ be the pedal points of $X \in \mathcal{P}_h$. Then, X is uniquely defined as the point of intersection between the orthogonal bisector plane σ_{GH} of G and H and the planes orthogonal to g and h through the respective points G and H. Now, we focus on the top view of the scene in the [x, y]-plane (Figure 2.36):

The top view of the axis s of $\mathcal{P}_h$ is the common point of g' and h'. Since G and H are in equal distance to the [x, y]-plane, but on different sides, the bisecting plane σ_{GH} intersects the $[x,y]$-plane along the perpendicular bisector of the top views G' and H'. The Thales circle with diameter $X's'$ passes through G' and H', and also through the pedal points E' and F' of X' on the x- and y-axes, which are the vertex generators f_0 and e_0 of $\mathcal{P}_h$. Hence, E and F in the $[x,y]$-plane are the trace points of the two generators e and f of $\mathcal{P}_h$ passing through X; they span the tangent plane τ_X to $\mathcal{P}_h$ at X. On the other hand, since g' and h' are symmetric w.r.t. e_0 and f_0, points E' and F' bisect the two arcs bounded by G' and H' on the Thales circle. Therefore, the connecting line $[E',F']$ coincides with the traces of σ_{GH} and τ_X, which proves that these two planes must coincide.

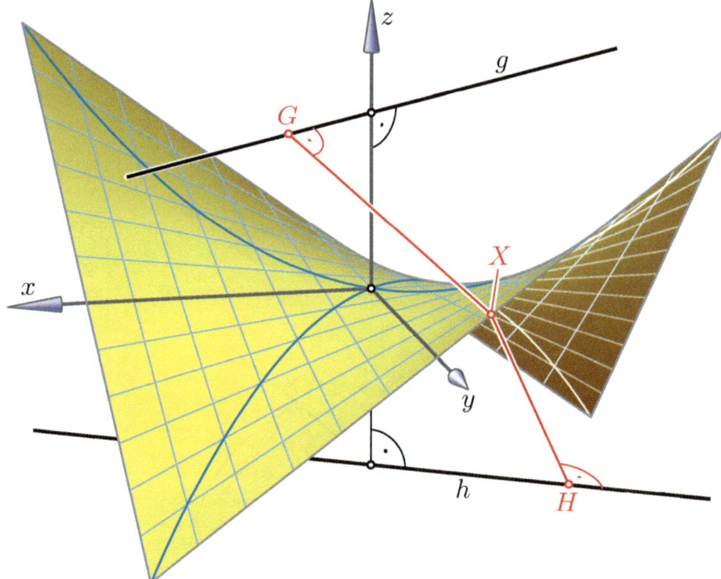

FIGURE 2.37. The bisector of the two lines g and h.

4. The reflection in σ_{GH} exchanges G with H and sends g to a line g_1 through H. We apply a second reflection in a bisecting plane of g_1 and h. This reflection keeps H fixed. Therefore, the product of these two reflections sends g to h and G to H. It acts like a rotation about the line r of intersection of σ_{GH} and the second mirror plane. Since each point of r remains fixed, it has equal distances to g and h. Consequently, r is a generator of $\mathcal{P}_h$.

The rotation about r also transforms the orthogonal plane of g through G onto the orthogonal plane of h through H. Both planes share on r a point which remains fixed under the rotation. Since this point also belongs to σ_{GH}, it coincides with X. Therefore, the generator r passes through X.

Conversely, if r is the axis of a rotation with $g \mapsto h$, then each point $X \in r$ has equal distances to g and h, hence $r \subset \mathcal{P}_h$. If we specify one orientation g, then rotations about axes taken from different reguli of $\mathcal{P}_h$ send g to opposite orientations of h (Figure 2.38). ■

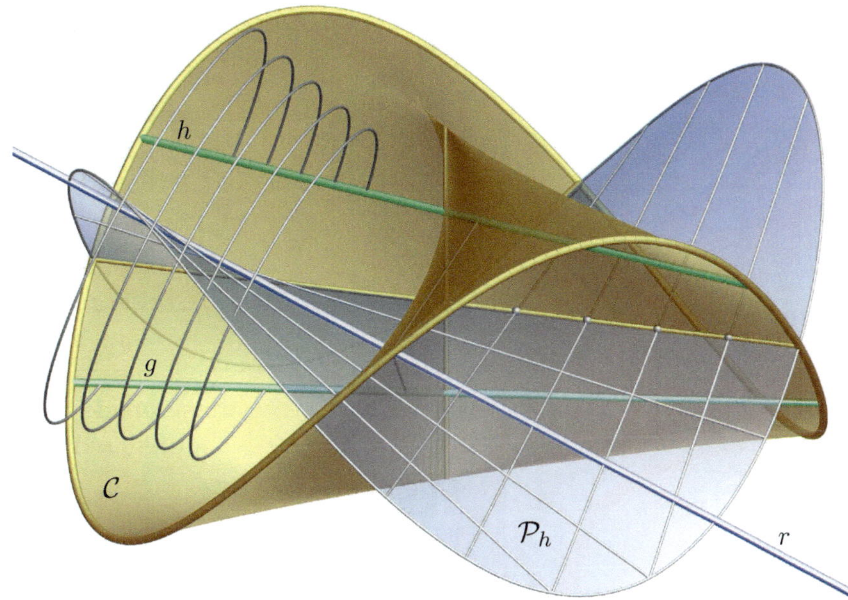

FIGURE 2.38. All pairs of skew lines (g, h) which share the bisecting orthogonal hyperbolic paraboloid $\mathcal{P}_h$ are located on a Plücker conoid $\mathcal{C}$. Generators r of $\mathcal{P}_h$ are axes of rotations with $g \mapsto h$.

Let us focus on the paraboloid $\mathcal{P}_h$ with the equation (2.54) and ask the following: Where are all pairs (g, h) of lines for which $\mathcal{P}_h$ is the bisector? The answer, as given below, was disclosed in [64].

Corollary 2.3.3 *All pairs of skew lines (g, h) which share the bisecting orthogonal hyperbolic paraboloid $\mathcal{P}_h$ are located on a Plücker conoid (cylindroid).*

Proof: Let the lines g and h be given in the same way as in Theorem 2.3.6. Then, the bisector $\mathcal{P}_h$ remains the same if the quotient $\sin 2\varphi/d$ does not change. Obviously, all points of g and h satisfy

$$(x^2 + y^2)z - \frac{2d}{\sin 2\varphi}xy = 0. \tag{2.56}$$

This equation defines a Plücker conoid $\mathcal{C}$, as introduced in (2.26) (see Figure 2.38). The cylindroid passes through the axes of the coordinate frame. The z-axis contains the cylindroid's double line (cf. Figure 2.20). ■

Remark 2.3.1 a) Surprisingly, there seems to be no close connection between the property of Plücker's conoid $\mathcal{C}$ mentioned in Theorem 2.2.4 (note Figure 2.19) and the statement above

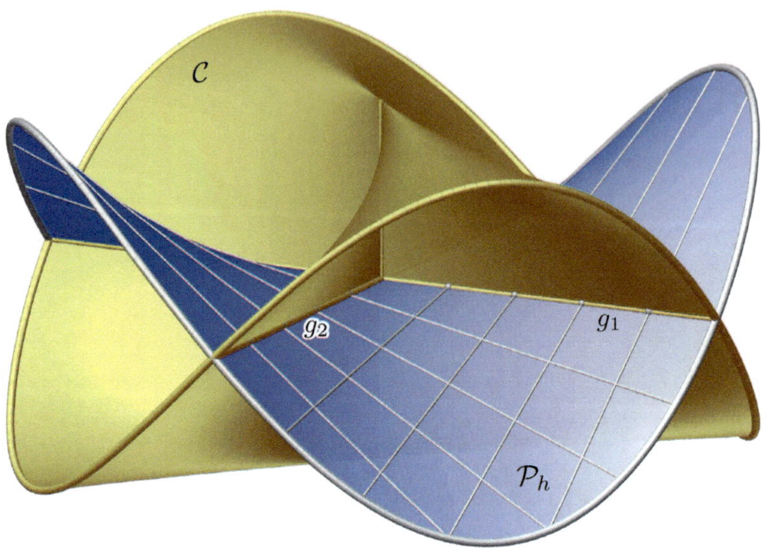

FIGURE 2.39. The surface normals of the Plücker conoid $\mathcal{C}$ along the two mean generators g_1, g_2 form an orthogonal hyperbolic paraboloid $\mathcal{P}_h$.

in Corollary 2.3.3, though both are related to one-sheeted hyperboloids of revolution. By the same token, another remarkable property of the cylindroid is reported in [117].

b) By virtue of item 2 in Theorem 2.3.6, the polarity in the orthogonal hyperbolic paraboloid $\mathcal{P}_h$ with Eq. (2.54) maps the cylindroid $\mathcal{C}$ from (2.56) onto itself.

c) The limit $g \mapsto h$ of the result stated in Corollary 2.3.3 reveals: The paraboloid $\mathcal{P}_h$ is, at the same time, the surface of normals of $\mathcal{C}$ along both vertex generators (see Figure 2.39).

• **Exercise 2.3.1** Projection of a skew quadrangle onto a parallelogram.

Prove the following statement: *For each skew quadrilateral with vertices $ABCD$, there is a direction for a parallel projection which maps the quadrilateral onto a parallelogram.*

• **Exercise 2.3.2** A statement equivalent to Theorem 2.3.5.

Prove that the hyperbolic paraboloid passing through the skew quadrilateral with vertices $\mathbf{a}\,\mathbf{b}\,\mathbf{c}\,\mathbf{d}$ is orthogonal if, and only if,

$$\langle(\mathbf{a}-\mathbf{b})\times(\mathbf{c}-\mathbf{d}),\ (\mathbf{a}-\mathbf{c})\times(\mathbf{b}-\mathbf{d})\rangle = 0.$$

• **Exercise 2.3.3** All points parametrized by (2.41) satisfy an equation of degree 2.

The parametrization $\mathbf{x}(u,v)$ in (2.41) of a hyperbolic paraboloid shows an affine combination of $\mathbf{a},\dots,\mathbf{d}$ on the right-hand side. This is invariant under affine transformations $\mathbf{a}\mapsto\mathbf{a}'$, ..., $\mathbf{d}\mapsto\mathbf{d}'$ which can be represented, in matrix form, as

$$\mathbf{x}\mapsto\mathbf{x}' = \mathbf{a}' + \big((\mathbf{b}'-\mathbf{a}')\ (\mathbf{c}'-\mathbf{a}')\ (\mathbf{d}'-\mathbf{a}')\big)\big((\mathbf{b}-\mathbf{a})\ (\mathbf{c}-\mathbf{a})\ (\mathbf{d}-\mathbf{a})\big)^{-1}(\mathbf{x}-\mathbf{a}),$$

where the involved (3×3)-matrices are represented by their three column vectors. Show that in the particular case

$$\mathbf{a}' = (1,0,1)^{\mathrm{T}}, \quad \mathbf{b}' = (0,1,-1)^{\mathrm{T}}, \quad \mathbf{c}' = (-1,0,1)^{\mathrm{T}}, \quad \mathbf{d}' = (0,-1,-1)^{\mathrm{T}}$$

the image points (x', y', z') satisfy the standard equation of a hyperbolic paraboloid.

• **Exercise 2.3.4** Filling hyperbolic paraboloid.

Verify for the hyperbolic paraboloid $\mathcal{P}_h$ passing through the sides of the skew quadrangle **abcd** (Corollary 2.3.1) that the axis has the direction of the vector $\mathbf{s}$ in (2.42) and the vertex has the position vector $\mathbf{x}(u,v)$ from (2.41) with parameters (u,v) listed in (2.43).

• **Exercise 2.3.5** Proof of Lemma 2.3.2.

Verify the equation (2.51) of the cone of tangents drawn from the apex $\mathbf{q}$ to the paraboloids $\mathcal{P}_e$ or $\mathcal{P}_h$.

Hint: Proceed in a similar manner to the proof of Lemma 2.1.2.

• **Exercise 2.3.6** Proof of Theorem 2.3.3.

The cone with apex S and tangent to the paraboloid $\mathcal{P}$ must be dual to an equilateral cone. This means that the trace of the cone's tangential equation must vanish. In order to obtain this equation similar to Exercise 2.1.4, we homogenize the dual equation (2.49) of $\mathcal{P}$ and translate the coordinate frame such that S becomes the new origin. Then, the trace of the tangential equation is $a^2 \pm b^2 + 2z_S$.

• **Exercise 2.3.7** Three mutually orthogonal tangents.

Prove the following theorem: *Given a paraboloid $\mathcal{P}$ satisfying one of the equations in* (2.39), *the locus of points Q from which three mutually orthogonal tangents of $\mathcal{P}$ can be drawn is a coaxial paraboloid $\mathcal{P}^*$ of revolution satisfying*

$$\pm(x^2 + y^2) - 2(b^2 \pm a^2)z = a^2 b^2,$$

which, in the case of an orthogonal paraboloid (lower sign, $b = a$), is empty and degenerate. Each point $X \in (\mathcal{P} \cap \mathcal{P}^*)$ is the meet of two orthogonal generators of $\mathcal{P}$.

Hint: By virtue of [46, p. 459ff], the cone with apex Q and tangent to $\mathcal{P}$ must be equilateral. In this case, the trace of the quadratic form (2.51) which is included in the Cartesian equation of the tangent cone, must vanish.

• **Exercise 2.3.8** Distribution parameter of vertex generators.

Prove that the two vertex generators of the orthogonal hyperbolic paraboloid in (2.39) with $b = a$ have the distribution parameters $\delta = \pm a^2$.

• **Exercise 2.3.9** Real and imaginary part of the complex function z^2.

Show that the real part and the imaginary part of the complex function $(x + \mathrm{i}y)^2$ represent two congruent orthogonal hyperbolic paraboloids. Their intersection splits into two parabolas in orthogonal planes (Figure 2.40).

• **Exercise 2.3.10** Generalization of Theorem 2.3.6.

Prove the following generalization of Theorem 2.3.6: The ruled quadrics can be defined as the set of points for which the ratio of the distances to two given skew lines is constant [6, p. 236].

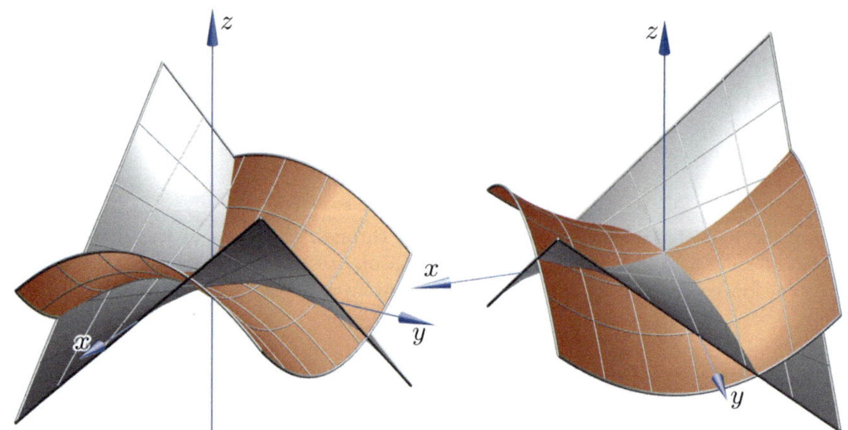

FIGURE 2.40. Visualizing the complex function $f(z) = z^2$.

2.4 Shared metric properties

Here, we present a couple of properties which hold for all types of quadrics simultaneously. The first result is a three-dimensional version of a result on conics (see [46, Theorem 4.2.3]).

Theorem 2.4.1 *Let $\mathcal{Q}$ be a quadric symmetric w.r.t. the plane σ. For each point $P \in \mathcal{Q}$ outside σ, we denote with P_σ its pedal point in σ and with P_n the point of intersection of σ with the surface normal n_P to $\mathcal{Q}$ at P. Then, $P_\sigma \mapsto P_n$ are corresponding in an affine transformation in σ, which keeps the axes of $\mathcal{Q}$ fixed.*

Proof: All central quadrics $\mathcal{Q}$ satisfy an equation of the type

$$\alpha x^2 + \beta y^2 + \gamma z^2 - 1 = 0$$

with $\alpha\beta\gamma \neq 0$. The tangent plane at the point $P = (x_1, y_1, z_1) \in \mathcal{Q}$ satisfies

$$\tau_P\colon\ \alpha x_1 x + \beta y_1 y + \gamma z_1 z - 1 = 0.$$

Therefore, the surface normal at P to $\mathcal{Q}$ can be parametrized as

$$n_P\colon \begin{pmatrix} x_1 \\ y_1 \\ z_1 \end{pmatrix} + t \begin{pmatrix} \alpha x_1 \\ \beta y_1 \\ \gamma z_1 \end{pmatrix}, \quad t \in \mathbb{R}.$$

Without restricting generality, we choose σ as the plane $z = 0$. Then, we obtain point P_n for $t = -1/\gamma$. Hence,

$$P_\sigma = (x_1, y_1, 0) \mapsto P_n = \left(\frac{\gamma - \alpha}{\gamma} x_1,\ \frac{\gamma - \beta}{\gamma} y_1,\ 0\right)$$

is an affine transformation in σ, which fixes the x- and y-axis.

Figure 2.41 shows the surface normals of an ellipsoid along the ellipse $c = \mathcal{Q} \cap \varepsilon$ as well as their respective intersections c_{xy} and c_{yz} with the $[x, y]$- and the $[y, z]$-plane, which are again ellipses. In Figure 8.25, the plane spanned by the ellipse e is orthogonal to the plane of symmetry σ. Therefore, the surface normals n_P along e intersect the plane σ along a line d.

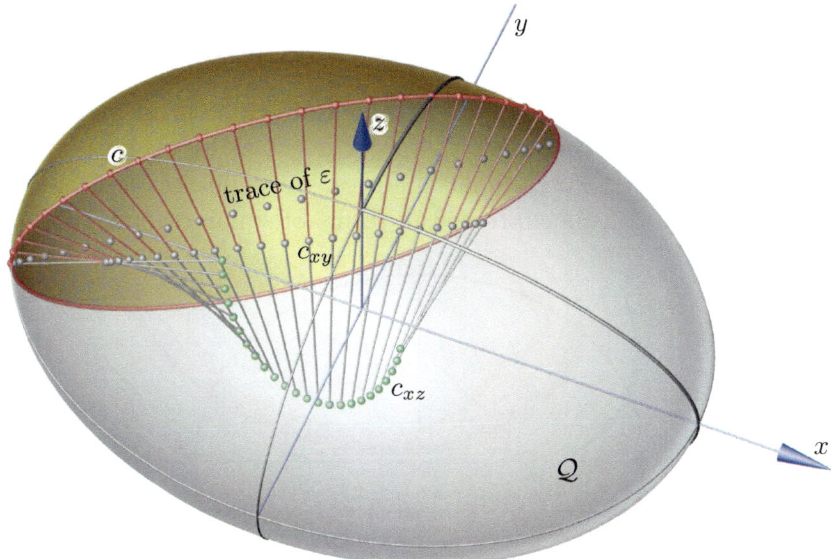

FIGURE 2.41. The surface normals of the quadric $\mathcal{Q}$ along the conic $c = \mathcal{Q} \cap \varepsilon$ intersect the $[x, y]$- and $[y, z]$-plane along curves c_{xy} and c_{yz}, respectively, which are related to c in affine mappings.

In the case of a paraboloid $\mathcal{P}$, we set up the quadric's equation as

$$\mathcal{P}\colon\ \alpha x^2 + \beta y^2 - 2z = 0.$$

Then, the tangent plane at $P \in \mathcal{P}$ satisfies

$$\tau_P\colon\ \alpha x_1 x + \beta y_1 y - (z + z_1) = 0.$$

We choose $\sigma\colon x = 0$ and obtain the mapping

$$P_\varepsilon = (0,\, y_1,\, z_1) \;\mapsto\; P_n = \left(0,\ \frac{\alpha - \beta}{\alpha}\, y_1,\ z_1 + \frac{1}{\alpha}\right).$$

This is the composition of a translation parallel to the z-axis and a scaling of y-coordinates. For quadrics of revolution, all points P^n lie on the axis; the affine transformation $P_\sigma \mapsto P_n$ is singular. ■

There is another theorem concerning the surface normals of a quadric $\mathcal{Q}$.

Theorem 2.4.2 *Given any quadric $\mathcal{Q}$, for any two points $P_1, P_2 \in \mathcal{Q}$ outside of any plane of symmetry, there is a similarity between the respective surface normals n_1 and n_2, mapping P_1 onto P_2 and, for each symmetry plane σ of $\mathcal{Q}$, the point of intersection $n_1 \cap \sigma$ onto $n_2 \cap \sigma$.*

This statement follows from properties of confocal central quadrics and confocal paraboloids. Therefore, the proof is left as an exercise in Chapter 7 (Exercise 7.1.4). For central quadrics, the affine ratio of any three points in question can be expressed in terms of squared semiaxes.

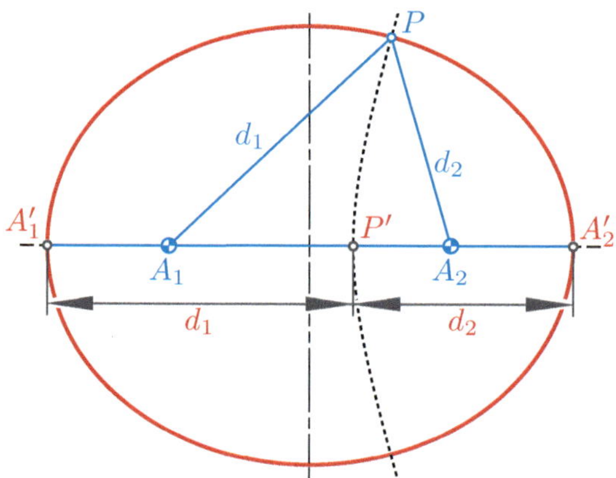

FIGURE 2.42. Two-dimensional version of JACOBI's focal property $\overline{PA_i} = \overline{P'A_i'}$.

The following theorem addresses one of the so-called *Jacobi focal properties*[13] (see [127, p. 208]). It generalizes the standard definitions of ellipses and hyperbolas (Figure 2.42).

Theorem 2.4.3 *Let $A_1A_2A_3$ and $A_1'A_2'A_3'$ be two non-congruent triangles in a plane σ. Then, the locus of points P in $\mathbb{E}^3$, whose distance to A_i for $i = 1, 2, 3$ is equal to the distance of a point $P' \in \sigma$ to A_i', i.e.,*

$$\overline{PA_1} = \overline{P'A_1'}, \quad \overline{PA_2} = \overline{P'A_2'}, \quad \overline{PA_3} = \overline{P'A_3'},$$

is a regular or singular quadric, symmetric w.r.t. σ. Conversely, each quadric can be generated this way.

[13] CARL GUSTAV JACOB JACOBI (1804–1851), German mathematician.

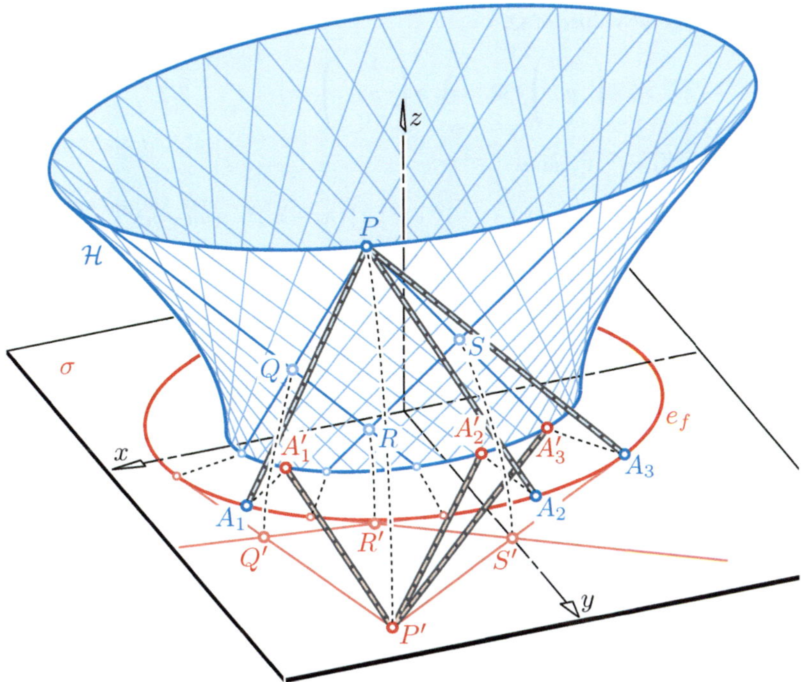

FIGURE 2.43. Three-dimensional version of JACOBI's focal property: $\overline{PA_i} = \overline{P'A'_i}$ for $i = 1, 2, 3$.

Proof: In Chapter 7, we will learn that the statement in question is a direct consequence of Ivory's theorem in 3-space. However, we can verify Theorem 2.4.3 in an analytic way, but we confine ourselves to two particular cases.

(a) Let $\mathcal{Q}$ be a one-sheeted hyperboloid with the standard equation

$$\mathcal{Q}\colon\ \frac{x^2}{a^2} + \frac{y^2}{b^2} - \frac{z^2}{c^2} = 1,$$

as depicted in Figure 2.43. The plane σ of symmetry is chosen as $[x, y]$-plane. In σ, we focus on the focal ellipse

$$e_f\colon\ \frac{x^2}{a^2 + c^2} + \frac{y^2}{b^2 + c^2} = 1, \quad z = 0,$$

which passes through the focal points of the principal sections of $\mathcal{Q}$ in the $[x, z]$- and $[y, z]$-planes. There is an affine mapping

$$\alpha_1\colon \begin{pmatrix} x \\ y \\ z \end{pmatrix} \mapsto \begin{pmatrix} \lambda x \\ \mu y \\ 0 \end{pmatrix} \text{ with } \lambda := \frac{\sqrt{a^2 + c^2}}{a} \text{ and } \mu := \frac{\sqrt{b^2 + c^2}}{b},$$

which sends points $P, Q, \ldots \in \mathcal{Q}$ to points $P', Q', \ldots \in \sigma$ in the closed exterior of the focal ellipse e_f. We specify three mutually different points A_1, A_2, A_3 on e_f with their α_1-preimages

A_1', A_2', A_3' on the gorge ellipse of $\mathcal{Q}$, *i.e.*,

$$A_i' = \begin{pmatrix} x_i \\ y_i \\ 0 \end{pmatrix} \in \mathcal{Q}, \quad A_i = \begin{pmatrix} \lambda x_i \\ \mu y_i \\ 0 \end{pmatrix} \in e_f, \quad i = 1, 2, 3.$$

Then, we obtain $\overline{PA_i} = \overline{P'A_i'}$, since

$$\begin{aligned} \overline{PA_i}^2 - \overline{P'A_i'}^2 &= \left[(x - \lambda x_i)^2 + (y - \mu y_i)^2 + z^2\right] - \left[(\lambda x - x_i)^2 + (\mu y - y_i)^2\right] \\ &= \left[x^2\left(1 - \tfrac{a^2+c^2}{a^2}\right) + y^2\left(1 - \tfrac{b^2+c^2}{b^2}\right) + z^2\right] + \left[x_i^2\left(\tfrac{a^2+c^2}{a^2} - 1\right) + y_i^2\left(\tfrac{b^2+c^2}{b^2} - 1\right)\right] \\ &= -c^2\left[\tfrac{x^2}{a^2} + \tfrac{y^2}{b^2} - \tfrac{z^2}{c^2}\right] + c^2\left[\tfrac{x_i^2}{a^2+c^2}\,\tfrac{a^2+c^2}{a^2} + \tfrac{y_i^2}{b^2+c^2}\,\tfrac{b^2+c^2}{b^2}\right] = -c^2 + c^2 = 0 \end{aligned}$$

because of $P = (x, y, z) \in \mathcal{Q}$ and $A_i = (\lambda x_i,\ \mu y_i,\ 0) \in f_e$. In addition, Figure 2.43 illustrates that the affine transformation α preserves distances along the generators of $\mathcal{Q}$, *i.e.*, $\overline{PQ} = \overline{P'Q'}$.

(b) In the case of a paraboloid $\mathcal{Q}$ with focal parabola f_p, satisfying

$$\mathcal{Q}\colon \frac{x^2}{a^2} + \frac{y^2}{b^2} - 2z = 0, \quad f_p\colon \frac{x^2}{a^2 - b^2} - 2z + b^2 = 0, \quad \sigma\colon y = 0,$$

we use the affine mapping

$$\alpha_2\colon \begin{pmatrix} x \\ y \\ z \end{pmatrix} \mapsto \begin{pmatrix} \lambda x \\ 0 \\ z + \frac{b^2}{2} \end{pmatrix} \text{ with } \lambda := \frac{\sqrt{a^2 - b^2}}{a}.$$

It sends points $P \in \mathcal{Q}$ to points $P' \in \sigma$ in the closed exterior of the focal parabola f_p. For any three mutually different points $A_1, A_2, A_3 \in f_p$ with their α_2-preimages A_1', A_2', A_3' on the principal section $\mathcal{Q} \cap \sigma$, *i.e.*,

$$A_i' = \begin{pmatrix} x_i \\ 0 \\ z_i \end{pmatrix} \in \mathcal{Q}, \quad A_i = \begin{pmatrix} \lambda x_i \\ 0 \\ z_i + \frac{b^2}{2} \end{pmatrix} \in f_p, \quad i = 1, 2, 3,$$

we obtain $\overline{PA_i} = \overline{P'A_i'}$, since

$$\begin{aligned} \overline{PA_i}^2 - \overline{P'A_i'}^2 &= \left[(x - \lambda x_i)^2 + y^2 + (z - z_i - \tfrac{b^2}{2})^2\right] - \left[(\lambda x - x_i)^2 + (z + \tfrac{b^2}{2} - z_i)^2\right] \\ &= \left[x^2\left(1 - \tfrac{a^2-b^2}{a^2}\right) + y^2 - 2zb^2\right] - \left[x_i^2\left(1 - \tfrac{a^2-b^2}{a^2}\right) + 2z_i b^2\right] \\ &= b^2\left[\tfrac{x^2}{a^2} + \tfrac{y^2}{b^2} - 2z\right] - b^2\left[\tfrac{x_i^2}{a^2}\,\tfrac{a^2-b^2}{a^2-b^2} - 2\left(z_i + \tfrac{b^2}{2}\right) + b^2\right] = 0, \end{aligned}$$

because of $P = (x, y, z) \in \mathcal{Q}$ and $A_i = (\lambda x_i,\ 0,\ z_i + \frac{b^2}{2}) \in f_p$.

(c) Conversely, it is sufficient to prove that the affine transformation defined by the two triangles $A_1'A_2'A_3' \mapsto A_1A_2A_3$ can always be expressed in one of the two ways given above. This follows from the singular-value-decomposition of the induced linear map. For the proof of the three-dimensional analogue, see Chapter 7. ■

The result below generalizes what is depicted in Figure 2.4. In fact, this result has a projective background. A proof can be found in Chapter 4 (note Exercise 4.2.5).

Theorem 2.4.4 *Let c and d be two different conics on the regular quadric $\mathcal{Q}$. If the line of intersection between the planes of c and d is tangent to $\mathcal{Q}$, then there exists one quadratic cone or cylinder connecting c and d. Otherwise, there are two quadratic cones or cylinders passing simultaneously through these two conics.*

Given a quadric $\mathcal{Q}$, let $\mathcal{C}_P$ and $\mathcal{C}_Q$ be tangent cones with different apices P and Q. If the line $[P, Q]$ contacts $\mathcal{Q}$, then the two tangent cones share a single conic. Otherwise, the intersection $\mathcal{C}_P \cap \mathcal{C}_Q$ splits into two conics.

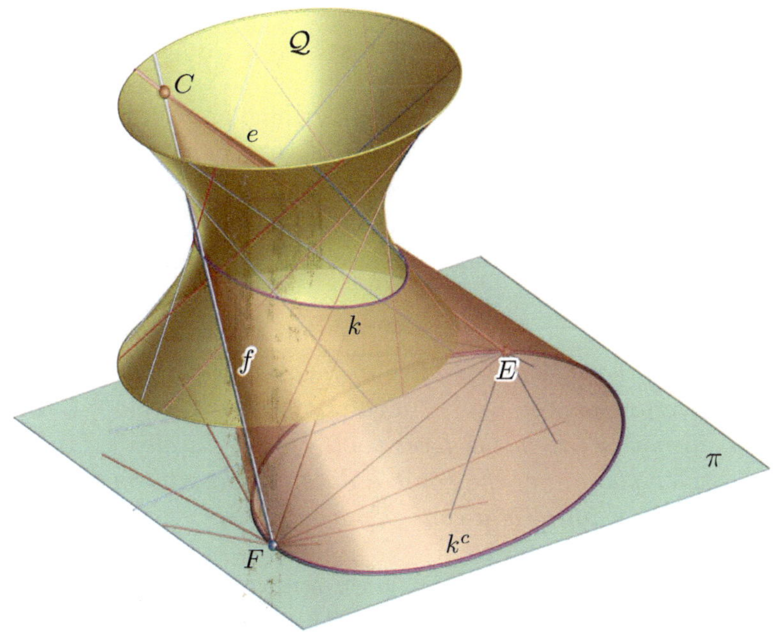

FIGURE 2.44. Stereographic projection $\mathcal{Q} \to \pi$.

Finally, we want to emphasize that the *stereographic projection* is not only defined in the standard position, from a sphere into a plane parallel to the sphere's tangent plane at the center C of projection. Any quadric $\mathcal{Q}$ can be projected from a point $C \in \mathcal{Q}$ into any plane π where $C \notin \pi$. If $\mathcal{Q}$ is a ruled quadric, as depicted in Figure 2.44, then all planar sections of $\mathcal{Q}$ are projected onto conics in π passing through the trace points E, F of the generators e, f through C. The only exceptions are sections with planes through C.

Suppose that two curves $c_1, c_2 \subset \mathcal{Q}$ meet at a point $P \neq C$, and t_1, t_2 are the respective tangents at P. Then, the tangents t_1, t_2 and the two concurrent generators $e_P, f_P \subset \mathcal{Q}$ form a cross ratio δ, which remains the same under stereographic projection, where the generators are depicted as lines through E and F, respectively. In the standard case of the stereographic projection, the generators of $\mathcal{Q}$ are isotropic, and the trace points E, F are the absolute circle points of π. Therefore, conics are mapped to circles, and the mapping is conformal, due to LAGUERRE's formula which relates the cross ratio $\delta = \operatorname{cr}(t_1, t_2, e_P, f_P)$ to the angle measure α between t_1 and t_2 as $\delta = e^{2i\alpha}$.

Quadrics of revolution

We conclude this section with a few results which are valid for all quadrics of revolution. In this case, all intersections with planes through the axis of revolution, called *meridians*, are congruent. Circles located in planes orthogonal to the axis s of revolution are called *parallel circles* of the quadric. Meridians and parallel circles form an orthogonal net on the quadric.

Theorem 2.4.5 *Let $\mathcal{Q}$ be a quadric of revolution with a real focal point F of the meridians on the axis of revolution. If a conic $e \subset \mathcal{Q}$ is located in a plane through F, then the focal point F of $\mathcal{Q}$ is also a focal point of e. If the carrier plane ε of the conic e does not pass through F, then the cone connecting e with F is a cone of revolution.*

Proof: The quadrics of revolution $\mathcal{Q}$ with real focal points on the axis s are either central and hence prolate ellipsoids of two-sheeted hyperboloids, or elliptic paraboloids. The statement of Theorem 2.4.5 is trivial when the cutting plane ε is orthogonal to the axis s. In all remaining cases, there is a unique plane π orthogonal to ε and passing through s. Then, the orthogonal projection into π shows an edge view of ε and, likewise, of all planes orthogonal to s. Below, we confine ourselves to an inspection of such orthogonal views in π (note Figures 2.45 and 2.46).

(a) To begin with, we assume that $\mathcal{Q}$ is a prolate ellipsoid of revolution $\mathcal{E}$ with focal points F_1 and F_2. The principal vertices A, B of $e = \mathcal{Q} \cap \varepsilon$ satisfy

$$\overline{F_1A} + \overline{F_2A} = \overline{F_1B} + \overline{F_2B} = k = \text{const.} \quad \text{and hence} \quad \overline{F_1A} - \overline{F_1B} = \overline{F_2B} - \overline{F_2A}.$$

As a result, the focal points F_1, F_2 are points on different branches of a hyperbola $h \subset \pi$ with focal points A and B (Figure 2.45, left). In order to show that h is the focal hyperbola of $e = \mathcal{E} \cap \varepsilon$, we recall (note [46, p. 143]) that all points X of the focal ellipse of h have the same sum of distances $\overline{F_1X} + \overline{F_2X}$, and the choice $X = A$ reveals that this sum equals the constant k. On the other hand, all points in 3-space with this constant sum lie on the ellipsoid $\mathcal{E}$.

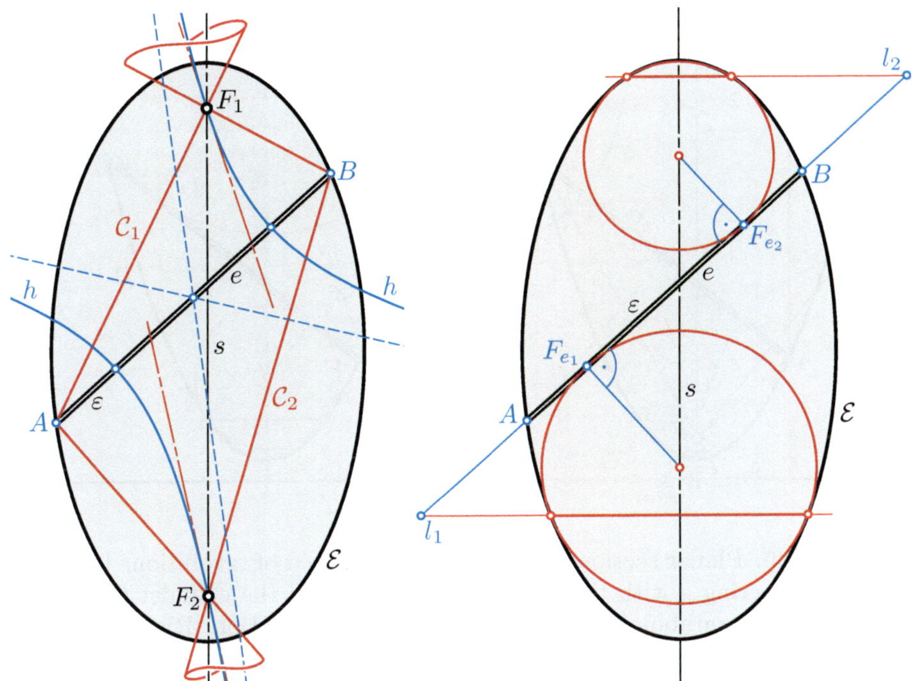

FIGURE 2.45. Intersection e of a prolate ellipsoid $\mathcal{E}$ of revolution with the plane ε: Left: The cones $\mathcal{C}_1, \mathcal{C}_2$ connecting e with the focal points F_1, F_2 of $\mathcal{E}$ are cones of revolution. Right: The focal points F_{e_1}, F_{e_2} of e are points of contact of ε with inscribed spheres.

Since the points F_1 and F_2 are located on the focal hyperbola of e, their connecting cones $\mathcal{C}_1$ and $\mathcal{C}_2$ with e are cones of revolution. In the limiting case $F_1 \subset \varepsilon$, the point F_1 must be a focal point of e [46, Theorem 4.2.1].

As an exercise, readers are invited to consider the case of a two-sheeted hyperboloid of revolution.

(b) Let $\mathcal{Q}$ be a paraboloid $\mathcal{P}$ of revolution with the focal point F. Then, the ideal point of the axis serves as a kind of second focal point. Hence, one of the connecting cones in question is replaced by the cylinder $\mathcal{C}_2$ through e with generators parallel to the axis s (see Figure 2.46, left). In order to prove that this cylinder is a surface of revolution, we choose an arbitrary point $X \in e \subset \mathcal{P}$.

According to the standard definition of parabolas, the point X is the center of a sphere which passes through F and contacts the director plane λ of $\mathcal{P}$ at a point X_0. The sphere in question which is centered in ε, also passes through the image $\widetilde{F}$ of F under reflection in the plane ε. Let M_0 denote the point of intersection between λ and the line $[F, \widetilde{F}]$. Then, the power of M_0 with respect to the sphere (note [46, p. 49]) equals

$$\overline{M_0X_0}^2 = \overline{M_0F} \cdot \overline{M_0\widetilde{F}},$$

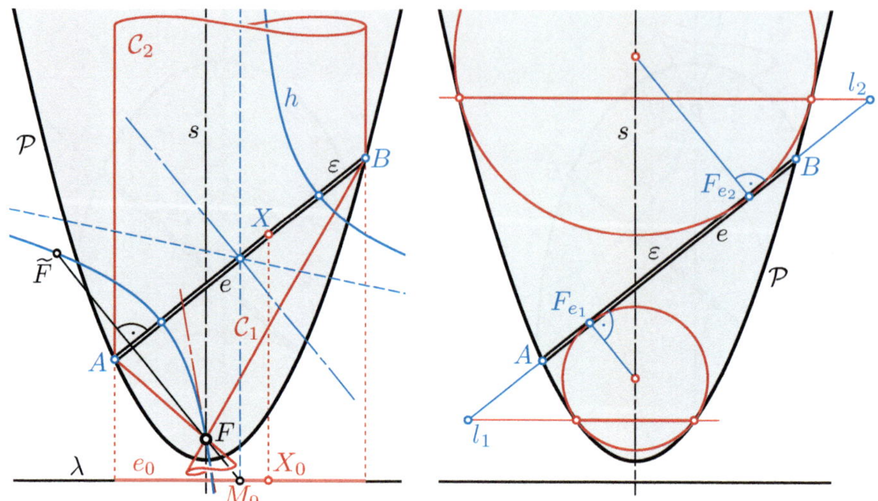

FIGURE 2.46. Planar section $e = \mathcal{P} \cap \varepsilon$ of a paraboloid of revolution: Left: The cone $\mathcal{C}_1$ connecting e with the focal point F of $\mathcal{P}$ and the cylinder $\mathcal{C}_2$ through e parallel to the paraboloid's axis s are surfaces of revolution. Right: The focal points F_{e_1}, F_{e_2} of e are points of contact of ε with spheres inscribed into $\mathcal{P}$.

and this is constant for all $X \in e$. Therefore, the orthogonal projection of e into the plane λ yields a circle e_0, and we obtain a right cylinder $\mathcal{C}_2$.

Next, we prove that the hyperbola h with focal points A, B and passing through F has the axis of $\mathcal{C}_2$ for an asymptote. It follows from

$$\overline{AF} = \overline{A\lambda}, \quad \overline{BF} = \overline{B\lambda}, \quad \text{and hence} \quad \overline{BF} - \overline{AF} = \overline{B\lambda} - \overline{A\lambda},$$

and this, as a limit of the standard definition of hyperbolas, characterizes s as a line parallel to one asymptote of h. The asymptote itself passes through the midpoint of AB and coincides with the axis of $\mathcal{C}_2$. Since $\mathcal{C}_2$ is a cylinder of revolution, e is the focal ellipse of h and C_1 is a cone of revolution, too.

(c) We conclude with an alternative proof, which is valid for all quadrics but uses the complex extension of the Euclidean 3-space. The tangent cone of $\mathcal{Q}$ with apex F is an isotropic cone $\mathcal{C}_F$ [46, p. 457]. The intersection $\mathcal{C}_F \cap \varepsilon$ must contact the conic $c = \mathcal{Q} \cap \varepsilon$ in points of the polar plane π_F of F w.r.t. $\mathcal{Q}$. This means for planes ε through F that the isotropic lines through F in ε are tangents of e, which characterizes F as a focus of e [46, p. 274]. For planes ε not passing through F, the quadratic cone connecting e with the vertex F is regular [46, Lemma 4.3.1]. Its intersection with ε contacts e at two points. Consequently, the isotropic cone with apex F contacts $\mathcal{C}_F$ along two generators, which characterizes $\mathcal{C}_F$ as a cone of revolution. ■

Remark 2.4.1 At each point X of the conic e, the tangent plane τ_X to the quadric of revolution $\mathcal{Q}$ bisects the angle between the lines $[X, F_1]$ and $[X, F_2]$. Therefore, the reflection in $\mathcal{Q}$ along e exchanges the two cones. Since the cone connecting X with the focal hyperbola h of e is also a cone of revolution, the tangent to e at X encloses congruent angles with the lines $[X, F_1]$ and $[X, F_2]$ (see [46, Fig. 4.18]).

The following theorem, depicted in Figures 2.45, right, and 2.46, right, generalizes Dandelin spheres as used in [46, p. 129ff] for planar sections of right cones or cylinders.

Theorem 2.4.6 *Let $\mathcal{Q}$ be a quadric of revolution and the conic e be the curve of intersection with a plane ε. If there is a sphere tangent to ε and contacting $\mathcal{Q}$ along a parallel circle, then the point F of contact between $\mathcal{S}$ and ε is a focal point of e.*

Proof: Again, we use the complex extension of the Euclidean space. Each sphere inscribed into $\mathcal{Q}$ along a parallel circle c intersects ε along a circle which can be real, or empty, or a null circle. The points of contact lie on the line of intersection between ε and the carrier plane of the parallel circle c. For the sphere mentioned in Theorem 2.4.6, the circle is a null circle, *i.e.*, it splits into two isotropic lines through the point F. Both lines are tangent to e; hence, F is a focal point of e. The points of tangency lie on the polar line l of F w.r.t. e, which lies in the polar plane of F w.r.t. $\mathcal{Q}$, a director plane (note the lines l_1 and l_2 in Figures 2.45, right, and 2.46, right).

Note that, contrary to the case of right cones or cylinders, there need not exist such contacting spheres for each planar section e of $\mathcal{Q}$. A necessary and sufficient condition is a non-empty intersection of $\mathcal{Q}$ with the plane passing through the polar line l (directrix) of F w.r.t. e and orthogonal to the axis of $\mathcal{Q}$. ■

- **Exercise 2.4.1** Surface of normals along a generator.

Prove the following statement (compare with Figure 9.7 on page 398): *The surface normals along a generator of any ruled quadric $\mathcal{Q}$ form an orthogonal hyperbolic paraboloid $\mathcal{N}$ which is tangent to the planes of symmetry of $\mathcal{Q}$.* In accordance with Footnote 8 on page 34, the first part of this statement is valid for all skew generators of any ruled surface.

Hint: Note Theorem 2.4.1. A plane τ contacts a hyperbolic paraboloid $\mathcal{N}$ if, and only if, the intersection $\tau \cap \mathcal{N}$ is reducible.

- **Exercise 2.4.2** Affine transformation preserves distances.

Verify that both affine transformations α_1 and $\alpha_2 : \mathcal{Q} \to \sigma$, as used in the proof of Theorem 2.4.3, preserve distances along the generators of the corresponding quadrics $\mathcal{Q}$.

Hint: Two points $P, Q \in \mathcal{Q}$ belong to the same generator if, and only if, they are conjugate.

- **Exercise 2.4.3** Planar sections of a two-sheeted hyperboloid of revolution.

Modify the proof of Theorem 2.4.5, as given in (a), for the case that the quadric in question is a two-sheeted hyperboloid.

- **Exercise 2.4.4** Planar section of a quadric of revolution.

Prove the following statement: On a given quadric of revolution $\mathcal{Q}$ with a real focal point F let $c \subset \mathcal{Q}$ be a conic, not coplanar with F (compare with Figures 2.45, left, and 2.46, left). Then, the pole of the plane of c w.r.t. $\mathcal{Q}$ lies on the axis of the cone of revolution connecting c with F (note Theorem 2.4.5).

2.5 Flexible models of quadrics

For more than 100 years, two types of physical models of quadrics habe been known which are also flexible [37, pp. 258, 261]. These models were designed in order to visualize quadrics and the variation of their shapes.[14]

- Type 1 is based on a flexible grid of parallel sections. Each section is manufactured from cardbord or wire. The design of such models is attributed to OLAUS HENRICI.[15] First physical models of this type were produced in 1874 by ALEXANDER VON BRILL[16] (see [127, p. 182]).
- Type 2 consists of a framework of generators with spherical joints at each crossing point. These models were also invented in 1874 by HENRICI [127, p. 208]. Later, their design was improved by H. WIENER[17].

Both types of models admit flat limiting poses.

Type 1, sliceform models

Figure 2.48 shows an example where the sections of the surface with two discrete sets of parallel planes are realized as slotted cardboard pieces, one as "slot-from-the-top", the other as "slot-from-the-bottom" (Figure 2.47).[18] Thus, the whole structure is made flexible, as the angle between the planes can vary simultaneously. We are going to prove that the variation of the angle acts as an affine transformation on the represented quadric.

---

[14]These models were distributed by the German companies M. SCHILLING, Halle a. d. Saale, and B.G. TEUBNER, Leipzig (see, *e.g.*, the catalogues [112, pp. 111–114], [138] or [139, pp. 13–18]). HENRICI's flexing hyperboloid is depicted, *e.g.*, in [73, Abb. 7] and [56, Figs. 23a,b]. Models made from cardboard have the advantage that they can be produced by students or teachers. Such models are also shown in [71, Fig. 500, p. 441], [118], or online, in collections of mathematical models, *e.g.*, at `http://www.geometrie.tuwien.ac.at/modelle/`.

[15]OLAUS HENRICI (1840–1918), German mathematician, was professor at University College London and director of the Laboratory of Mechanics.

[16]ALEXANDER WILHELM VON BRILL (1842–1935), German mathematician, was professor at TH Darmstadt and TH Munich.

[17]HERMANN LUDWIG GUSTAV WIENER (1857–1939), German mathematician, was professor for Geometry at TH Darmstadt.

[18]Additional stickers mounted at the slots might be necessary to prevent the slices from sliding along the slots, and thus, falling apart.

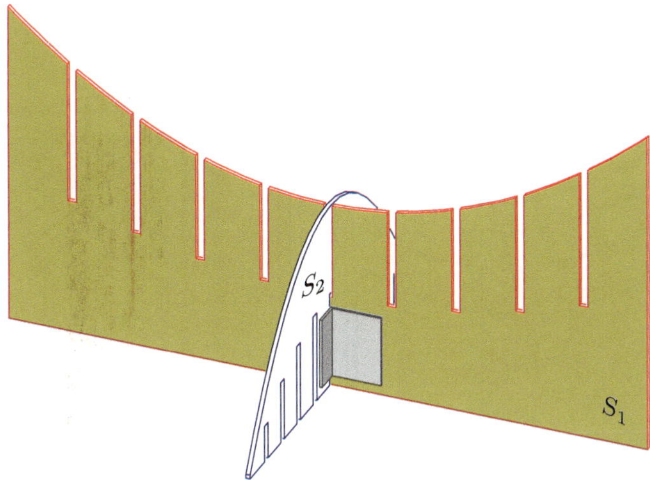

FIGURE 2.47. Two slices S_1, S_2 of a sliceform model with a revolute joint in between.

Lemma 2.5.1 *The restriction of the affine transformation*

$$\alpha : \begin{pmatrix} x \\ y \\ z \end{pmatrix} \mapsto \begin{pmatrix} x' \\ y' \\ z' \end{pmatrix} = \begin{pmatrix} \lambda x \\ \mu y \\ z \end{pmatrix} \tag{2.57}$$

on the planes with equations $mx \pm ny = 0$ *is an isometry if, and only if,*

$$(\lambda^2 - 1)n^2 + (\mu^2 - 1)m^2 = 0 .$$

Proof: It is necessary and sufficient that

$$x'^2 + y'^2 + z'^2 = x^2 + y^2 + z^2 \quad \text{for} \quad mx \pm ny = 0 .$$

This is equivalent to

$$(\lambda^2 - 1) : (\mu^2 - 1) = -y^2 : x^2 = -m^2 : n^2,$$

and hence to $n^2\lambda^2 + m^2\mu^2 = m^2 + n^2$. ■

The model depicted in Figure 2.48, left, is based on the mutually congruent sections parallel to the principal planes of a hyperbolic paraboloid. These curves belong to a net of translation curves on the paraboloid, as stated in Theorem 2.3.2. Figure 2.48, right, shows the same surface, but this time with the two reguli. By virtue of Theorem 2.3.1, the generators of each regulus are located in mutually parallel planes.

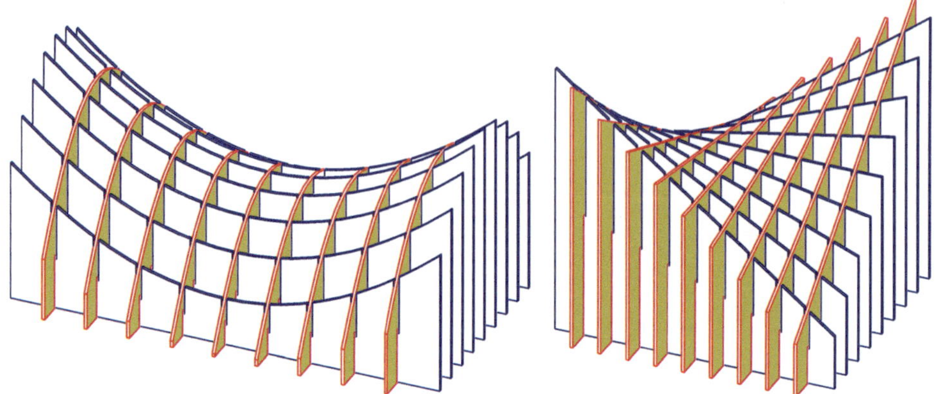

FIGURE 2.48. Flexing models of a hyperbolic paraboloid.

A similar physical model can be produced with parabolas opening to the same side, thus representing an elliptic paraboloid.

In the model shown in Figure 2.48 (right), we have $m : n = b : a$. According to Lemma 2.5.1, the scaling factors λ for x-coordinates and μ for y-coordinates obey

$$e^2 := a^2\lambda^2 + \mu^2 b^2 = a^2 + b^2.$$

Therefore, the flexions of this sliceform model show the two reguli of hyperbolic paraboloids with the equations

$$\frac{x'^2}{e^2\cos^2 t} - \frac{y'^2}{e^2\sin^2 t} - z = 0, \quad 0 < t < \frac{\pi}{2}, \tag{2.58}$$

with flat terminating poses for $t \to 0$ or $t \to \pi/2$. All paraboloids of the family (2.58) intersect any plane $z = \text{const.} \neq 0$ in confocal hyperbolas.[19]

We continue with sliceform models obtained from circular sections of quadrics (note [56, Figs. 25a,b]). The triaxial ellipsoid $\mathcal{E}_0$ with semiaxes $a > c > b$ has its circular sections, by virtue of (2.4), in planes parallel to

$$\frac{\sqrt{a^2 - b^2}}{ab}\, x \pm \frac{\sqrt{c^2 - b^2}}{bc}\, y = 0.$$

[19] After an appropriate translation of each single paraboloid along the axis, hyperbolic paraboloids can be brought into confocal positions, in the sense of Definition 7.1.1 (see Chapter 7).

In view of Lemma 2.5.1, we set

$$m : n = \frac{\sqrt{a^2 - c^2}}{a} : \frac{\sqrt{c^2 - b^2}}{b}$$

and obtain that all affine transformations α from (2.57) satisfying

$$\frac{a^2(c^2 - b^2)}{c^2(a^2 - b^2)}\,\lambda^2 + \frac{b^2(a^2 - c^2)}{c^2(a^2 - b^2)}\,\mu^2 = 1$$

induce isometries of all circular sections and keep the three coordinate planes fixed.

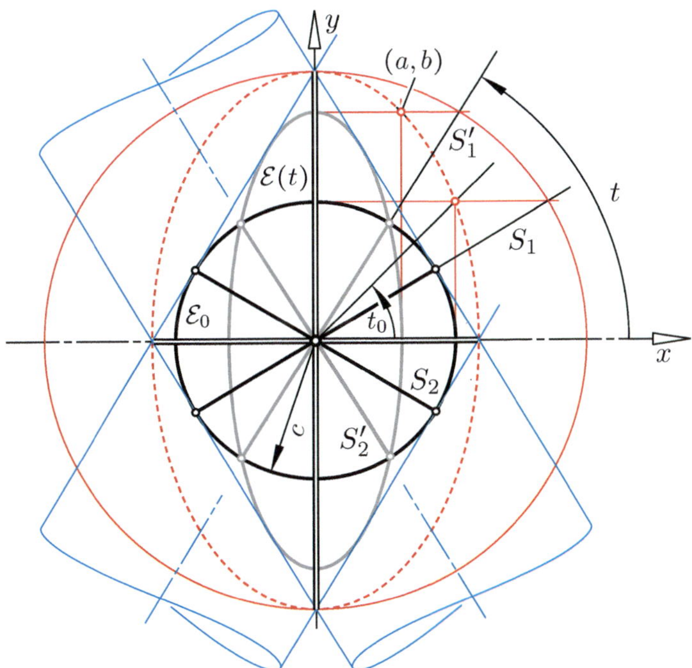

FIGURE 2.49. Top views of a sphere $\mathcal{E}_0$ and an ellipsoid $\mathcal{E}(t)$, produced with the same sliceform model.

This gives rise to a one-parameter family $\alpha(t)$ of affine transformations with

$$\lambda(t) := \frac{c}{a}\sqrt{\frac{a^2 - b^2}{c^2 - b^2}}\,\cos t\,, \quad \mu(t) := \frac{c}{b}\sqrt{\frac{a^2 - b^2}{a^2 - c^2}}\,\sin t,$$

which we can restrict to $\lambda, \mu \geq 0$, to say $0 \leq t \leq \frac{\pi}{2}$. Therefore, the semiaxes of the affine image $\mathcal{E}(t) = \alpha(\mathcal{E}_0)$ are

$$a(t) = \lambda(t)a = c\sqrt{\frac{a^2 - b^2}{c^2 - b^2}}\cos t, \quad b(t) = \mu(t)b = c\sqrt{\frac{a^2 - b^2}{a^2 - c^2}}\sin t, \quad c(t) = c.$$

The circular sections of $\mathcal{E}(t)$ match the proportion

$$m(t) : n(t) = \sin t : \cos t.$$

Therefore, $2t$ equals the angle between any two non-parallel circular sections of $\mathcal{E}(t)$ (see top view in Figure 2.49).

For variable t, the points $(a(t), b(t), 0)$ trace ellipses with the semiaxes $a(0)$ and $b(\frac{\pi}{2})$ in the plane $z = 0$, displayed as dotted curves in Figure 2.49. This confirms that, in the top view, each circular section of $\mathcal{E}$ performs an elliptic motion [46, p. 54] during the deformation. This also holds for the tangent planes at the umbilical points (2.5), which also trace ellipses.

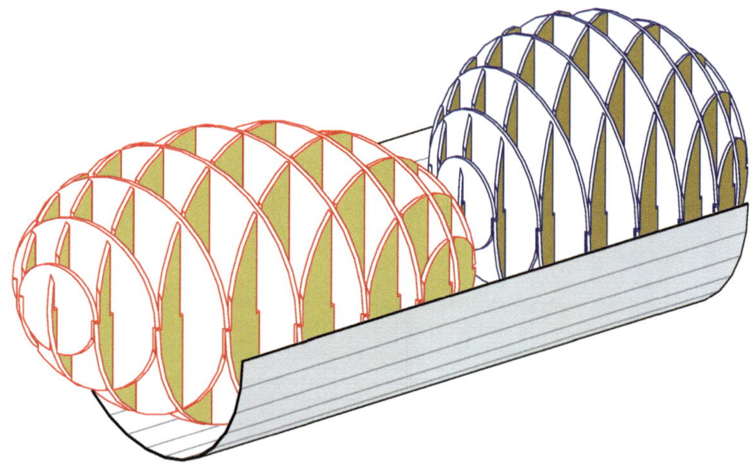

FIGURE 2.50. Sliceform model of an ellipsoid consisting of circular sections.

The limiting cases of $\mathcal{E}(t)$ are flat elliptic disks

$$t = 0: \qquad a(0) = c\sqrt{\frac{a^2 - b^2}{c^2 - b^2}}, \quad b(0) = 0, \quad c(0) = c$$

$$t = \tfrac{\pi}{2}: \qquad a(\tfrac{\pi}{2}) = 0, \quad b(\tfrac{\pi}{2}) = c\sqrt{\frac{a^2 - b^2}{a^2 - c^2}}, \quad c(\tfrac{\pi}{2}) = c.$$

There is always a sphere included satisfying $a(t_0) = b(t_0) = c$, because

$$\cos t_0 = \sqrt{\frac{c^2 - b^2}{a^2 - b^2}} \iff \sin t_0 = \sqrt{\frac{a^2 - c^2}{a^2 - b^2}}\,.$$

Therefore, without loss of generality, the initial ellipsoid $\mathcal{E}_0$ can be replaced by a sphere together with an arbitrary angle $2t_0$ between distinguished planar sections satisfying $0 < t_0 < \frac{\pi}{2}$. Then, the semiaxes of the set of ellipsoids $\mathcal{E}(t)$ read

$$a(t) = a(0)\cos t = \frac{c\,\cos t}{\cos t_0}, \quad b(t) = b(\tfrac{\pi}{2})\sin t = \frac{c\,\sin t}{\sin t_0}, \quad c(t) = c\,. \tag{2.59}$$

Note that for $t > t_0$ we get $b(t) > a(t)$ in contrast to the conditions stated above.

Theorem 2.5.1 *When the planes of symmetry of the considered slice-form model of a triaxial ellipsoid are kept fixed, then all flexions including the flat elliptic disks are in line contact with two cylinders of revolution.*

Proof: By virtue of (2.8), the dual equation of $\mathcal{E}(t)$ is $a(t)^2u^2 + b(t)^2v^2 + c^2w^2 - 1 = 0$, and hence

$$\left[a(0)^2u^2 + (c^2w^2 - 1)\right]\cos^2 t + \left[b\left(\tfrac{\pi}{2}\right)^2 v^2 + (c^2w^2 - 1)\right]\sin^2 t = 0\,. \tag{2.60}$$

This is a linear combination of two terms. Each of them, if set to equal zero, is the condition for any plane to be tangent to one of the two limiting ellipses, seen as 'flat' ellipsoids $\mathcal{E}(0)$ and $\mathcal{E}(\frac{\pi}{2})$. Hence, each common tangent plane of these two ellipses is also tangent to all flexions $\mathcal{E}(t)$. These common planes envelop two cylinders, which must be cylinders of revolution since they are also circumscribed to the sphere $\mathcal{E}(t_0)$. ■

The two cylinders mentioned in Theorem 2.5.1, can be seen in Figure 2.49. One of these two cylinders is also depicted in Figure 2.50, but there the ellipsoid $\mathcal{E}(t)$ is no longer concentric with $\mathcal{E}_0$ but translated in direction of the cylinder's generators.[20]

There are similar models generated by circular sections of elliptic paraboloids and one-sheeted or two-sheeted hyperboloids. When, again, half of the angle between two non-parallel section planes is used as a parameter t for $0 \le t \le \frac{\pi}{2}$, then the deformed quadrics can be represented

[20]These cylinders are also depicted in [71, p. 441, Fig. 501].

as

$$\begin{aligned}
&\text{elliptic paraboloid:} && \frac{x^2}{\cos^2 t} - \frac{cy}{\sin t} + z^2 = 0\,,\\
&\text{one-sheeted hyperboloid:} && \frac{\cosh^2 t_0}{\cos^2 t}\,x^2 - \frac{\sinh^2 t_0}{\sin^2 t}\,y^2 + z^2 = c^2\,,\\
&\text{two-sheeted hyperboloid:} && \frac{\sinh^2 t_0}{\cos^2 t}\,x^2 - \frac{\cosh^2 t_0}{\sin^2 t}\,y^2 - z^2 = c^2
\end{aligned}$$

with any constant $c \neq 0$ and $t_0 > 0$. The respective tangential equations are

elliptic paraboloid:

$$c\cos^2 t\, u^2 - 4\sin t\, v + c\,w^2 = 0\,,$$

one-sheeted hyperboloid:

$$\left[\frac{c^2\,u^2}{\cosh^2 t_0} + c^2w^2 - 1\right]\cos^2 t - \left[\frac{c^2\,v^2}{\sinh^2 t_0} - c^2w^2 + 1\right]\sin^2 t = 0\,,$$

two-sheeted hyperboloid:

$$\left[\frac{c^2\,u^2}{\sinh^2 t_0} - c^2w^2 - 1\right]\cos^2 t - \left[\frac{c^2\,v^2}{\cosh^2 t_0} + c^2w^2 + 1\right]\sin^2 t = 0.$$

These equations reveal that only the hyperboloids belong to a range of quadrics. In the case of paraboloids, there is no range, even when translations in direction of the y-axis were permitted.

Nets of translation curves on paraboloids

Let two parabolas in the symmetric planes $x\sin t \mp y\cos t = 0$ for $0 < t < \frac{\pi}{2}$ be given with the z-axis as the axis of symmetry and the origin as vertex. The parameters of the parabolas are denoted by p and q, respectively.

These two parabolas serve as translation curves of a paraboloid $\mathcal{P}(t)$. In affine coordinates (ξ, η, ζ) based on the vectors

$$\mathbf{b}_1 = \begin{pmatrix} \cos t \\ \sin t \\ 0 \end{pmatrix}, \quad \mathbf{b}_2 = \begin{pmatrix} \cos t \\ -\sin t \\ 0 \end{pmatrix}, \quad \mathbf{b}_3 = \begin{pmatrix} 0 \\ 0 \\ 1 \end{pmatrix},$$

i.e.,

$$\begin{pmatrix} x \\ y \\ z \end{pmatrix} = \begin{pmatrix} \mathbf{b}_1 & \mathbf{b}_2 & \mathbf{b}_3 \end{pmatrix} \begin{pmatrix} \xi \\ \eta \\ \zeta \end{pmatrix},$$

the paraboloid $\mathcal{P}(t)$ has the equation

$$\mathcal{P}(t):\ \zeta = \frac{1}{2p}\xi^2 + \frac{1}{2q}\eta^2.$$

After transformation into Cartesian coordinates, we obtain the representation

$$\mathcal{P}(t):\quad 8pqz = \frac{p+q}{\cos^2 t}x^2 + \frac{2(q-p)}{\sin t\cos t}xy + \frac{p+q}{\sin^2 t}y^2 \tag{2.61}$$

for the set of paraboloids that can be generated with a cardboard model of the type presented in Figure 2.48, left. The corresponding tangential equation reads

$$\widehat{\mathcal{P}}(t):\ 2w = (p+q)\cos^2 t\,u^2 + 2(p-q)\sin t\cos t\,uv + (p+q)\sin^2 t\,v^2\,. \tag{2.62}$$

Special cases:

a) $q = p$: Then, $\mathcal{P}(\frac{\pi}{4})$ is a paraboloid of revolution. The dual equation

$$\widehat{\mathcal{P}}(t):\ (pu^2 - w)\cos^2 t + (pv^2 - w)\sin^2 t = 0$$

reveals that all produced paraboloids belong to a tangential pencil. Each common tangent plane of this range matches $u^2 - v^2 = 0$. Therefore, the common developable consists of two parabolic cylinders. Each cylinder has line contact with each paraboloid $\mathcal{P}(t)$, even with the two parabolas, which are singular surfaces in the range.

b) $q = -p$: The corresponding equation

$$\mathcal{P}(t):\ z = \frac{xy}{2p\sin t\cos t}$$

shows that all regular surfaces of this family are orthogonal hyperbolic paraboloids sharing the x- and y-axis as vertex-generators. Both degenerated flat forms consist of translated parabolas through the origin.

Type 2, Henrici's flexible ruled quadrics

Let two one-sheeted hyperboloids $\mathcal{P}$ and $\mathcal{P}'$ with confocal principal sections be given. Such hyperboloids are called *confocal* (see Definition 7.1.1). We set

$$\begin{aligned}\mathcal{P}&:\ \frac{x^2}{a^2} + \frac{y^2}{b^2} - \frac{z^2}{c^2} = 1,\\ \mathcal{P}'&:\ \frac{x^2}{a^2-k^2} + \frac{y^2}{b^2-k^2} - \frac{z^2}{c^2+k^2} = 1,\end{aligned}$$

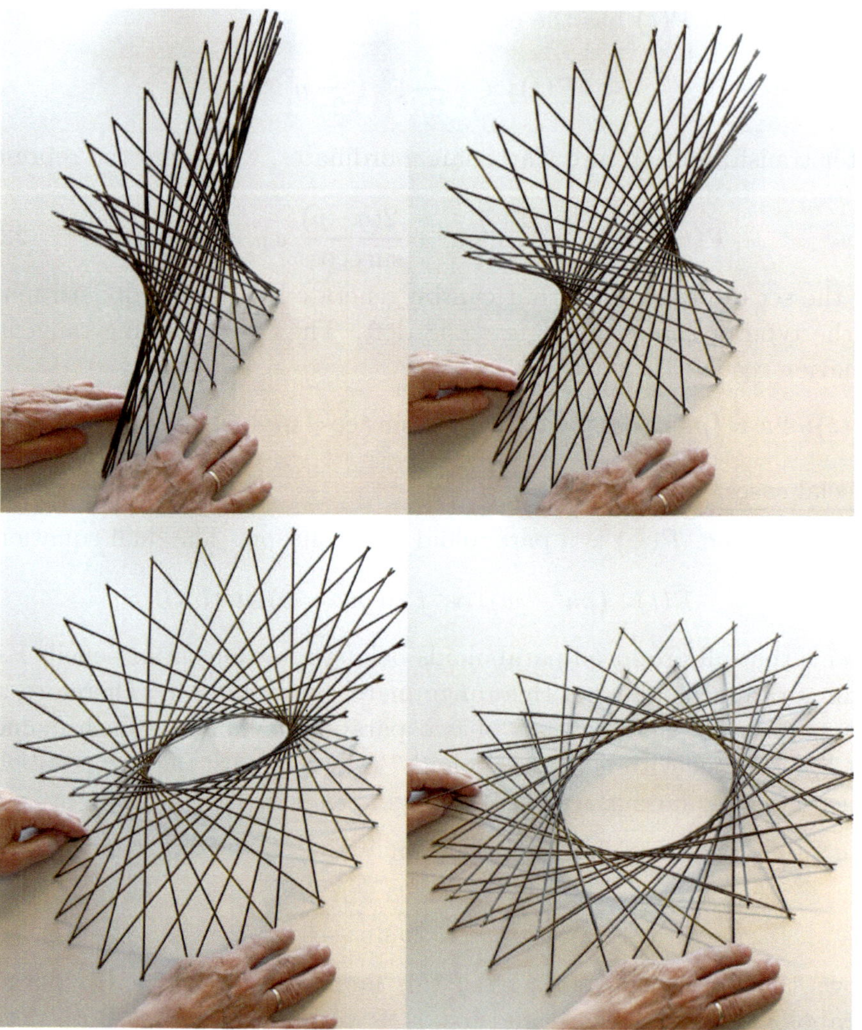

FIGURE 2.51. Confocal hyperboloids between flat limiting positions: The generators on the surfaces are materialized as rods with particular spherical joints.

where $a > b > k > 0$ and $c \neq 0$. After multiplication with k^2 and the substitutions

$$\frac{1}{1-\lambda^2} := \frac{a^2}{k^2}, \quad \frac{1}{1-\mu^2} := \frac{b^2}{k^2}, \quad \frac{-1}{1-\nu^2} := \frac{c^2}{k^2},$$

we obtain $\nu^2 > 1 > \lambda^2 > \mu^2 > 0$ and

$$\begin{aligned} \mathcal{P}&: (1-\lambda^2)x^2 + (1-\mu^2)y^2 + (1-\nu^2)z^2 = k^2, \\ \mathcal{P}'&: \frac{1-\lambda^2}{\lambda^2}x^2 + \frac{1-\mu^2}{\mu^2}y^2 + \frac{1-\nu^2}{\nu^2}z^2 = k^2. \end{aligned} \tag{2.63}$$

The affine transformation

$$\alpha: \begin{pmatrix} x \\ y \\ z \end{pmatrix} \mapsto \begin{pmatrix} x' \\ y' \\ z' \end{pmatrix} = \begin{pmatrix} \lambda x \\ \mu y \\ \nu z \end{pmatrix} \quad \text{with } \mathcal{P} \mapsto \mathcal{P}' \tag{2.64}$$

preserves the distance of two points $X_i = (x_i, y_i, z_i) \in \mathcal{P}$, $i = 1, 2$ if, and only if,

$$\begin{aligned} &(x_1 - x_2)^2 + (y_1 - y_2)^2 + (z_1 - z_2)^2 = (x_1' - x_2')^2 + (y_1' - y_2')^2 + (z_1' - z_2')^2 \\ \iff &(1-\lambda^2)(x_1 - x_2)^2 + (1-\mu^2)(y_1 - y_2)^2 + (1-\nu^2)(z_1 - z_2)^2 = 0\,. \end{aligned}$$

Because of $X_1, X_2 \in \mathcal{P}$, this is equivalent to

$$(1-\lambda^2)x_1x_2 + (1-\mu^2)y_1y_2 + (1-\nu^2)z_1z_2 = k^2,$$

which means that X_1 and X_2 are conjugate w.r.t. $\mathcal{P}$ (compare the proof given in [56], pp. 26–28). Only in this case, X_1 and X_2 are located on the same generator of $\mathcal{P}$. This well-known property[21] holds also for confocal hyperbolic paraboloids (see (2.58)).

The foregoing results, which were first proved by Arthur Cayley (1879), are summarized below:

Theorem 2.5.2 *Let α be an affine transformation that maps the ruled quadric $\mathcal{P}$ onto a confocal quadric $\mathcal{P}'$, while the common planes of symmetry are fixed. Then, the restrictions of α onto the generators of $\mathcal{P}$ are isometries.*

Therefore, when some generators of $\mathcal{P}(0)$ are materialized by rods with spherical joints[22] (see Figure 2.52) at each intersection between two rods,

[21] The quadrics $\mathcal{P}$ and $\mathcal{P}'$ can be embedded into a differentiable family of quadrics such that each pair of points corresponding under α belongs to the same orthogonal trajectory. Now, a well-known theorem by Gauss (see, *e.g.*, [129, p. 217]) concerning geodesic coordinate systems implies that α is length-preserving for each (geodesic) generator of $\mathcal{P}$.

[22] In the catalogues [138] and [139], this type of joint is proudly advertised as "H. Wieners geschränktes Verbindungsgelenk".

this structure is flexible. In the limiting cases $t = b$ and $t = 0$, we obtain flat poses; the generators are tangent to the focal conics of this set (see Chapter 7). This is the famous model of confocal hyperboloids designed by OLAUS HENRICI. This design works even for hyperboloids of revolution, where the focal hyperbola degenerates into the axis.

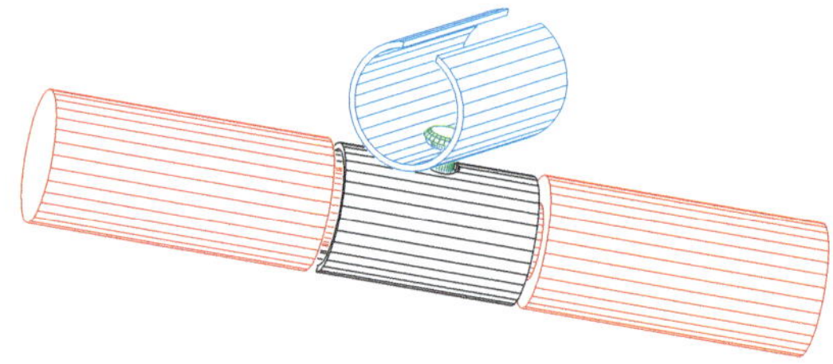

FIGURE 2.52. H. WIENER's design of the spherical joints.

The analogous version for confocal hyperbolic paraboloids

$$\mathcal{P}_h(t)\colon\ \frac{x^2}{a^2-t} - \frac{y^2}{b^2+t} = 4(z-t) \quad \text{with} \quad -b^2 \le t \le a^2 \tag{2.65}$$

(compare with (2.58)) is displayed in Figure 2.53. A cardboard model is shown in Figure 2.48.

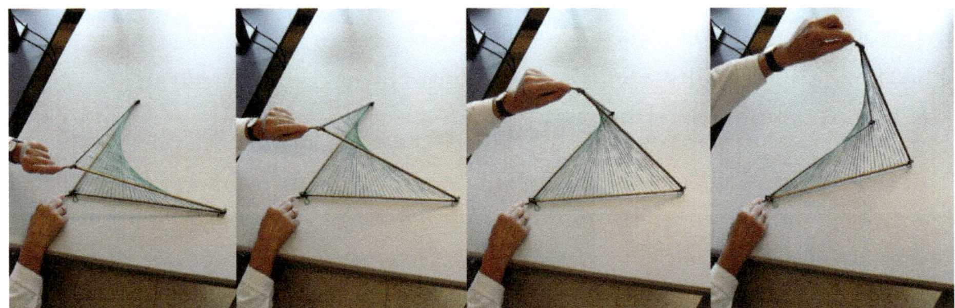

FIGURE 2.53. A flexible model of hyperbolic paraboloids out of a confocal family. The rulings within the bounding quadrangle are materialized by strings which remain taut while bending the quadrangle.

3 Linear algebraic approach to quadrics

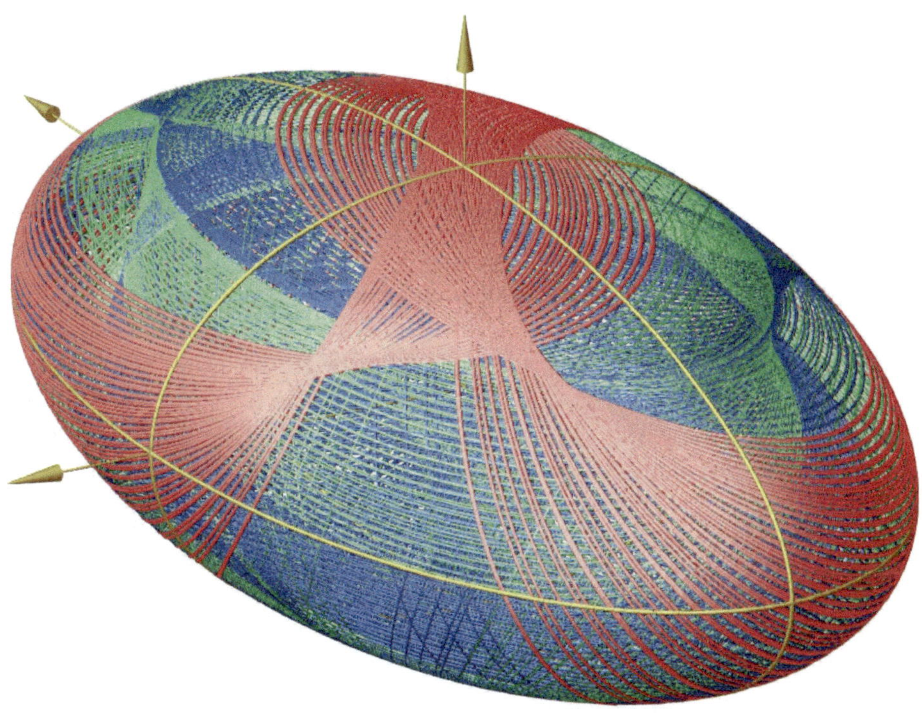

The picture shows a triaxial ellipsoid together with some ellipses which can move on the ellipsoid (courtesy: F. GRUBER).

B. Odehnal et al., *The Universe of Quadrics*,
https://doi.org/10.1007/978-3-662-61053-4_3

3.1 Principal-axes transformation in $\mathbb{E}^n$

Following the terminology of Linear Algebra, a *quadric* is synonymous with a *surface of degree two* in the Euclidean space $\mathbb{E}^n$, $n \geq 2$, and defined as the set of points whose Cartesian coordinates $(x_1, x_2, \dots, x_n)$ satisfy a quadratic equation

$$\begin{aligned} &a_{11}x_1^2 + 2a_{12}x_1x_2 + \dots + 2a_{1n}x_1x_n + a_{22}x_2^2 + 2a_{23}x_2x_3 + \dots + a_{nn}x_n^2 \\ &+ 2a_1x_1 + 2a_2x_2 + \dots + 2a_nx_n + a = 0 \end{aligned} \tag{3.1}$$

with $a_{ij}, a_k, a \in \mathbb{R}$. Only for formal reasons, we included the factor 2 at all mixed quadratic terms and all linear terms on the left-hand side in the polynomial of degree two.

Since quadrics are defined as sets of points, we can modify the underlying Cartesian coordinate frame, *i.e.*, its origin and the orthonormal basis of $\mathbb{R}^n$. Below, we prove that for each quadric there is an appropriate coordinate frame where the quadric's equation has a simple form, a so-called *normal form*. We can eliminate all mixed terms and almost all linear terms, so that finally, beside a constant, the purely quadratic terms and at most one linear term remain. This will be the starting point for a classification of quadrics.

By the same token, in this chapter we find the most general definition of quadrics. It refers to all dimensions $n \geq 2$, includes regular and singular ones and, of course, also the conics for $n = 2$.

The n-dimensional Euclidean space $\mathbb{E}^n$

Following the analytic approach, Euclidean geometry takes place in the vector space $\mathbf{V} = \mathbb{R}^n$ equipped with the *scalar product* for any two vectors $\mathbf{u}, \mathbf{v} \in \mathbb{R}^n$, which is defined as a positive definite bilinear form. There exist orthonormal bases $B = (\mathbf{b}_1, \mathbf{b}_2, \dots, \mathbf{b}_n)$ with $\langle \mathbf{b}_i, \mathbf{b}_j \rangle = \delta_{ij}$, such that, for $\mathbf{u} = u_1\mathbf{b}_1 + \dots + u_n\mathbf{b}_n$ and $\mathbf{v} = v_1\mathbf{b}_1 + \dots + v_n\mathbf{b}_n$, the scalar product has the standard form

$$\langle \mathbf{u}, \mathbf{v} \rangle = u_1v_1 + u_2v_2 + \dots + u_nv_n.$$

Vectors $\mathbf{v}$ with $\langle \mathbf{v}, \mathbf{v} \rangle = 1$ are called *unit vectors.* Two vectors $\mathbf{u}, \mathbf{v}$ with $\langle \mathbf{u}, \mathbf{v} \rangle = 0$ are called *orthogonal.* In this sense, an orthonormal basis of $\mathbb{R}^n$ consists of n mutually orthogonal unit vectors.

In addition, the Euclidean space $\mathbb{E}^n$ also has an affine structure. It is the union of *affine subspaces* $A = \mathbf{u} + \mathbf{U}$ with $\mathbf{u} \in \mathbb{R}^n$ and with $\mathbf{U}$ as a subspace of the vector space $\mathbb{R}^n$. In the case $d = \dim \mathbf{U}$, the affine subspace $\mathbf{u}+\mathbf{U}$ is called d-dimensional. Two affine subspaces $A_i = \mathbf{u}_i + \mathbf{U}_i$, $i = 1,2$ are called *totally orthogonal* if any two vectors $\mathbf{v}_1 \in \mathbf{U}_1$ and $\mathbf{v}_2 \in \mathbf{U}_2$ are orthogonal.

The one-dimensional affine subspaces are called *lines*, those of dimension two are called *planes*. Affine subspaces of dimension zero are *points*. Instead of the point $\{\mathbf{a}\}$, we write *point* $\mathbf{a}$ for short. At the beginning, it might be confusing that points and vectors are denoted in the same way. However, in most of the cases it will be clear what is meant. We usually prefer the symbols $\mathbf{a}, \mathbf{c}, \ldots, \mathbf{p}, \mathbf{x}$ for points and $\mathbf{b}, \mathbf{u}, \mathbf{v}, \ldots$ for vectors. Sometimes, we speak of the *position vector* $\mathbf{a}$ of the point A in order to avoid any confusion with a *direction vector* of a line.

In order to define a *Cartesian coordinate frame* in $\mathbb{E}^n$, we have to specify an orthonormal basis $B = (\mathbf{b}_1, \mathbf{b}_2, \ldots, \mathbf{b}_n)$ of $\mathbb{R}^n$ and a point $\mathbf{o}$ as the origin. Then, the point X with position vector

$$\mathbf{x} = \mathbf{o} + x_1\mathbf{b}_1 + \cdots + x_n\mathbf{b}_n$$

receives the Cartesian coordinates $(x_1, \ldots, x_n)$ w.r.t. the coordinate frame $(\mathbf{o}; B)$. We often write $X = (x_1, \ldots, x_n)$ for short and call the column vector $\mathbf{x} = (x_1, \ldots, x_n)$ the *coordinate vector* of X w.r.t. $(\mathbf{o}; B)$, occasionally denoted by $\mathbf{x}_{(\mathbf{o};B)}$. In this sense, the original position vector $\mathbf{x}$ of X is the coordinate vector w.r.t. the *canonical* coordinate frame in $\mathbb{R}^n$ with the zero-vector $\mathbf{0} = (0, \ldots, 0)$ as origin and the vectors $\mathbf{e}_1 = (1, 0, \ldots, 0)$, $\ldots$, $\mathbf{e}_n = (0, \ldots, 0, 1)$ of the canonical basis E of $\mathbb{R}^n$.

If we replace the coordinate frame $(\mathbf{o}; B)$ by another coordinate frame $(\mathbf{o}'; B')$, then the original coordinates $(x_1, \ldots, x_n)$ of point X have to be replaced by new coordinates $(x_1', \ldots, x_n')$ such that

$$\mathbf{o} + x_1\mathbf{b}_1 + \cdots + x_n\mathbf{b}_n = \mathbf{o}' + x_1'\mathbf{b}_1' + \cdots + x_n'\mathbf{b}_n'.$$

Each new basis vector $\mathbf{b}_i'$ can be expressed as a linear combination of the original basis vectors as $\mathbf{b}_i' = t_{1i}\,\mathbf{b}_1 + \cdots + t_{ni}\,\mathbf{b}_n$ with certain coefficients $t_{ji} \in \mathbb{R}$. Furthermore, we can set $\mathbf{o}' - \mathbf{o} = t_1\mathbf{b}_1 + \cdots + t_n\mathbf{b}_n$. Then, the comparison of coefficients yields

$$\begin{pmatrix} x_1 \\ \vdots \\ x_n \end{pmatrix} = \begin{pmatrix} t_1 \\ \vdots \\ t_n \end{pmatrix} + \begin{pmatrix} t_{11} & \ldots & t_{1n} \\ \vdots & & \vdots \\ t_{n1} & \ldots & t_{nn} \end{pmatrix} \begin{pmatrix} x_1' \\ \vdots \\ x_n' \end{pmatrix},$$

or, in matrix form, as the relation between the new coordinate vector and the original one for each point $X \in \mathbb{E}^n$,

$$\mathbf{x}_{(\mathbf{o};B)} = \mathbf{t} + \mathbf{T}_B^{B'} \mathbf{x}_{(\mathbf{o}',B')}. \tag{3.2}$$

We note that the vector $\mathbf{t}$ contains the coordinates of the new origin $\mathbf{o}'$ w.r.t. the original coordinate frame $(\mathbf{o}, B)$, *i.e.*, $\mathbf{t} = \mathbf{o}'_{(\mathbf{o};B)}$. Similarily, the columns in the $n \times n$ *transformation matrix* $\mathbf{T}_B^{B'}$ consist of the original B-coordinates of the new basis vectors $\mathbf{b}'_1, \dots, \mathbf{b}'_n$. Therefore, the column vectors are linearly independent, and furthermore, they are mutually orthogonal unit vectors. Thus, the transformation matrix is orthogonal, which means that $\left(\mathbf{T}_B^{B'}\right)^{-1} = \left(\mathbf{T}_B^{B'}\right)^{\mathrm{T}}$.

Sometimes, it is useful to combine the column vector $\mathbf{t}$ and the orthogonal $n \times n$ matrix $T_B^{B'}$ in a single $(n+1) \times (n+1)$ matrix of the form

$$T_B^{B'*} = \left(\begin{array}{c|c} 1 & \mathbf{0}^{\mathrm{T}} \\ \hline \mathbf{t} & \mathbf{T}_B^{B'} \end{array}\right) = \left(\begin{array}{c|ccc} 1 & 0 & 0 & 0 \\ \hline t_1 & t_{11} & \dots & t_{1n} \\ \vdots & \vdots & & \vdots \\ t_n & t_{n1} & \dots & t_{nn} \end{array}\right).$$

We call $T_B^{B'*}$ the *extended transformation matrix.* This matrix is again invertible. If we also extend the coordinate vectors $\mathbf{x}_{(\mathbf{o};B)} \in \mathbb{R}^n$ of the points X to vectors $\mathbf{x}^*_{(\mathbf{o};B)} \in \mathbb{R}^{n+1}$ by inserting 1 in the first row, we obtain the *extended matrix representation* of the coordinate transformation

$$\mathbf{x}^*_{(\mathbf{o};B)} = \begin{pmatrix} 1 \\ \mathbf{x}_{(\mathbf{o};B)} \end{pmatrix} = \begin{pmatrix} 1 & \mathbf{0}^{\mathrm{T}} \\ \mathbf{t} & \mathbf{T}_B^{B'} \end{pmatrix} \begin{pmatrix} 1 \\ \mathbf{x}_{(\mathbf{o}';B')} \end{pmatrix} = \mathbf{T}_B^{B'*} \, \mathbf{x}^*_{(\mathbf{o}';B')}. \tag{3.3}$$

The advantage of this representation becomes apparent when different coordinate transformations have to be applied repeatedly. Then, the transformation matrix of the composition is just the product of the extended matrices of the single transformations. Moreover, the inversion of a coordinate transformation is more transparent.

Quadratic functions

In Linear Algebra, a *bilinear form* on any vector space $\mathbf{V}$ over the symmetric field $\mathbb{F}$ is defined as a mapping

$$\sigma\colon\ \mathbf{V} \times \mathbf{V} \to \mathbb{F}, \quad (\mathbf{u}, \mathbf{v}) \mapsto \sigma(\mathbf{u}, \mathbf{v})$$

which satisfies, for all $\mathbf{u}, \mathbf{u}', \mathbf{v}, \mathbf{v}' \in \mathbf{V}$ and $\lambda \in \mathbb{F}$,

$$\begin{aligned} \sigma(\mathbf{u}+\mathbf{u}', \mathbf{v}) &= \sigma(\mathbf{u}, \mathbf{v}) + \sigma(\mathbf{u}', \mathbf{v}), & \sigma(\mathbf{u}, \mathbf{v}+\mathbf{v}') &= \sigma(\mathbf{u}, \mathbf{v}) + \sigma(\mathbf{u}, \mathbf{v}'), \\ \sigma(\lambda\mathbf{u}, \mathbf{v}) &= \lambda\sigma(\mathbf{u}, \mathbf{v}), & \sigma(\mathbf{u}, \lambda\mathbf{v}) &= \lambda\sigma(\mathbf{u}, \mathbf{v}). \end{aligned}$$

A bilinear form σ which satisfies $\sigma(\mathbf{v}, \mathbf{u}) = \sigma(\mathbf{u}, \mathbf{v})$ for all $\mathbf{u}, \mathbf{v} \in \mathbf{V}$ is called *symmetric*.

When the bilinear form is restricted to the case $\mathbf{v} = \mathbf{u}$, we obtain a *quadratic form*

$$\varrho\colon\ \mathbf{V} \to \mathbb{F}, \quad \mathbf{u} \mapsto \varrho(\mathbf{u}) = \sigma(\mathbf{u}, \mathbf{u}).$$

It satisfies

$$\begin{aligned} \varrho(\mathbf{u}+\mathbf{v}) &= \sigma(\mathbf{u}+\mathbf{v},\ \mathbf{u}+\mathbf{v}) \\ &= \sigma(\mathbf{u},\ \mathbf{u}) + \sigma(\mathbf{v},\ \mathbf{v}) + \sigma(\mathbf{u}, \mathbf{v}) + \sigma(\mathbf{v}, \mathbf{u}) \\ &= \varrho(\mathbf{u}) + \varrho(\mathbf{v}) + \sigma'(\mathbf{u}, \mathbf{v}) \end{aligned} \tag{3.4}$$

with $\sigma'(\mathbf{u}, \mathbf{v}) = \sigma(\mathbf{u}, \mathbf{v}) + \sigma(\mathbf{v}, \mathbf{u})$ as a symmetric bilinear form. Conversely, this can be used for defining a quadratic form ϱ over the vector space $\mathbf{V}$, independently from any bilinear form, as a map $\varrho\colon \mathbf{V} \to \mathbb{F}$ which firstly satisfies $\varrho(\lambda\mathbf{u}) = \lambda^2\varrho(\mathbf{u})$ and where secondly the mapping

$$\widetilde{\sigma}\colon\ (\mathbf{u}, \mathbf{v}) \ \mapsto\ \varrho(\mathbf{u}+\mathbf{v}) - \varrho(\mathbf{u}) - \varrho(\mathbf{v})$$

is a bilinear form, and symmetric by definition. The restriction of $\widetilde{\sigma}$ to the case $\mathbf{v} = \mathbf{u}$ is a quadratic form $\widetilde{\varrho}$ with

$$\widetilde{\varrho}(\mathbf{u}) = \varrho(2\,\mathbf{u}) - 2\,\varrho(\mathbf{u}) = 2\,\varrho(\mathbf{u}).$$

From now on we exclude the case that the field $\mathbb{F}$ is of characteristic 2. Then, we can state that for each quadratic form ϱ there exists a symmetric bilinear form

$$\sigma_1 = \tfrac{1}{2}\widetilde{\sigma}\colon\ (\mathbf{u}, \mathbf{v}) \ \mapsto\ \tfrac{1}{2}\left[\varrho(\mathbf{u}+\mathbf{v}) - \varrho(\mathbf{u}) - \varrho(\mathbf{v})\right],$$

which yields again ϱ, when restricted to $\mathbf{v} = \mathbf{u}$. This bilinear form is called a *polar form* of ϱ. It is easy to show that σ_1 is unique, because from $\sigma_2(\mathbf{u}, \mathbf{u}) = \sigma_1(\mathbf{u}, \mathbf{u})$ for all $\mathbf{u} \in \mathbf{V}$ follows $\sigma_1(\mathbf{u}+\mathbf{v},\ \mathbf{u}+\mathbf{v}) = \sigma_2(\mathbf{u}+\mathbf{v},\ \mathbf{u}+\mathbf{v})$, and hence,

$$\sigma_1(\mathbf{u}, \mathbf{u}) + 2\sigma_1(\mathbf{u}, \mathbf{v}) + \sigma_1(\mathbf{v}, \mathbf{v}) = \sigma_2(\mathbf{u}, \mathbf{u}) + 2\sigma_2(\mathbf{u}, \mathbf{v}) + \sigma_2(\mathbf{v}, \mathbf{v})$$

and, consequently, $\sigma_1(\mathbf{u},\mathbf{v}) = \sigma_2(\mathbf{u},\mathbf{v})$.

Now, we combine a quadratic form ϱ with a linear form $\varphi\colon \mathbf{V} \to \mathbb{F}$ and add a constant $a \in \mathbb{F}$. This results in the mapping

$$\psi\colon \mathbf{V} \to \mathbb{F}, \quad \mathbf{x} \mapsto \psi(\mathbf{x}) = \varrho(\mathbf{x}) + 2\varphi(\mathbf{x}) + a,$$

which we call a *quadratic function* on $\mathbf{V}$. We will see that it makes sense to understand quadratic functions on $\mathbb{R}^n$ as mappings of points into $\mathbb{R}$, where $\psi(X) = \psi(\mathbf{x})$, provided that $\mathbf{x}$ is the position vector of point X.

Let us return to the Euclidean n-space, *i.e.*, to the case $\mathbb{F} = \mathbb{R}$ and $\mathbf{V} = \mathbb{R}^n$, equipped with a Cartesian coordinate frame $(\mathbf{o}; B)$ where $B = (\mathbf{b}_1, \dots, \mathbf{b}_n)$ is a basis of $\mathbb{R}^n$. In order to express $\psi(X)$ in terms of the $(\mathbf{o}; B)$-coordinates $(x_1, \dots, x_n)$ of the point X, we recall that the basis B assigns to the polar form σ of the quadratic form ϱ a symmetric matrix

$$\mathbf{A} = \begin{pmatrix} a_{11} & \cdots & a_{1n} \\ \vdots & & \vdots \\ a_{n1} & \cdots & a_{nn} \end{pmatrix} \quad \text{with} \quad a_{ij} = \sigma(\mathbf{b}_i, \mathbf{b}_j) = a_{ji},$$

and hence,

$$\varrho(x_1\mathbf{b}_1 + \dots + x_n\mathbf{b}_n) = \sum_{i=1}^{n} a_{ii}x_i^2 + 2\sum_{i=1}^{n-1}\sum_{j=i+1}^{n} a_{ij}x_ix_j.$$

Analogously, the basis B assigns to the linear form φ the row vector

$$\mathbf{a}^{\mathrm{T}} = (a_1, \dots, a_n), \quad \text{where} \quad a_i = \varphi(\mathbf{b}_i),$$

and hence,

$$\varphi(x_1\mathbf{b}_1 + \dots + x_n\mathbf{b}_n) = \sum_{i=1}^{n} a_ix_i.$$

However, we must also pay attention to the origin $\mathbf{o}$ of our coordinate frame $(\mathbf{o}; B)$. Since the coordinate vector $(x_1, \dots, x_n)$ of the point X w.r.t. $(\mathbf{o}; B)$ originates from the representation $\mathbf{x} = \mathbf{o} + x_1\mathbf{b}_1 + \dots + x_n\mathbf{b}_n$, we take (3.4) into account, which yields

$$\varrho\left(\mathbf{o} + \sum_i^n x_i\mathbf{b}_i\right) = \varrho\left(\sum_i^n x_i\mathbf{b}_i, \sum_i^n x_i\mathbf{b}_i\right) + \sigma\left(\mathbf{o}, \sum_i^n x_i\mathbf{b}_i\right) + \varrho(\mathbf{o}).$$

On the right-hand side, we have, beside the quadratic form, a linear form and a constant. This shows that the choice of the origin affects the vector $\mathbf{a}^{\mathrm{T}}$ of the linear form as well as the constant a in the coordinate representation of the given quadratic function ψ. Only the matrix $\mathbf{A}$, which represents the corresponding quadratic form, remains unchanged.

The same can be concluded from the matrix representation of quadratic functions, as shown below. This representation is based on a fixed Cartesian coordinate frame $(\mathbf{o}; B)$, which assigns to the points X of the Euclidean n-space the coordinates $(x_1, \ldots, x_n)$. For the sake of brevity, we denote the column vectors $\mathbf{x}_{(\mathbf{o};B)}$ of coordinates by $\mathbf{x}$ only, but we will not forget that they depend on the choice of the coordinate frame. Thus, we can set up any given quadratic function $\psi\colon \mathbb{R}^n \to \mathbb{R}$ in matrix form as

$$\begin{aligned} \psi(X) &= \mathbf{x}^{\mathrm{T}} \mathbf{A}\, \mathbf{x} + 2\, \mathbf{a}^{\mathrm{T}} \mathbf{x} + a = \\ &= (x_1, \ldots, x_n) \begin{pmatrix} a_{11} & \cdots & a_{1n} \\ \vdots & & \vdots \\ a_{n1} & \cdots & a_{nn} \end{pmatrix} \begin{pmatrix} x_1 \\ \vdots \\ x_n \end{pmatrix} + 2(a_1, \ldots, a_n) \begin{pmatrix} x_1 \\ \vdots \\ x_n \end{pmatrix} + a \end{aligned} \tag{3.5}$$

with a symmetric matrix $\mathbf{A} \in \mathbb{R}^{n \times n}$, a vector $\mathbf{a} \in \mathbb{R}^n$, and an additional constant $a \in \mathbb{R}$, which is the image $\psi(\mathbf{o})$ of the origin of the underlying coordinate frame. It is easy to see that the explicit form of the polynomial on the right-hand side is exactly the same as the one written on the left-hand side in the quadric's equation (3.1), shown at the beginning of this chapter. According to our definition, even the zero-function with $\psi(X) = 0$ for all points X is a quadratic function.

Sometimes, it is advantageous to combine the n^2 coefficients a_{ij} of $\mathbf{A}$ with the n components $(a_1, \ldots, a_n)$ of the row vector $\mathbf{a}^{\mathrm{T}}$ and the constant a in a symmetric $(n+1) \times (n+1)$ matrix in the form

$$\mathbf{A}^* = \left(\begin{array}{c|c} a & \mathbf{a}^{\mathrm{T}} \\ \hline \mathbf{a} & \mathbf{A} \end{array} \right) = \left(\begin{array}{c|ccc} a & a_1 & \ldots & a_n \\ \hline a_1 & a_{11} & \ldots & a_{1n} \\ \vdots & \vdots & & \vdots \\ a_n & a_{n1} & \ldots & a_{nn} \end{array} \right).$$

We call this symmetric matrix $\mathbf{A}^* \in \mathbb{R}^{(n+1) \times (n+1)}$ the *extended* $(\mathbf{o}; B)$-*matrix* of the quadratic function ψ. If we also use the extended coordinate vectors $\mathbf{x}^* \in \mathbb{R}^{n+1}$ of points X, as introduced in (3.3), we obtain the extended matrix representation of the quadratic function as follows:

$$\psi(X) = \mathbf{x}^{*\mathrm{T}} \mathbf{A}\, \mathbf{x}^{*} = (1, x_1, \ldots, x_n) \begin{pmatrix} a & a_1 & \ldots & a_n \\ a_1 & a_{11} & \ldots & a_{1n} \\ \vdots & \vdots & & \vdots \\ a_n & a_{n1} & \ldots & a_{nn} \end{pmatrix} \begin{pmatrix} 1 \\ x_1 \\ \vdots \\ x_n \end{pmatrix}. \tag{3.6}$$

Example 3.1.1 Extended matrix of a quadratic function.

The quadratic function $\psi\colon \mathbb{R}^3 \to \mathbb{R}$, given in canonical coordinates, with

$$\psi(\mathbf{x}) = 2x_1^2 - x_2^2 + 4x_1x_3 - 6x_2 - 2x_3 + 5$$

can be written in matrix form as

$$\psi(\mathbf{x}) = (x_1, x_2, x_3) \begin{pmatrix} 2 & 0 & 2 \\ 0 & -1 & 0 \\ 2 & 0 & 0 \end{pmatrix} \begin{pmatrix} x_1 \\ x_2 \\ x_3 \end{pmatrix} + (0,\ -6,\ -2) \begin{pmatrix} x_1 \\ x_2 \\ x_3 \end{pmatrix} + 5.$$

Therefore, the extended matrix of ψ w.r.t. the canonical coordinate frame is

$$\mathbf{A}^{*} = \left(\begin{array}{r|rrr} 5 & 0 & -3 & -1 \\ \hline 0 & 2 & 0 & 2 \\ -3 & 0 & -1 & 0 \\ -1 & 2 & 0 & 0 \end{array}\right).$$

Definition 3.1.1 If $\psi\colon \mathbb{R}^n \to \mathbb{R}$ is a quadratic function other than the zero-function, then the set of zeros

$$\mathcal{Q}(\psi) = \psi^{-1}(0) = \{\, \mathbf{x} \mid \psi(\mathbf{x}) = 0 \,\}$$

is called a *quadric* of the Euclidean n-space $\mathbb{E}^n$.

When the condition $\psi(\mathbf{x}) = 0$ is expressed in Cartesian coordinates, we call this the *equation* of $\mathcal{Q}(\psi)$ w.r.t. the underlying coordinate frame $(\mathbf{o}; B)$. In the case of the canonical coordinate frame, we speak of the *canonical equation*.

Example 3.1.2 Selected quadrics.

1. In $\mathbb{E}^2$, the quadric given by the canonical equation $x_1^2 + x_2^2 - 1 = 0$ is the unit circle. In $\mathbb{E}^3$, the same equation defines a circular cylinder of revolution.

2. In $\mathbb{E}^2$, the equation $x_1^2 - x_2^2 = 0$ defines the pair of lines $x_1 \pm x_2 = 0$. In $\mathbb{E}^3$, the same equation describes a pair of planes intersecting in the x_3-axis.

3. The set of zeros of $\psi(\mathbf{x}) = x_1^2$ in $\mathbb{E}^2$ is a single line, namely the x_2-axis $x_1 = 0$, but with multiplicity two in the language of algebraic geometry.

4. In the case $\psi(\mathbf{x}) = x_1^2 + x_2^2$, the quadric $\mathcal{Q}(\psi)$ in $\mathbb{E}^2$ consists of a single point, the origin of the cooordinate frame. In the complex extension, we obtain a pair of conjugate complex lines $x_1 \pm \mathrm{i}x_2 = 0$. In $\mathbb{E}^3$, the equation $\psi(\mathbf{x}) = 0$ defines a pair complex conjugate planes through the (real) x_3-axis.

These examples reveal that the quadratic function ψ determines the quadric $\mathcal{Q}(\psi)$ uniquely. In the converse direction, this does not hold. For example, the two quadratic functions ψ_1 and ψ_2 with the canonical representations

$$\psi_1(\mathbf{x}) = x_1^2 + x_2^2 \quad \text{and} \quad \psi_2(\mathbf{x}) = 2x_1^2 + x_2^2$$

share the sets of zeros $\mathcal{Q}(\psi_1) = \mathcal{Q}(\psi_2) = \{\mathbf{o}\}$ in $\mathbb{E}^2$. However, if we extend the underlying field $\mathbb{R}$ to $\mathbb{C}$, then the quadric defines the corresponding function uniquely, up to a scaling factor, except for the case that $\psi(\mathbf{x})$ is the square of a linear function, since the sets of zeros of $\psi_1(\mathbf{x}) = x_1$ and $\psi_2(\mathbf{x}) = x_1^2$ are the same.

Simplifying quadratic functions by coordinate transformations

Our goal is to classify the quadrics in $\mathbb{E}^n$. We achieve this by developing an algorithm which modifies the given coordinate frame in such a way that the given quadratic function is transformed into a normal form.

- **Example 3.1.3** An ellipse and the normal form of its equation.

The quadratic function $\psi\colon \mathbb{E}^2 \to \mathbb{R}$ with the canonical representation

$$\psi(\mathbf{x}) = 9x_1^2 - 4x_1x_2 + 6x_2^2 - 32x_1 - 4x_2 + 24$$

defines as its set of zeros $\mathcal{Q}(\psi)$ an ellipse with the semiaxes $\sqrt{2}$ and 1 (Figure 3.1). This follows from a transformation to an appropriate coordinate frame, where the new coordinates (x_1', x_2') are related to the original coordinates by

$$\begin{pmatrix} x_1' \\ x_2' \end{pmatrix} = \frac{1}{\sqrt{5}}\begin{pmatrix} -3 \\ -4 \end{pmatrix} + \frac{1}{\sqrt{5}}\begin{pmatrix} 2 & -1 \\ 1 & 2 \end{pmatrix}\begin{pmatrix} x_1 \\ x_2 \end{pmatrix}$$

(compare with (3.2)). Then, the set $\mathcal{Q}(\psi)$ satisfies the equation

$$x_1'^2 + \tfrac{1}{2}x_2'^2 - 1 = 0,$$

which is written in normal form.

We can confirm this by straight forward computation. First, we look for the inverse coordinate transformation. Since the transformation matrix is orthogonal, we obtain this by transposition:

$$\begin{pmatrix} x_1 \\ x_2 \end{pmatrix} = \frac{1}{\sqrt{5}}\begin{pmatrix} 2 & 1 \\ -1 & 2 \end{pmatrix}\begin{pmatrix} x_1' + 3/\sqrt{5} \\ x_2' + 4/\sqrt{5} \end{pmatrix} = \begin{pmatrix} 2 \\ 1 \end{pmatrix} + \frac{1}{\sqrt{5}}\begin{pmatrix} 2 & 1 \\ -1 & 2 \end{pmatrix}\begin{pmatrix} x_1' \\ x_2' \end{pmatrix}.$$

We plug this into the given quadratic function $\psi(\mathbf{x})$ and obtain, after some computations, the new equation.

Let us analyse how a transformation of coordinates, by virtue of (3.2), affects the matrix representation (3.5) of a quadratic form. For the sake

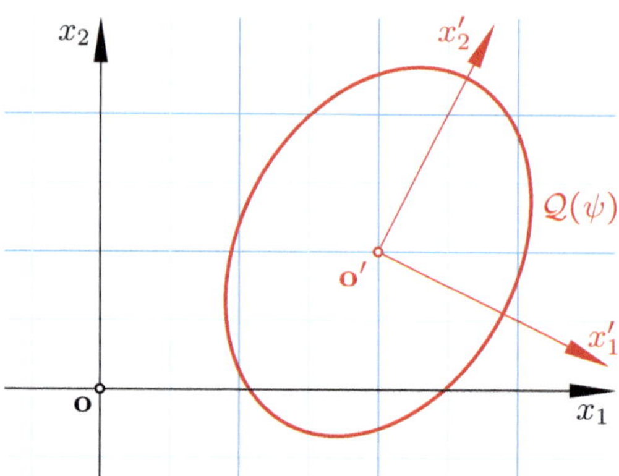

FIGURE 3.1. The quadric with the canonical equation $9x_1^2 - 4x_1x_2 + 6x_2^2 -32x_1 - 4x_2 + 24 = 0$ is an ellipse with the semiaxes $\sqrt{2}$ and 1.

of brevity, we write temporarily $\mathbf{x}$, $\mathbf{x}'$, and $\mathbf{T}$ instead of $\mathbf{x}_{(\mathbf{o};B)}$, $\mathbf{x}_{(\mathbf{o}';B')}$ and $\mathbf{T}_B^{B'}$, respectively. When plugging $\mathbf{x} = \mathbf{t} + \mathbf{T}\mathbf{x}'$ into

$$\psi(X) = \mathbf{x}^{\mathrm{T}}\mathbf{A}\mathbf{x} + 2\,\mathbf{a}^{\mathrm{T}}\mathbf{x} + a,$$

we obtain

$$\begin{aligned}\psi(X) &= (\mathbf{t}^{\mathrm{T}} + \mathbf{x}'^{\mathrm{T}}\mathbf{T}^{\mathrm{T}})\mathbf{A}(\mathbf{t} + \mathbf{T}\mathbf{x}') + 2\,\mathbf{a}^{\mathrm{T}}(\mathbf{t} + \mathbf{T}\mathbf{x}') + a \\ &= \mathbf{x}'^{\mathrm{T}}(\mathbf{T}^{\mathrm{T}}\mathbf{A}\,\mathbf{T})\,\mathbf{x}' + \left(\mathbf{t}^{\mathrm{T}}\mathbf{A}\,\mathbf{T}\mathbf{x}' + \mathbf{x}'^{\mathrm{T}}\mathbf{T}^{\mathrm{T}}\mathbf{A}\mathbf{t} + 2\,\mathbf{a}^{\mathrm{T}}\mathbf{T}\,\mathbf{x}'\right) \\ &\quad + \left(\mathbf{t}^{\mathrm{T}}\mathbf{A}\mathbf{t} + 2\,\mathbf{a}^{\mathrm{T}}\mathbf{t} + a\right).\end{aligned}$$

In the second row, the term which is linear in $\mathbf{x}'$ can still be modified. Because of $\mathbf{A}^{\mathrm{T}} = \mathbf{A}$, the real number $\mathbf{x}'^{\mathrm{T}}\mathbf{T}^{\mathrm{T}}\mathbf{A}\,\mathbf{t}$ can be rewritten as

$$\mathbf{x}'^{\mathrm{T}}\mathbf{T}^{\mathrm{T}}\mathbf{A}\,\mathbf{t} = \left(\mathbf{x}'^{\mathrm{T}}\mathbf{T}^{\mathrm{T}}\mathbf{A}\,\mathbf{t}\right)^{\mathrm{T}} = \mathbf{t}^{\mathrm{T}}\mathbf{A}\,\mathbf{T}\mathbf{x}'.$$

Thus, the linear term is

$$2\,\mathbf{t}^{\mathrm{T}}\mathbf{A}\,\mathbf{T}\mathbf{x}' + 2\,\mathbf{a}^{\mathrm{T}}\mathbf{T}\,\mathbf{x}' = 2\,(\mathbf{t}^{\mathrm{T}}\mathbf{A} + \mathbf{a}^{\mathrm{T}})\,\mathbf{T}\mathbf{x}'.$$

Lemma 3.1.1 *When the Cartesian coordinate frame* $(\mathbf{o}; B)$ *is replaced by* $(\mathbf{o}'; B')$*, where* $\mathbf{x}_{(\mathbf{o};B)} = \mathbf{t} + \mathbf{T}\,\mathbf{x}'_{(\mathbf{o}';B')}$*, then the quadratic function* ψ*, represented as* $\psi(X) = \mathbf{x}^{\mathrm{T}}_{(\mathbf{o};B)}\mathbf{A}\,\mathbf{x}_{(\mathbf{o};B)} + 2\,\mathbf{a}^{\mathrm{T}}\mathbf{x}_{(\mathbf{o};B)} + a$ *, gets the new matrix representation*

$$\psi(X) = \mathbf{x}^{\mathrm{T}}_{(\mathbf{o}';B')}\mathbf{A}'\mathbf{x}_{(\mathbf{o}';B')} + 2\,\mathbf{a}'^{\mathrm{T}}\mathbf{x}_{(\mathbf{o}';B')} + a'$$

where

$$\begin{aligned} \mathbf{A}' &= \mathbf{T}^{\mathrm{T}}\mathbf{A}\,\mathbf{T}, \\ \mathbf{a}' &= \mathbf{T}^{\mathrm{T}}(\mathbf{A}\,\mathbf{t}+\mathbf{a}), \\ a' &= \mathbf{t}^{\mathrm{T}}\mathbf{A}\,\mathbf{t}+2\mathbf{a}^{\mathrm{T}}\mathbf{t}+a=\psi(\mathbf{o}'). \end{aligned} \tag{3.7}$$

The transformation shown above becomes more transparent when extended matrices are used. The substitution

$$\mathbf{x}^* = \begin{pmatrix} 1 \\ \mathbf{x} \end{pmatrix} = \begin{pmatrix} 1 & \mathbf{0}^{\mathrm{T}} \\ \mathbf{t} & \mathbf{T} \end{pmatrix} \begin{pmatrix} 1 \\ \mathbf{x}' \end{pmatrix} = \mathbf{T}^*\mathbf{x}'^*$$

in (3.6) transforms $\mathbf{x}^{*\mathrm{T}}\mathbf{A}^*\mathbf{x}^*$ into

$$\mathbf{x}'^{*\mathrm{T}}\mathbf{A}'^*\mathbf{x}'^*, \text{ where } \mathbf{A}'^* = \mathbf{T}^{*\mathrm{T}}\mathbf{A}^*\mathbf{T}^*.$$

Now we start with the simplification of the quadric's equation w.r.t. a given Cartesian coordinate frame $(\mathbf{o}; B)$,

$$\mathbf{x}^{\mathrm{T}}\mathbf{A}\mathbf{x}+2\mathbf{a}^{\mathrm{T}}\mathbf{x}+a=0,$$

by an appropriate choice of a new basis B' and a new origin $\mathbf{o}'$. We proceed in two steps.

Step 1: We diagonalize the polar form of the corresponding quadratic form. In this way, we can eliminate all mixed terms $a_{ij}x_ix_j$ $(i \neq j)$ in the quadrics equation (3.1).

A well-known theorem from Linear Algebra states that, for each symmetric bilinear form σ over $\mathbb{R}^n$, there exists an orthonormal basis B' such that $\sigma(\mathbf{b}'_i, \mathbf{b}'_j) = 0$ for all (i, j) with $i \neq j$. With respect to the given basis B, the vectors $\mathbf{b}'_i$ are eigenvectors of the representing matrix $\mathbf{A} = (a_{ij})$ with $a_{ij} = \sigma(\mathbf{b}_i, \mathbf{b}_j)$. We recall that if $\mathbf{b}'_i$ is an eigenvector of $\mathbf{A}$ with eigenvalue λ_i, then

$$\mathbf{A}\,\mathbf{b}'_i = \lambda_i \mathbf{b}'_i.$$

Therefore, for all $i, j \in \{1, \dots, n\}$,

$$\begin{aligned} \sigma(\mathbf{b}'_i, \mathbf{b}'_j) &= \mathbf{b}'^{\mathrm{T}}_i\mathbf{A}\,\mathbf{b}'_j = \lambda_j\left(\mathbf{b}'^{\mathrm{T}}_i\mathbf{b}'_j\right) = \lambda_j\left\langle \mathbf{b}'_i, \mathbf{b}'_j\right\rangle \\ &= \left(\mathbf{A}^{\mathrm{T}}\mathbf{b}'_i\right)^{\mathrm{T}}\mathbf{b}'_j = \lambda_i\mathbf{b}'^{\mathrm{T}}_i\mathbf{b}'_j = \lambda_i\left\langle \mathbf{b}'_i, \mathbf{b}'_j\right\rangle. \end{aligned}$$

This implies

- $\sigma(\mathbf{b}'_i, \mathbf{b}'_i) = \lambda_i \|\mathbf{b}'_i\|^2 = \lambda_i$, and
- in the case $\lambda_i \neq \lambda_j$, $\langle \mathbf{b}'_i, \mathbf{b}'_j \rangle = 0$, which means that the vectors $\mathbf{b}'_i$ and $\mathbf{b}'_j$ are orthogonal.

Thus, we obtain the transformation matrix

$$\mathbf{T}_B^{B'} = (\mathbf{b}'_1 \ \dots \ \mathbf{b}'_n)$$

with the B-coordinates of the eigenvectors $\mathbf{b}'_1, \dots, \mathbf{b}'_n$ as column vectors, which transforms the matrix $\mathbf{A}$ representing σ w.r.t. B into the new matrix

$$\mathbf{A}' = \begin{pmatrix} \lambda_1 & & 0 \\ & \ddots & \\ 0 & & \lambda_n \end{pmatrix},$$

in diagonal form. We write $\mathbf{A}' = \operatorname{diag}(\lambda_1, \dots, \lambda_n)$ for short.

Another result from Linear Algebra states that, for symmetric bilinear forms over $\mathbb{R}^n$, the eigenvalues show the same distribution of signs, regardless of the used basis B. If we have p positive eigenvalues and σ has the rank r, then there are $r - p$ negative eigenvalues, and 0 is an $(n - r)$-fold eigenvalue. We call the triple $(p,\ r - p,\ n - r)$ the *signature*[1] of the symmetric bilinear form σ and of its representing matrices.

We reorder the eigenvalues with the corresponding vectors of our new basis B' such that we have p positive eigenvalues at the beginning, followed by $r - p$ negative eigenvalues, and finally $n - r$ zeros. Furthermore, we replace the positive eigenvalues λ_i by $1/\alpha_i^2$ and the negative ones with $-1/\alpha_i^2$, where $\alpha_i > 0$. As a result, the quadratic terms in the equation of $\mathcal{Q}(\psi)$ obtain the form

$$\mathbf{x}'^{\mathrm{T}} \mathbf{A}' \, \mathbf{x}' = \lambda_1 x_1'^2 + \dots + \lambda_r x_r'^2 = \frac{x_1'^2}{\alpha_1^2} + \dots + \frac{x_p'^2}{\alpha_p^2} - \frac{x_{p+1}'^2}{\alpha_{p+1}^2} - \dots - \frac{x_r'^2}{\alpha_r^2},$$

where $0 \le p \le r = \operatorname{rk}(\mathbf{A}) \le n$.

Elimination of linear terms in the quadric's equation

We continue to simplify the coordinate representation of a given quadratic function ψ on $\mathbb{R}^n$. We learned already that the modification of the origin

[1] In the literature, often the pair $(p,\ r - p)$ of natural numbers is called *signature* of σ, and also the symbol $(+\cdots+-\cdots-)$ with p plus and $r - p$ minus signs can be found.

does not affect the quadratic form, which is included in ψ. This means that we will not alter the diagonalization of the bilinear form σ, when we modify the origin such that most of the linear terms $a_i x_i$ in the equation of $\mathcal{Q}(\psi)$ will vanish. Hence, before we start with Step 2, we assume that in the equation w.r.t. $(\mathbf{o}; B)$,

$$\mathbf{x}^{\mathrm{T}} \mathbf{A} \mathbf{x} + 2\mathbf{a}^{\mathrm{T}} \mathbf{x} + a = 0$$

with $\mathbf{a} \neq \mathbf{0}$, we have already a diagonal matrix $\mathbf{A} = \mathrm{diag}(\lambda_1, \dots, \lambda_n)$, where $\lambda_i = 1/\alpha_i^2$ for $i = 1, \dots, p$, $\lambda_i = -1/\alpha_i^2$ for $i = p+1, \dots, r$, and finally $\lambda_i = 0$ for $i = r+1, \dots, n$.

Step 2: In order to eliminate the linear terms, our goal is to obtain a new Cartesian coordinate frame $(\mathbf{o}'; B)$ where the vector $\mathbf{a}'$ vanishes. By virtue of (3.7), we have $\mathbf{a}' = \mathbf{T}^{\mathrm{T}}(\mathbf{A}\,\mathbf{t} + \mathbf{a})$. Since the transformation matrix $\mathbf{T}$ is invertible, it is necessary and sufficient that the $(\mathbf{o}; B)$-coordinates of the new origin $\mathbf{o}'$, *i.e.*, the vector $\mathbf{t} = \mathbf{o}'_{(\mathbf{o};B)}$, fulfills the system of n inhomogeneous linear equations

$$\mathbf{A}\,\mathbf{t} = -\mathbf{a} \tag{3.8}$$

There are two cases to distinguish.

Case 2a: *The system* (3.8) *is solvable:*
In this case the rank of the matrix $\mathbf{A}$ must not increase if we extend the matrix $\mathbf{A}$ by the column $-\mathbf{a}$ to a matrix $\in \mathbb{R}^{n\times(n+1)}$. Consequently,

$$\mathrm{rk}(\mathbf{A}' \,|\, \mathbf{a}) = \mathrm{rk}(\mathbf{A}) = r,$$

hence, $a_{r+1} = \cdots = a_n = 0$, provided that $r < n$. Each vector $\mathbf{m}$ which satisfies (3.8) is the position vector w.r.t. $(\mathbf{o}; B)$ of a *center* of the quadric. The space of centers is $(n-r)$-dimensional and called *center space.* Quadrics with a center are called *central quadrics.*

In the Cartesian coordinate frame $(\mathbf{m}; B)$ the equation of the quadric $\mathcal{Q}(\psi)$ has the form

$$\frac{x_1'^2}{\alpha_1^2} + \cdots + \frac{x_p'^2}{\alpha_p^2} - \frac{x_{p+1}'^2}{\alpha_{p+1}^2} - \cdots - \frac{x_r'^2}{\alpha_r^2} + a' = 0. \tag{3.9}$$

The set $\mathcal{Q}(\psi)$ of points with coordinate vectors satisfying this equation will not change if we multiply the equation with any factor other than

0. In the case $a' \neq 0$, we multiply with $-1/a'$ which yields the constant -1. In the case $a' = 0$, we can assume that $p \geq r - p$, *i.e.*, there are at least as many positive summands as negative ones, since otherwise we can multiply the equation with -1.

In our simplified equation (3.9), only the squares of the $(\mathbf{m}; B)$-coordinates $x'_1, \ldots, x'_n$ of points X show up. Therefore, with $\psi(\mathbf{x}') = 0$, $\psi(-\mathbf{x}') = 0$ also holds. This means that each center $\mathbf{m}$ is a center of symmetry of the quadric $\mathcal{Q}(\psi)$.

If $r = \mathrm{rk}(\mathbf{A}) < n$, there exists a $(n-r)$-dimensional space of centers. Then, beside $\mathbf{m}$, each vector $\mathbf{m}' \in \mathbf{m} + \mathbf{U}$ satisfies (3.8), provided that $\mathbf{U}$ is the eigenspace to the eigenvalue 0, *i.e.*, it solves the system of homogeneous equations $\mathbf{Ax} = \mathbf{0}$.

In the case $a' = 0$, the quadratic polynomial on the left-hand side of (3.9) is homogeneous, and each center $\mathbf{m}$ is a point of the quadric. Then, the quadric is called *conical* or *quadratic cone*, since with each point $\mathbf{x} \in \mathcal{Q}(\psi)$, the full line connecting $\mathbf{m}$ and $\mathbf{x}$ belongs to the quadric, due to

$$\sum_{i=1}^{r} \lambda_i (t x'_i)^2 = t^2 \sum_{i=1}^{r} \lambda_i x'^2_i = 0 \quad \text{for all} \quad t \in \mathbb{R}.$$

Case 2b: *The system* (3.8) *has no solution, the quadric has no center:*
In this case with $r = \mathrm{rk}(\mathbf{A}) < \mathrm{rk}(\mathbf{A}\,|\,\mathbf{a}) \leq n$, not all linear terms $a_i x_i$ in the quadric's equation can be eliminated. As will be revealed, we can achieve that only the last term remains. This follows from a decomposition of the vector $\mathbf{a}$ in a sum of two orthogonal components, as stated below.

Lemma 3.1.2 *When the linear system* $\mathbf{A}\mathbf{x} = -\mathbf{a}$ *in* (3.8) *has no solution, we can represent the vector* $\mathbf{a}$ *as a sum of orthogonal vectors* $\mathbf{a}_0 + \mathbf{a}_1$ *such that* $\mathbf{a}_0 \neq \mathbf{0}$ *is an eigenvector of* $\mathbf{A}$ *with eigenvalue* 0 *and, on the other hand, the system* $\mathbf{A}\mathbf{x} = -\mathbf{a}_1$ *is solvable.*

Proof: We refer to the sequence of eigenvectors $\mathbf{b}_1, \ldots, \mathbf{b}_n$ of $\mathbf{A}$ in the same order as used before in Step 1. The r-dimensional space of vectors $\{\mathbf{A}\mathbf{x}\,|\,\mathbf{x} \in \mathbb{R}^n\}$ is spanned by the first r eigenvectors $\mathbf{b}_1, \ldots, \mathbf{b}_r$ of $\mathbf{A}$, because of $\mathbf{A}\mathbf{b}_i = \lambda_i \mathbf{b}_i$, while the remaining eigenvectors $\mathbf{b}_{r+1}, \ldots, \mathbf{b}_n$ span the eigenspace to the eigenvalue 0. Hence, in order to split $\mathbf{a}$ into the requested sum, it is sufficient to represent $\mathbf{a}$ as a linear combination of the basis vectors $\mathbf{b}_1, \ldots, \mathbf{b}_n$ and to decompose the sum into

$$\mathbf{a}_1 := \langle \mathbf{a}, \mathbf{b}_1 \rangle \, \mathbf{b}_1 + \cdots + \langle \mathbf{a}, \mathbf{b}_r \rangle \, \mathbf{b}_r \quad \text{and} \quad \mathbf{a}_0 := \langle \mathbf{a}, \mathbf{b}_{r+1} \rangle \, \mathbf{b}_{r+1} + \cdots + \langle \mathbf{a}, \mathbf{b}_n \rangle \, \mathbf{b}_n.$$

According to our assumption, we have $\mathbf{a}_0 \neq \mathbf{0}$, since otherwise the system $\mathbf{A}\mathbf{x} = -\mathbf{a}$ would be solvable. ■

When specifying a new origin $\mathbf{o}'$, *i.e.*, when replacing $\mathbf{x}$ with $\mathbf{x}' + \mathbf{o}'$, then, due to (3.7), the coordinate vector $\mathbf{a}$ of the linear form w.r.t. $(\mathbf{o}; B)$ has to be replaced by

$$\mathbf{a}' = \mathbf{A}\,\mathbf{o}' + \mathbf{a} = \mathbf{A}\,\mathbf{o}' + \mathbf{a}_1 + \mathbf{a}_0 .$$

By virtue of Lemma 3.1.2, there exists a vector $\mathbf{o}'$ which solves the system $\mathbf{A}\,\mathbf{x} = -\mathbf{a}_1$. Consequently, in the quadric's equation, the linear form $\mathbf{a}_0^{\mathrm{T}}\mathbf{x}'$, or explicitly,

$$\langle \mathbf{a}_0, \mathbf{b}_{r+1} \rangle\, x'_{r+1} + \cdots + \langle \mathbf{a}_0, \mathbf{b}_n \rangle\, x'_n$$

remains. We recall that, according to Step 1, the vectors $\mathbf{b}_{r+1}, \ldots, \mathbf{b}_n$ have to form an orthonormal basis of the $(n-r)$-dimensional eigenspace of $\mathbf{A}$ to the eigenvalue 0. Moreover, the vector $\mathbf{a}_0 \neq \mathbf{0}$ belongs to this space. Hence, we will not spoil the obtained diagonal form of the matrix $\mathbf{A}$ when we choose, within this subspace, a new orthonormal basis with $\mathbf{b}_n$ in the direction of $\mathbf{a}_0$, *i.e.*, with $\mathbf{a}_0 = a'_n \mathbf{b}_n$ and $a'_n \neq 0$. This reduces the coordinate representation of the linear form to the last term $2a'_n x'_n$. On the other hand, because of $\mathrm{rk}(\mathbf{A}) = r < n$, there is no term with x'^2_n in the quadric's equation.

Finally, we can eliminate the constant a', when we replace the origin $\mathbf{o}'$ by $\mathbf{o}'' = \mathbf{o}' + \mu\, \mathbf{b}_n$ with an appropriate $\mu \in \mathbb{R}$. This means that we substitute $x'_n = x''_n - \lambda$. Thus, we obtain, on the left-hand side of the quadrics equation,

$$\sum_{i=1}^{r} \lambda_i\, x_i^2 + 2\, a'_n (x''_n - \lambda) + a,$$

and we can set $\lambda = a/2a'_n$.

We continue our efforts of simplifying the quadric's equation. We can assume that, in the particular case in question, among the quadratic terms the number p of positive summands is not smaller than that of negative summands, at least after multiplying the equation with -1. After the division of the equation by $|a'_n|$, the last coefficient is reduced to 2 or -2. In the first case, we can reverse the direction of $\mathbf{b}_n$ and hence replace x''_n by $-x''_n$.

We now summarize the forgoing results: Due to an appropriate choice of a Cartesian coordinate frame for each given quadric $\mathcal{Q}(\psi)$, we can reduce the equation to one of three normal forms, as given below.

Theorem 3.1.1 (Classification of quadrics in $\mathbb{E}^n$) *There are three types of real quadrics to be distinguished. Their normal forms are as follows:*

Type 1, conical quadrics: $0 \le p \le r \le n,\ \ p \ge r-p,$
$\mathrm{rk}(\mathbf{A}^*) = \mathrm{rk}(\mathbf{A}\,|\,\mathbf{a}) = \mathrm{rk}(\mathbf{A}) = r:$

$$\frac{x_1^2}{\alpha_1^2} + \cdots + \frac{x_p^2}{\alpha_p^2} - \frac{x_{p+1}^2}{\alpha_{p+1}^2} - \cdots - \frac{x_r^2}{\alpha_r^2} = 0.$$

Type 2, central quadrics: $0 \le p \le r \le n,$
$\mathrm{rk}(\mathbf{A}^*) > \mathrm{rk}(\mathbf{A}\,|\,\mathbf{a}) = \mathrm{rk}(\mathbf{A}) = r:$

$$\frac{x_1^2}{\alpha_1^2} + \cdots + \frac{x_p^2}{\alpha_p^2} - \frac{x_{p+1}^2}{\alpha_{p+1}^2} - \cdots - \frac{x_r^2}{\alpha_r^2} - 1 = 0.$$

Type 3, parabolic quadrics: $0 \le p \le r < n,\ \ p \ge r-p,$
$\mathrm{rk}(\mathbf{A}\,|\,\mathbf{a}) > \mathrm{rk}(\mathbf{A}) = r:$

$$\frac{x_1^2}{\alpha_1^2} + \cdots + \frac{x_p^2}{\alpha_p^2} - \frac{x_{p+1}^2}{\alpha_{p+1}^2} - \cdots - \frac{x_r^2}{\alpha_r^2} - 2\,x_n = 0.$$

Central and conical quadrics with $r < n$ and parabolic quadrics with $r < n-1$ are also called cylindrical.

Now, we are able to define the term ‘regular quadric’ for all dimensions. We have already used this term in Chapter 2.

Definition 3.1.2 A quadric is called *regular* if its extended matrix $\mathbf{A}^*$ has full rank. Otherwise, we speak of a *singular* quadric.

In (3.3), it was shown that each coordinate transformation acts on the extended matrix of a quadric in the form of a right-multiplication with a regular matrix. This reveals that the term ‘regular quadric’ is invariant w.r.t. coordinate transformations. We conclude from the classification in Theorem 3.1.1 that regular quadrics are either central or parabolic, but never conical.

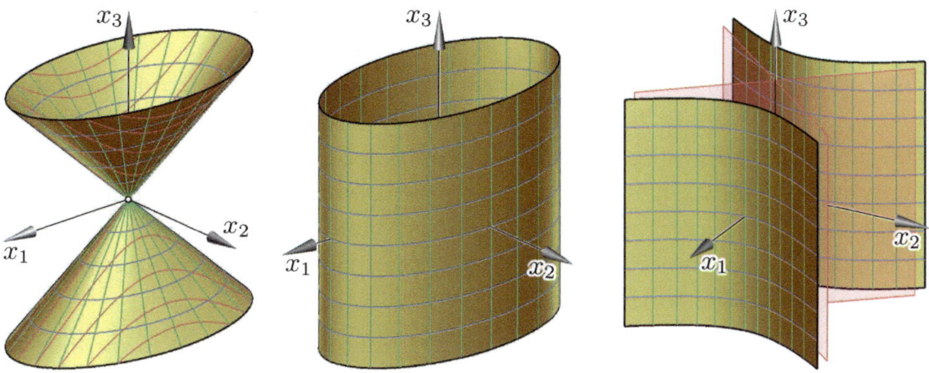

FIGURE 3.2. Examples of singular quadrics in $\mathbb{E}^3$: quadratic cone (left) with hyperbolas, elliptic cylinder (middle) and hyperbolic cylinder (right) with asymptotic planes.

Remark 3.1.1 It should be pointed out that, in order to figure out the type of any given quadric, it is not necessary to determine all eigenvalues of $\mathbf{A}$ explicitly. Firstly, the characteristic numbers p and r in these normal forms follow from the signature of the included quadratic form, and this can be found by elementary operations on rows and columns of the representing matrix $\mathbf{A}$. Secondly, the decision between the three types can be made by comparing the ranks of the matrices $\mathbf{A}$, $(\mathbf{A}\,|\,\mathbf{a})$ and the extended matrix $\mathbf{A}^*$, as listed above in Theorem 3.1.1.

In Chapter 2, we learned that one-sheeted hyperboloids and hyperbolic paraboloids are the only regular quadrics which are ruled surfaces. A generalization for higher dimensions will be presented in Section 4.3: The maximum dimension of linear spaces on quadrics in the real projective n-space depends on the signature of the representing quadratic form. For the sake of completeness, we anticipate this result w.r.t. the normal forms in $\mathbb{E}^n$.

Theorem 3.1.2 *Referring to the normal forms of quadrics $\mathcal{Q}$, as given in Theorem 3.1.1, the maximum dimension d of linear spaces on $\mathcal{Q}$ is as given below:*

Type 1, conical quadrics: $d = n - p$.

Type 2, central quadrics: *for* $2p \geq r + 1$: $d = n - p$,
for $2p < r + 1$: $d = p - 1 + n - r$.

Type 3, parabolic quadrics: $d = n - p - 1$.

$d < 0$ *means that* $\mathcal{Q} = \{\}$.

Principal axes and vertices

The normal form of a given quadric $\mathcal{Q}$ is always related to a particular Cartesian coordinate frame $(\mathbf{o}; B)$ with B as a basis of eigenvectors of the matrix $\mathbf{A}$. However, this frame need not be unique. In the case of an eigenvalue of multiplicity $k \geq 2$, there is a free choice of the orthonormal basis within the corresponding k-dimensional eigenspace. If the center space of $\mathcal{Q}$ has a dimension $(n-r) \geq 1$, then the coordinate frame can be translated parallel to this space.

Whenever a coordinate axis of a normalizing frame $(\mathbf{o}; B)$ is an axis of symmetry of the quadric $\mathcal{Q}$, it is called a *principal axis*. Hence, for central or conical quadrics $\mathcal{Q}$, all coordinate axes of the frame $(\mathbf{o}; B)$ are principal axes. For parabolic quadrics $\mathcal{Q}$, only the axis in direction of the last basis vector $\mathbf{b}_n$ is an axis of symmetry, since the mapping $(x_1, \dots, x_{n-1}, x_n) \mapsto (-x_1, \dots, -x_{n-1}, x_n)$ transforms $\mathcal{Q}$ onto itself.

A quadric $\mathcal{Q}$ of type 2, in the sense of Theorem 3.1.1, intersects the axis $\mathbf{m} + \mathbf{b}_j\mathbb{R}$, $1 \leq j \leq p$ at points with coordinates $(0, \dots, 0, x_j, 0, \dots, 0)$ where $x_j^2/\alpha_j^2 - 1 = 0$, and hence $x_j = \pm\alpha_j$. These points are called *vertices* of $\mathcal{Q}$, and the corresponding distance α_j to the center is a *semiaxis* of $\mathcal{Q}$. The principal axes $\mathbf{m} + \mathbf{b}_j\mathbb{R}$ with $p+1 \leq j \leq r$ are also called *secondary axes* of $\mathcal{Q}$. Their points of intersection with $\mathcal{Q}$ have an imaginary coordinate $x_j = \pm\, i\sqrt{-\alpha_j}$.

Quadrics of type 3 are called parabolic, because their intersections with planes spanned by the axes x_n and x_j for $j \leq r$ are parabolas satisfying the equations $x_j^2/\alpha_j \pm 2x_n = 0$ (note the Figures 3.4 and 3.7).

Corollary 3.1.3 *Let* $\mathbf{x}^{\mathrm{T}}\mathbf{A}\,\mathbf{x} + 2\,\mathbf{a}^{\mathrm{T}}\mathbf{x} + a = 0$ *be the equation of the quadric* $\mathcal{Q}$ *in* $\mathbb{E}^n$ *w.r.t. any coordinate frame* $(\mathbf{o}; B)$. *If* $\mathcal{Q}$ *is a central quadric, then the space of centers* $\mathbf{m}$ *satisfies the system of linear equations* $\mathbf{A}\mathbf{x} = -\mathbf{a}$. *In the case of a conical quadric, this system defines the* $(n-r)$*-dimensional apex, provided that* $\mathrm{rk}(\mathbf{A}) = r$.

The proof is left as an exercise for the readers (Exercise 3.1.1).

▪ **Example 3.1.4** Principal axis transformation of a conic.

We apply the algorithm, as presented above, to the ellipse depicted in Figures 3.1 and 3.3 with the canonical equation

$$\psi(\mathbf{x}) = 9x_1^2 - 4x_1x_2 + 6x_2^2 - 32x_1 - 4x_2 + 24 = 0.$$

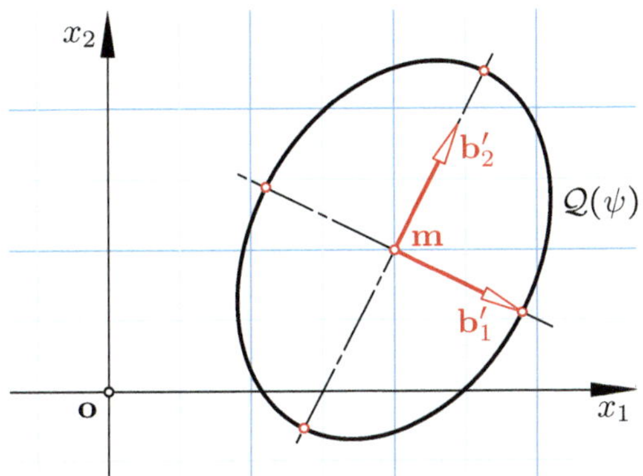

FIGURE 3.3. The principal axes of the ellipse with the canonical equation $9x_1^2 - 4x_1x_2 + 6x_2^2 - 32x_1 - 4x_2 + 24 = 0$ (Example 3.1.1).

In the notation of (3.5), we have

$$\mathbf{A} = \begin{pmatrix} 9 & -2 \\ -2 & 6 \end{pmatrix}, \quad \mathbf{a} = \begin{pmatrix} -16 \\ -2 \end{pmatrix} \quad \text{and} \quad a = 24.$$

Step 1: The characteristic polynomial of $\mathbf{A}$,

$$\det(\mathbf{A} - \lambda \mathbf{I}_2) = \det \begin{pmatrix} 9-\lambda & -2 \\ -2 & 6-\lambda \end{pmatrix} = \lambda^2 - 15\lambda + 50, \text{ where } \mathbf{I}_2 = \begin{pmatrix} 1 & 0 \\ 0 & 1 \end{pmatrix},$$

has the zeros

$$\lambda_1 = 10, \text{ and } \lambda_2 = 5.$$

These are the eigenvalues of $\mathbf{A}$. Corresponding eigenvectors satisfy the system $(\mathbf{A} - \lambda_i \mathbf{I}_2)\,\mathbf{x} = \mathbf{0}$. Thus, we obtain an orthonormal basis of eigenvectors

$$\mathbf{b}_1' = \frac{1}{\sqrt{5}} \begin{pmatrix} 2 \\ -1 \end{pmatrix}, \quad \mathbf{b}_2' = \frac{1}{\sqrt{5}} \begin{pmatrix} 1 \\ 2 \end{pmatrix}.$$

Step 2: According to Corollary 3.1.3, the center $\mathbf{m}$ satisfies

$$\mathbf{A}\,\mathbf{x} = -\mathbf{a}.$$

Its unique solution is $\mathbf{m} = (2, 1)$. The transformation of the given canonical coordinates $(\mathbf{o}; E)$ to the principal axes $(\mathbf{m}; B')$ can be expressed as

$$\begin{pmatrix} x_1 \\ x_2 \end{pmatrix} = \mathbf{m} + \mathbf{T}_E^{B'} \begin{pmatrix} x_1' \\ x_2' \end{pmatrix} \quad \text{with} \quad \mathbf{T}_E^{B'} = (\mathbf{b}_1' \ \mathbf{b}_2') = \frac{1}{\sqrt{5}} \begin{pmatrix} 2 & 1 \\ -1 & 2 \end{pmatrix}.$$

This generates the simplified equation

$$\lambda_1 \, x_1'^2 + \lambda_2 \, x_2'^2 + \psi(\mathbf{m}) = 10\, x_1'^2 + 5\, x_2'^2 - 10 = 0.$$

After division by 10, we obtain the normal form

$$x_1'^2 + \tfrac{1}{2}\, x_2'^2 - 1 = 0$$

with the semiaxes $\alpha_1 = 1$ and $\alpha_2 = \sqrt{2}$. The vertices of this ellipse are marked in Figure 3.3 and have the coordinates

$$\left(2 \pm \frac{2}{\sqrt{5}},\ 1 \mp \frac{1}{\sqrt{5}}\right), \quad \left(2 \pm \sqrt{\frac{2}{5}},\ 1 \pm 2\sqrt{\frac{2}{5}}\right).$$

The next example shows the algorithm in the case of a parabolic quadric of $\mathbb{E}^3$.

■ **Example 3.1.5** Principal axis transformation of a parabolic quadric.

Let the quadratic function $\psi\colon\ \mathbb{R}^3 \to \mathbb{R}$ be given by its coordinate representation w.r.t. any given Cartesian coordinate frame $(\mathbf{o}; B)$ as

$$\psi(\mathbf{x}) = x_1^2 + 2x_1x_2 + x_2^2 + 4x_1 + 2x_2 - 4x_3 - 3 = 0.$$

A comparison with the matrix representation $\mathbf{x}^{\mathrm{T}}\mathbf{A}\,\mathbf{x} + 2\,\mathbf{a}^{\mathrm{T}}\mathbf{x} + a$ results in

$$\mathbf{A} = \begin{pmatrix} 1 & 1 & 0 \\ 1 & 1 & 0 \\ 0 & 0 & 0 \end{pmatrix}, \quad \mathbf{a} = \begin{pmatrix} 2 \\ 1 \\ -2 \end{pmatrix}, \quad a = -3.$$

Step 1: The characteristic polynomial

$$\det(\mathbf{A} - \lambda\,\mathbf{I}_3) = -\lambda^2(\lambda - 2)$$

yields as zeros the eigenvalue $\lambda_1 = 2$ with multiplicity 1 and the two-fold eigenvalue $\lambda_2 = 0$.

$$\mathbf{b}_1' = \frac{1}{\sqrt{2}}\begin{pmatrix} 1 \\ 1 \\ 0 \end{pmatrix}$$

is a normalized eigenvector to λ_1. It spans the eigenspace $\mathbf{U}_1 := \{\mathbf{A}\mathbf{x} \mid \mathbf{x} \in \mathbb{R}^3\}$, which is the image space of the linear mapping $\mathbb{R}^3 \to \mathbb{R}^3$ with $\mathbf{x} \mapsto \mathbf{A}\mathbf{x}$. This eigenspace is orthogonal to the two-dimensional kernel of this mapping, the space $\mathbf{U}_2 := \{\mathbf{x} \mid \mathbf{A}\mathbf{x} = \mathbf{0}\}$ as the eigenspace of $\lambda_2 = 0$.

Step 2: Before we specify an orthonormal basis of $\mathbf{U}_2$, we check the system $\mathbf{A}\mathbf{x} = -\mathbf{a}$, given in Corollary 3.1.3.

Obviously, $\mathbf{a}$ is not contained in $\mathbf{U}_1$, *i.e.*, not a scalar multiple of $\mathbf{b}_1'$. According to Step 2b in our algorithm given above, we have to decompose $\mathbf{a}$ into a sum of two orthogonal components $\mathbf{a}_0 + \mathbf{a}_1$, where $\mathbf{a}_1 \in \mathbf{U}_1$ and $\mathbf{a}_0$ in the eigenspace $\mathbf{U}_2$ of $\lambda_2 = 0$. This yields

$$\mathbf{a}_1 = \langle \mathbf{a}, \mathbf{b}_1' \rangle \mathbf{b}_1' = \frac{3}{2}\begin{pmatrix} 1 \\ 1 \\ 0 \end{pmatrix} \quad \text{and} \quad \mathbf{a}_0 = \mathbf{a} - \mathbf{a}_1 = \frac{1}{2}\begin{pmatrix} 1 \\ -1 \\ -4 \end{pmatrix}.$$

Normalization of $\mathbf{a}_0$ yields the third vector of our orthonormal basis of eigenvectors, namely

$$\mathbf{b}_3' = \frac{1}{3\sqrt{2}}\begin{pmatrix} 1 \\ -1 \\ -4 \end{pmatrix}, \quad \text{and} \quad \mathbf{b}_2' = \mathbf{b}_3' \times \mathbf{b}_1' = \frac{1}{3}\begin{pmatrix} 2 \\ -2 \\ 1 \end{pmatrix}.$$

For the origin $\mathbf{o}'$ we choose, as a particular solution of $\mathbf{A}\,\mathbf{x} = -\mathbf{a}_1$, for example

$$\mathbf{o}' = \begin{pmatrix} -3/2 \\ 0 \\ 0 \end{pmatrix}.$$

The coordinate transformation

$$\begin{pmatrix} x_1 \\ x_2 \\ x_3 \end{pmatrix} = \mathbf{o}' + \mathbf{T}_B^{B'}\begin{pmatrix} x_1' \\ x_2' \\ x_3' \end{pmatrix} \quad \text{with} \quad \mathbf{T}_B^{B'} = (\mathbf{b}_1', \mathbf{b}_2', \mathbf{b}_3')$$

yields, by virtue of (3.7), the new matrix representation with

$$\mathbf{A}' = \begin{pmatrix} 2 & 0 & 0 \\ 0 & 0 & 0 \\ 0 & 0 & 0 \end{pmatrix}, \quad \mathbf{a}' = \begin{pmatrix} 0 \\ 0 \\ 3/\sqrt{2} \end{pmatrix}, \quad a' = -\frac{27}{4} = \psi(\mathbf{o}').$$

Consequently, the simplified equation of our quadric is

$$2x_1'^2 + 3x_3'\sqrt{2} - 27/4 = 0. \tag{3.10}$$

We combine the last two summands as

$$3\sqrt{2}\left(x_3' - \frac{27}{12\sqrt{2}}\right) = 3\sqrt{2}\left(x_3' - \frac{9}{4\sqrt{2}}\right) := 3\sqrt{2}\, x_3''$$

Thus, we replace the origin $\mathbf{o}'$ by

$$\mathbf{o}'' = \mathbf{o}' + \frac{9}{4\sqrt{2}}\mathbf{b}_3' = \begin{pmatrix} -9/8 \\ -3/8 \\ -3/2 \end{pmatrix}.$$

Finally, we multiply the equation (3.10) with $2/(3\sqrt{2})$ and reverse the direction of $\mathbf{b}_3'$. In order to avoid a left-handed coordinate frame, we simultaneously replace $\mathbf{b}_2'$ by $-\mathbf{b}_2'$. The basis $B'' = (\mathbf{b}_1', -\mathbf{b}_2', -\mathbf{b}_3')$ and the origin $\mathbf{o}''$ define a coordinate frame, w.r.t. which our quadric obtains the normal form

$$\frac{4}{3\sqrt{2}}x_1''^2 - 2x_3'' = 0.$$

This is the equation of a parabolic cylinder (see Figure 3.7 on page 117). The coordinate transformation between the original coordinate frame $(\mathbf{o}; B)$ and $(\mathbf{o}''; B'')$ can be expressed by

$$\begin{pmatrix} x_1 \\ x_2 \\ x_3 \end{pmatrix} = -\frac{3}{8}\begin{pmatrix} 3 \\ 1 \\ 4 \end{pmatrix} + \frac{1}{3\sqrt{2}}\begin{pmatrix} 3 & -2\sqrt{2} & -1 \\ 3 & 2\sqrt{2} & 1 \\ 0 & -\sqrt{2} & 4 \end{pmatrix}\begin{pmatrix} x_1'' \\ x_2'' \\ x_3'' \end{pmatrix},$$

or, in the inverse version,

$$\begin{pmatrix} x_1'' \\ x_2'' \\ x_3'' \end{pmatrix} = \frac{1}{4\sqrt{2}}\begin{pmatrix} 6 \\ -4\sqrt{2} \\ 7 \end{pmatrix} + \frac{1}{3\sqrt{2}}\begin{pmatrix} 3 & 3 & 0 \\ -2\sqrt{2} & 2\sqrt{2} & -\sqrt{2} \\ -1 & 1 & 4 \end{pmatrix}\begin{pmatrix} x_1 \\ x_2 \\ x_3 \end{pmatrix}.$$

• **Exercise 3.1.1** Center of a given quadric.

Prove in matrix form the statement of Corollary 3.1.3: If the quadric $\mathcal{Q}$ has the equation $\mathbf{x}^{\mathrm{T}}\mathbf{A}\mathbf{x} + 2\mathbf{a}^{\mathrm{T}}\mathbf{x} + a = 0$, then each vector $\mathbf{m}$ satisfying the system $\mathbf{A}\mathbf{x} = -\mathbf{a}$ is the position vector of a center of symmetry of $\mathcal{Q}$, *i.e.*, the reflection in $\mathbf{m}$ maps $\mathcal{Q}$ onto itself.

3.2 Quadrics in the Euclidean plane and 3-space

We continue with a detailed discussion of Theorem 3.1.1 in the form of enumerations of the quadrics in the Euclidean spaces $\mathbb{E}^2$ and $\mathbb{E}^3$. According to the underlying definition of quadrics in this Chapter, this results in a list of curves and surfaces of degree two (note Figures 3.4 and 3.7).

Regular and singular quadrics in $\mathbb{E}^2$

Type 1, conical quadrics: The conditions $0 \le p \le r \le 2$ and $p \ge r - p$ for the characteristic numbers (r, p) admit three cases:

1a) $(r,p) = (2,2)$, $\psi(\mathbf{x}) = \frac{x_1^2}{\alpha_1^2} + \frac{x_2^2}{\alpha_2^2}$, $\mathcal{Q}(\psi) = \{\mathbf{0}\}$,

1b) $(r,p) = (2,1)$, $\psi(\mathbf{x}) = \frac{x_1^2}{\alpha_1^2} - \frac{x_2^2}{\alpha_2^2}$, $\mathcal{Q}(\psi)$ consists of two lines,

1c) $(r,p) = (1,1)$, $\psi(\mathbf{x}) = x_1^2$, $\mathcal{Q}(\psi)$ is a single line (with multiplicity two).

In the case 1c, but also in the case 1b, the quadratic function $\psi(\mathbf{x})$ is *reducible* over $\mathbb{R}$, since

$$\frac{x_1^2}{\alpha_1^2} - \frac{x_2^2}{\alpha_2^2} = \left(\frac{x_1}{\alpha_1} + \frac{x_2}{\alpha_2}\right)\left(\frac{x_1}{\alpha_1} - \frac{x_2}{\alpha_2}\right).$$

Type 2, central quadrics: There are five cases:

2a) $(r,p) = (2,2)$, $\psi(\mathbf{x}) = \frac{x_1^2}{\alpha_1^2} + \frac{x_2^2}{\alpha_2^2} - 1$, $\mathcal{Q}(\psi)$ is an *ellipse*,

2b) $(r,p) = (2,1)$, $\psi(\mathbf{x}) = \frac{x_1^2}{\alpha_1^2} - \frac{x_2^2}{\alpha_2^2} - 1$, $\mathcal{Q}(\psi)$ is a *hyperbola*,

2c) $(r,p) = (2,0)$, $\psi(\mathbf{x}) = -\frac{x_1^2}{\alpha_1^2} - \frac{x_2^2}{\alpha_2^2} - 1$, $\mathcal{Q}(\psi) = \{\}$,

2d) $(r,p) = (1,1)$, $\psi(\mathbf{x}) = \frac{x_1^2}{\alpha_1^2} - 1$, $\mathcal{Q}(\psi)$ consists of two parallel lines,

2e) $(r,p) = (1,0)$, $\psi(\mathbf{x}) = -\frac{x_1^2}{\alpha_1^2} - 1$, $\mathcal{Q}(\psi) = \{\}$.

An ellipse with $\alpha_1 = \alpha_2$ is a circle with radius α_1. For an ellipse with $\alpha_1 > \alpha_2$, the first coordinate axis is called *major axis* with the *major semiaxis* α_1. The other is the *minor axis* with the *minor semiaxis* α_2.

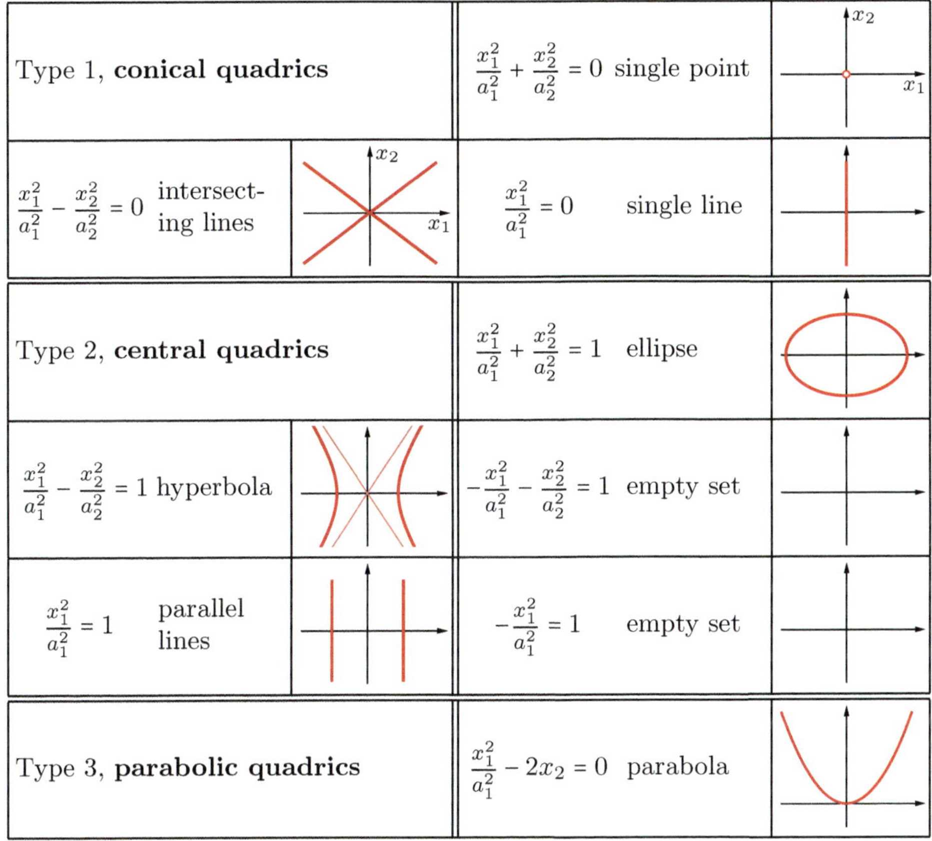

FIGURE 3.4. Quadrics in the Euclidean plane $\mathbb{E}^2$.

In the case of a hyperbola, the first coordinate axis is sometimes called *principal axis*, while the axis with non-real vertices is the *secondary axis*. The two lines through the origin with the equation $\frac{x_1}{\alpha_1} = \pm\frac{x_2}{\alpha_2}$ are the *asymptotes* of the hyperbola. They form a quadric of type 1b.

Type 3, parabolic quadrics: There are two normal forms to be distinguished.

3a) $(r,p) = (1,1)$, $\psi(\mathbf{x}) = \frac{x_1^2}{\alpha_1^2} - 2x_2$, $\mathcal{Q}(\psi)$ is a *parabola* with the *vertex* at the origin,

3b) $(r,p) = (0,0)$, $\psi(\mathbf{x}) = -x_2$, $\mathcal{Q}(\psi)$ is a single line.

FIGURE 3.5. Parabolic water trajectories, illustrating Exercise 3.2.1.

The constant α_1^2 in the parabola's equation is called the *parameter.*

According to Definition 3.1.2, the only regular quadrics in $\mathbb{E}^2$ are ellipses, parabolas, and hyperbolas. In the complex extension, the empty quadrics (type 2c) have to be included, too. Among the various common properties of the three types of conics, we pick out their Apollonian definition [46, p. 13ff] and their role as orbits of particles solving the two-body problem [46, p. 63ff].

Regular and singular quadrics in $\mathbb{E}^3$

There are many more cases to be distinguished in $\mathbb{E}^3$, as depicted in Figure 3.7 on the pages 116–117. The irreducible quadrics are printed in italics.

Type 1, conical quadrics: The numbers (r, p) are restricted by $0 \le p \le r \le 3$ and $p \ge r - p$. This yields five cases:

1a') $(r,p) = (3,3)$, $\psi(\mathbf{x}) = \dfrac{x_1^2}{\alpha_1^2} + \dfrac{x_2^2}{\alpha_2^2} + \dfrac{x_3^2}{\alpha_3^2}$, $\mathcal{Q}(\psi) = \{\mathbf{0}\}$,

1b') $(r,p) = (3,2)$, $\psi(\mathbf{x}) = \dfrac{x_1^2}{\alpha_1^2} + \dfrac{x_2^2}{\alpha_2^2} - \dfrac{x_3^2}{\alpha_3^2}$, $\mathcal{Q}(\psi)$ is a *quadratic cone,*

1c') $(r,p) = (2,2)$, $\psi(\mathbf{x}) = \dfrac{x_1^2}{\alpha_1^2} + \dfrac{x_2^2}{\alpha_2^2}$, $\mathcal{Q}(\psi)$ is a single line,

FIGURE 3.6. The cupola of St. Charles Church Vienna ("Karlskirche") has the shape of an ellipsoid with an elliptic base (ratio of semiaxes approx. 3 : 2). The depicted contour is an ellipse with the ratio of semiaxes width : height ≈ 1.03.

1d') $(r,p) = (2,1)$, $\psi(\mathbf{x}) = \dfrac{x_1^2}{\alpha_1^2} - \dfrac{x_2^2}{\alpha_2^2}$, $\mathcal{Q}(\psi)$ consists of two planes,

1e') $(r,p) = (1,1)$, $\psi(\mathbf{x}) = x_1^2$, $\mathcal{Q}(\psi)$ is a single plane.

Type 2, central quadrics: Here we have nine cases:

2a') $(r,p) = (3,3)$, $\psi(\mathbf{x}) = \dfrac{x_1^2}{\alpha_1^2} + \dfrac{x_2^2}{\alpha_2^2} + \dfrac{x_3^2}{\alpha_3^2} - 1$, $\mathcal{Q}(\psi)$ is an *ellipsoid,*

2b') $(r,p) = (3,2)$, $\psi(\mathbf{x}) = \dfrac{x_1^2}{\alpha_1^2} + \dfrac{x_2^2}{\alpha_2^2} - \dfrac{x_3^2}{\alpha_3^2} - 1$, $\mathcal{Q}(\psi)$ is a *one-sheeted hyperboloid,*

2c') $(r,p) = (3,1)$, $\psi(\mathbf{x}) = \dfrac{x_1^2}{\alpha_1^2} - \dfrac{x_2^2}{\alpha_2^2} - \dfrac{x_3^2}{\alpha_3^2} - 1$, $\mathcal{Q}(\psi)$ is a *two-sheeted hyperboloid,*

2d') $(r,p) = (3,0)$, $\psi(\mathbf{x}) = -\dfrac{x_1^2}{\alpha_1^2} - \dfrac{x_2^2}{\alpha_2^2} - \dfrac{x_3^2}{\alpha_3^2} - 1$, $\mathcal{Q}(\psi) = \{\}$,

Equation	Name	Equation	Name
Type 1, **conical quadrics**		$\frac{x_1^2}{a_1^2}+\frac{x_2^2}{a_2^2}+\frac{x_3^2}{a_3^2}=0$	single point
$\frac{x_1^2}{a_1^2}+\frac{x_2^2}{a_2^2}-\frac{x_3^2}{a_3^2}=0$	quadratic cone	$\frac{x_1^2}{a_1^2}+\frac{x_2^2}{a_2^2}=0$	single line
$\frac{x_1^2}{a_1^2}-\frac{x_2^2}{a_2^2}=0$	intersecting planes	$x_1^2=0$	single plane
Type 2, **central quadrics**		$\frac{x_1^2}{a_1^2}+\frac{x_2^2}{a_2^2}+\frac{x_3^2}{a_3^2}=1$	ellipsoid
$\frac{x_1^2}{a_1^2}+\frac{x_2^2}{a_2^2}-\frac{x_3^2}{a_3^2}=1$	one-sheeted hyperboloid	$\frac{x_1^2}{a_1^2}-\frac{x_2^2}{a_2^2}-\frac{x_3^2}{a_3^2}=1$	two-sheeted hyperboloid
$-\frac{x_1^2}{a_1^2}-\frac{x_2^2}{a_2^2}-\frac{x_3^2}{a_3^2}=1$	empty set	$\frac{x_1^2}{a_1^2}+\frac{x_2^2}{a_2^2}=1$	elliptic cylinder
$\frac{x_1^2}{a_1^2}-\frac{x_2^2}{a_2^2}=1$	hyperbolic cylinder	$-\frac{x_1^2}{a_1^2}-\frac{x_2^2}{a_2^2}=1$	empty set
$\frac{x_1^2}{a_1^2}=1$	parallel planes	$-\frac{x_1^2}{a_1^2}=1$	empty set

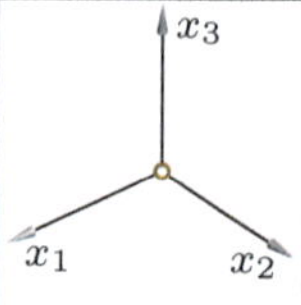

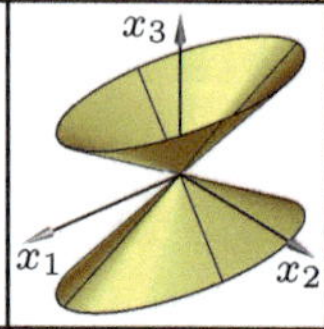

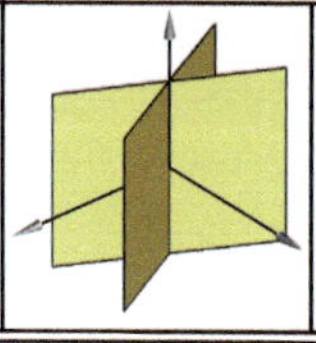

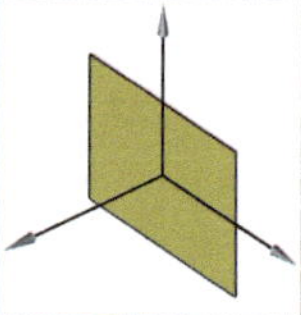

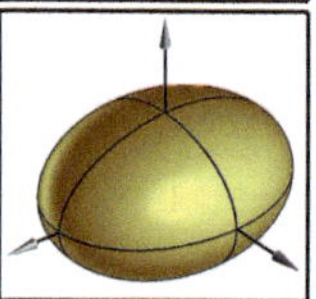

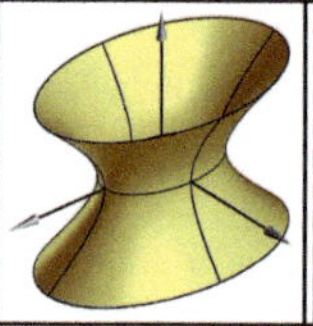

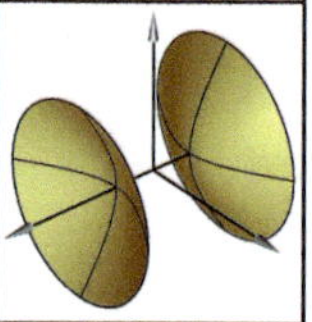

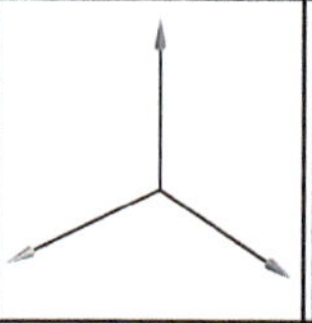

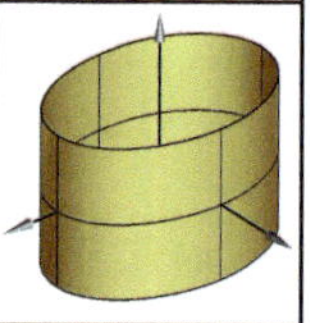

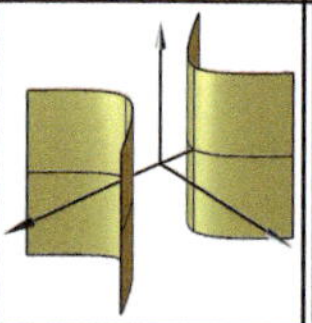

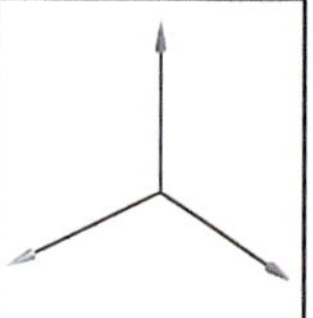

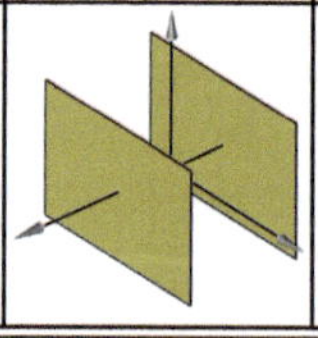

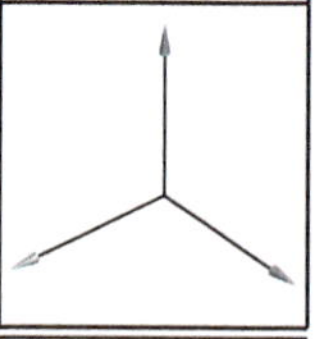

FIGURE 3.7. Quadrics in the Euclidean 3-space $\mathbb{E}^3$, part 1.

Type 3, **parabolic quadrics**	$\frac{x_1^2}{a_1^2} + \frac{x_2^2}{a_2^2} - 2x_3 = 0$ elliptic paraboloid 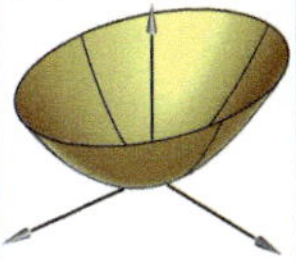
$\frac{x_1^2}{a_1^2} - \frac{x_2^2}{a_2^2} - 2x_3 = 0$ hyperbolic paraboloid	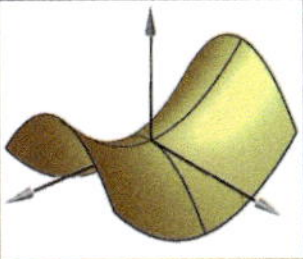 $\frac{x_1^2}{a_1^2} - 2x_3 = 0$ parabolic cylinder 

FIGURE 3.7. Quadrics in the Euclidean 3-space $\mathbb{E}^3$, ctd.

2e') $(r,p) = (2,2)$, $\psi(\mathbf{x}) = \frac{x_1^2}{\alpha_1^2} + \frac{x_2^2}{\alpha_2^2} - 1$, $\mathcal{Q}(\psi)$ is an *elliptic cylinder*,

2f') $(r,p) = (2,1)$, $\psi(\mathbf{x}) = \frac{x_1^2}{\alpha_1^2} - \frac{x_2^2}{\alpha_2^2} - 1$, $\mathcal{Q}(\psi)$ is a *hyperbolic cylinder*,

2g') $(r,p) = (2,0)$, $\psi(\mathbf{x}) = -\frac{x_1^2}{\alpha_1^2} - \frac{x_2^2}{\alpha_2^2} - 1$, $\mathcal{Q}(\psi) = \{\}$,

2h') $(r,p) = (1,1)$, $\psi(\mathbf{x}) = x_1^2 - 1$, $\mathcal{Q}(\psi)$ consists of two planes,

2i') $(r,p) = (1,0)$, $\psi(\mathbf{x}) = -x_1^2 - 1$, $\mathcal{Q}(\psi) = \{\}$.

The two hyperboloids (types 2b' and 2c') have *asymptotic cones* with the respective equations $\frac{x_1^2}{\alpha_1^2} \pm \frac{x_2^2}{\alpha_2^2} - \frac{x_3^2}{\alpha_3^2} = 0$. These cones are of type 1b'.

The regular quadrics in $\mathbb{E}^3$ are ellipsoids, one- and two-sheeted hyperboloids, and elliptic and hyperbolic paraboloids. In the complex extension, the empty quadric of type 2d' is also regular. Figure 3.2 shows quadratic cones (compare with [46, p. 449]) and cylinders as examples of singular, but irreducible quadrics.

Type 3, parabolic quadrics: There remain the following normal forms:

3a') $(r,p) = (2,2)$, $\psi(\mathbf{x}) = \frac{x_1^2}{\alpha_1^2} + \frac{x_2^2}{\alpha_2^2} - 2x_3$, $\mathcal{Q}(\psi)$ is an *elliptic paraboloid*,

3b') $(r,p) = (2,1)$, $\psi(\mathbf{x}) = \frac{x_1^2}{\alpha_1^2} - \frac{x_2^2}{\alpha_2^2} - 2x_3$, $\mathcal{Q}(\psi)$ is a *hyperbolic paraboloid*,

3c') $(r,p) = (1,1)$, $\psi(\mathbf{x}) = \frac{x_1^2}{\alpha_1^2} - 2x_3$, $\mathcal{Q}(\psi)$ is a *parabolic cylinder*.

• **Exercise 3.2.1** Parabolic water trajectory.

When neglecting air resistance, we can set up the water trajectory which is displayed in Figure 3.5 in parametrized form in $\mathbb{E}^2$ as

$$\begin{pmatrix} x_1(t) \\ x_2(t) \end{pmatrix} = \begin{pmatrix} p_1 \\ p_2 \end{pmatrix} + vt \begin{pmatrix} \cos\alpha \\ \sin\alpha \end{pmatrix} - \frac{g}{2} t^2 \begin{pmatrix} 0 \\ 1 \end{pmatrix}.$$

Herein, t designates the time, v the initial velocity, α the initial slope angle, and g the gravitational acceleration on Earth $\approx 9.80665\,\mathrm{m/s}^2$. Find an equation of this parabola and its normal form. What is the parameter of this parabola, as defined on page 114?

• **Exercise 3.2.2** Principal axis transformation.

Find the types, normal forms and, in non-parabolic cases, the center space of the following quadrics.

(a) $\psi(\mathbf{x}) = 5x_1^2 - 4x_1x_2 + 8x_2^2 + 4\sqrt{5}\,x_1 - 16\sqrt{5}\,x_2 + 4$,

(b) $\psi(\mathbf{x}) = 8x_1^2 + 4x_1x_2 - 4x_1x_3 - 2x_2x_3 + 2x_1 - x_3$,

(c) $\psi(\mathbf{x}) = 13x_1^2 - 10x_1x_2 + 13x_2^2 + 18x_3^2 - 72$,

(d) $\psi(\mathbf{x}) = x_1^2 - 4x_1x_2 + 2\sqrt{3}\,x_2x_3 - 2\sqrt{3}\,x_1 + \sqrt{3}\,x_2 + x_3$,

(e) $\psi(\mathbf{x}) = 4x_1^2 + 8x_1x_2 + 4x_2x_3 - x_3^2 + 4x_3$.

Solutions:

(a) $\psi(\mathbf{x}') = \frac{1}{4}x_1'^2 + \frac{1}{9}x_2'^2 - 1$; center $(0, \sqrt{5})$, direction of principal axes: $(2, 1)$ and $(-1, 2)$.

(b) $\mathcal{Q}(\psi)$ is conical (type 1) with a line as center space $G = (t, -\frac{1}{2} - 2t, 2t)$, $t \in \mathbb{R}$. Since $\psi(\mathbf{x}) = (2x_1 - x_3)(4x_1 + 2x_2 + 1)$, $\mathcal{Q}(\psi)$ consists of two planes through G.

(c) $\psi(\mathbf{x}') = \frac{x_1'^2}{9} + \frac{x_2'^2}{4} + \frac{x_3'^2}{4} - 1 = 0$; quadric $\mathcal{Q}(\psi)$ is an oblate ellipsoid of revolution.

(d) $\psi(\mathbf{x}') = 3x_1'^2 + (\sqrt{2}-1)x_2'^2 - (\sqrt{2}+1)x_3'^2 - \frac{11}{6} = 0$; quadric $\mathcal{Q}(\psi)$ is a one-sheeted hyperboloid.

(e) $\psi(\mathbf{x}') = \frac{\sqrt{6}(3 + \sqrt{105})x_1'^2}{8} - \frac{\sqrt{6}(\sqrt{105} - 3)x_2'^2}{8} + 2x_3' = 0$; quadric $\mathcal{Q}(\psi)$ is a hyperbolic paraboloid.

4 Projective and affine quadrics

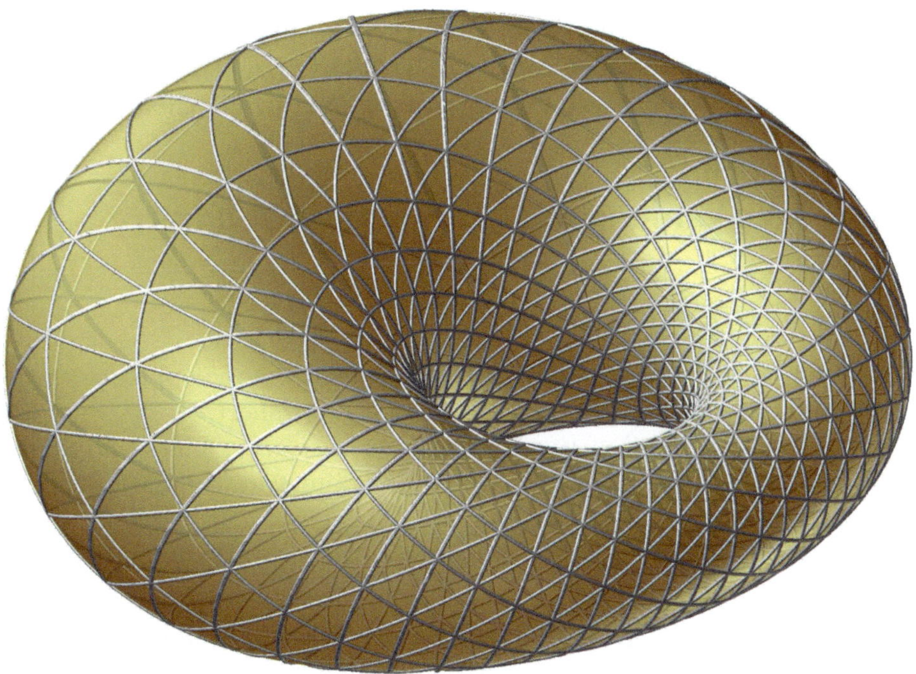

Dupin ring cyclide with a three-web consisting of Villarceau circles of both kinds and of isogonal trajectories of the circular curvature lines. In a conformal model of elliptic geometry, the cyclide represents a Clifford surface, where each kind of Villarceau circles is a family of Clifford parallels, while the isogonal trajectories play the role of coaxial helices.

B. Odehnal et al., *The Universe of Quadrics*,
https://doi.org/10.1007/978-3-662-61053-4_4

The projective space is the proper setting for many properties of quadrics and associated polarities. It will be shown that, in the real projective 3-space, there are only two types of quadrics to be distinguished, ruled and non-ruled quadrics. With regard to polarities, there are three, apart from null polarities. The affine classification of quadrics can be done by analyzing their relative position to the plane at infinity.

In this chapter, we expect the reader to be already familiar with the basics of Projective Geometry in the plane as presented in [46, chapters 5–7].

4.1 Three-dimensional Projective Geometry

Models of projective spaces

To begin with, we give a short overview of the axiomatic approach to the projective 3-space $\mathbb{P}^3$, called projective space for short. It consists of a set $\mathcal{P}$ of *points*, a set $\mathcal{L}$ of *lines*, and a set $\mathcal{E}$ of *planes*. Lines and planes are always considered as sets of points. Sometimes we emphasize this by calling them a *range of points*. Similarly, planes are considered as sets of points and lines, and we speak sometimes of a *field of points* or *field of lines*.

The set of planes through some line g is called *pencil of planes* $\mathcal{E}_g$ with the *carrier* or *axis* g. The set of lines through some point V is called *bundle* or *star of lines* $\mathcal{L}_V$, and the set of planes through V is the *bundle* or *star of planes* $\mathcal{E}_V$ with the *carrier* or *vertex* V. For a *pencils of lines*, *i.e.*, the set of lines through some point V in some plane ε, we use the notation $\mathcal{L}_{V,\varepsilon}$. Again, the point V is the *vertex* or *carrier* and ε is the *carrier plane* of this pencil. We use the symbols $\mathcal{P}_\varepsilon$ and $\mathcal{L}_\varepsilon$ for the sets of points and lines in the plane ε, respectively.

If, for some point $P \in \mathcal{P}$ and some line $l \in \mathcal{L}$, the relation $P \in l$ holds, we say that P *is contained in* the line l, P *is incident with* l, P *lies on* l, l *goes through* P, or l *passes through* P. A similar wording is used for the *incidence* $P \in \varepsilon$ of the point P with the plane ε and for the incidence $l \subset \varepsilon$ of the line l with ε. Any set of points on the same line is called *collinear* or *aligned*. Lines or planes through the same point are called *concurrent*, points and lines in the same plane are *coplanar*, and planes through the same line are *coaxial*.

The following list of *axioms* makes the triple $(\mathcal{P}, \mathcal{L}, \mathcal{E})$ a *three-dimensional projective space* or *projective 3-space* $\mathbb{P}^3$.

(A1) Any two different points $P \neq Q$ can be joined by a unique line $l = [P, Q]$, the *connecting* or *spanned* line.

(A2) Any point P and any line l with $P \notin l$ can be joined with a unique plane $[P, l]$, the *connecting* or *spanned* plane.

(A3) For any line l and plane ε with $l \not\subset \varepsilon$, there exists a unique common point $l \cap \varepsilon$, the *point of intersection* or *piercing point.*

(A4) There exist five points such that no four of them are coplanar. We call this quintuple a *fundamental figure* of $\mathbb{P}^3$.

Two projective spaces $\mathbb{P}^3 = (\mathcal{P}, \mathcal{L}, \mathcal{E})$ and $\mathbb{P}^{3\prime} = (\mathcal{P}', \mathcal{L}', \mathcal{E}')$ are called *isomorphic* if there exist bijective mappings $\kappa : \mathcal{P} \to \mathcal{P}'$, $\kappa^* : \mathcal{L} \to \mathcal{L}'$, and $\widehat{\kappa} : \mathcal{E} \to \mathcal{E}'$ such that each incidence $P \in l$, $P \in \varepsilon$, or $l \subset \varepsilon$ in $\mathbb{P}^3$ holds if, and only if, the respective incidence $\kappa(P) \in \kappa^*(l)$, $\kappa(P) \in \widehat{\kappa}(\varepsilon)$, or $\kappa^*(l) \subset \widehat{\kappa}(\varepsilon)$, holds in $\mathbb{P}^{3\prime}$. We also say that in this case the triples $(\mathcal{P}, \mathcal{L}, \mathcal{E})$ and $(\mathcal{P}', \mathcal{L}', \mathcal{E}')$ are *models* of the same abstract projective 3-space.

Below we present two three-dimensional projective spaces. As will be shown, the first is isomorphic to a particular case of the second.

(i) **The projectively closed Euclidean 3-space.** Similarly to the projective closure of the Euclidean plane in [46, p. 180], we begin with the Euclidean 3-space $\mathbb{E}^3$. We assign to each class of parallel lines in $\mathbb{E}^3$ a unique *ideal point* or *point at infinity* and define that each line out of this class passes through this ideal point. The set of all ideal points of the 3-space is called *ideal plane* or *plane at infinity.* If we concentrate only on lines parallel to a given plane ε, the set of associated ideal points is called *ideal line* or *line at infinity* of the plane ε. Obviously, parallel planes share their line at infinity.

Now, we define $\mathcal{P}$ as the union of the set of points of $\mathbb{E}^3$ and the set of all ideal points. $\mathcal{L}$ is the union of the set of lines of $\mathbb{E}^3$ with the set of ideal lines. Finally, we obtain $\mathcal{E}$ as the set of planes of $\mathbb{E}^3$, which is extended by the plane at infinity. It is easy to verify that the triple $(\mathcal{P}, \mathcal{L}, \mathcal{E})$ satisfies all axioms (A1) to (A4). The extended space is called the *projectively closed Euclidean 3-space* or the *real projective space*, and we denote it by $\mathbb{P}^3(\mathbb{R})$. Points, lines, and planes belonging to $\mathbb{E}^3$ are called *finite* or *proper* in the projective closure $\mathbb{P}^3(\mathbb{R})$.

Connecting the finite point P with an ideal point Q means drawing a parallel through P to some line of the class associated with Q. In order to obtain the connection of two different ideal points P and Q, we choose any finite point V. Then, $[P,Q]$ is the ideal line of the plane spanned by the lines $[V,P]$ and $[V,Q]$. The point of intersection of any finite line l with the plane at infinity is the ideal point of l. The line of intersection of any finite plane ε with the plane at infinity is the ideal line of ε.

(ii) **The vector space model of a projective space.** Let $\mathbb{F}$ be an arbitrary (commutative) field and $\mathbb{F}^4$ a four-dimensional vector space over $\mathbb{F}$. It can be shown that the set $\mathcal{P}$ of one-dimensional subspaces, the set $\mathcal{L}$ of two-dimensional subspaces, and the set $\mathcal{E}$ of three-dimensional subspaces of $\mathbb{F}^4$ defines a projective 3-space. We call it the *vector space model* of a projective space and denote it by $\mathbb{P}^3(\mathbb{F})$.

If point P is spanned by the non-zero vector $\mathbf{p} \in \mathbb{F}^4$, then, instead of $P = \{\lambda\mathbf{p} \,|\, \lambda \in \mathbb{F}\}$, we write $P = \mathbf{p}\mathbb{F}$ or sometimes only $P = \mathbf{p}$ for short. The line l spanned by the points $\mathbf{p}\mathbb{F}$ and $\mathbf{q}\mathbb{F}$ is written briefly as $l = \mathbf{p}\mathbb{F} + \mathbf{q}\mathbb{F}$, which means $l = \{\lambda_1\mathbf{p} + \lambda_2\mathbf{q} \,|\, (\lambda_1, \lambda_2) \in \mathbb{F}^2\}$, provided that $\mathbf{p}$ and $\mathbf{q}$ are linearly independent, which means that the two points are different. Similarly, for three linearly independent vectors $\{\mathbf{p}, \mathbf{q}, \mathbf{r}\}$, the plane spanned by the three corresponding points is denoted $\varepsilon = \mathbf{p}\mathbb{F} + \mathbf{q}\mathbb{F} + \mathbf{r}\mathbb{F}$.

Any basis $(\mathbf{b}_0, \dots, \mathbf{b}_3)$ of $\mathbb{F}^4$ gives rise to coordinates for each vector in $\mathbb{F}^4$ and also *homogeneous coordinates* for each point $P = \mathbf{p}\mathbb{F}$ in $\mathbb{P}^3(\mathbb{F})$, according to the rule

$$P = (x_0 : x_1 : x_2 : x_3) \iff \mathbf{p} = x_0\mathbf{b}_0 + \dots + x_3\mathbf{b}_3.$$

Apparently, the coordinates of points are only unique up to a common non-vanishing scalar.

Now, we show that the choice of a basis yields also homogeneous coordinates for each plane: The set of solutions of any linear equation

$$u_0x_0 + u_1x_1 + u_2x_2 + u_3x_3 = 0 \quad \text{with} \quad (u_0, \dots, u_3) \neq (0, \dots, 0)$$

is three-dimensional, and therefore, a plane of $\mathbb{P}^3(\mathbb{F})$. Conversely, a point $\mathbf{x}\mathbb{F}$ belongs to the plane $\varepsilon = \mathbf{p}\mathbb{F} + \mathbf{q}\mathbb{F} + \mathbf{r}\mathbb{F}$ if, and only if, the vectors $\{\mathbf{p}, \mathbf{q}, \mathbf{r}, \mathbf{x}\}$ are linearly dependent. In this case, the system of linear equations

$$\begin{aligned} u_0 p_0 + \ \dots \ + u_3 p_3 &= 0 \\ u_0 q_0 + \ \dots \ + u_3 q_3 &= 0 \\ u_0 r_0 + \ \dots \ + u_3 r_3 &= 0 \\ u_0 x_0 + \ \dots \ + u_3 x_3 &= 0 \end{aligned}$$

with the unknowns $(u_0, \dots, u_3)$ has a one-dimensional solution, and $\mathbf{x}$ must satisfy the last linear equation. We call the coefficients $(u_0 : \dots : u_3)$ *homogeneous plane coordinates* of ε and write also $\varepsilon = \mathbf{u}\mathbb{F}$, if $\mathbf{u} := (u_0, \dots, u_3)$.[1] Using the notation $\langle \cdot , \cdot \rangle$ for the standard scalar product, we can state:

Lemma 4.1.1 *Point* $\mathbf{x}\mathbb{F}$ *lies in the plane* $\mathbf{u}\mathbb{F}$ *if, and only if,* $\langle \mathbf{u}, \mathbf{x} \rangle = 0$.

A change of the basis in $\mathbb{F}^4$ yields new coordinates for each point. Due to a standard result of Linear Algebra, there is a regular 4×4 *transformation matrix* $\mathbf{T}$ such that for each point the *old* coordinates $(x_0 : \dots : x_3)$ and the *new* ones $(x'_0 : \dots : x'_3)$ are related in matrix form by

$$\begin{pmatrix} x'_0 \\ \vdots \\ x'_3 \end{pmatrix} = \mathbf{T} \begin{pmatrix} x_0 \\ \vdots \\ x_3 \end{pmatrix}.$$

Because of the homogeneity of the point coordinates, the transformation matrix $\mathbf{T}$ is unique only up to a non-vanishing scalar. Note that we still retain the convention introduced in Chapter 3 of writing coordinate vectors as columns.

Instead of choosing a basis of $\mathbb{F}^4$ for coordinatizing the projective space $\mathbb{P}^3(\mathbb{F})$, we can proceed in a more geometric way by specifying any fundamental figure $(A_0, \dots, A_3, E)$ in $\mathbb{P}^3(\mathbb{F})$:
If $A_i = \mathbf{a}_i\mathbb{F}$ for $i = 0, \dots, 3$, then $\{\mathbf{a}_0, \dots, \mathbf{a}_3\}$ is a basis of $\mathbb{F}^4$. This yields for the fifth point $E = \mathbf{e}\mathbb{F}$ a representation

$$\mathbf{e} = \lambda_0 \mathbf{a}_0 + \dots + \lambda_3 \mathbf{a}_3 \quad \text{with} \quad \lambda_0 \dots \lambda_3 \neq 0.$$

Now, we use $\mathbf{a}'_i = \lambda_i \mathbf{a}_i$ as the new coordinate vector of A_i for each $i \in \{0, \dots, 3\}$. The new basis $\{\mathbf{a}'_0, \dots, \mathbf{a}'_3\}$ brings about a coordinatization of $\mathbb{P}^3(\mathbb{F})$ with the following properties.

---

[1] In the language of Linear Algebra, planes are one-dimensional spaces in the dual space of $\mathbb{F}^4$, which can also be introduced in a coordinate-free way as the space of linear forms over $\mathbb{F}^4$.

Theorem 4.1.1 *For each fundamental figure* $(A_0, \dots, A_3, E)$ *in* $\mathbb{P}^3(\mathbb{F})$, *there is a coordinate frame such that* $A_0 = (1:0:0:0)$, ..., $A_3 = (0:0:0:1)$ *and* $E = (1:1:1:1)$. *The points* $A_0, \dots, A_3$ *are called base points of this frame, and point* E *is the unit point.*

Finally, we show that, in the projectively closed Euclidean 3-space, we can introduce homogeneous coordinates as well. This will confirm that this projective space is isomorphic to the projective 3-space over the field of real numbers, which justifies the introduced notation $\mathbb{P}^3(\mathbb{R})$.

It is quite natural to start with a Cartesian coordinate frame in the Euclidean 3-space $\mathbb{E}^3$, though the same procedure could be performed with affine coordinates: If (x_P, y_P, z_P) are the coordinates of any (finite) point P, we extend them by adding 1 as the first coordinate, and we permit that these four coordinates can be multiplied with any factor $\lambda \in \mathbb{R}\setminus\{0\}$.[2] If Q is the point at infinity of a line with direction vector (x_Q, y_Q, z_Q), then we insert 0 as additional coordinate at the beginning. Of course, the direction vector is unique only up to a non-vanishing scalar. Thus, we assigned to each point in the projective extension of $\mathbb{E}^3$ four homogeneous coordinates.

Conversely, from any given coordinates $(x_0 : \dots : x_3)$ with $(x_0, \dots, x_3) \neq (0, \dots, 0)$, we can retrieve the original point: In the case $x_0 \neq 0$, we get

$$(x, y, z) = \left(\frac{x_1}{x_0}, \frac{x_2}{x_0}, \frac{x_3}{x_0}\right).$$

In the case $x_0 = 0$, the vector $(x_1, x_2, x_3) \neq (0,0,0)$ gives the direction of the point at infinity.

We call these particular homogeneous coordinates *homogeneous Cartesian coordinates*. In the language of $\mathbb{P}^3(\mathbb{R})$, the base points of this coordinate frame are the origin and the ideal points of the coordinate axes. The unit point has the Cartesian coordinates $(x, y, z) = (1, 1, 1)$.

First consequences of the axioms

From the four axioms of projective 3-spaces (A1)–(A4) we can deduce a sequence of statements, as listed below.

[2]The same extension of coordinates was already used in Chapter 3 when extended coefficient matrices of linear mappings and quadrics were introduced.

• If $P, Q \in \varepsilon$, then $[P, Q] \subset \varepsilon$, and vice versa.

• For any three non-collinear points P, Q, R, *i.e.*, for each *triangle*, there exists a unique plane ε with $P, Q, R \in \varepsilon$, which is denoted by $[P, Q, R]$ and called the *connecting* or *spanned* plane.

• Each plane contains at least a *quadrangle*, *i.e.*, four coplanar points such that no three of them are collinear.

• Any two different planes ε, ψ have exactly one line in common, the *line of intersection* $l = \varepsilon \cap \psi$.

• If two different lines g, h share a point, then they are coplanar, hence spanning a plane $[g, h]$. Conversely, any two different but coplanar lines have exactly one point in common, the *point of intersection*.

Thus, we can state:

• For each plane ε in $\mathbb{P}^3$, the incident points and lines, *i.e.*, the sets $\mathcal{P}_\varepsilon$ and $\mathcal{L}_\varepsilon$, form a projective plane (see, *e.g.*, [46, Sec. 5.1]).

Each set of four non-coplanar points $\{O, P, Q, R\}$ is called a *tetrahedron* with the six *edges* $[O, P], \ldots, [Q, R]$. Let ε denote the plane $[P, Q, R]$. Then, the mapping

$$\pi\colon\ \mathbb{P}^3 \setminus \{O\} \to \mathcal{P}_\varepsilon,\ X \mapsto \pi(X) = [X, O] \cap \varepsilon,$$

is called a *projection* with *center* O and *image plane* ε.

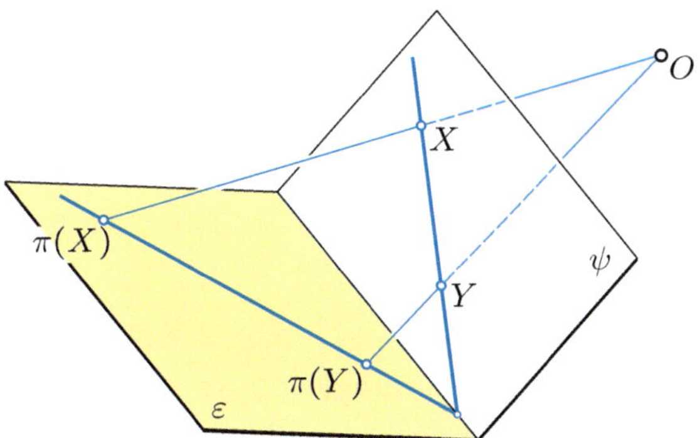

FIGURE 4.1. The projection $\pi\colon X \mapsto \pi(X)$ with center O and image plane ε.

The restriction of the projection π to any plane ψ not passing through O is bijective and preserves the collinearity of points (Figure 4.1). Hence, this is a collineation $\mathcal{P}_\psi \to \mathcal{P}_\varepsilon$. As a consequence, for any two planes ε, ψ

in $\mathbb{P}^3$, the induced projective planes are isomorphic in the sense of [46, p. 179].

• Conversely, if for two different planes ε, ψ in $\mathbb{P}^3$ there is a collineation $\lambda\colon\ \mathcal{P}_\varepsilon \to \mathcal{P}_\psi$, where each point of the line of intersection $\varepsilon \cap \psi$ remains fixed, then λ is the restriction of a projection.

• There exist pairs of *skew* lines in $\mathbb{P}^3$, *i.e.*, lines g, h having no point in common and being non-coplanar.

In each projective space $\mathbb{P}^3$, there exist five planes such that no four share a point. For example, given a fundamental figure $\{P_0, \dots, P_4\}$, the five planes $[P_0, P_1, P_2]$, $[P_1, P_2, P_3]$, $\dots$, $[P_4, P_0, P_1]$ have this property. Hence, similar to projective planes [46, p. 182], for each three-dimensional projective space $\mathbb{P}^3$, we can state the

Principle of Duality. *Any valid theorem formulated in terms of points, lines, planes, connection, intersection, and incidence remains true if we interchange the words points and planes, intersection and connection, but keep lines as lines and all incidences between points, lines, or planes.*

In other words, if $\mathbb{P}^3 = (\mathcal{P}, \mathcal{L}, \mathcal{E})$ is a projective space, then $\widehat{\mathbb{P}}^3 = (\mathcal{E}, \mathcal{L}, \mathcal{P})$ is also a projective space provided that we switch $\in$ and $\ni$ as well as $\subset$ and $\supset$. The space $\widehat{\mathbb{P}}^3$ is called the *dual space* of $\mathbb{P}^3$. When any statement is dualized, the obtained new result is called the *dual*.

The proof of the principle of duality consists in verifying that, from the axioms (A1) to (A4), all respective dual statements can be concluded, and this has already been done above.

In this spatial duality, the fields of points correspond to stars of planes, and the fields of lines (or ruled planes) to stars of lines. Therefore, for all points V, the pair $(\mathcal{E}_V, \mathcal{L}_V)$ is a projective plane, as well as $(\mathcal{L}_V, \mathcal{E}_V)$.

We recall that a projective plane is called a *Fano plane* if the axiom of FANO holds, *i.e.*, for each quadrangle the three diagonal points are not collinear. Similarly, we call a projective 3-space a *Fano space* if it contains a Fano plane.

Since any two planes are isomorphic, all quadrangles in a Fano space have a diagonal triangle. The principle of duality in the plane guarantees that, for the dual of a quadrangle, a *quadrilateral*, the three diagonals are never concurrent. It can be shown that $\mathbb{P}^3(\mathbb{F})$ is a Fano plane if, and only if, the characteristic of the field $\mathbb{F}$ is different from two (note [46, p. 184]. Otherwise, we speak of an *Anti-Fano plane.*

In this book, we will mainly focus on Fano planes without specifying so. Only in very exceptional cases, we will assume that this axiom does not hold.

For the axiomatic approach to projective planes, two other axioms are important, the axiom of DESARGUES and the axiom of PAPPUS (note [18], [46, p. 185–187], or [57]).

The first one states that if two triangles are perspective to a point, then their sides are perspective to a line. In other words, if, for two given triangles $P_1P_2P_3$ and $Q_1Q_2Q_3$, the points P_i, Q_i are collinear with a point (*perspector*) Z for $i = 1, 2, 3$, then the pairs of sides $([P_1, P_2], [Q_1, Q_2])$, $([P_2, P_3], [Q_2, Q_3])$, and $([P_3, P_1], [Q_3, Q_1])$ are concurrent with a line (*perspectrix*) a.

The axiom of Desargues guarantees the existence of perspective collineations in the plane. Furthermore, it can be proved that any Desarguesian plane is isomorphic to a projective plane $\mathbb{P}^2(\mathbb{F})$ over a field $\mathbb{F}$ which needs not be commutative (see [18]).

The situation in projective spaces is quite different:

- In each plane of a projective 3-space, the axiom of Desargues holds.

Proof: Suppose that the two triangles $\overline{P}_1\overline{P}_2\overline{P}_3$ and $Q_1Q_2Q_3$ in different planes are perspective w.r.t. $\overline{Z}$ (Figure 4.2). Then, the points $\{\overline{Z}, \overline{P}_1, \overline{P}_2, Q_1, Q_2\}$ are coplanar. Therefore, the sides $[\overline{P}_1, \overline{P}_2]$ and $[Q_1, Q_2]$ share a point A_3 on the line of intersection $a = [\overline{P}_1, \overline{P}_2, \overline{P}_3] \cap [Q_1, Q_2, Q_3]$. A cyclic permutation of the subscripts reveals that the sides of the two triangles are perspective w.r.t. a.

If two triangles $P_1P_2P_3$ and $Q_1Q_2Q_3$ belong to the same plane π and are perspective w.r.t. a point Z, we can confine ourselves to the case that the seven points $P_1, \ldots, Q_1, \ldots, Z$ are mutally different. Now, we interpret the scene as a projection of the spatial situation. This means that we choose a point $O \notin \pi$ and, on the line $[O, Z]$, another point $\overline{Z} \neq O, Z$. Then, there exist the points $\overline{P}_i = [O, P_i] \cap [\overline{Z}, Q_i]$ for $i = 1, 2, 3$, which do not belong to π. This results in a triangle $\overline{P}_1\overline{P}_2\overline{P}_3$ which is perspective to $Q_1Q_2Q_3$ w.r.t. $\overline{Z}$. Due to the spatial version of Desargues's theorem, the sides of these two triangles are perspective w.r.t. $a = [\overline{P}_1, \overline{P}_2, \overline{P}_3] \cap \pi$. The point of intersection $[P_1, P_2] \cap [\overline{P}_1, \overline{P}_2]$ must coincide with $A_3 = [P_1, P_2] \cap [Q_1, Q_2]$, which, after cyclic permutation of the subscripts, proves the original planar version. ∎

Finally, we call a projective space a *Pappian space,* if it contains a Pappian plane. Again, the isomorphy between any two planes guarantees that each plane is Pappian.

A standard result from the theory of projective planes states that each Pappian plane is isomorphic to a projective plane $\mathbb{P}^2(\mathbb{F})$ over a commu-

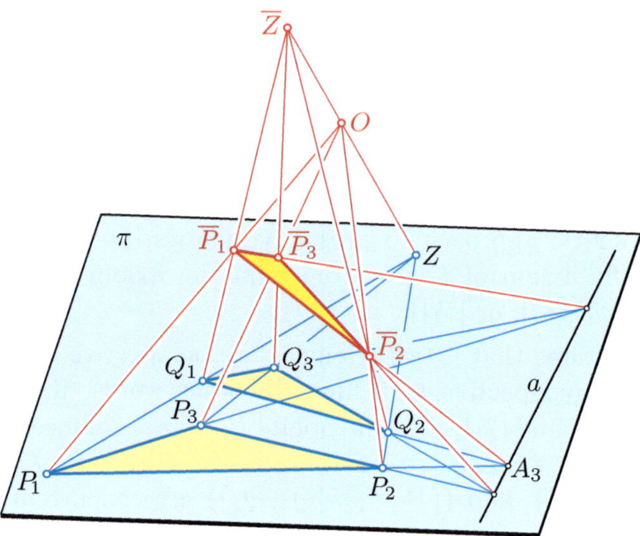

FIGURE 4.2. Proof of Desargues's theorem for the perspective triangles $P_1P_2P_3$ and $Q_1Q_2Q_3$ via the spatial version with the triangle $\overline{P}_1\overline{P}_2\overline{P}_3$.

tative field $\mathbb{F}$ (see [18]). Hence, the projective space $\mathbb{P}^3(\mathbb{F})$, as introduced above, is Pappian.

Collineations, perspectivities, and projectivities

Definition 4.1.1 A *collineation* $\kappa: \mathbb{P}^3 \to \mathbb{P}^{3\prime}$ between two projective spaces is a mapping that sends points to points and satisfies the following two conditions:

(1) Collinear points are mapped to collinear points.

(2) The mapping is one-to-one and onto.

Apparently, each isomorphism of $\mathbb{P}^3 = (\mathcal{P}, \mathcal{L}, \mathcal{E})$ to $\mathbb{P}^{3\prime} = (\mathcal{P}', \mathcal{L}', \mathcal{E}')$ defines a collineation $\kappa\colon \mathcal{P} \to \mathcal{P}'$. However, the converse is also true:

• Each collineation $\mathbb{P}^3 \to \mathbb{P}^{3\prime}$, defined as a mapping $\kappa\colon \mathcal{P} \to \mathcal{P}'$, induces an isomorphism between the two spaces, and hence the bijections $\kappa^*: \mathcal{L} \to \mathcal{L}'$ and $\widehat{\kappa}: \mathcal{E} \to \mathcal{E}'$. Each incidence in $\mathbb{P}^3$ is equivalent to the incidence between the corresponding elements in $\mathbb{P}^{3\prime}$.

For a proof, the reader is referred to the literature, *e.g.*, to [18].

This result reveals that the restriction of any collineation $\kappa\colon \mathbb{P}^3 \to \mathbb{P}^{3\prime}$ to each plane ε of $\mathbb{P}^3$ is a (planar) collineation $\mathcal{P}_\varepsilon \to \mathcal{P}_{\widehat{\kappa}(\varepsilon)}$. The duality guarantees that the same is true for corresponding stars $\mathcal{E}_S$ and $\mathcal{E}_{\kappa(S)}$.

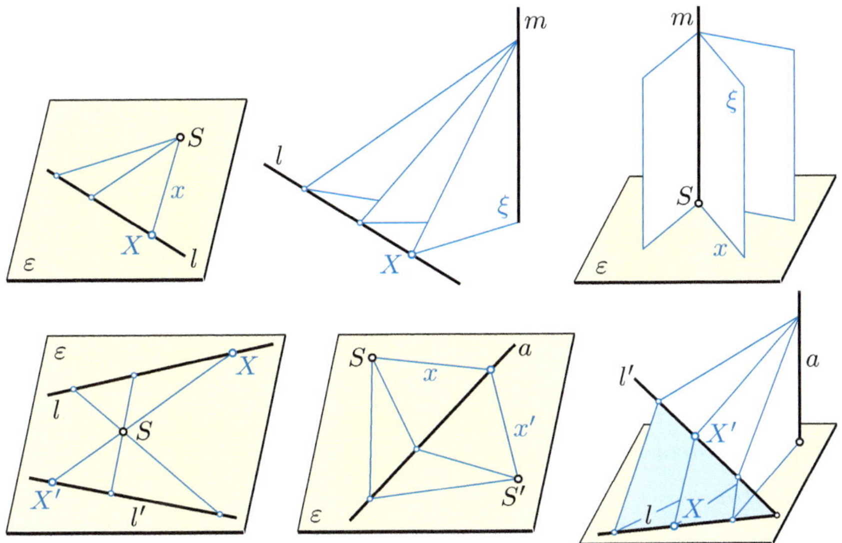

FIGURE 4.3. Perspectivities in the projective space: First row, from left to right: $l(X) \barwedge S_\varepsilon(x)$, $l(X) \barwedge m(\xi)$, and $S_\varepsilon(x) \barwedge m(\xi)$; second row, from left to right: $l(X) \overset{S}{\doublebarwedge} l'(X')$, $S_\varepsilon(x) \overset{a}{\doublebarwedge} S'_\varepsilon(x')$, and $l(X) \overset{a}{\doublebarwedge} l'(X')$.

Now, we are going to extend perspectivities and projectivities from the projective plane to the projective 3-space.

Definition 4.1.2 Let $\mathcal{P}_l$, $\mathcal{L}_{S,\varepsilon}$ and $\mathcal{E}_m$ be, respectively, a range of points, a pencil of lines, and a pencil of planes. Any bijection between any two of these sets is called a *perspectivity* if corresponding elements are incident (Figure 4.3). We denote them for short $l \barwedge S_\varepsilon$, $l \barwedge m$, and $S_\varepsilon \barwedge m$, respectively, or more extensively, by indicating corresponding elements, in the form

$$l(X) \barwedge S_\varepsilon(x), \quad l(X) \barwedge m(\xi), \quad S_\varepsilon(x) \barwedge m(\xi),$$

or in opposite order.

Any composition of finitely many perspectivities is called a *projectivity* and usually expressed with the symbol $\barwedge$.

If, in particular, the lines l and l' are coplanar with S, we call the composition $l(X) \barwedge S_\varepsilon(x) \barwedge l'(X')$ a perspectivity $l \overset{S}{\barwedge} l'$ with *center* S. Similarly, we speak of a perspectivity $S_\varepsilon \overset{a}{\barwedge} S'_\varepsilon$ with the *axis* a for two coplanar pencils or pencils with a common carrier, when either $S_\varepsilon(x) \barwedge a(X) \barwedge S'_\varepsilon(x')$ or $S_\varepsilon(x) \barwedge a(\xi) \barwedge S_{\varepsilon'}(x')$, respectively. Finally, we speak of a perspectivity between two pencils of planes $l \overset{a}{\barwedge} l'$ with the axis a when $l(\xi) \barwedge a \barwedge l'(\xi')$ and the carriers l and l' are coplanar.

The perspectivities, as defined above in the projective 3-space, are extensions of those related to the projective plane. Therefore, the set of projectivities between ranges of points or pencils of lines seems to be larger than that of the planar counterparts. However, a detailed case analysis reveals that each 3-space projectivity between coplanar ranges of points or pencils of lines is also a projectivity in the sense of planar Projective Geometry. This justifies keeping the same notations. Furthermore, the *fundamental theorem* of plane geometry is still valid. This means (note [46, p. 190]):

Lemma 4.1.2 *In a Pappian projective space, any projectivity between ranges of points, or pencils of lines or planes, is uniquely defined by three pairs of corresponding elements.*

The following lemma is also an extension of a result from plane geometry.

Lemma 4.1.3 *A projectivity between two different ranges of points with coplanar carriers is a perspectivity if, and only if, the common element remains fixed.*

The same holds for projectivities between different pencils of lines that are either in a common plane or share a common center. Finally, a similar statement is valid for projectivities between two different pencils of planes with intersecting carriers.

On the other hand, we can adopt, from the vector model $\mathbb{P}^3(\mathbb{F})$, the statement (note [46, Theorem 5.4.1]):

Lemma 4.1.4 *Each projectivity between ranges of lines in $\mathbb{P}^3(\mathbb{F})$ can be described by a bijective linear mapping between the corresponding two-dimensional subspaces of $\mathbb{F}^4$.*

In this sense, the group $\mathrm{PGL}(l)$ of projective mappings of any line l onto itself in the projective space is not bigger than that in the projective plane. It equals the factor group $\mathrm{GL}(\mathbb{F},2)/\mathrm{D}(\mathbb{F})$ of the general linear group, where $\mathrm{D}(\mathbb{F})$ is the group of dilations, represented by all scalar multiples of the unit matrix $\mathbf{I}_2$ (note [46, Theorem 5.4.2]).

Definition 4.1.3 We call a collineation *projective* if its restriction to a range of points or to a pencil of lines or planes is a projectivity.

It is easy to verify that the restriction of a projective collineation to all ranges of points or pencils of lines or planes is a projectivity. Hence, the restriction to any plane is projective, too, and the projective collineations $\mathbb{P}^3 \to \mathbb{P}^3$ form a group.

From the projective plane, we recall that under projective collineations all *cross ratios* remain unchanged ([46, Theorem 5.3.1]), which in the vector model can be defined in the following way:

If, on the line $l = [A,B] = \mathbf{a}\mathbb{F} + \mathbf{b}\mathbb{F}$, two more points are given as $C = (\lambda_1\mathbf{a} + \lambda_2\mathbf{b})\mathbb{F}$ and $D = (\mu_1\mathbf{a} + \mu_2\mathbf{b})\mathbb{F}$, then

$$\mathrm{cr}(A,B,C,D) := \frac{\lambda_2}{\lambda_1}\,\frac{\mu_1}{\mu_2}. \tag{4.1}$$

We call $(\mu_1 : \mu_2)$ *projective coordinates* of the point D w.r.t. the coordinate frame with the *base points* A and B and the *unit point* $(\mathbf{a} + \mathbf{b})\mathbb{F}$.

Similarly, a quadrangle $A_0A_1A_2E$ defines in its spanned plane a projective coordinate frame with the *base points* $A_i = \mathbf{a}_i\mathbb{F}$, $i = 0,1,2$, and the *unit points* $E = (\mathbf{a}_0 + \mathbf{a}_1 + \mathbf{a}_2)\mathbb{F}$, which assigns to any point $P = (x_0\mathbf{a}_0 + x_1\mathbf{a}_1 + x_2\mathbf{a}_2)\mathbb{F}$ the homogeneous triplet $(x_0 : x_1 : x_2)$ as coordinates.

A fundamental theorem of Projective Geometry over commutative fields states:

Theorem 4.1.2 *Each projective collineation $\mathbb{P}^3(\mathbb{F}) \to \mathbb{P}^3(\mathbb{F})$ of a Pappian projective space onto itself can be described by a bijective linear mapping $f\colon \mathbb{F}^4 \to \mathbb{F}^4$. This mapping is unique up to a scalar.*

Conversely, any linear mapping $f : \mathbb{F}^4 \to \mathbb{F}^4$ induces a projective collineation $\kappa\colon \mathbb{P}^3(\mathbb{F}) \to \mathbb{P}^3(\mathbb{F})$ with $\mathbf{p}\mathbb{F} \mapsto f(\mathbf{p})\mathbb{F}$.

We learn from the real projective plane that, in the real projective space $\mathbb{P}^3(\mathbb{R})$, each collineation is projective.

By the same token, all collineations in $\mathbb{P}^3(\mathbb{F}) \to \mathbb{P}^3(\mathbb{F})$ are induced by bijective semilinear mappings $\mathbb{F}^4 \to \mathbb{F}^4$, *i.e.*, by the compositions of linear mappings with automorphisms of $\mathbb{F}$ (see, *e.g.*, [18, 57]). A standard example of a non-projective collineation in the complex extension $\mathbb{P}^3(\mathbb{C})$ of the real projective space $\mathbb{P}^3(\mathbb{R})$ is the *complex conjugation* ι, which replaces each coordinate by its complex conjugate value. This conjugation is involutive and exchanges each non-real point P, line g, or plane ε with its respectively *complex conjugate* element $\overline{P} = \iota(P)$, $\overline{l} = \iota^*(l)$, or $\overline{\varepsilon} = \widehat{\iota}(\varepsilon)$. Elements which remain fixed under ι are called *real*, as they are already included in $\mathbb{P}^3(\mathbb{R})$. Otherwise, we call them *non-real* or *imaginary*. Therefore, $P = \mathbf{p}\mathbb{F}$ is imaginary if, and only if, the vectors $\{\mathbf{p}, \overline{\mathbf{p}}\}$ are linearly independent. In this case, the connection of P and $\overline{P}$ is real. For the same reason, two different complex conjugate planes ε and $\overline{\varepsilon}$ contain a single real line, the line of intersection $\varepsilon \cap \overline{\varepsilon}$.

The situation for lines is slightly different: If a line contains two different real points, then it is real. However, the converse is also true: If the line $l = \mathbf{p}\mathbb{F} + \mathbf{q}\mathbb{F}$ remains fixed under the conjugation, which means that $l = \overline{\mathbf{p}}\mathbb{F} + \overline{\mathbf{q}}\mathbb{F}$, where both pairs of vectors $(\mathbf{p}, \overline{\mathbf{p}})$ and $(\mathbf{q}, \overline{\mathbf{q}})$ are linearly independent, then the two points $R = (\mathbf{p} + \overline{\mathbf{p}})\mathbb{F}$ and $S = (\mathbf{p} - \overline{\mathbf{p}})\mathbb{F}$ on l are real. Moreover, the two points R and S must be different.

For imaginary lines l and $\overline{l}$, there are two possibilities:

- Either, they intersect each other. Then, they share a real point and span a real plane. We call them *imaginary of the first kind.*
- Or, the lines l and $\overline{l}$ are skew. Then, l neither includes a real point nor does it belong to a real plane. Line l is called *imaginary of the second kind.*

Perspective collineations

Suppose that a collineation κ in the projective space $\mathbb{P}^3$ keeps all points of a plane ζ fixed. Then, this collineation is projective by Definition 4.1.3. Moreover we can prove:

Lemma 4.1.5 *If under a collineation κ: $\mathbb{P}^3 \to \mathbb{P}^3$ all points of a plane ζ remain fixed, then there exists a point Z with the property that all lines and planes through Z remain fixed, too, and vice versa.*

Proof: If a point $Z \notin \zeta$ remains fixed, *i.e.*, $\kappa(Z) = Z$, then each line g through Z is a connection of Z with another fixed point $g \cap \zeta$, and therefore, a fixed line. Furthermore, each plane through Z is the connection of two fixed lines.

There remains the case that, for all points $P \notin \zeta$, we have $\kappa(P) \neq P$. Then, the line $p = [P, \kappa(P)]$ is fixed, since its image $\kappa^*(p)$ is the connection of $\kappa(P)$ with the fixed point $p \cap \zeta$. For any other point $Q \notin p, \zeta$ the fixed line $q = [Q, \kappa(Q)]$ is coplanar with p, since $g = [P, Q]$ and $\kappa^*(g) = [\kappa(P), \kappa(Q)]$ meet at the point $g \cap \zeta$. The point $Z := p \cap q$ remains fixed as the intersection of two fixed lines. Hence, $Z \in \zeta$.

Each line x through Z must be fixed, too, since, in the case $x \not\subset \zeta$ and $x \neq p$, it contains any point $Q \notin \zeta, p$, and, by the same argument as used above, x equals the fixed line $[Q, \kappa(Q)] = q$.

The converse statement holds due to duality. ∎

Definition 4.1.4 A collineation κ which keeps each point of a plane ζ and each plane through a point Z fixed is called *perspective collineation* with the *center* Z and the *axis* or *axial plane* ζ.

In the case $Z \in \zeta$, the collineation κ is called *elation* and otherwise *homology.*

All perspective collineations which share the center Z and the axis ζ form a group. We denote it by $\mathrm{PK}(Z, \zeta)$. A standard result of Projective Geometry states that the commutativity of this group is equivalent to the validity of the Pappus theorem.

Suppose that a perspective collineation $\kappa \in \mathrm{PK}(Z, \zeta)$, *i.e.*, with center Z and axis ζ, has another fixed point $P \neq Z$, $P \notin \zeta$. Then, each point $Q \notin [P, Z]$ remains fixed as the intersection of two fixed lines $[Q, Z]$ and $[Q, P]$. However, each point R on the fixed line $[P, Z]$ is also the intersection with another fixed line $[R, Q]$, and therefore, fixed. Hence, κ is the identity mapping $\mathrm{id}_{\mathbb{P}^3}$.

Lemma 4.1.6 *Any perspective collineation with center Z and axis ζ is uniquely defined by a pair of corresponding points P, P' collinear with Z and $P, P' \notin \zeta$.*

Proof: If there were two perspective collineations $\kappa, \lambda \in \mathrm{PK}(Z, \zeta)$ with $P \to P'$, then the product $\lambda^{-1} \circ \kappa$ keeps P fixed. This implies $\lambda^{-1} \circ \kappa = \mathrm{id}_{\mathbb{P}^3}$ and, consequently, $\kappa = \lambda$. ∎

Homologies in the plane have a *characteristic cross ratio* defined by the pairs of corresponding points, the center, and the collinear point on the axis. The analogue in 3-space is also valid: For all points $X \neq Z$, $X \notin \zeta$, the cross ratio

$$\delta = \operatorname{cr}\left(Z, X_\zeta, X, \kappa(X)\right), \quad \text{where} \quad X_\zeta = \zeta \cap [Z, X],$$

is the same, since the restriction to any plane through Z has a characteristic cross ratio, and any two planes through Z share a line through Z (note Figure 4.4, left). A homology is involutive if, and only if, its characteristic cross ratio is -1. In this case, the pairs $(X, \kappa(X))$ are harmonic w.r.t. Z and the axis ζ; we speak of a *harmonic homology*.

Suppose that in a Pappian projective space $\mathbb{P}^3$, a projective collineation κ has a fundamental figure of fixed points $P_0, \ldots, P_4$. Then, the restriction to the plane $\varepsilon = [P_0, P_1, P_2]$ has a quadrangle of fixed points, since $\varepsilon \cap [P_3, P_4]$ also remains fixed. Thus, this restriction is the identity, and the collineation κ is perspective with two more fixed points outside the axis ε. This yields $\kappa = \mathrm{id}_{\mathbb{P}^3}$. An immediate consequence is formulated below.

Theorem 4.1.3 *In Pappian projective spaces, for any two fundamental figures $P_0, \ldots, P_4$ and $P_0', \ldots, P_4'$, there exists exactly one projective collineation with $P_i \mapsto P_i'$ for $i = 0, \ldots, 4$.*

In other words, projective collineations act sharply transitive on fundamental figures. Since Pappian projective spaces are isomorphic to projective spaces over commutative fields, this result can also be concluded from Theorem 4.1.2.

After introducing a coordinate frame in $\mathbb{P}^3(\mathbb{F})$, each projective collineation $\kappa\colon (x_0 : \cdots : x_3) \mapsto (x_0' : \cdots : x_3')$ can be written in matrix form as

$$\kappa: \begin{pmatrix} x_0' \\ \vdots \\ x_3' \end{pmatrix} = \mathbf{M} \begin{pmatrix} x_0 \\ \vdots \\ x_3 \end{pmatrix}, \tag{4.2}$$

where $\mathbf{M}$ is a regular 4×4 matrix. Suppose that the image point lies in the plane with coordinates $(u_0' : \cdots : u_3')$, which can be written in matrix form as $\mathbf{u}'^{\mathrm{T}}\mathbf{x}' = 0$. The coordinates $(u_0 : \cdots : u_3)$ of the preimage satisfy the linear equation $\left(\mathbf{u}'^{\mathrm{T}}\mathbf{M}\right)\mathbf{u} = 0$. Therefore, the associated transformation of planes is given by

$$\widehat{\kappa}: \begin{pmatrix} u_0' \\ \vdots \\ u_3' \end{pmatrix} = \left(\mathbf{M}^{-1}\right)^{\mathrm{T}} \begin{pmatrix} u_0 \\ \vdots \\ u_3 \end{pmatrix}. \tag{4.3}$$

A point $(x_0 : \ldots : x_3)$ remains fixed under κ if, and only if, there exists $\lambda \in \mathbb{R} \setminus \{0\}$ with

$$\lambda \begin{pmatrix} x_0 \\ \vdots \\ x_3 \end{pmatrix} = \mathbf{M} \begin{pmatrix} x_0 \\ \vdots \\ x_3 \end{pmatrix}, \quad \text{and hence} \quad (\mathbf{M} - \lambda \mathbf{I}_4) \begin{pmatrix} x_0 \\ \vdots \\ x_3 \end{pmatrix} = \begin{pmatrix} 0 \\ \vdots \\ 0 \end{pmatrix}$$

with $\mathbf{I}_4$ as unit matrix. This reveals that λ must be an eigenvalue of the matrix $\mathbf{M}$. Once an eigenvalue λ is found, it depends on the rank of the matrix $(\mathbf{M}-\lambda\mathbf{I}_4)$ whether we get a fixed point, a line, or a plane consisting of fixed points only. A k-fold eigenvalue effects a rank deficiency of at least 1 and at most k.

Note that, for each eigenvalue λ of $\mathbf{M}$, the reciprocal $1/\lambda$ is an eigenvalue of $\mathbf{M}^{-1}$, since $\det(\mathbf{M} - \lambda\mathbf{I}_4) = 0$ is equivalent to $\det[\mathbf{M}^{-1}(\mathbf{M} - \lambda\mathbf{I}_4)] = \det(\mathbf{I}_4 - \lambda\mathbf{M}^{-1}) = 0$, which defines fixed planes of $\widehat{\kappa}^{-1}$ and $\widehat{\kappa}$.

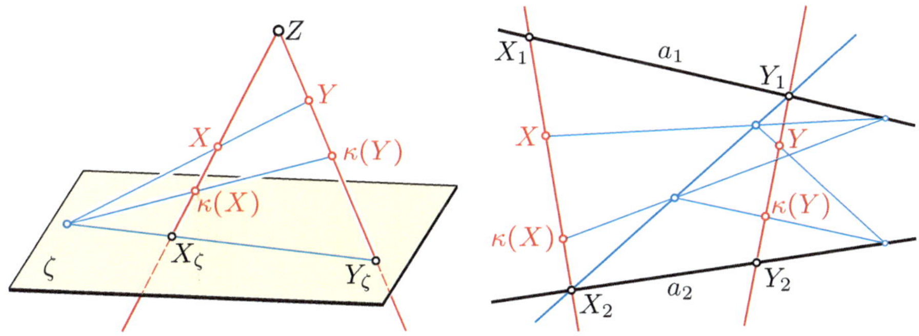

FIGURE 4.4. Left: characteristic cross ratio $\mathrm{cr}(Z, X_\zeta, X, \kappa(X))$ of a homology; Right: characteristic cross ratio $\mathrm{cr}(X_1, X_2, X, \kappa(X))$ of a biaxial collineation with axes a_1 and a_2.

In this way, we can classify projective collineations at hand of their configuration of fixed points. An appropriate tool, provided in Linear Algebra, is the Jordan normal form (see, *e.g.*, [53, 108]). For an algebraically closed field $\mathbb{F}$, there are already 13 different types to be distinguished, together with the identity $\mathrm{id}_{\mathbb{P}^3}$ (compare with (5.6) in Chapter 5). Otherwise, we obtain much more types. For example, in the real projective space, there are projective collineations without any fixed point, since the spectrum of $\mathbf{M}$ may consist of two pairs of complex conjugate eigenvalues.

Below, we concentrate on three types of projective collineations.

(i) The fixed points of a projective collineation κ can form a tetrahedron $A_0, \ldots, A_3$. In this case, the restriction to each edge of the tetrahedron is a hyperbolic projectivity. If the fixed points are chosen as base points of a coordinate frame, we obtain a diagonal matrix $\mathbf{M}$ as analytic representation of κ.

(ii) If a projective collineation $\kappa \neq \mathrm{id}_{\mathbb{P}^3}$ has, besides a fixed tetrahedron, another fixed point F, then F must be located either in a face or on an edge. In the first case, κ is perspective. In the second case, κ has a line consisting of fixed points only, and the collineation is called *uniaxial.*

(iii) It is even possible that κ has two skew lines a_1 and a_2 with fixed points. We call such a collineation *biaxial* with the axes a_1 and a_2. Here all transversals, meeting both axes, remain fixed, and the restriction to each line is a hyperbolic projectivity. The restriction to any plane through one of the axes is a homology, and any plane through a_1 shares with any plane through a_2 a transversal. Therefore, the characteristic cross ratios in all these planes and on all transversals are the same. We call

$$\delta := \mathrm{cr}\,(X_1, X_2, X, \kappa(X)) \text{ with } X_i \in a_i,\ i = 1, 2, \tag{4.4}$$

the *characteristic cross ratio* of the biaxial collineation (Figure 4.4, right). The involutive case with $\delta = -1$ is sometimes called a *skew involution.*

• **Exercise 4.1.1** Projective collineations are products of perspective collineations.

Prove the following theorem: *In Pappian projective spaces, each projective collineation is the product of four perspective collineations at most.*

Hint: Reduce it to the two-dimensional analogue.

• **Exercise 4.1.2** Perspective collineation in vector form.

Prove that any perspective collineation κ in $\mathbb{P}^3(\mathbb{F})$ with center $\mathbf{z}\mathbb{F}$ and axis $\mathbf{u}\mathbb{F}$ can be represented as $\kappa\colon \mathbf{x}\mathbb{F} \mapsto \mathbf{x}'\mathbb{F}$, where $\mathbf{x}' = \mathbf{x} + \langle \mathbf{u}, \mathbf{x} \rangle\, \mathbf{z}$, provided that $1 + \langle \mathbf{u}, \mathbf{z} \rangle \neq 0$.

4.2 Polarities

We proceed with a synthetic approach to polarities. From now on, the symbols $\mathbb{P}^3$ and $\mathbb{P}^{3\prime}$ stand for Pappian projective spaces. Later, the underlying fields (with characteristic $\neq 2$) will be mainly restricted to $\mathbb{R}$ and $\mathbb{C}$.

Correlations and polarities

Let κ be a collineation of $\mathbb{P}^3 = (\mathcal{P}, \mathcal{L}, \mathcal{E})$ to the dual space $\widehat{\mathbb{P}}^{3\prime}$ of $\mathbb{P}^{3\prime} = (\mathcal{P}', \mathcal{L}', \mathcal{E}')$. Then, $\kappa\colon \mathcal{P} \to \mathcal{E}'$ is called a *correlation* $\mathbb{P}^3 \to \mathbb{P}^{3\prime}$.

According to Definition 4.1.1, κ is bijective and sends collinear points to coaxial planes. Due to previous results, it induces an isomorphism $\mathbb{P}^3 \to \widehat{\mathbb{P}}^{3\prime}$. This means that when a point P runs along a line l, then the plane $\kappa(P)$ forms a pencil with the carrier $\kappa^*(l) \in \mathcal{L}'$. When P runs in a plane ε, the corresponding planes $\kappa(P)$ pass through the point $\widehat{\kappa}(\varepsilon) \in \mathcal{P}'$. Note that $\widehat{\kappa}^{-1}\colon \mathcal{P}' \to \mathcal{E}$ is again a correlation, but in the opposite direction. We call $\widehat{\kappa}^{-1}$ the *adjoint correlation* of κ.

Definition 4.2.1 1. Let κ be a correlation $\mathbb{P}^3 \to \mathbb{P}^{3\prime}$. Then, each point Q' in the plane $\kappa(P)$ is called *conjugate* to P w.r.t. κ. Similarly, each plane φ' through the point $\widehat{\kappa}(\varepsilon)$ is called *conjugate* to ε.

2. In the case of identical spaces $\mathbb{P}^{3\prime} = \mathbb{P}^3$, the correlation is called *self-adjoint* or *involutive* if the adjoint correlation $\widehat{\kappa}^{-1}$ coincides with the original correlation κ, *i.e.*, $\widehat{\kappa} \circ \kappa = \mathrm{id}_{\mathcal{P}}$. The correlation κ is called *projective* if its restriction to any range of points is a projectivity.

Apparently, if the correlation κ is projective, then its restrictions to all ranges of points or pencils of lines or planes are projective. Moreover, there exists exactly one projective correlation mapping a given fundamental figure onto a given dual of a fundamental figure.

Lemma 4.2.1 *Each of the following two statements is equivalent to the self-adjointness of a correlation* κ.

1. The induced mapping $\kappa^*\colon \mathcal{L} \to \mathcal{L}$ *is involutive, i.e.,* ${\kappa^*}^2 = \mathrm{id}_{\mathcal{L}}$, *while* $\kappa^* \neq \mathrm{id}_{\mathcal{L}}$.

2. The conjugacy of points is a symmetric relation which means $P \in \kappa(Q) \Leftrightarrow Q \in \kappa(P)$.

Proof: 1. For each correlation $\kappa\colon \mathcal{P} \to \mathcal{E}$

$$P \in g \iff \kappa(P) \supset \kappa^*(g) \iff \widehat{\kappa} \circ \kappa(P) \in \kappa^{*2}(g)$$

holds. Therefore, if $\widehat{\kappa} \circ \kappa = \mathrm{id}_{\mathcal{P}}$, then $P \in g$ is equivalent to $P \in \kappa^{*2}(g)$. This means that κ^{*2} keeps all lines fixed. Conversely, $\kappa^{*2} = \mathrm{id}_{\mathcal{L}}$ implies the equivalence $P \in g \Leftrightarrow \widehat{\kappa} \circ \kappa(P) \in g$ for all P, and hence $\widehat{\kappa} \circ \kappa = \mathrm{id}_{\mathcal{P}}$.

The mapping κ^* cannot be the identity, since it maps stars of lines onto coplanar lines.

2. If κ is self-adjoint, then $P \in \kappa(Q)$ implies $\kappa(P) \ni \widehat{\kappa} \circ \kappa(Q) = Q$.

Conversely, there exists a triangle $Q_1Q_2Q_3$ in the plane $\kappa(P)$. This implies $\kappa(Q_i) \ni \widehat{\kappa} \circ \kappa(P)$ for $i = 1, 2, 3$, where the three planes $\kappa(Q_1)$, $\kappa(Q_2)$, and $\kappa(Q_3)$ cannot be coaxial, and therefore, meet at a single point. If the conjugacy of points is symmetric, then $P \in \kappa(Q_i)$ for $i = 1, 2, 3$, and therefore, $P = \widehat{\kappa} \circ \kappa(P)$ for all P. ■

Definition 4.2.2 A correlation $\kappa\colon \mathbb{P}^3 \to \mathbb{P}^3$ is called *polarity* if it is self-adjoint and projective.[3]

From now on we use the symbol π for polarities. The following notations are restricted to polarities $\pi\colon \mathbb{P}^3 \to \mathbb{P}^3$.

Definition 4.2.3 1. The image $\pi(P)$ of any point P is called *polar plane* of P, and conversely, P is called *pole* of the plane $\pi(P)$. Moreover, the line $\kappa^*(g)$ is called *polar line* of g w.r.t. π.

2. A point P is called *self-conjugate* w.r.t. π if $P \in \pi(P)$. Similarly, a plane ε is called *self-conjugate* w.r.t. π if $\widehat{\pi}(\varepsilon) \in \varepsilon$.

3. A line g is called *self-polar* w.r.t. the polarity π if $g = \pi^*(g)$.

If a point P is self-conjugate w.r.t. π, *i.e.*, $P \in \pi(P)$, then the plane $\pi(P)$ is also self-conjugate, because it contains its pole $\widehat{\pi} \circ \pi(P) = P$. Conversely, the poles of self-conjugate planes are self-conjugate, too.

Lemma 4.2.2 *A polarity π and two different points P, Q are given. Then, the line $[P, Q]$ is self-polar if, and only if, the points P, Q are mutually conjugate and both are self-conjugate w.r.t. π.*

Proof: From the right-hand side, *i.e.*, from $P, Q \in \pi(P), \pi(Q)$ follows $\pi^*([P, Q]) = \pi(P) \cap \pi(Q) = [P, Q]$. Conversely, $P, Q \in g$ and $g = \pi^*(g)$ implies $g \subset \pi(P), \pi(Q)$, and therefore, $P, Q \in \pi(P), \pi(Q)$. ■

It turns out that there are different polarities to be distinguished.

[3] This definition is valid in all dimensions (see, [18]), particularly in the plane.

Definition 4.2.4 1. A polarity $\pi\colon \mathbb{P}^3 \to \mathbb{P}^3$ is called *elliptic* if there are no self-conjugate points, *i.e.*, $P \notin \pi(P)$ for all points P.

2. A polarity π is called *hyperbolic* if there are self-conjugate points as well as points which are not self-conjugate.

3. A polarity π is called *null polarity* if all points are self-conjugate.

Remark 4.2.1 Contrary to projective Fano planes, there exist null polarities in $\mathbb{P}^3$. It can be shown that, given a fundamental figure $A_0, \dots, A_4$, the correlation with $A_i \mapsto [A_{i-1}, A_i, A_{i+1}]$ (indices mod 5) is a null polarity. However, in what follows we focus only on elliptic and hyperbolic polarities.

Theorem 4.2.1 *Let π be an elliptic or hyperbolic polarity in $\mathbb{P}^3$.*

1. If the plane ε is not self-conjugate, then the mapping

$$\pi_\varepsilon\colon \mathcal{P}_\varepsilon \to \mathcal{L}_\varepsilon \text{ with } X \mapsto x := \pi(X) \cap \varepsilon$$

is a polarity in the plane, called the induced polarity. Any two points $X, Y \in \varepsilon$ are conjugate w.r.t. π if, and only if, they are conjugate w.r.t. π_ε.

2. When the polar lines g and $\pi^*(g)$ are skew, the mapping

$$\pi_g\colon \mathcal{P}_g \to \mathcal{P}_g \text{ with } X \mapsto X' := \pi(X) \cap g$$

is an involution and called induced involution or involution of conjugate points on g.

3. Let a point A be a self-conjugate point. Then, for each line t in the pencil $\mathcal{L}_{A,\pi(A)}$, the polar line $t' := \pi^*(t)$ belongs to the same pencil, and the mapping $t \mapsto t'$ is an involution in the pencil of lines. Fixed lines $t = \pi^*(t)$ are self-polar w.r.t. π.

Proof: 1. From $X \in \varepsilon$ follows $\pi(X) \in \overline{\pi}(\varepsilon)$. Since the pole $\overline{\pi}(\varepsilon)$ of ε does not belong to ε, the mapping $\pi_\varepsilon\colon X \mapsto x = \pi(X) \cap \varepsilon$ of points to lines is bijective and projective, and hence an elliptic or hyperbolic polarity in ε. Null polarities are impossible in the planes of $\mathbb{P}^3$.

Let X, Y be two points in ε. If they are conjugate w.r.t. π_ε, then Y is a point of $\pi(X) \cap \varepsilon$, and therefore, also conjugate to X w.r.t. π. The converse is also true.

2. From $X \in g$ follows $\pi^*(g) \in \pi(X)$. The mapping $\pi_g\colon X \mapsto X' := \pi(X) \cap g$ is a projectivity, since $g(X) \barwedge \pi^*(g)\,(\pi(X)) \barwedge g(X')$. The point X' is conjugate to X w.r.t. π. Since conjugacy is symmetric, the induced projectivity π_g maps X' back to X. Hence, π_g is either an involution or the identity. Below we show that the latter is impossible.

Suppose that all points $P \in g$ are self-conjugate. Then, all points $Q \in \pi^*(g)$ must also be self-conjugate, because $Q \in \pi^*(g)$ implies $\pi(Q) \in g$. If $Q \notin \pi(Q)$, then there is an induced polarity in $\pi(Q)$ where all points P of the line g are self-conjugate. This contradicts the properties of planar polarities.

Thus, we have a pair of skew polar lines g and $g^* := \pi^*(g)$ which contain only self-conjugate points. Now, through each point $A \notin g, g^*$, we can draw a transversal, which meets g at P and g^* at Q. By virtue of Lemma 4.2.2, this transversal $[P,Q]$ is self-polar. Hence, $A \in [P,Q]$ must be self-conjugate, too. The conclusion is that π must be a null polarity, which contradicts our assumption.

3. The pencil of lines $t \in \mathcal{L}_{A,\pi(A)}$ is mapped onto itself by π, since $A \in t$ and $t \subset \pi(A)$ imply $t' := \pi^*(t) \in \pi(A)$ and $\widehat{\pi} \circ \pi(A) = A \in \pi^*(t) = t'$.

Let g be any line in the polar plane $\pi(A)$, but not passing through the pole A. Then, the polar $\pi^*(g)$ of g passes through A, but does not lie in $\pi(A)$. Thus, the two polar lines g and $\pi^*(g)$ are skew, and the pairs of conjugate points on g form an involution π_g.

For $X = t \cap g$, the polar plane $\pi(X)$ passes through $t' = \pi^*(t)$. Therefore, the point of intersection $X' := t' \cap g$ is conjugate to X. Since X and X' are pairs of the involution π_g, the relation between the two mutually polar lines t and t' is also an involution. ■

Polarities and quadrics

The next theorem refers to the vector space model $\mathbb{P}^3(\mathbb{F})$.

Theorem 4.2.2 *Let the projective correlation $\kappa\colon \mathbb{P}^3 \to \mathbb{P}^3$ with $\mathbf{x}\mathbb{F} \mapsto \mathbf{u}\mathbb{F}$ be represented by the matrix equation $\mathbf{u} = \mathbf{M}\mathbf{x}$ with a regular 4×4 matrix $\mathbf{M}$.*

1. Then, κ is an elliptic or hyperbolic polarity if, and only if, the matrix $\mathbf{M}$ is symmetric, i.e., $\mathbf{M}^{\mathrm{T}} = \mathbf{M}$.

2. κ is a null polarity if, and only if, the matrix $\mathbf{M}$ is skew symmetric, i.e., $\mathbf{M}^{\mathrm{T}} = -\mathbf{M}$.

Proof: If by (4.2) the projective correlation $\kappa\colon \mathbf{x}\mathbb{F} \mapsto \mathbf{u}'\mathbb{F}$ is represented by $\mathbf{u}' = \mathbf{M}\mathbf{x}$, then by (4.3) we have $\widehat{\kappa}\colon \mathbf{v}\mathbb{F} \mapsto \mathbf{y}'\mathbb{F}$ with $\mathbf{y}' = (\mathbf{M}^{\mathrm{T}})^{-1}\mathbf{v}$. This results in

$$\widehat{\kappa} \circ \kappa\colon\ \mathbf{y}' = (\mathbf{M}^{\mathrm{T}})^{-1}\mathbf{M}\mathbf{x}.$$

κ is self-adjoint if, and only if, $\widehat{\kappa} \circ \kappa = \mathrm{id}_{\mathbb{P}^3}$, hence

$$(\mathbf{M}^{\mathrm{T}})^{-1}\mathbf{M} = \lambda \mathbf{I}_4 \text{ with } \lambda \in \mathbb{F} \setminus \{0\}.$$

From $\mathbf{M} = \lambda \mathbf{M}^{\mathrm{T}}$ follows $\mathbf{M}^{\mathrm{T}} = \lambda \mathbf{M}$ and $\mathbf{M} = \lambda^2 \mathbf{M}$ with the two possibilities $\lambda = +1$ and $\lambda = -1$. Both are sufficient for self-adjointness of the projective correlation π, hence for characterizing π as a polarity.

A point $\mathbf{x}\mathbb{F}$ is self-conjugate w.r.t. κ if, and only if, $\langle \mathbf{u}, \mathbf{x}\rangle = \langle \mathbf{M}\mathbf{x}, \mathbf{x}\rangle = 0$. In matrix form, this condition is equivalent to $(\mathbf{M}\mathbf{x})^{\mathrm{T}}\mathbf{x} = 0$ or

$$\mathbf{x}^{\mathrm{T}}\mathbf{M}^{\mathrm{T}}\mathbf{x} = 0. \tag{4.5}$$

Let a_{ik} be the entries of matrix $\mathbf{M}$ and $(x_0, \dots, x_3)$ be the coordinates of $\mathbf{x}$. Then, to be more precise, the condition (4.5) means

$$\sum_{i,k=0}^{3} a_{ik}x_i x_k = \sum_{i<k} (a_{ik} + a_{ki}) x_i x_k + \sum_{i=0}^{3} a_{ii} x_i^2 = 0.$$

For a skew symmetric matrix $\mathbf{M}$, we have $a_{ik} = -a_{ki}$, in particular $a_{ii} = 0$, which shows that the quadratic form on the left-hand side is the nullform. Each point is self-conjugate; the polarity is a null polarity.

Conversely, if π is a null polarity, then each point must be self-conjugate. When we substitute successively $\mathbf{x} = (1,0,0,0)$, $(0,1,0,0)$, ..., $(0,0,0,1)$ in (4.5), we obtain $a_{11} = \cdots = a_{33} = 0$. Then, setting $\mathbf{x} = (1,1,0,0)$ and so on, we obtain $a_{01} + a_{10} = 0$ etc., which shows that only a skew symmetric matrix represents a null polarity.

In the symmetric case, the set of solutions of (4.5) is either empty or a regular quadric with $\mathbf{M}$ as the extended coefficient matrix. The latter follows from Definition 3.1.2. ■

Thus, we can formulate as a corollary what is often called VON STAUDT's *definition of a quadric.*

Corollary 4.2.3 *Given a hyperbolic polarity π in $\mathbb{P}^3$, the set of self-conjugate points w.r.t. π is a regular quadric in $\mathbb{P}^3$. Conversely, each regular quadric $\mathcal{Q}$ is the set of self-conjugate points of a hyperbolic polarity $\pi_{\mathcal{Q}}$.*

For the rest of this section, quadrics $\mathcal{Q}$ are always assumed to be regular, without specifying so. Moreover, instead of referring to pole, polar plane, and conjugate elements 'w.r.t. the polarity w.r.t. $\mathcal{Q}$', we will use the shortened form 'w.r.t. $\mathcal{Q}$'.

Two points $\mathbf{p}\mathbb{F}$ and $\mathbf{q}\mathbb{F}$ are conjugate w.r.t. $\mathcal{Q}$ if $\langle \mathbf{Mp}, \mathbf{q} \rangle = 0$, and hence

$$\mathbf{p}^{\mathrm{T}}\mathbf{M}\,\mathbf{q} = 0. \tag{4.6}$$

The left-hand side here is exactly the polar form of the left-hand side of (4.5).

Now, we apply Theorem 4.2.1 to the hyperbolic polarity $\pi_{\mathcal{Q}}$ with the quadric $\mathcal{Q}$ as a set of self-conjugate points. The following definitions, originating from the theory of polarities, are obviously compatible with the terms introduced in Chapter 2: Any self-conjugate plane $\mathcal{Q}$ is called *tangent plane* of $\mathcal{Q}$. It includes its pole, the *point of contact.* The dual of Corollary 4.2.3 means that the self-conjugate planes of any hyperbolic polarity in $\mathbb{P}^3$ are the tangent planes of a quadric.

Tangent planes of $\mathcal{Q}$, given by (4.5), are self-conjugate, and therefore, by virtue of (4.3), their coordinates satisfy

$$\mathbf{u}^{\mathrm{T}}\mathbf{M}^{-1}\,\mathbf{u} = 0. \tag{4.7}$$

This is called the *tangential equation* or *dual equation* of $\mathcal{Q}$.

We summarize:

Theorem 4.2.3 *If a regular quadric $\mathcal{Q}$ in $\mathbb{P}^3(\mathbb{F})$ is given by*

$$\sigma(\mathbf{x},\mathbf{x}) := \mathbf{x}^{\mathrm{T}}\mathbf{M}\,\mathbf{x} = 0 \text{ with } \mathbf{M}^{\mathrm{T}} = \mathbf{M},$$

then two conjugate points $\mathbf{p}\mathbb{F}$ and $\mathbf{q}\mathbb{F}$ w.r.t. $\mathcal{Q}$ are characterized by

$$\sigma(\mathbf{p},\mathbf{q}) = \mathbf{p}^{\mathrm{T}}\mathbf{M}\,\mathbf{q} = 0.$$

Tangent planes $\mathbf{u}\mathbb{F}$ of $\mathcal{Q}$ satisfy the tangential equation

$$\mathbf{u}^{\mathrm{T}}\mathbf{M}^{-1}\mathbf{u} = 0.$$

Item 1 in Theorem 4.2.1 means w.r.t. the polarity $\pi_{\mathcal{Q}}$ that the intersection of any non self-conjugate plane ε with $\mathcal{Q}$ is either empty or a conic. If, in the latter case, A is a point of this conic, then the points of its tangent t_A are conjugate to A, and therefore, located in the tangent plane $\pi_{\mathcal{Q}}(A)$. All lines through A and within the tangent plane $\pi_{\mathcal{Q}}(A)$ are called *tangents* of $\mathcal{Q}$. They are characterized as lines which are coplanar with their polars.

As a result of dualization, either the set of tangent planes passing through any point $P \notin \mathcal{Q}$ is empty or the envelopes a quadratic cone (note [46, Sect. 10]), the dual of a conic. This cone is called the *tangent cone* of $\mathcal{Q}$ with apex A. The points of contact belong to the conic $\pi_{\mathcal{Q}}(A) \cap \mathcal{Q}$.

Let g be any line being skew to its polar line $g^* = \pi_{\mathcal{Q}}^*(g)$. Then, item 2 of Theorem 4.2.1 means that the points of intersection between g and $\mathcal{Q}$ are the fixed points of the involution π_g. If they exist, they are the points of contact of the tangent planes through g^*.

From item 3 in Theorem 4.2.1, we can conclude: For $A \in \mathcal{Q}$, the intersection of the tangent plane $\pi_{\mathcal{Q}}(A)$ with $\mathcal{Q}$ consists either only of the point A or of a pair of lines, the fixed lines of the involution of polar tangents at A.

Definition 4.2.5 A quadric $\mathcal{Q}$ is called *ruled quadric* if there exist self-polar lines of the polarity $\pi_{\mathcal{Q}}$. These lines are called *generators* or *rulings* of $\mathcal{Q}$. Non-ruled quadrics are called *oval.*

Suppose that the line e is self-polar w.r.t. $\pi_{\mathcal{Q}}$, *i.e.*, $\pi_{\mathcal{Q}}^*(e) = e$. Then, each point $A \in e$ is self-conjugate since $A \in e \subset \widehat{\pi}_{\mathcal{Q}}(A)$. Therefore, the line e is a subset of $\mathcal{Q}$, which justifies the notation.

Moreover, each plane through e is self-conjugate, and therefore, a tangent plane of $\mathcal{Q}$. Since polarities are projective, we obtain a projectivity

$$\gamma_e \colon \mathcal{P}_e \to \mathcal{E}_e \text{ with } A \mapsto \pi_{\mathcal{Q}}(A). \tag{4.8}$$

This projectivity is called the *contact projectivity* along $e \subset \mathcal{Q}$ (note Figures 2.14, 2.16, and 9.7). We summarize:

Lemma 4.2.4 *For each generator e of a ruled quadric $\mathcal{Q}$, there is a projectivity between the points $A \in e$ and the corresponding tangent planes $\pi_{\mathcal{Q}}(A) \supset e$.*

If the field $\mathbb{F}$ of $\mathbb{P}^3$ is algebraically closed, then each quadric is ruled, since all involutions of mutually polar tangents at points $A \in \mathcal{Q}$ are hyperbolic.

The followings theorem provides information about the set of generators of a ruled quadric.

Theorem 4.2.4 *1. Ruled quadrics $\mathcal{Q}$ in $\mathbb{P}^3$ contain two families of generators, called reguli. Generators of the same regulus are mutually skew, while generators out of complementary reguli are intersecting. Each point $P \in \mathcal{Q}$ is the meet of two generators, which span the tangent plane at P. The lines of one regulus intersect any two lines of the complementary regulus at points corresponding in a projectivity.*

2. Conversely, let $e_1(X_1) \barwedge e_2(X_2)$ be a projectivity between two skew lines. Then, the set of lines connecting corresponding points X_1, X_2 is a regulus of a quadric. The dual statement is also valid: If there is a projectivity $e_1(\xi_1) \barwedge e_2(\xi_2)$ between two pencils of planes with skew carriers e_1 and e_2, the set of lines of intersection $\xi_1 \cap \xi_2$ between corresponding planes is a regulus, too.

Proof: 1. Per definition, each ruled quadric $\mathcal{Q}$ contains a generator e. All points $A \in e$ are self-conjugate, and in the tangent plane $\pi_{\mathcal{Q}}(A)$ of A, there lies a second generator being the second fixed line f of the involution mentioned in item 3 of Theorem 4.2.1. The f-lines through different points A are mutually skew since the respective tangent planes through e are different.

Let A_1, A_2 be two different points of e and let f_1, f_2 denote the respectively second generators passing through. For each point $X_1 \in f_1$, the tangent plane $\pi_{\mathcal{Q}}(X_i)$ intersects the line f_2 at a point X_2, and the composition

$$f_1(X_1) \barwedge \pi_{\mathcal{Q}}(X_1) \bar{\barwedge} f_2(X_2)$$

is a projectivity. The points $X_1 \in f_1$ and $X_2 \in f_2$ are self-conjugate and also mutually conjugate. By virtue of Lemma 4.2.2, the connecting line is self-polar, hence a generator of $\mathcal{Q}$. In particular, the point $X_1 = A_1$ corresponds to $X_2 = A_2$.

We call the set of lines $[X_1, X_2]$, which includes the given line e, the *e-regulus* of $\mathcal{Q}$, generated by the projectivity $f_1(X_1) \barwedge f_2(X_2)$. The same line $[X_1, X_2]$ is also the line of intersection of the tangent planes $\xi_1 = \pi_{\mathcal{Q}}(X_1)$ and $\xi_2 = \pi_{\mathcal{Q}}(X_2)$. Therefore, the e-regulus can also be generated by projective pencils of planes $f_1(\xi_1) \barwedge f_2(\xi_2)$ with skew axes.

Similarly, we can select two different generators e_1, e_2 of the e-regulus and prove that there are projectivities $e_1(Y_1) \barwedge e_2(Y_2)$ and $e_1(\eta_1) \barwedge e_2(\eta_2)$ between skew ranges of points and between two pencils of planes which generate the *complementary regulus*, the *f-regulus* of $\mathcal{Q}$.

2. We prove the second statement in an analytic way in the vector model. Let the projectivity $e_1 \barwedge e_2$ be given by three pairs of points $A_i \mapsto A_i'$, $i = 1, 2, 3$. Now, we introduce a coordinate frame with the base points A_1, A_2, A_1', A_2' and with the unit point E specified on $[A_3, A_3']$, but different from A_3 and A_3'. Thus, we have assigned to the given points the respective homogeneous coordinates

$$\begin{array}{rclrclrcl} A_1 &=& (1:0:0:0), & A_2 &=& (0:1:0:0), & A_3 &=& (1:1:0:0), \\ A_1' &=& (0:0:1:0), & A_2' &=& (0:0:0:1), & A_3' &=& (0:0:1:1). \end{array}$$

The projectivity between the two ranges of points with $A_i \mapsto A_i'$ for $i = 1, 2, 3$ can be represented as

$$\alpha\colon\ (\lambda_1 : \lambda_2 : 0 : 0) \mapsto (0 : 0 : \lambda_1 : \lambda_2), \ \text{where } (\lambda_1, \lambda_2) \in \mathbb{F}^2 \setminus \{(0,0)\}.$$

Points being collinear with any pair $(X, \alpha(X))$ of corresponding points have coordinates

$$(x_0, \ldots, x_3) \in [(\lambda_1, \lambda_2, 0, 0)\mu_1 + (0, 0, \lambda_1, \lambda_2)\mu_2]\,\mathbb{F} = (\lambda_1\mu_1,\ \lambda_2\mu_1,\ \lambda_1\mu_2\ \lambda_2\mu_2)\mathbb{F}$$

with $(\mu_1, \mu_2) \in \mathbb{F}^2 \setminus \{(0,0)\}$. These coordinates satisfy

$$0 = x_0x_3 - x_1x_2 = \frac{1}{2}\,(x_0\ x_1\ x_2\ x_3)\begin{pmatrix} 0 & 0 & 0 & 1 \\ 0 & 0 & -1 & 0 \\ 0 & -1 & 0 & 0 \\ 1 & 0 & 0 & 0 \end{pmatrix}\begin{pmatrix} x_0 \\ x_1 \\ x_2 \\ x_3 \end{pmatrix},$$

which is the equation of a quadric. ■

We summarize what has just been shown.

Theorem 4.2.5 *For each ruled quadric $\mathcal{Q}$, there exists a coordinate frame such that all points of $\mathcal{Q}$ satisfy the equation*

$$x_0x_3 - x_1x_2 = 0\,.$$

In this frame, the quadric can be parametrized by pairs of homogeneous parameters $(\lambda_1, \lambda_2), (\mu_1, \mu_2) \in \mathbb{F}^2 \setminus \{(0,0)\}$ in the form

$$(x_0 : x_1 : x_2 : x_3) = (\lambda_1\mu_1 : \lambda_2\mu_1 : \lambda_1\mu_2 : \lambda_2\mu_2).$$

The parameter lines $\lambda_1 : \lambda_2 = \text{const.}$ and $\mu_1 : \mu_2 = \text{const.}$ are the generators of the two reguli.

Finally, we note that the intersection of a ruled quadric $\mathcal{Q}$ with a plane ε can never be empty, since each generator e of $\mathcal{Q}$ has a point of intersection with ε. When ε happens to be a tangent plane of $\mathcal{Q}$, the curve of intersection degenerates into the two generators in ε.

As to the dual statement, the tangent cone from any point P to $\mathcal{Q}$ can never be empty, since each connecting plane $[P, e]$ with $e \subset \mathcal{Q}$ is a tangent plane of this cone. For $P \in \mathcal{Q}$, this cone consists of two pencils of planes with the two generators through P as carriers.

Corollary 4.2.5 *The set of transversals meeting three mutually skew lines f_1, f_2, f_3 is a regulus.*

Proof: In order to get any transversal, we choose any point $A \in f_3$. Then, the transversal e through A is the intersection of the two connecting planes $\xi_1 = [A, f_1]$ and $\xi_2 = [A, f_2]$. The composition of perspectivities

$$f_1(\xi_1) \,\bar{\wedge}\, f_3(A) \,\bar{\wedge}\, f_2(\xi_2)$$

gives a projectivity which, by Theorem 4.2.4, generates a regulus. ■

Self-polar tetrahedra

Definition 4.2.6 A tetrahedron $A_0 \dots A_3$ is called *self-polar* w.r.t. a projective correlation κ if κ maps each vertex onto the respectively opposite plane, *i.e.*, $\kappa(A_i) = [A_{i+1}, A_{i+2}, A_{i+3}]$, $i = 0, 1, 2, 3 \mod 4$.

Note that any two vertices of a self-polar tetrahedron[4] are mutually conjugate w.r.t. the correlation κ. In the literature, self-polar tetrahedra are also called *polar tetrahedra* w.r.t. any quadric.

Lemma 4.2.6 *A projective correlation κ is an elliptic or hyperbolic polarity if, and only if, there exists a tetrahedron which is self-polar w.r.t. κ.*

Proof: Given a self-polar tetrahedron $A_0 \dots A_3$ w.r.t. κ, each point $X \in [A_0, A_1]$ is mapped onto a plane $\kappa(X)$ through $\kappa(A_0) \cap \kappa(A_1) = [A_2, A_3]$. Let X' denote the point of intersection $[A_0, A_1] \cap \kappa(X)$. Then, we obtain a projectivity

$$[A_0, A_1](X) \,\wedge\, [A_2, A_3](\kappa(X)) \,\bar{\wedge}\, [A_0, A_1](X').$$

This projectivity interchanges the points A_0 and A_1. Consequently, it is an involution and maps X' back to X. Therefore, $\kappa(X') = [X, A_2, A_3]$ and

$$\widehat{\kappa} \circ \kappa(X) = \widehat{\kappa}([X', A_2, A_3]) = X.$$

[4]Sometimes self-polar tetrahedra are also called *polar tetrahedra* in the literature.

The same holds for all other edges. Thus, the projective collineation $\widehat{\kappa} \circ \kappa$ keeps all points on the edges of the tetrahedron $A_0 \dots A_3$ fixed. This characterizes $\widehat{\kappa} \circ \kappa$ as the identity and κ as a polarity.

Conversely, given an elliptic or hyperbolic polarity π, there exists a point $A_0 \notin \alpha_0 := \pi(A_0)$. Then, we find in α_0 a triangle $A_1 A_2 A_3$ self-polar w.r.t. the induced polarity π_{α_0}. Together with A_0, this yields a tetrahedron where any two vertices are conjugate w.r.t. π. ■

Theorem 4.2.6 *An elliptic or hyperbolic polarity is uniquely defined by a self-polar tetrahedron $A_0 \dots A_3$ and an additional pair $B \mapsto \beta$ where $A_0, \dots, A_3, B$ is a fundamental figure and $A_0, \dots, A_3 \notin \beta$.*

Proof: The five given pairs of corresponding elements define a projective correlation uniquely, and by Lemma 4.2.6, this correlation is a polarity. ■

Let the polarity π be given according to Theorem 4.2.6. Then, on each edge of the tetrahedron, the involution of conjugate points is fixed, *e.g.*, on $l := [A_0, A_1]$ the pairs $A_0 \mapsto A_1$ and $l \cap \beta \mapsto l \cap [B, A_2, A_3]$.

If, conversely, the induced involutions are given on three concurrent edges of the self-polar tetrahedron, then for any given plane β the pole is determined as the intersection of three planes. In the following, we use this method for defining a polarity.

Theorem 4.2.7 *Given an elliptic or hyperbolic polarity π, let A not be a self-conjugate point and g be a line which is skew to its polar line $g^* := \pi^*(g)$.*
Then, the harmonic homology γ_1 with center A and axis $\pi(A)$ as well as the skew involution γ_2 with axes g and g^* keep the polarity fixed, i.e., they send pairs consisting of a pole and its polar plane w.r.t. π again onto such a pair of corresponding elements.

Proof: The given involutive transformations γ_1 and γ_2 map polar elements X and $\pi(X)$ w.r.t. the polarity π onto elements $\gamma_i(X)$ and $\widehat{\gamma}_i \circ \pi(X)$ of the projective correlation $\pi' := \widehat{\gamma}_i \circ \pi \circ \gamma_i$. This correlation is again self-adjoint, and therefore, a polarity, since $\widehat{\pi}' = \gamma_i \circ \widehat{\pi} \circ \widehat{\gamma}_i$ and $\widehat{\pi}' \circ \pi' = \mathrm{id}_{\mathbb{P}^3}$. Two points X, Y are conjugate w.r.t. π if, and only if, $\gamma_i(X)$ and $\gamma_i(Y)$ are conjugate w.r.t. π'.

We prove $\pi' = \pi$ for $i = 1, 2$ and begin with the harmonic homology γ_1:

Since all points in the axial plane $\pi(A)$ remain fixed, the planar polarities induced by π and π' are equal. Let $A_1 A_2 A_3$ be a self-polar triangle of these planar polarities. Together with the center A, they form a tetrahedron, which is self-polar w.r.t. π and π'.

Now, it remains to be shown that, on the three edges $l_j := [A, A_j]$, $j = 1, 2, 3$, the induced involutions are equal: There are two involutions acting on l_j, the induced involution π_{l_j}

and the restriction of γ_1 onto l_j with fixed points A and A_j. Since these two fixed points are corresponding under π_{l_j}, the two involutions commute (see, *e.g.*, [46, Exercise 5.4.1]). Therefore, when the points $X, Y \in l_j$ are conjugate w.r.t. π, *i.e.*, $Y = \pi_{l_j}(X)$, then $\gamma_1(Y) = \gamma_1 \circ \pi_{l_j}(X) = \pi_{l_j} \circ \gamma_1(X)$. This means that $\gamma_1(X)$ and $\gamma_1(Y)$ are also conjugate w.r.t. π or, in other words, the induced involutions of π and π' are the same.

With regard to the skew involution γ_2 with axes l and $l^* := \pi^*(l)$, we choose two different, but mutually conjugate points A and A' on l and show that γ_2 is the product of the two harmonic homologies with centers A and A' and respective axes $\pi(A)$ and $\pi(A')$.

Under the product of the two homologies, all points on $\pi(A) \cap \pi(A') = l^*$ remain fixed, but so do all points on $l = [A, A']$, since the restrictions of both harmonic homologies to the line l are involutions with the same fixed points A and A'. Furthermore, the two harmonic homologies commute since the center of one homology lies in the axial plane of the other. Consequently, the product of the two homologies is involutive and biaxial, and therefore, the skew involution with the axes l and l^*. ■

In the particular case $\pi = \pi_Q$, we can state:

Corollary 4.2.7 *Each quadric $\mathcal{Q}$ remains fixed under all harmonic homologies with a point $A \notin \mathcal{Q}$ as center and its polar plane w.r.t. $\mathcal{Q}$ as axial plane.*

If l and its polar line l^* w.r.t. $\mathcal{Q}$ are skew, then the skew involution with axes l and l^* maps $\mathcal{Q}$ onto itself.

Collineations which keep a quadric $\mathcal{Q}$ fixed are called *automorphisms* of $\mathcal{Q}$.

Quadrics in the real projective space

Let $A_0A_1A_2A_3$ be a self-polar tetrahedron of an elliptic or hyperbolic polarity π in the real projective space $\mathbb{P}^3(\mathbb{R})$. Then, each of its triangles is self-polar w.r.t. the induced polarity in the spanned plane ε, and this planar polarity can be elliptic or hyperbolic. We recall (note [46, p. 266]):

Lemma 4.2.8 *If $A_1A_2A_3$ is a self-polar triangle of a polarity π_ε in a real projective plane ε, then, in the case of an elliptic polarity, the involution of conjugate points on each side of the triangle is elliptic, while for a hyperbolic polarity, exactly on two sides the induced involutions are hyperbolic.*

We denote these two types of self-polar triangles by (eee) and (ehh), where 'e' stands for an elliptic involution and 'h' for a hyperbolic one (see [46, p. 278, Fig. 7.14]).

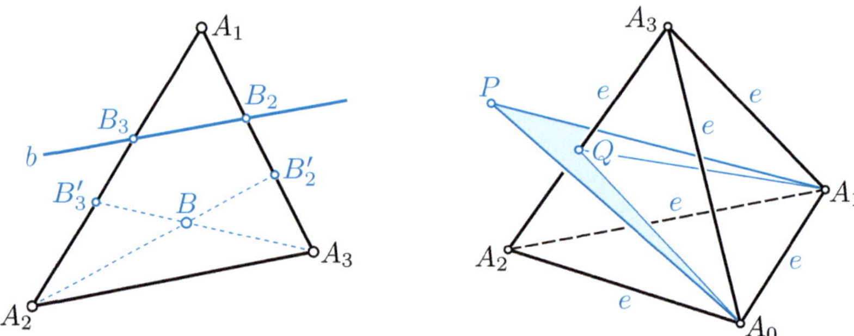

FIGURE 4.5. Left: Induced involutions of conjugate points on the sides of a self-polar triangle; Right: The point P is supposed to be self-conjugate.

Proof: We define π_ε by a self-polar triangle $A_1A_2A_3$ and an additional pair $B \mapsto b$ of pole and polar line, where $A_1, \dots, B$ is a quadrangle and $A_1, A_2, A_3 \notin b$. Then, on the side $[A_1, A_2]$, the involution of conjugate points maps A_1 onto A_2 and the intersection B_3 with b onto the point B'_3 being aligned with B and A_3 (Figure 4.5, left). A positive cross-ratio $\mathrm{cr}(A_1, A_2, B_3, B'_3)$ of the two pairs characterizes hyperbolic involutions, *i.e.*, real self-conjugate points on $[A_1, A_2]$. The same applies to the other sides. Due to the theorems of Ceva and Menelaos, the product of the cross-ratios on the three sides must be negative, which results in the two cases. ∎

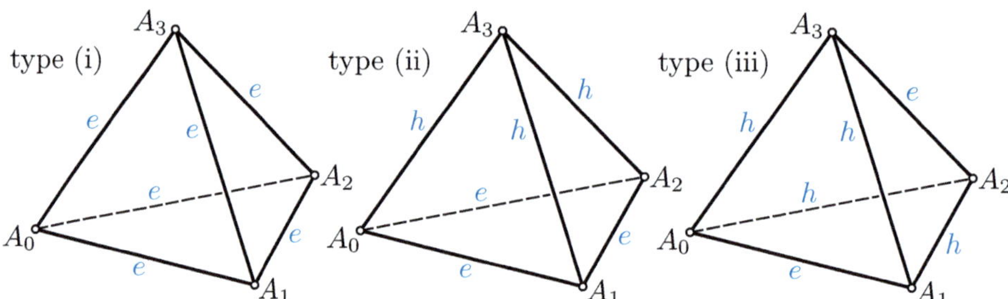

FIGURE 4.6. Induced involutions of conjugate points on the edges of a self-polar tetrahedron.

By virtue of Lemma 4.2.8, for the involutions induced by any elliptic or hyperbolic polarity π on the edges of a self-polar tetrahedron $A_0 \dots A_3$ in $\mathbb{P}^3(\mathbb{R})$, there remain three possibilities (see Figure 4.6):

type (i): all involutions are elliptic;

type (ii): the involutions are elliptic on exactly three coplanar edges;

type (iii): the involutions are elliptic on exactly two opposite edges.

If π is the polarity w.r.t. a ruled quadric $\mathcal{Q}$, then each induced polarity must be hyperbolic, since each generator of $\mathcal{Q}$ meets all planes at self-conjugate points. Therefore, each self-polar triangle must be of type (ehh), and consequently, all self-polar tetrahedra are of type (iii).

Conversely, any self-polar tetrahedron of type (iii) defines self-polar lines, because there exist self-conjugate points on opposite edges (l, l^*) and, by virtue of Lemma 4.2.2, the connection of a self-conjugate point $P \in l$ with a self-conjugate point $Q \in l^*$ is self-polar.

If π is elliptic, then the induced polarity must be elliptic in each plane, too. Therefore, each self-polar tetrahedron must be of type (i).

Conversely, the induced polarities are elliptic in the planes of any self-polar tetrahedron of type (i). Suppose that there is a self-polar point P outside the faces of $A_0 \dots A_3$ (see Figure 4.5, right). Then, the induced polarity in the plane $\varepsilon := [A_0, A_1, P]$ must be hyperbolic. Together with A_0 and A_1, the point of intersection $Q := \varepsilon \cap [A_2, A_3]$ forms a self-polar triangle of π_ε of type (ehh). This yields a self-conjugate point on the side $[Q, A_0]$ in the plane $[A_0, A_2, A_3]$, which contradicts our assumption.

What remains now is that all self-polar tetrahedra of oval quadrics must be of type (ii). We summarize:

Theorem 4.2.8 *All self-polar tetrahedra of ruled quadrics in $\mathbb{P}^3(\mathbb{R})$ are of type* (iii) *and those of oval quadrics are of type* (ii).

Let π be any elliptic or hyperbolic polarity in $\mathbb{P}^3(\mathbb{R})$ with a self-polar tetrahedron $A_0 \dots A_3$. When the vertices $A_0, \dots, A_3$ are used as base points of a coordinate frame, then the π representing matrix $\mathbf{M}$ already has diagonal form $\operatorname{diag}(a_{00}, \dots, a_{33})$ with $a_{00}, \dots, a_{33} \neq 0$, since $\pi_Q \colon (1, 0, 0, 0)\mathbb{F} \mapsto (1 : 0 : 0 : 0)\mathbb{F}$, and so on. By (4.6), the two points $X = (x_0 : x_1 : 0 : 0)$ and $X' = (x_0' : x_1' : 0 : 0)$ on $[A_0, A_1]$ are conjugate w.r.t. π if, and only if,

$$a_{00} x_0 x_0' + a_{11} x_1 x_1' = 0.$$

This symmetric bilinear form characterizes the induced involution on $[A_0, A_1]$, which, in the case $a_{00} a_{11} > 0$, is elliptic and otherwise hyperbolic. The same holds for all other edges of the tetrahedron.

If π is the polarity w.r.t. an oval quadric $\mathcal{Q}$, we can assume that the involutions on the edges through A_3 are hyperbolic. Now, we define the unit point such that the planes connecting one side of the triangle $A_0 A_1 A_2$ with E intersect the opposite edge at a self-conjugate point.

This means, *e.g.*, that the point $[A_0, A_1, E] \cap [A_2, A_3]$ with coordinates $(0:0:1:1)$ is self-conjugate, which implies that $a_{22} + a_{33} = 0$ and, similarly, $a_{00} + a_{33} = a_{11} + a_{33} = 0$.

If the induced involution on an edge through A_3, say on $[A_2, A_3]$, is elliptic, we specify the plane $[A_0, A_1, E]$ such that the conjugate point of $X = (0:0:1:1)$ is the fourth harmonic point w.r.t. A_2 and A_3, hence $X' = (0:0:1:-1)$, which implies $a_{22} - a_{33} = 0$, and so on. Finally, we may set $a_{33} = -1$, which gives rise to the following standard forms.

Theorem 4.2.9 *For each polarity π in $\mathbb{P}^3(\mathbb{R})$, there is a coordinate frame such that the representing matrix $\mathbf{M}$ has one of the following forms:*

$$\mathbf{M} = \begin{cases} \mathrm{diag}(1,1,1,1) & \Leftrightarrow \quad \pi \text{ is elliptic,} \\ \mathrm{diag}(1,1,1,-1) & \Leftrightarrow \quad \pi = \pi_Q \text{ and } Q \text{ is oval,} \\ \mathrm{diag}(1,1,-1,-1) & \Leftrightarrow \quad \pi = \pi_Q \text{ and } Q \text{ is ruled.} \end{cases}$$

For the sake of completeness, we shall mention that even a symmetric 4×4 matrix $\mathbf{M}$ with rank $r < 4$ defines a symmetric relation called *conjugacy* between two points $\mathbf{p}\mathbb{R}$ and $\mathbf{q}\mathbb{R}$ by the condition (4.6), $\mathbf{p}^{\mathrm{T}}\mathbf{M}\mathbf{q} = 0$. There is a $(3-r)$-dimensional space of *singular points* $\mathbf{s}\mathbb{R}$ satisfying $\mathbf{M}\mathbf{s} = \mathbf{0}$. Singular points are conjugate to each point in $\mathbb{P}^3(\mathbb{R})$. When conjugacy is restricted to an $(r-1)$-dimensional space of non-singular points, then for $r = 3$, we obtain pairs of conjugate points w.r.t. a planar polarity and for $r = 2$ we obtain pairs of an elliptic or hyperbolic involution.

Theorem 4.2.10 *The set of points in $\mathbb{P}^3(\mathbb{R})$ satisfying $\mathbf{x}^{\mathrm{T}}\mathbf{M}\mathbf{x} = 0$ with a symmetric 4×4 matrix $\mathbf{M}$ of rank $r \leq 3$ is a singular quadric Q.*
When $\mathbf{M}$ has rank deficiency 1, i.e., $r = 3$, this is a quadratic or imaginary cone with the apex at the singular point. In the case $r = 2$, we obtain either two real or two complex conjugate planes intersecting at the line of singular points. For $r = 1$, the set of singular points is a plane, which equals the set of self-conjugate points.
In the case of $r = 3$, it is the sets of tangent planes of a conic that are dual to the singular point quadrics; for $r = 2$, it is pairs of line bundles; and for $r = 1$, it is single line bundles.

Euclidean geometry in the projective setting

When the real projective space $\mathbb{P}^3(\mathbb{R})$ is introduced as the projective closure of the Euclidean 3-space $\mathbb{E}^3$, then it is endowed with additional structures. Below, we analyse some of them.

The orthogonality in $\mathbb{E}^3$ defines a bijectivity between bundles of parallel lines l and pencils of parallel planes ε orthogonal to l. This induces, in the plane ω at infinity, a bijective mapping $\nu\colon \mathcal{P}_\omega \to \mathcal{L}_\omega$ with $(l \cap \omega) \mapsto (\varepsilon \cap \omega)$ for $l \perp \omega$. If l runs through a pencil $\mathcal{L}_{A,\varphi}$ of lines, the corresponding planes $\varepsilon \perp l$ are parallel to lines f perpendicular to φ. Hence, ω is a correlation. Since the orthogonality $l \perp f$ between two lines is symmetric, the conjugacy between $l \cap \omega$ and $f \cap \omega$ is symmetric, too. Consequently, orthogonality in $\mathbb{E}^3$ induces a polarity ν in ω, which is called the Euclidean *absolute polarity*.

No real line can be orthogonal to itself. Therefore, the absolute polarity in ω is elliptic. In the complex extension, the self-conjugate points form the *absolute conic* j, which satisfies

$$x_0 = 0 \quad \text{and} \quad x_1^2 + x_2^2 + x_3^2 = 0, \tag{4.9}$$

when using homogeneous Cartesian coordinates $(x_0 : x_1 : x_2 : x_3) = (1 : x : y : z)$. Finite imaginary lines which meet j are called *isotropic* as well as finite planes tangent to j. Pairs of orthogonal lines in any pencil $\mathcal{L}_{A,\varepsilon}$ form an involution with fixed lines passing through the absolute circle points of ε. Hence, the absolute conic j comprises the absolute circle points of all finite planes ε in $\mathbb{E}^3$.

We recall LAGUERRE's formula from page 76: If l_1 and l_2 are two lines in the pencil $\mathcal{P}_{A,\varepsilon}$ and i and $\bar{i}$ are the two complex conjugate isotropic lines in this pencil, then the measure α of the angle enclosed by l_1 and l_2 is related to the cross ratio of the four lines by

$$\operatorname{cr}(l_1, l_2, i, \bar{i}) = e^{2i\alpha}, \text{ where } \alpha = \sphericalangle\, l_1 l_2. \tag{4.10}$$

The dual formula holds for two intersecting planes $\varepsilon_1, \varepsilon_2$ and the two isotropic planes $\iota, \bar{\iota}$ through the line $\varepsilon_1 \cap \varepsilon_2$ in the form

$$\operatorname{cr}(\varepsilon_1, \varepsilon_2, \iota, \bar{\iota}) = e^{2i\alpha}, \text{ where } \alpha = \sphericalangle\, \varepsilon_1 \varepsilon_2. \tag{4.11}$$

Collinear transformations κ which map the plane ω onto itself, are called *affine transformations.* If κ sends all pairs of orthogonal lines again to orthogonal lines, then κ is called a *similarity.* In this case, the restriction of κ to the plane ω keeps the absolute conic j fixed. Consequently, by virtue of LAGUERRE's formula, all angle measures between lines or planes remain fixed, too. Each triangle if mapped onto a similar triangle, *i.e.*, scaled with a certain factor, and this scaling factor is the same for all distances. In other words, for each similarity κ in $\mathbb{E}^3$, there exists a scaling factor f_κ with $\overline{\kappa(A)\kappa(B)} = f_\kappa \cdot \overline{AB}$. Similarities with scaling factor $f_\kappa = 1$ preserve all distances and are called *motions*, whether they are continuous transformations or discrete.

Important examples of motions are the *reflections in planes.* They are special harmonic homologies where the center is the ideal point orthogonal to the plane of the mirror. The *reflection in a line* g is a skew involution with axes g and g^*, where g^* is the ideal line of all planes orthogonal to g. Rotations about the axis g are examples of uniaxial collineations.

Affine and Euclidean classification of quadrics

From the viewpoint of the projective space $\mathbb{P}^3(\mathbb{F})$, the affine space $\mathbb{A}^3(\mathbb{F})$ can be defined by appointing any plane ω to be the 'plane at infinity' and focusing on elements outside ω, *i.e.*, on finite points, lines, and planes. Affine transformations are collineations which keep the plane at infinity fixed. In order to classify quadrics $\mathcal{Q}$ in the real affine space $\mathbb{A}^3(\mathbb{R})$, we have to study the relation of $\mathcal{Q}$ and $\pi_\mathcal{Q}$ to the plane ω.

If ω is self-conjugate, then the quadric $\mathcal{Q}$ is a *paraboloid.* Each finite line l through the point of contact with ω is called a *diameter line* of $\mathcal{Q}$. Its polar line l^* lies in ω, and the skew involution with the axes l and l^*, which maps $\mathcal{Q}$ onto itself, is an *affine reflection* in l. This characterizes l as an axis of affine symmetry. Each plane through l is the axial plane of an affine reflection which also leaves the paraboloid invariant.

Paraboloids can be oval or ruled and accordingly, they are called *elliptic* or *hyperbolic.* In the latter case, the paraboloid intersects ω along two lines, taken from different reguli. Therefore, all finite generators of each regulus of $\mathcal{Q}$ are parallel to a fixed plane.

If the pole M of ω is finite, then M is called the *center*, and the quadric is a *central quadric.* The harmonic homology with the center M and axis ω, which maps $\mathcal{Q}$ onto itself, is the reflection in M and shows that M is a center of symmetry. When $MA_1A_2A_3$ is a self-polar tetrahedron, then the edges through M form a *triplet of conjugate diameters.* The polar line of the diameter line $l_1 = [M, A_1]$ is $l_1^* = [A_2, A_3]$ at infinity. The affine reflection in l maps $\mathcal{Q}$ onto itself.

When the induced polarity in ω is elliptic, $\mathcal{Q}$ is an *ellipsoid.* The self-polar tetrahedra are of type (ii). Each line through M is an edge of a self-polar tetrahedron and carries two real points of $\mathcal{Q}$.

A hyperbolic polarity in ω belongs to a *hyperboloid* $\mathcal{Q}$. In this case, the tangent cone with apex M contacts the hyperboloid along its curve at infinity; this cone is called *asymptotic cone.* It depends on the type of the self-polar tetrahedra whether the hyperboloid $\mathcal{Q}$ is oval or ruled. In the first case, each triplet of conjugate diameters contains only one axis with real points of $\mathcal{Q}$, while the existence of a tetrahedron of type (iii) yields two axes with real points. Accordingly, the hyperboloids are two-sheeted or one-sheeted, respectively. Quadratic cones with an apex at infinity are, of course, quadratic *cylinders.*

It can be shown that, for each regular quadric in $\mathbb{A}^3(\mathbb{R})$, there exists a coordinate frame such that the equation can be reduced to one of the normal forms, as listed in Theorem 3.1.1, but where all coefficients α_i are equal to 1.

When classifying quadrics w.r.t. Euclidean transformations, we must additionally analyse the relation between the quadrics and the absolute polarity in ω.

In the case of a paraboloid $\mathcal{Q}$, let U denote its point of contact with ω. The absolute polar u of U has a polar line u^* w.r.t. $\mathcal{Q}$, and this is the *axis* of the paraboloid $\mathcal{Q}$, because the skew involution with axes u and u^* is the *reflection* in u^*, which maps the paraboloid onto itself. There are two cases to be distinguished:

(i) Either there are exactly two tangents t_1, t_2 at U which are polar w.r.t. $\mathcal{Q}$ and at the same time conjugate w.r.t. the absolute polarity. Then, the planes $[t_1, u^*]$ and $[t_2, u^*]$ are the only planes of symmetry. The reflection in each of these planes maps the paraboloid onto itself.

(ii) When the absolute polarity as well as $\pi_{\mathcal{Q}}$ induce the same involution in the pencil of lines $\mathcal{L}_{U,\omega}$ or, in other words, when the complex conjugate lines of intersection between an elliptic paraboloid and ω are tangents of the absolute conic, then each plane through u^* is a plane of symmetry, and $\mathcal{Q}$ is a paraboloid of revolution.

In the case of central quadrics $\mathcal{Q}$, the induced polarity in ω and the absolute polarity share at least one common self-polar triangle $A_1A_2A_3$ (see [46, p. 278]). Then, the diameter lines $[M, A_i]$ are axes of symmetry of $\mathcal{Q}$, and the self-polar tetrahedron is called the *principal self-polar tetrahedron* of $\mathcal{Q}$. In the following two cases, the axes of $\mathcal{Q}$ are not unique:

(i) When the quadric $\mathcal{Q}$, which can also be imaginary, has a two-point contact with the absolute conic, each real point A_1 on the common chord is a point of a principal self-polar tetrahedron, and $\mathcal{Q}$ is a surface of revolution, an ellipsoid or a hyperboloid.

(ii) When the induced polarity π_ω is the absolute polarity, then, in the case of a hyperbolic polarity $\pi_\mathcal{Q}$, the quadric $\mathcal{Q}$ is a *sphere.*

Finally, a few words about elliptic polarities π: On each line l through the pole M of ω, the induced involution is elliptic. Consequently, there exist two points $P, P' \in l$ which are conjugate w.r.t. π and symmetric w.r.t. M. The harmonic homology γ with center M and axial plane ω is the reflection in M and leaves π invariant, which means $\widehat{\gamma} \circ \pi \circ \gamma = \pi$, hence $\widehat{\gamma} \circ \pi = \pi \circ \gamma$. In this case, the product $\pi' := \pi \circ \gamma$ is also self-adjoint, since

$$\widehat{\pi}' \circ \pi' = (\widehat{\pi} \circ \widehat{\gamma}) \circ (\pi \circ \gamma) = \widehat{\pi} \circ \pi = \mathrm{id}_\mathcal{P}.$$

The point M is also the pole of ω w.r.t. this composed polarity π'. If, on any line l through M, the points P and P' are conjugate w.r.t. π and symmetric w.r.t. M, then $\pi' = \pi \circ \gamma$ maps P via P' to the plane $\pi(P')$ which passes through P. Hence, P is self-conjugate w.r.t. π'. This holds for each diameter line l. Therefore, the self-polar tetrahedra are of type (ii), and π' is the polarity w.r.t. an ellipsoid with center M. Thus, we proved

Theorem 4.2.11 *Each elliptic polarity in the projective extension of $\mathbb{E}^3$ is the commutative product of the polarity w.r.t. an ellipsoid $\mathcal{Q}$ and the reflection in the center of the ellipsoid. This composition is called the anti-polarity w.r.t. $\mathcal{Q}$.*

If the ellipsoid $\mathcal{Q}$ is given in standard form, then the respective symmetric matrices $\mathbf{M}_\mathcal{Q}$ of $\mathcal{Q}$ and $\mathbf{M}$ of the elliptic polarity are

$$\mathbf{M}_\mathcal{Q} = \operatorname{diag}\left(-1, \frac{1}{a^2}, \frac{1}{b^2}, \frac{1}{c^2}\right) \quad \text{and} \quad \mathbf{M} = \operatorname{diag}\left(1, \frac{1}{a^2}, \frac{1}{b^2}, \frac{1}{c^2}\right)$$

In the literature, the ellipsoid $\mathcal{Q}$ is sometimes called the *real representative* of the imaginary quadric, associated with the elliptic polarity.

Since the field of complex numbers is algebraically closed, in the complex extension of $\mathbb{P}^3(\mathbb{R})$, every polarity π with a real symmetric transformation matrix $\mathbf{M}$ defines a quadric as the set of self-conjugate points. With each complex point P, the complex conjugate $\overline{P}$ is also self-conjugate. In the case of an elliptic polarity π, we call this quadric *empty* or *imaginary*.

Moreover, in the complex extension, every quadric is ruled, and it contains pairs of complex conjugate generators. When the polarity π is elliptic, the generators must be complex of the second kind, since otherwise pairs of complex conjugate generators would intersect at a real self-conjugate point. This means that complex conjugate generators of an imaginary quadric belong to the same regulus. On the other hand, the generators of oval quadrics are imaginary of the first kind, and the pairs of complex conjugate generators belong to different reguli. In particular, the generators of a sphere are isotropic. The intersection of a sphere with any tangent plane is a null circle consisting of two complex conjugate isotropic lines of the first kind.

If a polarity is elliptic and the induced polarity in ω equals the absolute polarity, we call the corresponding imaginary quadric of self-conjugate points an *imaginary sphere*. Its generators are isotropic of the second kind.

• **Exercise 4.2.1** Equations of a ruled quadric.

Find a coordinate transformation which transforms the equation $x_0x_3 - x_1x_2 = 0$ of a ruled quadric, as given in Theorem 4.2.5, to the standard form $x_0^2 + x_1^2 - x_2^2 - x_3^3 = 0$ given in Theorem 4.2.9.

• **Exercise 4.2.2** Polarity w.r.t. a ruled quadric.

Prove the following statements:

(i) *Let an arbitrary line l intersect a generator e of the ruled quadric $\mathcal{Q}$. Then, also the polar line l^* of l w.r.t. $\mathcal{Q}$ intersects the generator e.*

(ii) *If the four sides of the (non-planar) quadrangle $A_1 \dots A_4$ are generators of $\mathcal{Q}$, then the two diagonals $[A_1, A_3]$ and $[A_2, A_4]$ are polar w.r.t. $\mathcal{Q}$.*

• **Exercise 4.2.3** Two pairs of polar lines.

Prove the following statement concerning the complex extension of $\mathbb{P}^3(\mathbb{R})$: *Let* (g, g^*) *and* (h, h^*) *be two pairs of polar lines. Then, the four lines* g, g^*, h, h^* *belong to a regulus if, and only, if they intersect two different self-polar lines* e_1 *and* e_2.

• **Exercise 4.2.4** Reducible intersection of two quadrics.

Prove the following statement: *If two quadrics* $\mathcal{Q}_1, \mathcal{Q}_2$ *in* $\mathbb{P}^3(\mathbb{R})$ *contact each other at two points* P, Q, *and the common chord* $l = [P, Q]$ *and the intersection* $l^* = \tau_P \cap \tau_Q$ *of the corresponding tangent planes are skew, then all common points of the quadrics are located in two planes through* l, *which can also coincide or be complex conjugate. This is true also when one or both quadrics are quadratic cones.*

Hint: Choose a coordinate frame with the first two base points at P and Q and the two remaining base points on l^*. Common points of $\mathcal{Q}_1$ and $\mathcal{Q}_2$ satisfy all linear combinations of the quadrics' equations. There is a linear combination of the representing symmetric matrices with zeros in the first two lines and rows (see also Theorem 6.9.2).

• **Exercise 4.2.5** Proof of Theorem 2.4.4.

Prove the statements of Theorem 2.4.4 concerning cones (or cylinders) passing through two different conics c and d of a regular quadric $\mathcal{Q}$.

Hint: Let l denote the line of intersection between the planes of c and d, and select any point $C \in l$ in the exterior of $\mathcal{Q}$ as the center of a projection into the polar plane γ of C. Then in the image plane γ, there are one or two centers V of planar harmonic homologies which map the contour of $\mathcal{Q}$ onto itself and exchange the contour points of c with those of d. Each homology can be continued to a harmonic homology in the 3-space which preserves $\mathcal{Q}$ and sends c to d. This confirms that its center V is the apex of a quadratic cone through c and d.

The last statement in Theorem 2.4.4 follows by dualization.

• **Exercise 4.2.6** Projective generation of quadrics in projective 3-space.

Given two different points P, Q in $\mathbb{P}^3(\mathbb{F})$, let $\kappa\colon \mathcal{L}_P \to \mathcal{E}_Q$ be a projective correlation between the bundles with respective carriers P and Q. Then, the set of points of intersection $X = x \cap \kappa(x)$ is a regular or singular quadric. Conversely, each quadric can be generated in this way.

Hint: Choose a coordinate frame with $P = (1:0:0:0)$ and $Q = (0:0:0:1)$. Note that for describing a quadric in the form $\mathbf{x}^\mathrm{T} \mathbf{A}\, \mathbf{x} = 0$, the matrix A need not be symmetric.

4.3 The projective n-space

Let V be a vectorspace of dimension $n+1 \geq 4$ over a commutative field $\mathbb{F}$ with $\operatorname{Char}\mathbb{F} \neq 2$. Then, we obtain the analytic model[5] of the *projective n-space* $\mathbb{P}^n(\mathbb{F})$ in the following way:

The one-dimensional subspaces $\mathbf{x}\mathbb{F}$ of V with $\mathbf{x} \neq \mathbf{0}$ are the *points* X of $\mathbb{P}^n(\mathbb{F})$. More general, the $(d+1)$-dimensional subspaces U_{d+1} of the vectorspace V, $1 \leq d < n$, are the d-dimensional *projective subspaces* or *linear spaces* S_d of $\mathbb{P}^n(\mathbb{F})$, when seen as point sets. In the particular cases $d = 1$, 2, and $n-1$ we speak of *lines*, *planes*, and *hyperplanes* of $\mathbb{P}^n(\mathbb{F})$.

The intersection $S_1 \cap S_2$ of two projective subspaces is either a projective subspace of dimension $d \geq 0$ or it is empty, when for the corresponding sub-vectorspaces holds $U_1 \cap U_2 = \{\mathbf{0}\}$. In the latter case, we speak of *skew* spaces S_1, S_2. The *span* or *connecting space* $S_1 \vee S_2$ corresponds to the vectorspace $U_1 + U_2$. The well-known dimension formula for vectorspaces yields the *projective dimension formula*

$$\dim S_1 + S_2 = \dim(S_1 \cap S_2) + \dim(S_1 \vee S_2), \tag{4.12}$$

where $\dim(S_1 \vee S_2) \leq n$ and $\dim(S_1 \cap S_2) \geq -1$. Here, -1 characterizes an empty intersection.

A set of $d+1$ points $\{\mathbf{p}_0\mathbb{F}, \dots, \mathbf{p}_d\mathbb{F}\}$ of $\mathbb{P}^n$ is called *projectively independent* if the points span a d-dimensional space, which means that the vectors $\mathbf{p}_0, \dots, \mathbf{p}_d$ are linearly independent. The set of $d+1$ points $\{\mathbf{p}_0\mathbb{F}, \dots, \mathbf{p}_d\mathbb{F}\}$ is called a *simplex* or *n-simplex* in the spanned d-dimensional space S_d. A set of $d+2$ points in S_d, consisting of a simplex $\{\mathbf{p}_0\mathbb{F}, \dots, \mathbf{p}_d\mathbb{F}\}$ and an additional point $\mathbf{q}\mathbb{F}$, is called a *fundamental figure* of S_d if, and only if, the point $\mathbf{q}\mathbb{F}$ is projectively independent of any d points out of the simplex.

Any basis $(\mathbf{b}_0, \dots, \mathbf{b}_d)$ of the $(d+1)$-dimensional sub-vectorspace of V, $0 < d < n$, defines *homogeneous coordinates* of the points X in the corresponding d-dimensional projective space S_d. Conversely, for a given fundamental figure $A_0, \dots, A_d, E$ in S_d, there exists a *projective coordinate frame* in S_d, which assigns to the *base points* $A_0, \dots, A_d$ the respective

[5] For an axiomatic approach to projective spaces, defined as sets of points and lines, readers are referred to [18].

coordinates $(1:0:\ldots:0)$, …, $(0:\ldots:0:1)$, and to the *unit point* E the coordinates $(1:\ldots:1)$.

Hyperplanes in $\mathbb{P}^n(\mathbb{F})$ are the sets of points X with coordinates $(x_0 : \ldots : x_n)$ satisfying a linear equation $u_0x_0 + \cdots + u_nx_n = 0$ with coefficients $\mathbf{u} := (u_0, \ldots, u_n) \neq (0, \ldots, 0)$. The coefficients serve as homogeneneous coordinates $(u_0 : \ldots : u_n)$ of the hyperplane, which will be denoted by $\mathbf{u}\mathbb{F}$ for short. Consequently, the point $\mathbf{x}\mathbb{F}$ lies in the hyperplane $\mathbf{u}\mathbb{F}$ if, and only if, $\langle \mathbf{u}, \mathbf{x} \rangle = 0$. If hyperplanes serve as basic elements in the projective n-space, we call this the *dual space* $\widehat{\mathbb{P}}^n(\mathbb{F})$ of $\mathbb{P}^n(\mathbb{F})$.

Collineations, correlations, polarities

A *collineation* $\kappa\colon \mathbb{P}^n(\mathbb{F}) \to \mathbb{P}'^n(\mathbb{F})$ between two projective n-spaces over $\mathbb{F}$ is a bijective mapping that preserves the collinearity of points. This implies that d-dimensional subspaces are mapped to d-dimensional subspaces, and this is one-to-one for d between 0 and n. A collineation is called *projective*, if the restriction to a line is a projectivity.

The fundamental theorem states that each projective collineation $\mathbb{P}^3(\mathbb{F}) \to \mathbb{P}^3(\mathbb{F})$ is induced by a bijective linear mapping $f\colon V \to V$ of the underlying vectorspace, and conversely, any linear mapping $f : V \to V$ induces a projective collineation $\kappa\colon \mathbb{P}^n(\mathbb{F}) \to \mathbb{P}^n(\mathbb{F})$ with $\mathbf{x}\mathbb{F} \mapsto f(\mathbf{x})\mathbb{F}$.

A *projective correlation* $\kappa\colon \mathbb{P}^n(\mathbb{F}) \to \mathbb{P}'^n(\mathbb{F})$ is a projective collineation of $\mathbb{P}^n(\mathbb{F})$ to the dual space $\widehat{\mathbb{P}}'^n(\mathbb{F})$ with $\kappa\colon \mathbf{x}\mathbb{F} \mapsto \mathbf{u}'\mathbb{F}$, where $\mathbf{u}' = \mathbf{M}\mathbf{x}$ with a regular matrix $\mathbf{M} \in \mathbb{F}^{(n+1)\times(n+1)}$. Points $\mathbf{y}'\mathbb{F} \in \kappa(\mathbf{x}\mathbb{F})$ are called *conjugate* to $\mathbf{x}\mathbb{F}$ w.r.t. the correlation κ.

A projective correlation of the projective n-space $\mathbb{P}^n(\mathbb{F})$ onto itself is called a *polarity* if, and only if, conjugacy is a symmetric relation. The n-dimensional analogue of Theorem 4.2.2 states that κ is a polarity if, and only if, the matrix $\mathbf{M}$ is either symmetric or skew symmetric. In the latter case, κ is a *null polarity* with the property that all points X are *self-conjugate, i.e.*, $X \in \kappa(X)$. In this case, we conclude from

$$\det \mathbf{M} = \det(\mathbf{M}^{\mathrm{T}}) = \det(-\mathbf{M}) = (-1)^{n+1} \det \mathbf{M},$$

that null polarities exist only in projective spaces of odd dimension.

In the case of a symmetric matrix $\mathbf{M}$, the self-conjugate points satisfy the equation $\mathbf{x}^{\mathrm{T}}\mathbf{M}\mathbf{x} = 0$. We call this set a regular quadric $\mathcal{Q}$ in $\mathbb{P}^n(\mathbb{F})$ and κ the polarity $\pi_{\mathcal{Q}}$ w.r.t. $\mathcal{Q}$, even when the polarity is *elliptic* and

$\mathcal{Q}$ is an empty set (Definition 4.2.4). The hyperplane $\kappa(X)$ is called the *polar hyperplane* of the point X, and vice versa, X is the *pole* of the hyperplane. More general, the polarity in $\mathcal{Q}$ sends the points of a d-dimensional subspace S_d to the hyperplanes through an $(n-d-1)$-dimensional space, called the *polar space* $\pi_{\mathcal{Q}}(S_d)$ of S_d. Each point $X \in S_d$ is conjugate to all points $Y \in \pi_{\mathcal{Q}}(S_d)$, and vice versa.

A subspace S_d is called *self-polar* if $\pi_{\mathcal{Q}}(S_d) = S_d$. With regard to the dimensions, we obtain the necessary condition $2d = n - 1$.

If S_d and its polar space $\pi_{\mathcal{Q}}(S_d)$ are skew, then for each point $X \in S_d$ the mapping $X \mapsto \pi_{\mathcal{Q}}(X) \cap S_d$ is a polarity in S_d, which is called the polarity *induced* by $\pi_{\mathcal{Q}}$. In S_d, the self-conjugate points w.r.t. the induced polarity equal the subquadric $\mathcal{Q} \cap S_d$. In the particular case $d = 1$, this induced polarity is an involution.

The polar plane $\pi_{\mathcal{Q}}(X)$ of points $X \in \mathcal{Q}$ is called the *tangent hyperplane* T_X to $\mathcal{Q}$ at X. Its intersection with $\mathcal{Q}$ is a quadratic cone with apex X, because each other self-conjugate point $Y \in T_X$, $Y \neq X$, is conjugate to X. Now, we can apply Lemma 4.2.2, which is valid also in higher dimensions, and conclude that, with $Y \in \mathcal{Q}$, the full line $[X, Y]$ is contained in $T_X \cap \mathcal{Q}$.

Lemma 4.3.1 *Every two points* $\mathbf{p}\mathbb{F}$ *and* $\mathbf{q}\mathbb{F}$ *of a regular quadric* $\mathcal{Q}$ *in* $\mathbb{P}^n(\mathbb{F})$ *are projectively equivalent, i.e., there exists a projective collineation* κ *which maps* $\mathcal{Q}$ *onto itself and interchanges the two points.*

If $\mathbf{p}\mathbb{F}$ *and* $\mathbf{q}\mathbb{F}$ *are not conjugate w.r.t.* $\mathcal{Q}$*, then a harmonic homology* κ *with the center* $\mathbf{z}\mathbb{F}$ *and the axial hyperplane* $\pi_{\mathcal{Q}}(\mathbf{z}\mathbb{F})$ *has the requested properties.*

Proof: Let $\mathbf{p}\mathbb{F}$ and $\mathbf{q}\mathbb{F}$ be two different and not conjugate points of $\mathcal{Q}$. Then, they are the points of intersection between $\mathcal{Q}$ and the connecting line $\mathbf{p}\mathbb{F} + \mathbf{q}\mathbb{F}$. Any other point $\mathbf{z}\mathbb{F}$ on this line is not self-conjugate and can be represented by $\mathbf{z} = \mathbf{p} + \mathbf{q}$. In particular, the point $\mathbf{z}^* := \mathbf{p} - \mathbf{q}$ is conjugate to $\mathbf{z}\mathbb{F}$, because the two equations $\mathbf{p}^{\mathrm{T}}\mathbf{M}\mathbf{p} = \mathbf{q}^{\mathrm{T}}\mathbf{M}\mathbf{q} = 0$ imply that

$$(\mathbf{p}+\mathbf{q})^{\mathrm{T}}\mathbf{M}(\mathbf{p}-\mathbf{q}) = -\mathbf{p}^{\mathrm{T}}\mathbf{M}\mathbf{q} + \mathbf{q}^{\mathrm{T}}\mathbf{M}\mathbf{p} = -\mathbf{p}^{\mathrm{T}}\mathbf{M}\mathbf{q} + \left(\mathbf{q}^{\mathrm{T}}\mathbf{M}\mathbf{p}\right)^{\mathrm{T}} = -\mathbf{p}^{\mathrm{T}}\mathbf{M}\mathbf{q} + \mathbf{p}^{\mathrm{T}}\mathbf{M}\mathbf{q} = 0.$$

Note that the four points $\mathbf{p}\mathbb{F}$, $\mathbf{q}\mathbb{F}$, $\mathbf{z}\mathbb{F}$, and $\mathbf{z}^*\mathbb{F}$ form a harmonic quadruple.

On the other hand, when $\mathbf{p}\mathbb{F} \in \mathcal{Q}$ and $\mathbf{z}\mathbb{F} \notin \mathcal{Q}$, let $\mathbf{q}\mathbb{F}$ be the remaining point of intersection between the line $\mathbf{z}\mathbb{F} + \mathbf{p}\mathbb{F}$ and $\mathcal{Q}$. There are two cases: If $\mathbf{q}\mathbb{F} = \mathbf{p}\mathbb{F}$, then $\mathbf{z}\mathbb{F}$ lies in the tangent hyperplane $\pi_{\mathcal{Q}}(\mathbf{p}\mathbb{F})$ and, vice versa, $\mathbf{p}\mathbb{F}$ in the polar hyperplane $\pi_{\mathcal{Q}}(\mathbf{z}\mathbb{F})$ of $\mathbf{z}\mathbb{F}$. If otherwise $\mathbf{p}\mathbb{F} \neq \mathbf{q}\mathbb{F}$, then $\mathbf{p}\mathbb{F}$ and $\mathbf{q}\mathbb{F}$ are harmonic conjugates w.r.t. $\mathbf{z}\mathbb{F}$ and $\mathbf{z}^*\mathbb{F}$ on $\mathbf{p}\mathbb{F} + \mathbf{q}\mathbb{F}$.

In both cases, the harmonic homology with the center $\mathbf{z}\mathbb{F}$ and the axial hyperplane $\pi_{\mathcal{Q}}(\mathbf{z}\mathbb{F})$ sends $\mathbf{p}\mathbb{F}$ to the point $\mathbf{q}\mathbb{F} \in \mathcal{Q}$. Hence, this harmonic homology maps $\mathcal{Q}$ onto itself.

If the two given points $\mathbf{p}\mathbb{F}$ and $\mathbf{q}\mathbb{F}$ of $\mathcal{Q}$ are conjugate, *i.e.*, the connecting line $\mathbf{p}\mathbb{F}+\mathbf{q}\mathbb{F}$ belongs to $\mathcal{Q}$, then we use another point $\mathbf{r}\mathbb{F} \in \mathcal{Q}$ which is not conjugate to $\mathbf{p}\mathbb{F}$ or $\mathbf{q}\mathbb{F}$. In this case, a composition of two harmonic homologies sends $\mathbf{p}\mathbb{F}$ via $\mathbf{r}\mathbb{F}$ to $\mathbf{q}\mathbb{F}$, while $\mathcal{Q}$ remains fixed. It is possible to specify two conjugate points on the lines $\mathbf{p}\mathbb{F}+\mathbf{r}\mathbb{F}$ and $\mathbf{r}\mathbb{F}+\mathbf{q}\mathbb{F}$ as centers of the two homologies. In this case, the product of the two homologies is also involutive. ■

The second part of Lemma 4.3.1 is an n-dimensional analogue of Theorem 4.2.7. The original proof of this theorem can easily be adapted to prove also for elliptic polarities that each non-null polarity π in $\mathbb{P}^n(\mathbb{F})$ commutes with harmonic homologies whose centers and axial hyperplanes are polar w.r.t. π.

A simplex $A_0 \dots A_n$ in $\mathbb{P}^n(\mathbb{F})$ is called *self-polar* w.r.t. to a regular quadric $\mathcal{Q}$ if the points are mutually conjugate. This means that for each point A_i of the simplex, the polar hyperplane $\pi_{\mathcal{Q}}(A_i)$ is the span of the remaining n points. For defining a self-polar simplex, we can select any non self-conjugate point A_0. The given polarity induces a polarity in the polar plane $\pi_{\mathcal{Q}}(A_0)$, and there, we specify any non self-polar point A_1, and so on. If the points $A_0, \dots, A_n$ are chosen as base points of a projective coordinate frame, then the regular matrix $\mathbf{M}$ of $\mathcal{Q}$ must have diagonal form $\operatorname{diag}(\lambda_0, \dots, \lambda_n)$ with $\lambda_i \in \mathbb{F}$.

A change of the unit point E causes a coordinate transformation of the type $(x_0, \dots, x_n) \mapsto (x'_0, \dots, x'_n) = (\alpha_0 x_0, \dots, \alpha_n x_n)$ with $\alpha_0, \dots, \alpha_n \neq 0$. Consequently, the matrix $\mathbf{M}$ is replaced by $\mathbf{M}' = \operatorname{diag}(\lambda_0/\alpha_0^2, \dots, \lambda_n/\alpha_n^2)$. Now, it depends on the underlying field $\mathbb{F}$ whether other simplifications are possible:

In the case $\mathbb{F} = \mathbb{C}$, each λ_i is a square, and therefore, we obtain for all regular quadrics the normal form $\mathbf{M} = \operatorname{diag}(1, \dots, 1)$. This means that in the complex projective n-space $\mathbb{P}^n(\mathbb{C})$, any two regular quadrics are equivalent. The case $\mathbb{F} = \mathbb{R}$ follows below.

The real projective n-space

For achieving normal forms of quadrics in the real projective n-space $\mathbb{P}^n(\mathbb{R})$, we can reorder the base points of any self-polar simplex such that, in the diagonal form of the coefficient matrix $\mathbf{M}$, the entries $\lambda_0, \dots, \lambda_p$ are positive and $\lambda_{p+1}, \dots, \lambda_n$ are negative. Then, after an appropriate change of the unit point E, the coefficient matrix $\mathbf{M}$ of $\mathcal{Q}$ gets the form $\mathbf{M} = \operatorname{diag}(1, \dots, 1, -1, \dots, -1)$ with $p+1$ plus and $n-p$ minus. Since we can multiply the equation with -1, it is no restriction to require $p+1 \geq n-p$, *i.e.*, $2p \geq n-1$.

Due to SYLVESTER's *law of inertia*, the number p is independent of the choice of the self-polar simplex. However, the invariance of p has also a geometric background:

Theorem 4.3.1 *For each regular quadric $\mathcal{Q}$ in the real projective n-space, there exist coordinate frames such that $\mathcal{Q}$ has the equation*

$$\mathbf{x}^{\mathrm{T}}\mathbf{M}\mathbf{x} = x_0^2 + \cdots + x_p^2 - x_{p+1}^2 - \cdots - x_n^2 = 0 \text{ with } p \geq \frac{n-1}{2}\,.$$

The number p is the same for all possible coordinate frames, because $d := n-p-1$ is the maximum dimension of linear spaces on $\mathcal{Q}$.

In the literature, the maximum dimension d is sometimes called the *index* of the quadric $\mathcal{Q}$ (cf. [18, I, p. 190]), and the d-dimensional linear spaces of $\mathcal{Q}$ are called *generators*. Theorem 4.3.1 is the projective version of Theorem 3.1.2, as far as regular central quadrics are concerned.

Proof: All points $X \in \mathbb{P}^n(\mathbb{R})$ whose coordinates $(x_0 : \ldots : x_n)$ satisfy the system of $p+1$ linearly independent homogeneous equations

$$x_0 = x_{p+1},\ x_1 = x_{p+2},\ \ldots,\ x_{n-p-1} = x_n,\ x_{n-p} = \cdots = x_p = 0,$$

belong to $\mathcal{Q}$ and form an $(n-p-1)$-dimensional linear space S_1. On the other hand, no real point of the p-dimensional subspace

$$S_2\colon\ x_{p+1} = x_{p+2} = \cdots = x_n = 0$$

satisfies the given equation of $\mathcal{Q}$. Consequently, the quadric $\mathcal{Q}$ cannot contain a linear space of dimension greater than $n-p-1$, because otherwise it would intersect S_2 in at least one point by virtue of (4.12), and this yields a contradiction.

We recall from Chapter 3 that the pair of numbers $(p+1,\, n-p)$ is called the signature of the quadratic form $\mathbf{x}^{\mathrm{T}}\mathbf{M}\mathbf{x}$. ■

By virtue of Theorem 4.3.1, in real projective spaces of even dimension $n = 2k$, $k \in \mathbb{N}$, there exist k different types of non-empty quadrics. Here, the maximum dimension of linear spaces on quadrics is $d = k-1$ and the corresponding quadrics have the signature $(k+1,\, k)$.

For odd dimension $n = 2k+1$, there are $k+1$ different types of non-empty quadrics. The quadrics with the maximum dimension $d = k$ of linear spaces have the signature $(k+1,\, k+1)$.

Note that, in the complex extension of $\mathbb{P}^n(\mathbb{R})$, all regular quadrics are equivalent. Therefore, all of them carry linear spaces of the maximum dimension d, as given before. In this sense, the following theorem (cf. [21, p. 232]) is also valid for all regular quadrics in $\mathbb{P}^n(\mathbb{C})$.

Theorem 4.3.2 *1. In the real projective space $\mathbb{P}^n(\mathbb{R})$ of odd dimension $n = 2k+1$, the quadrics $\mathcal{Q}$ with signature $(k+1,\, k+1)$ carry two $\frac{k(k+1)}{2}$-parameter families of k-dimensional linear spaces. Each automorphic harmonic homology of $\mathcal{Q}$ interchanges the two families.*
For two different k-dimensional subspaces S_k, S_k' of $\mathcal{Q}$ holds:

$$\begin{aligned} S_k,\ S_k' \text{ from different families} &\Longrightarrow \dim(S_k \cap S_k') \in \{k-1,\, k-3, \dots\}, \\ S_k,\ S_k' \text{ from the same family} &\Longrightarrow \dim(S_k \cap S_k') \in \{k-2,\, k-4, \dots\}. \end{aligned}$$

2. In the real projective space of even dimension $n = 2k$, the quadrics with signature $(k+1,\, k)$ carry a $\frac{k(k+1)}{2}$-parameter family of $(k-1)$-dimensional linear spaces. The dimension of the intersection between any two of them varies between $k-2$ and -1.

Proof: 1. By virtue of Lemma 4.3.1, it is no restriction to focus on the particular point $P = (1 : 0 : \ldots : 0 : 1)$ of the quadric $\mathcal{Q}$ satisfying $x_0^2 + \cdots + x_k^2 - x_{k+1}^2 - \cdots - x_{2k+1}^2 = 0$. Its tangent hyperplane $T_P\colon x_0 - x_{2k+1} = 0$ intersects $\mathcal{Q}$ in a quadratic cone which connects the apex P with the subquadric $\mathcal{Q}'\colon x_1^2 + \cdots + x_k^2 - x_{k+1}^2 - \cdots - x_{2k}^2 = 0$ of signature (k,k) in the $(2k-1)$-dimensional subspace $x_0 = x_{2k+1} = 0$. The connection of each $(k-1)$-dimensional linear space of $\mathcal{Q}'$ with the apex P is a k-dimensional subspace of $\mathcal{Q}$.

From now on, we use the induction on k: For $k = 2$, the subquadric $\mathcal{Q}'$ in the three-dimensional subspace is ruled and has two reguli. The connection with P defines two one-parameter families of planes on the quadratic cone $T_P \cap \mathcal{Q}$. Planes from different families intersect along lines, planes from the same family share only the apex P.

Let us assume, as induction hypothesis for $k-1$, that the subquadric $\mathcal{Q}'$ contains two families of $(k-1)$-dimensional linear spaces with dimensions of mutual intersections, as stated in Theorem 4.3.2, when k is replaced by $k-1$. After connection with P, we observe that we can split the k-dimensional spaces through P on $\mathcal{Q}$ into two families such that the stated dimensions of mutual intersections are fulfilled.

Now, we extend the separation of k-dimensional spaces to all other points of $\mathcal{Q}$. By virtue of Lemma 4.3.1, for each point $Q \notin T_P$, there is an automorphic homology γ of $\mathcal{Q}$ which switches P and Q. The $(n-2)$-dimensional intersection $T_P \cap T_Q$ between the tangent hyperplanes remains fixed under γ and intersects $\mathcal{Q}$ in a subquadric $\mathcal{Q}'$ of the same type as studied before.

Let us define that two k-dimensional subspaces $S_k \subset T_P$ and $S_k' \subset T_Q$ belong to the same family on $\mathcal{Q}$ if, and only if, their intersections with $T_P \cap T_Q$ belong to different families of the subquadric $\mathcal{Q}'$.

This means for S_k and S_k' in the same family that their intersection equals the intersection between complementary subspaces of $\mathcal{Q}'$, and by our induction hypothesis, the dimension is $\in \{k-2,\, k-4, \dots\}$.

Two spaces S_k and S_k' taken from different families share either the $(k-1)$-dimensional intersection with $\mathcal{Q}'$, if they are corresponding under γ. Or their intersection equals that between subspaces of the same family of $\mathcal{Q}'$, and consequently, it is of a dimension $\in \{k-3,\, k-5, \dots\}$.

Since the same holds for any two non-conjugate points $P, Q \in \mathcal{Q}$, we obtained a partition of the set of k-dimensional linear spaces on $\mathcal{Q}$ which satisfies all conditions claimed with regard to the dimensions of mutual intersections. Therefore, the family assigned to each S_k must be unique, because we only need to check the intersection of S_k with a single distinguished k-space.

Each harmonic homology γ with $\mathcal{Q} \to \mathcal{Q}$ maps S_k onto another space S'_k, where $S_k \cap S'_k$ equals the $(k-1)$-dimensional intersection with the axial hyperplane of γ. Therefore, S_k and S'_k belong to different families. Note that no S_k can be fixed under γ, because all points in S_k and all hyperplanes through S_k are self-conjugate, and therefore, they are not usable as center or axis of any automorphic harmonic homology.

This proves that the statement in the induction hypothesis is also valid for k. The remaining statement that we obtain two $\dfrac{k(k+1)}{2}$-parameter families of generators on $\mathcal{Q}$, can be proved by induction, too.

2. In contrast to the first case, the $(k-1)$-dimensional spaces S_{k-1} on the regular quadric $\mathcal{Q}$ in $\mathbb{P}^{2k}(\mathbb{R})$ are not self-polar. There exist hyperplanes through S_{k-1} which are not self-conjugate, and therefore, available as axial hyperplanes of harmonic homologies which keep S_{k-1} fixed. Consequently, we cannot split the set of $(k-1)$-dimensional spaces of $\mathcal{Q}$ into two families which switch under harmonic homologies.

In order to prove that, for the intersection of two linear spaces S_d, S'_d of maximum dimension (cf. Theorem 4.3.1), all dimensions between -1 and d are possible, we confine ourselves to one example: The following two spaces of $\mathcal{Q}$,

$$\begin{aligned} S_d\colon\ & x_0 - x_{p+1} = x_1 - x_{p+2} = \cdots = x_{n-p-1} - x_n = 0,\ x_{n-p} = \cdots = x_p = 0 \text{ and} \\ S'_d\colon\ & x_0 + x_{p+1} = x_1 + x_{p+2} = \cdots = x_{n-p-1} + x_n = 0,\ x_{n-p} = \cdots = x_p = 0, \end{aligned}$$

are skew, hence $\dim(S_d \cap S'_d) = -1$. As an exercise, readers are invited to prepare examples with other dimensions of $S_d \cap S'_d$. ■

Quadrics on given points

For defining a symmetric matrix $\mathbf{M}$ with $n+1$ rows, $\frac{1}{2}(n+1)(n+2)$ entries a_{ij} have to be given. This means that, for defining a quadric by single points, $\frac{1}{2}n(n+3)$ points are necessary and sufficient, provided that the corresponding system of homogeneous linear equations for the entries a_{ij} has maximal rank (note the Veronese mapping on page 180).

In the literature, there are graphical constructions with ruler and compass for determining, in the real projective 3-space, a quadric given by nine points (see, *e.g.*, [127, no. 49] or [77]). Still open is the question, how to characterize in a geometric way the spatial position of ten points which belong to the same quadric. This would be a spatial analogue of the well-known theorem of Pappus [46, Sect. 6.2] which characterizes six points of a conic in the projective plane in form of a linear construction.

• Exercise 4.3.1 Quadrics passing through seven vertices of a box.

Prove the following statement: *If, in the projective 3-space, a quadric $\mathcal{Q}$ passes through seven vertices of a projectively transformed box, then $\mathcal{Q}$ contains the eighth vertex, too* (see [12, p. 99]).

Hint: If a quadrangle is inscribed in a conic c, then the diagonal triangle is a polar triangle, *i.e.*, self-polar w.r.t. c (see, *e.g.*, [46, p. 294]). This defines corresponding elements of the polarity w.r.t. $\mathcal{Q}$, and moreover, harmonic homologies mapping $\mathcal{Q}$ onto itself. The eight vertices are common points of three quadrics with a common self-polar tetrahedron.

• Exercise 4.3.2 Induced involutions on the edges of a self-polar simplex in $\mathbb{P}^n(\mathbb{R})$.

With regard to the types displayed in Figure 4.5 (left), enumerate the different types of self-polar simplices in $\mathbb{P}^n(\mathbb{R})$, based on the classification of quadrics in Theorem 4.3.1.

• Exercise 4.3.3 Projective generation of generators of quadrics in $\mathbb{P}^n(\mathbb{R})$.

For given skew linear spaces S_d and S'_d of maximal dimension d, according to Theorem 4.3.1, let $\kappa\colon S_d \to S'_d$ be a projective correlation. Then, the connecting spaces $X \vee \kappa(X)$ belong to a regular quadric.

Conversely, if the regular quadric $\mathcal{Q}$ contains the two skew spaces S_d and S'_d, then the polarity $\pi_{\mathcal{Q}}$ induces a correlation $\kappa\colon S_d \to S'_d$, and the connecting spaces $X \vee \kappa(X)$ belong to $\mathcal{Q}$.

4.4 Projective models of non-Euclidean geometries

The axioms of the hyperbolic 3-space $\mathbb{H}^3$ and the Euclidean 3-space $\mathbb{E}^3$ differ only in the parallel postulate: In $\mathbb{E}^3$, for every given plane ε, there is a unique parallel plane through any point P. In $\mathbb{H}^3$, the opposite holds true: Through each point $P \notin \varepsilon$, there pass (at least) two planes, which do not meet the plane ε. For more information on Hyperbolic Geometry, see [31, 44]).

Cayley-Klein model of Hyperbolic Geometry

We refer to the *projective* or CAYLEY-KLEIN *model* of $\mathbb{H}^3$: Let an oval quadric Ω be given, which is called the *absolute quadric* in the projective space $\mathbb{P}^3$. Then, *points* of $\mathbb{H}^3$ are the points in the interior of Ω, *lines* of $\mathbb{H}^3$ are secants of Ω, and *planes* of $\mathbb{H}^3$ are those which intersect Ω along (non-empty) conics. The polarity π_Ω w.r.t. Ω is called the *absolute polarity.* Points of Ω are called *absolute points* and do not belong to $\mathbb{H}^3$. The same holds for so-called *absolute* lines and planes which are tangent to Ω.

We prefer the standard version of the CAYLEY-KLEIN model of $\mathbb{H}^3$. It is embedded into the projective extension of the Euclidean space $\mathbb{E}^3$. We use homogeneous Cartesian coordinates $(x_0 : x_1 : x_2 : x_3) = (1 : x : y : z)$ and the unit sphere as the absolute quadric Ω. Then, the polar form with respect to Ω can be written as

$$\langle \mathbf{p}, \mathbf{q} \rangle_h := p_0 q_0 - p_1 q_1 - p_2 q_2 - p_3 q_3. \tag{4.13}$$

Points P in $\mathbb{H}^3$ have coordinate vectors $\mathbf{p}$ where $\langle \mathbf{p}, \mathbf{p} \rangle_h > 0$. The hyperbolic distance $d_h(P, Q)$ between $P = \mathbf{p}\mathbb{R}$ and $Q = \mathbf{q}\mathbb{R}$ is defined by

$$\begin{aligned} \cosh d_h(P,Q) &= \left| \frac{\langle \mathbf{p}, \mathbf{q} \rangle_h}{\sqrt{\langle \mathbf{p}, \mathbf{p} \rangle_h \langle \mathbf{q}, \mathbf{q} \rangle_h}} \right| \quad \text{or} \\ d_h(P,Q) &= \frac{1}{2} \ln \operatorname{cr}(P, Q, U_1, U_2), \end{aligned} \tag{4.14}$$

where U_1 and U_2 are the absolute points of the line $[P, Q]$. There are similar formulas for the hyperbolic measures of angles between intersecting lines or planes, based on cross ratios between the given elements and the absolute elements in the spanned pencils. A line l is *hyperbolic orthogonal* to a plane ε if l passes through the absolute pole of ε. Two planes are *hyperbolic orthogonal* if they are conjugate w.r.t. Ω.

Hyperbolic motions are the restrictions to $\mathbb{H}^3$ of projective transformations in $\mathbb{P}^3$ which map the absolute quadric Ω onto itself. Due to the second formula in (4.14), hyperbolic motions preserve all hyperbolic distances. From the invariance of cross ratios, we also infer the invariance of hyperbolic angle measures under hyperbolic motions.

Hyperbolic reflections are, according to Theorem 4.2.7, harmonic homologies of Ω, which are restricted to the interior of Ω. Hence, depending on whether the center E or the axial plane $\varepsilon = \pi_\Omega(E)$ of the harmonic homology is an element of $\mathbb{H}^3$, we obtain the hyperbolic reflection in the point E or in the plane ε.

It can be proved that each hyperbolic motion can be represented as a product of four reflections in points or planes at most. Now, we refer to our standard model, embedded into the Euclidean 3-space: The stereographic projection of Ω in a plane transforms the restrictions of harmonic homologies to Ω into inversions or line reflections. This reveals that the group of hyperbolic motions is isomorphic to the Möbius group in the plane.

Let us classify the products of two reflections σ_1, σ_2 in planes, *i.e.*, the products of two harmonic homologies with centers A_1 and A_2 in the exterior of Ω. We denote the line of intersection between the axial planes $\pi_\Omega(A_1)$ and $\pi_\Omega(A_2)$ with a. Then, a is the absolute polar line of $a^* = [A_1, A_2]$. Under the product of the two harmonic homologies, each point of a remains fixed, as well as each plane through a^*. Three cases need to be distinguished:

(i) The line $a = \pi_\Omega(A_1) \cap \pi_\Omega(A_2)$ is a secant of Ω. In this case, the product $\sigma_2 \circ \sigma_1$ is a *hyperbolic rotation* about the axis a. Points $X \in \mathbb{H}^3 \backslash a$ remain on hyperbolic circles centered on a and within planes orthogonal to a. Only in the case of absolute conjugate points A_1 and A_2, the two axial planes are orthogonal and the two reflections commute. Then, the product is involutive and a *hyperbolic reflection in a*.

(ii) The line a is a tangent of Ω. In this case, the product is called a *horolation*. In each plane through a^*, the points X move on concentric horocircles.

(iii) The line a lies in the exterior of Ω, *i.e.*, $a^* = [P, Q]$ is a secant of Ω. Then, the product $\sigma_Q \circ \sigma_P$ is a *hyperbolic translation* along a^*. Points $X \notin a^*$ remain in planes through a^* and move on hypercircles with the axis a^*.

In all three cases, the motion can be embedded into a one-parameter subgroup of hyperbolic motions, namely a rotation, horolation, or translation. The corresponding orbits of points are hyperbolic circles, horocircles, or hypercircles.

Clifford surfaces and hyperbolic screw motions

Under rotations about an axis a or translations along the same axis a, the hyperbolic distance of points X to the axis a remains fixed. The geometric locus of points $X \in \mathbb{H}^3$ with a constant distance $d_h(X, a)$ is called a *Clifford surface*, named after W.K. CLIFFORD[6]. Clifford surfaces can be generated by rotating hypercircles with the axis a about a. Figure 4.7 (left) shows the case where the axis is chosen as a Euclidean diameter of Ω. Then, the hyperbolic rotation about a is also a Euclidean rotation, and the corresponding Clifford surfaces are Euclidean ellipsoids of revolution. This reveals that Clifford surfaces with the axis a share with Ω the absolute generators through the absolute points B_1, B_2 of a (Figure 4.7, left).

Clifford surfaces are the hyperbolic counterparts of Euclidean cylinders of revolution. In the coordinate frame, as used in (4.13), we can introduce *hyperbolic cylinder coordinates* (r_h, φ, z_h), where r_h is the hyperbolic distance to the z-axis a, z_h the signed hyperbolic length along a, and φ the polar angle, measured from the x-axis. Then, the normalized coordinate vectors of points in $\mathbb{H}^3$ can be expressed as

$$\mathbf{p} = (\cosh r_h \cosh z_h,\ \sinh r_h \cos\varphi,\ \sinh r_h \sin\varphi, \cosh r_h \sinh z_h), \quad (4.15)$$

where $\langle \mathbf{p}, \mathbf{p} \rangle_h = 1$. For $r_h = \text{const.}$, we obtain a normalized parametrization of the Clifford surface $\mathcal{C}_{r_h}$ with the hyperbolic radius r_h (Figure 4.7, right). When setting $u := \varphi \sinh r_h$ and $v := z_h \cosh r_h$, we obtain

$$\begin{aligned}\mathbf{x}(u,v) = \Big(&\cosh r_h \cosh \frac{v}{\cosh r_h}\,,\ \sinh r_h \cos \frac{u}{\sinh r_h}\,,\\ &\sinh r_h \sin \frac{u}{\sinh r_h}\,,\ \cosh r_h \sinh \frac{v}{\cosh r_h}\Big)\end{aligned}$$

[6]WILLIAM KINGDON CLIFFORD (1845–1879), English mathematician and philosopher.

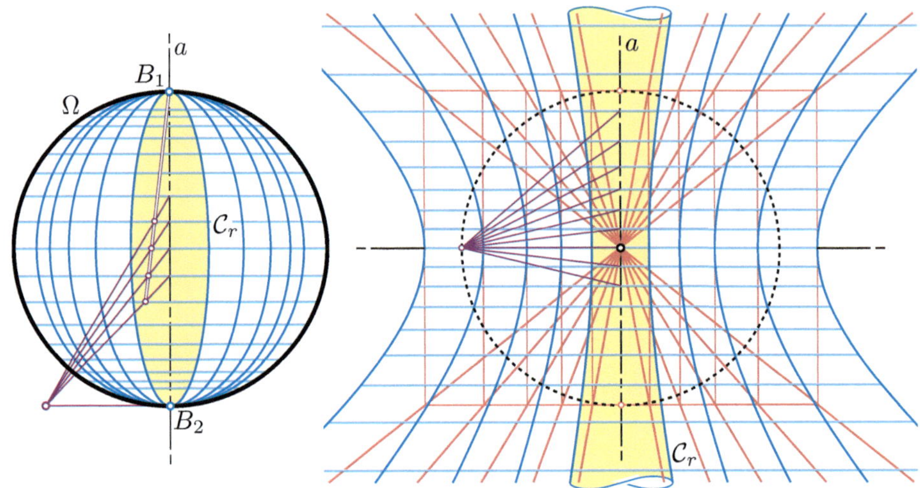

FIGURE 4.7. Clifford surfaces with radii $r, 2r, 3r, \ldots$ and iteratively translated normal planes in the hyperbolic space (left) and in the elliptic space (right).

for $0 \le u < 2\pi \sinh r_h$ and $v \in \mathbb{R}$.

It turns out that the hyperbolic metric induces on $\mathcal{C}_{r_h}$ a Euclidean metric, since the coefficients of the first fundamental form are

$$E = -\langle \mathbf{x}_u, \mathbf{x}_u \rangle = 1, \quad F = -\langle \mathbf{x}_u, \mathbf{x}_v \rangle = 0, \quad G = -\langle \mathbf{x}_v, \mathbf{x}_v \rangle = 1.$$

Thus, there is an isometry $\mathbf{x}(u,v) \mapsto (u,v)$ of the Clifford surface $\mathcal{C}_{r_h}$ in $\mathbb{H}^3$ to a parallel strip in the Euclidean (u,v)-plane.

The composition of rotations about the axis a through φ and a translation along a through a proportional length $z_h = p \cdot \varphi$ with $p =$ const. is a hyperbolic *helical motion* or *screw motion* with the *pitch* or *screw parameter* p.

The trajectories of points are called *helices* or *screws.* Since hyperbolic motions preserve distances and angle measures, along each helix the enclosed angle with the parallel circles remains constant. The isometry mentioned above sends screws to straight lines.

The hyperbolic screw motion about a induces a group of projective automorphisms of Ω. In our standard model, the trajectories of absolute points $Q \in \Omega$ must enclose a constant Euclidean angle with the parallel circles on the sphere. This is a consequence of (4.10), since along the trajectory the cross ratio of the tangents to the trajectory and the parallel circle and the

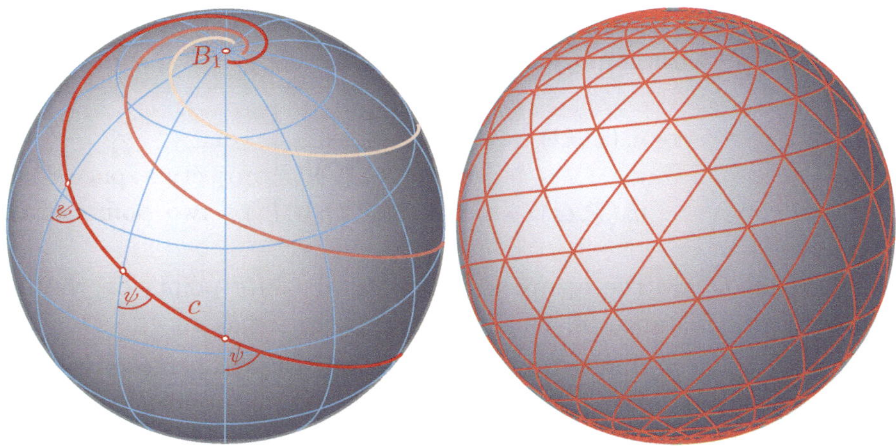

FIGURE 4.8. Left: Spherical loxodromes play the role of hyperbolic helices. Right: A 3-web of spherical loxodromes.

isotropic generators of the Euclidean sphere Ω remains constant. Hence, these isogonal trajectories are *spherical loxodromes* (Figure 4.8, left). The isometry to the plane can be used to transfer, conversely, a three-web built from three families of parallel lines to a three-web of spherical loxodromes (Figure 4.8, right).

For a deeper insight into hyperbolic motions, see, *e.g.*, [31] or [73].

Cayley-Klein model of Elliptic Geometry

The projective model of the *elliptic 3-space* is based on an elliptic polarity in $\mathbb{P}^3(\mathbb{R})$, called the *absolute polarity.* This means that the set of self-conjugate points is an empty quadric Ω consisting only of imaginary *absolute points.* The generators of Ω are imaginary of the second kind (note page 132). Tangent lines and planes of Ω are called *absolute lines* and *absolute planes*, respectively.

We refer to the standard model which is embedded in the projective closure of the Euclidean 3-space with Ω as an imaginary unit sphere. Then, in terms of homogeneous Cartesian coordinates, the polar form w.r.t. Ω can be expressed as

$$\langle \mathbf{p}, \mathbf{q} \rangle_e := p_0q_0 + p_1q_1 + p_2q_2 + p_3q_3. \tag{4.16}$$

All points in $\mathbb{P}^3(\mathbb{R})$ are admitted, and the *elliptic distance* $d_e(P,Q)$ of two points P,Q with homogeneous position vectors $\mathbf{p}$ and $\mathbf{q}$ is defined by

$$\begin{aligned} \cos d_e(P,Q) &= \left| \frac{\langle \mathbf{p}, \mathbf{q} \rangle_e}{\sqrt{\langle \mathbf{p}, \mathbf{p} \rangle_e \langle \mathbf{q}, \mathbf{q} \rangle_e}} \right| \quad \text{or} \\ d_e(P,Q) &= \frac{1}{2\mathrm{i}} \ln \operatorname{cr}(P,Q,U,\overline{U}), \end{aligned} \tag{4.17}$$

where U and $\overline{U}$ are the complex conjugate absolute points on the line $[P,Q]$. Elliptic distances d_e can be restricted by $0 \le d_e < \pi$.

There are similar formulas for the elliptic measures of angles between coplanar lines and pairs of planes, expressed in terms of cross ratios with the absolute lines or planes which belong to the spanned pencils, respectively. Two lines g,h are called *elliptic orthogonal* if they are intersecting and g meets the absolute polar h^* of h, too. Two planes are called *elliptic orthogonal* if they are conjugate w.r.t. Ω.

When $\mathbb{P}^3(\mathbb{R})$ is embedded as a projectively closed hyperplane $x_4 = -1$ in the Euclidean 4-space, then $d_e(P,Q)$ equals the Euclidean angle between the lines connecting P and Q with the origin O in $\mathbb{E}^4$. This star of lines gives the four-dimensional *star model* of the elliptic 3-space. Since the connections of absolute elements in $\mathbb{P}^3(\mathbb{R})$ with the point O are isotropic in $\mathbb{E}^4$, the elliptic measures in $\mathbb{P}^3(\mathbb{R})$ equal Euclidean measures in the four-dimensional star model.

Elliptic motions are projective transformations in $\mathbb{P}^3(\mathbb{R})$ which commute with the elliptic absolute polarity, *i.e.*, which map the absolute quadric Ω onto itself. Due to the second formula in (4.17), elliptic motions preserve all elliptic distances and angle measures. Since all elliptic measures equal Euclidean measures in the isomorphic four-dimensional star model, the group of elliptic motions in $\mathbb{P}^3(\mathbb{R})$ is isomorphic to the orthogonal group $\mathrm{O}(4)$.

Elliptic reflections are automorphic harmonic homologies of Ω, according to Theorem 4.2.7. In contrast to the hyperbolic space, they always have a center and an axial plane, polar w.r.t. Ω. Each elliptic reflection in a point is at the same time an elliptic reflection in a plane. The product of any two different reflections is an *elliptic rotation* about an axis a as well as an *elliptic translation* along the absolute polar a^* of a. The rotations about a and the translations along a form one-parameter groups. If the ratio of the length z_e of translation, measured along a, and the angle φ of the rotation is kept constant, we obtain a group of elliptic screw motions about a with the pitch $p = z_e/\varphi$.[7]

Elliptic screw motions and Clifford translations

As an analogy to the hyperbolic space, the locus of points at constant elliptic distance r_e to any line a is called an elliptic *Clifford surface* with the axis a and elliptic radius r_e. All points of this Clifford surface also have a constant distance $\frac{\pi}{2} - r_e$ to the absolute polar a^* of a. Clifford surfaces remain fixed unter all screw motions about the axes a and a^*.

In our standard model with the axis a as a diameter of Ω, the Clifford surfaces $\mathcal{C}_{r_e}$ are one-sheeted hyperboloids of revolution (see Figure 4.7, right). In terms of *elliptic cylinder coordinates* (r_e, φ, z_e), we can parametrize the Clifford surface with radius r_e as

$$\mathbf{p} = (\cos r_e \cos z_e, \sin r_e \cos\varphi, \sin r_e \sin\varphi, \cos r_e \sin z_e)\,, \tag{4.18}$$

where $\langle \mathbf{p}, \mathbf{p}\rangle_e = 1$. The inhomogeneous coordinates (x, y, z) of the points of $\mathcal{C}_{r_e}$ satisfy

$$\frac{x^2}{\tan^2 r_e} + \frac{y^2}{\tan^2 r_e} - z^2 = 1 \quad \text{or} \quad \frac{1}{\sin^2 r_e}\,(x^2 + y^2) - (x^2 + y^2 + z^2 + 1) = 0.$$

This confirms that the Clifford surfaces with a common axis belong together with Ω to a pencil of quadrics (see Chapter 5). All quadrics in this pencil share the two pairs of complex conjugate generators in the planes $z = \pm\mathrm{i}$.

When setting $u := \varphi \sin r_e$ and $v := z_e \cos r_e$, we obtain

$$\mathbf{x}(u, v) = \left(\cos r_e \cos \frac{v}{\cos r_e},\ \sin r_e \cos \frac{u}{\sin r_e},\ \sin r_e \sin \frac{u}{\sin r_e},\ \cos r_e \sin \frac{v}{\cos r_e}\right)$$

for $-\pi \sin r_e \le u < \pi \sin r_e$ and $-\frac{\pi}{2} \cos r_e \le u < \frac{\pi}{2} \cos r_e$, which defines an isometry of the Clifford surface $\mathcal{C}_{r_e}$ to a rectangle in the (u, v)-plane. Helices on $\mathcal{C}_{r_e}$ correspond to straight line segments.

Quite contrary to the hyperbolic space, in the elliptic 3-space, the Clifford surfaces $\mathcal{C}_{r_e}$ are ruled quadrics. Along each generator f of $\mathcal{C}_{r_e}$, the distance

[7] The sign of the pitch depends on the specified orientations for measuring φ and z_e. Note that the screw motion about a is at the same time a screw motion about a^* with z_e as the angle of rotation and φ as the length of translation.

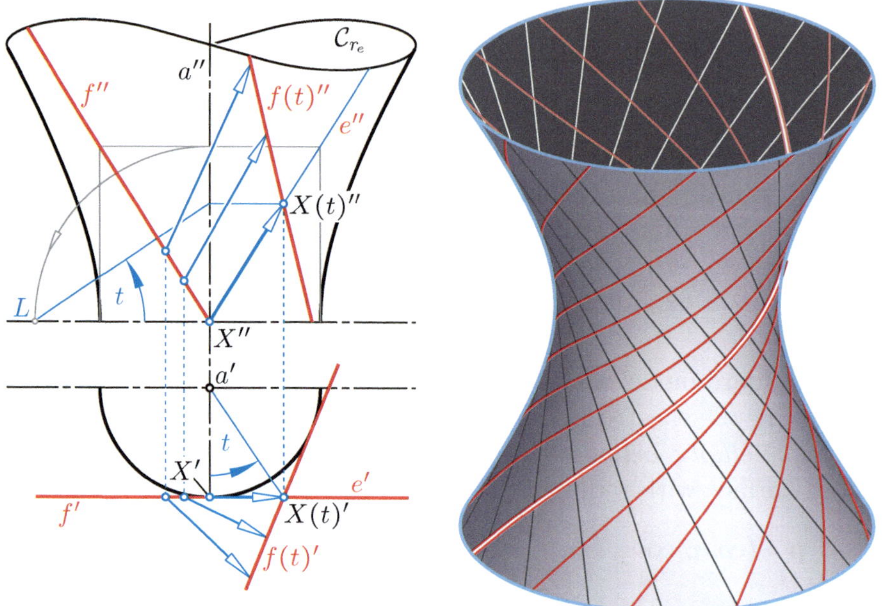

FIGURE 4.9. Left: Clifford translation acting on the Clifford surface $\mathcal{C}_{r_e}$. Right: Quartics of the second kind as elliptic helices with pitch $p = 1/2$. The picture shows the particular case with $\cot r_e = \sqrt{2}$, where the helices are elliptic orthogonal trajectories of one regulus on $\mathcal{C}_{r_e}$.

to the axis a remains constant and there exist infinitely many common perpendiculars of f and a. Hence, the two lines f and a behave like parallel lines, and indeed, we speak of a *Clifford parallelism*[8].

The generators of each regulus on $\mathcal{C}_{r_e}$ serve as trajectories of an elliptic screw motion with pitch $p = \pm 1$. This follows from (4.18): When setting $\varphi = t$ and $z_e = \pm t$ and switching to inhomogeneous coordinates, the particular helix is represented as

$$\mathbf{s}(t) = (\tan r_e,\ \tan r_e \tan t, \pm \tan t)\,.$$

[8]According to W.K. Clifford [28], two distinct lines f and g are said to be *Clifford parallel*, if f, g, and their absolute polars f^* and g^* belong to the same regulus, which implies that the complementary regulus consists of common normals of f and g. According to F. Klein [74, p. 241], parallel lines f and g meet the same pair of complex conjugate generators of the absolute quadric (note Exercise 4.2.3). Depending on the kind of generators of the Clifford surfaces, one can speak of *right parallel* or *left parallel* lines. For further information about Clifford parallelism see, *e.g.*, [31], [44, Kap. 12], or [54].

These particular screw motions are called *Clifford translations.* In the particular vertical position of the axis a, as chosen in Figure 4.9 (left), the angle t shows up in the top view as the angle of rotation. However, t shows up also in the front view, when the length of translation, measured along a, is projected from a Laguerre point L (see [46, p. 252]) of the elliptic involution of absolute conjugate points on a.

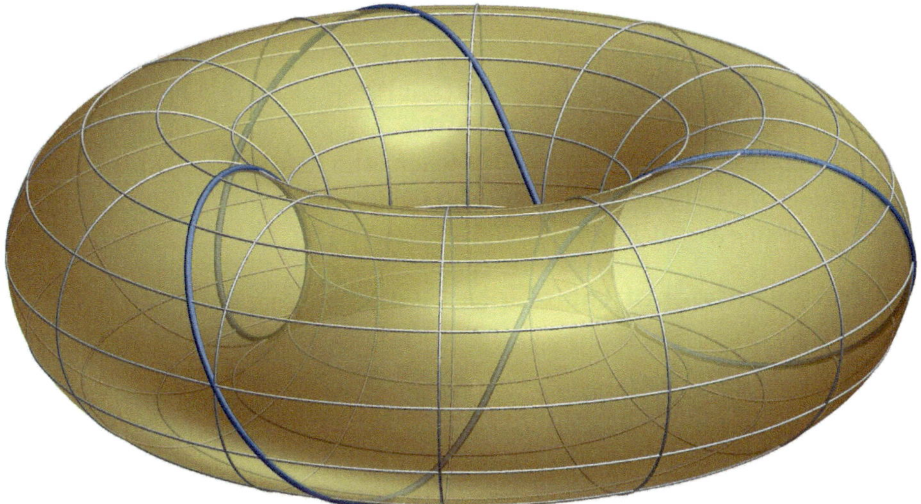

FIGURE 4.10. In the conformal model of elliptic geometry, loxodromes on a torus are helical curves. The depicted curve is an elliptic helix with pitch $p = 1/3$.

In the projective setting, Clifford translations are biaxial collineations with complex conjugate generators of Ω and $\mathcal{C}_{r_e}$ serving as axes. By virtue of (4.10), the angle t is related to the characteristic cross ratio of the collineation. All lines meeting the two axes of this collineation form an elliptic linear line congruence (see Figure 10.4) and are mutually left- or right parallel in the sense of Clifford parallelism. Once the orientations for angle measures about a and lengths of translations along a are fixed, we can differ between *left* translations with pitch $p = -1$ and *right* translations with pitch $p = +1$.[9]

[9] Another approach to Clifford translations is presented in [11] and [91]: When the quadruples of homogeneous coordinates $(x_0, \dots, x_3)$ are identified with Hamiltonian quaternions, then

Since the group generated by rotations about and translations along a is commutative, the product of Clifford right and left translations through the same length t is a rotation about a through $\varphi = 2t$. More generally, the composition of continuous right translations through $c_1 t$ and left translations through $c_2 t$ yields screw motions with the pitch

$$p = \frac{c_1 + c_2}{c_1 - c_2}.$$

For rational p we can assume $c_1, c_2 \in \mathbb{Z}$ with $(c_1, c_2) = 1$. Then, the helices are algebraic and of degree $|c_1| + |c_2|$ (see [128]). Figure 4.9 (right) shows quartics of the second kind, obtained for $c_1 = 3$ and $c_2 = -1$. Depicted is the particular case with $\cot r_e = \sqrt{2}$, where the helices are elliptic orthogonal trajectories of one regulus on the Clifford surface $\mathcal{C}_{r_e}$ (compare with the Euclidean case in Figure 9.15).

Conformal model of the elliptic 3-space

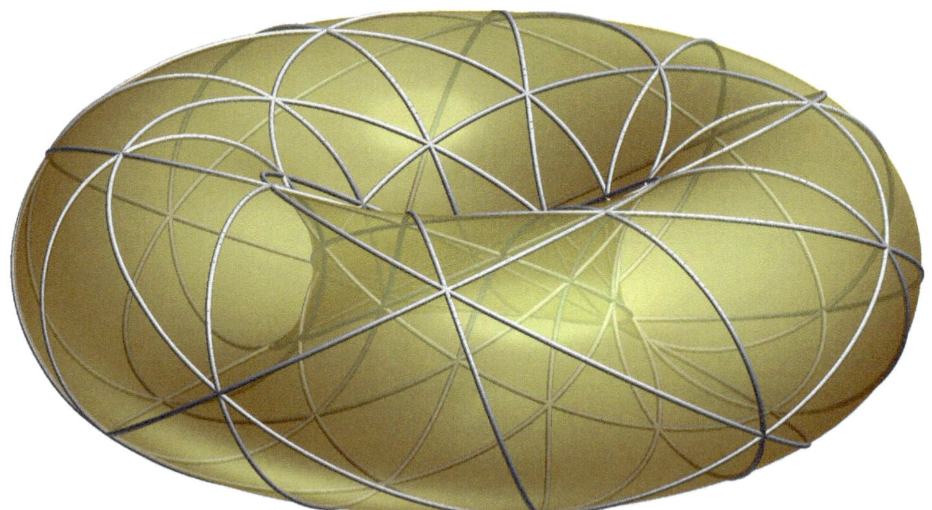

FIGURE 4.11. A 3-web consisting of loxodromes and Villarceau circles on the torus.

---

the multiplication with fixed quaternions, either from the left or from the right, represent exactly the two types of Clifford translations in the elliptic 3-space. This is also the basis for the kinematic mapping of the elliptic space to the group SO(3) of orientation preserving spherical motions in $\mathbb{E}^3$ (note Theorem 10.3.1).

After connecting the points P, Q of the elliptic 3-space with the exterior point O in $\mathbb{E}^4$, we obtain the star model, where the elliptic distance $d_e(P, Q)$ shows up as the Euclidean angle enclosed by the lines $[P, O]$ and $[Q, O]$. Now, we intersect the lines with the unit sphere centered at O and project it stereographically back to $\mathbb{P}^3(\mathbb{R})$. Then, each point P is sent to a pair of points (P_c, P_c'); we can achieve that P_c and P_c' are inverse w.r.t. the empty unit sphere Ω.In this way, intersecting straight lines g, h of the projective model are, in general, mapped to self-inverse circles g_c and h_c, and the elliptic angle between g and h equals the Euclidean angle between g_c and h_c. Consequently, we speak of a *conformal model* of the elliptic 3-space.

It turns out that, in the conformal model, tori and Dupin ring cyclides $\mathcal{C}_{r_e}$ (note Figure 10.18) take over the role of Clifford surfaces. Elliptic helices of the projective model are sent to loxodromes on tori (Figure 4.10) or, more generally, to isogonal trajectories of the curvature lines on Dupin ring cyclides. The reguli on $\mathcal{C}_{r_e}$ correspond to Villarcau circles on tori and cyclides.

This tranference was used in [142] to study loxodromes of the torus. In the same way, three-webs of helices can be transferred to three-webs on a torus (Figure 4.11) and on Dupin cyclides (see page 119).

5 Pencils of quadrics

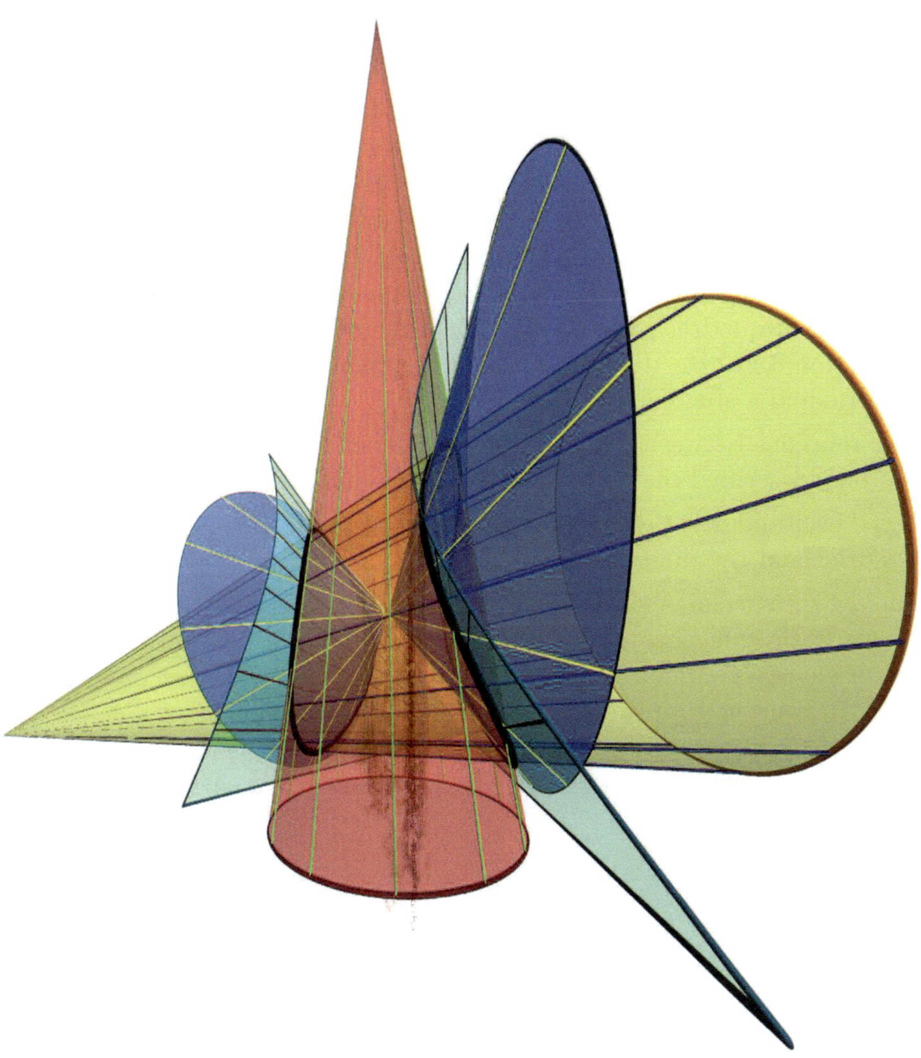

Pencils of quadrics can be classified by means of the number, type, and configuration of singular quadrics in the pencil. The generic pencil of quadrics contains four singular quadrics which are quadratic cones. The thirteen types of pencils will be discussed in detail in this section.

B. Odehnal et al., *The Universe of Quadrics*,
https://doi.org/10.1007/978-3-662-61053-4_5

We have studied pencils of conics in [46, Ch. 7]. Since conics are quadrics in the plane, the five types of pencils of conics are precisely the pencils of quadrics in a projective plane.

In this chapter, we start with an overview of pencils of quadrics in general, and especially in projective 3-space. We describe the classification of pencils of quadrics in projective 3-space and provide normal forms of the equations. A complete classification of pencils of quadrics in an affine 3-space is far beyond the scope of this chapter and book. The classification of pencils of quadrics in projective 3-space is based on the theory of *characteristics* and *elementary divisors* (cf. [87]).

The notion of principal points and common polar tetrahedra is of interest when we are looking for a favorable or simple coordinate representation of a pair of quadrics. Especially, the principal axes transform described in Section 3.1 is the determination of a common polar simplex of a given quadric and a singular polarity.

The relations between the quadrics in a pencil and a generic line or plane are best described by Desargues's theorem and induced polarities.

Since each quadric defines a dual quadric, we can define pencils of dual quadrics and the dual of a pencil of quadrics.

Among the special pencils, we discuss singular pencils which consist either of quadratic cones projecting a pencil of conics or of quadratic cones that touch along a common generator. Pencils of spheres play an outstanding role: In Section 10.2, we shall see that parabolic pencils of spheres correspond to straight lines in Lie's quadric L_2^4. The pencils of spheres give rise to a construction of systems of triply orthogonal surfaces.

5.1 Definition of pencils, basics, invariants

Let us start in an arbitrary projective space $\mathbb{P}^n(\mathbb{F})$ over some commutative field $\mathbb{F}$ with $\operatorname{char}\mathbb{F} \neq 2$ and let us describe points with homogeneous coordinates. We shall exclude the case $\operatorname{char}\mathbb{F} = 2$, although this has a charm of its own. Now, we may assume that $\mathbf{A}, \mathbf{B} \in \mathbb{F}^{(n+1)\times(n+1)}$ are two symmetric matrices which are not each other's scalar multiples. However, $\mathbf{A}$ and $\mathbf{B}$ need not be regular. With $\mathbf{x} = (x_0, \dots, x_n)$, the equations of two quadrics $\mathcal{A}$ and $\mathcal{B}$ are

$$\mathcal{A}:\ \mathbf{x}^{\mathrm{T}}\mathbf{A}\mathbf{x} = 0 \quad \text{and} \quad \mathcal{B}:\ \mathbf{x}^{\mathrm{T}}\mathbf{B}\mathbf{x} = 0. \tag{5.1}$$

We use the following:

Definition 5.1.1 Let $\mathcal{A}$ and $\mathcal{B}$ be two quadrics in a projective space $\mathbb{P}^n(\mathbb{F})$ with equations (5.1), then the equations of all quadrics in the *pencil* $\mathcal{P}$ *of quadrics* spanned by $\mathcal{A}$ and $\mathcal{B}$ are given by

$$\mathcal{P}:\ \mathbf{x}^{\mathrm{T}}(\alpha\mathbf{A} + \beta\mathbf{B})\mathbf{x} = 0 \quad (\alpha, \beta) \in \mathbb{F}^2 \setminus \{(0,0)\}. \tag{5.2}$$

It is clear that each point $B = \mathbf{b}\mathbb{F}$ that is common to $\mathcal{A}$ and $\mathcal{B}$ is also common to all quadrics in the pencil: Since $B \in \mathcal{A}$ and $B \in \mathcal{B}$ causes $\mathbf{b}^{\mathrm{T}}\mathbf{A}\mathbf{b} = 0$ and $\mathbf{b}^{\mathrm{T}}\mathbf{B}\mathbf{b} = 0$, the linear combination $\lambda\mathbf{b}^{\mathrm{T}}\mathbf{A}\mathbf{b} + \mu\mathbf{b}^{\mathrm{T}}\mathbf{B}\mathbf{b} = \lambda\cdot 0 + \mu\cdot 0 = 0$ is annihilated. Therefore, it makes sense to call each point $B \in \mathcal{A} \cap \mathcal{B}$ a *base point* of the pencil $\mathcal{P}$.

Moreover, through each point $P = \mathbf{p}\mathbb{F} \subset \mathbb{P}^n(\mathbb{F})$, there exists a unique quadric $\mathcal{Q}_P \subset \mathcal{P}$ of the pencil: Inserting $\mathbf{p}$ for $\mathbf{x}$ in the equation (5.2) of the pencil, yields

$$\lambda\mathbf{p}^{\mathrm{T}}\mathbf{A}\mathbf{p} + \mu\mathbf{p}^{\mathrm{T}}\mathbf{B}\mathbf{p} = 0,$$

i.e., a single linear homogeneous equation in the unknown $\lambda : \mu$. Its solution

$$\lambda : \mu = -\mathbf{p}^{\mathrm{T}}\mathbf{B}\mathbf{p} : \mathbf{p}^{\mathrm{T}}\mathbf{A}\mathbf{p}$$

defines a unique quadric in the pencil, unless $P \notin \mathcal{A} \cap \mathcal{B}$.

According to our definition, a pencil of quadrics is a linear one-parameter manifold of quadrics. Definition 5.1.1 goes well with the definition of pencils of conics given in [46, Def. 7.3.1.].

In the literature, one can find an alternative way of defining pencils of quadrics in an arbitrary projective space $\mathbb{P}^n$: We recall that the coefficient matrix of a quadrics's equation in $\mathbb{P}^n(\mathbb{F})$ has $(n+1)\times(n+1)$ entries from $\mathbb{F}$, is symmetric, and only the ratio of these matters. Therefore, a quadric in $\mathbb{P}^n(\mathbb{F})$ is defined by prescribing

$$(n+1)+(n)+(n-1)+\ldots+2+1-1=\frac{1}{2}n(n+3)$$

points in *admissible position*. Hence, one point less should define a pencil.

Lemma 5.1.1 *In a projective space $\mathbb{P}^n(\mathbb{F})$ (of n dimensions), the set of all quadrics that pass through $\frac{1}{2}(n^2+3n-2)$ points in admissible position is a pencil.*

Proof: The determination of a quadric $\mathcal{Q}: \mathbf{x}^{\mathrm{T}}\mathbf{Q}\mathbf{x}=0$ that passes through the maximum number $\frac{1}{2}n(n+3)$ of points that can be prescribed is equivalent to solving a system of that many linear equations in the unknown coefficients q_{ij} with $i,j\in\{0,\ldots,n\}$ and $q_{ij}=q_{ji}$. Using one point less drops the rank of the coefficient matrix of the system of linear equations about one. Consequently, the solutions for q_{ij} can be expressed with the help of homogeneous parameters $\alpha:\beta\neq 0:0$. ■

The equations in the system of linear equations used in the proof of Lemma 5.1.1 arise by substituting the coordinates of prescribed points of x_i in the equation $\mathbf{x}^{\mathrm{T}}\mathbf{Q}\mathbf{x}=0$ of $\mathcal{Q}$. Then, the monomials x_ix_j evaluate to numbers in $\mathbb{F}$ and the coefficients q_{ij} remain undetermined and are thus the variables. Basically, each homogeneous coordinate vector $\mathbf{x}=(x_0,\ldots,x_n)\mathbb{F}^{n+1}\setminus\{\mathbf{o}\}$ is mapped to a new vector

$$\mathbf{v}(\mathbf{x})=(x_0^2:x_0x_1:\ldots:x_0x_n:x_1^2:x_1x_2:\ldots:x_n^2)\in\mathbb{F}^{\frac{(n+1)(n+2)}{2}}\setminus\{\mathbf{o}\}. \quad (5.3)$$

The mapping $v: \mathbf{x}\mapsto\mathbf{v}(\mathbf{x})$ is called the *Veronese*[1] *mapping*. The manifold of points parametrized by (5.3) is called *Veronese variety* V_2^n. The hyperplanar intersections of this manifold correspond to quadrics in $\mathbb{P}^n(\mathbb{F})$ and the coefficients of the hyperplane's equation equal the coefficient of the quadric's equations. In the case $n=2$, *i.e.*, conics in a projective plane, the Veronese variety V_2^2 turned out to be very useful for the study of conics, pencils of conics, and the quadratic Cremona transformations induced by linear two-parameter manifolds of conics. For details on Veronese varieties and related varieties that frequently show up in algebraic and higher dimensional geometry, see [21].

[1] Giuseppe Veronese (1854–1917) was an Italian mathematician who contributed to algebraic geometry and projective geometry in higher dimensions. In his later life, he turned to politics and became a senator of the Italian Republic.

From the basic definition of a pencil of quadrics, the following is clear:

Theorem 5.1.1 *A pencil of quadrics is projectively equivalent to a projective line. The group of projective transformations acting on a projective line acts in the same way as on a pencil of lines.*

Proof: The matrices $\mathbf{A}$ and $\mathbf{B}$ in (5.2) can be written as vectors $\mathbf{a}, \mathbf{b} \in \mathbb{F}^{\frac{1}{2}(n+1)(n+2)}$. Since scalar multiples of the equations describe the same geometric object, $\mathbf{a}$ and $\mathbf{b}$ can be interpreted as homogeneous coordinates of points in a projective space of $\frac{1}{2}(n+1)(n+2)-1$ dimensions. Then, by virtue of (5.2), $\mathcal{P}$ is a projective line spanned by $\mathbf{a}\mathbb{F}$ and $\mathbf{b}\mathbb{F}$.

We can describe projective mappings on a projective line in terms of (the) homogeneous coordinates $\alpha : \beta \neq 0 : 0$. According to [46, Theorem 5.4.1], a projective mapping on a projective line is described by a linear mapping $\mathbb{F}^2 \to \mathbb{F}^2$. ■

The regularity of the coefficient matrix of a quadric is equivalent to the regularity of the quadric. So, it is clear that the singular quadrics in a pencil correspond to the zeros of the form

$$\Delta(\alpha, \beta) = \det(\alpha\mathbf{A} + \beta\mathbf{B}). \tag{5.4}$$

Therefore, we can expect up to $n+1$ different singular quadrics in a pencil of quadrics.

The following example shows that the maximum number of singular quadrics does not necessarily occur:

■ **Example 5.1.1** Number of singular quadrics in a pencil.

The coefficient matrices $\mathbf{Q} = \alpha\mathbf{A} + \beta\mathbf{B}$ of pencils of conics (*i.e.*, quadrics in a plane) of all five kinds can be given as

$$\mathbf{Q}_1 = \begin{pmatrix} \alpha+\beta & 0 & 0 \\ 0 & \alpha & 0 \\ 0 & 0 & \beta \end{pmatrix}, \quad \mathbf{Q}_2 = \begin{pmatrix} 0 & 0 & \alpha \\ 0 & \beta & 0 \\ \alpha & 0 & -\beta \end{pmatrix}, \quad \mathbf{Q}_3 = \begin{pmatrix} -\alpha & 0 & 0 \\ 0 & \alpha & 0 \\ 0 & 0 & \beta \end{pmatrix},$$

$$\mathbf{Q}_4 = \begin{pmatrix} 0 & 0 & \beta \\ 0 & -2\beta & \alpha \\ \beta & \alpha & -2\alpha \end{pmatrix}, \quad \mathbf{Q}_5 = \begin{pmatrix} 0 & 0 & \beta \\ 0 & -2\beta & 0 \\ \beta & 0 & 2\alpha \end{pmatrix}.$$

The determinants $\Delta_i = \det \mathbf{Q}_i$ are (up to non-zero factors)

$$\Delta_1 = \alpha\beta(\alpha+\beta), \quad \Delta_2 = \Delta_3 = \alpha^2\beta, \quad \Delta_4 = \Delta_5 = \beta^3.$$

It is clear that we can find three singular conics (pairs of lines) in a pencil of the first kind (cf. [46, p. 285]).

The pencils of the second kind and third kind as well as the pencils of the fourth and fifth kind cannot be distinguished by means of the determinant Δ.

In the case of the pencil of the second and third kind, there are two different zeros of Δ: $\alpha : \beta = 0 : 1$ and $\alpha : \beta = 1 : 0$ with the multiplicities $\mu(0,1) = 2$ and $\mu(1,0) = 1$. For the pencil of the second kind, we find

$$\text{rk}(0 \cdot \mathbf{A} + 1 \cdot \mathbf{B}) = \text{rk}(0 \cdot \mathbf{A} + 1 \cdot \mathbf{B})^2 = 2, \;\; \text{rk}(1 \cdot \mathbf{A} + 0 \cdot \mathbf{B}) = \text{rk}(1 \cdot \mathbf{A} + 0 \cdot \mathbf{B})^2 = 2,$$

while in the case of the pencil of the third kind, we have

$$\text{rk}(0 \cdot \mathbf{A} + 1 \cdot \mathbf{B}) = 2, \;\; \text{rk}(0 \cdot \mathbf{A} + 1 \cdot \mathbf{B})^2 = 1, \;\; \text{rk}(1 \cdot \mathbf{A} + 0 \cdot \mathbf{B}) = 2, \;\; \text{rk}(1 \cdot \mathbf{A} + 0 \cdot \mathbf{B})^2 = 1,$$

which shows that the geometric multiplicity and algebraic multiplicity of the zeros of Δ_2 and Δ_3 differ. Clearly, the number of singular conics in both pencils equals two.

The same phenomenon can be observed with pencils of the fourth and fifth kind. The determinant Δ is a complete cube in both cases, and thus, there is only one zero with algebraic multiplicity three. However, the geometric multiplicity in the case of the pencil of the fourth kind equals two, while the geometric multiplicity in the case of the pencil of the fifth kind equals one.

Following the Example 5.1.1, we see that pencils of quadrics can roughly be characterized by the multiplicities of the zeros of the determinant (5.4) of the pencil. Moreover, we have learned that the multiplicities alone are not sufficient in order to distinguish between a pencil of the second and a pencil of the third kind.

The sum of multiplicities of the solutions $\alpha_i : \beta_i$ of the quartic form (5.4) always equals four. There are five different ways to build four as the sum of positive integers:

$$[1111], \quad [211], \quad [22], \quad [31], \quad [4]. \tag{5.5}$$

These five groups of summands – called the *characteristics* of $\mathcal{Q}$ – are the first step towards the classification of the pencils of quadrics, cf. [87, 127].

The fine-tuning now involves the *elementary divisors* of the matrix $\mathcal{Q}$. If the multiplicity μ_i of a zero of $\alpha_i : \beta_i$ of (5.4) is greater or equal to two, then it can be a zero of multiplicity 2 or 1 of the determinant of the *characteristic matrix*

$$\mathbf{C}_i = \alpha_i \mathbf{A} + \beta_i \mathbf{B}$$

of $\mathcal{Q}$. If this is the case, we break up the characteristic [211] into [(11)11], expressing the fact that the first zero of (5.4) is a double root of Δ and a root of multiplicity one of $\det \mathbf{C}_1 = 0$.

Each integer $\mu \geq 2$ displaying a multiplicity in the five characteristics given above can now be a sum of smaller integers. Splitting these multiplicities into lower integers takes into account all possible combinations of multiplicities of zeros of (5.4) and the determinants of its elementary matrices.

Hence, the first coarse classification is refined by breaking up the larger integers, and we arrive at the following list of characteristics:

$$\begin{array}{r|llll} [4] & [(31)] & [(22)] & [(211)] \\ [31] & [(21)1] & [(111)1] & \\ [22] & [2(11)] & [(11)(11)] & \\ [211] & [(11)11] & & \\ [1111] & & & \end{array} \tag{5.6}$$

This allows us to formulate:

Theorem 5.1.2 *In a projective 3-space $\mathbb{P}^3(\mathbb{F})$ with* char$\mathbb{F} \neq 2, 3$[2]*, there exist 13 different types of pencils of quadrics.*

In the cases char$\mathbb{F} = 2, 3$, the characteristic polynomials of pencils of quadrics reduce to lower degree polynomials. Formulated in geometrical terms, a projective line in a finite projective plane of order 2 or 3 carries exactly 3 or 4 points, cf. [46, Example 6.4.8]. So does each pencil of quadrics and we cannot expect to find more singular quadrics in a pencil than there are points on a line. Further, the multiplicities of zeros (of polynomials) over fields with low characteristics cannot grow arbitrarily high.

■ **Example 5.1.2** Pencils of quadrics in $\mathbb{P}^2$.

Example 5.1.1 taught us to distinguish between different types of pencils of conics in some projective plane. According to the procedure described prior to Theorem 5.1.2, we can write down the characteristics for the pencils of conics for all five types as follows:

Following what we have learned about the five types of pencils of conics in [46, Section 7.3], we can say: For the pencils of conics of the fourth and fifth kind, the multiplicity of the one and only zero of Δ from (5.4) equals 3. The pencil of the fifth kind has the characteristic [3], while the pencil of the fourth kind has the characteristic [(21)]. In the case of the pencils of second and third kind, there are only two singular conics contained, *i.e.*, the list of multiplicities equals [21], which equals the characteristic of the pencil of the third kind and becomes the

---

[2]In the case of char$\mathbb{F} = 2$, each polynomial of degree $d \geq 2$ can be reduced to a polynomial whose degree is at most two. If char$\mathbb{F} = 2$, the set of variable values as well as the set of values of any polynomial is restricted to $\{0, 1\}$. The set of quadratic polynomial equals $\{x^2 + x + 1, x^2 + x, x^2 + 1, x^2\}$ and reduces (in the same order) to $\{1, 0, x^2 + 1, x^2\}$, due to the values they achieve. Since $f: x \mapsto x^3$, sends 0 to 0 and 1 to 1, the function f agrees with $\mathrm{id}_{\mathbb{F}}$, and thus, $x^3 \cong x$. Similar reasoning works if char$\mathbb{F} = 3$.

characteristic of the pencil of the second kind by splitting 2 into (11) and returns [(11)1]. It is not at all surprising that [111] characterized the pencil of the first kind.

• **Exercise 5.1.1** Pencils of quadrics in $\mathbb{P}^4$ and $\mathbb{P}^5$.

Try to solve the classification problem for pencils of quadrics in a projective four space $\mathbb{P}^4$. Do the same in a projective five space $\mathbb{P}^5$. The latter exercise is not a mere training. It also discloses the number of different quadratic complexes of lines (cf. Section 10.1).

The simultaneous diagonalization of **A** and **B** is equivalent to the computation of a common polar tetrahedron. If we set

$$\mathbf{B} = \begin{pmatrix} 0 & 0 & 0 & 0 \\ 0 & -1 & 0 & 0 \\ 0 & 0 & -1 & 0 \\ 0 & 0 & 0 & -1 \end{pmatrix}$$

then the computation of the polar tetrahedron common to $\mathcal{A}$ and the singular quadric $\mathcal{B}$ is nothing but the determination of the principal axes, for

$$\mathcal{B}: \ x_1^2 + x_2^2 + x_3^2 = 0$$

is the absolute conic of Euclidean geometry.

For all types of pencils of quadrics in projective 3-space, we can give normal forms of the equations of the two quadrics $\mathcal{A}$ and $\mathcal{B}$ that span the pencil (see [127]):

(1) In the case [4], the coefficient matrices of $\mathcal{A}$ and $\mathcal{B}$ can be given as

$$\mathbf{A} = \begin{pmatrix} 0 & 0 & 1 & \lambda_1 \\ 0 & 1 & \lambda_1 & 0 \\ 1 & \lambda_1 & 0 & 0 \\ \lambda_1 & 0 & 0 & 0 \end{pmatrix} \quad \text{with } \lambda \neq 0, \quad \mathbf{B} = \begin{pmatrix} 0 & 0 & 0 & 1 \\ 0 & 0 & 1 & 0 \\ 0 & 1 & 0 & 0 \\ 1 & 0 & 0 & 0 \end{pmatrix}.$$

Common to all quadrics in this pencil is a cubic space curve c together with one of its tangents t (see Figure 5.1). For details on cubic and quartic space curves, we refer to Chapter 6.

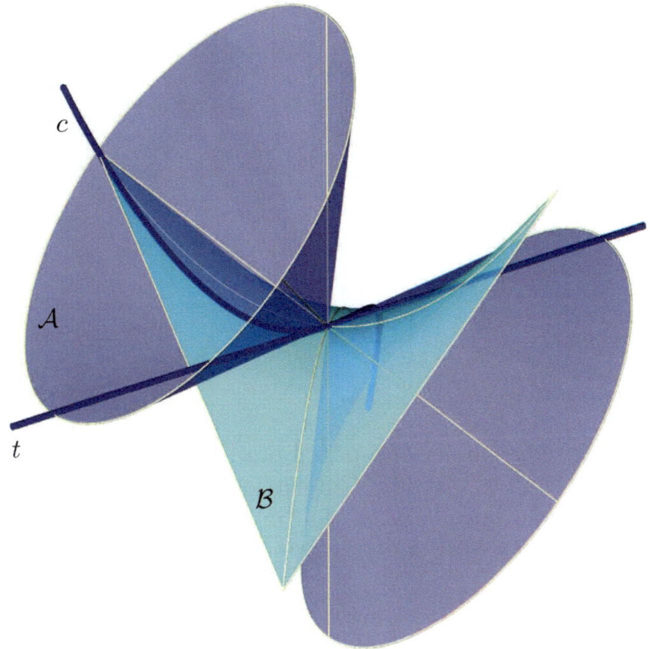

FIGURE 5.1. Quadrics that share a cubic space curve c plus a tangent t of c constitute the pencil of type 1 with characteristic [4].

(2) For the case [31], we have the normal forms

$$\mathbf{A} = \begin{pmatrix} \lambda_1 & 0 & 0 & 0 \\ 0 & 0 & 1 & \lambda_2 \\ 0 & 1 & \lambda_2 & 0 \\ 0 & \lambda_2 & 0 & 0 \end{pmatrix} \quad \text{with } \lambda_1 : \lambda_2 \neq 0 : 0, \quad \mathbf{B} = \begin{pmatrix} 1 & 0 & 0 & 0 \\ 0 & 0 & 0 & 1 \\ 0 & 0 & 1 & 0 \\ 0 & 1 & 0 & 0 \end{pmatrix}.$$

The carrier of this pencil is either

(2.1) a quartic space curve q with a cusp (cf. Definition 6.9.1 in Chapter 6) or

(2.2) a regular conic c with a pair (l, m) of intersecting straight lines where $l \cap m \in c$.

Both cases are illustrated in Figure 5.2.

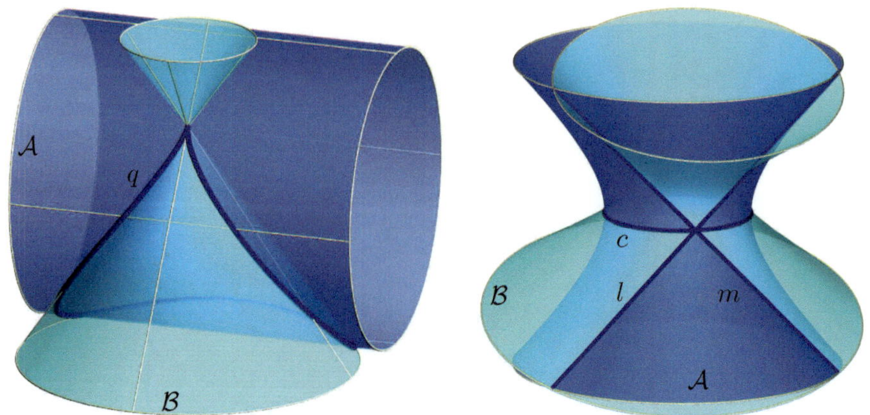

FIGURE 5.2. Left: The carrier of a pencil of quartics of type (2.1) is a quartic space curve q with a cusp. Right: The carrier of type (2.2) is the union of a conic c and two straight lines l and m that intersect each other on c. The latter configuration can be viewed as a degenerate quartic space curve.

We will meet the quartic curve shown in Figure 5.2 again in Section 6.9, especially in Example 6.9.3.

(3) The case [22] with two double solutions has the normal forms

$$\mathbf{A} = \begin{pmatrix} 1 & \lambda_1 & 0 & 0 \\ \lambda_1 & 0 & 0 & 0 \\ 0 & 0 & 1 & \lambda_2 \\ 0 & 0 & \lambda_2 & 0 \end{pmatrix} \quad \text{with } \lambda_1 : \lambda_2 \neq 0:0, \quad \mathbf{B} = \begin{pmatrix} 0&1&0&0 \\ 1&0&0&0 \\ 0&0&0&1 \\ 0&0&1&0 \end{pmatrix}.$$

The carrier of this particular pencil is either

(3.1) a cubic space curve c together with one of its chords l or

(3.2) a straight line d with multiplicity 2 and a pair (l, m) of skew straight lines both intersecting d.

Examples of carriers of these two types of pencils are shown in Figure 5.3.

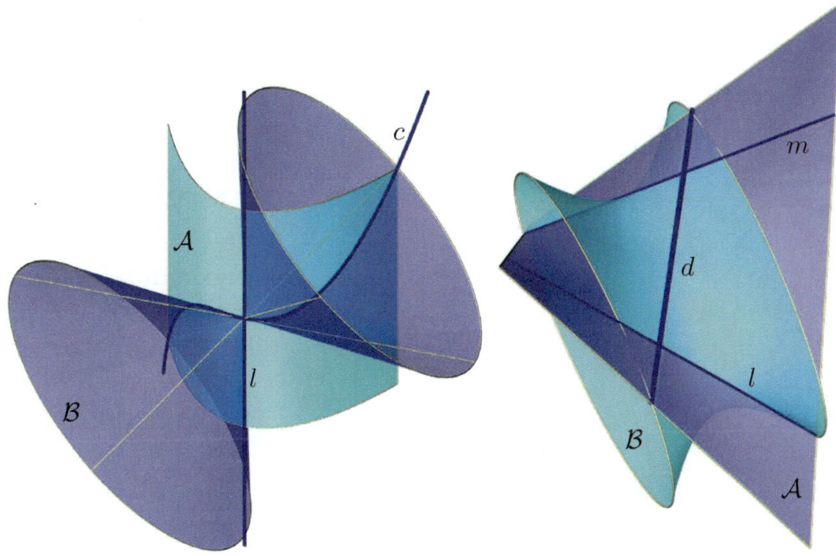

FIGURE 5.3. Left: A cubic space curve c together with one of its chords l can be considered the base of a pencil of quadrics of type (3.1). Right: The pencil of type (3.2) consists of all quadrics on a double line d together with two skew straight lines l that intersect d. The double line d takes over the role of the tangent, and $d \cup l \cup m$ is a degenerate cubic space curve.

(4) If the pencil's characteristic equals [211], the normal forms read

$$\mathbf{A} = \begin{pmatrix} \lambda_1 & 0 & 0 & 0 \\ 0 & \lambda_2 & 0 & 0 \\ 0 & 0 & 1 & \lambda_3 \\ 0 & 0 & \lambda_3 & 0 \end{pmatrix} \quad \text{with } \lambda_1 : \lambda_2 : \lambda_3 \neq 0 : 0 : 0, \; \mathbf{B} = \begin{pmatrix} 1 & 0 & 0 & 0 \\ 0 & 1 & 0 & 0 \\ 0 & 0 & 0 & 1 \\ 0 & 0 & 1 & 0 \end{pmatrix}.$$

The curves of intersection of the quadrics in this pencil are determined by the parameters λ_i and can be either

(4.1) a quartic space curve q with a double point,

(4.2) a regular conic c and a pair (l, m) of straight lines that intersect c and each other (but not on c),

(4.3) a pair (c_1, c_2) of touching conics, or

(4.4) a pair (d_1, d_2) of intersecting lines each of which has multiplicity two.

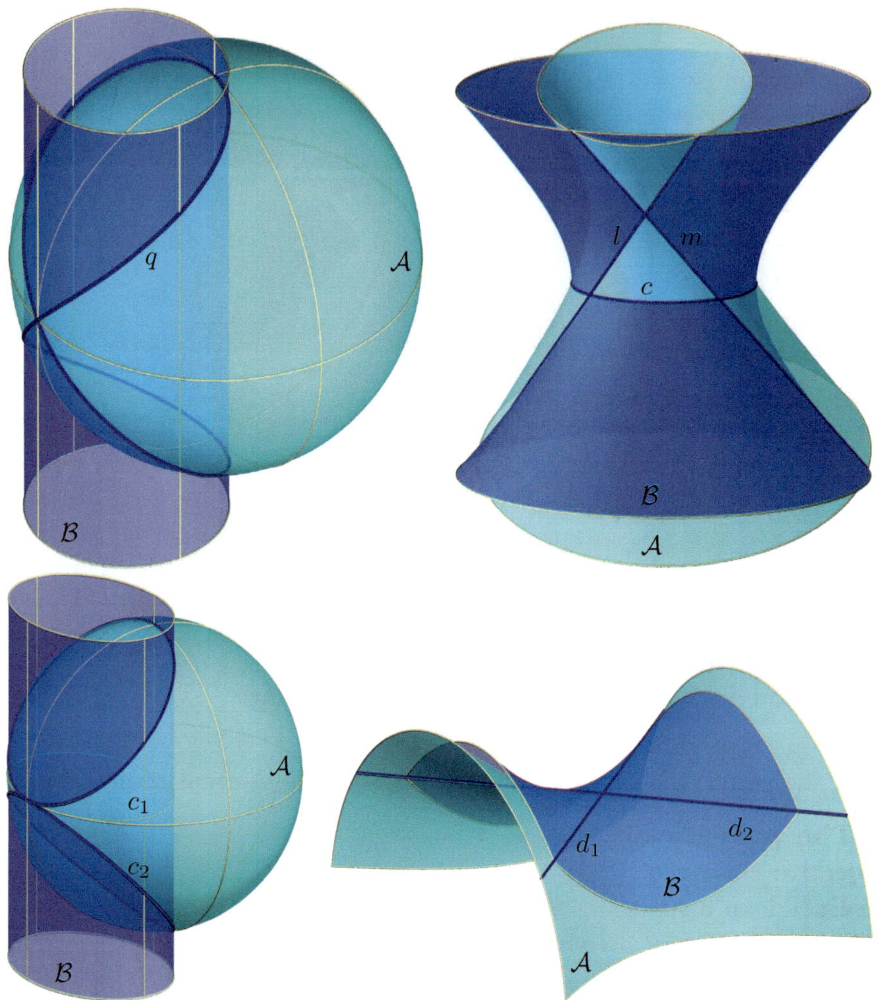

FIGURE 5.4. Top row: The pencil of type (4.1) consists of all quadrics that pass through a quartic space curve q with a double point (left). The carrier of the pencil of quadrics of type (4.2) is a degenerate form of quartic space curve and consists of a conic c together with a pair (l, m) of lines that intersect in a point not on c (right). Bottom row: The quartic q of type (4.1) can also decompose into two conics c_1 and c_2 that touch in a point which characterizes the pencil of type (4.3), (left). The most degenerate appearance of the quartic shows up as the carrier of the pencil of type (4.4). It is a pair (d_1, d_2) of double lines.

Figure 5.4 shows possible configurations and shapes of pairs of quadrics spanning pencils of type (4.1) to (4.4).

(5) Finally, we give the normal forms for the pencil with the multiplicities [1111]

$$\mathbf{A} = \begin{pmatrix} \lambda_1 & 0 & 0 & 0 \\ 0 & \lambda_2 & 0 & 0 \\ 0 & 0 & \lambda_3 & 0 \\ 0 & 0 & 0 & \lambda_4 \end{pmatrix} \text{ with } \lambda_1 : \lambda_2 : \lambda_3 : \lambda_4 \neq 0:0:0:0, \mathbf{B} = \begin{pmatrix} 1 & 0 & 0 & 0 \\ 0 & 1 & 0 & 0 \\ 0 & 0 & 1 & 0 \\ 0 & 0 & 0 & 1 \end{pmatrix}.$$

Depending on the choice of the parameters λ_i, the quadrics in the pencil intersect along

(5.1) a quartic space curve q of the first kind (without singularities; cf. Definition 6.9.1 in Chapter 6), along

(5.2) two conics c_1 and c_2 with two points D_1 and D_2 of intersection, along

(5.3) a skew quadrilateral (l_1, l_2, l_3, l_4), or along

(5.4) a regular conic d with multiplicity two.

In Figure 5.5, we can see some examples of pairs of quadrics spanning pencils of type (5.1) to (5.4).

• **Exercise 5.1.2** Various pencils of quadrics.

The following pairs $(\mathcal{A}, \mathcal{B})$ of quadrics span pencils $\mathcal{P}$ of quadrics that appear frequently in geometry. Determine the type of pencil and compute the singular quadrics therein.

1. Let $\mathcal{A}$ be a sphere with the equation
$$x^2 + y^2 + z^2 = r^2$$
with the radius $r > 0$, and let $\mathcal{B}$ be the coaxial cylinder of revolution with the equation
$$x^2 + y^2 = 1.$$
Depending on the choice of r, different types of pencils of quadrics appear. What happens if we drop the condition $r > 0$ and allow $r = 0$?
2. What is the type of pencil spanned by two concentric spheres?
3. Two coaxial cylinders of revolution clearly span a singular pencil. What is the type number of this pencil? Is there a difference if we allow the axis of the two cylinders to be skew?
4. Let $\mathcal{A}$ be the sphere with the equation
$$x^2 + y^2 + z^2 = 1$$
and $\mathcal{B}$ be the cone emanating from the sphere's center with the equation(s)
$$x^2 + y^2 + \varepsilon z^2 = 0.$$
Discuss the two pencils corresponding to $\varepsilon = +1$ and $\varepsilon = -1$.

5. Which types of pencils of quadrics occur if we choose $\mathcal{A}$ as the triaxial ellipsoid

$$\frac{x^2}{a^2} + \frac{y^2}{b^2} + \frac{z^2}{c^2} = 1$$

with $a > b > c$ and the concentric spheres $\mathcal{B}:\ x^2 + y^2 + z^2 = r^2$? Note the cases $r = a$, $r = b$, and $r = c$.

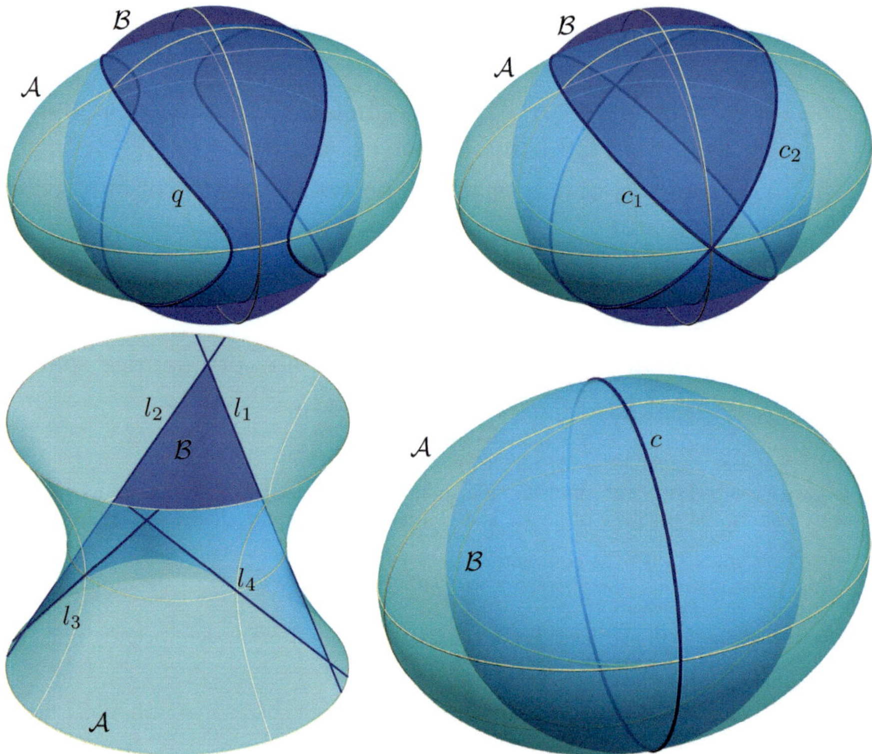

FIGURE 5.5. Top row: The pencil of type (5.1) consists of all quadrics through a quartic space curve q without singularities (left). The quartic q in the pencil of type (5.1) can split into a pair (c_1, c_2) of conics with two common points D_1 and D_2 (right). Bottom row: In the pencil of type (5.3), the common quartic of all quadrics in the pencil splits into four straight lines $l_1, \ldots, l_4$ forming a skew quadrilateral (left). If the two conics c_1 and c_2 in the pencil of type (5.2) coincide, we have a pencil of type (5.4) consisting of all quadrics that touch along a regular conic (right).

5.2 Principal points and common polar tetrahedron

We have seen that a pencil of quadrics in $\mathbb{P}^n(\mathbb{F})$ contains up to $n+1$ singular quadrics that may be quadratic cones (with vertices of arbitrary dimensions), pairs of hyperplanes, or double hyperplanes.

Consequently, in three-dimensional projective space, a pencil $\mathcal{P}$ of quadrics contains four singular quadrics at most, provided that the spanning quadrics $\mathcal{A}$ and $\mathcal{B}$ are regular.

The polar planes of a point P w.r.t. all quadrics in a pencil of quadrics either form a pencil of planes or coincide. If the latter happens, P is a singular point of a singular quadric contained in the pencil.

For the moment, we shall assume that the pencil $\mathcal{P}$ in question has four different singular quadrics. The singular points of a singular surface in the pencil, *i.e.*, points of the vertices of the quadratic cones, are called *principal points* of the pencil. The polar planes of a principal point w.r.t. all quadrics in the pencil coincide in a so-called *principal plane* of the pencil.

For any pair $\alpha_i : \beta_i \neq \alpha_j : \beta_j$ of different zeros of (5.4), the corresponding principal points are harmonic w.r.t. all quadrics in the pencil. If (5.4) has four different zeros, there are four different principal points and principal planes in the pencil. In this case, the four principal planes form a tetrahedron that is self-polar w.r.t. quadrics in the pencil.

5.3 Desargues's involution theorem

A pencil of quadrics and a plane

We continue the study of pencils of quadrics in three-dimensional projective space $\mathbb{P}^3(\mathbb{F})$ by intersecting the pencil (5.2) with a plane. This yields the following result:

Theorem 5.3.1 *The quadrics in a pencil of quadrics* (5.2) *intersect a generic plane* ε *in conics of a pencil, provided that* ε *is not contained in the pencil (as a part of a singular cone).*

Proof: It means no restriction if we assume that the plane ε is given by the equation $x_0 = 0$. Then, all terms involving x_0 in (5.2) disappear, and we arrive at the equation

$$\mathcal{P}_\varepsilon : \underline{\mathbf{x}}^{\mathrm{T}}(\alpha\underline{\mathbf{A}} + \beta\underline{\mathbf{B}})\underline{\mathbf{x}} \tag{5.7}$$

with $\underline{\mathbf{x}} = (x_1, x_2, x_3)$ and $\underline{\mathbf{A}}, \underline{\mathbf{B}} \in \mathbb{F}^{3\times 3}$ obtained from $\mathcal{A}$ and $\mathcal{B}$ with the first rows and columns removed.

Now, it is obvious that (5.7) is the algebraic description of a pencil of quadrics, since the coefficient matrix depends linearly on $\alpha : \beta \neq 0 : 0$ (see [46, Thm. 7.3.1, p. 289]).

Let us assume that the plane ε is a part of at least one quadric in the pencil, say $\varepsilon \subset \mathcal{B}$. Then, the quadratic form (5.2) is constant, *i.e.*, $\underline{\mathbf{x}}^{\mathrm{T}}(\alpha\underline{\mathbf{A}})\underline{\mathbf{x}} = 0$, since the factor $\alpha \in \mathbb{F} \setminus \{0\}$ can be cancelled. Now, the intersection of the pencil $\mathcal{P}$ with ε is one conic at most, not a pencil of conics. ■

Theorem 5.3.1 holds in projective spaces of any dimension. The term *plane* needs to be replaced by the term *hyperplane*, and the pencil in the hyperplane is a pencil of hyperquadrics in the hyperplane. From the theory of polarities and induced polarities in Section 4, Theorem 5.3.1 is obvious.

The conics in a pencil of conics meet a line that does not pass through any base point of the pencil in pairs of points which correspond to each other in an involution (cf. [46, Thm. 7.4.1, p. 309]). This involution is called *Desargues involution*.

A pencil of quadrics and a straight line

A straight line l can be contained in a quadric $\mathcal{Q}$ in a projective space $\mathbb{P}^n$ if $n \geq 3$. Then, for any pair (P, Q) of different points on l, any linear combination $\mathbf{g} = \lambda\mathbf{p} + \mu\mathbf{q}$ of the respective homogeneous coordinate vectors with $(\lambda, \mu) \in \mathbb{F}^2 \setminus \{\mathbf{o}\}$ annihilates the quadric's equation $\mathbf{x}^{\mathrm{T}}\mathbf{Q}\mathbf{x} = 0$. A straight line l is a tangent of $\mathcal{Q}$ if $\mathbf{g}^{\mathrm{T}}\mathbf{Q}\mathbf{g} = 0$ has a double solution $\lambda : \mu$, *i.e.*, if l and $\mathcal{Q}$ have exactly one point in common.

For all lines that are neither tangent to a quadric $\mathcal{Q}$ nor contained in it, we can formulate:

Theorem 5.3.2 *If the points S_1 and S_2 are the points of intersection of a quadric $\mathcal{Q}$ of any given pencil of quadrics and a given line l, then S_1 and S_2 are corresponding in an involution δ_l, provided that l does not pass through a base point of the pencil and is not a component of any quadric in the pencil.*

The proof of Theorem 5.3.2 does not differ from the proof of the similar theorem for pencils of conics in a projective plane (see [46, p. 309 f.]).

Moreover, by analogy with a theorem given in [46, Thm. 7.4.2], we can clarfiy the meaning of fixed points of the Desargues involution:

Theorem 5.3.3 *Any fixed point of the Desargues involution δ_l on the line l is either the point of contact of l with a quadric out of the pencil or a singular point of a singular quadric contained in the pencil.*

5.4 Dual pencils

Let $\mathcal{A}$ and $\mathcal{B}$ be two regular quadrics as given in (5.1), *i.e.*, their (symmetric) coefficient matrices $\mathbf{A}$ and $\mathbf{B}$ are regular. Each of the given quadrics has a dual and these shall be labelled with $\mathcal{A}^\star$ and $\mathcal{B}^\star$. If $\mathbf{u}\mathbb{F}$ be the homogeneous coordinates of the polar plane of $X = \mathbf{x}\mathbb{F}$ w.r.t. a quadric, say $\mathcal{A}$, then $\mathbf{A}\mathbf{x} = \mathbf{u}$ and $\mathbf{x} = \mathbf{A}^{-1}\mathbf{u}$. For all points $X = \mathbf{x}\mathbb{F} \in \mathcal{A}$, we have

$$\langle \mathbf{x}, \mathbf{u} \rangle = \langle \mathbf{x}, \mathbf{A}\mathbf{x} \rangle = \mathbf{x}^{\mathrm{T}}\mathbf{A}\mathbf{x} = 0.$$

Therefore,

$$\mathcal{A}^\star : \ \mathbf{x}^{\mathrm{T}}\mathbf{u} = \mathbf{u}^{\mathrm{T}}\mathbf{A}^{-1}\mathbf{u} = 0,$$

since $\mathbf{A} = \mathbf{A}^{\mathrm{T}}$ and $\mathbf{A}^{-1} = (\mathbf{A}^{-1})^{\mathrm{T}}$ hold. This is the *dual* or *tangential equation* of $\mathcal{A}$.

Now, the equations of the quadrics in the pencil $\mathcal{P}^\star$ of dual quadrics spanned by $\mathcal{A}^\star$ and $\mathcal{B}^\star$ can be given as

$$\mathcal{P}^\star : \ \mathbf{u}^{\mathrm{T}}(\alpha\mathbf{A}^{-1} + \beta\mathbf{B}^{-1})\mathbf{u} = 0. \tag{5.8}$$

In the literature, $\mathcal{P}^\star$ is also called a *range of quadrics.* Note that this is an equation in terms of homogeneous coordinates for (hyper)planes in $\mathbb{P}^n(\mathbb{F})$.

Especially in $\mathbb{P}^3(\mathbb{F})$, the singular dual quadrics are conics, as duals of quadratic cones. It is worth mentioning that a common tangent hyperplane of $\mathcal{A}^\star$ and $\mathcal{B}^\star$ is a common tangent hyperplane of all (dual) quadrics in the range $\mathcal{P}$.

In [46, p. 340], we have seen that ranges of confocal conics can be defined as ranges of dual conics that touch four given isotropic lines.

The representation of the quadrics in the pencil as given in (5.8) depends linearly on a homogeneous parameter $\alpha : \beta \neq 0$ and is independent of the dimension of $\mathbb{P}$.

Let us focus on the dual equation of the regular quadrics in the original pencil spanned by $\mathcal{A}$ and $\mathcal{B}$. The dual equation of the quadrics in the pencil can be obtained by directly inverting $\mathbf{Q} = \alpha\mathbf{A} + \beta\mathbf{B}$ using a special form of the Woodbury matrix identity (cf. [47]): Let $\mathbf{M}$ and $\mathbf{N}$ be two $n \times n$-matrices where $\mathbf{M}$ and $\mathbf{I}_n + \mathbf{M}^{-1}\mathbf{N}$ are regular. Then, the inverse of the sum of the two matrices $\mathbf{M}$ and $\mathbf{N}$ equals

$$(\mathbf{M} + \mathbf{N})^{-1} = \mathbf{M}^{-1} - (\mathbf{I}_n + \mathbf{M}^{-1}\mathbf{N})^{-1}\mathbf{M}^{-1}\mathbf{N}\mathbf{M}^{-1}. \tag{5.9}$$

Thus, the dual equations of the quadrics in the pencil have the equations $\mathbf{u}^{\mathrm{T}}\mathbf{Q}^{-1}\mathbf{u} = 0$ and for the inverse of $\mathbf{Q} = \alpha\mathbf{A} + \beta\mathbf{B}$, we have either

$$\begin{aligned} \mathbf{Q}^{-1} &= \tfrac{1}{\alpha}\mathbf{A}^{-1} - \left(\mathbf{I}_n + \tfrac{\beta}{\alpha}\mathbf{A}^{-1}\mathbf{B}\right)^{-1}\mathbf{A}^{-1}\mathbf{B}\mathbf{A}^{-1}, \quad \text{or} \\ \mathbf{Q}^{-1} &= \tfrac{1}{\beta}\mathbf{B}^{-1} - \left(\mathbf{I}_n + \tfrac{\alpha}{\beta}\mathbf{B}^{-1}\mathbf{A}\right)^{-1}\mathbf{B}^{-1}\mathbf{A}\mathbf{B}^{-1} \end{aligned} \tag{5.10}$$

depending on the order of the summands of $\mathbf{Q}$. Especially in $\mathbb{P}^3(\mathbb{F})$, we can infer from (5.10) that the coefficient matrix $\mathcal{Q}^{-1}$ equals

$$\mathcal{Q}^{-1} = \alpha^3\mathbf{A}^{-1} + \alpha^2\beta\mathbf{C}_1 + \alpha\beta^2\mathbf{C}_2 + \beta^3\mathbf{B}^{-1}$$

up to non-zero scalar multiples. This seems to be a cubic parametrization of the dual pencil of quadrics. However, the coefficient matrices $\mathbf{A}^{-1}$ and $\mathbf{B}^{-1}$ of α^3 and β^3 are the coefficient matrices of the equations of the dual quadrics $\mathcal{A}^\star$ and $\mathcal{B}^\star$.

The planes of the dual quadrics $\mathcal{C}_i$ given by $\mathbf{x}^{\mathrm{T}}\mathbf{C}_i\mathbf{x} = 0$, with $i = 1, 2$, intersect the two base quadrics along conics c_i such that there exist polar triangles inscribed into the first conic and circumscribed to the second, and *vice versa* (cf. [127, p. 213]).

From (5.10), we can immediately infer:

Theorem 5.4.1 *In a pencil of quadrics in $\mathbb{P}^n(\mathbb{F})$, there are up to $n-1$ quadrics that touch a generic hyperplane.*

Proof: Assume that the generic plane v has the homogeneous coordinate vector $\mathbf{u} \neq \mathbf{o}$. Then, we choose one of the equations given in (5.10) and observe that

$$\mathbf{u}^{\mathrm{T}}\mathbf{Q}^{-1}\mathbf{u} = 0$$

is a binary form of degree $n-1$ in the homogeneous variable $\lambda : \mu$. It can have up to $n-1$ zeros $\alpha_i : \beta_i \neq 0 : 0$, which correspond to planes in $\mathcal{P}^\star$, and thus, to tangent planes of quadrics in $\mathcal{P}$. ■

Theorem 5.4.1 covers the case of pencils of conics. We have learned that the fixed points of Desargues's involution induced by a pencil on a line (that does not contain a base point of the pencil) are the points of contact of conics in the pencil (see [46, Thm. 7.4.2]). These fixed points correspond to the zeros of a binary quadratic form, so that there exist two of them. In 3-space we have up to three possible points of contact between a plane and certain quadrics in a pencil. These are the singular points of the singular conics in the pencil of intersection curves.

5.5 Special pencils

Singular pencils of quadrics

There are two distinct types of *singular pencils of quadrics*:

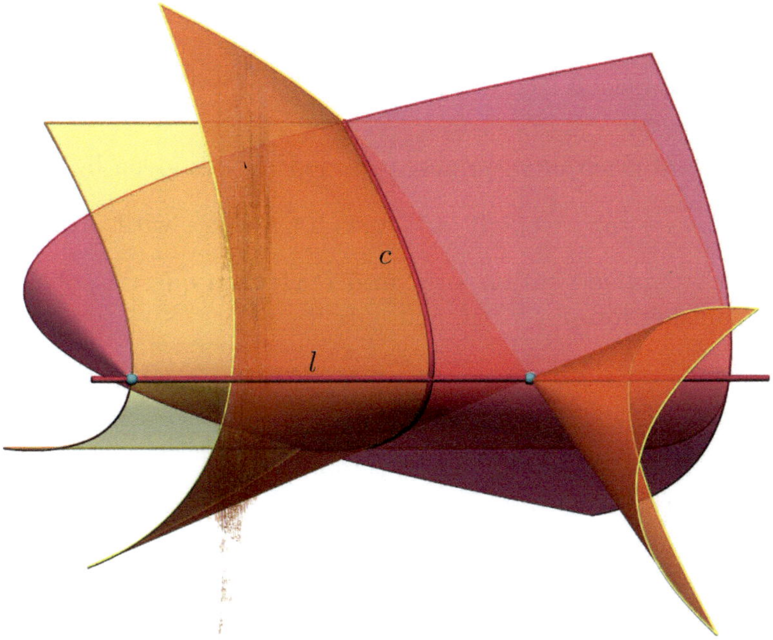

FIGURE 5.6. A singular pencil of quadrics that consists of quadratic cones that are tangent to a common tangent plane along a common generator l and pass through a common regular conic c.

(1) The projection of a pencil $\mathcal{P}$ of conics in a projective plane $\mathbb{P}^2(\mathbb{F})$ from a point $V \notin \mathbb{P}^2(\mathbb{F})$ produces a pencil of quadratic cones. (2) Quadratic cones that touch along a common generator and share a further (regular) conic (shown in Figure 5.6).

(1) In the singular pencil that is obtained by projecting a pencil of conics in a plane, we find five different types of singular pencils of quadrics. Obviously, these pencils of quadrics contain, besides the regular quadratic cones joining the conics with V, singular quadratic cones which consist of pairs of planes or double planes, depending on whether the pencil of conics contains pairs of lines or double lines.

An analytic description of this kind of singular pencil of quadrics can be obtained in an easy way. We assume that the point V has the homogeneous coordinates $(1:0:0:0)$ and the conics lie in the plane $\omega:\ x_0 = 0$. Then, the pencil of conics in the plane ω has an equation of the form given in (5.7). Note that $\Delta = 0$ from (5.4) for all $\alpha:\beta \neq 0:0$.

(2) A pencil of quadrics that consists only of quadratic cones that touch along a common generator l and share a regular conic c can be parametrized by first assuming that $\mathcal{P}:\ \mathbf{x}^{\mathrm{T}}\mathbf{Q}\mathbf{x} = 0$ is the equation of all quadrics in the pencil. The entries $q_{ij} \in \mathbb{F}$ of $\mathbf{Q}$ with $i,j \in \{0,\dots,3\}$ are to be determined.

Without loss of generality, we may assume that c is given by

$$c:\ x_3 = x_0x_2 - x_1^2 = 0 \quad \text{or} \quad \mathbf{c} = (1,t,t^2,0), \quad \text{with } \ t \in \mathbb{F}.$$

Inserting c's parametrization for $\mathbf{x}$ into $\mathcal{P}$ yields a quartic polynomial in t which vanishes for all $t \in \mathbb{F}$ if, and only if,

$$q_{00} = q_{01} = q_{12} = q_{22} = 0, \quad 2q_{02} + q_{11} = 0.$$

Further, it is admissible to assume that l is the intersection of $x_1 = x_2 = 0$ which can also be given in parametric form as $\mathbf{l}(t) = (1,0,0,t)$. Again, we insert the parametrization of l into $\mathcal{P}$, which causes

$$q_{00} = q_{03} = q_{33} = 0.$$

Since all the cones in this pencil shall be tangent to one plane π along l, this plane is uniquely determined by l and the tangent to c at $c \cap l$. Therefore, $\pi:\ x_1 = 0$, and finally, we find

$$q_{02} = 0.$$

Now, we are able to write down $\mathcal{Q}$ as

$$\mathcal{Q} = \begin{pmatrix} 0 & 0 & \alpha & 0 \\ 0 & -2\alpha & 0 & 0 \\ \alpha & 0 & 0 & \beta \\ 0 & 0 & \beta & 0 \end{pmatrix} \tag{5.11}$$

where $q_{02} = \alpha$ and $q_{23} = \beta$. Again, we observe $\Delta(\alpha, \beta) = 0$ for all $\alpha : \beta \neq 0 : 0$.

The matrix (5.11) is of rank 3 as long as $\alpha \neq 0$. In the case $\alpha = 0$, the rank of $\mathcal{Q}$ equals 2, and thus, the only singular cone in the pencil consists of the two planes $x_2 = x_3 = 0$.

Quadratic cones that osculate or hyperosculate along a common generator belong to the pencils of the first type.

Pencils of spheres

Analytic description of pencils of spheres

A pencil of spheres is a pencil of quadrics that contains only spheres (including planes and points as null spheres). Any two spheres, say $\mathcal{A}$ and $\mathcal{B}$, given by their homogeneous equations

$$\mathcal{A}: \ \langle \mathbf{x} - \mathbf{a}, \mathbf{x} - \mathbf{a} \rangle = a_0^2 x_0^2, \quad \mathcal{B}: \ \langle \mathbf{x} - \mathbf{b}, \mathbf{x} - \mathbf{b} \rangle = b_0^2 x_0^2,$$

where $\mathbf{x} = (x_1, x_2, x_3)$, intersect the plane at infinity $\omega: \ x_0 = 0$ along the absolute circle

$$\langle \mathbf{x}, \mathbf{x} \rangle = x_1^2 + x_2^2 + x_3^2 = 0$$

of Euclidean geometry. On the other hand, we can eliminate the quadratic form $\langle \mathbf{x}, \mathbf{x} \rangle$ from the inhomogeneous equations by simply subtracting them. This yields the equation

$$\varrho: \ \langle \mathbf{a} - \mathbf{b}, \mathbf{x} \rangle = \langle \mathbf{a}, \mathbf{a} \rangle - \langle \mathbf{b}, \mathbf{b} \rangle - a_0^2 + b_0^2$$

of the *radical plane* of the spheres $\mathcal{A}$ and $\mathcal{B}$. The plane ϱ is the radical plane of all spheres in the pencil. If the radii a_0, b_0, and the distance $d = \|\mathbf{a} - \mathbf{b}\|$ between the spheres' centers satisfy

$$a_0 + b_0 > d,$$

then $c := \varrho \cap \mathcal{A} = \varrho \cap \mathcal{B}$ is a real circle which lies on all spheres of the pencil. However, c is a circle even if $a_0 + b_0 < d$, but it has no real points. If $\mathcal{A}$ and

$\mathcal{B}$ are in contact, then $a_0 + b_0 = d$ and $\mathcal{A}$ and $\mathcal{B}$ share a pair of complex conjugate (and isotropic) generators. Families of concentric spheres are in contact along the absolute circle of Euclidean geometry and form the fourth kind of pencils of spheres.

From pencils of circles to pencils of spheres

In [46, p. 320], we have learned that there exist four different types of pencils of circles: hyperbolic, elliptic, parabolic pencils, or a pencil of concentric circles.

We can distinguish between as many pencils of spheres as we can distinguish pencils of circles.

The classification of pencils of circles can be done by means of the number and reality of base points of these special forms of pencils of conics or by means of the Desargues involution induced by the circles of the pencil on the line joining the centers. The classification with the help of the Desargues involution fails in the case of a parabolic pencil or a pencil of concentric circles. The elliptic pencil has two real base points, while the hyperbolic pencil of circles has no real base point. The parabolic pencil of circles is the one-parameter family of circles that touch a common line element. Clearly, the circles of a concentric pencil share the center.

The pencils of spheres can easily be generated from the pencils of circles. We apply a rotation about the line a joining the centers of any two different circles in the pencil. The line a shall be called the *axis of the pencil.* During the rotation, each circle sweeps a sphere and each base point, whether real or not, traces a circle c that is common to all spheres in the pencil (see Figure 5.7),

The plane ϱ spanned by the circle c is the *radical plane* of the spheres in the pencil. For example, let

$$S: \ (x-m)^2 + y^2 + z^2 = r^2$$

be the equation of a sphere in the elliptic pencil. Assume that all spheres in the pencil share the circle

$$c: \ x^2 + y^2 - R^2 = x = 0, \quad \varrho \in \mathbb{R} \setminus \{0\}.$$

Then, r, m, and R are subject to

$$r^2 = R^2 + m^2.$$

In this case, the radical plane ϱ has the equation $x = 0$. With the latter relation, we obtain a parametrization of the elliptic pencil of spheres as

$$\mathcal{P}_e:\ x^2 - 2mx + y^2 + z^2 = R^2.$$

The smallest sphere in the elliptic pencil is that of radius R, which is sometimes called *central sphere*. The *largest sphere* in the pencil is the radical plane, *i.e.*, a sphere with infinitely large radius.

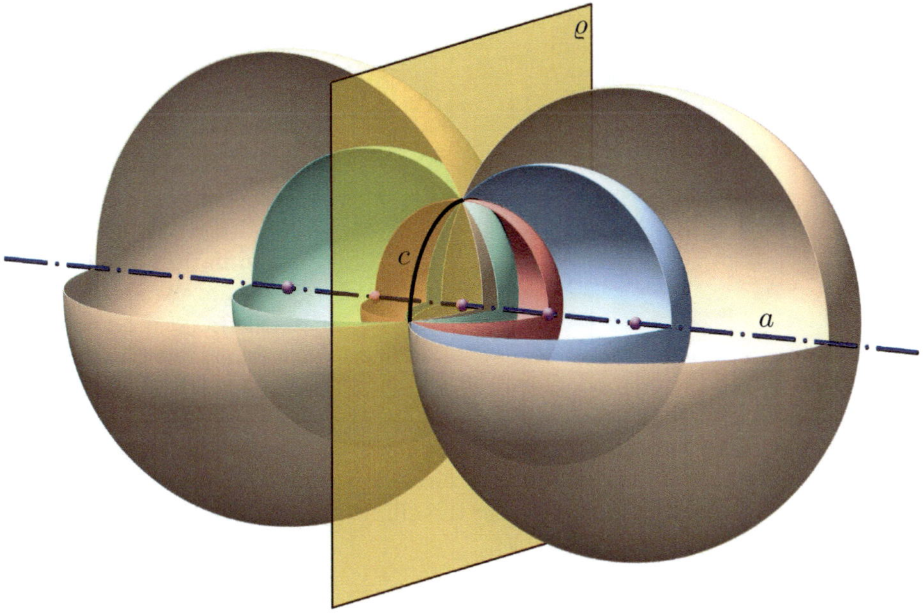

FIGURE 5.7. An elliptic pencil of spheres is uniquely determined by prescribing a real circle c through which all spheres of the pencil have to pass. The Desargues involution on the axis a of the pencil is elliptic.

A hyperbolic pencil is obtained by replacing the abovementioned circle c by a circle with a purely imaginary radius, *i.e.*, the trace of the complex conjugate pair of base points of a hyperbolic pencil of circles. Even though the circle c does not carry a real point, its carrier plane ϱ is real and is the radical plane of all spheres in the hyperbolic pencil. The axis of the hyperbolic pencil of spheres joins the centers of all spheres and carries the two *null spheres* N_1 and N_2, two points considered as spheres of radius 0. Speaking in terms of projective geometry, N_1 and N_2 are the fixed points of the Desargues involution determined by the hyperbolic pencil of spheres on the axis a (see Figure 5.8).

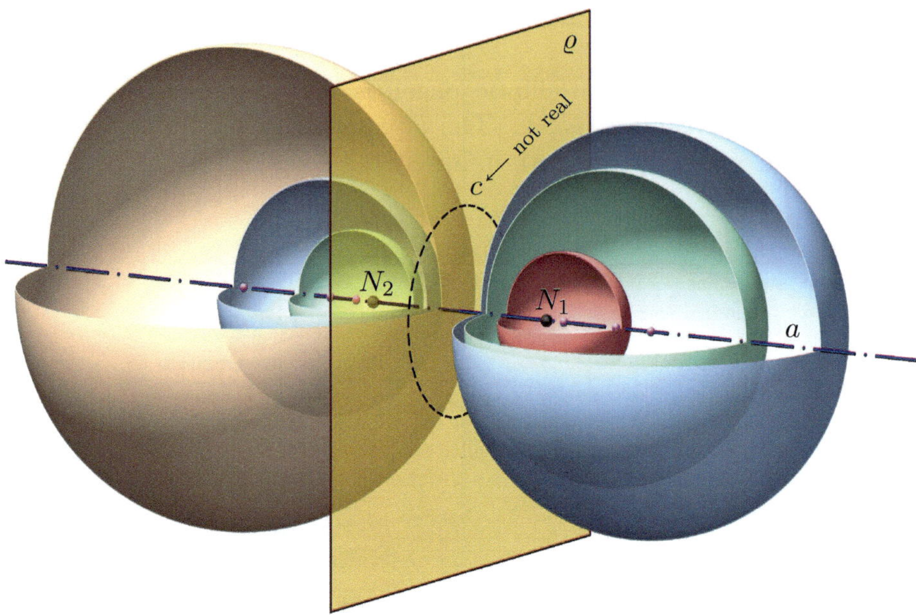

FIGURE 5.8. The hyperbolic pencil of spheres consists of the one-parameter family of spheres that pass through a circle c without any real point in the common (real) radical plane ϱ. Unlike the elliptic pencil, the hyperbolic pencil contains two null spheres N_1 and N_2.

The parabolic pencil of spheres consists of all spheres that touch a plane ϱ at a point P (see Figure 5.9). It is clear that ϱ is the radical plane of any pair of spheres taken from the pencil. The axis a of the pencil coincides with the common normal of all spheres. The range of points cut out from the spheres by the axis a is no longer an involution in the case of a parabolic pencil. The radical plane can be viewed as a sphere with infinitely large radius and belongs to the pencil.

The pencil of concentric spheres has infinitely many axes which are all common diameters of the spheres in the pencil. The hyperbolic pencil of spheres consists of all spheres that share a complex circle in a real plane. All spheres of an elliptic pencil pass through a real circle. Rotating a parabolic pencil of circles results in a parabolic pencil of spheres, *i.e.*, the

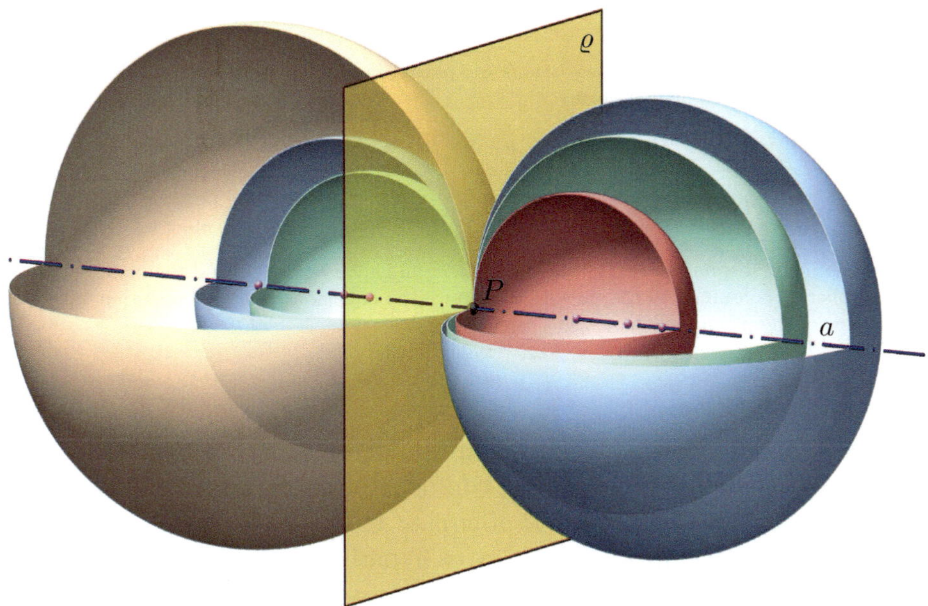

FIGURE 5.9. The parabolic pencil of spheres is formed by the one-parameter family of spheres that touch a plane ϱ at a common point P. The plane ϱ is the common radical plane of all spheres in the pencil and it is also a sphere of the pencil. The point P is a null sphere.

one-parameter family of spheres that touch a real plane in a certain real point.

Finally, a range of concentric spheres can be obtained by rotating concentric circles about a common diameter. Any two spheres in the pencil of concentric spheres share only the absolute circle of Euclidean geometry. Moreover, they are in line contact along the absolute circle. Each spherical motion with its fixed point in the common center of all spheres leaves the concentric pencil fixed as a whole.

Thus, we can classify the pencils of spheres as pencils of quadrics in the following way:

Theorem 5.5.1 *The elliptic and hyperbolic pencils of spheres are pencils of quadrics of type (5.1). The parabolic pencils of spheres are pencils of quadrics of type (4.2). The pencils of concentric spheres are pencils of quadrics of type (5.4).*

• **Exercise 5.5.1** Pencils of spheres.

Show that a normal form of the equations of spheres in the four kinds of pencils of spheres can be written in the following forms

$$\begin{array}{llll} x^2+y^2+z^2-2mx-\varrho^2=0, & \varrho\in\mathbb{R}^+,\ m\in\mathbb{R} & \text{elliptic pencil,} \\ x^2+y^2+z^2-2mx+\varrho^2=0, & \varrho\in\mathbb{R},\ \ m\in\mathbb{R} & \text{hyperbolic pencil,} \\ x^2+y^2+z^2-2mx=0, & m\in\mathbb{R} & \text{parabolic pencil,} \\ x^2+y^2+z^2-m^2=0, & m\in\mathbb{R} & \text{concentric spheres.} \end{array} \tag{5.12}$$

Determine the null spheres in all those cases where null spheres exist. Find the planes (viewed as flat spheres) contained in the pencils. Compute the equations of all spheres that intersect all spheres of the given pencils at right angles.

Triply orthogonal systems of spheres

From pencils of spheres, we can create triply orthogonal families of surfaces comparable to the system of confocal quadrics. In [46, Thm. 7.4.5], we have seen that each pencil $\mathcal{P}$ of circles defines a pencil $\mathcal{P}^\perp$ of orthogonal circles. If $\mathcal{P}$ is elliptic, parabolic, or hyperbolic, then $\mathcal{P}^\perp$ is hyperbolic, parabolic, or elliptic. The ortho circles of a pencil of concentric circles is the set of diameters common to all circles in the pencil.

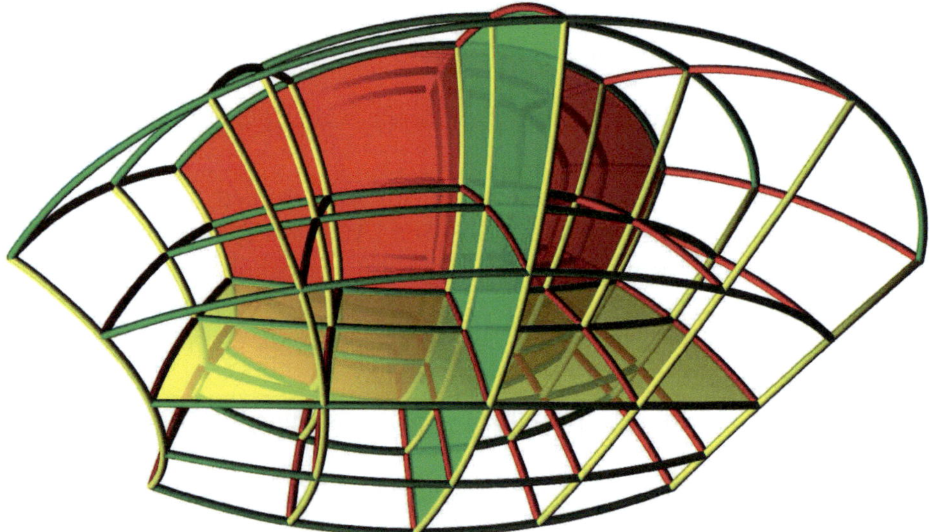

FIGURE 5.10. A triply orthogonal system of surfaces can be obtained from the simple triply orthogonal system built by a pencil of concentric spheres and two pencils of common meridian planes by inversion. The parameter curves are circles. The parameter surfaces are spheres.

In the previous section, we have generated pencils of spheres by rotating pencils of circles about their axes. Not only do we apply the generating rotation to the initial pencil $\mathcal{P}$, but we also rotate the orthogonal pencil $\mathcal{P}^{\perp}$. This yields, besides the pencil of spheres, a one-parameter family of tori (ring-, thorn-, or spindle-shaped).

FIGURE 5.11. A system of triply orthogonal surfaces consisting of the Möbius transforms of spheres from an elliptic pencil (yellow), spheres as the transforms of the common meridian planes (red), and Dupin cyclides (green).

The spheres of the pencil and the tori share the meridian planes, *i.e.*, the pencil of planes through the axis of the sphere pencil. The planes, the spheres, and the tori comprise the three independent one-parameter families of surfaces that form the triply orthogonal system of surfaces.

Each sphere $\mathcal{A}$ in the pencil intersects each torus $\mathcal{B}$ along a circle at right angles. Both the spheres and the tori intersect each meridian μ along circles at right angles. Thus, the grid of intersection lines between the surfaces is a grid of orthogonally intersecting circles, as illustrated in Figure 5.10.

Since intersection at right angles along circles is invariant with respect to Möbius transformations, we can generate the more general forms of families of triply orthogonal surfaces by applying a Möbius transformation, *i.e.*, a combination of a finite number of translations, equiform transformation, and inversions in spheres.

Then, the surfaces in the first family are spheres (including planes). The surfaces in the second family transform to Dupin cyclides (ring-, spindle-, thorn-shaped, depending on the shape of the tori), and the meridian planes become spheres (including planes).

Figure 5.11 shows a Möbius transformation of some spheres from an elliptic pencil together with the transforms of some tori whose meridian circles stem from the orthogonal hyperbolic pencil of the spheres' meridians. The meridian planes are bent to spheres.

As we shall see in Chapter 7, confocal quadrics also constitute a triply orthogonal system of quadrics. The quadrics in such systems of surfaces also intersect along lines of curvature at right angles, cf. Theorem 7.1.2 and Corollary 7.1.1.

6 Cubic and quartic space curves as intersections of quadrics

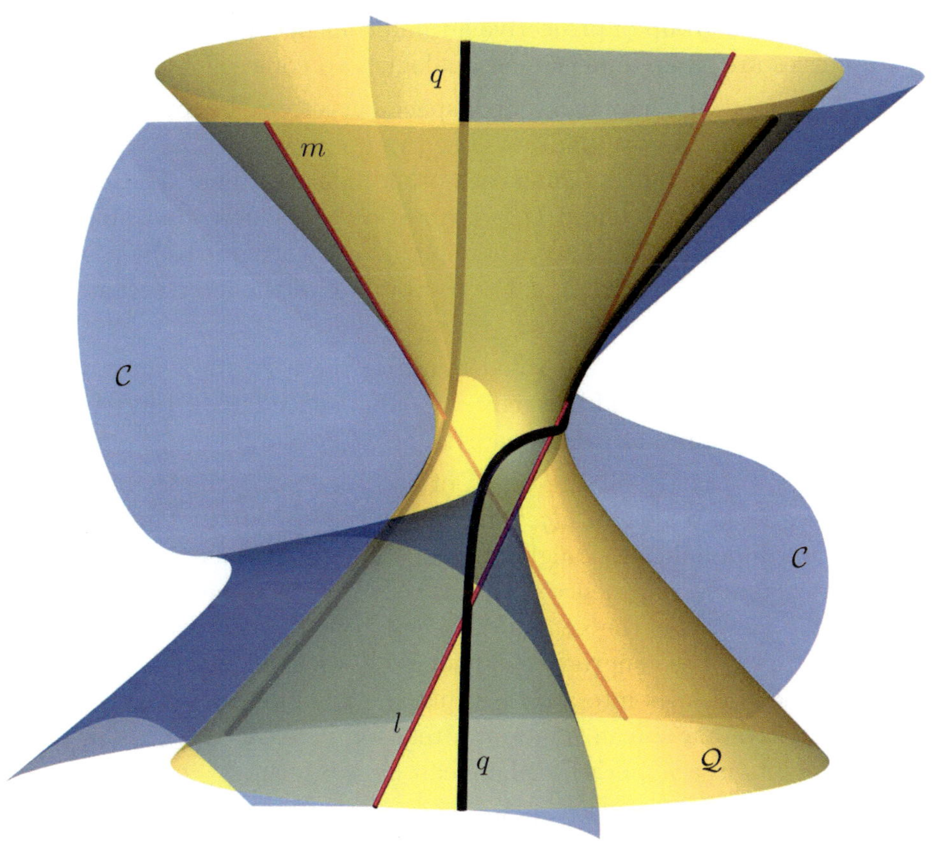

Quartic space of curves q of the second kind are part of the intersection of a quadric $\mathcal{Q}$ and a cubic surface $\mathcal{C}$ which share two further straight lines l and m. Unlike quartic space curves of the first kind, there exists only one quadric through a quartic space curve of the second kind.

B. Odehnal et al., *The Universe of Quadrics*,
https://doi.org/10.1007/978-3-662-61053-4_6

In this chapter, we discuss cubic space curves. From the view point of projective geometry, there exist several types of *planar* cubics, but only one type of *cubic space curve.* Unlike planar cubics, cubic space curves are always rational normal curves, *i.e.*, collinear images of the Veronese variety V_1^3, and thus, they can always be parametrized over the projective line and admit a group of projective automorphisms. In some sense, cubic space curves are three-dimensional analogues to conics.

As we have seen in Chapter 5, cubic space curves can be found as a part of the intersection of two quadrics. The two quadrics must have a straight line in common without touching each other along this line. The complete intersection of the two quadrics is a quartic space curve that splits into the cubic and the straight line. Therefore, it will turn out to be necessary to impose a further condition on the coordinates of the intersection of two quadrics in order to describe only the cubic part.

As we will see soon, on any ruled quadric, there exists a huge variety of cubic space curves; and through any cubic space curve, there exists a two-parameter family of ruled quadrics.

The chords of a conic fill the plane of the conic; the chords of a cubic space curve form a special *algebraic congruence of lines.* This is a two-dimensional manifold of straight lines in the three-dimensional ambient space. The cubic's congruence of chords contains the tangents of the cubic space curve as well as the generators of all (ruled) quadrics that contain the cubic curve. The congruence of chords allows us to define a projection from the space of lines into a certain plane, and composed with a certain quadratic Cremona transformation, a relation between the lines in space and the points in a hyperbolic plane can be established.

Moreover, it turns out that cubics have a projective generation which mirrors Steiner's construction of conics (cf. [46, p. 218, Definition 6.1.1]). A dual version of this projective generation can be given.

Finally, we study quartic space curves which arise in a natural way as the intersection of two quadrics, cf. Chapter 5. The quartics obtained in this way are called *quartic space curves of the first kind.* We shall give normal forms of their equations and show some examples. It turns out that there are *quartic space of curves of the second kind*, *i.e.*, quartic space curves which lie on a single ruled quadric and are cut out by a cubic surface sharing two straight lines with the quadric. Some appearances of such quartics of the second kind shall also be discussed in this chapter.

6.1 Standard cubic

In the following, we shall use homogeneous coordinates in projective 3-space $\mathbb{P}^3(\mathbb{F})$. The underlying field $\mathbb{F}$ shall not be of characteristic 2 or 3, cf. the footnote on page 183. A cubic space curve c can be given by a parametrization over the projective line $\mathbb{P}^1(\mathbb{F})$ in the form

$$\mathbf{c}(t_0,t_1) = \begin{pmatrix} a_{00} & a_{01} & a_{02} & a_{03} \\ a_{10} & a_{11} & a_{12} & a_{13} \\ a_{20} & a_{21} & a_{22} & a_{23} \\ a_{30} & a_{31} & a_{32} & a_{33} \end{pmatrix} \begin{pmatrix} t_0^3 \\ t_0^2 t_1 \\ t_0 t_1^2 \\ t_1^3 \end{pmatrix}, \quad t_0 : t_1 \neq 0 : 0. \tag{6.1}$$

Note that the four coordinate functions are cubic forms depending on the homogeneous parameter $t_0 : t_1 \neq 0 : 0$. So, it is clear that any plane $\varepsilon = \mathbf{e}\mathbb{F}$ shares at most three points with c.

We treat only those cases where the matrix $\mathbf{A} := (a_{ij})$ (with $i,j \in \{0,\dots,3\}$) is regular. Otherwise, (6.1) describes a planar and rational cubic if $\mathrm{rk}\mathbf{A} = 3$. The cases $\mathrm{rk}\mathbf{A} \leq 2$ correspond to lines and points, or even the empty set. Planar cubics can also be elliptic, but these cubics, *i.e.*, non-rational cubics, cannot be described by (6.1).

The following result is helpful and will simplify many problems by tracing them back to an apparently simpler parametrization and representation of a cubic space curve. In [46, p. 239, Theorem 6.4.1], we have shown that there exists only one conic in a projective plane. There is a similar result for cubic space curves:

Theorem 6.1.1 *In a projective space* $\mathbb{P}^3(\mathbb{F})$ *with* $\mathrm{char}\mathbb{F} \neq 2$, *there exists only one type of cubic space curve, up to projective collineations. In an appropriately chosen coordinate system, the cubic can be written as*

$$\mathbf{c}(t_0,t_1) = (t_0^3 : t_0^2 t_1 : t_0 t_1^2 : t_1^3), \quad (t_0,t_1) \in \mathbb{F}^2 \setminus \{\mathbf{0}\}. \tag{6.2}$$

Proof: The matrix $\mathbf{A}$ in (6.1) describes the linear mapping that transforms the canonical basis of $\mathbb{F}^4$ into an arbitrary one, since we have assumed that $\mathrm{rk}\mathbf{A} = 4$. Consequently, a suitable change of coordinates transforms $\mathbf{A}$ into a multiple of the unit matrix $\mathbf{I}_4 = \mathrm{diag}(1,1,1,1)$, and (6.1) changes to (6.2). ■

A straight line l is called a *tangent* of a cubic c at a point P if it meets the curve at P with multiplicity two. An equivalent definition of the tangent uses differential calculus: For that approach, we assume that

$\mathbf{c}(\tau): I \subset \mathbb{F} \to \mathbb{F}^4$ is a parametrization of c with an affine parameter τ. $\mathbb{F}$ can be either the real or complex number field. Here, and in the following, a dot shall indicate the differentiation w.r.t. the parameter. The tangent l of c at $P = \mathbf{c}(\tau_0)$ is spanned by P and the first derivative $\dot{\mathbf{c}}(\tau_0)$ of the parametrization at τ_0 which, in terms of homogeneous coordinates, reads

$$l = [\mathbf{c}(\tau_0), \dot{\mathbf{c}}(\tau_0)],$$

provided that $\mathbf{c}(\tau_0)$ and $\dot{\mathbf{c}}(\tau_0)$ are linearly independent, *i.e.*, $\mathbf{c}(\tau_0)$ is a *regular point* of c.

The differential geometric definition of the tangent to a curve is, indeed, independent of parametrizations and *normalizations* of the curve's parametric representation. If

$$\widetilde{\mathbf{c}} = \varrho(\upsilon(\tau)) \cdot \mathbf{c}(\upsilon(\tau)) \quad \text{with} \quad \varrho(\upsilon(\tau)): I \to \mathbb{F}$$

is a reparametrization (combined with renormalization $\varrho: I \to \mathbb{F}$) of c, then

$$\dot{\widetilde{\mathbf{c}}} = \frac{\mathrm{d}\varrho}{\mathrm{d}\upsilon}\frac{\mathrm{d}\upsilon}{\mathrm{d}\tau} \cdot \mathbf{c} + \varrho \cdot \frac{\mathrm{d}\mathbf{c}}{\mathrm{d}\tau}\frac{\mathrm{d}\upsilon}{\mathrm{d}t}$$

and the tangent remains unchanged:

$$[\widetilde{\mathbf{c}}, \dot{\widetilde{\mathbf{c}}}] = \left[\varrho \cdot \mathbf{c},\ \tfrac{\mathrm{d}\varrho}{\mathrm{d}\upsilon}\tfrac{\mathrm{d}\upsilon}{\mathrm{d}\tau} \cdot \mathbf{c} + \varrho \cdot \tfrac{\mathrm{d}\mathbf{c}}{\mathrm{d}\upsilon}\tfrac{\mathrm{d}\upsilon}{\mathrm{d}\tau}\right] = [\mathbf{c}, \dot{\mathbf{c}}] = l.$$

■ **Example 6.1.1** Cubics in the real affine space.

In the case of conics, the number of common points of the conic with the ideal line determines the affine type and name. If now $\mathbb{F} = \mathbb{R}$, then we can distinguish the following cases:

1. A *cubic hyperbola* has three real points at infinity.
2. A *cubic parabolic hyperbola* has exactly two points at infinity.
3. A *cubic ellipse* has one real point and a pair of complex conjugate points at infinity. If the complex conjugate pair of points equals a pair of absolute points of Euclidean geometry, then the cubic is called a *cubic circle*.
4. A *cubic parabola* has exactly one real point at infinity. The intersection multiplicity with the plane at infinity equals three.

The cubics depicted in Figure 6.1 are a cubic hyperbola h (left) and a cubic parabolic hyperbola p (right). The curves are parametrized by

$$\begin{aligned}\mathbf{h}(t_0, t_1) &= (t_1(t_0^2 - t_1^2) : t_0^3 : t_0^2 t_1 : t_0 t_1^2),\\ \mathbf{p}(t_0, t_1) &= (t_0 t_1^2 : t_0^3 : t_0^2 t_1 : t_1^3),\end{aligned}$$

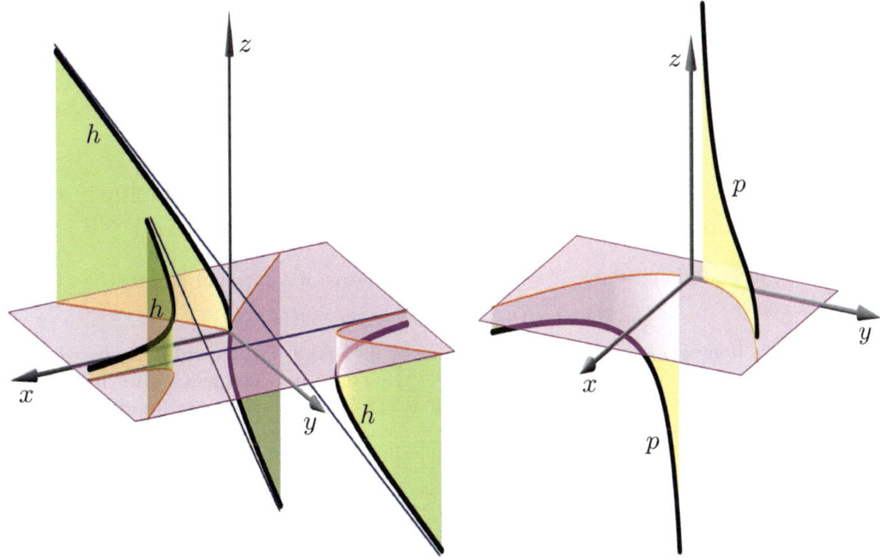

FIGURE 6.1. Left: A cubic hyperbola h has three points at infinity, and thus, three asymptotes. Right: A cubic parabolic hyperbola p has exactly two points at infinity.

in terms of homogeneous coordinates depending on a homogeneous parameter $t_0 : t_1 \neq 0 : 0$. In terms of affine (or even Cartesian) coordinates and with an affine parameter $t = t_1 t_0^{-1}$, we have

$$\mathbf{h}(t) = \left(\tfrac{1}{t(1-t^2)}, \tfrac{1}{1-t^2}, \tfrac{t}{1-t^2}\right), \quad t \in \mathbb{R} \setminus \{0, \pm 1\},$$
$$\mathbf{p}(t) = \left(\tfrac{1}{t^2}, \tfrac{1}{t}, t\right), \qquad t \in \mathbb{R}^\star.$$

The cubic hyperbola h has the three ideal points

$$\mathbf{f}_1 = (0:1:0:0), \quad \mathbf{f}_{2,3} = (0:1:\pm 1:1)$$

with the tangents (asymptotes)

$$\begin{array}{rcccc} \mathbf{a}_1 & = & (u_0 : u_1 : u_0 : 0) & \Leftrightarrow & x_3 = x_0 - x_2 = 0, \\ \mathbf{a}_{2,3} & = & (2u_0 : u_1 : \pm u_1 - u_0 : u_1 \mp 2u_0) & \Leftrightarrow & x_0 \mp 2x_1 + 2x_2 = x_0 \mp x_1 \pm x_3 = 0 \end{array}$$

with $u_0 : u_1 \neq 0 : 0$.

On the cubic parabolic hyperbola p, we find two points at infinity

$$\mathbf{f}_1 = (0:0:0:1), \quad \mathbf{f}_2 = (0:1:0:0)$$

with the tangents

$$\begin{array}{rcccc} \mathbf{a}_1 & = & (0 : u_0 : u_1 : 0) & \Leftrightarrow & x_0 = x_3 = 0, \\ \mathbf{a}_2 & = & (u_0 : 0 : 0 : u_1) & \Leftrightarrow & x_1 = x_2 = 0 \end{array}$$

with $u_0 : u_1 \neq 0 : 0$.

Figure 6.6 shows a cubic circle (left) and a cubic parabola (right).

Rational planar cubics

It is well-known (see, for example, [19], [8, p. 232–250], or [22, p. 17–29]) that planar cubic curves with a singularity are rational, *i.e.*, they admit a rational parametrization. We can state:

Theorem 6.1.2 *Any planar rational cubic $c' \subset \mathbb{P}^2(\mathbb{F})$ with* $\operatorname{char}\mathbb{F} \neq 2$ *can be obtained as a linear image of the standard cubic* (6.2) *in* $\mathbb{P}^3(\mathbb{F})$.

Proof: Each rational planar cubic c' can be given in terms of a rational parametrization

$$\mathbf{c}'(t_0, t_1) = (\gamma_0(t_0 : t_1) : \gamma_1(t_0 : t_1) : \gamma_2(t_0 : t_1)) \tag{6.3}$$

depending on a homogeneous parameter $t_0 : t_1 \neq 0 : 0$. Herein, γ_i are cubic forms which can be written in the monomial basis $\{t_0^3, t_0^2 t_1, t_0 t_1^2, t_1^3\}$. Then, it is elementary to find a matrix $\mathbf{P} \in \mathbb{F}^{3\times 4}$ that maps (6.2) to (6.3). Clearly, $\ker \mathbf{P}$ is the homogeneous coordinate vector of the center of the projection that maps c to c'. ∎

■ **Example 6.1.2** Rational planar cubics as images of the standard cubic.

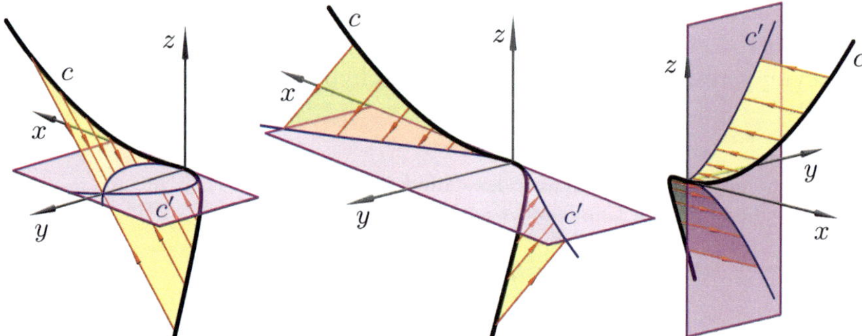

FIGURE 6.2. The three types of planar rational cubics are linear images c' of the standard cubic c (6.2): Left: the cubic with a node; in the middle: with an isolated double point; right: the semi-cubic parabola.

There are three different types of planar cubics with singularities:

1. The cubic c' has an *ordinary node*, and thus, its equation in normal form reads $x_0 x_1^2 = x_2(x_2 - x_0)^2$ (in the plane $x_3 = 0$). It can be parametrized by $\mathbf{c}'(t_0, t_1) = (t_0^3 : t_1(t_0^2 - t_1^2) : t_0 t_1^2 : 0)$ with $t_0 : t_1 \neq 0 : 0$. Obviously, c' is the image of c (given in (6.2)) under the projection from the point $(0:1:0:1) \in \mathbb{P}^3(\mathbb{F})$ onto the plane $x_3 = 0$.
2. If the planar cubic c' has an isolated double point $\mathbf{n} = (1:0:-1:0)$, then its equation can be given in the normal form $x_1^2 = x_2(x_2 + x_0)^2$ (in the plane $x_3 = 0$). The projection of c (from (6.2)) with center $(0:1:0:-1)$ onto the plane $x_3 = 0$ yields $\mathbf{c}'(t_0, t_1) = (t_0^3 : t_1(t_0^2 + t_1^2) : t_0 t_1^2 : 0)$, which is a parametrization of c'. The isolated double point is the image of a projecting chord that carries the complex conjugate pair of points $(1 : \pm\mathrm{i} : -1 : \mp\mathrm{i})$ which corresponds to the parameter values $t_0 : t_1 = 1 : \pm\mathrm{i}$ on the cubic.

3. In the case of a semi-cubic parabola c', *i.e.*, the prototype of a cusped (planar) cubic, we write the normal form of the equation of c' as $x_2^2 = x_0x_3^3$ (in the plane $x_1 = 0$). A parametrization of c' is $\mathbf{c}'(t_0, t_1) = (t_0^3 : 0 : t_0t_1^2 : t_1^3)$ and is obtained from (6.2) by the projection with center $(0:1:0:0)$ onto the plane $x_1 = 0$.

The projections of the standard cubic to the rational planar curves are shown in Figure 6.2.

Remark 6.1.1 In [46, Chapter 7.5], we have seen that rational planar cubics can be obtained as the images of conics under quadratic Cremona transformations. The relation between the resolution of singularities on planar curves and quadratic Cremona transformations is obvious.

Remark 6.1.2 The resolution of a singularity (on a cubic) is somehow the reconstruction of a (rational cubic) space curve c plus the projection that maps c to the planar curve with the desired singularity. From the case of the semi-cubic parabola, we can deduce that a projection in the direction of a tangent of a curve (whether cubic or not) results in an image curve with a cusp. If a fiber of a projection hits a curve twice, the image curve has double point.

• **Exercise 6.1.1** Three points in a generic plane.

The curve c (6.2) is a *true* space curve, *i.e.*, there are no four different coplanar points on the curve c. This can easily be shown if we choose four different parameter values $t_0 : t_i \neq 0 : 0$ (with $i \in \{1, \dots, 4\}$) and build the 4×4-matrix

$$V(t_0, \dots, t_3) = \begin{pmatrix} t_0^3 & t_0^2t_1 & t_0t_1^2 & t_1^3 \\ \vdots & & & \vdots \\ t_0^3 & t_0^2t_4 & t_0t_4^2 & t_4^3 \end{pmatrix}$$

whose rows contain the homogeneous coordinates of the four corresponding points on the curve c. The matrix $V(t_0, \dots, t_3)$ is known as the *Vandermonde matrix*, and its determinant equals the product

$$\det V(t_0, \dots, t_3) = \prod_{i<j}(t_i - t_j).$$

Therefore, $\det V(t_0, \dots, t_3)$ vanishes if, and only if, $t_i = t_j$ for at least one pair of indices. This would be the case if two points are identical, or equivalently, two corresponding parameter values are proportional. This holds true even if one of the four points is replaced by c's only point at infinity $(0:0:0:1)$.

In this case the parameters $t_0 : t_i$ are scaled in such way that the first coordinate equals t_0 for *all four points*. This is admissible as long as no homogeneous parameter equals $(0:\tau)$.

Cubic Bézier curves

The homogeneous parameter $t_0 : t_1$ in (6.2) and in any other cubic can be replaced by an inhomogeneous (affine) parameter $t = t_1t_0^{-1}$. Thus, the coordinate functions become univariate polynomials and can be written either in the monomial basis $\{1, t, t^2, t^3\}$ or in the geometrically favorable Bernstein basis $\{(1-t)^3, 3t(1-t)^2, 3t^2(1-t), t^3\}$, see [38]. Consequently, we can state:

Theorem 6.1.3 *Each rational cubic space curve and each rational planar cubic is a rational cubic Bézier curve.*

6.2 Quadrics containing a cubic

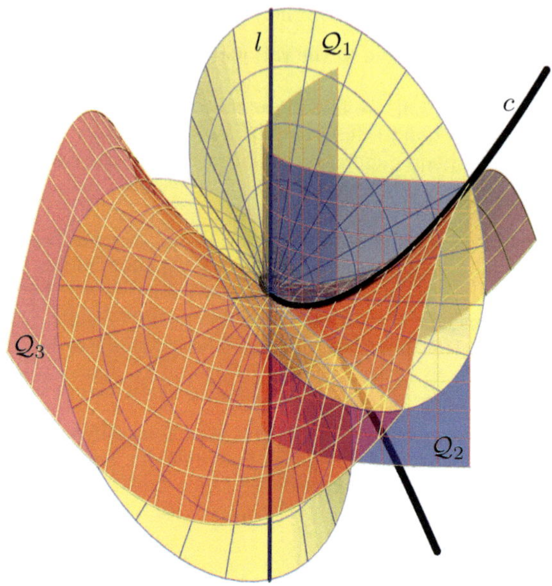

FIGURE 6.3. The prototype of a cubic space curve is obtained as the intersection of the three quadrics $\mathcal{Q}_1$, $\mathcal{Q}_2$, and $\mathcal{Q}_3$.

In the following, almost all results on cubic space curves can be deduced from the standard cubic (6.2), according to Theorem 6.1.1. Now, we shall have a look at the coordinate functions of (6.2) and ask for relations among them. Therefore, we let $x_i = t_0^{3-i}t_1^i$ (with $i \in \{0,1,2,3\}$) and find

$$\begin{gathered}\mathcal{Q}_1:\ x_2^2 - x_1x_3 = 0, \quad \mathcal{Q}_2:\ x_1^2 - x_0x_2 = 0,\\ \mathcal{Q}_3:\ x_1x_2 - x_0x_3 = 0.\end{gathered} \tag{6.4}$$

The various possible choices of commutative fields $\mathbb{F}$ may change the appearance and properties of the quadrics as well as those of the cubic.

The quadrics $\mathcal{Q}_1$ and $\mathcal{Q}_2$ are quadratic cones with vertices $(1:0:0:0)$ and $(0:0:0:1)$, respectively. The third quadric $\mathcal{Q}_3$ is a ruled quadric (a hyperbolic paraboloid if we choose $x_0 = 0$ as the plane at infinity).

The set of common points of $\mathcal{Q}_1$ and $\mathcal{Q}_2$ is the union of the straight line l with the parametrization

$$\mathbf{l}(t_0,t_1) = (t_0:0:0:t_1), \quad t_0:t_1 \neq 0:0 \tag{6.5}$$

and the standard cubic c given in (6.2). Therefore, two quadratic equations out of the three (6.4) are not sufficient in order to describe only the cubic curve c.

This fits well with Bézout's theorem, according to which the intersection of two quadrics has to be an algebraic curve of degree four. In this case, the curve of degree four splits into the straight line l and the cubic curve c. We observe that the tangent plane of $\mathcal{Q}_1$ along the line l equals $x_1 = 0$ and that of $\mathcal{Q}_2$ along l is $x_2 = 0$. So, the two cones do not touch each other along l. At this point we note:

Corollary 6.2.1 *The quartic intersection of two quadratic cones that are in line contact along a generator l consists of l with multiplicity two and a further planar curve of degree two, i.e., a conic.*

Proof: Without loss of generality, we may assume that the two quadratic cones are given by

$$\Gamma_0 : \overline{\mathbf{x}}^{\mathrm{T}} \mathbf{A} \overline{\mathbf{x}} = 0 \quad \text{and} \quad \Gamma_3 : \underline{\mathbf{x}}^{\mathrm{T}} \mathbf{B} \underline{\mathbf{x}} = 0, \tag{6.6}$$

with symmetric and regular matrices $\mathbf{A}, \mathbf{B} \in \mathbb{F}^{4\times 4}$, $\overline{\mathbf{x}} = (0, x_1, x_2, x_3)$, and $\underline{\mathbf{x}} = (x_0, x_1, x_2, 0)$.

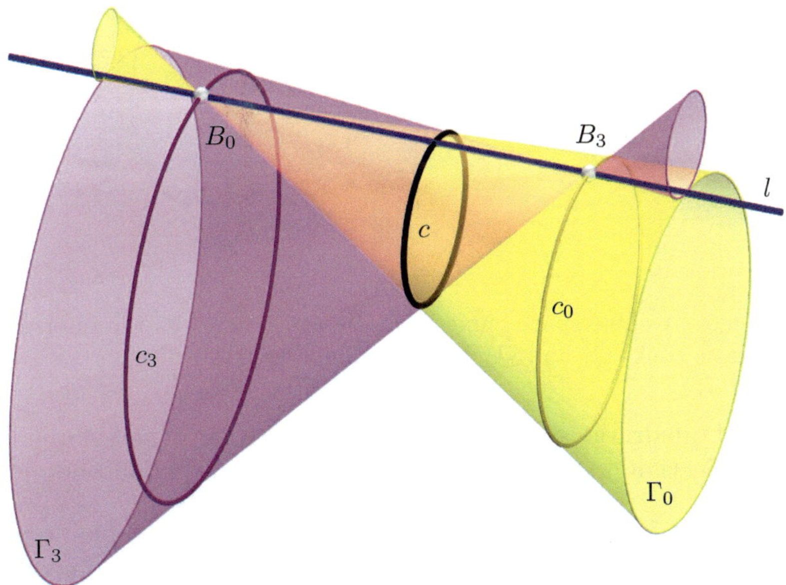

FIGURE 6.4. The two quadratic cones Γ_0 and Γ_3 are in contact along a common generator l share a conic c.

The vertex of the cone Γ_0 is the point $B_0 = (1:0:0:0)$; the vertex of Γ_3 is the point $B_3 = (0:0:0:1)$. We denote the entries of $\mathbf{A}$ and $\mathbf{B}$ by a_{ij} and b_{ij}. If the line $l = [B_0, B_3]$

is the common generator of the cones, then $b_{00} = a_{33} = 0$. Furthermore, the homogeneous coordinates of the tangent planes of Γ_0 and Γ_3 along l are $\tau_0 = (0 : a_{13} : a_{23} : 0)$ and $\tau_3 = (0 : b_{01} : b_{02} : 0)$ and have to be linearly dependent if the cones are in line contact along l. This results in the simple condition $a_{13}b_{02} - a_{23}b_{01} = 0$, or equivalently,

$$b_{02} = wa_{23}, \quad b_{01} = wa_{13} \tag{6.7}$$

with some $w \in \mathbb{F}^\star$. Inserting (6.7) into (6.6), we can solve for x_i ($i \in \{0,1,2,3\}$) and obtain, besides a parametrization of l, a parametrization of the remaining set of common points

$$\mathbf{c}(t_0, t_1) = (b_{11}t_1^2 + 2b_{12}t_1t_2 + b_{22}t_2^2 : -2w(a_{13}t_1 + a_{23}t_2)t_1 : \\ : -2w(a_{13}t_1 + a_{23}t_2)t_2 : w(a_{11}t_1^2 + 2a_{12}t_1t_2 + a_{22}t_2^2))$$

with $t_1 : t_2 \neq 0 : 0$, which is a planar curve of degree two, *i.e.*, a conic, cf. Figure 6.4. ■

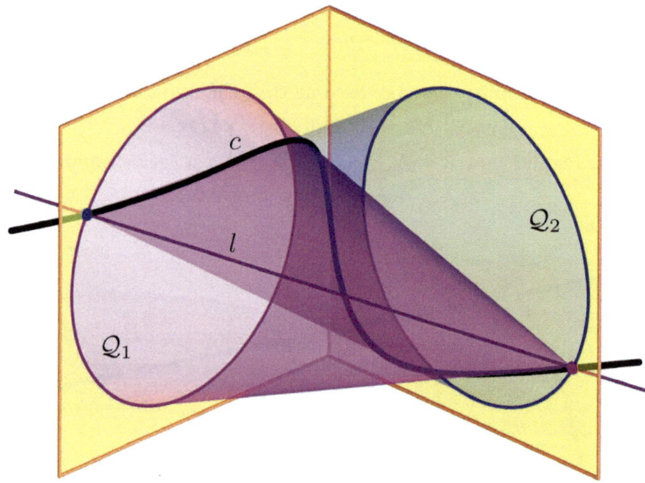

FIGURE 6.5. A cubic space curve c can be constructed as the intersection of two quadratic cones $\mathcal{Q}_1$ and $\mathcal{Q}_2$ with a common generator l.

The pair of quadratic cones in (6.6) spans a singular pencil of quadrics as defined in Section 5.5. Corollary 6.2.1 gives rise to the following result:

Theorem 6.2.1 *A cubic space curve can be obtained as the intersection of two quadratic cones with different vertices sharing a generator l (straight line) without touching each other along l.*

Proof: Assume that the two quadratic cones with different vertices share a generator l, as depicted in Figure 6.5. They cannot share a further generator, otherwise the two vertices would coincide. Therefore, the straight line l splits off from the quartic intersection curve, and a cubic curve c remains. The cubic curve c is, indeed, a space curve, since planar intersections of quadratic cones are always of degree two. ■

The two quadrics $\mathcal{Q}_1$ and $\mathcal{Q}_2$ from (6.4) span a pencil $\mathcal{P}$ of quadrics. Let $\mathbf{Q}_1, \mathbf{Q}_2 \in \mathbb{F}^{4\times 4}$ be the coefficient matrices of the respective quadrics $\mathcal{Q}_1$ and $\mathcal{Q}_2$ in (6.4). Then, the quadrics in the pencil $\mathcal{P}$ have the equations

$$\mathbf{x}^{\mathrm{T}}(\lambda\mathbf{Q}_1 + \mu\mathbf{Q}_2)\mathbf{x} = 0, \quad \lambda : \mu \neq 0 : 0. \tag{6.8}$$

Any quadric in this pencil passes through the intersection $\mathcal{Q}_1 \cap \mathcal{Q}_2 = l \cap c$. There, l is the straight line given in (6.5) and c is the cubic from (6.2). It is easy to see that the quadric $\mathcal{Q}_3$ given in (6.4) is not contained in the pencil $\mathcal{P}$, but passes through the cubic c. In order to construct the cubic space curve c as the intersection of (quadratic) surfaces, we have to add a third quadric which is not contained in the pencil $\mathcal{P}$. For instance, we add $\mathcal{Q}_3$. We can summarize this in

Theorem 6.2.2 *Each cubic space curve can be found as the complete intersection of three quadrics.*

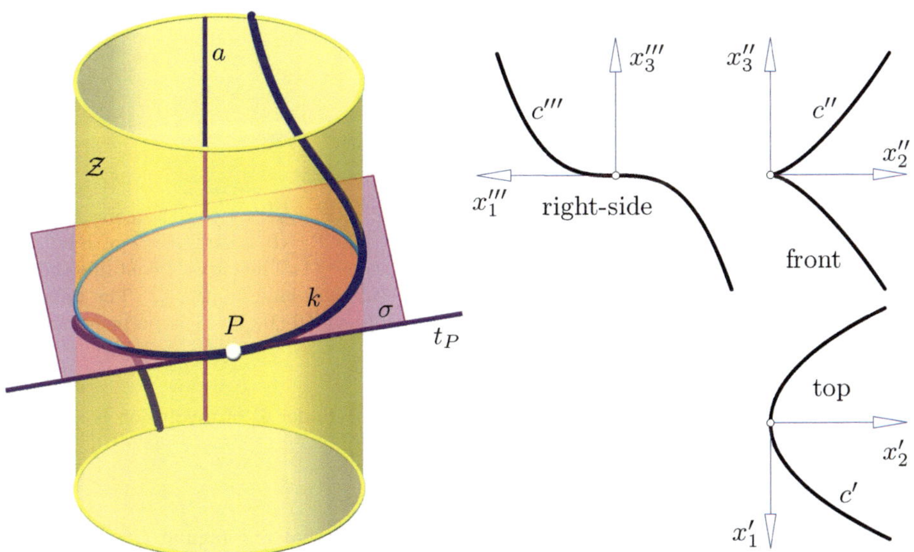

FIGURE 6.6. Left: A straight cubic circle k with asymptote a is located on a cylinder of revolution $\mathcal{Z}$. Right: The three principal views (orthogonal projections of a generic cubic space curve onto the osculating plane at P [top], the normal plane [front], the rectifying plane [right side]) of the canonical expansion show a parabola, a (planar) cubic with a cusp (semi-cubic parabola), and a (planar) cubic with a point of inflection.

■ **Example 6.2.1** Rational Pythagorean-hodograph curve.

The cubic space curve c admitting the parametrization

$$\mathbf{c}(t) = \begin{pmatrix} 3t \\ 3t^2 \\ 2t^3 \end{pmatrix}$$

with the affine parameter $t \in \mathbb{R}$, written in terms of Cartesian coordinates, is a *rational Pythagorean-hodograph curve*, *i.e.*, it has a rational parametrization such that the tangent unit vector $\dot{\mathbf{c}}\|\dot{\mathbf{c}}\|^{-1}$ is a rational curve on the unit sphere. A small part of $\mathbf{c}$ in a vicinity of $\mathbf{c}(0)$ is shown in Figure 6.6 (right). Rational Pythagorean-hodograph curves play an important role in computer aided geometric design (CAGD) and in computer graphics (see [39]).

Moreover, it is a curve of constant slope $\sigma = 45°$, since the curvature $\kappa(t)$ and torsion $\tau(t)$ of c satisfy $\kappa(t)/\tau(t) \equiv 1 =$ const. (for all $t \in \mathbb{R}$), which is characteristic for curves of constant slope w.r.t. the direction $(1, 0, 1)$.

■ **Example 6.2.2** Straight cubic circle.

Now, we move over to Euclidean space and use Cartesian coordinates. The cubic space curve k with the parametrization

$$\mathbf{k}(\tau) = \begin{pmatrix} \cos\tau \\ \sin\tau \\ \tan\frac{\tau}{2} \end{pmatrix} \quad \text{with} \quad \tau \in]-\pi, \pi[$$

can also be parametrized by rational functions as

$$\mathbf{k}(t) = \frac{1}{1+t^2}\begin{pmatrix} 1-t^2 \\ 2t \\ t(1+t^2) \end{pmatrix} \quad \text{with} \quad t \in \mathbb{R}.$$

The curve k is located on the cylinder of revolution $\mathcal{Z} : x^2 + y^2 = 1$, on the cone $\Gamma : (x - 1)^2 + y^2 - 2yz = 0$, and passes through the absolute points $I = (0, 1, \mathrm{i}, 0)$, $J = \overline{I}$ of Euclidean geometry. It is called a *straight cubic circle* for two reasons: (1) The curve k contains the two absolute points $(0 : 1 : \pm\mathrm{i} : 0)$ of Euclidean geometry (in the plane $z = 0$). (2) The only real asymptote, *i.e.*, k's tangent at its only real point at infinity $(0 : 0 : 0 : 1)$ is perpendicular to the generators of $\mathcal{Z}$.

The planes $z =$ const. intersect $\mathcal{Z}$ and Γ along circles.

The curve k together with its real asymptote a and the cylinder $\mathcal{Z}$ of revolution is shown in Figure 6.6 (left).

■ **Example 6.2.3** Canonical expansion of a space curve.

From the Differential Geometry of curves, it is well-known that any regular C^3 space curve c can be expanded locally (around any regular, non-inflection, and non-handle point) in a Taylor series such that

$$\widetilde{\mathbf{c}}(s) \approx \left(s - \frac{\kappa^2}{6}s^3, \frac{\kappa}{2}s^2 + \frac{\dot{\kappa}}{6}s^3, \frac{\kappa\tau}{6}s^3\right)$$

(with s being the arc length parameter and ignoring terms of higher order) agrees with c in derivatives up to the order three and the Frenet frame is chosen as the coordinate frame. This expansion is frequently called the *canonical expansion* of c at some regular point $c_0 = c(t_0)$, and it obviously describes a cubic space curve which is in third-order contact with c. The curve parametrized by $\widetilde{\mathbf{c}}$ is a cubic parabola and agrees with the initial curve c in derivatives up to order three, *i.e.*, it hyperosculates the curve c. Therefore, it can be viewed as the three-dimensional analogue of the hyperosculating parabola of a planar curve (cf. [46, p. 100]). From this expansion, the local behavior and the appearance of the initial curve c in the orthogonal projections onto the planes of the Frenet frame can be read off, and locally, the curve looks as shown in Figure 6.6 (right).

Regular quadrics through a cubic

Since a quadric $\mathcal{Q}$ is an algebraic surface of degree two and a cubic c is an algebraic curve of degree three, we can expect that $\#\mathcal{Q} \cap c \leq 6$. The bound on the number of common points is sharp if, and only if, $\mathcal{Q}$ and c are located in a projective space $\mathbb{P}^3(\mathbb{F})$ with an algebraically closed field $\mathbb{F}$. Even in the real projective space $\mathbb{P}^3(\mathbb{R})$, the number of six common points of a quadric and a cubic can be achieved (see Figure 6.7).

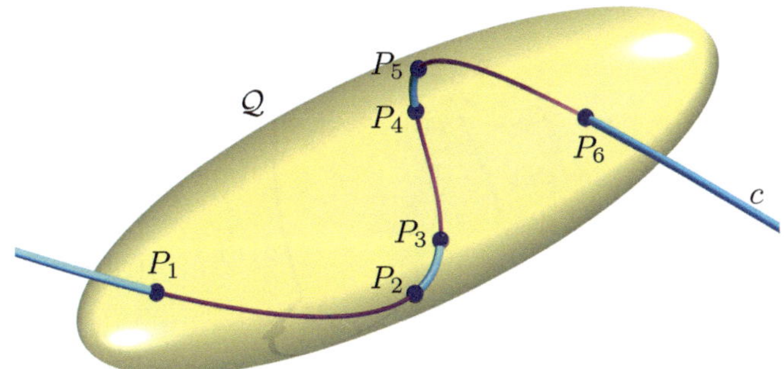

FIGURE 6.7. In the generic case, a quadric $\mathcal{Q}$ meets a cubic space curve c in six different points. Those segments of the cubic that lie in the interior of $\mathcal{Q}$ are drawn with a thin line.

A cubic space curve c is completely contained in a quadric $\mathcal{Q}$ if the quadric and the cubic share at least seven points. This fact allows us to determine the totality of quadrics through a cubic space curve:

Corollary 6.2.2 *There exists a linear two-parameter family of quadrics through any cubic space curve.*

Proof: According to Chapter 4, a quadric is uniquely defined by nine points in admissible position. In order to force a quadric $\mathcal{Q}$ to carry a given cubic (space curve) c completely, it needs to have seven points with $\mathcal{Q}$ in common. Thus, we have to determine $\mathcal{Q}$ such that it carries seven arbitrary (but mutually different) points $P_0, \ldots, P_6$ of c. This results in solving a system of seven homogeneous linear equations in the ten unknown coefficients of $\mathcal{Q}$'s equation. The solutions gather in a three-dimensional subspace of $\mathbb{F}^{10}$, and due to the homogeneity, we end up with a two-parameter family of quadrics. ■

It is elementary to solve the system of linear equations mentioned in the proof of Corollary 6.2.2. It is sufficient to compute the equations of the quadrics $\mathcal{Q}$ passing through the standard cubic c. This yields

$$\mathcal{Q}:\ \mathbf{x}^{\mathrm{T}}\begin{pmatrix} 0 & 0 & u_0 & -u_1 \\ 0 & -2u_0 & u_1 & u_2 \\ u_0 & u_1 & -2u_2 & 0 \\ -u_1 & u_2 & 0 & 0 \end{pmatrix}\mathbf{x} = 0, \quad u_0 : u_1 : u_2 \neq 0:0:0 \tag{6.9}$$

The quadric (6.9) is regular if the determinant of the coefficient matrix

$$\delta := (u_0u_2 - u_1^2)^2 \tag{6.10}$$

does not vanish.

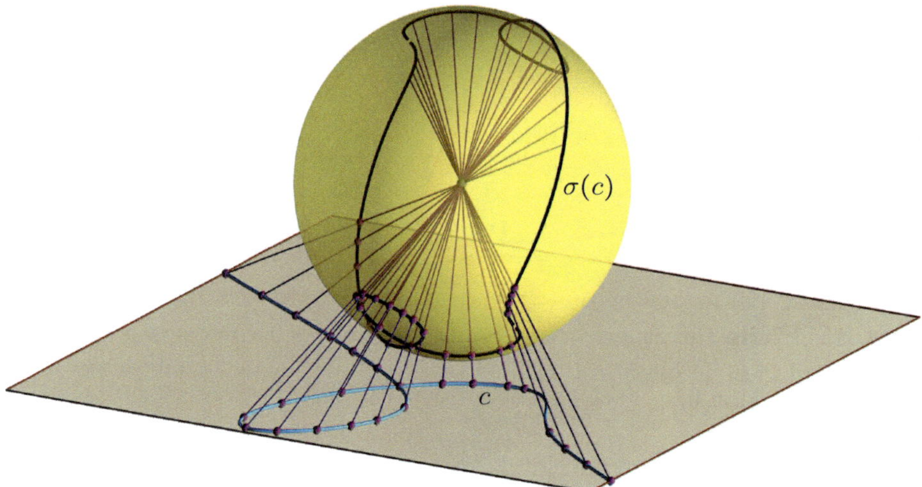

FIGURE 6.8. The curve $\sigma(c)$ is a spherical cubic obtained by the gnomonic projection of a planar cubic c onto a sphere. As such it is actually a sextic curve as the intersection of the sphere with the cubic cone projecting c onto the sphere. There are no real cubics on oval quadrics.

In the real projective 3-space $\mathbb{P}^3(\mathbb{R})$, the quadrics carrying a cubic space curve are never oval:

Lemma 6.2.3 *The quadrics through cubic space curves given in* (6.9) *are either quadratic cones or ruled quadrics.*

Proof: An oval quadric carries no real line that could eventually split off from the intersection curve of the oval quadric with another quadric. ■

Figure 6.8 shows a spherical cubic $\sigma(c)$ obtained as the gnomonic projection of a planar cubic c. (The center of the gnomic projection is the sphere's center.) The spherical cubic is a sextic curve. Only the cone projecting the planar cubic onto the sphere is a cubic variety in the star of lines about the center of the projection. This fits with the definition of spherical conics which are curves of degree four, but the cone projecting them from the center of the sphere is a quadratic cone (cf. [46, Ch. 10.1]).

Figure 6.9 shows the cubic curve c with the parametrization

$$\mathbf{c}(t) = \frac{1}{1+t^2}\left(ct(1+t^2), bt(3-t^2), a(1-3t^2)\right), \quad t \in \mathbb{R}$$

on the (regular) ruled quadric

$$\mathcal{H}: \; -\frac{x^2}{c^2} + \frac{y^2}{b^2} + \frac{z^2}{a^2} = 1.$$

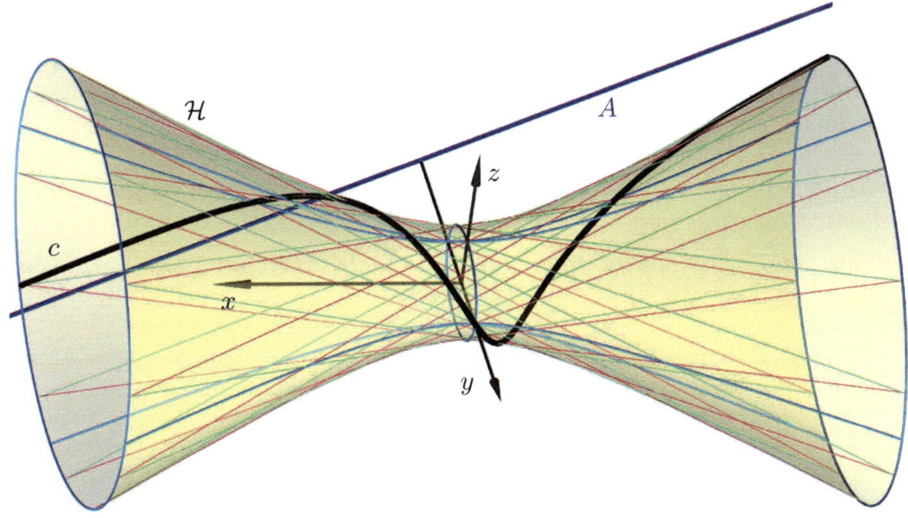

FIGURE 6.9. The cubic c winds about a one-sheeted hyperboloid $\mathcal{H}$. It is a cubic ellipse with one real ideal point $(0:0:-b:a)$ and the asymptote A (blue).

Quadratic cones through cubics

The singular quadrics through a cubic space curve can be obtained from (6.9) by imposing the (in fact) quadratic condition $\delta = 0$ (cf. (6.10)). The parameters u_i can be replaced by

$$u_0 = v_1^2, \quad u_1 = v_0 v_1, \quad u_2 = v_0^2$$

with $u_0 : u_1 \neq 0 : 0$. Consequently, (6.9) changes to

$$\Gamma : \mathbf{x}^{\mathrm{T}} \begin{pmatrix} 0 & 0 & v_1^2 & -v_0 v_1 \\ 0 & -2v_1^2 & v_0 v_1 & v_0^2 \\ v_1^2 & v_0 v_1 & -2v_0^2 & 0 \\ -v_0 v_1 & v_0^2 & 0 & 0 \end{pmatrix} \mathbf{x} = 0 \tag{6.11}$$

and yields a parametric representation of all quadratic cones (singular quadrics) through the standard cubic. The coefficient matrix in (6.11) is of rank three if $v_0^2 + v_1^2 \neq 0$ and of rank two otherwise.

When the homogeneous parameters $(u_0 : u_1 : u_2)$ of the quadrics through the cubic c given by (6.9) are interpreted as homogeneous coordinates in a projective plane, then the cones determine conics in the plane. The two-parameter family of quadrics through a cubic is equivalent to a projective plane.

The following fact has important consequences for the linear images of cubic space curves:

Corollary 6.2.4 *1. The vertex of any quadratic cone through a cubic is located on the cubic.*

2. From any of its points, the cubic is projected through a quadratic cone.

Proof:

1. The radicals of the quadratic forms (6.11) are given by $(v_0^3 : v_0^2 v_1 : v_0 v_1^2 : v_1^3)$ with $v_0 : v_1 \neq 0 : 0$ which parametrizes (6.2), and thus, the cones' vertices are located on the cubic. The cones have a point-shaped vertex for any admissible homogeneous parameter, *i.e.*, for any $v_0 : v_1$ as long as $v_0 : v_1 \neq 0 : 0$.
2. Assume that $V = \mathbf{v}\mathbb{F}$ is the cone's vertex and $v_0 : v_1 \neq 0 : 0$ is the corresponding homogeneous parameter on the curve (6.2). Then,

$$\mathbf{q}(\lambda, \mu) = \lambda \mathbf{v} + \mu \mathbf{c}$$

with $\lambda : \mu \neq 0 : 0$ is a homogeneous parametrization of the cone. Identifying the components of $\mathbf{q}$ with the homogeneous coordinates $x_0 : \ldots : x_3$, we can eliminate the parameters $\lambda : \mu$

and $t_0 : t_1$. This yields the implicit equation of the cone Γ_V as given in (6.11), provided that

$$t_0 v_1 - t_1 v_0 \neq 0 \quad \text{and} \quad v_0 x_1 - v_1 x_0 \neq 0.$$

The first inequality expresses the fact that we cannot expect a unique result when connecting C with itself. At the vertex, the cubic's tangent equals the cone's generator. ■

Consequently, Corollary 6.2.4 leads to:

Theorem 6.2.3 *The image c' of a cubic c under a central projection with the center C on the cubic c is a conic c' when adding c's tangent at C to the projecting cone.*

Proof: Assume that $C \in c$ is the center of a central projection. According to Corollary 6.2.4, the cone joining C with all points on c is a quadratic cone Γ_c, and the image c' of c is the intersection of Γ_C with an image plane. ■

Theorem 6.2.3 applies to parallel projections as well as to central projections. Figure 6.6 shows a parabola as the top view of a cubic parabola. In this case, the parabola is the orthogonal projection of the standard cubic (6.2) onto the xy-plane, *i.e.*, the parallel projection from the ideal point of the z-axis. Since the ideal point of the z-axis is a point of the cubic c, the image c' is (according to Theorem 6.2.3) a conic.

As a consequence of the proof of the first part of Corollary 6.2.4, there exist up to three quadratic cylinders carrying a cubic space curve c, see Example 6.2.4.

■ **Example 6.2.4** Three quadratic cylinders through a cubic space curve.

We shall determine all quadratic cylinders through the cubic space curve c (cubic hyperbola, shown in Figure 6.10) contained in the projectively extended Euclidean 3-space parametrized by

$$\mathbf{c}(t) = \frac{1}{t(t^2-1)} \begin{pmatrix} 1+t^2 \\ 2t \\ t^2-1 \end{pmatrix}, \quad t \in \mathbb{R}.$$

According to Corollary 6.2.4, the cones projecting c from any of its points are quadratic. The quadratic cones are cylinders if their vertices are points at infinity. Therefore, $t(t^2-1) = 0$ yields the parameter values $t = -1, 0, 1$ corresponding to c's ideal points, which are

$$\mathbf{u}_1 = \begin{pmatrix} 0 \\ 1 \\ -1 \\ 0 \end{pmatrix}, \quad \mathbf{u}_2 = \begin{pmatrix} 0 \\ 1 \\ 0 \\ 1 \end{pmatrix}, \quad \mathbf{u}_3 = \begin{pmatrix} 0 \\ 1 \\ 1 \\ 0 \end{pmatrix}.$$

From the parametrizations $\mathbf{z}_i = \mathbf{c} + \lambda \mathbf{v}_i$ ($i \in \{1, 2, 3\}$) where $\mathbf{v}_i \in \mathbb{R}^3$ are the vectors aiming in the directions defined by $\mathbf{u}_i$, we can find the implicit equations of the cylinders. The equations are

$$\mathcal{Z}_1: \; xz + yz - z^2 - x - y - z = 0,$$
$$\mathcal{Z}_2: \; 4x^2 - 8xz - 4y^2 + 4z^2 + 8y = 0,$$
$$\mathcal{Z}_3: \; xz - yz - z^2 + x - y + z = 0.$$

All three cylinders are hyperbolic.

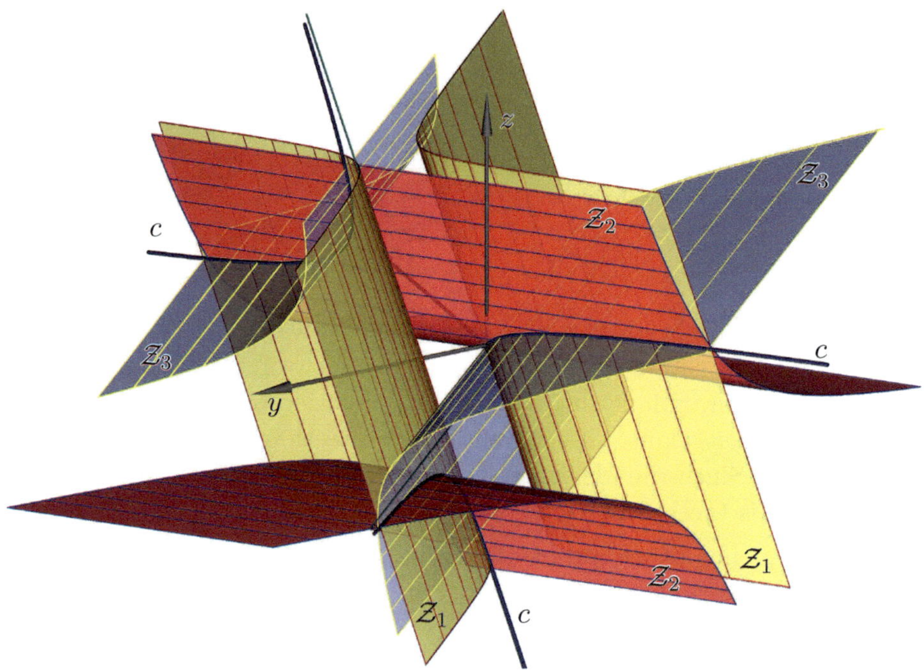

FIGURE 6.10. The three quadratic cylinders $\mathcal{Z}_1$, $\mathcal{Z}_2$, and $\mathcal{Z}_3$ pass through the cubic hyperbola c from Example 6.2.4.

The unique cubic on six points

In [46, p. 219, Theorem 6.1.2], we have seen that a conic in a projective plane is uniquely defined by five given points in admissible position (*i.e.*, no three points are collinear). It is reasonable to ask for the number of points that define a cubic space curve:

Theorem 6.2.4 *A cubic space curve is uniquely determined by prescribing six points $P_0, \ldots, P_5$ with any five out of these forming a projective frame (i.e., no four points are coplanar).*

Proof: A cubic space curve can be constructed as the intersection of two quadratic cones Γ_0 and Γ_5 (vertices P_0 and P_5) having a common generator $g = [P_0, P_5]$ but different tangent planes along f. (It means no restriction to assume that P_0 and P_5 are the cones' vertices.)

There is a uniquely determined quadratic cone on the five lines $l_i = [P_0, P_i]$ (with $i \in \{1, \ldots, 5\}$): The lines l_i meet an arbitrary plane ε_0 in five points on which a unique conic c_0 is defined, provided that ε_0 contains neither l_i. Then, the quadratic cone is obtained by joining all points of c_0 with P_0, *i.e.*, $\Gamma_0 = P_0 \vee c_0$ (see Figure 6.11).

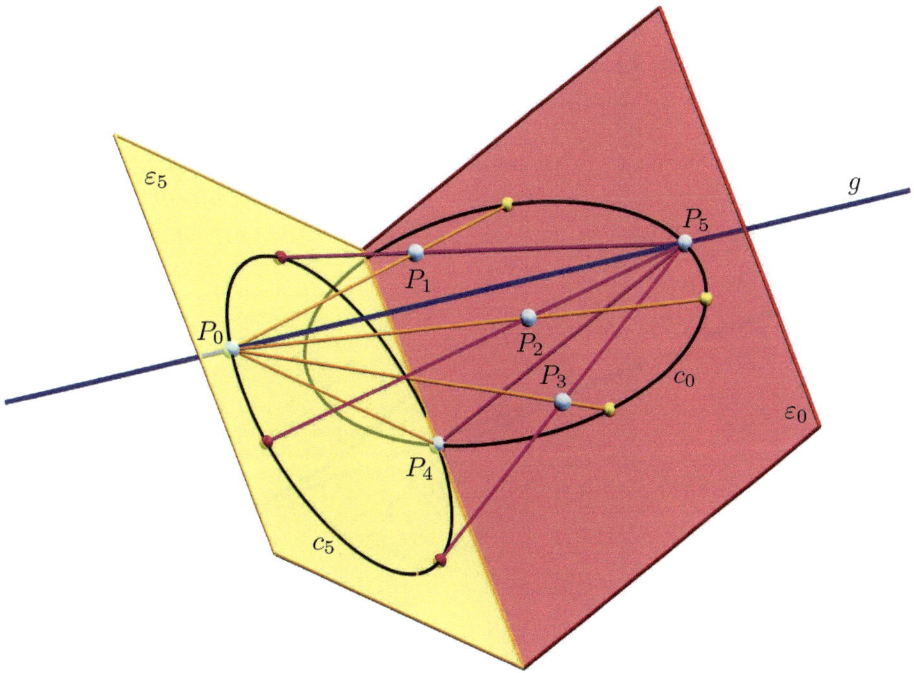

FIGURE 6.11. Construction of the cubic on six points $P_0, \ldots P_5$: Two quadratic cones can be defined by five generators each. The generators emerge at two chosen (given) points P_0 and P_5.

In the same way, we proceed with the lines $m_i = [P_5, P_i]$ (with $i \in \{0, \ldots, 4\}$) and obtain the cone Γ_5.

From the construction it is clear that $g \subset \Gamma_0 \cap \Gamma_5$. Provided that Γ_0 and Γ_5 have different tangent planes along g, the curve $c = \Gamma_0 \cap \Gamma_5 \setminus \{g\}$ is a cubic space curve (cf. Figure 6.12). ■

Five points in a projective plane that are in an admissible position determine a conic, even if two points come *infinitely close together*. In such a case, the two points become a *line element*, *i.e.*, a pair (P, t_P) of a line t_P incident with a point, and t_P is then considered to be the conic's tangent at P. The same holds true for cubics. Therefore, cubics can be constructed on

1. four points $P_0, \ldots, P_3$ plus a line element (P_4, t_4);
2. two points P_0, P_1 plus two line elements (P_2, t_2) and (P_3, t_3);
3. tree line elements (P_i, t_i) (with $i \in \{1,2,3\}$),

provided that the points and lines are in admissible position. In these three cases, each line element (X, t_X) defines a generator x and the tangent plane τ_x (along x) of a quadratic cone through the cubic. Consequently, a conic in the auxiliary planes is then determined by points and line elements.

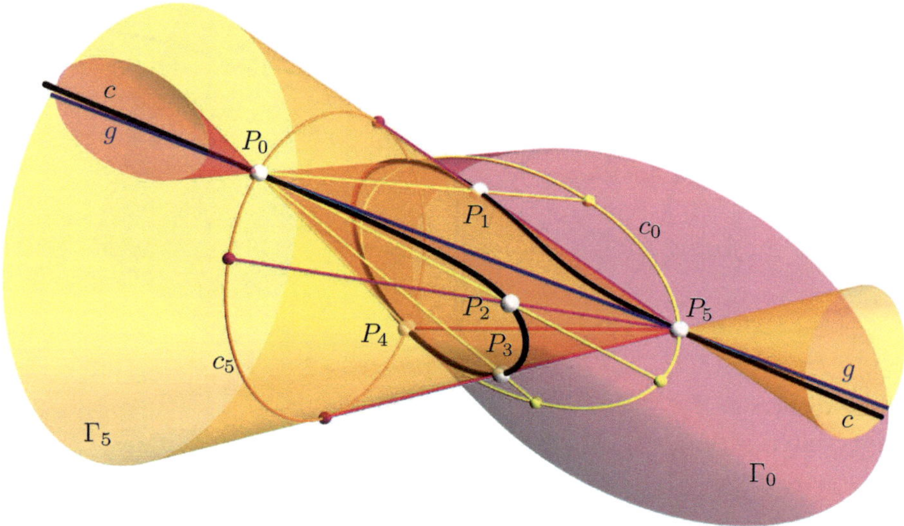

FIGURE 6.12. The construction of the unique cubic on six points $P_0, \ldots P_5$: The intersection of the cones Γ_0 and Γ_5 contains the straight line g and the cubic c.

Cubics, conics, and lines – equally many points.

The stereographic projection from a conic to a straight line establishes a bijective mapping between them (cf. [46, p. 231]) and leads to the finding that there are as many points on a conic as there are on a line (see [46, p. 232, Corollary 6.3.4]). With a little bit more effort, we can show

Corollary 6.2.5 *There are as many points on a cubic space curve as there are on a projective line (or on a projective conic).*

Proof: According to Corollary 6.2.4, from any point P on the cubic c, a quadratic cone $\Gamma_P = P \vee c$ emanates (cf. Figure 6.13). The intersection c' of Γ_P with a plane $\varepsilon \not\ni P$ is a conic. The generators of Γ_P establish a bijective mapping between the points of c and those of c'. The point P is mapped to the intersection of the cubic's tangent t_P with ε. Consequently, the cubic c carries as many points as the conic c'. Since conics and lines are of equal cardinality (cf. [46, p. 232, Corollary 6.3.4]), the cubic c and a projective line are of equal cardinality. ■

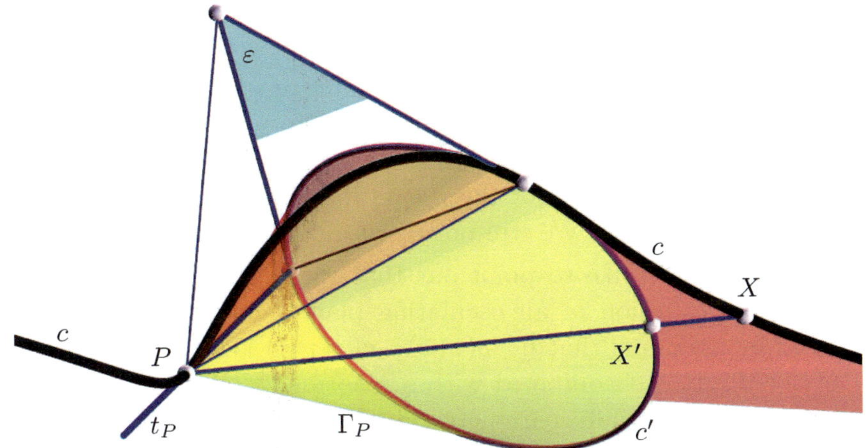

FIGURE 6.13. Some kind of stereographic projection from a cubic space curve c onto a plane ε with center $P \in c$ establishes a one-to-one and onto mapping $X \mapsto X'$ between a conic c' and a cubic c in order to show that a projective line, a conic, and a cubic are of the same cardinality.

• **Exercise 6.2.1** Cardinality of a rational normal curve.

Show that a rational normal curve $\mathbf{n}(t_0, t_1) = (t_0^n : t_0^{n-1} t_1 : \ldots : t_0 t_1^{n-1} : t_1^n)$ of degree $n \geq 3$ has as many points as a projective line. Use the proof of Corollary 6.2.5 and induction over n.

6.3 Osculating planes of a cubic

We defined the tangent of a cubic as a line intersecting the cubic with multiplicity two, *i.e.*, it is the limit of a chord where one of the two points of intersection converges towards the other. In what follows, we shall have a look at three points and make use of

Definition 6.3.1 A plane $\omega \subset \mathbb{P}^3(\mathbb{F})$ is called an *osculating plane* of c at the point P if it intersects c at P with multiplicity three.

In the case $\mathbb{F} = \mathbb{R}$, it is easy to show that a limit procedure yields the same osculating plane. Any three sufficiently close points on a cubic space curve and indeed, on any C^2 space curve define a plane which converges towards the osculating plane at one of the points if the remaining two converge towards the third point. We will not discuss this here in detail.

Moreover, we would like to point out that the algebraic and the differential geometric notion of an osculating plane coincide. This is easily confirmed by assuming that $\widetilde{\mathbf{c}} = \varrho(\upsilon(\tau)) \cdot \mathbf{c}(\upsilon(\tau)))$ is a reparametrization $\tau \mapsto \upsilon(\tau) : I \subset \mathbb{F} \to \mathbb{F}$ combined with a renormalization $\varrho : I \subset \mathbb{F} \to \mathbb{F}$ of the cubic curve c initially parametrized by $\mathbf{c}(\tau) : I \subset \mathbb{F} \to \mathbb{F}^4$ over some interval I. We assume that $\mathbb{F} = \mathbb{C}$ or $\mathbb{F} = \mathbb{R}$. Again, dots shall indicate differentiation. The osculating plane at $P = \mathbf{c}(\tau_0)$ is spanned by P and the first two derivatives $\dot{\mathbf{c}}(\tau_0)$ and $\ddot{\mathbf{c}}(\tau_0)$. The invariance of the definition of the osculating plane under renormalizations and reparametrizations is shown in a manner similar to that of the tangent (see p. 208). We just have to show that $[\mathbf{c}, \dot{\mathbf{c}}, \ddot{\mathbf{c}}] = [\widetilde{\mathbf{c}}, \dot{\widetilde{\mathbf{c}}}, \ddot{\widetilde{\mathbf{c}}}]$. In order to achieve, that we compute the derivatives of the alternative representation of c and observe that these are linear combinations of $\mathbf{c}$, $\dot{\mathbf{c}}$, and $\ddot{\mathbf{c}}$.

Finally, it is worth mentioning that there is a certain kind of tetrahedron that can be attached to a cubic (in many ways). Let c be a cubic space curve. A tetrahedron $A_0A_1A_2A_3$ is called an *osculating tetrahedron* of c if $A_0, A_3 \in c$, $[A_0, A_1]$ and $[A_2, A_3]$ are tangents, and $[A_0, A_1, A_2]$ and $[A_1, A_2, A_3]$ are osculating planes of c at A_0 and A_3. Note that the base points of the coordinate frame which represents c in the standard form (6.2) forms an osculating tetrahedron. Conversely, if the base points of a projective coordinate frame form an osculating tetrahedron and the unit point belongs to c, then c can be parametrized in the standard form.

The null polarity associated with the cubic and the dual cubic

FIGURE 6.14. A cubic space curve with some null planes and some lines of the pencils of null lines.

An interesting relation between cubics and special types of correlations can be disclosed:

Theorem 6.3.1 *For each cubic c in projective 3-space $\mathbb{P}^3(\mathbb{F})$ (with* $\operatorname{char}\mathbb{F} \neq 2, 3$*), there exists a null polarity mapping the points of c to its osculating planes.*

Proof: Let us assume that $\varepsilon \subset \mathbb{P}^3(\mathbb{F})$ is a plane with the homogeneous coordinates $\mathbf{e} = (e_0, e_1, e_2, e_3)$ with $e_0 : \ldots : e_3 \neq 0 : \ldots : 0$. Further, we assume that the coordinate system is chosen such that the parametrization $\mathbf{c}$ of the cubic equals that given in (6.2). The points of intersection of the cubic c and the plane ε correspond to the solutions of the cubic form

$$\langle \mathbf{e}, \mathbf{c} \rangle = e_0 t_0^3 + e_1 e_0^2 t_1 + e_2 t_0 t_1^2 + e_3 t_1^3 = 0. \tag{6.12}$$

A plane ε is an osculating plane if there exists only one point of intersection (with multiplicity 3), and thus, the cubic form has to be the *complete cube* $(\alpha_0 t_0 + \alpha_1 t_1)^3$ of a linear form $\alpha_0 t_0 + \alpha_1 t_1$. Hence, we have

$$\alpha_0 : \alpha_1 = -t_1 : t_0$$

and the homogeneous coordinates $e_0 : \ldots : e_3$ of ε are

$$e_0 = -t_1^3, \quad e_1 = t_1^2 t_0, \quad e_2 = -t_1 t_0^2, \quad e_3 = t_0^3.$$

This shows that the homogeneous coordinates $\mathbf{e}$ of the osculating planes of c and the corresponding points $\mathbf{c}$ on the cubic are related via a linear mapping $\nu:\ \mathbb{F}^4 \to \mathbb{F}^4$ which reads

$$\mathbf{e} = \underbrace{\begin{pmatrix} 0 & 0 & 0 & -1 \\ 0 & 0 & 3 & 0 \\ 0 & -3 & 0 & 0 \\ 1 & 0 & 0 & 0 \end{pmatrix}}_{=:\mathbf{N}} \mathbf{c}. \tag{6.13}$$

The matrix $\mathbf{N}$ that describes ν is skew symmetric, and therefore, ν is a null polarity (cf. Theorem 4.2.2). ■

According to (6.13), the family of osculating planes can be given in terms of a one-parameter family of implicit linear equations as

$$\mathcal{O}(t_0, t_1):\ t_1^3 x_0 - 3t_0 t_1^2 x_1 + 3t_0^2 t_1 x_2 - t_0^3 x_3 = 0. \tag{6.14}$$

At each point P of the cubic c, the osculating plane of c contains a pencil of lines through P. Among these lines, we find the tangent t of c at P. The lines in the pencil are self-polar w.r.t. the null polarity, and therefore, called *null lines*.

Figure 6.14 shows a cubic curve with some of its osculating planes. In each of the osculating planes, some lines of the pencil of null lines are shown.

• **Exercise 6.3.1** Null polarities in higher dimensions.

A rational normal curve $n \subset \mathbb{P}^n(\mathbb{F})$ in an n-dimensional projective space with $\mathbb{F} = \mathbb{R}$ or $\mathbb{F} = \mathbb{C}$ can be given in terms of homogeneous point coordinates

$$\mathbf{n}(t_0, t_1) = (t_0^n : t_0^{n-1} t_1 : \ldots : t_0 t_1^{n-1} : t_1^n)$$

parametrized by the homogeneous parameter $t_0 : t_1 \neq 0 : 0$.

Show that the mapping

$$\nu:\ \mathbb{P}^n(\mathbb{F}) \to \mathbb{P}^{n\star}(\mathbb{F})$$

that assigns to each point $P \in n$ the osculating hyperplane $\mathcal{H}$ is a null polarity, provided that n is odd.

The homogeneous coordinates of the hyperosculating hyperplane at the point $\mathbf{n}(t_0, t_1)$ are

$$\mathbf{h}(t_0, t_1) = \left(t_1^n : -\binom{n}{1} t_1^{n-1} t_0 : \ldots : (-1)^{n+1} \binom{n}{k} t_1^{n-k} t_0^k : \ldots : (-1)^{n-1} \binom{n}{n-1} t_1 t_0^{n-1} : (-1)^n t_0^n\right)$$

What kind of mapping is ν if n is even?

Consider also finite fields with characteristic 2 or 3.

The dual construction

The construction of a cubic c as the intersection of two quadratic cones Γ_A and Γ_B with a common generator $l := [A, B]$ (along which the cones are not in line contact along l) can be dualized (see Figure 6.15).

The dual counterparts to the quadratic cones Γ_A and Γ_B are two conics $a \subset \alpha$ and $b \subset \beta$ in different planes α and β. The line of intersection $l = \alpha \cap \beta$ of the conics' planes is the only common tangent of both conics, since the cones Γ_A and Γ_B are assumed not to be in line contact along their common generator. The common tangent l of a and b corresponds to the common generator of Γ_A and Γ_B. The conics a and b do touch l in different points $A \neq B$, since the cones Γ_A and Γ_B of the original construction do not share the tangent plane along the common generator.

As shown in Figure 6.15, the planes of the dual cubic can be found as the common tangent planes of the conics a and b. To each tangent t_a of a there exists exactly one tangent $t_b \neq l$ of b that intersects t_a in a point on l, and therefore, t_a and t_b are coplanar. The plane $\tau = [t_a, t_b]$ is a plane of the dual cubic. According to Theorem 6.3.1, all these planes τ are the osculating planes of a cubic space curve, provided that $\mathrm{char}\mathbb{F} \neq 2, 3$. Their envelope is a cubic developable.

Tangent planes of a cubic curve

The osculating plane σ of a curve c at the point P contains c's tangent t at P. Among the planes through t, the osculating plane is somewhat distinct, though all these planes touch the cubic c. Indeed, the manifold $\mathcal{T}$ of *tangent planes of the curve* is the two-parameter family of planes through the curve's tangents. Since c is algebraic, the manifold of tangent planes is, too. This holds particularly true for cubic curves.

We aim at an algebraic description of this two-parameter family $\mathcal{T}$ of planes. For that purpose, we assume that (6.12) is the product of the full square of a linear form $\alpha_0 t_0 + \alpha_1 t_1$ with a linear form $\beta_0 t_0 + \beta_1 t_1$. We compare the coefficients and find

$$e_0 = \alpha_0^2 \beta_0, \quad e_1 = 2\alpha_0\alpha_1\beta_0 + \alpha_0^2\beta_1, \quad e_2 = 2\alpha_0\alpha_1\beta_1 + \alpha_1^2\beta_0, \quad e_3 = \alpha_1^2\beta_1.$$

The elimination of the α_i and β_i results in a polynomial condition on the homogeneous coordinates e_i of the tangent planes

$$\mathcal{T}: \ 27e_0^2e_3^2 - 18e_0e_1e_2e_3 + 4e_0e_2^3 + 4e_1^3e_3 - e_1^2e_2^2 = 0. \tag{6.15}$$

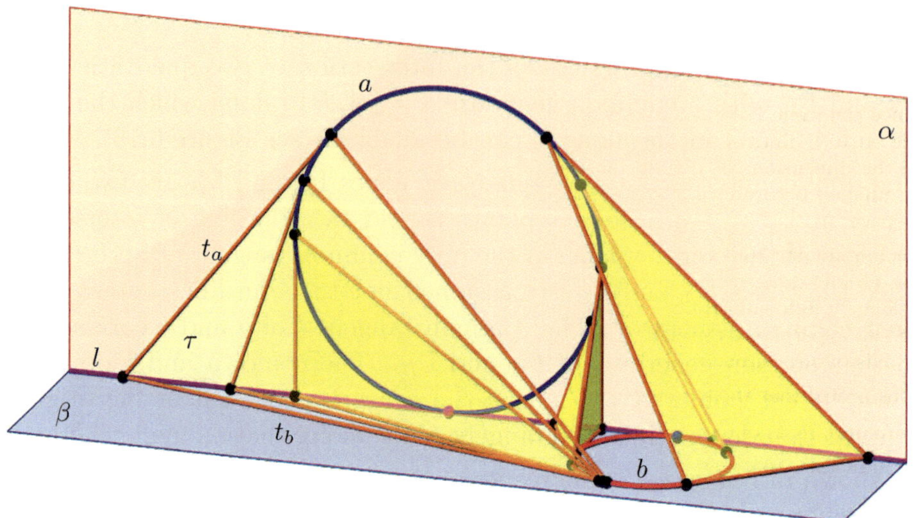

FIGURE 6.15. The constructive approach to the dual cubic: The dual cubic comprises the set of planes that are tangent to a pair of conics a and b in different planes α and β, but with a common tangent $l = \alpha \cap \beta$.

This dual surface contains the dual curve of the osculating planes (6.14).

Cubic developable

The tangents of each C^1 space curve form a developable ruled surface $\mathcal{D}$ sometimes called the *tangent developable* or *tangent surface*. This is also true for cubic space curves. The example shown in Figure 6.16 is the tangent developable of the straight cubic circle presented in Example 6.2.2. The curve c whose tangents sweep the developable ruled surface $\mathcal{D}$ is called the *curve of regression* on this ruled surface. The curve c is a sharp ridge separating the two branches of $\mathcal{D}$. In any case, the tangent developable agrees with the envelope of the osculating planes of the space curve.

In the case of the cubic (6.2), we shall give the equation of the surface $\mathcal{D}$ in terms of homogeneous point coordinates. The equations $\mathcal{O}(t_0, t_1)$ of the osculating planes (6.14) of the standard cubic c (given in (6.2)) depend smoothly on the homogeneous parameter $t_0 : t_1 \neq 0 : 0$. For the sake of simplicity, we switch to an affine parameter by letting $t_1 = t$ and $t_0 = 1$. Then, we compute the derivative $\dot{\mathcal{O}}$ of $\mathcal{O}$ w.r.t. the parameter t

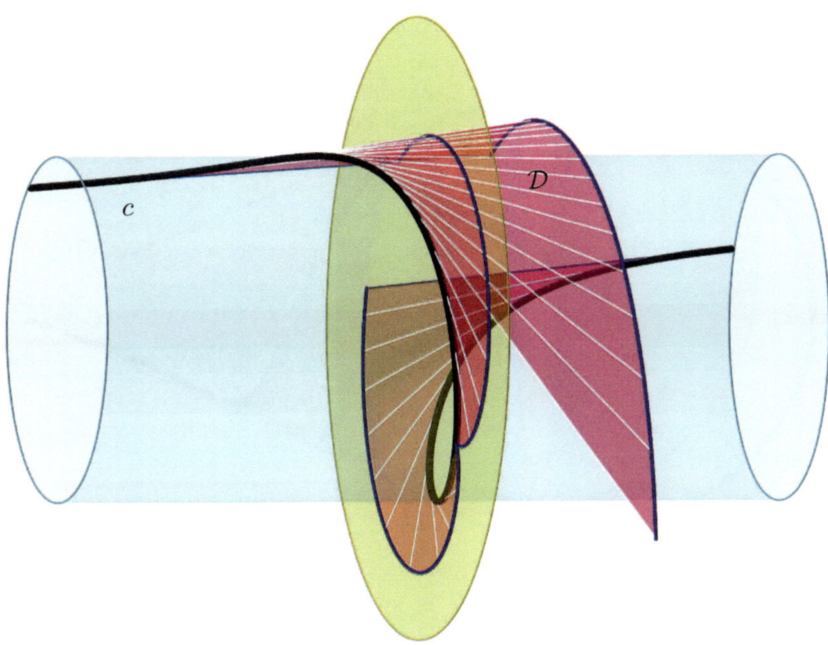

FIGURE 6.16. The surface $\mathcal{D}$ swept by the tangents of a cubic space curve c is a cubic developable. The tangents of c comprise the set of rulings (generators) of $\mathcal{D}$.

and intersect with $\mathcal{O}$ by eliminating t, which yields

$$\mathcal{D}: \; x_0^2x_3^2 - 6x_0x_1x_2x_3 + 4x_0x_2^3 + 4x_1^3x_3 - 3x_1^2x_2^2 = 0. \tag{6.16}$$

This is obviously a quartic surface and agrees with (6.15) up to a collineation from the projective space $\mathbb{P}^3$ to its dual $\mathbb{P}^{3\star}$, *i.e.*, a correlation.

• **Exercise 6.3.2** Finding a correlation.

Show that (6.15) and (6.16) are collinear images of each other. Assume that $x_i = \alpha_i e_i$ for $i \in \{0,1,2,3\}$ with $\alpha_i \in \mathbb{R}$. A comparison of coefficients yields $\alpha_0 : \alpha_1 : \alpha_2 : \alpha_3 = 1 : \frac{1}{3\sqrt{3}} : 1 : \sqrt{3}$.

From (6.16), we can find a one-parameter family of conics on $\mathcal{D}$:

Theorem 6.3.2 *The tangent planes of $\mathcal{D}$ (osculating planes of c) intersect $\mathcal{D}$ along conics (together with the twofold generators of $\mathcal{D}$). The conics osculate c.*

Proof: Each osculating plane $\mathcal{O}$ of c touches $\mathcal{D}$ along a tangent l of c, since $\mathcal{D}$ is the envelope of c's osculating planes, and therefore, l is of multiplicity two in $\mathcal{D} \cap \mathcal{O}$. Considered as a set of

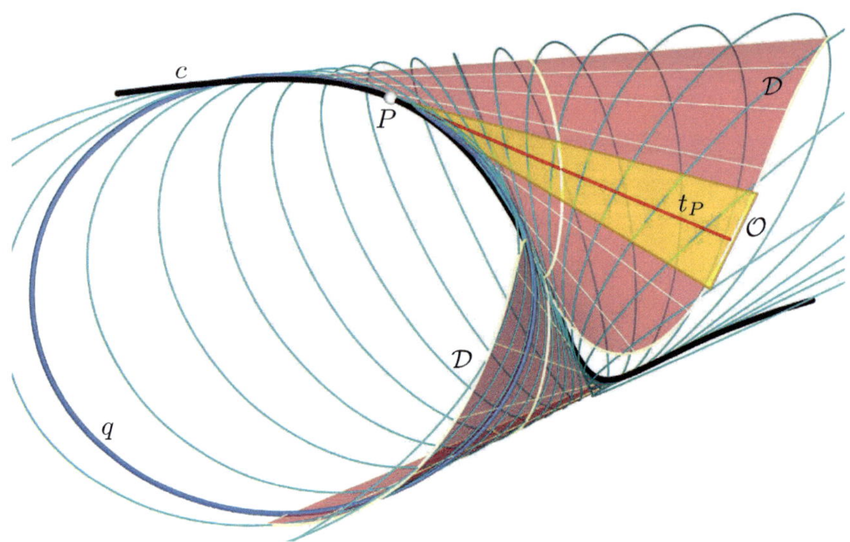

FIGURE 6.17. The developable $\mathcal{D}$ is swept by c's tangents t_P. The one-parameter family of tangent planes of $\mathcal{D}$, *i.e.*, the osculating planes of c intersect $\mathcal{D}$ along the tangents of c (with multiplicity two) and conics q osculating c.

points, $\mathcal{D}$ is of algebraic degree four, and thus, $\mathcal{D} \cap \mathcal{O}$ is a planar quartic containing a double line. Consequently, a quadratic part q remains (cf. Figure 6.17).

We have to show that q is indeed a conic and does not degenerate any further. The two branches of $\mathcal{O} \cap \mathcal{D}$ can be given explicitly as

$$\mathbf{q}(t_0, t_1) = \left(4w_0^2 t_0^3, 4w_0 t_0^3 w_1, t_0(t_0 w_1 + t_1 w_0)(3t_0 w_1 - t_1 w_0), t_1(3t_0 w_1 - t_1 w_0)^2\right),$$
$$\mathbf{l}(t_0, t_1) = \left(w_0 t_0^3, t_0^3 w_1, t_0 t_1(2t_0 w_1 - t_1 w_0), t_1^2(3t_0 w_1 - 2t_1 w_0)\right)$$

where $t_0 : t_1 \neq 0 : 0$ is the homogeneous parameter on c and $w_0 : w_1 \neq 0 : 0$ is that on either branch of $\mathcal{O} \cap \mathcal{D}$. It is clear that q describes a conic, for it carries only regular points.

Finally, we have to show that each q osculates c. This is done by introducing an affine parameter on each curve and showing that q and c agree in their derivatives up to second order. ■

Figure 6.17 shows the conics on the tangent developable of the straight cubic circle from Example 6.2.2.

Projective generation of the dual cubic

In [46, p. 226], we have learned that a projective mapping between two lines generates a dual conic. The lines joining corresponding points are the lines of the thus generated dual conic, or the tangents of the associated point conic. The projective mapping is not allowed to be a perspectivity.

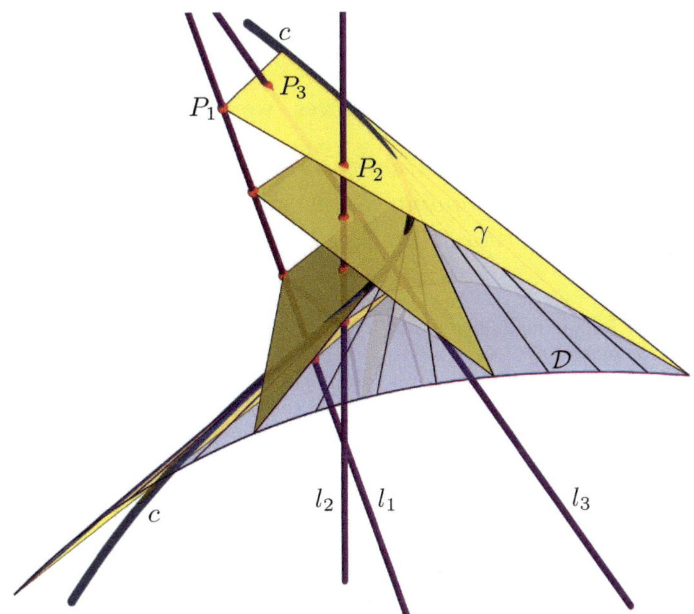

FIGURE 6.18. Projective generation of a dual cubic: Given $l_1(P_1) \barwedge l_2(P_2) \barwedge l_3(P_3)$, the projective range of planes $\gamma = [P_1, P_2, P_3]$ is the set of tangent planes of a cubic developable $\mathcal{D}$ with the space curve c as the curve of regression.

We need the notion of a projective mapping between three mutually skew straight lines l_1, l_2, l_3 in a projective 3-space. A projective mapping π_{12} between two lines l_1 and l_2 assigns to each point $P_1 \in l_1$ exactly one point $P_2 \in l_2$ and preserves cross ratios. A further projective mapping $\pi_{13}: l_1 \to l_3$ performs $P_1 \mapsto P_3$. In this way, P_1 is related to two points P_2 and P_3. We assume that the span $\gamma = [P_1, P_2, P_3]$ of the three points $P_i \in l_i$ (with $i \in \{1,2,3\}$) is a plane in any case, otherwise we end up with a regulus, which shall not be the case here. In total, each point $P_1 \in l_1$ is mapped to a uniquely defined plane γ. Furthermore, since the lines l_i are joined by projective mappings (one-to-one and onto), the points P_2 and P_3 on the respective lines define the same plane γ as, for example, the span

$$[P_2, \pi_{12}^{-1}(P_2), \pi_{13}(\pi_{12}^{-1}(P_2))].$$

Since cross ratios are preserved in any of the mappings $\pi_{1j}: l_1 \to l_j$ (with $j \in \{2,3\}$), homogeneous coordinates on any of the three lines can serve as coordinates on all three lines simultaneously by an appropriate choice of the coordinate system on each.

Relating the latter to cubic space curves, we can state:

Theorem 6.3.3 *Let three mutually skew straight lines $l_1, l_2, l_3 \subset \mathbb{P}^3(\mathbb{F})$ be joined by a projective mapping between them. Then, the set of planes γ joining three corresponding points $P_i \in l_i$ (with $i \in \{1,2,3\}$) are the planes of a dual cubic.*

Proof: We identify four independent points $B_0, \ldots, B_3$ in projective 3-space $\mathbb{P}^3(\mathbb{F})$ with the canonical basis vectors of $\mathbb{F}^4$. Now, we let $l_1 = [B_0, B_1]$, $l_2 = [B_2, B_3]$, and $l_3 = [E, F]$ with $E = \mathbf{e}\mathbb{F}$ and $F = \mathbf{f}\mathbb{F}$ as the unit points in the planes $[B_0, B_1, B_2]$ and $[B_1, B_2, B_3]$. If now $P_1 = t_1\mathbf{b}_0 + t_0\mathbf{b}_1$, then we can always choose the coordinate frames on l_2 and l_3 such that $\pi_{1i} : l_1 \to l_i$ with $i \in \{2,3\}$ and the homogeneous coordinates $\mathbf{p}_i$ (with $i \in \{1,2,3\}$) of P_i are related via

$$\pi_{12}(\mathbf{p}_1) = \mathbf{p}_2 = t_1\mathbf{b}_2 + t_0\mathbf{b}_3 \quad \text{and} \quad \pi_{13}(\mathbf{p}_1) = \mathbf{p}_3 = t_1\mathbf{e} + t_0\mathbf{f}.$$

Thus, the homogeneous coordinates of the plane γ are

$$\gamma(t_0 : t_1) = (t_0^3 : -t_0^2 t_1 : t_0 t_1^2 : -t_1^3),$$

which is obviously a representation of the standard cubic (6.2) up to a collineation. ■

Figure 6.18 illustrates the contents of Theorem 6.3.3. Four planes out of the projective range are shown together with the envelope $\mathcal{D}$ and the cubic c as the curve of regression.

6.4 An analogue to Steiner's generation

Projective mappings that map lines from one pencil to lines in another pencil of a projective plane generate conics. The points of the conic are the points of intersection of lines corresponding in the projective mapping (see [46, p. 218]). In analogy to STEINER's generation of conics, we provide:

Definition 6.4.1 Let $\kappa : S \to T$ be a collineation from the star of lines about S to the star of lines about T where no plane is mapped to itself. Then, we call the set c of all points $P = s \cap t$ with $s \ni S$, $t \ni T$, and $\kappa(s) = t$ a *cubic curve*. The set $\mathcal{C}$ of all lines $l = \sigma \cap \tau$ with $\sigma \ni S$, $\tau \ni T$, and $\kappa(\sigma) = \tau$ the *congruence of chords* of the cubic. A line $g \in \mathcal{C}$ is called a tangent of c at P if g meets c only in P. A plane ω is called an *osculating plane* of c at P if it meets c only in P.

Definition 6.4.1 indeed leads to the cubic curves we already know:

Theorem 6.4.1 *The cubic curves c from Definition 6.4.1 are exactly the cubics obtained as intersections of quadratic cones with a common generator l which are not in line contact along l.*

Proof: It means no restriction to assume that the points S and T have the homogeneous coordinates $\mathbf{s} = (1:0:0:0)$ and $\mathbf{t} = (0:0:0:1)$. The collineation $\kappa: S \to T$ can also be described by its action from the plane $\alpha: x_0 = 0$ to the plane $\beta: x_3 = 0$. It is well-known (see [46, p. 237]) that the collineation $\kappa: \alpha \to \beta$ can be written in *equal coordinates*. Any point $X = \mathbf{x}\mathbb{F} \in \alpha$ with coordinates $\mathbf{x} = (x_0 : x_1 : x_2 : 0)$ is mapped to a unique image point $\kappa(X) = Y = \mathbf{y}\mathbb{F}$ with coordinates $\mathbf{y} = (0 : x_0 : x_1 : x_2) \in \beta$. Then, the line $s = [S, X]$ is mapped to $t = \kappa(s) = [T, Y]$. These lines intersect if, and only if, $\mathbf{s}$, $\mathbf{x}$, $\mathbf{t}$, $\mathbf{y}$ are linearly dependent, *i.e.*,

$$\det\begin{pmatrix} 1 & 0 & 0 & 0 \\ 0 & x_0 & x_1 & x_2 \\ x_0 & x_1 & x_2 & 0 \\ 0 & 0 & 0 & 1 \end{pmatrix} = x_0x_2 - x_1^2 = 0 \quad \text{with} \quad x_0 : x_1 : x_2 \neq 0:0:0,$$

which gives the second equation of (6.4) and describes the quadratic cone $\mathcal{Q}_2$ with vertex T. The lines $[S, X]$ that are the κ-preimages of $[T, Y]$ (the generators of $\mathcal{Q}_1$) are now parametrized by $(\lambda : \mu u_0^2 : \mu u_0 u_1 : \mu u_1^2)$ with $\lambda : \mu \neq 0:0$ and $u_0 : u_1 : u_2 \neq 0:0:0$. The intersection points of corresponding lines are obtained by inserting the latter parametrization into $\mathcal{Q}_2$'s equation. So, we find $(u_0^3 : u_0^2 u_1 : u_0 u_1^2 : u_1^3)$ with $u_0 : u_1 \neq 0:0$, which parametrizes the cubic (6.2). ■

• **Exercise 6.4.1** Images and pre-images of tangents and osculating planes of a cubic.

In [46, Def. 6.1.1], we have learned that a projective mapping π from a pencil of lines about a point S to a pencil of lines about a point $T \neq S$ generates a conic c. Thereby, the projectivity π maps the line $[S, T]$ to the tangent t_T of c at T and the pre-image of $[S, T]$ equals the tangent t_S at S, *i.e.*, $\pi^{-1}([S, T]) = t_S$.

Show that for the cubic space curve as defined in Definition 6.4.1 we have in analogy the following relations:

(1) The π-image of $[S, T]$ equals the cubic's tangent at T, *i.e.*, $\pi([S, T]) = t_T$.

(2) The π-image of $[S, t_T]$ equals c's osculating plane at T, *i.e.*, $\pi([S, t_T]) = \mathcal{O}_T$.

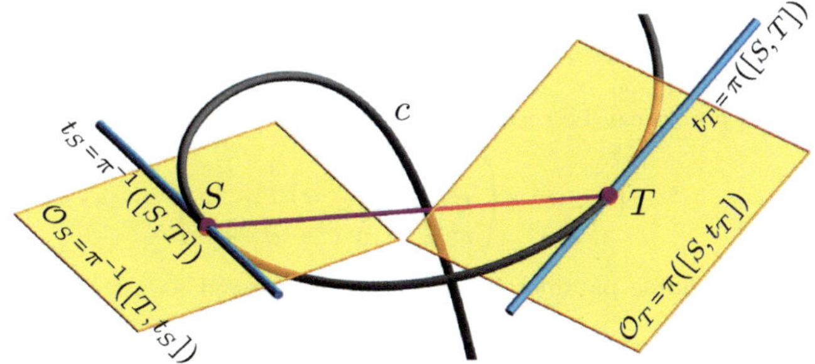

FIGURE 6.19. STEINER's generation of a cubic c by a projective collineation between two stars of lines gives a synthetic definition of the tangents and osculating planes of c (cf. Exercise 6.4.1).

6.5 Chords of a cubic

The set of chords of a conic fills the conic's plane and makes it a ruled plane. Loosely speaking, in three-dimensional space, there is *more space.* What does the family of all chords of a cubic look like? Does it *fill* the space? It is helpful to clarify this by

Definition 6.5.1 Let c be a cubic space curve in a projective space $\mathbb{P}^3(\mathbb{F})$ over an algebraically closed field $\mathbb{F}$. A line that joins two points of the cubic is called a *bisecant.*

We are able to show:

Theorem 6.5.1 *The bisecants of a cubic space curve are chords in the sense of Definition 6.4.1. All chords of a cubic form a congruence of lines with the cubic as the only (degenerate) focal surface. The congruence of chords of a cubic space curve is a fibration of* $\mathbb{P}^3(\mathbb{F})$.

Proof: Let $l = [P_1, P_2]$ be a bisecant with $P_i = s_i \cap t_i$. Then, $l = [s_1, s_2] \cap [t_1, t_2]$, where $[t_1, t_2] = \kappa([s_1, s_2])$. We use the standard cubic (6.2) and assume that $t_0 : t_1 \neq 0 : 0$ and $u_0 : u_1 \neq 0 : 0$ are independent homogeneous parameters of two different points on the cubic. Algebraically, this can be expressed as

$$\det \begin{pmatrix} t_0 & t_1 \\ u_0 & u_1 \end{pmatrix} = t_0 u_1 - t_1 u_0 \neq 0.$$

We compute the Plücker coordinates of c's bisecants by joining

$$\mathbf{c}(t_0, t_1) = (t_0^3 : t_0^2 t_1 : t_0 t_1^2 : t_1^3) \quad \text{and} \quad \mathbf{c}(u_0, u_1) = (u_0^3 : u_0^2 u_1 : u_0 u_1^2 : u_1^3),$$

which is described in more detail in Section 10.1 (we refer especially to (10.5) and (10.6)):

$$\begin{aligned} \mathcal{B} = (t_0^2 u_0^2,\ -t_0 u_0 (t_0 u_1 + t_1 u_0),\ t_0^2 u_1^2 + t_0 t_1 u_0 u_1 + t_1^2 u_0^2; \\ t_1^2 u_1^2,\ -t_1 u_1 (t_0 u_1 + t_1 u_0),\ t_0 t_1 u_0 u_1). \end{aligned} \tag{6.17}$$

The focal surfaces of the congruence $\mathcal{B}$ are easier to find if we replace the homogeneous parameters $u_0 : u_1$ and $t_0 : t_1$ by affine ones, say $u = u_1 u_0^{-1}$ and $t = t_1 t_0^{-1}$. This means no restriction and ideal points can be reached by a further reparametrization. Then,

$$\mathbf{c}(t, u, w) = \begin{pmatrix} t \\ t^2 \\ t^3 \end{pmatrix} + w \begin{pmatrix} 1 \\ t + u \\ t^2 + ut + u^2 \end{pmatrix}$$

with $(t, u) \in \mathbb{R}^2$ as an affine parametrization of $\mathcal{B}$ (considered as set of points). The focal polynomial

$$F(w) = \det \left(\frac{\mathrm{d}\mathbf{c}}{\mathrm{d}t}, \frac{\mathrm{d}\mathbf{c}}{\mathrm{d}u}, \frac{\mathrm{d}\mathbf{c}}{\mathrm{d}w} \right) = w(u - t)(t - u + w)$$

yields two zeros for w, which fix the singular points of the mapping $\mathbf{c} \colon \mathbb{R}^3 \to \mathbb{R}^3$ on each line. These are $w = 0$ and $w = u - t$. Inserting them into the parametrization of the congruence shows that the set of singular points of the congruence, *i.e.*, the focal surfaces coincide and shrink to the cubic.

In order to show that the congruence is a fibration, we have to show that there exists a unique chord through each point in space. Therefore, we insert the components of (6.17) into a skew symmetric 4×4-matrix and obtain the incidence condition of a point $X = \mathbf{x}\mathbb{F}$ and a line in the congruence

$$\begin{pmatrix} 0 & t^2u^2 & -ut(t+u) & tu \\ -t^2u^2 & 0 & t^2+tu+u^2 & -t-u \\ tu(t+u) & -t^2-tu-u^2 & 0 & 1 \\ -tu & t+u & -1 & 0 \end{pmatrix} \begin{pmatrix} x_0 \\ x_1 \\ x_2 \\ x_3 \end{pmatrix} = \mathbf{0}, \tag{6.18}$$

which can be compared to that given in Section 10.1 (see Exercise 10.1.10). The equations for given homogeneous coordinates $x_0 : x_1 : x_2 : x_3$ of a point X have a unique solution for t and for u. These parameters correspond to two different points on the cubic whose join is the unique chord through X. ■

In the real projective space $\mathbb{P}^3(\mathbb{R})$, the system of equations given in (6.18) may also have pairs of complex conjugate solutions. These also correspond to real chords of the cubic.

6.6 The cubic of coincidence – points with coinciding images

In Descriptive Geometry, pairs of mappings are frequently used in order to create images of Euclidean 3-space that also allow us to reconstruct points in three-dimensional (Euclidean) space. A well-known and wide-spread example is the pair consisting of top view and front view. However, it is also possible to combine a perspective image with an auxiliary perspective top view.

Returning to the pair of top view and front view, we can describe the involved mappings in terms of Cartesian coordinates. In order to get the complete picture, we perform the projective closure of $\mathbb{R}^3$ and use homogeneous coordinates whenever necessary.

We assume that points $X \in \mathbb{R}^3$ have the coordinates $\mathbf{x} = (x, y, z)$. The top view $X^1 = (x, y, 0)$ of X is the image of X under the orthogonal projection $\pi_1 : \mathbb{R}^3 \to \mathbb{R}^2$ onto the plane $\sigma_1 : z = 0$, *i.e.*, the projection of X from the ideal point Z_u of the z-axis of the initially chosen Cartesian frame. By the same token, the front view $X^2 = (0, y, z)$ of X is obtained as the orthogonal projection $\pi_2 : \mathbb{R}^3 \to \mathbb{R}^2$ onto the plane $\sigma : x = 0$, which is nothing but the projection of X from the ideal point X_u of the x-axis from the projective point of view.

The two views defined so far (cf. Figure 6.20, left) would be useful if they were drawn in one plane. Usually, the images showing up in σ_1 and σ_2 are mapped to one plane via the two collineations:

$$X^1 \mapsto \iota_1(X^1) := X' = (y, -x) \quad \text{and} \quad X^2 \mapsto \iota_2(X^2) := X'' = (y, z)$$

as illustrated in Figure 6.20, described in a Cartesian coordinate system in the image plane (drawing plane) such that the first axis equals the y'-axis ($= y''$-axis) and the second axis coincides with the z''-axis ($= x'$-axis).

We can easily see that X' and X'' are points with equal first coordinates. Therefore, the top view and the front view of any point X are aligned on a vertical line, the *ordering line*. Now, it is a fairly simple task to find those points $X \in \mathbb{R}^3$ that have coinciding top and front views. Since the first coordinate of X' and X'' agree in any case, only the second coordinates have to match, which yields

$$\kappa : \; x + z = 0.$$

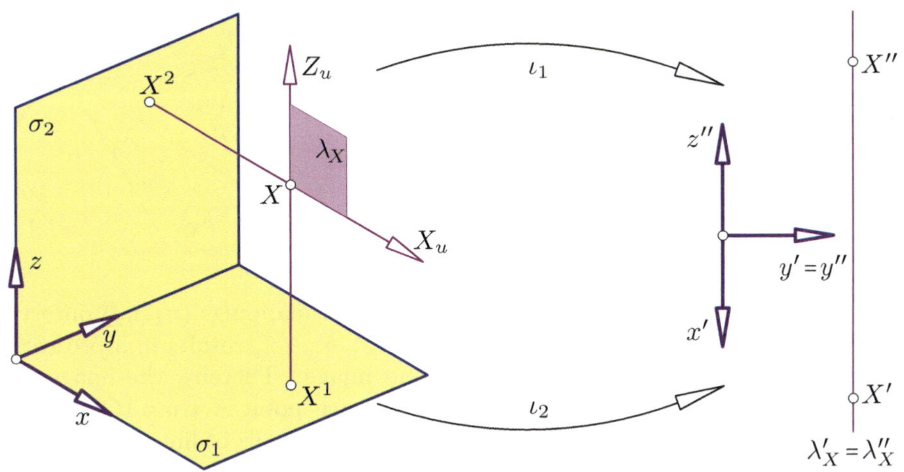

FIGURE 6.20. Top and front view are orthogonal projections in orthogonal directions (left). The situation on the screen is obtained by applying two regular collinear mappings ι_1, ι_2 to the results of the projection. This arranges the two views in a useful way (right).

The set κ is called the *plane of coincidence.*

The simple case of top and front view is very special for two reasons: Firstly, the projections π_1 and π_2 are orthogonal projections, and in addition, they project to a pair of orthogonal planes. Secondly, the projective closure of Euclidean 3-space makes clear that either center of the projections is contained in the other image plane, which implies $\pi_1 \circ \pi_2 = \pi_2 \circ \pi_1$.

We shall have a look at a more general setting. Assume that the image planes σ_1 and σ_2 are two different planes in a projective 3-space $\mathbb{P}^3$. Further, let the centers of the projections $\pi_i : \mathbb{P}^3 \to \sigma_i$ be two points O_1 and O_2 such that $o = [O_1, O_2] \notin \sigma_i$. The projections π_i are followed by two (regular) collinear transformations $\iota_i : \sigma_i \to \pi_i$ mapping each image to the final image plane π (which is comparable to a sheet of paper or the screen), so

$$X' = \iota_1 \circ \pi_1(X) \quad \text{and} \quad X'' = \iota_2 \circ \pi_2(X).$$

Then, the points $X \in \mathbb{P}^3$ with $X' = X''$ are the fixed points of the collineation

$$\beta := \iota_2 \circ \varphi_2 \circ \pi_1^{-1} \circ \iota_1^{-1} : \sigma_1 \to \sigma_2,$$

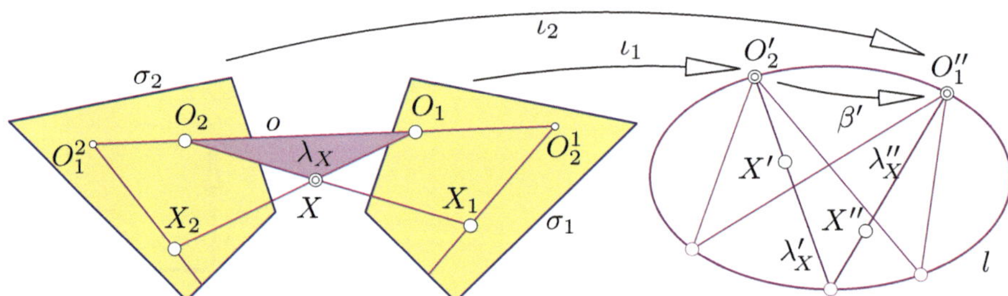

FIGURE 6.21. The most general form of a pair of mappings (π_1, π_2) onto two planes σ_1, σ_2 followed by collinear mappings $\iota_i : \sigma_i \to \pi_i$ results in a projective correspondence between the first and second image. Thereby, the images λ_X' and λ_X'' of the doubly projecting plane λ_X of each point $X \notin o = [O_1, O_2]$ are related in a projectivity $\beta : O_2' \to O_1''$ between two pencils of lines.

i.e., the locus of intersections of lines joining the two centers O_i with points in σ_i related in a collineation $\beta : \sigma_1 \to \sigma_2$. According to Theorem 6.4.1, the points X with $X' = X''$ form a cubic space curve c. This cubic curve is the *cubic of coincidence*, all of whose points have coinciding first and second image.

Figure 6.21 shall give an idea of what the general setting may look like. After the images in the planes σ_i have been transferred via the collinear mappings ι_i to the image plane π, the projective mapping $\beta' : O_2' \to O_1''$ joins the pencils of ordering lines. Each of these lines is the image of a doubly projecting plane $\lambda_X = [O_1, O_2, X]$ through the points X. In general, the projective mapping β' between the pencils of ordering lines generates a conic l (see [46, p. 218 ff.]).

In the case of the pair (top view, front view) as illustrated in Figure 6.20, the centers of the pencils of ordering lines become one, *i.e.*, $O_2' = O_1''$ and β' is the identity mapping in the pencil.

6.7 Projection with the chords of a cubic

The chords of a cubic c can be used to project the points of the projective 3-space $[c]$ spanned by c onto a plane π. This results in interesting relations between the points and lines in space and conics in the image plane π. We follow an idea by H. BRAUNER[1] (see [16, 17]).

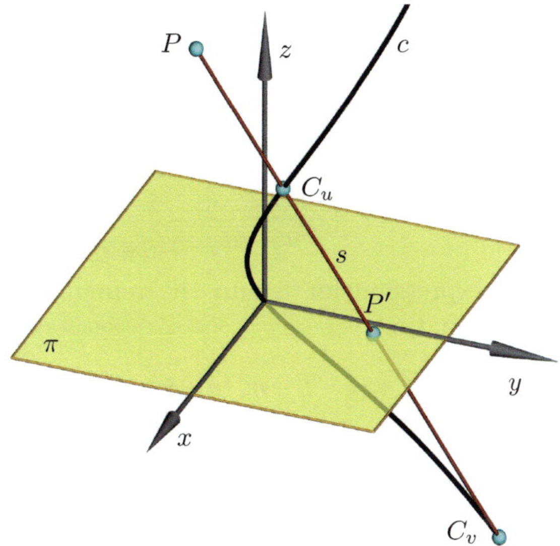

FIGURE 6.22. The unique chord s of c through $P \notin c$ meets the image plane π in the point P'.

Although any cubic in 3-space $\mathbb{P}^3(\mathbb{F})$ would serve as the center of the projection with the chords, we choose a special curve. Figure 6.22 shows the action of the projection with the chords of a cubic circle. Therefore, we assume that c is parametrized by

$$\mathbf{c}(t) = (1 + t^2 : t : t^2 : t + t^3), \quad t \in \mathbb{R}. \tag{6.19}$$

This cubic passes through the absolute points

$$I_1 = (0 : 1 : \mathrm{i} : 0) \quad \text{and} \quad I_2 = (0 : 1 : -\mathrm{i} : 0)$$

[1] HEINRICH BRAUNER (1928–1990) was an Austrian mathematician and geometer who worked on various fields of geometry, such as Line Geometry, the theory of linear mappings, and Differential Geometry.

of the Euclidean plane $z = 0$ and through the point $B_0 = (1:0:0:0)$. The corresponding parameter values are $\mathbf{c}(0) = B_0$ and $\mathbf{c}(\pm\mathrm{i}) = I_{1,2}$.

Let us now map a point P with Cartesian coordinates $\mathbf{p} = (\xi, \eta, \zeta)$, or in terms of homogeneous coordinates $(1:\xi:\eta:\zeta)$, to the corresponding image point $P' \in \pi$, see Figure 6.22. For that purpose, we are looking for chords of c that pass through P. Let $u \neq v$ be two different parameter values determining two different points C_u and C_v on c. The chord $s = [C_u, C_v]$ contains the point P if, and only if, the two vectors $\mathbf{c}(u) - \mathbf{p}$ and $\mathbf{c}(v) - \mathbf{p}$ are linearly dependent, *i.e.*, their cross product equals the zero vector. This yields three conditions on u, v, and ξ, η, ζ, of which the third can be solved for v and yields

$$v = \frac{u\eta + \xi - \zeta}{u\xi - \eta}. \tag{6.20}$$

Plugging the latter expression for v into the remaining two conditions yields the quadratic equation

$$(\xi^2 + \eta^2 - \eta)u^2 + (\zeta - \xi - \eta\zeta)u + \xi^2 + \eta^2 - \xi\zeta = 0. \tag{6.21}$$

This makes clear that for each point P with coordinates $\mathbf{p} = (\xi, \eta, \zeta)$, there are two parameter values u_1, u_2 fixing those points S_1 and S_2 on the cubic such that $[S_1, S_2] = s$ is the chord through P. In any case, the chord is real: If both u_1 and u_2 are real, then the points on the cubic are real and so is their join s. If u_1 and u_2 are not real, then they form a complex conjugate pair, and so do S_1 and S_2. Hence, $s = [S_1, S_2]$ is real. The case of a real double solution makes s a tangent of c.

The point of intersection of the chord of c with the image plane $z = 0$ has the coordinates

$$\mathbf{s}(u, v) = \left(\frac{uv(u+v)}{(u^2+1)(v^2+1)}, \frac{uv(uv-1)}{(u^2+1)(v^2+1)}, 0 \right). \tag{6.22}$$

In order to find the image point of P, we eliminate u and v with (6.20) and (6.21) from (6.22) and find

$$P' = \left(\frac{(\xi^2 + \eta^2 - \xi\zeta)(\eta\zeta + \xi - \zeta)}{(\eta - \xi\zeta)^2 + (\xi - \zeta + \eta\zeta)^2}, \frac{(\xi^2 + \eta^2 - \xi\zeta)(\eta - \xi\zeta)}{(\eta - \xi\zeta)^2 + (\xi - \zeta + \eta\zeta)^2} \right), \tag{6.23}$$

neglecting the third coordinate, which is zero anyway. The mapping $P \mapsto P'$ is quartic in the coordinates of P. A subsequent quadratic Cremona transformation (cf. [46, p. 329 ff.]) allows us to simplify the coordinate representation of the field of image points. We use the standard inversion in the Euclidean unit circle S^1. Therefore, (6.23) can be simplified to

$$P'' = \left(\frac{\eta\zeta + \xi - \zeta}{\xi^2 + \eta^2 - \xi\zeta}, \frac{\eta - \xi\zeta}{\xi^2 + \eta^2 - \xi\zeta} \right). \tag{6.24}$$

We see that the mapping $P \mapsto P''$ is quadratic in the coordinates of the points in 3-space.

Now, we shall apply the mapping with the cubic's chords to a straight line $l \subset \mathbb{P}^3$. We assume that $(l_1, l_2, l_3; l_4, l_5, l_6)$ are the Plücker coordinates (cf. Section 10.1, especially (10.1) and (10.5)) of the line l. An affine parametrization of l involving the Plücker coordinates reads

$$\mathbf{l}(t) = \begin{pmatrix} -l_5 l_3^{-1} \\ l_4 l_3^{-1} \\ 0 \end{pmatrix} + t \begin{pmatrix} l_1 \\ l_2 \\ l_3 \end{pmatrix}, \quad t \in \mathbb{R}.$$

Inserting this into (6.24) yields a parametrization of a planar curve l''. The latter is a conic, since its homogenized coordinate representation consists of three non-proportional quadratic polynomials. The elimination of the parameter t results in an equation of the image curve l'' of l. Using the relation (10.2), which has to be fulfilled for Plücker coordinates of a line, the equation of l'' reads

$$l'' : \; l_6 x^2 + (l_4 + l_6) y^2 + l_5 xy + (l_5 - l_2) x + (l_3 + l_4 - l_1) y + l_3 = 0. \tag{6.25}$$

The coefficients of the conic's equation are linear in the Plücker coordinates of l. We can summarize this as follows:

Theorem 6.7.1 *The composition of the projection with the chords of a cubic circle c and the inversion in the Euclidean unit circle results in a linear mapping from the manifold of lines in 3-space onto the manifold of conics in a plane.*

Figure 6.23 shows the action of the mapping $P \mapsto P''$. The image of a straight line l is a rational and planar quartic curve l'. The inversion in the unit circle S^1 yields a conic l''.

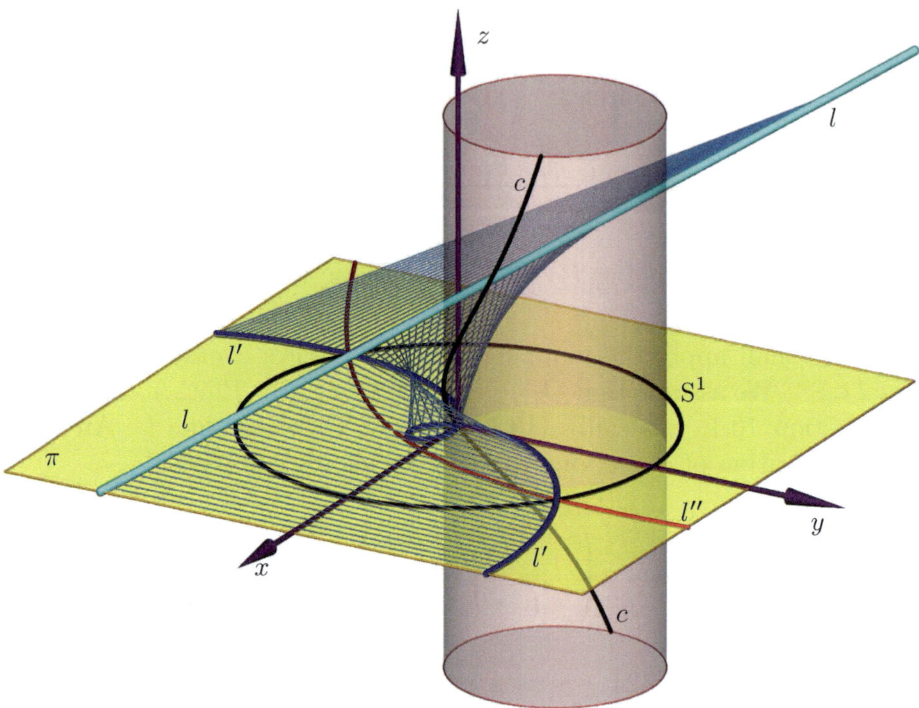

FIGURE 6.23. The image of a line l is a quartic curve $l' \subset \pi$ which is inverse to a conic l'' w.r.t. the unit circle S^1.

There are various other approaches to mappings from the manifold of lines in a projective 3-space to points in a plane or points in a 3-space, see, *e.g.*, [51, 52, 92]. These mappings are related with kinematic mappings and admit constructive realizations in the classical sense.

6.8 Cross ratios and projective automorphisms

The cross ratio is a value that can be assigned to ordered quadruples of points on a line, lines though a point, or even to four ordered points on a conic. In [46, p. 202, Equation (5.6)], we have seen that the cross ratio can easily be computed from the homogeneous coordinates of points or lines, and extending the definition to complex numbers, we can characterize concyclic points in the Euclidean plane (cf. [46, p. 209, Example 5.4.1]).

We can define cross ratios on cubic (space curves) as well. We recall Corollary 6.2.5. There, we used a stereographic projection from the cubic to a conic and, further, a stereographic projection from the conic to a straight line. In this way, we established a one-to-one and onto mapping between a straight line and cubic space curve. Thus, we can say:

Theorem 6.8.1 *A non-degenerate cubic space curve $c \subset \mathbb{P}^3(\mathbb{F})$ is projectively equivalent to a straight line and the group of projective automorphisms on the cubic is acting sharply transitive on pairs of ordered triplets of points, provided $\mathbb{F}$ is commutative and* $\operatorname{char}\mathbb{F} \neq 2, 3$.

Proof: The projective automorphisms on c are the projective automorphisms of a straight line lifted via stereographic projection to c. ■

Using a parametric representation of a cubic space curve c as given in (6.2), the homogeneous parameter $t_0 : t_1 \neq 0 : 0$ can be viewed as homogeneous coordinates on c. Consequently, the projective automorphisms on c can be written as linear mappings $\mathbb{F}^2 \to \mathbb{F}^2$ with regular transformation matrix $\mathbf{A} \in \mathbb{F}^{2\times 2}$. In terms of an affine parameter $t = t_1 t_0^{-1}$, the projective automorphisms are the linear rational transformations

$$t \to \frac{a_{00} + a_{01}t}{a_{10} + a_{11}t}.$$

Obviously, the projective mappings $c \to c$ can be extended to the ambient projective space.

In the same way, the definition of the cross ratio as given initially for collinear points carries over to cubic space curves. For four points $P_i \in c$ from (6.2) corresponding to the respective four affine parameters t_i with $i \in \{1, 2, 3, 4\}$, the cross ratio is simply

$$\operatorname{cr}(P_1, P_2, P_3, P_4) = \frac{(t_1 - t_3)(t_2 - t_4)}{(t_1 - t_4)(t_2 - t_3)}.$$

• **Exercise 6.8.1** Cross ratio of tangents of a cubic.

Assume P_i are four different points on the cubic (6.2) defined by four different affine parameters $t_i \in \mathbb{F} \cup \{\infty\}$ with $i \in \{1,2,3,4\}$. Assume that $\operatorname{cr}(P_1, P_2, P_3, P_4) = \delta$. Now, compute the Plücker coordinates of the four tangents T_i at P_i to c with (10.5) and (10.6) and show that the classical definition of the cross ratio of four straight lines in space,

$$\Delta := \operatorname{cr}(T_1, T_2, T_3, T_4) = \frac{\Omega(P_1, P_3)\Omega(P_2, P_4)}{\Omega(P_1, P_4)\Omega(P_2, P_3)},$$

yields $\Delta = \delta^4$, where $\Omega(P_i, P_j)$ equals the value of the bilinear form Ω from (10.10) on P_i and P_j.

• **Exercise 6.8.2** Cross ratio of chords of a cubic.

Assume that $C_1, \ldots, C_4$ are four chords of the cubic (6.2). Let each chord C_i be defined by two different points S_i and T_i determined by the affine parameters s_i and t_i (with $i \in \{1,2,3,4\}$.

Use the formula for the cross ratio of four lines given in Exercise 6.8.1 and show that the cross ratio of the four chords is related via

$$\operatorname{cr}(C_1, C_2, C_3, C_4) = \operatorname{cr}(S_1, S_2, S_3, S_4)\operatorname{cr}(T_1, T_2, T_3, T_4)\operatorname{cr}(S_1, S_2, T_3, T_4)\operatorname{cr}(T_1, T_2, S_3, S_4).$$

to the cross ratios between the points $S_1, \ldots, S_4$ and $T_1, \ldots, T_4$.

Discuss the case if one, two, or three chords are replaced by tangents of the cubic.

6.9 Quartic space curves

We have seen that cubic space curves can be obtained as the intersection of two quadrics $\mathcal{A}$ and $\mathcal{B}$, *i.e.*, as the intersection of two algebraic surfaces of degree two if they share a straight line l (without touching along l). In general, two quadrics will not share a straight line, and thus, the intersection will not split into lower degree parts. An algebraic space curve of degree four is defined as a curve that has four common points with a generic plane. Clearly, the multiplicities have to be taken into account, and, in general, an algebraic closure has to be performed. According to Bézout's theorem, a quartic curve meets a generic quadric in eight points. If a quadric contains nine points of a quartic space curve, then it carries the entire curve.

There are two types of quartic space curves. If the quartic curve is the intersection of two quadrics, then it is called a quartic space curve of the *first kind*. In this case, there exists a pencil of quadrics through the quartic curve.

If there exists only one quadric through the quartic curve, then we shall call it a quartic space curve of the *second kind*.

The next sections are dedicated to the study of quartic space curves of the first kind. Later, we shall add some material dealing with quartics of the second kind.

Singular quadrics through a quartic space curve

Let $\mathbf{A}$ and $\mathbf{B}$ be two symmetric 4×4-matrices with entries from an arbitrary commutative field $\mathbb{F}$ with $\operatorname{char}\mathbb{F} \neq 2, 3$ and let further $\mathbf{x} = (x_0, x_1, x_2, x_3)$ be homogeneous coordinates of points in $\mathbb{P}^3(\mathbb{F})$.

Then, we assume that $\mathcal{A}:\ \mathbf{x}^{\mathrm{T}}\mathbf{A}\mathbf{x} = 0$ and $\mathcal{B}:\ \mathbf{x}^{\mathrm{T}}\mathbf{B}\mathbf{x} = 0$ are two quadrics which do not share a straight line. The quadrics in the pencil of quadrics defined by $\mathcal{A}$ and $\mathcal{B}$ have the equations

$$\mathbf{x}^{\mathrm{T}}(\alpha\mathbf{A} + \beta\mathbf{B})\,\mathbf{x} = 0, \quad \alpha : \beta \neq 0 : 0. \tag{6.26}$$

The curve $q = \mathcal{A} \cap \mathcal{B}$ is a quartic space curve of the first kind. Since the coordinates of the points of q satisfy both equations $\mathbf{x}^{\mathrm{T}}\mathbf{A}\mathbf{x} = 0$ and $\mathbf{x}^{\mathrm{T}}\mathbf{B}\mathbf{x} = 0$, each linear combination of these is annihilated by the coordinates of all points on q. Therefore, we have

Lemma 6.9.1 *A quartic space curve of the first kind that is the intersection of two quadrics $\mathcal{A}$ and $\mathcal{B}$ lies on each quadric in the pencil spanned by $\mathcal{A}$ and $\mathcal{B}$.*

From Chapter 5, we know that a pencil of quadrics in $\mathbb{P}^3(\mathbb{F})$ contains at most four singular quadrics. Thus, we can state

Lemma 6.9.2 *Through a quartic space curve $q \subset \mathbb{P}^3$ of the first kind, there exist at most four singular quadrics.*

The upper bound of singular quadrics through a quartic space curve given in Lemma 6.9.2 can, indeed, be achieved, as illustrated for an affine special case in Figure 6.24.

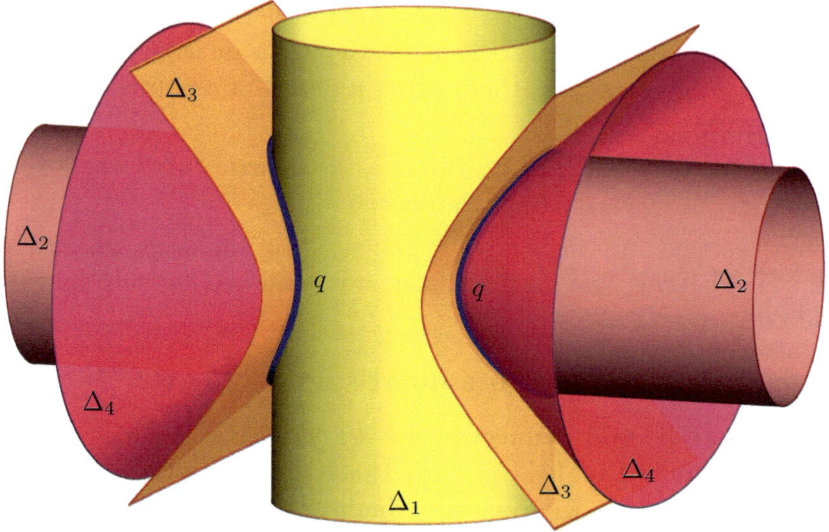

FIGURE 6.24. Through a quartic q of the first kind, we can find up to four real singular quadrics. Here, q lies on two cylinders Δ_1, Δ_2 of revolution, on a hyperbolic cylinder Δ_3, and on a quadratic cone of revolution Δ_4.

Quartic space curves of the first kind

In the following, we assume that two of the carrier quadrics, say $\mathcal{A}$ and $\mathcal{B}$, are symmetric w.r.t. to a plane σ. Consequently, the quartic space curve $q = \mathcal{A} \cap \mathcal{B}$ in question is also symmetric w.r.t. σ.

We use Cartesian coordinates and let σ be the $[x,z]$-plane. Then, $\mathcal{A}$ and $\mathcal{B}$ have the inhomogeneous equations

$$\begin{aligned}\mathcal{A}:\ \mathbf{x}^{\mathrm{T}}\begin{pmatrix} a_{11} & 0 & a_{13}\\ 0 & a_{22} & 0\\ a_{13} & 0 & a_{33}\end{pmatrix}\mathbf{x}+2(a_{01},0,a_{03})^{\mathrm{T}}\mathbf{x}+a_{00}=0,\\ \mathcal{B}:\ \mathbf{x}^{\mathrm{T}}\begin{pmatrix} b_{11} & 0 & b_{13}\\ 0 & b_{22} & 0\\ b_{13} & 0 & b_{33}\end{pmatrix}\mathbf{x}+2(b_{01},0,b_{03})^{\mathrm{T}}\mathbf{x}+b_{00}=0\end{aligned}\tag{6.27}$$

with $\mathbf{x}=(x,y,z)^{\mathrm{T}}$, where terms linear in y do not show up due to the quadrics' symmetry w.r.t. σ. These equations can be simplified further by replacing $\mathcal{A}$ and $\mathcal{B}$ with singular quadrics in the pencil. To that end, we return to the homogeneous equations of the quadrics in the pencil and the coefficient matrices read

$$\mathbf{P}(\alpha,\beta)=\begin{pmatrix}\alpha a_{00}+\beta b_{00} & \alpha a_{01}+\beta b_{01} & 0 & \alpha a_{03}+\beta b_{03}\\ \alpha a_{10}+\beta b_{10} & \alpha a_{11}+\beta b_{11} & 0 & \alpha a_{13}+\beta b_{13}\\ 0 & 0 & \alpha a_{22}+\beta b_{22} & 0\\ \alpha a_{30}+\beta b_{30} & \alpha a_{31}+\beta b_{31} & 0 & \alpha a_{33}+\beta b_{33}\end{pmatrix},\tag{6.28}$$

where $\alpha:\beta\neq 0:0$. The matrix $\mathbf{P}$ in (6.28) is singular if, and only if, $\det\mathbf{P}=0$, which is the case especially for

$$\alpha:\beta=-b_{22}:a_{22}.\tag{6.29}$$

The homogeneous parameter (6.29) determines a cylinder Δ_1 orthogonal to σ with an equation of the form

$$\Delta_1:\ c_{11}x^2+2c_{12}xz+c_{22}z^2+2c_{01}x+2c_{02}z+c_{00}=0.\tag{6.30}$$

It seems obvious to assume that this cylinder does not degenerate any further in order to obtain a non-degenerate quadric.

The remaining singular quadrics in the pencil correspond to the zeros of the cubic form arising as the determinant of the 3×3-matrix that is obtained from (6.28) by removing the third column and the third row.

A cubic form always has at least one real root. Let the singular quadric Δ_2, which corresponds to a real root be given by

$$\Delta_2:\ y^2=(x,z)^{\mathrm{T}}\mathbf{R}^{\star}\begin{pmatrix}x\\ z\end{pmatrix}+2(x,z)^{\mathrm{T}}\mathbf{x}+r_{00}\tag{6.31}$$

with

$$\mathbf{R}^\star = \begin{pmatrix} r_{11} & r_{12} \\ r_{21} & r_{22} \end{pmatrix}, \quad \mathbf{r} = \begin{pmatrix} r_{01} \\ r_{02} \end{pmatrix}, \quad \text{and} \quad \mathbf{R} = \left(\begin{array}{c|c} r_{00} & \mathbf{r}^{\mathrm{T}} \\ \hline \mathbf{r} & \mathbf{R}^\star \end{array}\right).$$

For the singularity of Δ_2, it is necessary and sufficient that

$$\det \mathbf{R} = 0. \tag{6.32}$$

As shown in Chapter 3, it is a matter of elementary computations to show that the equation admits further simplifications without changing (6.30). These simplifications are coordinate transformations aiming at a translation that removes the linear terms in x and z from (6.31). There are three cases to be distinguished:

(1) It is possible to translate the Cartesian coordinate system such that the coefficients of x and z vanish. In this case, $\det \mathbf{R}^\star \neq 0$ has to hold and (6.31) is the equation of a quadratic cone with vertex

$$\mathbf{v}_2 = -\mathbf{R}^{\star^{-1}} \mathbf{r}. \tag{6.33}$$

Note that $\mathbf{v}_2 = (v_x, v_z)$ are local coordinates in the $[x, z]$-plane. In the global coordinate system, the cone's vertex has the coordinate vector $(v_x, 0, v_z)$. Choosing the latter as the origin of the coordinate system transforms (6.31) into

$$\Delta_2 : \; y^2 = s_{11}x^2 + 2s_{12}xz + s_{22}z^2 \quad \text{with} \quad s_{11}s_{22} - s_{12}^2 \neq 0.$$

(2) If $\mathbf{R}^\star$ is not regular, *i.e.*, $\det \mathbf{R}^\star = 0$, then there is no such translation. Without loss of generality, we can assume that at least $r_{11} \neq 0$ in (6.31). From $\det \mathbf{R}^\star = r_{11}r_{22} = r_{12}^2 = 0$ and with $\det \mathbf{R} = 0$, we find

$$2r_{01}r_{02}r_{12} - r_{01}^2 r_{22} - r_{02}^2 r_{11} = 0. \tag{6.34}$$

Therefore, we can immediately infer

$$r_{01}(r_{02}r_{12} - r_{01}r_{22}) = r_{02}(r_{02}r_{11} - r_{01}r_{12}),$$

which is equivalent to

$$r_{01} \det \begin{pmatrix} r_{12} & r_{01} \\ r_{22} & r_{02} \end{pmatrix} = r_{02} \det \begin{pmatrix} r_{11} & r_{01} \\ r_{12} & r_{02} \end{pmatrix}. \tag{6.35}$$

Multiplying (6.34) with r_{11}, we find

$$r_{02}^2 r_{11}^2 - 2r_{01}r_{02}r_{11}r_{12} + r_{01}^2 \underbrace{r_{12}^2}_{r_{11}r_{22}} = \left(\det\begin{pmatrix} r_{12} & r_{01} \\ r_{22} & r_{02} \end{pmatrix}\right)^2 .$$

By virtue of (6.35), the latter determinant is equal to zero, and thus, the vertices (6.33) trace the line

$$y = 0, \quad r_{11}x + r_{12}z + r_{02} = 0.$$

Hence, the singular quadric Δ_2 is either an elliptic or a hyperbolic cylinder with the equation

$$\Delta_2: \; y^2 = s_{11}x^2 + 2s_{12}xz + s_{22}z^2 + s_{00},$$

depending on whether the right-hand side describes an ellipse or a hyperbola in the $[x, z]$-plane.

(3) Finally, we have to discuss the case $\det \mathbf{R}^\star = 0$ with $r_{11} = 0$. Consequently, $r_{12} = 0$, and from (6.35), it follows that $r_{01}^2 r_{22} = 0$.

In the first subcase, $r_{22} \neq 0$, we obtain $r_{01} = 0$. Then, (6.31) has the equation $y^2 = r_{22}z^2 + 2r_{02}z + r_{00}$, which can be simplified by applying a translation. Its equation then reads $\Delta_2: \; y^2 = s_{22}z^2 + s_{00}$, which describes an elliptic or a hyperbolic cylinder and coincides with a previous case.

So, we have to consider the second subcase where $r_{22} = 0$. This implies $y^2 = 2r_{01}x + 2r_{02}z + r_{00}$, where r_{01} and r_{02} cannot vanish simultaneously. Again, a suitable translation yields a simplified equation of Δ_2, and we have

$$\Delta_2: \; y^2 = 2s_{01}x + 2s_{02}z,$$

which is the equation of a parabolic cylinder.

We summarize our results in

Theorem 6.9.1 *The equations of the quartic space curves of the first kind symmetric w.r.t. the $[x, z]$-plane of a Cartesian coordinate system can be transformed into three canonical forms. Let $\mathbf{x} = (x, z)^{\mathrm{T}}$ and $\mathbf{A}, \mathbf{B} \in \mathbb{R}^{2\times 2}$. Then, the normal forms are:*

$$\begin{array}{lll} 1. & \mathbf{x}^{\mathrm{T}}\mathbf{A}\mathbf{x} = 0, \; y^2 = \mathbf{x}^{\mathrm{T}}\mathbf{B}\mathbf{x} & \text{with } \det\mathbf{A} \neq 0, \; \det\mathbf{B} \neq 0, \\ 2. & \mathbf{x}^{\mathrm{T}}\mathbf{A}\mathbf{x} = 0, \; y^2 = \mathbf{x}^{\mathrm{T}}\mathbf{B}\mathbf{x} + b_{00} & \text{with } \det\mathbf{A} = 0, \; \det\mathbf{B} \neq 0, \\ 3. & \mathbf{x}^{\mathrm{T}}\mathbf{A}\mathbf{x} = 0, \; y^2 = 2b_{01}x + 2b_{02}z & \text{with } \det\mathbf{A} = 0, \; b_{01}^2 + b_{02}^2 \neq 0. \end{array} \tag{6.36}$$

Figure 6.25 illustrates the first type of quartic space curve of the first kind as listed in Theorem 6.9.1. In $\mathbf{A} = (a_{ij})_{i,j\in\{0,1,2\}}$, we have chosen $a_{11} = 4$, $a_{22} = -a_{00} = 1$, $a_{01} = a_{02} = a_{12} = 0$, and in $\mathbf{B} = (b_{ij})_{i,j\in\{1,2\}}$, we have set $b_{11} = 1$, $b_{22} = 9$, and $b_{12} = 0$. The quartic $q = \Delta_1 \cap \Delta_2$ consists of two separate parts symmetric w.r.t. $y = 0$.

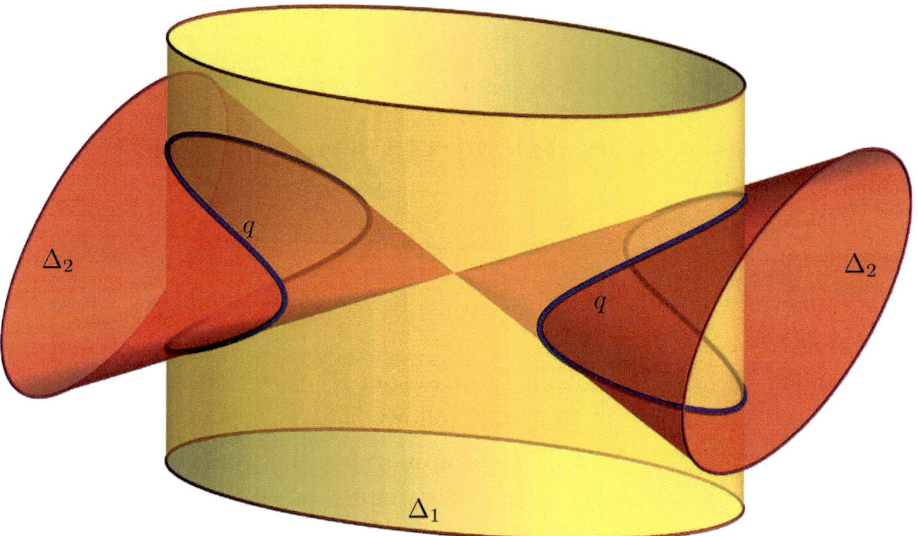

FIGURE 6.25. The first type of a quartic space curve q of the first kind according to Theorem 6.9.1.

Additionally to Figure 6.25, we show the three principal views (top, front, and right side view) of the quartic in Figure 6.26, the orthogonal projections onto three mutually orthogonal planes. These planes are the planes of symmetry of q and coincide with the coordinate planes.

The curve q is (by assumption) symmetric w.r.t. the three coordinate planes. Thus, each point on one of the images of q is the image of two points on the space curve. Therefore, each projection of q is at the same the time image of a projecting cylinder through q. Consequently, the degree of all three images of q is reduced to two. In the front view (cf. Figure 6.26), we observe another phenomenon. According to the chosen values for a_{ij} and b_{ij}, the points on the curve q are determined by

$$q = \Delta_1 \cap \Delta_2, \quad \Delta_1:\ 4x^2 + y^2 = 1, \quad \Delta_2:\ x^2 + 9z^2 = y^2.$$

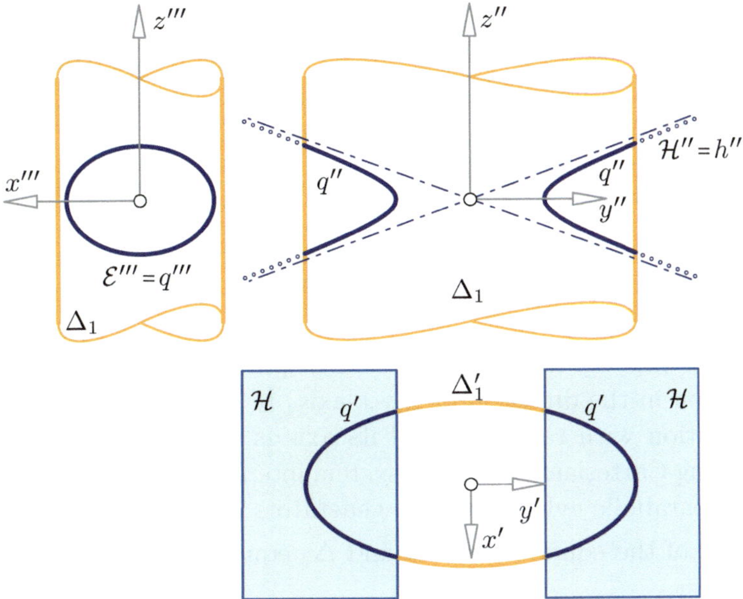

FIGURE 6.26. Top, front, and side view of the quartic q of the first kind shown in Figure 6.25. Since q is symmetric w.r.t. the coordinate planes, each point on the curve's image is the image of two points in space and either image of q coincides with the image of a (projecting) singular quadric through q. In the front view, the parasitic branches of the curve are shown (dotted parts).

We can eliminate x from these equations in order to find an equation of the front view. This yields

$$h: \ 5y^2 - 36z^2 = 1,$$

which is, of course, the equation of a hyperbola when interpreted as the equation of a planar curve in the $[y, z]$-plane. The equation of h can also be interpreted as the equation of a hyperbolic cylinder $\mathcal{H}$ in 3-space. $\mathcal{H}$ is one of the singular quadrics through q. Only that part of the hyperbola h that lies between the contour generators of the cylinder Δ_1 is in fact the front view of the quartic q. The points outside (dotted) are so-called *parasitic points*. These points lie on the hyperbola h that carries the front view q'' of q.

The right side view of q (cf. Figure 6.26) shows an ellipse q''' as the image of the quartic. This ellipse is also the image of the horizontal elliptic cylinder $\mathcal{E}: \ 5x^2 + 9z^2 = 1$, whose equation can be obtained from those of

Δ_1 and Δ_2 by eliminating y. With the cone Δ_2 and the cylinders Δ_1, $\mathcal{H}$, and $\mathcal{E}$, the list of singular quadrics through q is completed.

An example of the third type of a quartic space curve of the first kind (as classified in Theorem 6.9.1) is illustrated in Figure 6.27. We have chosen the singular quadrics according to the second type mentioned in Theorem 6.9.1 (cf. (6.36)) as

$$\Delta_1: \; x^2 + z^2 - x = 0, \quad \Delta_2: \; y^2 - 1 + x = 0.$$

Note that this agrees with the settings prior to Theorem 6.9.1 up to a translation in the direction of the x-axis. It is clear that Δ_1 is a cylinder of revolution with radius $\frac{1}{2}$, and its axis is parallel to the y-axis of the underlying Cartesian coordinate system intersecting the x-axis at $(\frac{1}{2}, 0, 0)$. Δ_2 is a parabolic cylinder whose generators are parallel to the z-axis.

The sum of the equations of Δ_1 and Δ_2 equals

$$\Sigma: \; x^2 + y^2 + z^2 = 1, \tag{6.37}$$

which is the equation of the Euclidean unit sphere. Since Σ is among the quadrics of the pencil spanned by Δ_1 and Δ_2, the quartic curve $q = \Delta_1 \cap \Delta_2$ (which is common to all quadrics in the pencil) also lies on Σ and is, therefore, a spherical curve. Up to a non-vanishing factor, the characteristic polynomial of the pencil reads

$$\alpha\beta(\alpha + \beta)^2.$$

The zeros $\alpha = 0$ and $\beta = 0$ correspond to the given cylinders Δ_2 and Δ_1. The zero $\alpha : \beta = -1 : 1$ with multiplicity two corresponds to the quadratic cone

$$\Gamma: \; (x - 1)^2 - y^2 + z^2 = 0,$$

whose equation is simply the difference of the equations of the given cylinders. Γ's vertex D has the Cartesian coordinates $\mathbf{d} = (1, 0, 0)$. At D, the cylinders Δ_1 and Δ_2 share the tangent plane $x = 1$, *i.e.*, the surfaces are in contact there. Therefore, $q = \Delta_1 \cap \Delta_2$ has a double point at D, as can be seen in Figure 6.27.

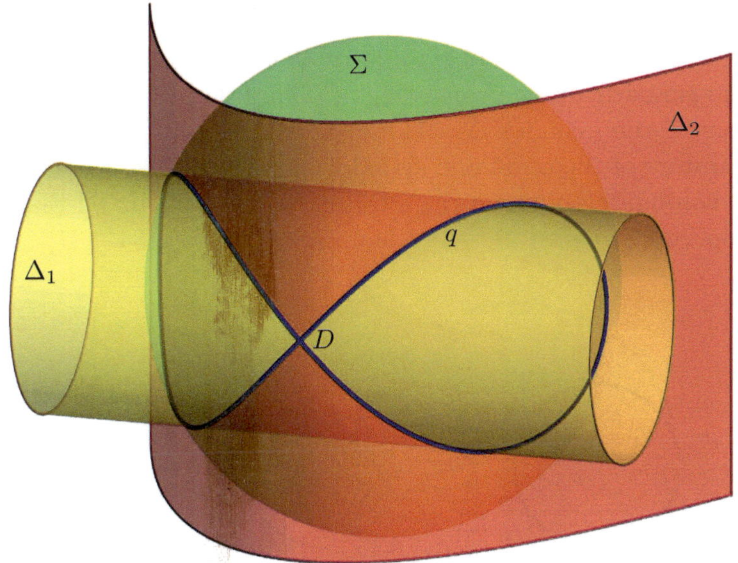

FIGURE 6.27. Viviani's curve is spherical, rational, and a quartic space curve of the first kind.

The curve q is known as *Viviani's curve* or *Viviani's window*.[2] Viviani's curve admits a rational parametrization: Inserting

$$x = \frac{1}{2}(1 + \cos\tau) \quad \text{and} \quad z = \frac{1}{2}\sin\tau$$

into any of the equations of Δ_1, Δ_2, or Σ, we find

$$y = \sqrt{\frac{1 - \cos\tau}{2}} = \cos\frac{\tau}{2}.$$

Then, we reparametrize by letting $t = \frac{\tau}{2}$, and Viviani's curve achieves the simple parametrization

$$\mathbf{q}(t) = (\cos^2 t, \sin t, \sin t\cos t) \quad t \in [0, 2\pi[. \tag{6.38}$$

A conversion into a rational parametrization can easily be done by replacing $\cos t = \frac{1-u^2}{1+u^2}$ and $\sin t = \frac{2u}{1+u^2}$.

[2] VINCENZO VIVIANI (1622–1703) was an Italian mathematician, a pupil of EVANGELISTA TORRICELLI, and a disciple of Galileo Galilei.

Figure 6.28 shows the three principal views, *i.e.*, the orthogonal projections onto the three planes of symmetry of Viviani's curve. The top view shows the parasitic branch consisting of points that are the images of real generators of the cylinder Δ_1. However, these points are the images of real generators of the parabolic cylinder and chords of q carrying pairs of complex conjugate points on the quartic. The eight-loop q'' that appears in the top view is called Gerono's lemniscate (cf. [41, 84, 137]).

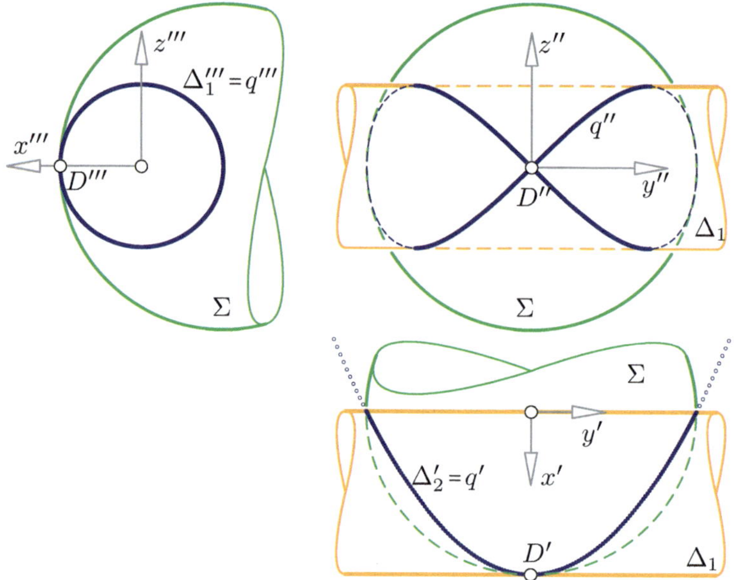

FIGURE 6.28. The three principal views of Viviani's curve. Top-left: The right side view q''' is a circle which is also the image of the cylinder Δ_1 of revolution. Top-right: Gerono's lemniscate (eight-loop). In the top view, we see points of the parabola q' outside the sphere $\Sigma \supset q$. These blue dots are parasitic points: They are the images of real generators of the parabolic cylinder Δ_2 that carry pairs of complex conjugate points of q.

■ **Example 6.9.1** A torus and a Möbius strip through Viviani's curve.

In [79] it is shown that there exists a unique torus through Viviani's curve (6.38) (cf. Figure 6.29, left). This is a special result on spherical quartic curves: In general such curves lie simultaneously on three pairs of tori (cf. [59]). The torus $\mathcal{T}$ has the absolute circle of Euclidean geometry for its double curve. The sphere Σ (given by (6.37)) also intersects the plane at infinity along the absolute circle. From the curve of intersection of $\mathcal{T}$ and Σ, which is of degree 8 according to Bézout's theorem, the absolute circle splits off with multiplicity two and a spherical quartic remains.

Show that in the case of the particular curve q given by (6.38), the torus $\mathcal{T}$ carrying q can be parametrized by

$$\mathbf{t}(u,v) = \left(\left(\frac{3}{4} + \frac{1}{2\cos u}\right)\cos v + \frac{3}{4}, \left(\frac{3}{4} + \frac{1}{2\cos u}\right)\sin v, \frac{1}{2}\sin u \right), \quad (u,v) \in [0,2\pi[^2,$$

which can also be described by the implicit equation

$$\mathcal{T}:\ (4(x^2+y^2+z^2) - 6x - 1)^2 - 9(1-4z^2) = 0.$$

Surprisingly, the cubic ruled surface $\mathcal{R}$ of the topological type of the Möbius strip also passes through Viviani's curve (6.38). Figure 6.29 (right) shows Viviani's curve q together with the sphere Σ and the cubic ruled surface $\mathcal{R}$. The ruled surface $\mathcal{R}$ can be parametrized by

$$\mathbf{r}(u,v) = \begin{pmatrix} -\cos 2u \\ 0 \\ \sin 2t \end{pmatrix} + v \begin{pmatrix} -\cos 2t \cos t \\ \sin t \\ \sin 2t \cos t \end{pmatrix}, \quad (u,v) \in [0,\pi[\times\mathbb{R}.$$

An implicit equation of the ruled surface reads:

$$\mathcal{R}:\ 2y(x^2+z^2) - (x^2+y^2+z^2)z - 2xy + z = 0.$$

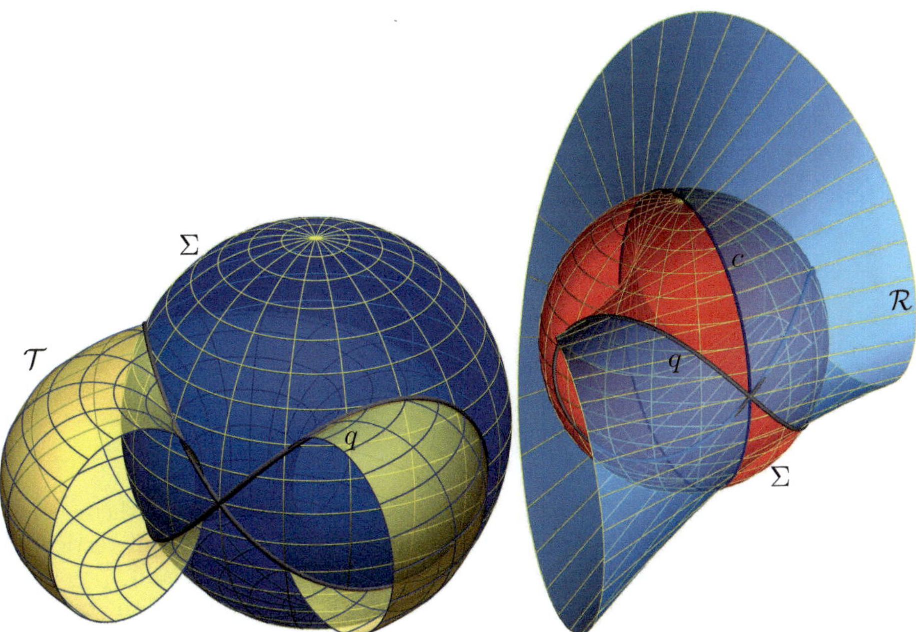

FIGURE 6.29. Left: $\mathcal{T}$ is a ring torus through Viviani's curve q. Right: The one-sided cubic ruled surface $\mathcal{R}$ intersects the sphere Σ along Viviani's curve q and a great circle c.

▪ **Example 6.9.2** Constructive approach to the intersection of two quadratic cones.

Figure 6.30 shows how to find the point of a quartic space curve q (of the first kind) that is the intersection of two quadratic cones Δ_1 and Δ_2. We use auxiliary planes ε through the line $v = [V_1, V_2]$ that joins the cones' vertices. Such a plane intersects either cone along two (possibly) real generators, say $a_1, a_2 \subset \Delta_1$ and $b_1, b_2 \subset \Delta_2$. Since these generators are coplanar, they intersect at the points Q_1, Q_2, Q_3, Q_4, which are common to both cones and belong to the quartic intersection curve.

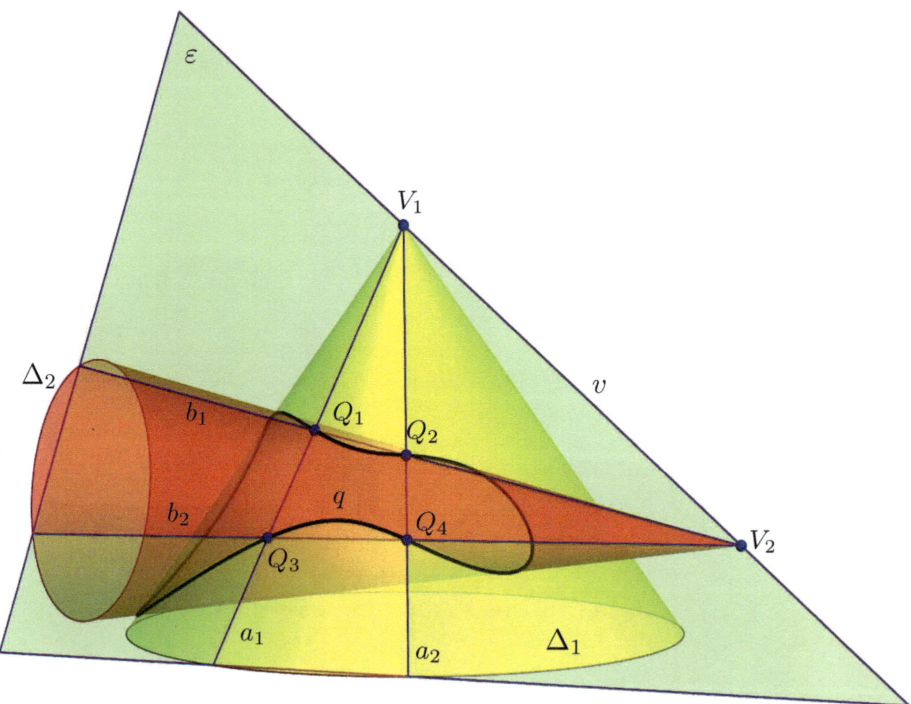

FIGURE 6.30. Constructive treatment of a quartic space curve q as the intersection of two quadratic cones Δ_1 and Δ_2: Each plane π through both vertices V_1 and V_2 meets both cones along (coplanar) generators $a_1, a_2 \subset \Delta_1$ and $b_1, b_2 \subset \Delta_2$. The generators intersect at the common points Q_1, Q_2, Q_3, Q_4 of the cones.

If one of the planes from the pencil about v is tangent to both cones, then q has a double point. When there are two common tangent planes of Δ_1 and Δ_2, the quartic has two double points and splits into two conics.

Degenerate quartic space curves of the first kind

In this section, we shall give a short overview on degenerate quartic space curves of the first kind. As we shall see, quartics frequently degenerate and fall into lower degree parts when two quadrics intersect.

We state:

Theorem 6.9.2 *If a quartic space curve q of the first kind degenerates, then it can only have one of the following shapes:*

(1) q is the union of a cubic space curve c and a straight line l where l is either a chord (meets c in two points) or l is a tangent of q.

(2) q is the union of two conics c_1 and c_2 sharing two points D_1 and D_2. (The two points can coincide, and c_1 and c_2 are then tangent to each other.)

(3) q is the union of a conic c and two straight lines l_1 and l_2 that meet in one point $L = l_1 \cap l_2$, but neither line lies in the plane of c.

(4) q is the union of four straight lines that form a skew quadrilateral.

Proof: Since quartic space curves of the first kind are the complete intersections of a pair of quadrics, we simply give pairs $(\mathcal{A}, \mathcal{B})$ of quadrics that intersect along degenerate quadrics.

(1) The hyperbolic paraboloid $\mathcal{A}: xy = z$ intersects the cylinder of revolution $\mathcal{B}: y^2 + z^2 = z$ along the cubic c with parametrization

$$\mathbf{c}(t) = \left(\frac{1+t}{1-t}, \frac{1-t^2}{2(1+t^2)}, \frac{(1+t)^2}{2(1+t^2)}\right) \quad \text{with} \quad t \in \mathbb{R} \setminus \{1\}$$

and the straight line l with $\mathbf{l}(t) = (t, 0, 0)$ (with $t \in \mathbb{R}$) since $\mathcal{A}$ and $\mathcal{B}$ do not touch each other along the common straight line l (cf. Figure 6.31, left).

The line l meets c at $(0,0,0)$ and at the ideal point $(0:1:0:0)$ of l, and is thus a chord.

Replacing $\mathcal{B}: y^2 + z^2 = z$ with the quadratic cone $\mathcal{C}: xz = y^2$ that emanates from $(0,0,0) \in c$, we have the same cubic component c. However, $\mathcal{A} \cap \mathcal{C} = c \cup m$ with the straight line m tangent to c. The tangent can be viewed as a limiting case of a chord (cf. Figure 6.31, right).

A degenerate quartic space curve $q = c \cup l$ of the first kind cannot be the union of a cubic space curve c and a straight line l that share only one point: Each of the quadrics through the (degenerate) quartic has to contain both components. According to Bézout's theorem, a cubic c has six points in common with a generic quadric $\mathcal{Q}$. In order to make c a cubic in $\mathcal{Q}$, it has to have at least seven points in common with $\mathcal{Q}$. For the straight line l, it is necessary to have three points of intersection with the quadric. Thus, the quadric $\mathcal{Q}$ has to carry ten points, which is, in general, impossible. For the same reason, a quartic space curve of the first kind cannot split into a cubic and a disjoint straight line.

(2) It is admissible to assume that the quadrics through q are singular. So, we let $\Delta_1: y^2 + z^2 = 1$ and $\Delta_2: x^2 + z^2 = 1$. The difference of the latter equations equals $\mathcal{P}: y^2 - z^2 = (y-z)(y+z) = 0$, which describes a pair of planes forming a further singular quadric in the pencil spanned by $\mathcal{A}$ and $\mathcal{B}$. Since all points of $q = \mathcal{A} \cap \mathcal{B}$ are also located in $\mathcal{P} = \varepsilon_1 \cup \varepsilon_2$, the quartic is the union of the two conics $c_1 = \varepsilon_1 \cap \Delta_1$ and $c_2 = \varepsilon_2 \cap \Delta_1$ (see Figure 6.32).

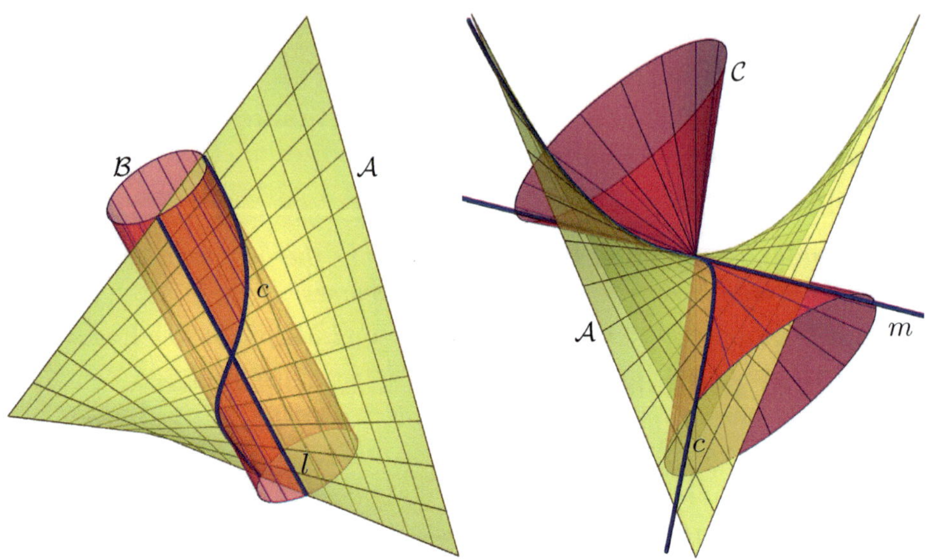

FIGURE 6.31. Degenerate quartic space curves of the first kind: a cubic c and a chord l (left), a cubic c with a tangent m (right) together with two quadrics spanning the pencil through q.

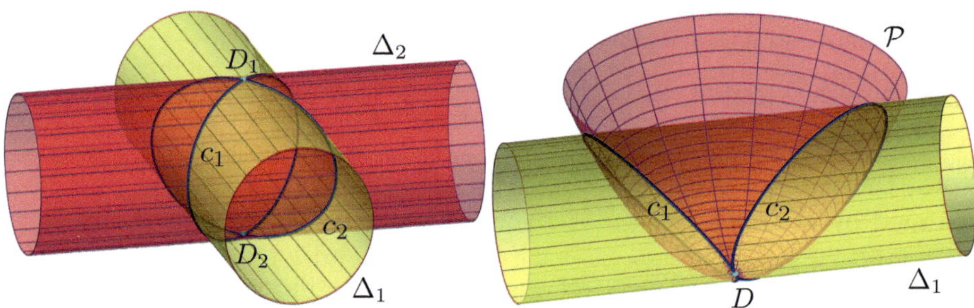

FIGURE 6.32. Quartic space curves of the first kind may also split into a pair of conics. Left: If a quartic of the first kind has two double points, then it is the union of two conics. Right: The two double points may coincide and the two quadratic branches of the quartic are in contact.

The case of a degenerate quartic space curve of the first kind that is the union of two conics touching each other once can be found as follows: Let $\mathcal{A}: x^2+y^2=2z$ and $\mathcal{B}: x^2+(z-1)^2=1$ span the pencil of quadrics. Then, $q=\mathcal{A}\cap\mathcal{B}$ splits into two planar components c_1 and c_2 located in the planes $x^2+y^2-2z-(x^2+(z-1)^2-1)=y^2-z^2=(y-z)(y+z)=0$. The intersection of $y=z$ and $y=-z$ is the x-axis of the underlying Cartesian frame which is tangent to $\mathcal{A}$ and $\mathcal{B}$, and thus, to q (note Figure 6.32).

(3) Now, we choose $\mathcal{A}: xy-z=0$ and $\mathcal{B}: xy-xz+yz-z=0$. Obviously, $xy-xz+yz-z-(xy-z)=yz-xz=(y-x)z$ is a singular quadric passing through $q=\mathcal{A}\cap\mathcal{B}$. Indeed, it is the only singular quadric in this pencil and consists of the two planes $z=0$ and $x=y$. Intersecting the first with $\mathcal{A}$, we find the straight lines $\mathbf{l}_1=(t,0,0)$ and $\mathbf{l}_2(t)=(0,t,0)$ with $t\in\mathbb{R}$ as the two straight parts of q. The plane $x=y$ meets $\mathcal{A}$ (and also $\mathcal{B}$) along the conic $\mathbf{c}=(t,t,t^2)$ (with $t\in\mathbb{R}$). In this case, $q=l_1\cap l_2\in c$ (cf. Figure 6.33, left).

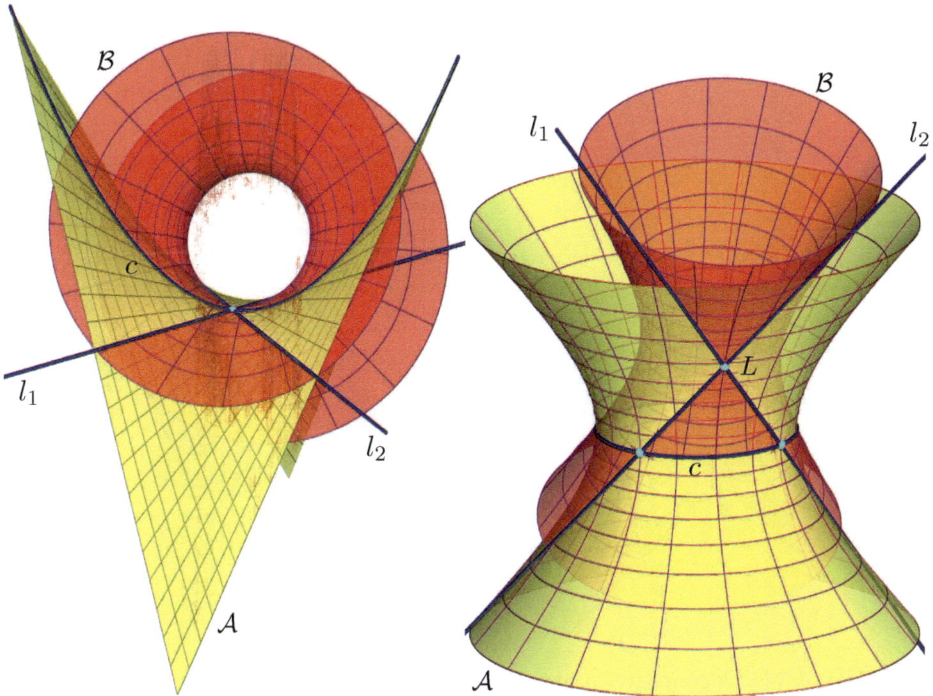

FIGURE 6.33. Left: The quadrics mentioned in the proof of Theorem 6.9.2 with the degenerate quartic $q=c\cup l_1\cup l_2$. Right: The quadrics given in Exercise 6.9.1 intersect in a degenerate quartic of the first kind. Here, the straight components l_1 and l_2 intersect in a point $L\notin c$.

It is possible to prescribe two quadrics as the carrier of a pencil such that $q=\mathcal{A}\cap\mathcal{B}$ is the union of a conic c and a pair of lines (l_1,l_2) such that $L=l_1\cap l_2\notin c$ (see Exercise 6.9.1 and Figure 6.33, right).

(4) We assume that the pencil is spanned by $\mathcal{A}:\ xy=z$ and $\mathcal{B}:\ xy=-z$. In order to get the complete picture, we perform the projective closure and homogenize the equations of both quadrics by substituting $x\to x_1x_0^{-1}$, $y\to x_2x_0^{-1}$, and $z\to x_3x_0^{-1}$. Then, we have $\mathcal{A}:\ x_1x_2-x_0x_3=0$ and $\mathcal{B}: x_1x_2+x_0x_3=0$. Now, $q=\mathcal{A}\cap\mathcal{B}$ is the union of the four lines $\mathbf{l}_1(t_0,t_1)=(t_0:t_1:0:0)$, $\mathbf{l}_2(t_0:t_1)=(0:t_0:0:t_1)$, $\mathbf{l}_3(t_0,t_1)=(0:0:t_0:t_1)$, and $\mathbf{l}_4(t_0,t_1)=(t_0:0:t_1:0)$ (parametrized by the homogeneous parameter $t_0:t_1\neq 0:0$). If B_i ($i\in\{0,1,2,3\}$) are the base points of the coordinate frame, then $l_1=[B_0,B_1]$, $l_2=[B_1,B_3]$,

$l_3 = [B_3, B_2]$, and $l_4 = [B_2, B_0]$, which is a skew quadrilateral. Figure 6.34 (left) illustrates the present case, whereas Figure 6.34 (right) shows that it is easy to realize a configuration of quadrics sharing four proper lines (in infinitely many ways).

The skew quadrilateral cannot open at a single vertex. Otherwise, the cubic component would be a triple (l_1, l_2, l_3) of mutually skew lines together with the line l_4, intersecting l_1 and l_3. This would define a unique singular quadric $\mathcal{Q}$ (consisting of two planes) on ten points through $q \bigcup_{i=1}^{4} l_i$ if l_4 meets at least one of the others. For similar reasons, the completely degenerate quartic space curve cannot open up at more than one point. ■

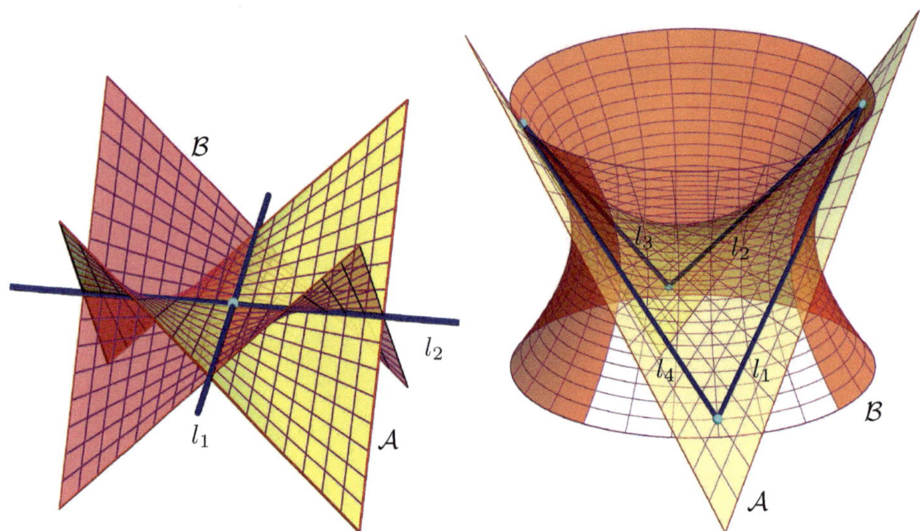

FIGURE 6.34. Degenerate quartics may consist of four lines carrying the edges of a skew quadrilateral. Left: The simplest version appears as the intersection of doubly touching hyperbolic paraboloids $\mathcal{A}$ and $\mathcal{B}$ and has two lines at infinity. Right: On a one-sheeted hyperboloid of revolution $\mathcal{B}$, we can find a smooth one-parameter family of skew quadrilaterals that allow a rotation about the hyperboloid's axis. Through each quadrilateral, there exists a unique hyperbolic paraboloid $\mathcal{A}$.

• **Exercise 6.9.1** A singular quartic as the union of a conic and two straight lines.

Show that the two quadrics (depicted in Figure 6.33 (right))

$$\mathcal{A}:\ x^2 + y^2 - yz - xz + z - 1 = 0 \quad \text{and} \quad \mathcal{B}:\ x^2 + y^2 - z^2 - 1 = 0$$

intersect in a degenerate quartic space curve q of the first kind. The components of q are a conic c and two straight lines l_1 and l_2 that intersect in a point L which is not contained in c. Derive parametrizations of all components of q.

Viviani's curve (6.38) was an example of a quartic space curve of the first kind with a double point. Indeed, the degenerate cases with a cubic and an intersecting line could be seen as limiting cases. In any case, a singular point S on a quartic space curve q of the first kind appears when the quadrics of the pencil through q touch at a certain point S. The following definition (illustrated in Figure 6.35) seems useful:

Definition 6.9.1 The singular point S on a quartic space curve q of the first kind is called, respectively, a *double point* or a *cusp* if the common tangent planes at S of two quadrics through q intersect the quadratic cone Γ_S emanating from S and passing through q along either two generators or along one generator.

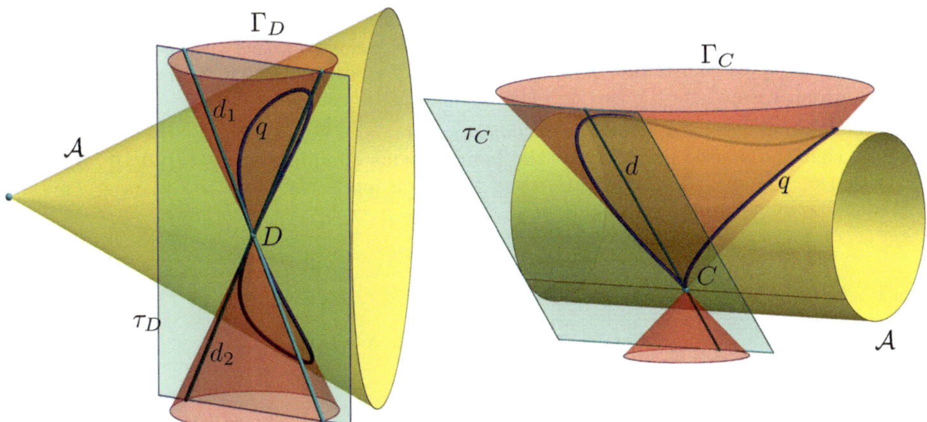

FIGURE 6.35. Left: At the double point D of a quartic space curve q, there are two tangents d_1 and d_2. They are the intersection of the cone Γ_D emanating from D containing q with the tangent plane τ_D of the quadric $\mathcal{A}$ at D. Right: At the cusp C of the quartic q, there is only one tangent d, *i.e.*, the intersection of the cone Γ_C and τ_C.

Definition 6.9.1 is a purely synthetical definition and can be extended to triple points and points with arbitrary multiplicity. If a point D of an intersection curve $q = \mathcal{A} \cap \mathcal{B}$ is a double point, then the tangents d_1 and d_2 can be constructed with the help of Dupin's indicatrix (for a precise definition of Dupin's indicatrix as a curvature diagram, see [46, p. 122]). The indicatrices of both surfaces $\mathcal{A}$ and $\mathcal{B}$ (drawn in the common tangent plane with the same scaling constant) intersect in pairs of points that are symmetric with regard to the common center D of both indicatrices. These opposite points define the tangents at the double points. Moreover, this approach as well as the synthetical definition make clear that at D there can be two real tangents (ordinary double point), one real tangent (cusp), or a pair of complex conjugate tangents (isolated double point).

Now, we are able to show:

Theorem 6.9.3 *Quartic space curves of the first kind have at most one singularity which is either an ordinary or isolated double point or a cusp. Quartics of the first kind with singularities admit a polynomial parametrization.*

Proof: If a quartic q has two double points, say $D_1 \neq D_2$, then each plane through the line $d = [D_1, D_2]$ meets q in no further point according to Bézout's theorem. There are two exceptional planes: Since D_1 is a double point, there exist two tangents $d_1 \neq d_2$ to q, and each of the exceptional planes $\delta_1 = [d, d_1]$ and $\delta_2 = [d, d_2]$ has to contain exactly one tangent of q at the other double point, again according to Bézout's theorem. Otherwise, $\#q \cap \delta_i = 5$, which contradicts $\deg q = 4$. Therefore, q splits into a pair of conics. The case with the cusp can be seen as a limit case where the two ordinary double points coincide.

A rational parametrization of a curve becomes polynomial if we use homogeneous coordinates. Assume that the double point D has the homogeneous coordinates $\mathbf{d} = (1 : 0 : 0 : 0)$. The projection of q from D is a quadratic cone Γ_1 which intersects the plane x_0 in a conic c_1. Without loss of generality, we may assume that c_1 is given by the equation

$$c_2 : \; x_0 = 0, \quad x_1 x_3 - x_2^2 = 0$$

which admits a rational parametrization

$$\mathbf{c}_2 = (0 : \lambda^2 : \lambda\mu : \mu^2) \quad \text{with} \quad \lambda : \mu \neq 0 : 0.$$

Now, the lines

$$\mathbf{l}(t_0, t_1) = t_0 \mathbf{d} + t_1 \mathbf{c}_2 \quad \text{with} \quad t_0 : t_1 \neq 0 : 0$$

intersect any other quadric through q in precisely one point that is different from D, which yields a polynomial parametrization of q. ■

▪ **Example 6.9.3** Rational quartic space curves.

We study the quartic intersection curve q of the cylinder of revolution

$$\mathcal{A} : \; (x+1)^2 + (z-1)^2 = 2$$

and the cone of revolution

$$\Gamma_C : \; x^2 + y^2 = z^2$$

shown in Figure 6.36 (right). In order to find a parametrization of q, we insert the parametrization

$$\mathbf{c}(u, v) = (u \cos v, u \sin v, u), \quad (u, v) \in \mathbb{R} \times [0, 2\pi[$$

of Γ_C into $\mathcal{A}$'s equation and find a relation between the surface coordinates u and v

$$u((1 + \cos^2 v)u + 2(\cos v - 1)) = 0.$$

The solution $u = 0$ returns the cusp $(0,0,0)$. Therefore, we solve the second factor for u and find

$$u = \frac{1 - 2\cos v}{1 + \cos^2 v},$$

which, together with the parametrization of Γ_C, yields the parametrization of q

$$\mathbf{q}(v) = \left(\frac{1 - 2\cos v}{1 + \cos^2 v}\cos v, \frac{1 - 2\cos v}{1 + \cos^2 v}\sin v, \frac{1 - 2\cos v}{1 + \cos^2 v}\right).$$

This parametrization is well-defined over the interval $[0, 2\pi[$. Naturally, the curve q allows a rational parametrization, which can be obtained from the trigonometric one by substituting $\cos v = \frac{1-t^2}{1+t^2}$ and $\sin v = \frac{2t}{1+t^2}$. This yields

$$\mathbf{q}(t) = \left(\frac{2t^2(1 - t^2)}{1 + t^4}, \frac{4t^3}{1 + t^4}, \frac{2t^2(1 + t^2)}{1 + t^4}\right).$$

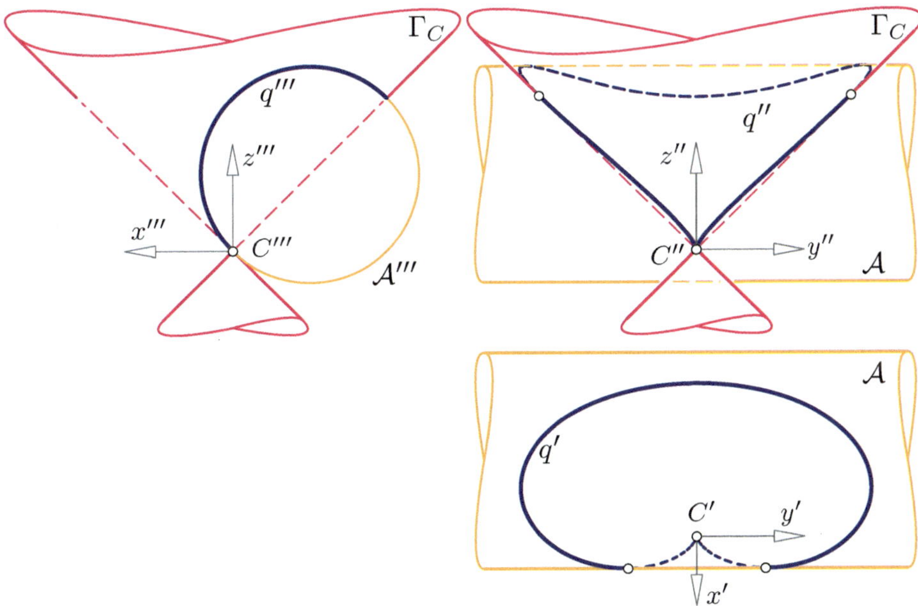

FIGURE 6.36. The quartic q of the first kind with a cusp as derived in Example 6.9.3 is depicted in the three principal views. The image q''' of q in the right-side view is a part of the circle $\mathcal{A}'''$. The top and the front view show a cusped quartic.

Figure 6.37 shows that any type of linear image – especially orthogonal projections – of a rational quartic space curve can result in rational planar quartic curves with different types of singularities on the image curve. On the one hand, such singularities can occur as the images of the singular point. On the other hand, singularities of the planar images such as the additional two cusps (cf. Figure 6.37, top-right), the two ordinary double points (bottom-left), or the triple point (bottom-right) are images of at least two regular points on the quartic space curve q.

In Figure 6.37 (top-left), there is hardly a singularity to be seen. However, the point C which is a cusp on q is mapped to a singular point where the local expansion of the image curve starts with (t^3, t^4). This sort of points on planar algebraic curves is called a cusp of the *third kind*.[3] Indeed, it is possible to find a projection (as shown in Figure 6.37, bottom-right) such that a triple point occurs on the image of q. This triple point C^4 is a so-called *composed singularity* and consists of a linear branch (image of the part locally around the handle point H) and a cusp of the first kind (which is the image of the cusp C on q).

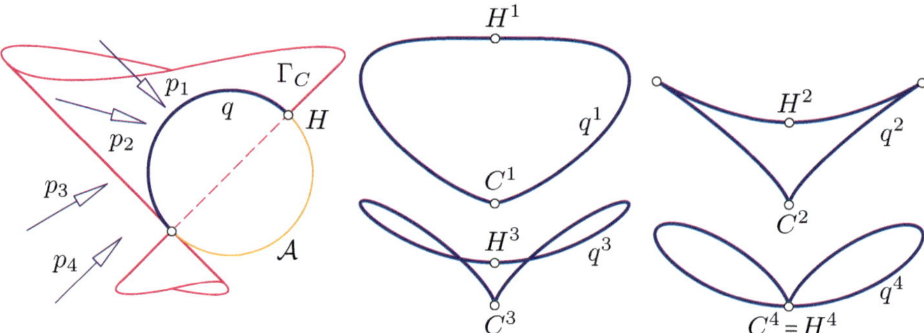

FIGURE 6.37. Four different projections p_i of a rational quartic space curve q of the first kind produce four different rational planar quartics q^i with different types of singularities. Top-left: C^1 is a cusp of the third kind on q^1 and the image H^1 of the handle point H is (occasionally) an undulation point on q^1. Top-right: C^2 is a cusp of the first kind on q^2. The additional two cusps are of the same type, but images of regular points on q. Bottom-left: C^3 is an ordinary cusp and each of the two (ordinary) double points is the image of two different regular points on q. Bottom-right: The projection p_4 is chosen such that the image of the handle point and the image of the cusp C coincide. This causes a composed singularity on q^4. The point $C^4 \in q^4$ is a triple point with a linear branch and a cusp sitting on it.

Quartic space curves of the second kind

A quartic space curve of the *second kind* can be defined as the intersection of a quadric $\mathcal{Q}$ and a cubic surface $\mathcal{C}$ that share two skew rulings with $\mathcal{Q}$. In the following, we aim to provide an analytic description of quartic space curves of the second kind. It is no loss of generality to assume that

[3] The cusp of the third kind has the local expansion (t^3, t^4), while the expansions at the cusps of the first and second kind start with (t^2, t^3) and (t^2, t^4) and are usually referred to as ordinary cusp and rhamphoid cusp. The German term for the cusp of the third kind is *Spitzpunkt*.

$\mathcal{Q}$ is given in the very simple form

$$\begin{gathered}\mathcal{Q}:\ x_0x_3 - x_1x_2 = 0 \quad \text{or}\\ \mathbf{q}=(\lambda_0\mu_0 : \lambda_0\mu_1 : \lambda_1\mu_0 : \lambda_1\mu_1) \ \text{ with } \lambda_0 : \lambda_1 : \mu_0 : \mu_1 \neq 0{:}0{:}0{:}0\end{gathered} \tag{6.39}$$

in terms of homogeneous coordinates and homogeneous parameters. The straight lines l_1 and l_2 on $\mathcal{Q}$ are determined by $\lambda_1 = 0$ ($\lambda_0 \neq 0$) and $\lambda_0 = 0$ ($\lambda_1 \neq 0$) and admit the parametric representation

$$\mathbf{l}_1 = (\mu_0 : \mu_1 : 0 : 0) \quad \mathbf{l}_2 = (0 : 0 : \mu_0 : \mu_1) \quad \mu_0 : \mu_1 \neq 0 : 0. \tag{6.40}$$

The second main ingredient is the family of cubic surfaces through l_1 and l_2. In terms of homogeneous coordinates, a generic cubic surface $\mathcal{C}$ in projective 3-space $\mathbb{P}^3(\mathbb{F})$ over an arbitrary commutative field (with $\operatorname{char}\mathbb{F} \neq 2, 3$) has an equation of the form

$$\mathcal{C}: \sum_{i,j,k=0,1,2,3} c_{ijk}x_ix_jx_k = 0 \tag{6.41}$$

where not all the coefficients $c_{ijk} \in \mathbb{F}$ vanish simultaneously. We assume that $\mathcal{C}$ is not degenerate, *i.e.*, it does not contain a linear factor. In order to find a quartic space curve $q = \mathcal{Q} \cap \mathcal{C} \setminus \{l_1, l_2\}$, we have to make sure that l_1 and l_2 split off from the sextic curve $\mathcal{Q} \cap \mathcal{C}$, and therefore, l_1 and l_2 have to be contained in $\mathcal{C}$. This is guaranteed if we insert the parametrizations (6.40) into (6.41) and it causes

$$c_{000} = c_{111} = c_{222} = c_{333} = c_{001} = c_{011} = c_{223} = c_{233} = 0.$$

Therefore, the totality of cubic surfaces through l_1 and l_2 is given by

$$\begin{gathered}\mathcal{C}:\ c_{002}x_0^2x_2 + c_{003}x_0^2x_3 + c_{022}x_0x_2^2c_{022} + c_{033}x_0x_3^2 +\\ +c_{112}x_1^2x_2 + c_{113}x_1^2x_3 + c_{122}x_1x_2^2 + c_{133}x_1x_3^2 + c_{012}x_0x_1x_2 +\\ +c_{013}x_0x_1x_3 + c_{023}x_0x_2x_3 + c_{123}x_1x_2x_3 = 0.\end{gathered} \tag{6.42}$$

The curve of intersection of $\mathcal{Q}$ and $\mathcal{C}$ is now determined by a condition imposed on the parameters λ_i and μ_i. This condition is found by inserting the parametric representation of $\mathcal{Q}$ from (6.39) into the equations of the cubic surfaces (6.42) passing through l_1 and l_2 and reads

$$\begin{gathered}(c_{002}\mu_0^3+c_{003}\mu_0^2\mu_1+c_{012}\mu_0^2\mu_1+c_{013}\mu_0\mu_1^2+c_{112}\mu_0\mu_1^2+c_{113}\mu_1^3)\lambda_0+\\ +(c_{022}\mu_0^3+c_{023}\mu_0^2\mu_1+c_{033}\mu_0\mu_1^2+c_{122}\mu_0^2\mu_1+c_{123}\mu_0\mu_1^2+c_{133}\mu_1^3)\lambda_1=0.\end{gathered} \tag{6.43}$$

The fact that (6.43) is linear in $\lambda_0 : \lambda_1$ and cubic in $\mu_0 : \mu_1$ gives rise to

Theorem 6.9.4 *The intersection of a regular ruled quadric $\mathcal{Q}$ and a cubic surface $\mathcal{C}$ that share two skew lines l_1 and l_2 (i.e., from the same regulus on $\mathcal{Q}$) consists of the two lines and a quartic space curve q of the second kind (as long as $\mathcal{Q} \not\subset \mathcal{C}$ and as long as q does not degenerate any further).*

The lines of the regulus $\mathcal{R}_1 \subset \mathcal{Q}$ that contains l_1 and l_2 are met by q in three points (algebraically counted), while the lines of the complementary regulus $\mathcal{R}_2$ are met only once by q.

Proof: The only thing that remains to be confirmed is the number of common points of q and the lines in either regulus.

For fixed $\lambda_0 : \lambda_1 \neq 0 : 0$, (6.43) is cubic in $\mu_0 : \mu_1$ and has three roots over the complex number field, and thus, there are as many points on the lines determined by $\lambda_0 : \lambda_1$ as is the case with l_1 and l_2.

On the other hand, for fixed $\mu_0 : \mu_1 \neq 0 : 0$, (6.43) is linear in $\lambda_0 : \lambda_1$, and therefore, there is exactly one point on each line in the second regulus. ■

The regulus $\mathcal{R}_1$ mentioned in Theorem 6.9.4 is usually referred to as the *regulus of trisecants*, since all its lines meet the quartic q in three points. The complementary regulus $\mathcal{R}_2$ is called *unisecant* or *monosecant regulus*, since these lines meet q only in one point. The quadric $\mathcal{Q}$ is usually called the *triceant quadric* of q.

Figure 6.38 shows a quartic space curve q of the second kind as the intersection of a quadric $\mathcal{Q}$ and a cubic surface $\mathcal{C}$, that share two straight lines l_1 and l_2 with $\mathcal{Q}$.

■ **Example 6.9.4** A closed rational quartic of the second kind.

We investigate the intersection of a one-sheeted hyperboloid $\mathcal{H}$ of revolution and a cubic conoidal ruled surface $\mathcal{C}$ (cf. Figure 6.39) with the respective equations

$$\mathcal{Q}:\ x^2 + z^2 - y^2 = 1 \quad \text{and} \quad \mathcal{C}:\ z(x^2+y^2) - 2xy = 0. \tag{6.44}$$

The surface $\mathcal{C}$ is a cubic ruled surface known as Plücker's conoid (cf. Sections 2.2, 10.1, and 11.4).

Eliminating z from both equations yields

$$q':\ (x^2+y^2)^2 - (x^2-y^2) = 0, \tag{6.45}$$

which is the equation of a planar curve that can be considered to be the top view q' of q. The curve (6.45) is known as *Bernoulli's lemniscate* (cf. [41, 84, 137] and also [46, p. 111 & p. 411]). It allows the rational parametrization

$$\mathbf{q}'(t) = \frac{1}{1+t^4}\begin{pmatrix} t(1+t^2) \\ t(t^2-1) \end{pmatrix},$$

which, inserted into $\mathcal{C}$'s equation, yields

$$z(t) = \frac{(t^2+1)(t^2-1)}{1+t^4}.$$

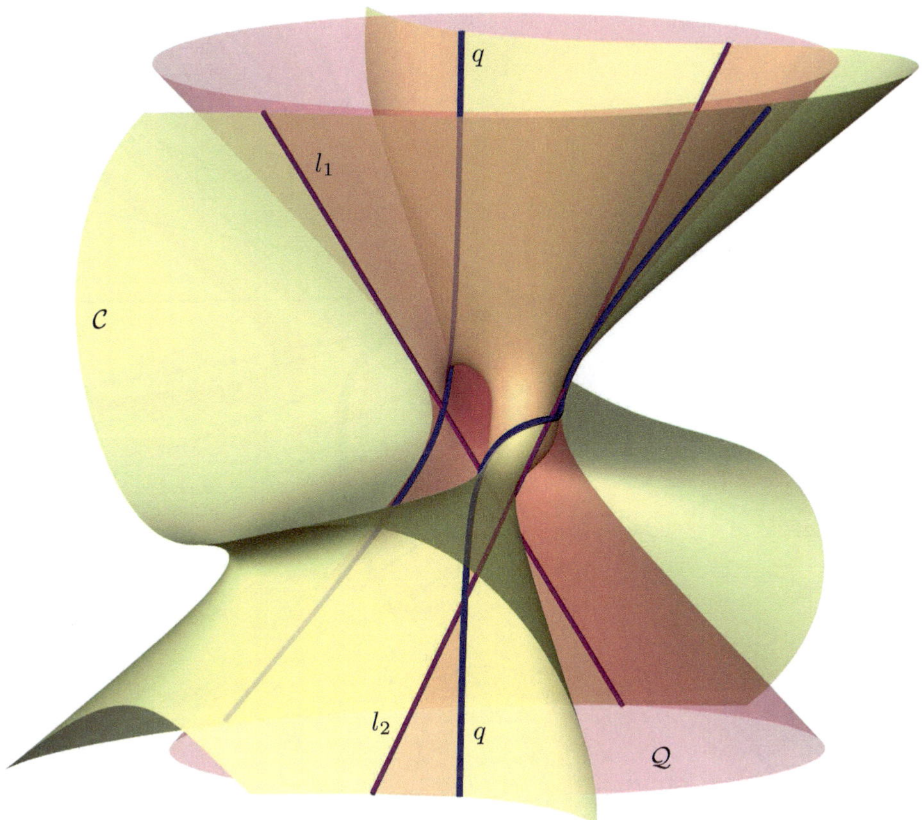

FIGURE 6.38. From the sextic intersection of the quadric $\mathcal{Q}$ and the cubic $\mathcal{C}$, two straight lines l_1 and l_2 split off. What remains is a quartic space curve q of the second kind.

Thus, we have the rational parametrization of q

$$\mathbf{q}(t) = \frac{1}{1+t^4}\begin{pmatrix} t(1+t^2) \\ t(t^2-1) \\ (t^2+1)(t^2-1) \end{pmatrix},$$

which can also be turned into a trigonometric one by substituting $t = \tan\frac{\tau}{2}$. This results in

$$\mathbf{q}(\tau) = \frac{1}{3+\cos 2\tau}\begin{pmatrix} 2\sin\tau \\ -\sin 2\tau \\ 4\cos\tau \end{pmatrix}, \quad \tau \in [0, 2\pi[. \tag{6.46}$$

Note that all linear images of q are rational curves. So are the principal projections shown in Figure 6.40.

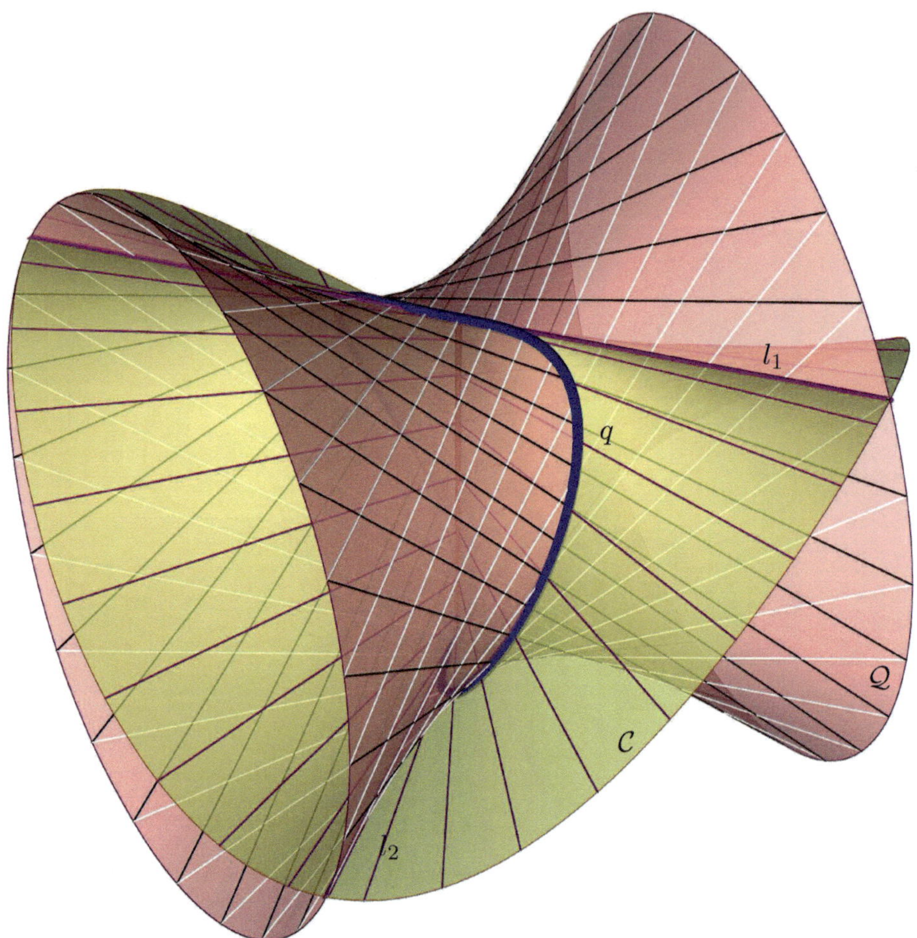

FIGURE 6.39. This closed and rational quartic curve q is the intersection of a one-sheeted hyperboloid $\mathcal{H}$ of revolution and a cubic conoidal ruled surface $\mathcal{C}$. The trisecant regulus (black) and the unisecant regulus (white) are also shown.

■ **Example 6.9.5** A regulus of trisecants.

Show that the quartic q (displayed in Figure 6.41) of the second kind parametrized by

$$\mathbf{q}(t) = \frac{1}{1+t^4} \begin{pmatrix} 2at \\ b(t^4 + 2t^2 - 1) \\ 2ct^3 \end{pmatrix}$$

lies on the one-sheeted hyperboloid

$$\mathcal{H}: \ \frac{x^2}{a^2} + \frac{y^2}{b^2} - \frac{z^2}{c^2} = 1.$$

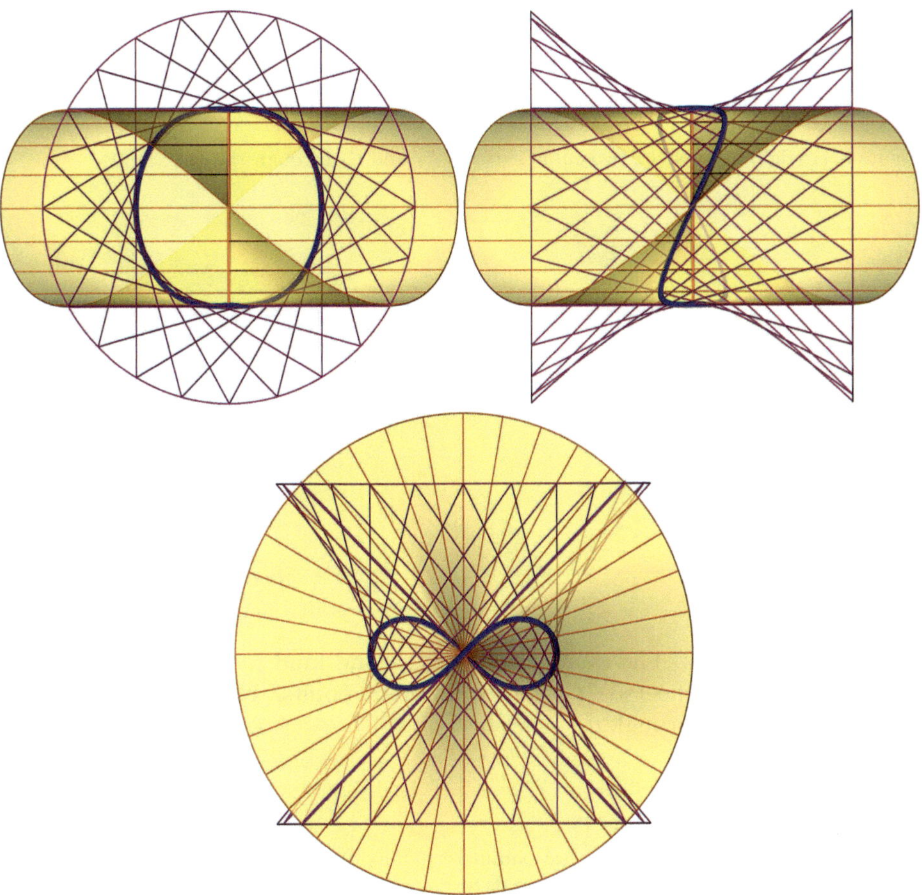

FIGURE 6.40. Top row: The right-side view (left) and the front view (right) of (6.46). The front view shows an eight-loop with two undulation points (top and bottom) and an inflection node in the center. Bottom: The top view of q is Bernoulli's lemniscate.

Show further that the lines of the regulus

$$\mathcal{R} = \begin{pmatrix} a\cos u \\ b\sin u \\ 0 \end{pmatrix} + v \begin{pmatrix} -a\sin u \\ b\cos u \\ -c \end{pmatrix}$$

intersect the curve in three different points. The curve q admits a trigonometric parametrization by substituting $t = \tan\frac{\tau}{2}$, which reads

$$\mathbf{q}(t) = \frac{1}{3+\cos 2t} \begin{pmatrix} a(2\sin\tau + \sin 2\tau) \\ b(1-\cos 2\tau - 4\cos\tau) \\ c(2\sin\tau - \sin 2\tau) \end{pmatrix}.$$

Further, q lies on the two cubic surfaces

$$\mathcal{C}_1: \ ay(c^2y^2 - 2b^2z^2) + bc(acy^2 + 2b^2xz) - ab^2c^2y - ab^3c^2 = 0,$$
$$\mathcal{C}_2: \ az(c^2y^2 - 2b^2z^2) + bc^2y(az + cz) - b^2c^2(2az - cx) = 0.$$

Figure 6.41 (right) shows the two cubic surfaces $\mathcal{C}_1$ and $\mathcal{C}_2$ through the quartic q.

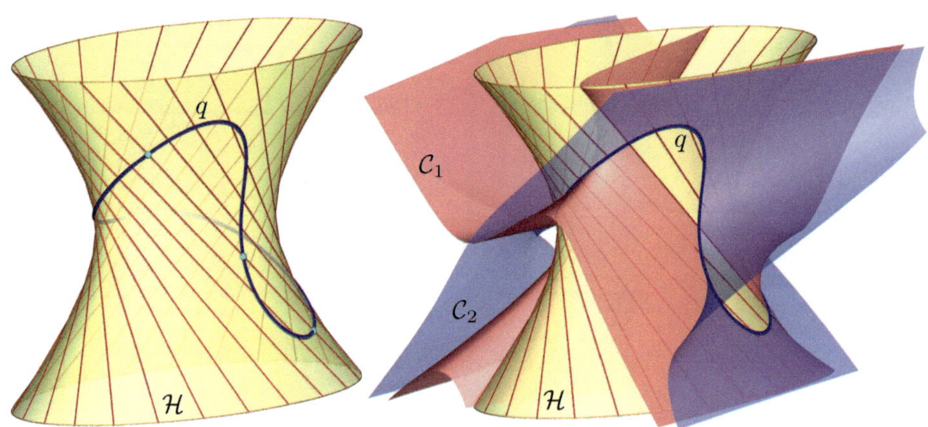

FIGURE 6.41. Left: The quartic q of the second kind meets each line of one regulus on the one-sheeted $\mathcal{H}$ in three different points. Right: The two cubic surfaces $\mathcal{C}_1$ and $\mathcal{C}_2$ through q.

■ **Example 6.9.6** Contours of Plücker's conoid.

On Plücker's conoid, we can find a two-parameter family of quartic space curves of the second kind. As we shall see soon, there are also degenerate quartics among them.

Let $\mathbf{l} = (l_1, l_2, l_3)$ with $\|\mathbf{l}\| = 1$ be a unit vector. A contour point C with coordinates $\mathbf{c} = (x, y, z)$ on Plücker's conoid (second equation in (6.44)) with regard to the parallel projection in the direction of $\mathbf{l}$ is characterized by having a surface normal (vector) $\mathbf{n}_C$ orthogonal to $\mathbf{l}$. Since

$$\operatorname{grad}\mathcal{P} = (2xz - 2y, 2yz - 2x, x^2 + y^2) = \mathbf{n_c}$$

yields the normal vector at C. The condition $\langle \mathbf{n_c}, \mathbf{l} \rangle = 0$, *i.e.*,

$$\mathcal{U}: \ 2(xz - y)l_1 + 2(yz - x)l_2 + (x^2 + y^2)l_3 = 0, \tag{6.47}$$

is the equation of a regular quadric as long as $(l_1 - l_2)^2(l_1 + l_2)^2 \neq 0$ and gives a relation between the coordinates x, y, and z of the contour points. So, the contour q of $\mathcal{P}$ w.r.t. the parallel projection is a part of the intersection of $\mathcal{U}$ and $\mathcal{P}$, which is expected to be a sextic. We can easily verify that the z-axis is contained in $\mathcal{U}$ and, provided that $l_2(l_1^2 + l_2^2) \neq 0$, the quadric always has a center whose coordinates are

$$\frac{1}{(l_1^2 + l_2^2)l_3} \begin{pmatrix} l_2(l_2^2 - l_1^2) \\ l_1(l_1^2 - l_2^2) \\ 2l_1l_2l_3 \end{pmatrix}. \tag{6.48}$$

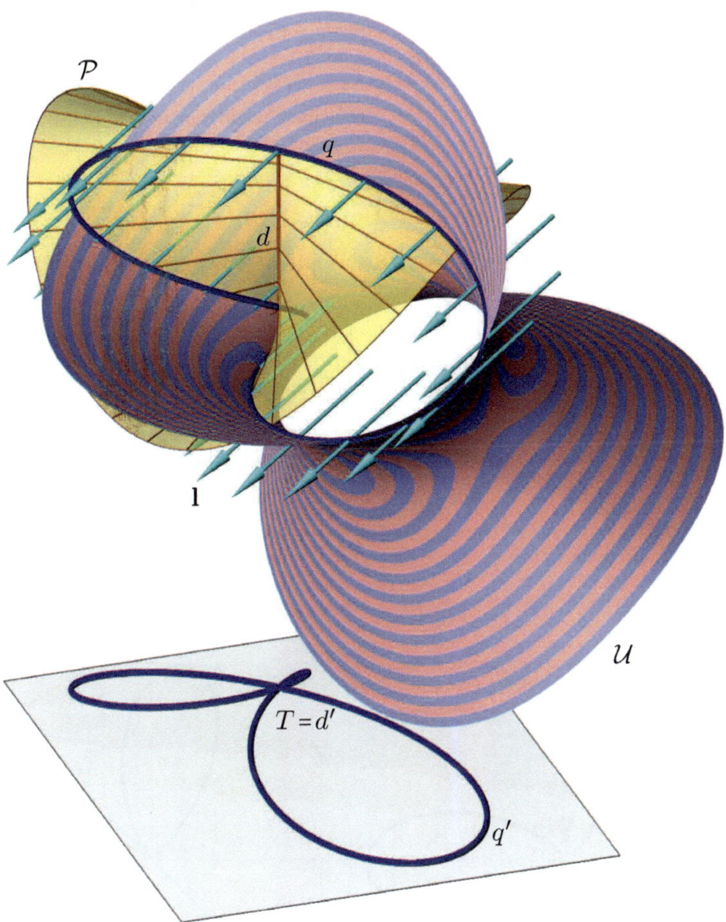

FIGURE 6.42. The contour q of Plücker's conoid $\mathcal{P}$ w.r.t. the parallel projection in the direction $\mathbf{l}$ is a quartic space curve of the second kind. The curve q is a part of the intersection of $\mathcal{P}$ with a ruled quadric $\mathcal{U}$ that contains the directrix $d \subset \mathcal{P}$. The orthogonal projection of q in the direction of d (top view) is a quartic curve q' with a triple point $T = d'$ that is the directrix's image.

Since the z-axis is a part of the double curve of $\mathcal{P}$, it splits off from $\mathcal{U} \cap \mathcal{P}$ with multiplicity two. Consequently, the remaining part is a quartic space curve of the second kind, an example of which is displayed in Figure 6.42.

We can describe this quartic more precisely. For that purpose, we eliminate z from the equations of $\mathcal{U}$ and of $\mathcal{P}$, which yields an equation of the top view $\mathbf{q}'$ of the contour curve:

$$q' : \; l_3(x^2 + y^2) + 2(x - y)(x + y)(l_1 y - l_2 x) = 0. \tag{6.49}$$

Since there are no terms of degree two or less, we see that the point $(0,0)$ is a triple point on q' with two tangents if $l_1 : l_2$ is either $1:1$ or $1:-1$. In any other case, there are three separate tangents at the triple point, which makes it an ordinary triple point, then.

In order to find a parametrization of q', we intersect it with the lines of the pencil about the triple point, *i.e.*, we insert $x = u$ and $y = tu$ with $t, u \in \mathbb{R}$ into (6.49) and find

$$\mathbf{q}'(t) = \frac{1}{l_3(1+t^2)^2}\begin{pmatrix} 2(t^2-1)(tl_1-l_2) \\ 2t(t^2-1)(tl_1-l_2) \end{pmatrix}.$$

Inserting the two coordinate functions of $\mathbf{q}'$ for x and y into the equation of $\mathcal{P}$ or of $\mathcal{U}$, we find

$$z(t) = \frac{2t}{1+t^2},$$

which, together with $\mathbf{q}'$, gives a parametrization of the quartic contour curve q. A trigonometric form can be given by substituting $t = \tan\frac{\tau}{2}$, which yields

$$\mathbf{q}(\tau) = \frac{1}{l_3}\begin{pmatrix} (l_2\cos\tau - l_1\sin\tau + l_2)\cos\tau \\ (l_1\cos\tau + l_2\sin\tau - l_1)\cos\tau \\ \sin\tau \end{pmatrix}.$$

The cases where $\mathcal{U}$ is singular or has no center are to be discussed separately. Only if $l_3 = 0$, the quadric $\mathcal{U}$ has no center (for $l_1^2+l_2^2 = 0$ cannot happen over the reals). Then, (6.49) becomes $(x-y)(x+y)(l_1y - l_2x) = 0$, and the quartic q becomes the union of three (mutually skew) generators of $\mathcal{P}$ together with the directrix $x = y = 0$ (intersecting them all).

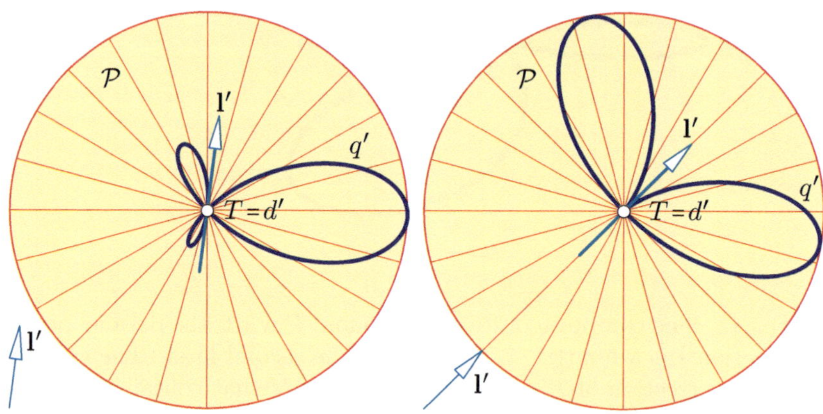

FIGURE 6.43. The top views q' of $\mathcal{P}$'s contour curve q w.r.t. two different parallel projections in the direction $\mathbf{l}$. Left: $\mathbf{l} = \frac{1}{9}(1,8,4)$. Right: $\mathbf{l} = \frac{1}{9}(4,4,7)$, *i.e.*, $l_1 = l_2$ where q' has only two tangents at the triple point T.

The quadric $\mathcal{U}$ is singular if, and only, if $l_1 = \pm l_2$. Then, (6.47) becomes a quadratic cone centered at one of $\mathcal{P}$'s pinch points, *i.e.*, whose vertex is either $(0,0,1)$ or $(0,0,-1)$, depending on whether $l_1 = l_2$ or $l_1 = -l_2$. The top view q' of q shows a triple point T with only two tangents (see Figure 6.43).

As can be guessed from Figure 6.43, the top view $\mathbf{l}'$ of the projection vector is parallel to one of the tangents of q' at the triple point.

The visual contour of Plücker's conoid $\mathcal{P}$ in an orthogonal view is simply the orthogonal projection of the quartic q onto a plane orthogonal to $\ker \mathbf{l}$. The visual contour of $\mathcal{P}$ is a planar rational quartic with three cusps, which is a collinear image of Steiner's hypocycloid (sometimes referred to as deltoid). Figure 6.44 (left) shows an orthogonal projection of Plücker's conoid. The visual contour is drawn in dark blue. The right-hand side of Figure 6.44 shows only the three-cusped curve.

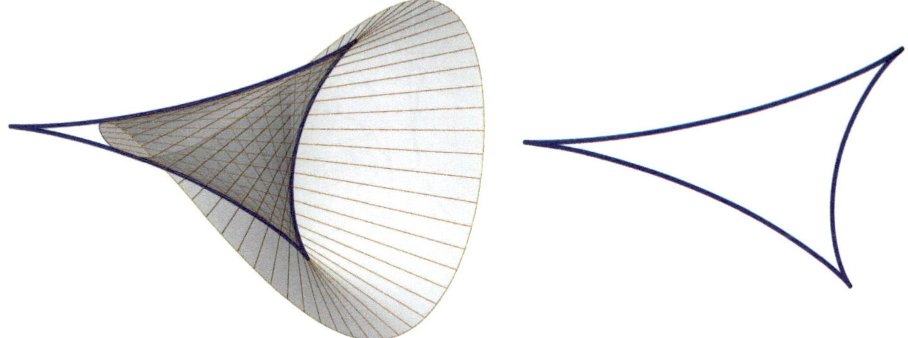

FIGURE 6.44. Left: Plücker's conoid with its visual contour (thick blue curve) as the envelope of the images of the rulings. Right: The visual contour as the orthogonal projection of the quartic q (of the second kind) onto the plane $\ker \mathbf{l}$.

■ **Example 6.9.7** The asymptotic lines of Plücker's conoid.

We can use either the second equation of (6.44) (implicit equation of Plücker's conoid) or the parametrization (10.22) in order to compute the curved asymptotic lines of Plücker's conoid. In (10.22), we replace t by v and w by u. Then, the second fundamental form reads

$$\mathrm{II} = 4((1-2\cos^2 v)\mathrm{d}u - u\sin 2v)\mathrm{d}v.$$

Setting II equal to zero yields two linear differential equations for the asymptotic lines on the conoid. The factor $\mathrm{d}v = 0$, or equivalently, $v = \text{const.}$ corresponds to the rulings, which are the straight asymptotic lines. The second family of asymptotic lines is the one-parameter family of integral curves of

$$((1-2\cos^2 v)\mathrm{d}u - u\sin 2v)\mathrm{d}v = 0,$$

which can be given as $k^2u^2 = -\cos 2v$ with an arbitrary constant of integration $k \in \mathbb{R}$. Thus, the asymptotic lines on Plücker's conoid can be parametrized by

$$\mathbf{a}(t) = \begin{pmatrix} k\sqrt{-\cos 2v}\cos v \\ k\sqrt{-\cos 2v}\sin v \\ \sin 2v \end{pmatrix}.$$

Some of these curves are shown in Figure 6.45 (left). Their top views (cf. Figure 6.45, right) are Bernoulli's lemniscates (cf. [41, 84, 137]).The latter fact can be verified by implicitizing the parametrization of the top view

$$\mathbf{a}'(t) = \begin{pmatrix} k\sqrt{-\cos 2v}\cos v \\ k\sqrt{-\cos 2v}\sin v \end{pmatrix}.$$

The curved asymptotic lines are quartic space curves of the second kind. What are the quadrics through the asymptotic lines?

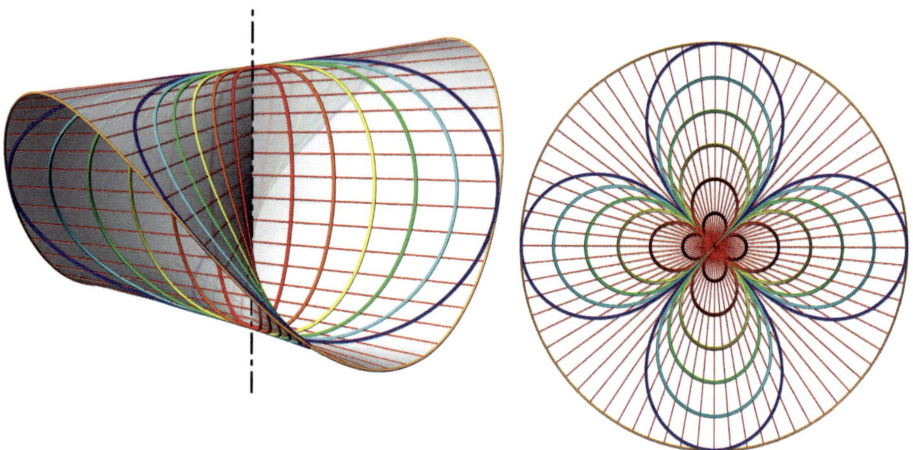

FIGURE 6.45. Left: The curved asymptotic lines of Plücker's conoid are quartic space curves of the second kind. Right: The top view (*i.e.*, orthogonal projections in the direction of the directrix (or double line) d shows the curved asymptotic lines as Bernoulli's lemniscates.

Degenerate quartics space curves of the second kind

In analogy to Theorem 6.9.2, we can describe the degenerate quartic space curves of the second kind:

Theorem 6.9.5 *A quartic space curve q of the second kind can degenerate in the following three ways:*

(1) the union of a rational cubic space curve c and a straight line l that intersects c in exactly one point P, and l does not touch c at P, or

(2) the union of a conic c and a pair of skew straight lines l_1 and l_2 such that each line meets c in precisely one point, neither line is contained in the plane spanned by c, or

(3) the union of three mutually skew straight lines l_1, l_2, l_3, and a fourth line intersecting all three lines l_i.

Proof: In order to verify the contents of Theorem 6.9.5, we only have to study all possible cases of degeneration of the form (6.43).

If (6.43) is independent from $\lambda_0 : \lambda_1$, then q is the union of a straight line l and a cubic space curve c. From Theorem 6.9.4, it is clear that c and l can have only one point in common.

If the cubic form has one real factor, then q becomes a conic and a pair of straight lines.

In the third case, the cubic form in $\mu_0 : \mu_1$ is the product of three homogeneous linear factors each of which determines a straight line from one regulus, and thus, they are mutually skew and meet the fourth line (which is from the complementary regulus). An example of this kind of degenerate quartic space curve of the second kind is given by the contour of Plücker's conoid in Example 6.9.6. ■

Double binary forms

On the ruled quadric (6.39), we have determined quartic curves on a ruled quadric $\mathcal{Q}$ as the intersections with cubic surfaces $\mathcal{C}$ given (6.42) that contain two rulings of the same regulus on $\mathcal{Q}$. We have found the condition (6.43) on the homogeneous surface parameters $\lambda_0 : \lambda_1 : \mu_0 : \mu_1$. It is easy to see that (6.43) is homogeneous in $\lambda_0 : \lambda_1$ and $\mu_0 : \mu_1$ separately. Therefore, this form is called a *double binary form.*

Moreover, (6.43) is of degree 1 in $\lambda_0 : \lambda_1$ and of degree 3 in $\mu_0 : \mu_1$. Thus, the quartic space curve defined by this double binary form is sometimes denoted by $C^{1,3}$.

• **Exercise 6.9.2** Quartic space curves of the first and second kind.

Use the parametric and implicit representation of a ruled quadric as given (6.39) and show that the algebraic curves $C^{1,2}$ are cubic space curves and $C^{2,2}$ are quartic space curves of the first kind.

Show further that the thus defined quartic space curves of the first kind lie on all quadrics in a pencil of quadrics.

7 Confocal quadrics

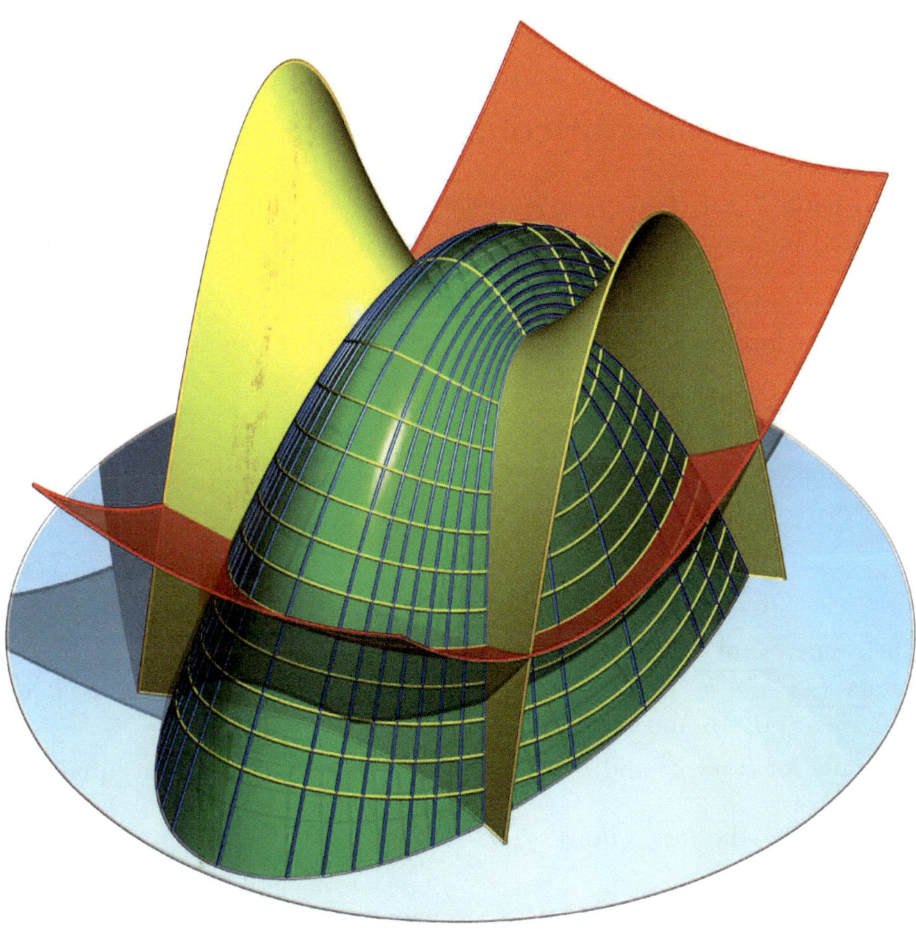

The picture shows an elliptic paraboloid with its curvature lines. These lines are the curves of intersection with confocal elliptic and hyperbolic paraboloids, and therefore, in general of degree four.

B. Odehnal et al., *The Universe of Quadrics*,
https://doi.org/10.1007/978-3-662-61053-4_7

Families of quadrics which intersect the planes of symmetry along confocal conics are called confocal. They belong to a range, *i.e.*, a dual pencil. It is interesting to see that confocal quadrics are connected with various geometric problems, and often the famous theorem of IVORY plays a central role.

7.1 The families of confocal quadrics

Definition 7.1.1 Two quadrics are called *confocal* if they have common axes and intersect each plane of symmetry along confocal conics.[1]

Let $\mathcal{E}$ be a triaxial ellipsoid with semiaxes a, b, and c. Then, the one-parameter family of quadrics being confocal with $\mathcal{E}$ is given as

$$\frac{x^2}{a^2+k}+\frac{y^2}{b^2+k}+\frac{z^2}{c^2+k}=1 \quad \text{for} \quad k\in\mathbb{R}\setminus\{-a^2,-b^2,-c^2\}. \tag{7.1}$$

In the case $a>b>c>0$, this family contains (see Figure 7.1)

$$\text{for}\ \begin{cases} -c^2<k<\infty & \text{triaxial ellipsoids,}\\ -b^2<k<-c^2 & \text{one-sheeted hyperboloids,}\\ -a^2<k<-b^2 & \text{two-sheeted hyperboloids.}\end{cases} \tag{7.2}$$

Their curves of intersections with the plane $z=0$ share the focal points $(\pm\sqrt{a^2-b^2},0,0)$. In the plane $y=0$, the common foci are $(\pm\sqrt{a^2-c^2},0,0)$, and in $x=0$ they are $(0,\pm\sqrt{b^2-c^2},0)$.
As limits for $k\to -c^2$ and $k\to -b^2$ (Figure 7.2), we obtain 'flat' quadrics,

$$\begin{aligned} &\text{the } \textit{focal ellipse } f_e\colon && \frac{x^2}{a^2-c^2}+\frac{y^2}{b^2-c^2}=1,\ z=0,\\ &\text{and the } \textit{focal hyperbola } f_h\colon && \frac{x^2}{a^2-b^2}-\frac{z^2}{b^2-c^2}=1,\ y=0. \end{aligned} \tag{7.3}$$

This pair of conics has already been studied under the notation 'focal conics' in [46, p. 137ff]: Each of these conics is the locus of apices of right cones passing through the other conic. However, this time these conics are embedded in a family of confocal quadrics.[2]

---

[1] M. BERGER uses in [5, 15.7.17] the term *homofocal* instead of confocal.

[2] Note that, for $k\to -c^2$, the right-hand limit via ellipsoids covers the interior of the focal ellipse f_e twice (Figure 7.3), while the left-hand limit via one-sheeted hyperboloids covers the exterior of f_e twice. A similar effect shows up at the focal hyperbola f_h (cf. Figure 7.4).

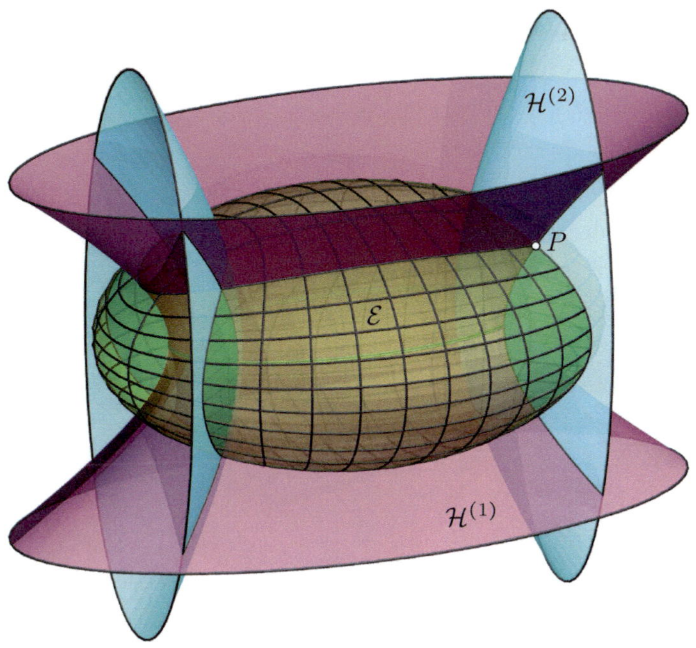

FIGURE 7.1. Confocal quadrics and the curves of intersection between the triaxial ellipsoid $\mathcal{E}$ and the confocal hyperboloids, which are curvature lines on $\mathcal{E}$.

The limit for $k \to -a^2$ gives the empty conic

$$\frac{y^2}{a^2 - b^2} + \frac{z^2}{a^2 - c^2} + 1 = 0, \quad x = 0. \tag{7.4}$$

In the sense of Theorem 7.1.1, this conic is the fourth singular dual quadric, besides the focal ellipse f_e, the focal hyperbola f_h, and the absolute conic, in the linear system of confocal quadrics.

The dual representation of the confocal family will reveal that the two focal conics have to be seen as a set of tangent planes, and then, they are included as singular dual quadrics (note the coming Theorem 7.1.1 on page 285). Due to this interpretation, all lines in space which meet a focal conic f in at least one point are *tangent lines* of f. It makes sense to specify that all lines in a plane of symmetry are also tangents of the included focal conic, even when the conic is empty or the points of intersection with the conic are complex conjugate. When an ordinary tangent line of the plane curve f is meant, we speak in the following of a *proper tangent line.*

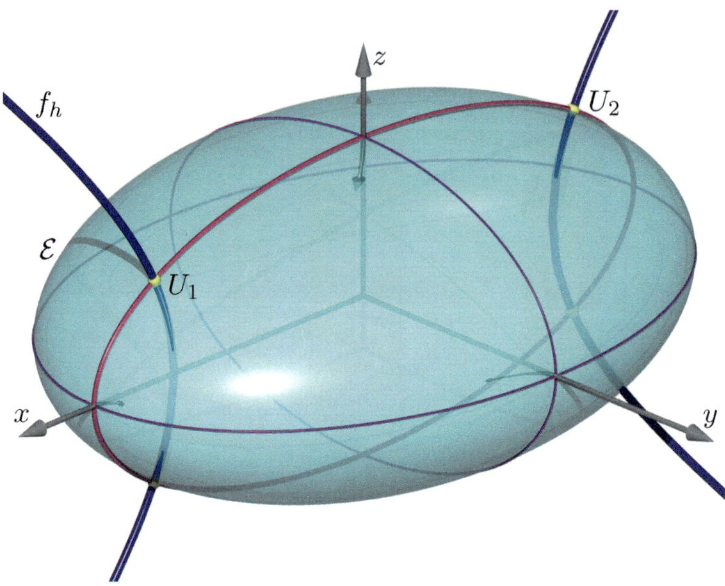

FIGURE 7.2. Ellipsoid $\mathcal{E}$ with umbilic points U_1, U_2 and focal hyperbola f_h.

If $a = b \neq c$, then (7.1) defines a family of confocal central quadrics of revolution. In this case, the meridians in the $[x, z]$-plane are confocal conics. The 'focal ellipse' is a circle, the focal hyperbola as the limit $k \to -b^2 = -a^2$ degenerates to the z-axis. This time, two-sheeted hyperboloids do not show up in the family; they can be replaced with meridian planes.

For $a = b = c$, we obtain concentric spheres.

The quadrics being confocal with an elliptic paraboloid with the standard equation (see Table 2.1) can be represented as

$$\frac{x^2}{a^2+k} + \frac{y^2}{b^2+k} - 2z - k = 0 \quad \text{for} \quad k \in \mathbb{R} \setminus \{-a^2, -b^2\}. \tag{7.5}$$

In the case $a > b > 0$, this one-parameter set contains

$$\text{for} \begin{cases} -b^2 < k < \infty & \text{elliptic paraboloids,} \\ -a^2 < k < -b^2 & \text{hyperbolic paraboloids,} \\ k < -a^2 & \text{elliptic paraboloids.} \end{cases} \tag{7.6}$$

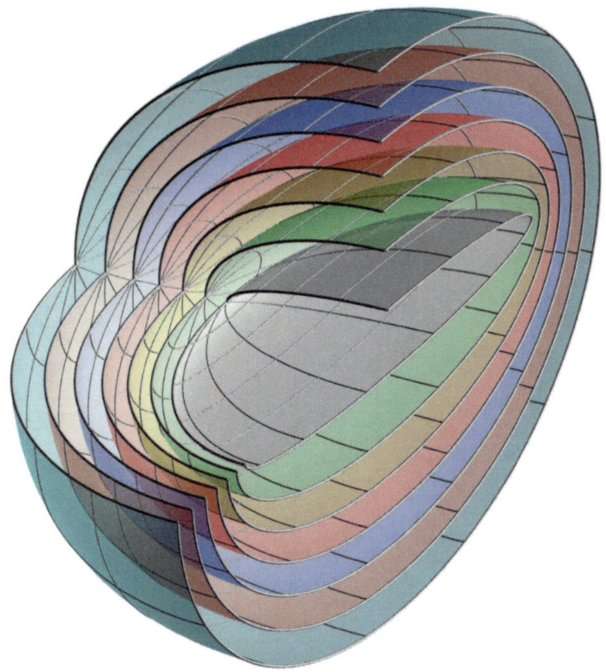

FIGURE 7.3. Confocal ellipsoids, where k tends to $-c^2$.

For each k, the vertex of the corresponding paraboloid has the coordinates $(0,0,-k/2)$. The point $(0,0,b^2/2)$ is the common focal point of the principal sections in the plane $x=0$, and $(0,0,a^2/2)$ is the analogue for the sections with $y=0$.

The limits for $k \to -b^2$ or $k \to -a^2$ define the pair of *focal parabolas* (Figure 7.5)

$$\begin{aligned} f_1 &: \frac{x^2}{a^2-b^2} - 2z + b^2 = 0, \quad y = 0\,, \\ f_2 &: \frac{y^2}{a^2-b^2} + 2z - a^2 = 0, \quad x = 0\,, \end{aligned} \tag{7.7}$$

within the family of confocal quadrics. The vertex of each focal parabola coincides with the focal point of the other parabola. Therefore, this pair is the same as shown in [46, Fig. 4.15]. The limits of the paraboloids for $k \to -a^2$ and for $k \to -b^2$ again show a difference between the right-hand and the left-hand version: In one case it is the interior of a focal parabola that is covered twice, and in the other case the exterior.

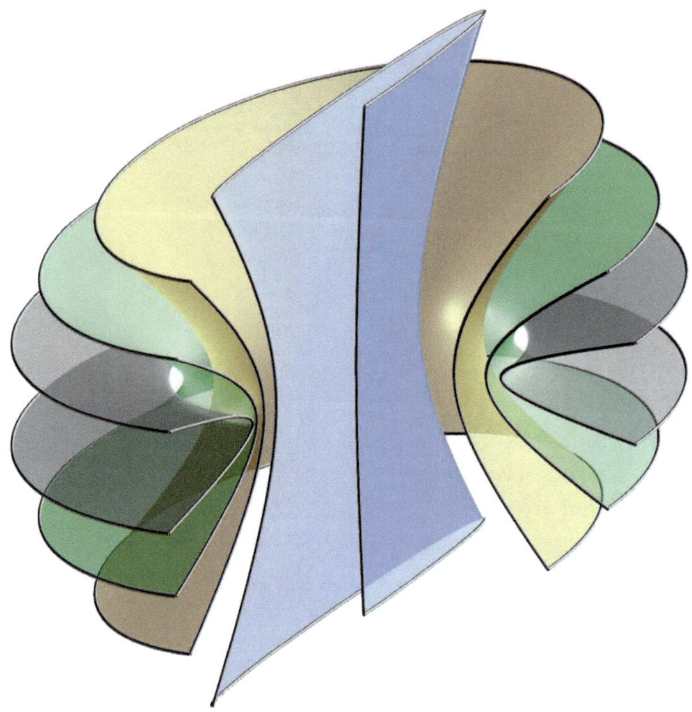

FIGURE 7.4. Confocal one-sheeted hyperboloids, where k tends to $-b^2$.

For the sake of completeness, we recall from [46, (10.12)] that, after a slight modification, *confocal quadratic cones* with their common apex at the origin satisfy

$$\frac{x^2}{a^2+k} + \frac{y^2}{b^2+k} - \frac{z^2}{c^2-k} = 0, \quad k \in \mathbb{R} \setminus \{-a^2, -b^2, c^2\}. \tag{7.8}$$

Their intersections with the unit sphere are confocal spherical conics. We assume $a > b > 0$. Then, for $k < -a^2$ and $k > c^2$, the cones do not contain real points other than the origin. The flat limit for $k \to -b^2$ is a sector bounded by the lines

$$\frac{x}{\sqrt{a^2-b^2}} \pm \frac{z}{\sqrt{b^2+c^2}} = 0 \tag{7.9}$$

in the plane $y = 0$. These lines are called *focal lines* of the cones since they pass through the focal points of the spherical conics. Another name is *focal axes* according to the general Definition 7.1.2, which follows below.

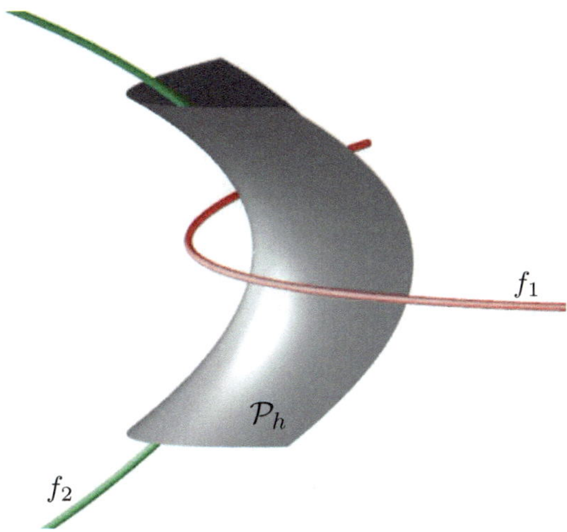

FIGURE 7.5. A hyperbolic paraboloid $\mathcal{P}_h$ together with its focal parabolas f_1 and f_2.

At the limits for $k \to -a^2$ and $k \to c^2$, the corresponding spherical conics become great circles in the common planes of symmetry.

Confocal quadrics define a range

Theorem 7.1.1 *Confocal quadrics belong to a range of dual quadrics, which includes the absolute conic as a singular surface. Therefore, confocal quadrics share their isotropic tangent planes.*
Also confocal quadratic cones, each seen as a set of tangent planes, belong to a range, which includes the isotropic cone with the common apex O.

Proof: In analogy to the dual equations of ellipsoids $\mathcal{E}$ (2.8) and hyperboloids $\mathcal{H}$ (2.30) with the respective standard equations (2.2) and (2.20), we can state that the plane given by

$$u_0 + u_1 x + u_2 y + u_3 z = 0$$

is tangent to any surface of the family of confocal quadrics (7.1) if, and only if,

$$(-u_0^2 + a^2 u_1^2 + b^2 u_2^2 + c^2 u_3^2) + k(u_1^2 + u_2^2 + u_3^2) = 0. \tag{7.10}$$

This is a linear combination of the homogenized dual equation of $\mathcal{E}$ and the equation of a singular quadric consisting of all isotropic planes.
The homogenized dual equation of confocal paraboloids satisfying (7.5) shows a similar result, namely

$$(a^2 u_1^2 + b^2 u_2^2 - 2u_0 u_3) + k(u_1^2 + u_2^2 + u_3^2) = 0.$$

Finally, the dual equation of confocal cones, as given in (7.8), reads

$$(a^2 u_1^2 + b^2 u_2^2 - c^2 u_3^2) + k(u_1^2 + u_2^2 + u_3^2) = 0, \quad u_0 = 0,$$

and it shows again a range, spanned by the given cone ($k = 0$) and the isotropic cone with its apex at the origin. ■

By virtue of Theorem 7.1.1, two confocal quadrics share their isotropic tangent planes. The confocal families given by (7.1) contain four rank-3 dual quadrics, the absolute conic, and the two focal conics (7.3), each seen as a set of tangent planes. The fourth singular quadric (7.4) is an imaginary (or empty) conic in the third plane of symmetry.

Theorem 7.1.2 *The cones or cylinders, drawn from any finite or ideal point P to the quadrics of a confocal family or connecting P with one of the focal conics, are confocal. For finite P, the common planes of symmetry of the confocal cones are tangent to one of the three quadrics passing through P. Therefore, confocal quadrics form a triply orthogonal system of surfaces.*

Proof: The tangent cones in question share all isotropic planes which are common to the confocal quadrics and pass through P. Therefore, these cones with apex P are confocal, too. This is a classical result attributed to C. G. J. Jacobi in 1834 [127, p. 204] and a special case of a theorem concerning ranges of surfaces of degree two (see Chapter 4).

When the quadric $\mathcal{Q}$ passes through the point P, the tangent cone from P to $\mathcal{Q}$, defined as a set of planes, splits into two pencils of planes with real or complex conjugate axes. These two axes are generators of $\mathcal{Q}$ and span the tangent plane τ_P to $\mathcal{Q}$ at P. On the other hand, the planes spanned by these pairs of axes are the common planes of symmetry of the confocal cones. This confirms that confocal quadrics form a triply orthogonal system of surfaces. ■

This result has a series of important consequences. We will consider four of them.

(1) A famous theorem by Dupin[3] states that the surfaces of a triply orthogonal system intersect each other along curvature lines (cf. [56, p. 187] or [129, p. 80]). This is equivalent to the statement that the surface normals to any surface $\mathcal{S}_1$ out of this family along the intersection curve $c = \mathcal{S}_1 \cap \mathcal{S}_2$ form a developable ruled surface, namely the developable which contacts $\mathcal{S}_2$ along c. We summarize the foregoing result for the case of confocal quadrics (note Figure 7.1 and that on page 279).

Corollary 7.1.1 *The quadrics of a confocal range intersect each other along their curvature lines. Therefore, the curvature lines of regular*

[3] Charles Dupin (1784–1873), French mathematician.

quadrics Q are algebraic curves. They appear as conics when projected orthogonally onto any plane of symmetry of Q.

(2) In the family of confocal quadrics, there are three quadrics passing through any given point P in space, outside the common planes of symmetry. In the case of central quadrics (see Figure 7.1), we obtain an ellipsoid, a one-sheeted hyperboloid, and a two-sheeted hyperboloid. In the case of paraboloids, two elliptic paraboloids, opening to opposite sides, and a hyperbolic paraboloid are meeting at P. Usually (see, *e.g.*, [73, p. 22], [129, p. 81], or [133, p. 64]), this is proved by showing that, after plugging the coordinates of P into (7.1) or (7.5), respectively, the equations have three zeros for the unknown parameter k: one in each of the parameter intervals listed in (7.2) and (7.6).

The parameters (k_1, k_2, k_3) of the three central quadrics passing through the point P are called *elliptic coordinates* of P (see, *e.g.*, [73, § 5] or [129, p. 81]). For the family given in (7.1), the Cartesian coordinates (x, y, z) of the point P and its elliptic coordinates with $k_1 > -c^2 > k_2 > -b^2 > k_3 > -a^2$ are related via

$$\begin{gathered} x^2 = \frac{(a^2+k_1)(a^2+k_2)(a^2+k_3)}{(a^2-b^2)(a^2-c^2)}, \quad y^2 = \frac{(b^2+k_1)(b^2+k_2)(b^2+k_3)}{(b^2-c^2)(b^2-a^2)}, \\ z^2 = \frac{(c^2+k_1)(c^2+k_2)(c^2+k_3)}{(c^2-a^2)(c^2-b^2)}. \end{gathered} \tag{7.11}$$

Apparently, eight points in space, symmetrically placed w.r.t. the coordinate frame, share their elliptic coordinates. The n-dimensional analogues of (7.11) are provided in [60, Lemma 1].

We retain the name *elliptic coordinates* also for the case of confocal paraboloids.[4] Then, for the family or quadrics (7.5), the relations between elliptic and Cartesian coordinates are

$$\begin{gathered} x^2 = -\frac{(a^2+k_1)(a^2+k_2)(a^2+k_3)}{(a^2-b^2)}, \quad y^2 = \frac{(b^2+k_1)(b^2+k_2)(b^2+k_3)}{(a^2-b^2)}, \\ z = -\frac{a^2+b^2+k_1+k_2+k_3}{2}, \quad \text{for} \quad k_1 > -b^2 > k_2 > -a^2 > k_3. \end{gathered} \tag{7.12}$$

Readers are invited to develop the proofs of (7.11) and (7.12) in Exercise 7.1.7.

[4] Staude uses the notation *parabolic coordinates* in [126].

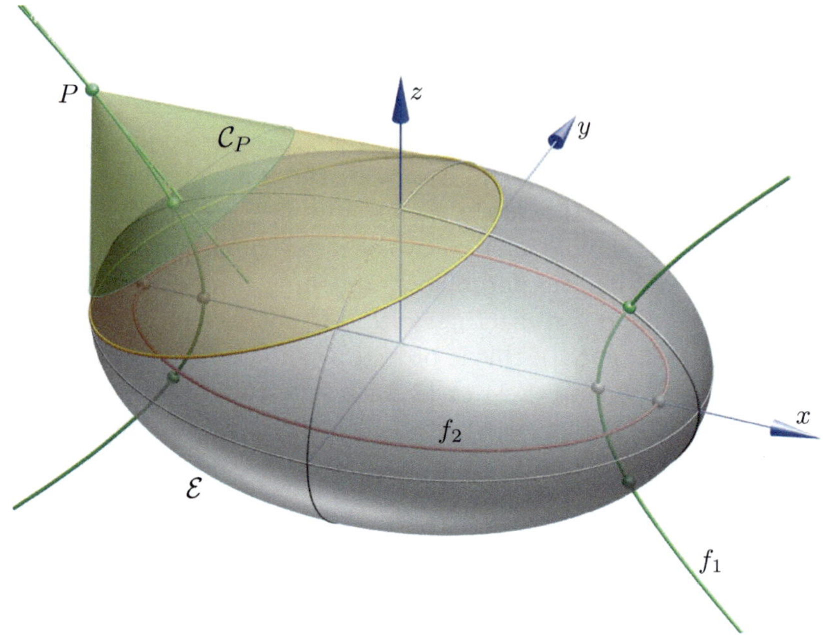

FIGURE 7.6. For an apex P on the focal hyperbola f_1 of the triaxial ellipsoid $\mathcal{E}$, the tangent cone $\mathcal{C}_P$ to $\mathcal{E}$ is a cone of revolution.

At any point P of a quadric $\mathcal{Q}$, the tangents t_1, t_2 at P to the curvature lines of $\mathcal{Q}$ are the surface normals of the other two confocal quadrics passing through P. They are axes of the Dupin indicatrix at P which, in the case of quadratic surfaces, is homothetic to planar sections of $\mathcal{Q}$ in planes parallel to the tangent plane τ_P to $\mathcal{Q}$ (note Section 8.2). This proves that t_1 and t_2 are *principal curvature tangents*, as defined for ellipsoids on page 13.

(3) Two confocal cones share the planes and axes of symmetry.

Corollary 7.1.2 *If tangents can be drawn from a finite or ideal point P to two confocal quadrics, then there exist two or four common tangents through P. They are symmetric w.r.t. three orthogonal planes through P.*

(4) Finally, we focus on confocal tangent cones in the case when the apex P is located on a focal conic.

Corollary 7.1.3 *Given a point P in the exterior of a quadric $\mathcal{Q}$, the tangent cone $\mathcal{C}_P$ from P to $\mathcal{Q}$ is a cone of revolution if, and only if, P is located on a focal conic f of $\mathcal{Q}$. The axis of $\mathcal{C}_P$ is the proper tangent of f at P.*

Proof: If P lies on a focal conic f_1 of a family of confocal quadrics, then the cone connecting P with the other focal conic f_2 is a cone of revolution. By virtue of Theorem 7.1.2, all tangent cones from P to quadrics confocal with f_2 are of the same type, and their common axis is the proper tangent line to f_1 at P (see Figure 7.6).

Conversely, f_1 is the locus of points P from which the connecting cone with f_2 is a cone of revolution, provided that P is not coplanar with f_2 (note [46, p. 139]). ■

Focal axes of quadrics

The following definition addresses a generalization of the focal points of conics [46, p. 254].

Definition 7.1.2 A (proper) line ℓ is called a *focal axis* of any quadric, quadratic cone, or cylinder if, and only if, the two isotropic planes through ℓ are tangent planes of the quadric, cone, or cylinder, respectively [127, pp. 205–206].

In the case of a quadratic cylinder, the focal axes are parallel to the cylinder's generators and pass through the focal points of the orthogonal sections. Therefore, elliptic or hyperbolic cylinders have two real and two complex conjugate focal axes. A parabolic cylinder has only one real focal axis.

A quadratic cone $\mathcal{C}$ has two real focal axes and two pairs of complex conjugate focal axes. The real ones pass through the focal points of the spherical conic, which is the intersection of $\mathcal{C}$ with the unit sphere centered at the apex of $\mathcal{C}$. If $\mathcal{C}$ satisfies (7.8) for any $k \neq -b^2$ in the open interval $(-a^2, c^2)$, the two real focal axes satisfy

$$y = x\sqrt{b^2 + c^2} \pm z\sqrt{a^2 - b^2} = 0,$$

in accordance with (7.9). The planes $x = 0$ and $z = 0$ contain pairs of complex conjugate focal axes satisfying, respectively,

$$z = x\sqrt{b^2 + c^2} \pm iy\sqrt{a^2 + c^2} = 0 \quad \text{or} \quad x = y\sqrt{a^2 + c^2} \pm iz\sqrt{a^2 - b^2} = 0.$$

With regard to the focal axes of regular quadrics, we can state

Lemma 7.1.4 *1. Each focal axis of a quadric $\mathcal{Q}$ is a focal axis of all quadrics being confocal with $\mathcal{Q}$.*

2. Each focal axis ℓ of a regular quadric $\mathcal{Q}$ is either a generator of a confocal ruled quadric or a proper tangent of a focal conic of $\mathcal{Q}$.

3. If ℓ is a focal axis of $\mathcal{Q}$ and P a point of ℓ, then ℓ is a focal axis of the cones drawn from P tangent to $\mathcal{Q}$ and to all confocal quadrics.

Proof: The first statement follows from Theorem 7.1.1. For each real plane, there exists in the confocal range exactly one regular or singular surface that is tangent to the plane. If ℓ is a focal axis, we choose any non-isotropic plane through ℓ. This plane is tangent to a quadric $\mathcal{Q}$ in the range. Since, together with the two isotropic planes, there already exist three planes through ℓ being tangent to $\mathcal{Q}$, the line ℓ must be a generator of a ruled quadric $\mathcal{Q}$ or a proper tangent of a focal conic. ■

Theorem 7.1.3 *Given a family of confocal quadrics, a real line ℓ is either a focal axis or it touches exactly two surfaces of the confocal family, and the tangent planes at the respective points of contact are orthogonal.*

Proof: We apply the spatial version of Desargues's involution theorem [46, p. 309] to the given range of quadrics. This theorem states that either the pairs of tangent planes through ℓ to the quadrics of the range form an involution or a plane through ℓ contacts all quadrics.

All common tangent planes of confocal quadrics are isotropic. So, either the real line ℓ is the intersection of two common tangent planes and ℓ or it is a focal axis or the two isotropic planes through ℓ are a pair of an involution. The latter implies that the two fixed planes of the involution are orthogonal. We obtain a fixed plane exactly in the case when the corresponding quadric is tangent to ℓ (note Exercise 7.1.11). ■

Remark 7.1.1 Concerning Theorem 7.1.3, the points of contact of the two tangent planes through ℓ with the quadrics need not be different. By virtue of Corollary 7.1.1, they coincide if, and only if, ℓ is tangent to a line of curvature on a quadric.

Theorem 7.1.3 is attributed to M. CHASLES.[5] Its n-dimensional version can be found in [133, p. 65].

Further, Theorem 7.1.3 gives rise to a characterization of geodesics on quadrics (cf. Figure 7.7), which dates back to JACOBI (1839) [68] and CHASLES [27].

Theorem 7.1.4 *On each quadric $\mathcal{Q}$, the geodesics are curves c with tangents contacting another fixed quadric $\mathcal{Q}' \neq \mathcal{Q}$ out of the confocal family.*

[5] MICHEL CHASLES (1793–1880), a French mathematician and professor at the École Polytechnique.

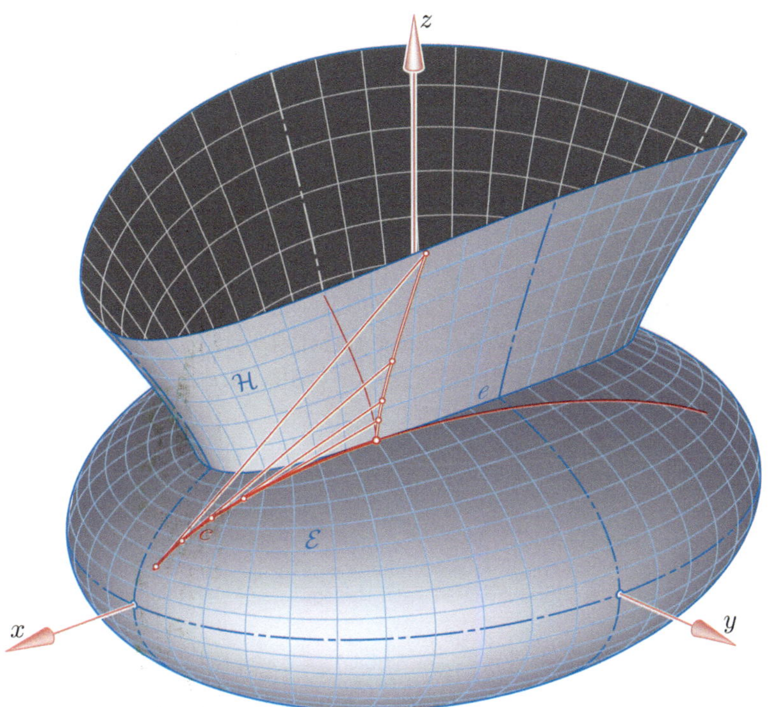

FIGURE 7.7. The tangent surface along any geodesic c on the ellipsoid $\mathcal{E}$ contacts a confocal quadric $\mathcal{H}$ (Theorem 7.1.4).

This theorem holds also for the intersections of $\mathcal{Q}$ with its planes of symmetry, provided that each line in a plane of symmetry is defined as a (not necessarily proper) tangent of the singular curve which lies in this plane. This is in agreement with the convention described on page 281.

Proof: Suppose that a smooth curve c on $\mathcal{Q}$ has the property that all its tangents t contact the confocal quadric $\mathcal{Q}'$, but c lies outside $\mathcal{Q}'$. Let t be the tangent to c at any point $P \in c$ and let $P' \neq P$ denote its point of contact with $\mathcal{Q}'$. Since the osculating plane of c at $P \in c$ is tangent to the tangential surface of c along t, it coincides with the tangent plane $\tau_{P'}$ to $\mathcal{Q}'$ at P'. By virtue of Theorem 7.1.3, the osculating plane of c at P is orthogonal to the tangent plane τ_P at P to $\mathcal{Q}$. This characterizes c as a geodesic. We can state that the *integral curves* on $\mathcal{Q}$, defined by the congruence of common tangents of the two confocal quadrics $\mathcal{Q}$ and $\mathcal{Q}'$, are geodesic lines on $\mathcal{Q}$, except the connected components e and $\overline{e}$ of $\mathcal{Q} \cap \mathcal{Q}'$ (Figure 7.7).

Geodesics oscillate between two antipodal curvature lines e and $\overline{e}$. In particular, geodesics passing through one umbilic point U of a central quadric $\mathcal{Q}$ must also pass through the antipode of U [7, p. 519].

Conversely, each geodesic c on the quadric Q is uniquely defined by an initial point and its tangent t [82, p. 143]. If t is a focal axis, then, by Lemma 7.1.4, t is a generator of $\mathcal{Q}$ and hence geodesic. Otherwise, by virtue of Theorem 7.1.3, t contacts exactly two surfaces of the family of confocal quadrics. One surface is $\mathcal{Q}$ itself. The other quadric $\mathcal{Q}'$ defines a congruence of common tangents with $\mathcal{Q}$, and, as a consequence of the first part of this proof, c must be one of the integral curves. Formulas expressing geodesics in terms of elliptic coordinates can be found in [73, p. 25–31] ■

In the literature, there are many results about properties of geodesics. We consider a theorem by F. JOACHIMSTHAL[6] [7, p. 517]: Let X be a point of a geodesic line with tangent t and tangent plane τ_X to the carrier quadric $\mathcal{Q}$. Then, along the geodesic the product of the distance of τ_X to the center and the length of the diameter parallel to t is constant. Geodesics of ellipsoids in $\mathbb{E}^n$ are studied, *e.g.*, in [76]. Geodesics of ellipsoids in the 3-dimensional Minkowski space are treated in [43]; geodesics in the n-dimensional Lorentz space are a topic in [72].

• **Exercise 7.1.1** Vertex generators of confocal paraboloids.
Prove the following statement [64]: For each family of confocal paraboloids satisfying (7.5), the vertex generators of the hyperbolic paraboloids and the proper vertex tangents of the focal parabolas are located on a Plücker conoid with the equation

$$(x^2+y^2)\left(z-\frac{a^2+b^2}{4}\right)+\frac{a^2-b^2}{4}(x^2-y^2)=0$$

(compare with the equations in (2.26) and (2.56)).

• **Exercise 7.1.2** Affine transformation $P_\varepsilon \mapsto P_n$ in Theorem 2.4.1.
Let the quadric $\mathcal{Q}$ intersect its plane ε of symmetry along the conic c_ε, while $f \subset \varepsilon$ is a focal conic of $\mathcal{Q}$. Show for the affine transformation $P_\varepsilon \mapsto P_n$, as explained in Theorem 2.4.1, that the polar of P_ε w.r.t. c_ε coincides with the polar of P_n w.r.t. f. Therefore, the affine transformation is the composition of two polarities: the first as a point-to-line transformation and the second as a line-to-point transformation.

• **Exercise 7.1.3** Polarity w.r.t. confocal quadrics in $\mathbb{E}^3$.
Prove the following assertions concerning a given family of confocal quadrics in $\mathbb{E}^3$:
1. The poles of a given finite plane ε w.r.t. the quadrics of the confocal range are located on a line orthogonal to ε. This line is the surface normal of the unique quadric which contacts ε.
2. For a given finite line ℓ outside the common planes of symmetry, the polar lines w.r.t. the quadrics of the confocal range form a hyperbolic paraboloid. This paraboloid is orthogonal if, and only if, ℓ is a focal axis.

• **Exercise 7.1.4** Proof of Theorem 2.4.2.

Prove the following statement concerning the surface normals n_P of any central or parabolic quadric $\mathcal{Q}$ in $\mathbb{E}^3$:

---

[6] FERDINAND JOACHIMSTHAL (1818–1861), German mathematician, professor in Halle and Wrozław (Breslau).

For any two given points $P_1, P_2 \in \mathcal{Q}$ outside of any plane of symmetry, there is a similarity between the respective surface normals n_1 and n_2, mapping P_1 onto P_2 and, for each symmetry plane σ of $\mathcal{Q}$, the point of intersection $n_1 \cap \sigma$ onto $n_2 \cap \sigma$.

Hint: By virtue of Exercise 7.1.3, the points of intersection of n_i with the planes of symmetry are the poles of the tangent plane τ_i w.r.t. the singular surfaces in the range of surfaces being confocal with $\mathcal{Q}$. These points of intersection are directly related to the fixed parameters k of the singular surfaces. In the case of central quadrics, these are the k-values $-a^2$, $-b^2$, and $-c^2$. For confocal paraboloids the k-values in question are $-a^2$ and $-b^2$, while P_i corresponds to the parameter $k = 0$ of the given paraboloid.

• **Exercise 7.1.5** Tangent plane and normal line of a quadric.

Prove the following spatial analogue of focal involutions of conics (see [46, p. 340]): Given any quadric $\mathcal{Q}$ with σ as a plane of symmetry. Then, for all points $P \in \mathcal{Q}$ outside of σ, the normal line n_P and the tangent plane τ_P to $\mathcal{Q}$ at P intersect σ in a point and its polar line w.r.t. the focal conic of $\mathcal{Q}$ which is located in σ. Why is the affine transformation of Exercise 7.1.2 the product of two polarities?

• **Exercise 7.1.6** Orthogonal views of confocal quadrics are confocal.

Prove the statement that, for every orthogonal projection, the visual contours of confocal quadrics are confocal. If the rays r of the orthogonal projection are parallel to a plane σ of symmetry of the quadric $\mathcal{Q}$, then the tangents to the focal conic f in σ that are parallel to r are projected onto the focal points of the visual contour of $\mathcal{Q}$.

• **Exercise 7.1.7** Relations between Cartesian and elliptic coordinates.

Prove the relations between the Cartesian coordinates (x, y, z) and the elliptic coordinates (k_1, k_2, k_3) of points, as listed in (7.11) and (7.12).

• **Exercise 7.1.8** Points of contact with parallel planes.

Verify that the points of contact between a pencil of parallel planes and the quadrics of a confocal family are located on an equilateral hyperbola [60, Statement 3].

• **Exercise 7.1.9** Involution related to the principal sections of a quadric.

Given a central quadric $\mathcal{Q}$, let ν_1, ν_2 be the planes of the two principal sections, *i.e.*, the normal planes through the principal curvature tangents, at a point $X \in \mathcal{Q}$. Then, for all points X outside the planes of symmetry, the two planes ν_1, ν_2 intersect any axis of symmetry of $\mathcal{Q}$ in pairs of an involution.

• **Exercise 7.1.10** An invariant along geodesics.

Confirm the following statement: For points $\mathbf{p}$ with tangent vectors $\mathbf{v} = (v_1, v_2, v_3)$ of a geodesic on the ellipsoid with semiaxes a, b, c, the quantity

$$k := \frac{\|\mathbf{v} \times \mathbf{p}\|^2 - v_1^2(b^2 + c^2) - v_2^2(c^2 + a^2) - v_3^2(a^2 + b^2)}{\|\mathbf{v}\|^2}$$

remains constant.

Hint: By virtue of Theorem 7.1.4, the tangents of any geodesic contact a confocal quadric. Show that k is the parameter of this quadric.

• **Exercise 7.1.11** Compute confocal quadrics which are tangent to a given line.

Confirm the following statement: If a line ℓ through a point $\mathbf{p} = (p_1, p_2, p_3)$ and with direction vector $\mathbf{v} = (v_1, v_2, v_3)$ contacts a quadric $\mathcal{Q}$ out of the confocal range spanned by the ellipsoid with standard equation (2.2), then the parameter k of $\mathcal{Q}$ is a root of the quadratic equation

$$\begin{aligned} \|\mathbf{v}\|^2 k^2 + \big(\|\mathbf{w}\|^2 - v_1^2(b^2+c^2) - v_2^2(c^2+a^2) - v_3^2(a^2+b^2)\big)\, k \\ + \big(a^2b^2v_3^2 + b^2c^2v_1^2 + c^2a^2v_2^2 - a^2w_1^2 - b^2w_2^2 - c^2w_3^2\big) = 0, \end{aligned}$$

where $\mathbf{w} = (w_1, w_2, w_3) = \mathbf{v} \times \mathbf{p}$. What characterizes ℓ, in accordance with Theorem 7.1.3, as a focal axis of the confocal family?

7.2 Ivory's Theorem and bipartite frameworks

Ivory's Theorem (see Figure 7.8), implicitly included in Sir James Ivory's paper [66] from 1809, deals with confocal surfaces and holds in Euclidean, hyperbolic, and elliptic spaces of all dimensions. The two-dimensional version, the spherical version in $\mathbb{E}^3$, as well as the versions in the Minkowski plane $\mathbb{M}^2$ and in the hyperbolic plane $\mathbb{H}^2$ were already addressed in [46, pp. 39, 454, 477, 479].

Here, we focus on Ivory's Theorem in $\mathbb{E}^n$, $n \geq 2$, together with a remarkable converse which illustrates the role of quadrics in the theory of rigidity. The converse in question deals with *bar frameworks* (*frameworks*, for short), *i.e.*, with graphs of any combinatorial type and related *configurations* (or *realizations*) in $\mathbb{E}^n$ where the knots as points in $\mathbb{E}^n$ and the edges are bars in the form of straight segments terminating in two knots.

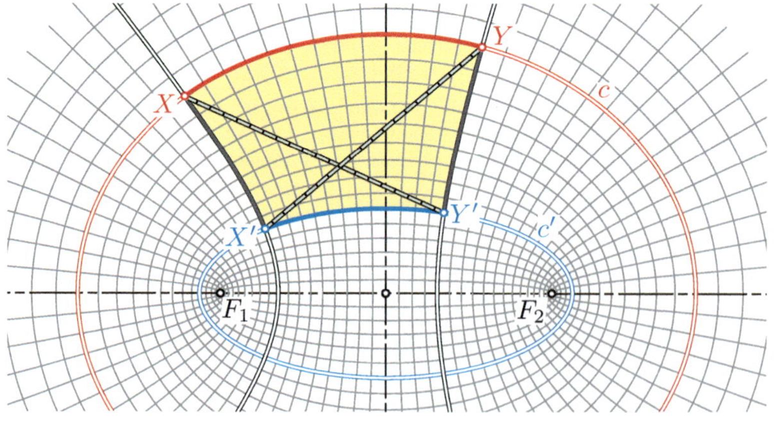

FIGURE 7.8. Ivory's Theorem in the Euclidean plane $\mathbb{E}^2$: $\overline{XY'} = \overline{X'Y}$.

Corresponding points of confocal quadrics

As a first step, we pay attention to the fact that, for any two confocal regular quadrics $\mathcal{Q}$, $\mathcal{Q}'$ of the same type in $\mathbb{E}^3$, *e.g.*, for two confocal ellipsoids, there is an axial scaling $\gamma\colon\ (x, y, z) \mapsto (\lambda_1 x,\ \lambda_2 y, \lambda_3 z)$ with coefficients $\lambda_i \in \mathbb{R}$, which maps $\mathcal{Q}$ onto $\mathcal{Q}'$, while the axes of the coordinate frame remain fixed. In the case of confocal parabolic quadrics, we have to combine the scaling of the x- and y-coordinates with an appropriate translation along the z-axis in order to obtain again an affine transformation γ with $\mathcal{Q} \mapsto \mathcal{Q}'$, which preserves the planes of symmetry.

We generalize to the Euclidean n-space, $n \geq 2$:

(i) Let $\mathcal{Q}$ be a regular central quadric in $\mathbb{E}^n$. By virtue of Theorem 3.1.1 and by analogy to (7.1), the family of confocal quadrics of $\mathcal{Q}$ can be represented as

$$F(\mathbf{x},\, k) := \sum_{i=1}^{n} \varepsilon_i \frac{x_i^2}{a_i^2 + \varepsilon_i k} - 1 = 0, \tag{7.13}$$

where the sequence of coefficients $\varepsilon_1, \ldots, \varepsilon_n$ with $\varepsilon_i \in \{+1, -1\}$ follows from the condition that $F(\mathbf{x}, 0) = 0$ is the equation of the given quadric $\mathcal{Q}$. If $F(\mathbf{x},\, k') = 0$ defines a confocal quadric $\mathcal{Q}'$ of the same type as $\mathcal{Q}$, the parameter k' must be sufficiently close to 0, or to be precise, $a_i^2 + \varepsilon_i k' > 0$ for all i. Then, the n-dimensional version of the axial scaling

$$\gamma\colon\ \mathbf{x} = (x_1, \ldots, x_n) \mapsto \mathbf{x}' = (x_1', \ldots, x_n')$$

is defined by

$$x_i'^2 = \frac{a_i^2 + \varepsilon_i k'}{a_i^2}\, x_i^2 \quad \text{for all } i = 1, \ldots, n. \tag{7.14}$$

This guarantees that $\mathbf{x} \in \mathcal{Q}$, *i.e.*, $F(\mathbf{x}, 0) = 0$, implies $\gamma(\mathbf{x}) \in \mathcal{Q}'$, hence $F(\mathbf{x}',\, k') = 0$.

(ii) In the case of a regular parabolic quadric $\mathcal{Q}$ in $\mathbb{E}^n$, the confocal family is given by

$$F_p(\mathbf{x}, k) := \sum_{j=1}^{n-1} \varepsilon_j \frac{x_j^2}{a_j^2 + \varepsilon_j k} - 2x_n - k = 0. \tag{7.15}$$

To obtain an affine transformation which fixes again all planes of symmetry, we set

$$x_j'^2 = \frac{a_j^2 + \varepsilon_j k'}{a_j^2}\, x_j^2 \text{ for } j = 1, \ldots, n-1 \text{ and } x_n' = x_n + \frac{k - k'}{2}\,. \tag{7.16}$$

(iii) Let $\mathcal{Q}$ be a conical quadric in $\mathbb{E}^n$ with a 0-dimensional apex at the origin. Then, the confocal family can be defined by

$$F_c(\mathbf{x}, k) := \sum_{i=1}^{n} \varepsilon_i \frac{x_i^2}{a_i^2 + \varepsilon_i k} = 0 \tag{7.17}$$

with $\mathcal{Q}$ as $F_c(\mathbf{x}, 0) = 0$. The confocal cone $\mathcal{Q}'$ of the same type satisfies $F(\mathbf{x}, k') = 0$. We define the affine transformation $\gamma\colon\ \mathcal{Q} \mapsto \mathcal{Q}'$ by (7.14).

It is easy to see that, for the definitions of confocal central, parabolic, and conical quadrics in $\mathbb{E}^n$, as given above, the Definition 7.1.1 is still valid for all planes spanned by two coordinate axes. More generally, we can state that the intersections of confocal quadrics with each subspace, spanned by k coordinate axes with $k < n$, are again confocal.

In each of the three cases listed above, the scaling γ is not unique since the coefficients λ_i with

$$\lambda_i^2 = \frac{a_i^2 + \varepsilon_i k'}{a_i^2}, \quad i = 1, \dots$$

need not be positive (Figure 7.9). Some λ_i can even be zero, *e.g.* in $\mathbb{E}^3$, when $\mathcal{Q}$ is an ellipsoid and $\mathcal{Q}'$ covers the closed interior of the focal ellipse.

We can also extend the definition of γ to the 'cylindric' cases, where the sums in (7.13) and (7.17) go from 1 to $r < n$ and those in (7.15) from 1 to $r - 1 < n - 1$. Then, the respective transformations of $x_1, \dots, x_r$, as given above, are completed by

$$x_k' = x_k \quad \text{for} \quad k = r + 1, \dots n. \tag{7.18}$$

The following definition refers to the affine transformations γ and originates from classical literature on this topic (see, *e.g.*, [66] or [35, p. 116]).

Definition 7.2.1 Let $\mathcal{Q}$ and $\mathcal{Q}'$ be two confocal quadrics of the same type in $\mathbb{E}^n$. The pointset $\mathcal{Q}'$ can even be an adjacent flat limit, related to a singular quadric in the confocal range.
If γ denotes a fixed affine transformation $\mathcal{Q} \mapsto \mathcal{Q}'$, as defined above in (7.14) and (7.16) with the extension (7.18) for cylindrical cases, then the points $X \in \mathcal{Q}$ and $X' := \gamma(X) \in \mathcal{Q}'$ are called *corresponding* points of the pair of confocal quadrics $(\mathcal{Q}, \mathcal{Q}')$ (Figure 7.9).

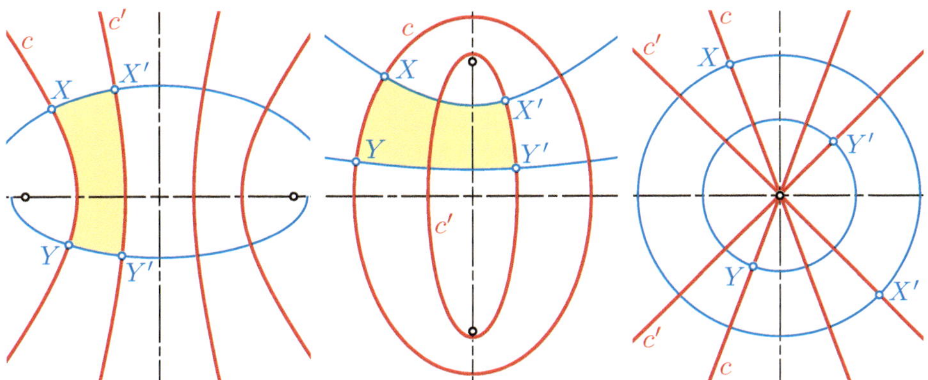

FIGURE 7.9. Examples of corresponding points on two confocal conics c, c'.

It means no restriction to exclude quadrics of revolution, *i.e.*, cases with equal semiaxes $a_i = a_j$ for $i \neq j$. This is so because otherwise γ preserves all hyperplanes of the pencil spanned by $x_i = 0$ and $x_j = 0$, and each of these $(n-1)$-dimensional 'meridian hyperplanes' shows congruent transformations $X \mapsto X'$ between corresponding points. The same happens in 'cylindric cases' $r < n$ in the hyperplanes from $x_{r+1} = \text{const.}$ to $x_n = \text{const.}$ In the following lemma, this case is also excluded.

Lemma 7.2.1 *Let X and X' be corresponding points of two confocal quadrics $\mathcal{Q}$ and $\mathcal{Q}'$ of the same type, where $\mathcal{Q}$ is neither cylindric nor a surface of revolution. Then, all other $n-1$ confocal quadrics, passing through X, also contain the point X'. In the conical case, the n-th quadric through X is a sphere centered at the origin.*

Proof: It is sufficient to show that the equations $F(\mathbf{x}, 0) = 0$ and $F(\mathbf{x}, k) = 0$ with $k \neq 0$ imply $F(\mathbf{x}', k) = 0$ for $\mathbf{x}' = \gamma(\mathbf{x})$ with $F(\mathbf{x}', k') = 0$.

(i) In the case of central quadrics, this follows from an identity which expresses $F(\mathbf{x}', k)$, defined in (7.13), as an affine combination

$$F(\mathbf{x}', k) = \frac{k'}{k}\,F(\mathbf{x}, 0) + \left(1 - \frac{k'}{k}\right)F(\mathbf{x}, k).$$

This identity can readily be verified by (7.14) as

$$\frac{1}{a_i^2 + \varepsilon_i k}\,x_i'^2 = \frac{a_i^2 + \varepsilon_i k'}{a_i^2(a_i^2 + \varepsilon_i k)}\,x_i^2 = \left[\frac{k'}{k a_i^2} + \frac{k - k'}{k(a_i^2 + \varepsilon_i k)}\right] x_i^2 \quad \text{for all } i = 1, \dots$$

(ii) It is easy to see that the affine combination, as given above, is also valid for the function F_p from (7.15), since the identity holds for the included non-quadratic terms.

(iii) Concerning the case of confocal cones, the same affine combination works for the function F_c. We note that corresponding points have the same distance to the common apex at the origin, because, by virtue of (7.14),

$$\sum_{i=1}^{n} x_i'^2 = \sum_{i=1}^{n} \frac{a_i^2 + \varepsilon_i k'}{a_i^2} x_i^2 = \sum_{i=1}^{n} x_i^2 + \varepsilon_i k' \sum_{i=1}^{n} \frac{x_i^2}{a_i^2} = \sum_{i=1}^{n} x_i^2,$$

due to $\mathbf{x}$ being a point of $\mathcal{Q}: F_c(\mathbf{x}, 0) = 0$.

All spheres centered at the apex, and therefore orthogonal to all cones in the family, complete the family of confocal cones to an n-fold orthogonal system of hypersurfaces in $\mathbb{E}^n$. Two of these spheres can also be seen as confocal quadrics $\mathcal{Q}, \mathcal{Q}'$, and points $\mathbf{x} \in \mathcal{Q}$ and $\mathbf{x}' \in \mathcal{Q}'$ on the same diameter are corresponding. ■

As a direct consequence of Lemma 7.2.1, we can state, in terms of elliptic coordinates:

Corollary 7.2.2 *If X and X' are corresponding points of two confocal quadrics $\mathcal{Q}$ and $\mathcal{Q}'$ in $\mathbb{E}^3$, then the related elliptic coordinates of X and X' differ in no more than one coordinate.*

This condition is not sufficient since, due to (7.11) and (7.12), the elliptic coordinates only depend on the squared Cartesian coordinates of the two points, apart from the last coordinate in the parabolic case.[7]

Any two scalings γ_1, γ_2 related to two confocal conics or quadrics out of a confocal family satisfy $\gamma_1 \circ \gamma_2 = \gamma_2 \circ \gamma_1$. Therefore, if $\gamma_1 : X \mapsto X', Y \mapsto Y'$, and $\gamma_2 : X \mapsto Y$, then also $\gamma_2 : X' \mapsto Y'$. This means for any Ivory quadrangle (see Figure 7.8): If (X, X') and (Y, Y') are corresponding w.r.t. two confocal conics or quadrics, and if the elliptic coordinates of X and Y differ in one coordinate at most, then (X, Y) and (X', Y') are pairs of corresponding points w.r.t. two other confocal conics or quadrics.

In the three-dimensional case, the Ivory quadrangles are replaced by curved rectangular boxes, so-called *Ivory boxes*, as displayed in Figure 7.10 with the vertices $A, \dots, D$ on the ellipsoid $\mathcal{E}$ and the corresponding points $A', \dots, D'$ on the confocal ellipsoid $\mathcal{E}'$. Ivory's Theorem implies $\overline{AC'} = \overline{A'C}$ and $\overline{BD'} = \overline{B'D}$. Furthermore, since the pairs (A, B) and (D', C') are corresponding points of two confocal one-sheeted hyper-

[7] Usually, a curved quadrangle $XX'Y'Y$ formed by two pairs of corresponding points is called *Ivory quadrangle*. For example, in the two-dimensional case, it is not sufficient to speak of a quadrangle bounded by two pairs of confocal conics. Ivory's theorem (Figure 7.8) is valid only for curvilinear quadrangles formed by *'arcs semblables'*, according to [26], which means that opposite arcs are corresponding under an affine transformation γ, as defined above.

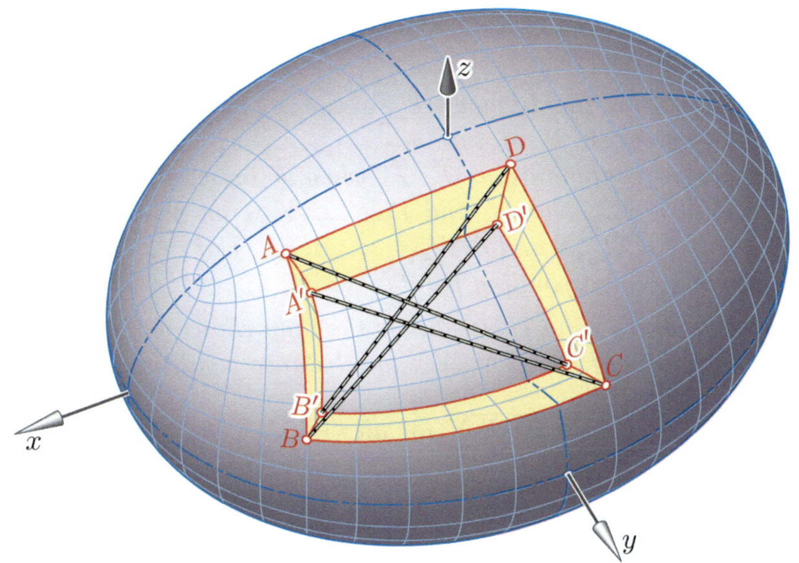

FIGURE 7.10. Curved Ivory box in $\mathbb{E}^3$ with its four diagonals of equal lengths.

boloids, we obtain $\overline{AC'} = \overline{BD'}$. Therefore, all four spatial diagonals are of equal lengths.

Configurations of bipartite frameworks

Theorem 7.2.1 (Configuration theorem of bipartite frameworks in $\mathbb{E}^n$) *Let $\mathcal{F}_0$ be a complete bipartite framework of combinatorial type $\mathcal{K}_{n+1,q+1}$ in $\mathbb{E}^n$, i.e., consisting of two disjoint sets of knots together with bars connecting every two knots from different classes. Suppose that this framework with given bar lengths admits two non-congruent configurations $\mathcal{F}$ and $\mathcal{F}'$ in $\mathbb{E}^n$. This means that the classes of knots $\{X_0, \dots, X_n\}$, $\{Y_0, \dots, Y_q\}$ of $\mathcal{F}$ and $\{X'_0, \dots, X'_n\}$, $\{Y'_0, \dots, Y'_q\}$ of $\mathcal{F}'$ satisfy the $(n+1)(q+1)$ quadratic equations*

$$l_{ik}^2 = \overline{X'_i Y'_k}^{\,2} = \overline{X_i Y_k}^{\,2} \text{ for all } i \in \{0, \dots, n\} \text{ and } k \in \{0, \dots, q\}. \tag{7.19}$$

If the knots $X_0, \dots, X_n$ are supposed to form a simplex in $\mathbb{E}^n$ and $q \geq 1$, then the following holds true:

1. *There is an appropriate motion $\beta\colon \mathbb{E}^n \to \mathbb{E}^n$ of the knots of $\mathcal{F}'$ such that, for all (i,k), the knots $X_i \mapsto \beta(X'_i)$ and $\beta(Y'_k) \mapsto Y_k$ are corre-*

sponding points of two confocal quadrics $\mathcal{X}$ and $\mathcal{Y}$ of the same type (see Figure 7.11).

2. *For any $r, s \in \mathbb{N}$, the framework $\mathcal{F}_0$ can be extended to a complete bipartite framework $\widetilde{\mathcal{F}}$ of type $\mathcal{K}_{n+r+1,\,q+s+1}$ which still admits two non-congruent configurations $\widetilde{\mathcal{F}}$, $\widetilde{\mathcal{F}}'$ when $\widetilde{\mathcal{F}}$ arises from $\mathcal{F}$ by adding knots $X_{n+1}, \dots, X_{n+r}$ on $\mathcal{X}$ and $Y_{q+1}, \dots, Y_{q+s}$ on $\mathcal{Y}$, while the knots $\beta(X'_{n+1}), \dots, \beta(X'_{n+r})$ and $\beta(Y'_{q+1}), \dots, \beta(Y'_{q+s})$ of $\beta(\widetilde{\mathcal{F}}')$ are specified as respectively corresponding points.*
3. *For knots of the second class, this is the only choice for extending $\mathcal{F}_0$, i.e., for any pair of points Y, Y', the distance conditions*

$$\overline{X'_iY'} = \overline{X_iY} \text{ for all } i \in \{0, \dots, n\}$$

imply $Y \in \mathcal{Y}$ and $Y' \in \beta^{-1}(\mathcal{X})$.

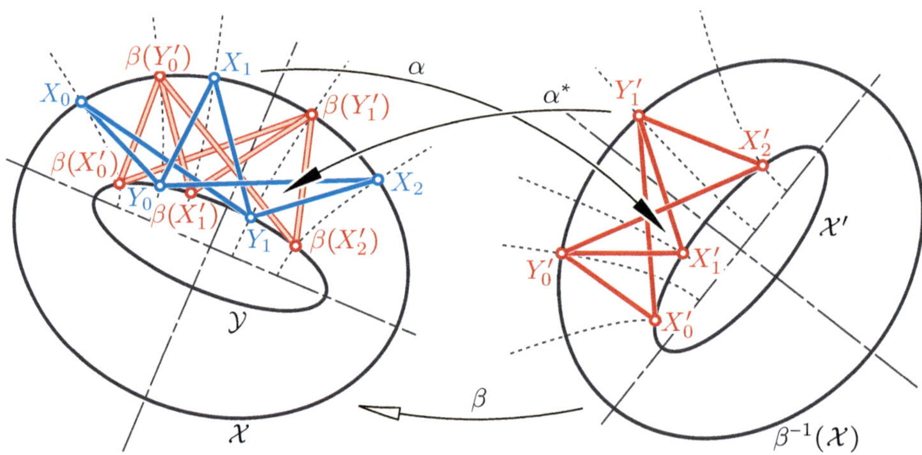

FIGURE 7.11. Illustrating Theorem 7.2.1 for dimension $n = 2$.

Proof: We use two Cartesian coordinate frames in $\mathbb{E}^n$, one for each configuration. Let $\mathbf{x}_i$ and $\mathbf{y}_k$ denote the respective position vectors of the knots X_i and Y_k of $\mathcal{F}$ and $\mathbf{x}'_i, \mathbf{y}'_k$ those of the knots X'_i, Y'_k of $\mathcal{F}'$. Then, the equations (7.19) are equivalent to

$$(\mathbf{x}_i - \mathbf{y}_k)^2 = (\mathbf{x}'_i - \mathbf{y}'_k)^2 \text{ for all } i \in \{0, \dots, n\} \text{ and } k \in \{0, \dots, q\}. \tag{7.20}$$

After subtracting from the equation for the indices $(0, k)$ that for the indices (i, k), we obtain

$$\mathbf{x}_0^2 - \mathbf{x}_i^2 + 2\langle \mathbf{x}_i - \mathbf{x}_0, \mathbf{y}_k \rangle = \mathbf{x}_0'^2 - \mathbf{x}_i'^2 + 2\langle \mathbf{x}'_i - \mathbf{x}'_0, \mathbf{y}'_k \rangle, \quad i = 0, \dots, n. \tag{7.21}$$

For each k, this can be seen as a fixed system of n linear equations for the unknown vector $\mathbf{y}_k$. For given $\mathbf{y}'_k$, this system has a unique solution, since $X_0, \dots, X_n$ is supposed to form a simplex. In order to express this solution in an appropriate form, we take into account that there is a unique affine transformation

$$\alpha\colon\ \mathbb{E}^n \to \mathbb{E}^n,\ \mathbf{x} \mapsto \alpha(\mathbf{x}) = \mathbf{a}' + l(\mathbf{x}) \text{ with } \mathbf{x}_i \mapsto \mathbf{x}'_i \text{ for all } i \in \{0,\dots,n\}. \tag{7.22}$$

Here, l denotes the induced linear map defined by $l(\mathbf{x}_i - \mathbf{x}_0) = \mathbf{x}'_i - \mathbf{x}'_0$ for $i = 1,\dots,n$. Let $l^*\colon \mathbb{R}^n \to \mathbb{R}^n$ be the *adjoint* map satisfying

$$\langle \mathbf{u},\, l^*(\mathbf{v}')\rangle = \langle l(\mathbf{u}),\, \mathbf{v}'\rangle \text{ for all } \mathbf{u}, \mathbf{v}' \in \mathbb{R}^n. \tag{7.23}$$

In the case $q \geq 1$, we subtract from (7.21) the equation for $k = 0$ and obtain

$$\langle \mathbf{x}_i - \mathbf{x}_0,\, \mathbf{y}_k - \mathbf{y}_0\rangle = \langle l(\mathbf{x}_i - \mathbf{x}_0),\, \mathbf{y}'_k - \mathbf{y}'_0\rangle \text{ for all } i \in \{1,\dots,n\} \text{ and } k \in \{1,\dots,q\}. \tag{7.24}$$

Then, we can fulfill all equations (7.24) by setting

$$\mathbf{y}_k - \mathbf{y}_0 = l^*(\mathbf{y}'_k - \mathbf{y}'_0) \text{ for all } k \in \{1,\dots,q\}.$$

This adjoint map l^*, together with one pair $\mathbf{y}'_0 \mapsto \mathbf{y}_0$ of points, defines the affine transformation

$$\alpha^*\colon\ \mathbb{E}^n \to \mathbb{E}^n,\ \mathbf{y}' \mapsto \alpha^*(\mathbf{y}') = \mathbf{b} + l^*(\mathbf{y}') \text{ with } \mathbf{y}'_k \mapsto \mathbf{y}_k \text{ for all } k \in \{0,\dots,q\}. \tag{7.25}$$

We call α and α^* *"adjoint"* affine transformations. This results in the necessary condition:

Lemma 7.2.3 *Let $\mathcal{F}$, $\mathcal{F}'$ in $\mathbb{E}^n$ be two non-congruent configurations of the bipartite framework $\mathcal{F}_0$, as defined in Theorem 7.2.1. Then, there are two adjoint affine transformations α, α^* with*

$$\alpha\colon X_i \mapsto X'_i \text{ for all } i = 0,\dots,n \text{ and } \alpha^*\colon Y_k \mapsto Y'_k \text{ for all } k = 0,\dots,q.$$

Neither α nor α^* is an isometry.

Proof: (Proof of the last statement.) Let us assume that α is an isometry. We can specify the coordinate frame of $\mathcal{F}'$ such that $\mathbf{x}'_i = \mathbf{x}_i$ holds for all $i = 0,\dots,n$. Then, (7.21) gives rise to a homogeneous system of linear equations

$$\langle \mathbf{x}_i - \mathbf{x}_0,\, \mathbf{y}_k - \mathbf{y}'_k\rangle = 0 \text{ for each } k \in \{0,\dots,q\},$$

which admits only the trivial solution $\mathbf{y}'_k = \mathbf{y}_k$. The two configurations $\mathcal{F}$, $\mathcal{F}'$ are congruent, thus contradicting the initial assumption. ∎

We continue the proof of Theorem 7.2.1 and replace $\mathbf{x}'_i$ by $\alpha(\mathbf{x}_i)$ and $\mathbf{y}_k$ with $\alpha^*(\mathbf{y}'_k)$ in the distance equations (7.20). Thus, we obtain, by virtue of (7.22) and (7.25),

$$\|\mathbf{x}_i\|^2 - 2\langle \mathbf{x}_i,\ \mathbf{b} + l^*(\mathbf{y}'_k)\rangle + \|\mathbf{b} + l^*(\mathbf{y}'_k)\|^2 = \|\mathbf{a}' + l(\mathbf{x}_i)\|^2 - 2\langle \mathbf{a}' + l(\mathbf{x}_i),\ \mathbf{y}'_k\rangle + \|\mathbf{y}'_k\|^2,$$

or, because of $\langle \mathbf{x}_i,\, l^*(\mathbf{y}'_k)\rangle = \langle l(\mathbf{x}_i),\, \mathbf{y}'_k\rangle$ due to (7.23),

$$\begin{aligned} &\|\mathbf{x}_i\|^2 - \|l(\mathbf{x}_i)\|^2 - 2\langle \mathbf{x}_i, \mathbf{b}\rangle - 2\langle l(\mathbf{x}_i), \mathbf{a}'\rangle + \|\mathbf{b}\|^2 \\ &= \|\mathbf{y}'_k\|^2 - \|l^*(\mathbf{y}'_k)\|^2 - 2\langle \mathbf{y}'_k, \mathbf{a}'\rangle - 2\langle l^*(\mathbf{y}'_k), \mathbf{b}\rangle + \|\mathbf{a}'\|^2. \end{aligned}$$

This equation holds for all $i \in \{0,\dots,n\}$ and all $k \in \{0,\dots,q\}$. Since the left-hand side depends on the index i and the right-hand side only on k, both sides must equal a constant C. This results in two quadratic functions

$$\begin{aligned} f(\mathbf{x}) &:= \|\mathbf{x}\|^2 - \|l(\mathbf{x})\|^2 - 2\langle \mathbf{x}, \mathbf{b} + l^*(\mathbf{a}')\rangle + \|\mathbf{b}\|^2 - C, \\ g'(\mathbf{y}') &:= \|\mathbf{y}'\|^2 - \|l^*(\mathbf{y}')\|^2 - 2\langle \mathbf{y}', \mathbf{a}' + l(\mathbf{b})\rangle + \|\mathbf{a}'\|^2 - C \end{aligned} \quad \text{and} \quad \begin{aligned} f(\mathbf{x}_i) &= 0\ \ \forall i = 0,\dots,n, \\ g'(\mathbf{y}'_k) &= 0\ \ \forall k = 0,\dots,q. \end{aligned} \tag{7.26}$$

Conversely, for all $\mathbf{x}, \mathbf{y}' \in \mathbb{R}^n$, the equations $f(\mathbf{x}) = g'(\mathbf{y}') = 0$ imply

$$\|\mathbf{x} - \alpha^*(\mathbf{y}')\| = \|\alpha(\mathbf{x}) - \mathbf{y}'\|.$$

In what follows, we will see that this is precisely Ivory's Theorem for corresponding points of confocal surfaces in $\mathbb{E}^n$ (see [127, p. 208]). The forgoing results are summarized below.

Lemma 7.2.4 *Let α and α^* be adjoint affine transformations according to (7.22) and (7.25). Then, for any $\mathbf{x}_0, \dots, \mathbf{x}_p, \mathbf{y}'_0, \dots, \mathbf{y}'_q \in \mathbb{R}^n$, the points*

$$\mathbf{x}_0, \dots, \mathbf{x}_p, \alpha^*(\mathbf{y}'_0), \dots, \alpha^*(\mathbf{y}'_q) \ \textit{and}\ \alpha(\mathbf{x}_0), \dots, \alpha(\mathbf{x}_p), \mathbf{y}'_0, \dots, \mathbf{y}'_q \ \textit{in}\ \mathbb{E}^n$$

form two configurations of a complete bipartite framework of combinatorial type $\mathcal{K}_{p+1,q+1}$, i.e.,

$$\|\mathbf{x}_i - \alpha^*(\mathbf{y}'_k)\| = \|\alpha(\mathbf{x}_i) - \mathbf{y}'_k\| \ \textit{for all}\ i \in \{0, \dots, p\} \ \textit{and}\ j \in \{0, \dots, q\},$$

if, and only if, there is a constant C such that for the quadratic functions f and g', as defined in (7.26), the following equations hold true:

$$f(\mathbf{x}_i) = 0, \ \forall i = 0, \dots, p \quad \textit{and} \quad g'(\mathbf{y}'_k) = 0, \ \forall k = 0, \dots, q.$$

In order to determine the geometric meaning of the equations $f(\mathbf{x}) = 0$ and $g'(\mathbf{y}') = 0$, we specify the coordinate systems associated with the two configurations $\mathcal{F}$ and $\mathcal{F}'$. It is well-known that the composition of the two linear maps $l^* \circ l \colon \mathbb{R}^n \to \mathbb{R}^n$ is self-adjoint since

$$\langle \mathbf{u}, l^*\circ l(\mathbf{v})\rangle = \langle l(\mathbf{u}), l(\mathbf{v})\rangle = \langle l^*\circ l(\mathbf{u}), \mathbf{v}\rangle \quad \text{for all } \mathbf{u}, \mathbf{v} \in \mathbb{R}^n.$$

There is an orthonormal basis of eigenvectors $\mathbf{e}_1, \dots, \mathbf{e}_n$ satisfying

$$l^*\circ l(\mathbf{e}_i) = \lambda_i^2 \mathbf{e}_i \text{ with } \lambda_i^2 = \langle l(\mathbf{e}_i), (\mathbf{e}_i)\rangle \geq 0,$$

because of

$$\langle l(\mathbf{e}_i), l(\mathbf{e}_j)\rangle = \langle l^*\circ l(\mathbf{e}_i), \mathbf{e}_j\rangle = \lambda_i^2 \langle \mathbf{e}_i, \mathbf{e}_j\rangle = \lambda_i^2 \delta_{ij}. \tag{7.27}$$

We assume

$$\lambda_1 = \dots = \lambda_s = 0,\ \lambda_{s+1}, \dots, \lambda_r \neq 0, 1,\ \lambda_{r+1} = \dots = \lambda_n = 1 \text{ with } 0 \leq s \leq r \leq n \text{ and } r \geq 1, \tag{7.28}$$

because of $l^*\circ l \neq \mathrm{id}\,\mathbb{R}^n$, according to Lemma 7.2.3. Obviously, $|\lambda_{s+1}|, \dots, |\lambda_n|$ are the *singular values* of the linear map l.

The eigenvectors $\mathbf{e}_{s+1}, \dots, \mathbf{e}_n$ have mutually orthogonal images $l(\mathbf{e}_{s+1}), \dots, l(\mathbf{e}_n)$ due to (7.27). Thus, there is also an orthonormal basis $\mathbf{e}'_1, \dots, \mathbf{e}'_n$ such that

$$l(\mathbf{e}_i) = \lambda_i \mathbf{e}'_i \text{ and } l^*(\mathbf{e}'_k) = \lambda_k \mathbf{e}_k \text{ for all } i, k \in \{1, \dots, n\}. \tag{7.29}$$

We use $\mathbf{e}_1, \dots, \mathbf{e}_n$ and $\mathbf{e}'_1, \dots, \mathbf{e}'_n$ as the basis vectors of the two Cartesian coordinate frames associated with $\mathcal{F}$ and $\mathcal{F}'$, respectively. With regard to an appropriate choice of the origin, we distinguish two cases:

Case 1: The affine transformation $\alpha^*\circ\alpha\colon\ \mathbb{E}^n\to\mathbb{E}^n$ keeps a point $\mathbf{f}$ fixed:

Then, we choose $\mathbf{f}$ as the origin of the coordinate frame for $\mathcal{F}$ and $\alpha(\mathbf{f})$ as the origin for $\mathcal{F}'$. Due to (7.29), we obtain

$$\begin{aligned}\alpha\colon\quad & \mathbf{x}=\mathbf{f}+\sum_{i=1}^{n}x_i\mathbf{e}_i \ \mapsto\ \alpha(\mathbf{x})=\alpha(\mathbf{f})+\textstyle\sum_{i=1}^{n}\lambda_i x_i\mathbf{e}'_i\,,\\ \alpha^*\colon\quad & \mathbf{y}'=\alpha(\mathbf{f})+\sum_{k=1}^{n}y'_k\mathbf{e}'_k \ \mapsto\ \alpha^*(\mathbf{y}')=\mathbf{f}+\textstyle\sum_{k=1}^{n}\lambda_k y'_k\mathbf{e}_k\,.\end{aligned}\tag{7.30}$$

A comparison with (7.22) and (7.25) yields $\mathbf{a}'=\mathbf{b}=\mathbf{0}$. Thus, because of (7.28), the coordinate representations of the quadratic functions in (7.26) is

$$f(\mathbf{x})=x_1^2+\cdots+x_s^2+(1-\lambda_{s+1}^2)x_{s+1}^2+\cdots+(1-\lambda_r^2)x_r^2-C=g'(\mathbf{x}).\tag{7.31}$$

The equation $f(\mathbf{x})=0$ defines a non-parabolic quadric $\mathcal{X}$. Its affine image $\mathcal{X}':=\alpha(\mathcal{X})$ is

$$\begin{aligned}&\text{for } s=0\colon\ f'(\mathbf{x}'):=\frac{1-\lambda_1^2}{\lambda_1^2}\,x_1'^2+\cdots+\frac{1-\lambda_r^2}{\lambda_r^2}\,x_r'^2-C=0,\\ &\text{for } s>0\colon\ x'_1=\cdots=x'_s=0 \text{ and } f'(\mathbf{x}'):=\frac{1-\lambda_{s+1}^2}{\lambda_{s+1}^2}\,x_{s+1}'^2+\cdots+\frac{1-\lambda_r^2}{\lambda_r^2}\,x_r'^2-C\le 0.\end{aligned}\tag{7.32}$$

In the "singular" case $s>0$, the symbol $\partial\mathcal{X}'$ denotes the boundary of $\mathcal{X}'=\alpha(\mathcal{X})$, *i.e.*, the set

$$\partial\mathcal{X}'=\{\mathbf{x}'=(0,\dots,0,x'_{s+1},\dots,x'_n)\mid f'(\mathbf{x}')=0\}.$$

Now we move the second configuration such that the coordinate frame associated with $\mathcal{F}'$ coincides with that of $\mathcal{F}$. This motion is denoted by β (see Figure 7.11). Then, because of

$$\frac{\lambda_i^2}{1-\lambda_i^2}=\frac{1}{1-\lambda_i^2}-1\quad\text{for } i=1,\dots,r,$$

the quadric $\mathcal{X}$ and the displaced affine image $\mathcal{Y}:=\beta\circ\alpha(\mathcal{X})$, or $\partial\mathcal{Y}$ if $s>0$, are *confocal* in $\mathbb{E}^n$. If $C\ne 0$, we obtain a central quadric $\mathcal{X}$, and the comparison with (7.13) reveals

$$\varepsilon_i a_i^2=\frac{C}{1-\lambda_i^2},\quad \varepsilon_i(a_i^2+\varepsilon_i k)=\frac{C\lambda_i^2}{1-\lambda_i^2}=\varepsilon_i a_i^2-C,\ \text{hence } k=-C.$$

For a conical quadric in the case $C=0$, we obtain, by virtue of (7.17),

$$\varepsilon_i a_i^2=\frac{1}{1-\lambda_i^2},\quad \varepsilon_i(a_i^2+\varepsilon_i k)=\frac{\lambda_i^2}{1-\lambda_i^2}=\varepsilon_i a_i^2-1,\ \text{hence } k=-1.$$

Case 2: There is *no fixed point* of

$$\alpha^*\circ\alpha\colon\ \mathbf{x}\ \mapsto\ \mathbf{b}+l^*(\mathbf{a}'+l(\mathbf{x}))=\mathbf{b}+l^*(\mathbf{a}')+l^*\circ l(\mathbf{x}).$$

This means that the system of linear equations

$$\left(l^*\circ l-\mathrm{id}\,\mathbb{R}^n\right)(\mathbf{x})=-\mathbf{b}-l^*(\mathbf{a}')$$

has no solution, *i.e.*, the rank r of $(l^*\circ l-\mathrm{id}\,\mathbb{R}^n)$ is smaller than n, and therefore, 1 is an eigenvalue of $l^*\circ l$ and

$$\mathbf{d}:=-\mathbf{b}-l^*(\mathbf{a}')\notin(l^*\circ l-\mathrm{id}\,\mathbb{R}^n)(\mathbb{R}^n)=[\mathbf{e}_1,\dots,\mathbf{e}_r].$$

We decompose

$$\mathbf{d}=\mathbf{d}_0+\mathbf{d}_1\ \text{such that}\ \mathbf{d}_0\in[\mathbf{e}_1,\dots,\mathbf{e}_r]\ \text{and}\ \mathbf{d}_1\in[\mathbf{e}_{r+1},\dots,\mathbf{e}_n]$$

with $\mathbf{d}_1 \neq \mathbf{0}$. We fix the origin $\mathbf{f}$ of the coordinate frame associated with $\mathcal{F}$ as a solution to

$$(l^* \circ l - \mathrm{id}\,\mathbb{R}^n)(\mathbf{f}) = \mathbf{d}_0, \quad i.e., \quad l^* \circ l(\mathbf{f}) = \mathbf{f} + \mathbf{d}_0.$$

Furthermore, we modify the basis vectors $\mathbf{e}_{r+1}, \ldots, \mathbf{e}_n$ in the eigenspace of 1 such that $\mathbf{d}_1 = -2a\,\mathbf{e}_n$ with $a \neq 0$. This implies

$$\alpha^* \circ \alpha(\mathbf{f}) = -\mathbf{d}_0 - \mathbf{d}_1 + l^* \circ l(\mathbf{f}) = -\mathbf{d}_1 + \mathbf{f} = 2a\,\mathbf{e}_n + \mathbf{f}.$$

Let $\mathbf{f}' := \alpha(\mathbf{f}) - a\,\mathbf{e}'_n$ be the origin of the coordinate frame associated with $\mathcal{F}'$. Then, we get with (7.29) and $\lambda_n = 1$

$$\alpha(\mathbf{f}) = \mathbf{f}' + a\,\mathbf{e}'_n \quad \text{and} \quad \alpha^*(\mathbf{f}') = \alpha^* \circ \alpha(\mathbf{f}) - a\,l^*(\mathbf{e}'_n) = \mathbf{f} + a\,\mathbf{e}_n,$$

and hence, $\mathbf{a}' = a\,\mathbf{e}'_n$ in (7.22), $\mathbf{b} = a\,\mathbf{e}_n$ in (7.25), $\mathbf{b} + l^*(\mathbf{a}') = 2a\,\mathbf{e}_n$. From (7.26), we obtain

$$f(\mathbf{x}) = x_1^2 + \cdots + x_s^2 + (1 - \lambda_{s+1}^2)x_{s+1}^2 + \cdots + (1 - \lambda_r^2)x_r^2 - 4ax_n + D = g'(\mathbf{x}) \tag{7.33}$$

with $D := a^2 - C$. The image $\mathcal{X}'$ of the quadratic surface $\mathcal{X}$: $f(\mathbf{x}) = 0$ under

$$\alpha\colon\ (x_1, \ldots, x_n) \mapsto (x'_1, \ldots, x'_n) = (0, \ldots, 0,\ \lambda_{s+1}x_{s+1}, \ldots, \lambda_r x_r,\ x_{r+1}, \ldots,\ x_{n-1},\ x_n + a)$$

satisfies

$$x'_1 = \cdots = x'_s = 0, \quad f'(\mathbf{x}') := \frac{1 - \lambda_{s+1}^2}{\lambda_{s+1}^2}\, x'^2_{s+1} + \cdots + \frac{1 - \lambda_r^2}{\lambda_r^2}\, x'^2_r - 4a(x'_n - a) + D \leq 0, \tag{7.34}$$

where equality holds in the regular case $s = 0$ only.

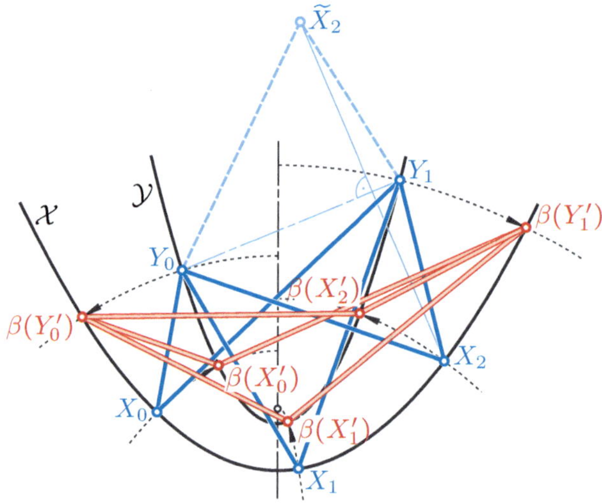

FIGURE 7.12. Theorem 7.2.1 in the parabolic case (Case 2).

Again, we apply to $\mathcal{F}'$ an isometry β which lets the two coordinate frames coincide. Then, a comparison with (7.15) reveals that, after replacing x_n with $\widetilde{x}_n = x_n - D/4a$, we obtain *confocal* parabolic quadrics with

$$\varepsilon_i a_i^2 = \frac{2a}{1 - a_i^2} \quad \text{and} \quad k = -2a,$$

(see Figure 7.12).

In the cases 1 and 2, the points $\mathbf{x}_i \mapsto \beta(\mathbf{x}'_i)$ and $\beta(\mathbf{y}'_j) \mapsto \mathbf{y}_j$ are corresponding in the sense of Definition 7.2.1. This proves

Lemma 7.2.5 *Let $\mathcal{F}$, $\mathcal{F}'$ in $\mathbb{E}^n$ be two non-congruent configurations as defined in Theorem 7.2.1. There is an isometry $\beta\colon \mathbb{E}^n \to \mathbb{E}^n$ such that the knots of $\mathcal{F}$ and $\beta(\mathcal{F}')$ are corresponding points of two confocal quadrics $\mathcal{X}$ and $\mathcal{Y}$ with*

$$X_0, \dots, X_n, \beta(Y_0'), \dots, \beta(Y_q') \in \mathcal{X} \text{ and } \beta(X_0'), \dots, \beta(X_n'), Y_0, \dots, Y_q \in \mathcal{Y}.$$

Remark 7.2.1 After applying β, the affine transformations α and α^* coincide, or more precisely,

$$\beta\circ\alpha = \alpha^*\circ\beta^{-1} = (\beta\circ\alpha)^*.$$

The last equation results from $\beta^{-1} = \beta^*$ for isometries and $(\beta\circ\alpha)^* = \alpha^*\circ\beta^*$. Therefore, the affine transformation $\beta\circ\alpha$ is self-adjoint (compare with [120]).

Thus, Theorem 7.2.1 is a consequence of Lemmas 7.2.3, 7.2.4, and 7.2.5. ∎

Remark 7.2.2 Item 3 in Theorem 7.2.1 is not valid when $\mathcal{F}$ and $\mathcal{F}'$ are exchanged and the span of $\{Y_0, \dots, Y_q\}$ has a dimension $< n$. The underlying reason is that the reflection in the span of these $q+1$ points sends X to a point $\widetilde{X}$, which again satisfies

$$\overline{\widetilde{X}Y_j} = \overline{XY_j} = \overline{X'Y_j'} \quad \text{for all} \quad j = 0, \dots, q,$$

though $\widetilde{X}$ need not be a point of $\mathcal{X}$ (note, *e.g.*, the point $\widetilde{X}_2$ in Figure 7.12). Of course, the reflection in question can be replaced by any other motion which keeps the span fixed.

Lemma 7.2.6 *Let $\mathcal{F}$, $\mathcal{F}'$ in $\mathbb{E}^n$ be two non-congruent configurations of a bipartite framework, as defined in Theorem 7.2.1. Two different knots X_i, X_j of $\mathcal{F}$ share their distance with the corresponding points of X_i', X_j' of $\mathcal{F}'$ if, and only if, the spanned line $[X_i, X_j]$ is a generator of the quadric $\mathcal{X}$, i.e.,*

$$\overline{X_iX_j} = \overline{X_i'X_j'} \iff [X_i, X_j] \subset \mathcal{X}.$$

Similarly, for $Y_i', Y_j' \in \mathcal{F}'$ with $i \neq j$ and the corresponding points $Y_i, Y_j \in \mathcal{F}$, we have

$$\overline{Y_i'Y_j'} = \overline{Y_iY_j} \iff [Y_i', Y_j'] \subset \beta^{-1}(\mathcal{X}).$$

Proof: The statement $[X_i, X_j] \subset \mathcal{X}$ is equivalent to $f(\lambda\mathbf{x}_i + (1-\lambda)\mathbf{x}_j) = 0$ for all $\lambda \in \mathbb{R}$. From (7.26) and $f(\mathbf{x}_i) = f(\mathbf{x}_j) = 0$, we get

$$\begin{aligned}
&f(\lambda\mathbf{x}_i + (1-\lambda)\mathbf{x}_j)\\
&= \|\lambda\mathbf{x}_i + (1-\lambda)\mathbf{x}_j\|^2 - \|l(\lambda\mathbf{x}_i + (1-\lambda)\mathbf{x}_j)\|^2 - 2\langle\lambda\mathbf{x}_i + (1-\lambda)\mathbf{x}_j,\ \mathbf{b} + l^*(\mathbf{a}')\rangle + \|\mathbf{b}\|^2 - C\\
&= \lambda^2[\|\mathbf{x}_i\|^2 - \|l(\mathbf{x}_i)\|^2] + (1-\lambda)^2[\|\mathbf{x}_j\|^2 - \|l(\mathbf{x}_j)\|^2] + 2\lambda(1-\lambda)\,[\langle\mathbf{x}_i,\,\mathbf{x}_j\rangle - \langle l(\mathbf{x}_i),\, l(\mathbf{x}_j)\rangle]\\
&\quad - \lambda[\|\mathbf{x}_i\|^2 - \|l(\mathbf{x}_i)\|^2] - (1-\lambda)[\|\mathbf{x}_j\|^2 - \|l(\mathbf{x}_j)\|^2]\\
&= \lambda(1-\lambda)\left[-\|\mathbf{x}_i\|^2 - \|\mathbf{x}_j\|^2 + \|l(\mathbf{x}_i)\|^2 + \|l(\mathbf{x}_j)\|^2 + 2\langle\mathbf{x}_i,\,\mathbf{x}_j\rangle - 2\langle l(\mathbf{x}_i),\, l(\mathbf{x}_j)\rangle\right]\\
&= \lambda(\lambda-1)\left[\|\mathbf{x}_i - \mathbf{x}_j\|^2 - \|l(\mathbf{x}_i) - l(\mathbf{x}_j)\|^2\right] = \lambda(\lambda-1)\left[(\mathbf{x}_i - \mathbf{x}_j)^2 - \|\alpha(\mathbf{x}_i) - \alpha(\mathbf{x}_j)\|^2\right],
\end{aligned}$$

which yields the equivalence

$$[X_i, X_j] \subset \mathcal{X} \iff \|\mathbf{x}_i - \mathbf{x}_j\| = \|\alpha(\mathbf{x}_i) - \alpha(\mathbf{x}_j)\|.$$

The analogous computation shows that $[Y_i', Y_j'] \subset \beta^{-1}(\mathcal{X})$ is equivalent to $\overline{Y_iY_j} = \overline{Y_i'Y_j'}$. This is also a consequence of statements 2 and 3 in Theorem 7.2.1, as $\overline{Y_iY_j} = \overline{Y_i'Y_j'}$ implies congruent triangles $Y_iY_jX_k = Y_i'Y_j'X_k'$ for each $k \in \{0, \ldots, n\}$. Therefore, for any point Y of the line $[Y_i, Y_j]$ and its image $Y' \in [Y_i', Y_j']$ under the isometry $[Y_i, Y_j] \to [Y_i', Y_j']$ with $Y_i \mapsto Y_i'$ and $Y_j \mapsto Y_j'$, we have $\overline{YX_k} = \overline{Y'X_k'}$, and therefore, $Y' \in \beta^{-1}(\mathcal{X})$. ■

In order to avoid confusion with the notation, one needs to note that, previously, pairs of corresponding points on confocal quadrics were denoted by $X \mapsto X'$, $Y \mapsto Y' \ldots$ with $X, Y \in \mathcal{Q}$ and $X', Y' \in \mathcal{Q}'$. With regard to bipartite frameworks, primes indicate the knots of the second configuration. Consequently, by virtue of the Configuration Theorem 7.2.1, we obtained corresponding knots of the first class in the order $X_i \mapsto X_i'$, while, for corresponding knots of the second class, the order was reversed as $Y_j' \mapsto Y_j$, where $X_i, Y_j' \in \mathcal{X}$ and $X_i', Y_j \in \mathcal{Y}$. In the upcoming subsection, we return to the original notation.

Ivory's Theorem

This famous theorem can be directly extracted from Theorem 7.2.1. Below, we combine it with Lemma 7.2.6.

Theorem 7.2.2 (Ivory's Theorem in $\mathbb{E}^n$)
Let (X, X') and (Y, Y') be two pairs of corresponding points of confocal quadrics $\mathcal{Q}$, $\mathcal{Q}'$ in the sense of Definition 7.2.1, i.e., with the affine transformation $\gamma\colon \mathcal{Q} \to \mathcal{Q}'$, $X \mapsto X'$, and $Y \mapsto Y'$. Then, we obtain equal distances

$$\overline{XY'} = \overline{X'Y}.$$

In addition, $\overline{XY} = \overline{X'Y'}$ for $X \neq Y$ if, and only if, the line $[X, Y]$ is a generator of $\mathcal{Q}$.

It is surprising that Ivory's Theorem is also valid on a quadric's surface, when, *e.g.* in $\mathbb{E}^3$, the conics are replaced by curvature lines and the straight diagonals with geodesics (Figure 7.13), while the vertices still form an Ivory quadrangle. The most general form of Ivory's Theorem on surfaces is valid for Liouville nets and, in the case of higher dimensions, for Stäckel nets. For further details and references, see [67, p. 21]. The same holds for GRAVES's construction (cf. planar version in [46, p. 45]). We revisit here GRAVES's *construction on ellipsoids*:

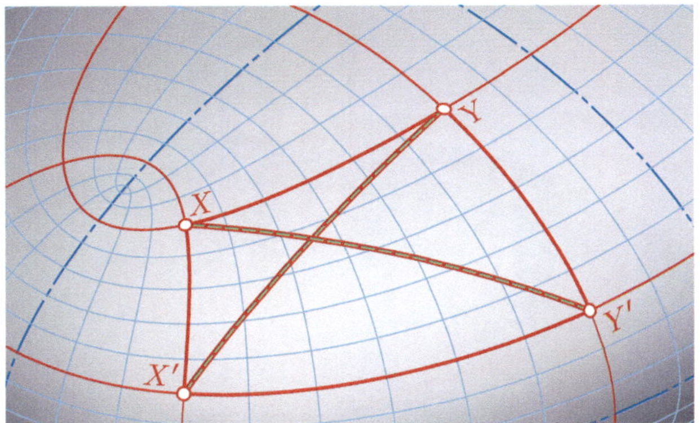

FIGURE 7.13. Ivory's Theorem is also valid on a quadric's surface.

Theorem 7.2.3 *Let a line of curvature e_0 on the ellipsoid $\mathcal{E}$ be given as well as a closed piece of string strictly longer than the perimeter of e_0, but sufficiently short. Then, the locus of a point P used to pull the string taut around e_0 is again a line of curvature e (Figure 7.14).*

When e_0 shrinks to the elliptic arc terminated by two adjacent umbilic points, GRAVES's construction turns into the "gardener's construction" [46, Fig. 1.8]. This reveals that the curvature lines are geodesic ellipses and hyperbolas, *i.e.*, along each line of curvature, either the sum or the difference of the geodesic distances to non-antipodal umbilic points U_1 and U_2 is constant (see, *e.g.*, [7, p. 521]).

Proof: We proceed similarly to the proof of the planar version in [46, p. 45]. In each pose, the string is composed of an arc along the curve e_0 terminated by points T_1 and T_2, and two contacting geodesics which connect T_1 and T_2, respectively, with the point P. By virtue of Theorem 7.1.4 and Corollary 7.1.2, the tangents to these two geodesics at P (note Figure 7.14) are symmetric w.r.t. the principal curvature tangents of $\mathcal{E}$ at P.

Let us cut the taut string in our imagination at P. First, we focus on the left-hand part, which follows, from an initial point on e_0, the line of curvature e_0 and, from point T_1 onward, the geodesic. We assume that the initial point of this left-hand part is attached to e_0, which prevents the string from sliding on e_0. If we keep the string under tension and try to move the point P, then T_1 varies on e_0, and consequently, the geodesic between T_1 and P varies on $\mathcal{E}$. However, it remains tangent to e_0. Since the arc length along the string is preserved, a point P, fixed on the string, traces an orthogonal trajectory of the geodesics, due to a theorem by GAUSS concerning geodesic parallel coordinates [82, p. 160]. Therefore, the tangent vector to this trajectory at P is orthogonal to the geodesic.

However, in GRAVES's construction, the point P varies also relative to the string. Therefore, w.r.t. the left-hand part, the absolute velocity vector $\mathbf{v}$ of P is the sum of the velocity vector $\mathbf{v}_1$ along the string and an orthogonal component.

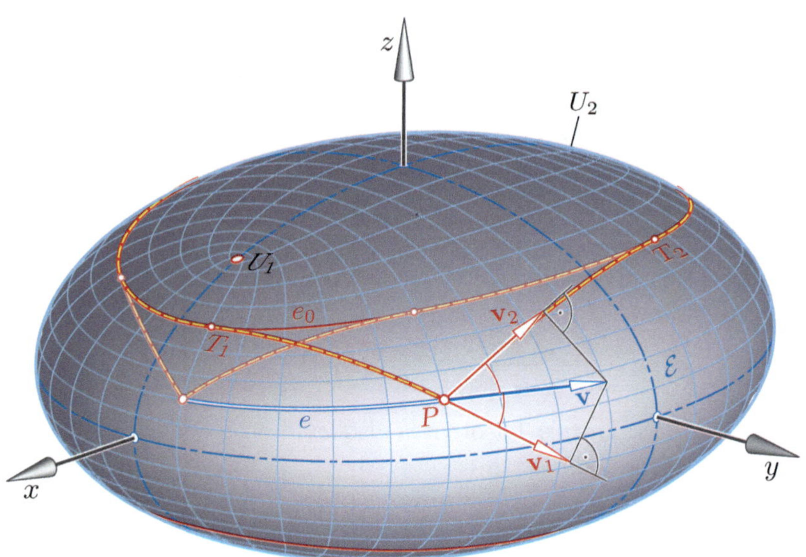

FIGURE 7.14. GRAVES's construction on the ellipsoid.

The same holds for the right-hand part of the string. Thus, we obtain a second decomposition of $\mathbf{v}$ into two orthogonal components. One of them, the vector $\mathbf{v}_2$, is tangent to the second geodesic through P. Because of the constant total length of the string, the vectors $\mathbf{v}_1$ and $\mathbf{v}_2$ are of equal lengths. Furthermore, if the length of one part increases, that of the other part must decrease by the same amount. The symmetry of the two decompositions of $\mathbf{v}$ reveals (see Figure 7.14) that $\mathbf{v}$ bisects the exterior angle between the geodesics. Thus, at each position the path of P contacts a principal curvature tangent. The path must be an integral curve of the set of principal curvature tangents on $\mathcal{E}$, and hence a line of curvature. ■

The same arguments can be used for proving GRAVES's construction for other regular quadrics, provided that e_0 is a closed line of curvature. Otherwise, on paraboloids, one has to replace the Graves condition by

$$\overline{PT_1} + \overline{PT_2} - \widetilde{T_1T_2} = \text{const.},$$

where the $\widetilde{T_1T_2}$ denotes the length of the arc along e_0 between T_1 and T_2, while $\overline{PT_i}$, $i = 1, 2$, denotes the lengths of the connecting geodesic arcs.

Remark 7.2.3 The line of curvature e_0 on the ellipsoid $\mathcal{E}$ can be the intersection either with a one-sheeted or a two-sheeted hyperboloid. It can happen that, for a sufficiently long string, the contact points T_1 and T_2 coincide in a particular position, and when moving further, the string slips over the ellipsoid. As a consequence, not every point on the ellipsoid $\mathcal{E}$ in the exterior of e_0 needs to be reached by GRAVES's construction with a given e_0. There is an upper bound for the length of closed string in order to prevent it from slipping away.

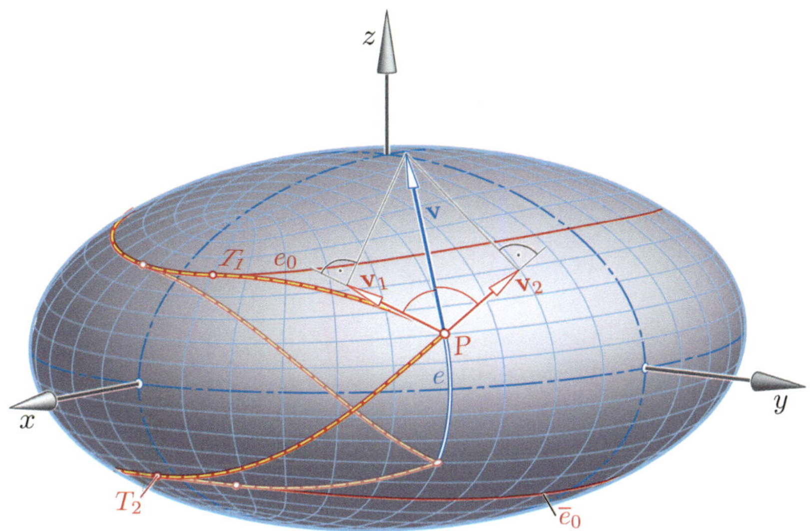

FIGURE 7.15. When a string is attached with both ends to a pair of opposite curvature lines and kept taut at P, then P traces again a line of curvature e.

With regard to GRAVES's construction on an ellipsoid $\mathcal{E}$, the hyperboloid through e_0 intersects $\mathcal{E}$ along a second curve, the antipode $\overline{e}_0$ of e_0. Now we see that the construction, as depicted in Figure 7.15, using a string with its ends fixed on e_0 and $\overline{e}_0$, respectively, produces a line of curvature e of the second family on $\mathcal{E}$. This follows since, by virtue of Theorem 7.1.4 and Corollary 7.1.2, the tangents to the two geodesics at P are again symmetric w.r.t. the principal curvature tangents.

Referring to Figure 7.15, the same trajectory e can be obtained by using the extensions of the geodesics through P on the right-hand side. The components $\mathbf{v}_1$ and $\mathbf{v}_2$ of the velocity vector $\mathbf{v}$ in direction of the extended geodesics remain the same. Thus, if the length of the string at the left-hand side decreases, that of the extension increases by the same amount. This results in the following statement:

A closed string of appropriate length that is placed in the form of a figure eight around e_0 and $\overline{e}_0$ enables the crossing point P to move freely on the ellipsoid, while the string does not slide on e_0 or $\overline{e}_0$ (Figure 7.16). However, in accordance with Remark 7.2.3, not each position in the zone between e_0 and $\overline{e}_0$ needs to be reachable.

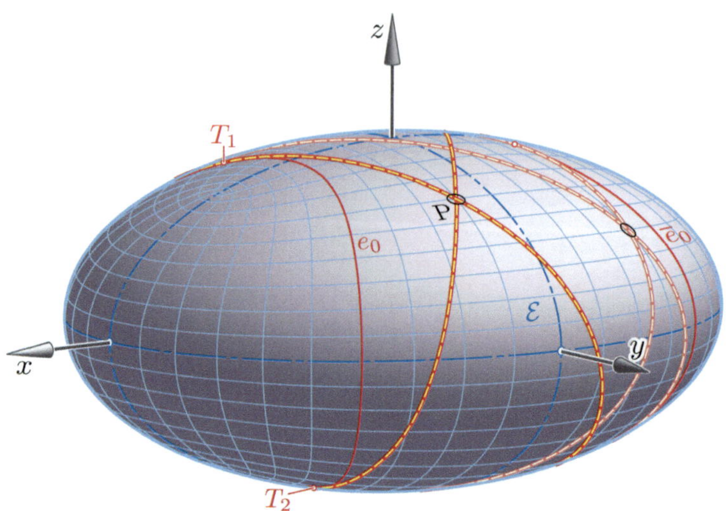

FIGURE 7.16. A closed string wrapped taut in the form of a figure eight around a pair of antipodal curvature lines $(e_0, \overline{e}_0)$ on the ellipsoid $\mathcal{E}$ allows for the free movement of the crossing point P on $\mathcal{E}$.

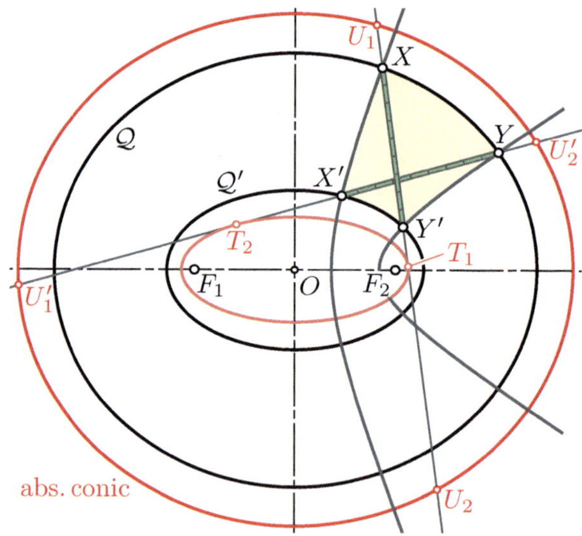

FIGURE 7.17. The extended diagonals of the Ivory quadrangle $XX'Y'Y$ contact the confocal conic.

As already mentioned, Ivory's Theorem is also valid in non-Euclidean spaces. A proof can be found in [122]. In the Cayley-Klein model $\mathbb{H}^n$ of hyperbolic geometry, two quadrics $\mathcal{Q}$ and $\mathcal{Q}'$ are *confocal* if, and only if, the spanned range includes also the absolute quadric of $\mathbb{H}^n$. Consequently, any two confocal quadrics $\mathcal{Q}, \mathcal{Q}'$ in $\mathbb{E}^n$, *i.e.*, in the Euclidean sense, are also confocal in the hyperbolic sense if a third confocal quadric $\mathcal{Q}''$ is specified as the absolute quadric. Of course, $\mathcal{Q}''$ must be regular.

According to [122], corresponding points X, X' of two confocal quadrics $\mathcal{Q}, \mathcal{Q}'$ in $\mathbb{H}^n$ are related by a self-adjoint endomorphism $\mathcal{Q} \to \mathcal{Q}'$. As noted in Remark 7.2.1, the affine mapping $\gamma = \beta \circ \alpha$ is self-adjoint. Therefore, points $X \in \mathcal{Q}$ and $X' \in \mathcal{Q}'$ corresponding in the Euclidean sense, are also corresponding in the hyperbolic sense when another confocal quadric $\mathcal{Q}''$ serves as absolute quadric. Let (Y, Y') be a second pair of corresponding points. According to Ivory's Theorem in $\mathbb{H}^n$, the absolute points U_1, U_2 on the line $[X, Y']$ and V_1, V_2 on the line $[X', Y]$ yield equal cross ratios

$$\operatorname{cr}\left(XY'U_1U_2\right) = \operatorname{cr}\left(X'YV_1V_2\right).$$

If the line $[X, Y']$ happens to be tangent to $\mathcal{Q}''$, then the cross ratio in question equals 1, *i.e.*, the line $[X', Y]$ also contacts $\mathcal{Q}''$ (Figure 7.17). This is still valid in the limiting case when $\mathcal{Q}''$ is singular.

Lemma 7.2.7 *Let (X, X') and (Y, Y') be two pairs of conjugate points of confocal quadrics $\mathcal{Q}$ and $\mathcal{Q}'$ in $\mathbb{E}^n$. Then, each quadric $\mathcal{Q}''$ which contacts the diagonal line $[X, Y']$ is also tangent to the other diagonal line $[X', Y]$ in the Ivory quadrangle $XX'Y'Y$.*

This result is cited in [13, p. 153] and proved in [2, p. 118]. However, most probably, it can already be found in the literature of the 19th century. An alternative proof of Lemma 7.2.7 is the subject of Exercise 7.2.1. Figure 7.18 illustrates this property even on an ellipsoid.

This means that, in the 3-dimensional case, all four diagonals in any Ivory box (Figure 7.10) contact the same two quadrics included in the confocal family. In general, these diagonals are mutually skew without any symmetry, and they do not belong to a regulus.

Finally, it must be noted that, due to (7.11) and (7.12), the linearity of the transformation between corresponding points of confocal conics $\mathcal{Q}, \mathcal{Q}'$ has some other consequences. For example, in the case of conics with center O, it is easy to confirm that the distances of the vertices of an

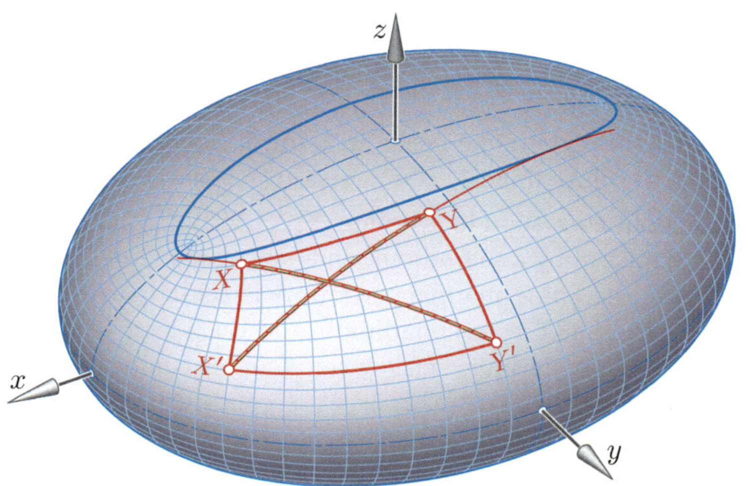

FIGURE 7.18. Even on the ellipsoid, the extended diagonals of the Ivory quadrangle $XX'Y'Y$ contact the same line of curvature.

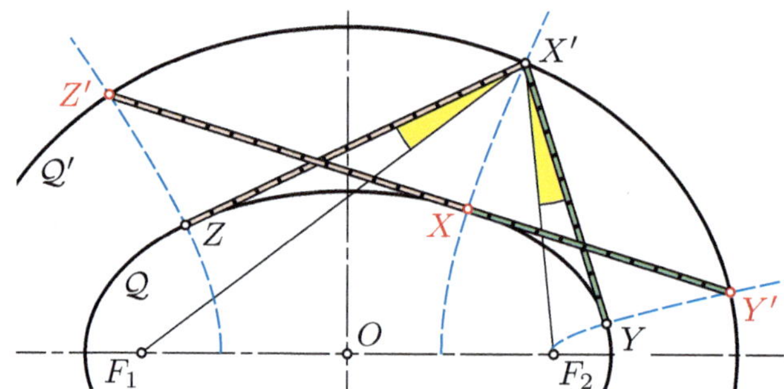

FIGURE 7.19. If $[X, Y']$ contacts $\mathcal{Q}$ at X, then $[X', Y]$ contacts $\mathcal{Q}$ at Y.

Ivory quadrangle $XX'Y'Y$ satisfy

$$\overline{OX}^2 + \overline{OY'}^2 = \overline{OX'}^2 + \overline{OY}^2.$$

Similarly, we obtain equal dot products for the vectors

$$\langle \overrightarrow{OX}, \overrightarrow{OY'} \rangle = \langle \overrightarrow{OX'}, \overrightarrow{OY} \rangle.$$

More generally, for any quadric $\mathcal{R}$ centered at O, we can state: Two points X and Y' are conjugate w.r.t. $\mathcal{R}$ if, and only if, X' and Y are conjugate w.r.t. $\mathcal{R}$. This means in the case $\mathcal{R} = \mathcal{Q}$: If the line $[X, Y']$ contacts $\mathcal{Q}$

at X, then $[X', Y]$ contacts $\mathcal{Q}$ at Y (Figure 7.19). In the 2-dimensional case, the tangents from X' to the conic $\mathcal{Q}$ have common angle bisectors with the lines connecting X' with the focal points (see, *e.g.*, [46, p. 42]).

More about bipartite frameworks

An immediate consequence of Theorem 7.2.1 is a statement about the flexibility of bipartite frameworks.

Corollary 7.2.8 *A complete bipartite framework of combinatorial type $\mathcal{K}_{p,q}$ in $\mathbb{E}^n$ with knots $\{X_1, \ldots, X_p\}$ of one class spanning $\mathbb{E}^n$ permits a continuous flexion $\mathbb{E}^n$ if, and only if, there is an at least one-parametric set of quadrics $\mathcal{X}$ passing through $X_1, \ldots, X_p$ such that a confocal quadric $\mathcal{Y}$ contains all knots $Y_1, \ldots, Y_q$ of the second class.*

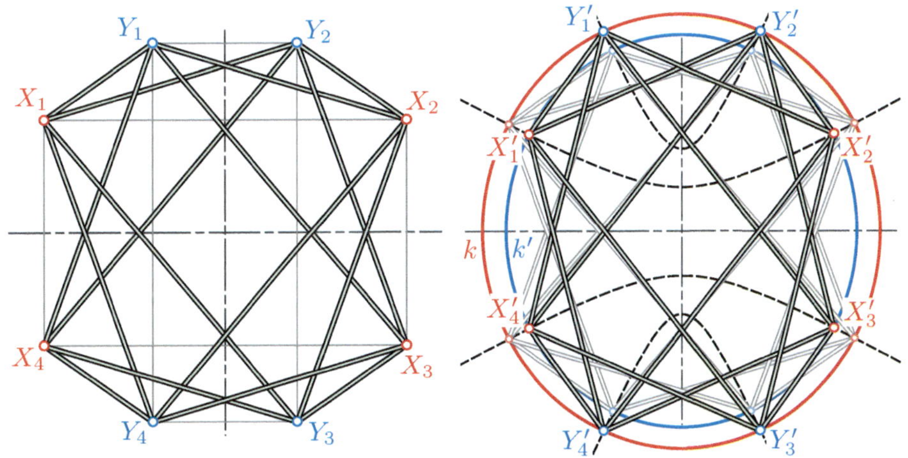

FIGURE 7.20. Dixon's flexible framework of type $\mathcal{K}_{4,4}$ in $\mathbb{E}^2$.

As a first example, we present a flexible framework attributed to DIXON in 1899.[8] The vertices of two rectangles with two common axes of symmetry constitute the classes of vertices. There is a one-parameter set of conics k passing through $X_1, \ldots, X_4$. They all share the axes of symmetry, and for each conic k, there exists a confocal conic k' through $Y_1, \ldots, Y_4$.

[8] Sir ALFRED CARDEW DIXON (1865–1936), English mathematician, professor at the Queen's University Belfast and president of the London Mathematical Society.

Thus, Corollary 7.2.8 guarantees its flexibility (Figure 7.20). According to DIXON in [36], there are only two continuously flexible bipartite frameworks of type $\mathcal{K}_{3,3}$ in the Euclidean plane, and the second one with knots $X_1, X_2, X_3, Y_1, Y_2, Y_3$ is a subset of the framework presented above.

The spherical version of Dixon's framework is also flexible. To prove this statement, we only have to replace the confocal conics in the presented proof by confocal spherical conics or confocal cones in $\mathbb{E}^3$ [46, p. 456]. This spherical framework is also known under the name *Bottema's 16-bar framework* [14, 85, 86, 116, 147].

Finally, an n-dimensional version is treated in a similar way: We specify the $\mathbf{x}$-knots as vertices of $n-1$ coaxial boxes in $\mathbb{E}^n$ and the Y-knots as vertices of another box with the same axes. Then, this yields a continuously flexible bipartite framework of combinatorial type $\mathcal{K}_{(n-1)2^n,2^n}$.

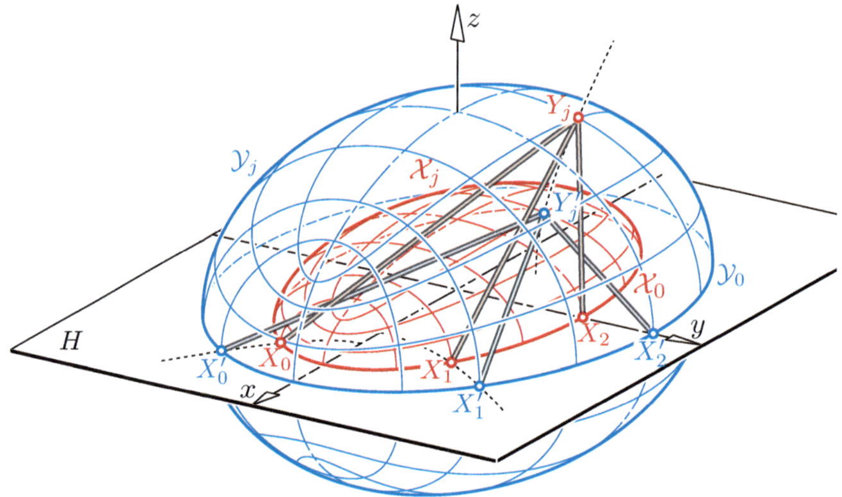

FIGURE 7.21. A bipartite framework in $\mathbb{E}^3$ with knots $\{X_1, X_2, \dots\}$ of one class located on a conic $\mathcal{X}_0$ is continuously flexible (Corollary 7.2.9).

As a second example, we focus on a particular bipartite framework of type $\mathcal{K}_{p,q}$ in $\mathbb{E}^n$, where the knots of one class are placed on a quadric within a hyperplane H. In this case as well, we obtain a continuously flexible framework. For the 3-dimensional version, this has first been proved by W. WUNDERLICH in [150].

Corollary 7.2.9 *A bipartite framework of type $\mathcal{K}_{p,q}$ in $\mathbb{E}^n$, $n \geq 3$, where the knots $X_1, \dots, X_q$ of one class are located on a quadric $\mathcal{X}_0$ within a hyperplane H, is continuously flexible wherever the knots $Y_1, \dots, Y_q$ of the second class are specified in $\mathbb{E}^n$ outside of H.*

Proof: In the hyperplane H, there is a one-parametric family of quadrics $\mathcal{Y}_0$ that are confocal with $\mathcal{X}_0$ and of the same type. The equations (7.13), (7.15), and (7.17) of confocal quadrics reveal that, through each point Y_j, $j = 1, 2 \dots$ outside H, there passes a quadric $\mathcal{Y}_j$ which is symmetric w.r.t. H and intersects H at $\mathcal{Y}_0$. Furthermore, there exists a confocal quadric $\mathcal{X}_j$ of the same type through $\mathcal{X}_0$. If $Y_j' \in \mathcal{X}_j$ corresponds to $Y_j \in \mathcal{Y}_j$ and $X_1', \dots, X_p' \in \mathcal{Y}_0$ have the given knots $X_1, \dots, X_p$ as corresponding points, then, by virtue of Ivory's Theorem, we have $\overline{X_i Y_j} = \overline{X_i' Y_j'}$ for all $i \in \{1, \dots, p\}$.

Note that in this case each Y-knot has its own quadric $\mathcal{Y}_j$. The three-dimensional version is depicted in Figure 7.21. ■

In [121, Theorem 2] it is shown that for $p > n$ the flexions of $\mathcal{F}$, as explained in the proof before, are the only possible ones, apart from reflections of single knots Y_j in the hyperplane H. Another result in [121] addresses flexions of a bipartite framework in $\mathbb{E}^n$ where the two classes of knots are chosen in two perpendicular hyperplanes. In this case as well, the proof can be based on Ivory's Theorem.

If two configurations of a framework are sufficiently close, then a physical model can be forced to change from one configuration into the other due to small deformations of bars and clearances at joints. This is called a *flipping* or *snapping* structure. By virtue of Theorem 7.2.1, a flipping bipartite framework can be produced by specifying the two quadrics $\mathcal{X}$ and $\mathcal{Y}$ through the respective classes of knots sufficiently close.

As the limit $\mathcal{Y} \to \mathcal{X}$, we obtain an infinitesimally flexible framework as defined below.

Definition 7.2.2 A framework is called *infinitesimally flexible* (or *first-order flexible*) if, and only if, a *velocity vector* $\mathbf{v}_i$ can be assigned to each knot $\mathbf{p}_i$ such that

1. for all bars $\mathbf{p}_i\mathbf{p}_j$ the *projection theorem* holds, *i.e.*(see Figure 7.22), $\langle \mathbf{p}_i - \mathbf{p}_j, \mathbf{v}_i - \mathbf{v}_j \rangle = 0$, hence $\langle \mathbf{p}_i - \mathbf{p}_j, \mathbf{v}_i \rangle = \langle \mathbf{p}_i - \mathbf{p}_j, \mathbf{v}_j \rangle$, and
2. the assignment is nontrivial, *i.e.*, the velocity vectors must not originate from a motion of the whole framework as a rigid body.

In the case of a flipping bipartite framework in the standard position (Figure 7.11 (left) or 7.12), for each knot the two positions X, X' are

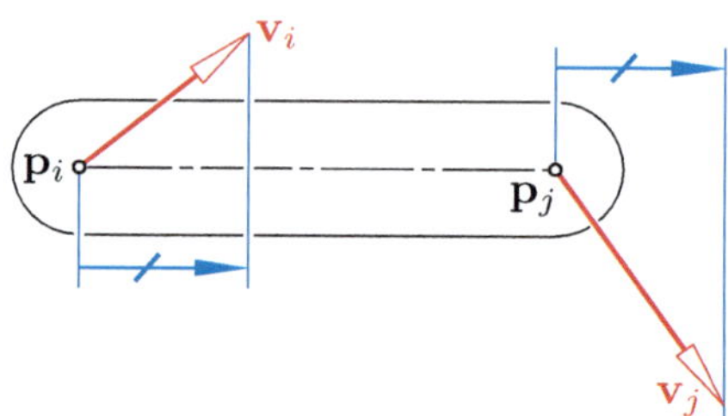

FIGURE 7.22. Geometric interpretation of the projection theorem in $\mathbb{E}^2$.

corresponding points of the confocal quadrics, and therefore, located on a common orthogonal trajectory (Theorem 7.1.2). For the limiting case of an infinitesimally flexible framework, the following theorem, which is attributed to WALTER WHITELEY (1984) (see [136, p. 1337]), holds true. The three-dimensional version was used earlier to characterize singular positions in satellite geodesy [116, 148].

Theorem 7.2.4 *A bipartite framework with knots placed on a quadric $\mathcal{Q}$ is infinitesimally flexible. There is an assignment of velocity vectors orthogonal to $\mathcal{Q}$ in such a way that, for knots of different classes, the vectors point to opposite sides of $\mathcal{Q}$.*

Conversely, suppose that in $\mathbb{E}^n$ a bipartite framework with both classes of knots spanning an affine n-space is infinitesimally flexible. Then, all knots must be placed on a quadric $\mathcal{Q}$.

Proof: Let $F(\mathbf{x}) = \mathbf{x}^{\mathrm{T}}\mathbf{A}\,\mathbf{x} + 2\mathbf{a}^{\mathrm{T}}\mathbf{x} + a$ be a quadratic function with $\mathbf{A}^{\mathrm{T}} = \mathbf{A}$ which vanishes for all knots $\mathbf{p}_i$ of the first class and $\mathbf{q}_j$ of the second class. Then, the assignment of velocities $\mathbf{p}_i \mapsto \mathbf{v}_i = \mathbf{A}\mathbf{p}_i + \mathbf{a}$ and $\mathbf{q}_j \mapsto \mathbf{w}_j = -\mathbf{A}\mathbf{q}_j - \mathbf{a}$ satisfies the projection theorem for all bars $\mathbf{p}_i\mathbf{q}_j$ since

$$\begin{aligned}(\mathbf{p}_i - \mathbf{q}_j)^{\mathrm{T}}(\mathbf{v}_i - \mathbf{w}_j) &= (\mathbf{p}_i - \mathbf{q}_j)^{\mathrm{T}}(\mathbf{A}\mathbf{p}_i + 2\mathbf{a} + \mathbf{A}\mathbf{q}_j)\\ &= \mathbf{p}_i^{\mathrm{T}}\mathbf{A}\,\mathbf{p}_i - \mathbf{q}_j^{\mathrm{T}}\mathbf{A}\,\mathbf{q}_j + \mathbf{p}_i^{\mathrm{T}}\mathbf{A}\,\mathbf{q}_j - \mathbf{q}_j^{\mathrm{T}}\mathbf{A}\,\mathbf{p}_i + 2\mathbf{p}_i^{\mathrm{T}}\mathbf{a} - 2\mathbf{q}_j^{\mathrm{T}}\mathbf{a}\\ &= -2\mathbf{a}^{\mathrm{T}}\mathbf{p}_i - a + 2\mathbf{a}^{\mathrm{T}}\mathbf{q}_j + a + 2\mathbf{p}_i^{\mathrm{T}}\mathbf{a} - 2\mathbf{q}_j^{\mathrm{T}}\mathbf{a} = 0,\end{aligned}$$

because of $F(\mathbf{p}_i) = F(\mathbf{q}_i) = 0$ and the symmetry of both the dot product and the matrix $\mathbf{A}$. A comparison with (2.8) and (2.48) reveals that the velocity vectors $\mathbf{v}_i$ and $\mathbf{w}_j$, as given above, are orthogonal to the tangent planes to $\mathcal{Q}$: $F(\mathbf{x}) = 0$ at the respective points $\mathbf{p}_i$ and $\mathbf{q}_j$ (see Figure 7.23).

For the proof of the converse, we use another theorem by WHITELEY, his 'averaging theorem' 1990 ([136, p. 1341]): Let $\mathbf{p}_1, \dots$ and $\mathbf{p}'_1, \dots$ be the knots of two non-congruent configurations of any framework $\mathcal{F}$. Then, the midpoints $\mathbf{m}_i = \frac{1}{2}(\mathbf{p}_i + \mathbf{p}'_i)$ of corresponding knots constitute an infinitesimally flexible framework $\widetilde{\mathcal{F}}$ of the same combinatorial type with velocity vectors $\mathbf{v}_i = \frac{1}{2}(\mathbf{p}_i - \mathbf{p}'_i)$, and vice versa. This principle of averaging can be proved in two lines.

The condition $\|\mathbf{p}_i - \mathbf{p}_j\|^2 - \|\mathbf{p}'_i - \mathbf{p}'_j\|^2 = 0$ is equivalent to

$$\langle \mathbf{p}_i - \mathbf{p}_j + \mathbf{p}'_i - \mathbf{p}'_j,\ \mathbf{p}_i - \mathbf{p}_j - \mathbf{p}'_i + \mathbf{p}'_j \rangle = \langle 2(\mathbf{m}_i - \mathbf{m}_j),\ 2(\mathbf{v}_i - \mathbf{v}_j)\rangle = 0,$$

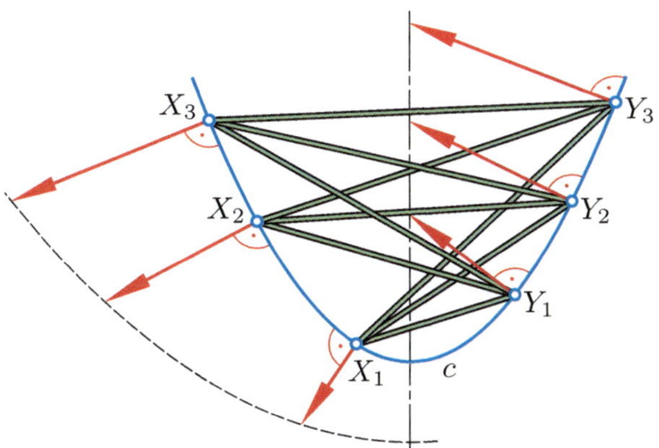

FIGURE 7.23. Infinitesimally flexible bipartite framework (Theorem 7.2.4).

and hence, to the condition of the projection theorem. It can be shown that a non-trivial assignment of velocity vectors implies the incongruence of the two configurations. Of course, all velocity vectors can be multiplied with a common scalar $\lambda \in \mathbb{R}$.

Given an infinitesimally flexible bipartite framework in $\mathbb{E}^n$, we select $n+1$ affinely independent knots $\mathbf{x}_1, \dots$ of the first class. Averaging produces two non-congruent configurations, and in both configurations, the first class knots must still be affinely independent, at least after scaling the velocity vectors with a sufficiently small λ_0. By virtue of the configuration theorem 7.2.1, all knots of the first configuration are located on two confocal quadrics, and this holds for all $\lambda < \lambda_0$, and hence also for the limit $\lambda = 0$.

When the knots of the second class are also affinely independent, we can extend the first class only with knots on the same quadric, while all distance relations are still fulfilled. ■

As an example, we deal with infinitesimally flexible octahedra in $\mathbb{E}^3$, *i.e.*, four-sided double pyramids where the basis $X_1 \dots X_4$ need not be planar (Figure 7.24). They also form a bipartite framework, with the apices Y_1, Y_2 of the pyramids as knots of the 2nd class. Since the sides $X_1X_2, \dots$ must preserve their length, Lemma 7.2.6 and Theorem 7.2.4 yield:

Corollary 7.2.10 *An octahedron in $\mathbb{E}^3$ is infinitesimally flexible if, and only if, there is a quadric $\mathcal{Q}$ through the four sides of the basis $A_1 \dots A_4$ which passes also through both apices B_1 and B_2.*

Equivalent characterizations are subject of Exercise 7.2.2.

• **Exercise 7.2.1** Alternative proof of Lemma 7.2.7.
Within a range of confocal quadrics, let $\mathcal{Q}_0$ be the regular quadric with parameter k. On the other hand, let (X, X') and (Y, Y') be two pairs of corresponding points on quadrics $\mathcal{Q}$ and $\mathcal{Q}'$ out of the range. Formulate a condition that the line $[X, Y']$ contacts $\mathcal{Q}_0$, and prove that it implies that $[X', Y]$ is also tangent to $\mathcal{Q}_0$.

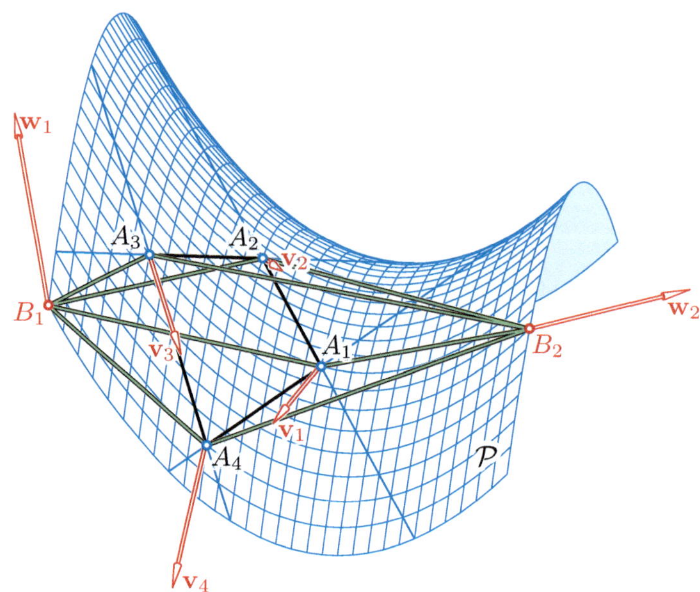

FIGURE 7.24. An infinitesimally flexible octahedron: The assigned velocity vectors $\mathbf{v}_i$ of A_i, $i = 1, \ldots, 4$, and $\mathbf{w}_j$ of B_j, $j = 1, 2$, are orthogonal to the hyperbolic paraboloid $\mathcal{P}$.

• **Exercise 7.2.2** Infinitesimally flexible octahedra.

Based on Corollary 7.2.10, prove the following statements (cf. [3, p. 316], [9], [143]):

(i) An octahedron is infinitesimally flexible if, and only if, the affine spans of four faces, of which no two share an edge, meet at a point C_1 or C_2.

(ii) An octahedron is infinitesimally flexible if, and only if, the tetrahedra $A_1A_3B_1C_1$ and $A_2A_4B_2C_2$ are of the Moebius type, *i.e.*, each vertex of one tetrahedron is incident with a face of the other tetrahedron.

7.3 String constructions of quadrics

The role of quadrics in $\mathbb{E}^3$ is analogous to that of conics in the plane. Therefore, it makes sense to ask for possibilities of string constructions of ellipsoids, as spatial analogues of the gardener's construction or of GRAVES's construction (cf. [46, Figs. 1.8 and 2.29]). The problem was solved in 1882 by STAUDE,[9] who also provided some generalizations in [126]. Here we partly follow the synthetic approach presented by W. BÖHM in [13], where historical remarks and generalizations are also included. For further references note [6, p. 236], [34, p. 11], [56, p. 19], [67], or [125, Theorem 4.3]. Historical models of string constructions are included in [112, p. 139].[10]

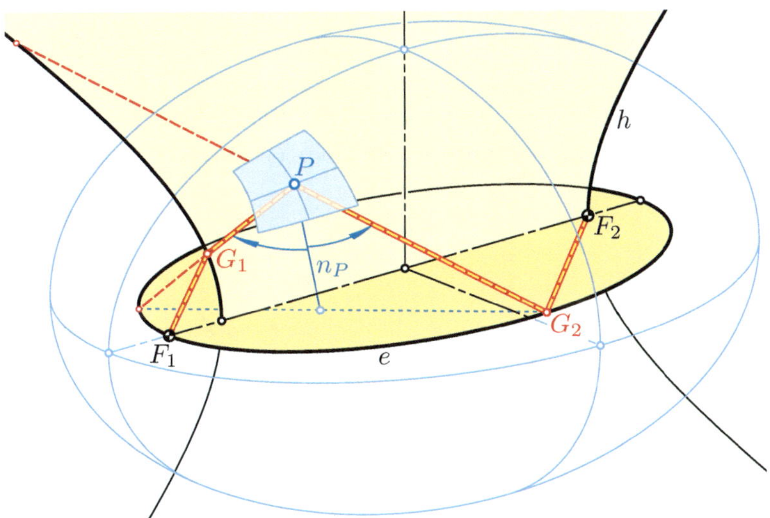

FIGURE 7.25. String construction of an ellipsoid.

Theorem 7.3.1 *Let the ellipse e and the hyperbola h be the focal conics within a range of confocal central quadrics. Let F_1 be a vertex of e and focal point of h, and point F_2 be the focal point of e at a greater distance*

[9] OTTO STAUDE (1857–1928), German mathematician.

[10] These models are displayed at *Digitales Archiv Mathematischer Modelle (DAMM)*, TU Dresden, `https://mathematical-models.org/index.php/models/view/279` and `https://mathematical-models.org/index.php/models/view/345`.

to F_1 and vertex of h. A string of a given length, fixed with one end at F_1, is passed behind the nearest branch of h and in front of e and finally attached to F_2 (see Figure 7.25).

If the string is stretched at a point P such that it forms a spatial polygon with vertices F_1, $G_1 \in h$, P, $G_2 \in e$, and F_2, then the point P traces a patch of an ellipsoid confocal with e and h.

The proof of Theorem 7.3.1, as presented below, is based on two lemmas.

Lemma 7.3.1 *Let a string with endpoints F_1 and F_2 be stretched over a curve c. Then, the corner-point $G \in c$ of the string is characterized by two conditions:*

(i) The tangent t_G to c at G subtends congruent angles with the straight segments F_1G and F_2G, and

(ii) the normal plane to c at G either passes through both endpoints or separates F_1 and F_2. In the latter case, the lines $[F_1, G]$ and $[F_2, G]$ are generators of a cone of revolution with apex G and axis t_G.

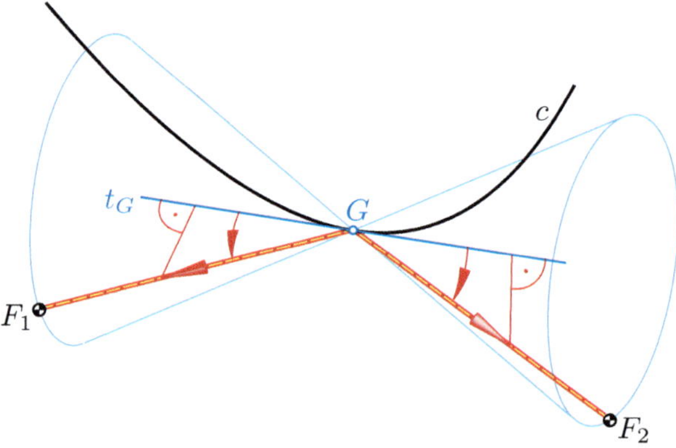

FIGURE 7.26. A string, with fixed endpoints F_1, F_2 and stretched over the curve c makes equal angles with the tangent to c at the vertex G (Lemma 7.3.1).

Proof: When the string has reached its equilibrium at $G \in c$, the stress along the string induces two forces of equal quantity, which act along the segments GF_1 and GF_2 and result in a force orthogonal to c. Therefore, the components of the two forces in direction of the tangent t_G must be opposite to each other, *i.e.*, they compensate each other (see Figure 7.26). This implies congruent angles between t_G and the two segments of the strengthened string. ■

Lemma 7.3.2 *Let a strengthened string of a given length with fixed endpoint F_1 be bent over a curve c while the second endpoint P traces any smooth curve p. Then, at each pose P, the curve p is orthogonal to the final segment of the string.*

It is noteworthy that G need not be unique. If, for example, the curve c lies on an ellipsoid with focal points F_1 and F_2, then each point $G \in c$ satisfies the claimed equilibrium condition. The sum of distances $\overline{GF_1}$ and $\overline{GF_2}$ is stationary (compare with [46, Fig. 4.17, p. 143]). A similar example plays a role below in connection with Corollary 7.3.3.

Proof: With respect to F_1 as the origin of a coordinate frame, the curve c can be parametrized as $\mathbf{c}(t) = \lambda(t)\mathbf{e}_1(t)$ with $\|\mathbf{e}_1(t)\| = 1$ for t in some interval J. After being bent over c, the upper segments form a ruled surface, and the position vector of the trajectory of F_2 can be written as

$$\mathbf{f}(t) = \mathbf{c}(t) + (k - \lambda(t))\,\mathbf{e}_2(t) \text{ with } \|\mathbf{e}_2(t)\| = 1 \text{ for } k = \text{const.}$$

The angle condition claimed in Lemma 7.3.1, implies for all $t \in J$

$$\langle\,\dot{\mathbf{c}}, \mathbf{e}_2\rangle = \langle\,\dot{\mathbf{c}}, \mathbf{e}_1\rangle, \text{ and hence } \langle\dot{\lambda}\mathbf{e}_1 + \lambda\,\dot{\mathbf{e}}_1,\, \mathbf{e}_2\rangle = \dot{\lambda}.$$

We differentiate $\mathbf{f}(t)$ w.r.t. t and obtain for the dot product with $\mathbf{e}_2$

$$\langle\,\dot{\mathbf{f}}, \mathbf{e}_2\rangle = \langle\dot{\lambda}\mathbf{e}_1 + \lambda\,\dot{\mathbf{e}}_1 - \dot{\lambda}\mathbf{e}_2 + (k-\lambda)\,\dot{\mathbf{e}}_2,\, \mathbf{e}_2\rangle = \dot{\lambda} + \langle-\dot{\lambda}\mathbf{e}_2 + (k-\lambda)\,\dot{\mathbf{e}}_2,\, \mathbf{e}_2\rangle = \dot{\lambda} - \dot{\lambda} = 0.$$

This proves the statement. ■

Proof: [Theorem 7.3.1] By virtue of Lemma 7.3.1, point $G_1 \in h$ is the apex of a cone of revolution which passes through F_1 and P and has the tangent t_{G_1} to h as its axis. We learned in [46, Theorem 4.2.1] that each conic out of a pair of focal conics is the locus of apices of cones of revolution which pass through the other conic, and the axes of these cones are tangents to the conics. Therefore, since the segment G_1F_1 meets the focal ellipse e, the same must hold for the extension of the segment G_1P.

Lines through the point P meeting e and h are common generators of two confocal cones (Theorem 7.1.2). Thus, if there exists one line, then there are four that are mutually symmetric w.r.t. the tangent planes to the three confocal quadrics through P. A 180° rotation about the surface normal n_P of the ellipsoid through P takes the line $[P, G_1]$ to a line $[P, G_2]$ which again meets the two focal conics e and h. The traces of the plane $[P, G_1, G_2]$ in the planes of e and h reveal that, starting from P, the line $[P, G_2]$ meets first e and then h. Due to the properties of a pair of focal conics, the bent portion PG_2F_2 of the string is in equilibrium because of Lemma 7.3.1.

Suppose that the point P moves such that the string remains strengthened. We are going to prove that in this case the tangents to all possible trajectories of P are orthogonal to n_P.

If the point P were fixed on the moving string, then, by Lemma 7.3.2, the tangent vector $\mathbf{v}_{t_1}$ of the point P is orthogonal to PG_1. Similarly, for the point P being fixed on the final portion of the string, the velocity vector $\mathbf{v}_{t_2}$ would be orthogonal to PG_2. Because of the constant total length of the string, the relative velocities of P with respect to the two sections of the string must be equal. When the length of the initial part increases, that of the final part must decrease about the same rate, and vice versa. This shows that the vector of the absolute velocity of P equals

$$\mathbf{v}_P = \mathbf{v}_{t_1} + \mathbf{v}_{r_1} = \mathbf{v}_{t_2} + \mathbf{v}_{r_2},$$

where $\mathbf{v}_{r_1}$ and $-\mathbf{v}_{r_2}$ are symmetric w.r.t. n_P. The orthogonal projection of the involved vectors into the plane $[P, G_1, G_2]$, as depicted in Figure 7.27, reveals that $\mathbf{v}_P$ must be orthogonal to n_P.

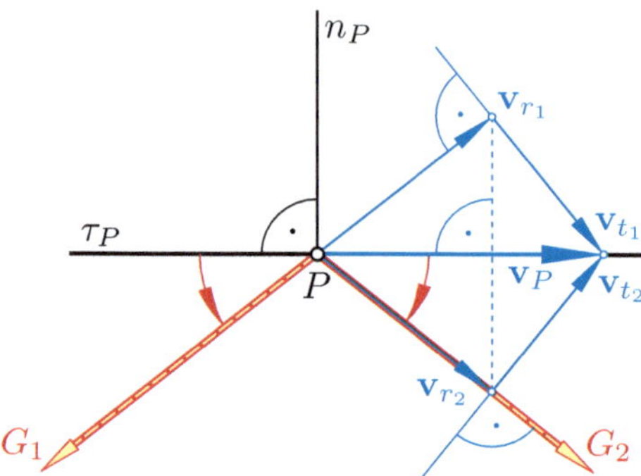

FIGURE 7.27. Decomposition of the velocity vector $\mathbf{v}_P$ at P, while the length of the strengthened string $F_1G_1PG_2F_2$ is kept fixed.

Consequently, at all poses, the point P moves tangentially to the ellipsoid. Now, with reference to theorems from the theory of partial differential equations, we conclude that P traces a patch of the unique ellipsoid through the given initial pose.

Conversely, if P remains on the ellipsoid, then $\mathbf{v}_P$ is orthogonal to n_P. This implies equal relative velocities $\|\mathbf{v}_{r_1}\| = \|\mathbf{v}_{r_2}\|$ in appropriate directions, and therefore, a constant length of the string. ■

We can figure out the total length L of the string by specifying the point P at one of the vertices of the ellipsoid. Thus, we obtain

$$L = 2a + a_e - a_h,$$

where a, a_e, and a_h are the respective principal semiaxes of the ellipsoid, the focal ellipse, and the focal hyperbola. It should be noted that W. BÖHM uses Ivory's Theorem in [13] to prove, for each pose of P, that L equals the string's total length, which hereby remains constant.

Remark 7.3.1 1. One cannot obtain the complete ellipsoid with this string construction, as described in Theorem 7.3.1, since the string, starting at F_1 and coming from behind, has to be bent around the hyperbola h. This does not work if point P also lies behind the plane spanned by h. With regard to the end of the string, the point P cannot lie under the plane of the ellipse e.

2. The same ellipsoid can be generated by using the remaining two common generators of the confocal cones which connect P with the pair of confocal conics. Similarly to Figure 7.16, the two strengthened strings could be bound together at P with a small ring through which the two strings can glide independently from each other.

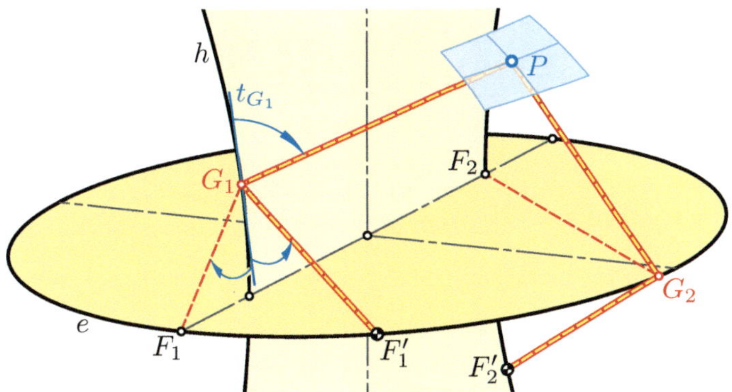

FIGURE 7.28. The fixed points F_1 and F_2 can be replaced by F_1' and F_2'.

Corollary 7.3.3 *The string construction, as described before, remains valid if the fixed endpoints F_1 and F_2 are replaced by other but sufficiently close points of the respective conics. This variation affects only the total length L of the string and the traced patch of the ellipsoid.*

Proof: The condition stated in Lemma 7.3.1 remains valid when F_1 is replaced by a sufficiently close point $F_1' \in e$ (Figure 7.28). On the other hand, since t_{G_1} forms congruent angles with G_1F_1 and G_1F_1', while F_1 and F_1' lie on the same side of the normal plane to t_{G_1} at G_1, the difference of distances

$$d := \overline{G_1F_1'} - \overline{G_1F_1}$$

remains constant. Thereofore, the distance d has to be added to the total length L of the string when P should remain on the same ellipsoid.

The same is valid for the other fixed endpoint $F_2 \in h$. Note that similar arguments were used in [46, Fig. 4.18]. ■

In [126], STAUDE presented another string construction, which is also documented as a historical model in [112, p. 139]. It generalizes the version of GRAVES's construction on an ellipsoid $\mathcal{E}_0$, as displayed in Figure 7.16.

Theorem 7.3.2 *Let a string of appropriate length with both ends be attached to a pair of antipodal curvature lines e_1, e_2 of an ellipsoid $\mathcal{E}_0$ and kept taut so that it follows a geodesic crossing from e_1 to e_2. If we elongate this string to a fixed length and keep it taut at a point P between the two curves e_1 and e_2, then P traces a patch of an ellipsoid $\mathcal{E}$ confocal with $\mathcal{E}_0$.*

Conversely, for a point P moving locally on $\mathcal{E}$, the length of the described taut string connecting e_1 via P with e_2 remains fixed.

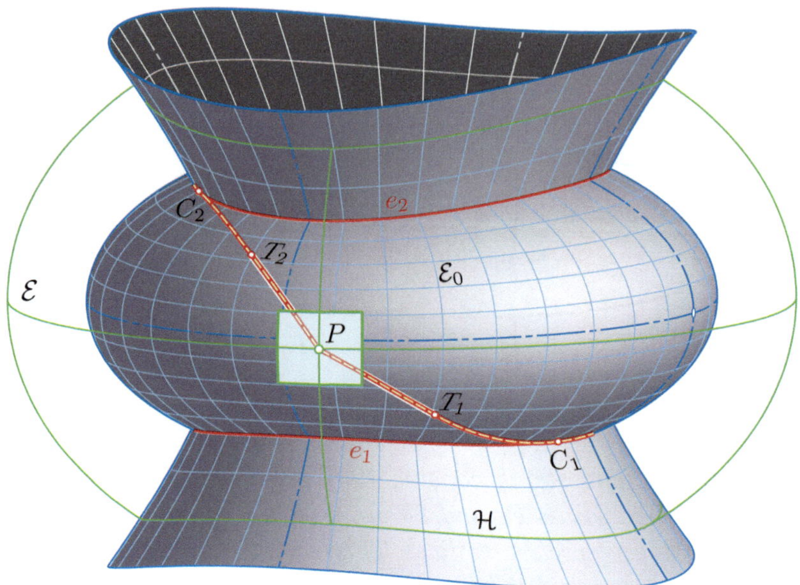

FIGURE 7.29. According to STAUDE's generalized string construction, the two ends of the string have to be attached to two antipodal curvature lines e_1, e_2 of an ellipsoid $\mathcal{E}_0$, while the point P moves on a confocal ellipsoid $\mathcal{E}$.

If the string is sufficiently short, then if follows, from the two antipodal curvature lines $e_1, e_2 \subset \mathcal{E}_0$ on, geodesic arcs and respective tangents meet at the point P (see Figure 7.29). By virtue of Theorem 7.1.4, the two tangents $[P, T_1]$ and $[P, T_2]$ contact $\mathcal{E}_0$ and a confocal hyperboloid $\mathcal{H}$. By Theorem 7.1.2, they are common to two confocal tangent cones with apex P, and consequently, symmetric w.r.t. the normal n_P at P to the confocal ellipsoid $\mathcal{E}$ passing through P. In order to prove Theorem 7.3.2, we need a statement similar to Lemma 7.3.2.

Lemma 7.3.4 *Let one end of a strengthened string of given length be attached to a line e of curvature, which is the intersection of an ellipsoid $\mathcal{E}_0$ with a confocal hyperboloid $\mathcal{H}$. Suppose that the string is a C^1-composition of three arcs. It begins along the curved edge e between $\mathcal{E}_0$ and $\mathcal{H}$. Then, it continues from a point $C \in e$, along a geodesic c of $\mathcal{E}_0$ or $\mathcal{H}$, until a point T. Finally, there is a straight segment on the tangent to c at T. Then, in which way ever, the second endpoint P of the string moves smoothly in space, its trajectory p is orthogonal to the straight segment TP.*

Proof: We proceed similar to the proof of GRAVES's Theorem 7.2.3 on the ellipsoid. Any point Q attached to the string between C and T will trace an orthogonal trajectory of the geodesic c. There is a local parametrization $\mathbf{x}(u,v)$, $(u,v) \in I \times J$, of $\mathcal{E}_0$ with the u-lines as geodesics tangent to c and the v-lines as orthogonal trajectories. By virtue of a theorem by GAUSS, u can be assumed as common arc length along the geodesics. This implies for the partial derivatives

$$\langle \mathbf{x}_u, \mathbf{x}_v \rangle = 0, \quad \langle \mathbf{x}_u, \mathbf{x}_u \rangle = 1, \quad \langle \mathbf{x}_u, \mathbf{x}_{uu} \rangle = \langle \mathbf{x}_u, \mathbf{x}_{uv} \rangle = \langle \mathbf{x}_v, \mathbf{x}_{uu} \rangle = 0$$

for all $(u,v) \in I \times J$. The equations $\langle \mathbf{x}_u, \mathbf{x}_{uu} \rangle = \langle \mathbf{x}_v, \mathbf{x}_{uu} \rangle = 0$ confirm that the u-lines are geodesics, since their osculating planes are orthogonal to the tangent plane.

If, under the motion of the string's endpoint P in space, the geodesic c remains fixed, *i.e.*, T remains on c, then the point P traces on the tangent surface of c an involute, which is an orthogonal trajectory of the generators and also a curvature line on the developable. Otherwise, the v-coordinate of T varies. Let T trace the curve on $\mathcal{E}_0$ given by $u = u(t)$ and $v = t$ for $t \in J$. This yields the parametrization

$$\mathbf{p}(t) = \mathbf{x}\,(u(t),\, t) + (k - u(t))\,\mathbf{x}_u(t) \quad \text{with} \quad k = \text{const.}$$

From

$$\dot{\mathbf{p}}(t) := \frac{\mathrm{d}\,\mathbf{p}(t)}{\mathrm{d}t} = \dot{u}\,\mathbf{x}_u + \mathbf{x}_v - \dot{u}\,\mathbf{x}_u + (k-u)(\dot{u}\,\mathbf{x}_{uu} + \mathbf{x}_{uv})$$

follows the stated orthogonality, since

$$\langle \mathbf{x}_u,\, \dot{\mathbf{p}} \rangle = \langle \mathbf{x}_u,\, \mathbf{x}_v \rangle + (k-u)\langle \mathbf{x}_u,\, \dot{u}\,\mathbf{x}_{uu} \rangle + (k-u)\langle \mathbf{x}_u,\, \mathbf{x}_{uv} \rangle = 0.$$

The same holds when we replace the ellipsoid $\mathcal{E}_0$ by the hyperboloid $\mathcal{H}$. ■

Proof: [Theorem 7.3.2]

(i) We first assume that the string is sufficiently short so that its path between e_1 and e_2 is composed of two geodesics and two line segments with the common endpoint P.

Based on Lemma 7.3.4, the proof is similar to that of Theorem 7.3.1. With respect to the part of the string attached to the line of curvature e_i, $i \in \{1,2\}$, a point P which is fixed on the moving string has a tangent vector $\mathbf{v}_{t_i}$ orthogonal to the segment PT_i. If, additionally, the point P is moving relative to the string with velocity vector $\mathbf{v}_{r_i}$ in direction of PT_i, we obtain for the absolute velocity of P the decomposition

$$\mathbf{v}_P = \mathbf{v}_{t_1} + \mathbf{v}_{r_1} = \mathbf{v}_{t_2} + \mathbf{v}_{r_2}.$$

When the total length of the string remains constant, the relative velocities $\|\mathbf{v}_{r_1}\|$ and $\|\mathbf{v}_{r_2}\|$ must be equal. This implies, as depicted in Figure 7.27, that $\mathbf{v}_P$ is orthogonal to the interior angle bisector of $\sphericalangle T_1 P T_2$ and tangent to the confocal ellipsoid $\mathcal{E}$ passing through P.

Conversely, if P remains on the ellipsoid $\mathcal{E}$, then we obtain equal relative velocities in appropriate directions, and therefore, a constant length of the string.

(ii) The arguments given above remain valid if, from e_1 on, a sufficiently long string also contacts the hyperboloid $\mathcal{H}$ before reaching the point P. In this case, the string also follows – after the straight segment between $\mathcal{E}$ and $\mathcal{H}$ – a geodesic arc on $\mathcal{H}$ and continues along a tangent line. Nevertheless, we can still claim that a point P fixed on the string will move instantaneously orthogonal to the final straight segment. ■

Remark 7.3.2 In the case of confocal paraboloids, each intersection between an elliptic and a hyperbolic paraboloid is a curvature line which consist of two simply connected components e_1 and e_2. The string construction, as explained in Theorem 7.3.2, works in this case as well, yielding string constructions of confocal elliptic paraboloids.

8 Special problems

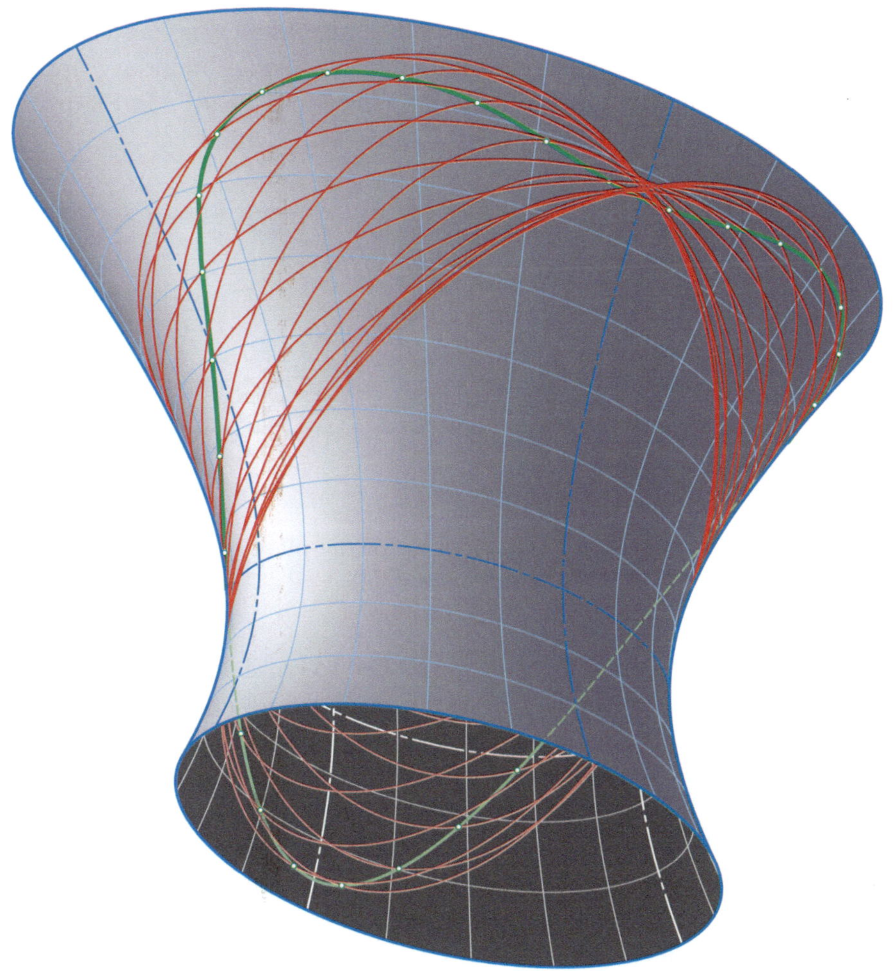

Patches of quadrics can be swept out by particular movements of conics. The picture shows poses of an ellipse moving on a one-sheeted hyperboloid, together with the trajectory of the principal vertices.

B. Odehnal et al., *The Universe of Quadrics*,
https://doi.org/10.1007/978-3-662-61053-4_8

8.1 Reflection in quadratic surfaces

The stimulus for this section is a photograph showing a coffee cup, which is made of ceramics and stands on a plate.[1] The cup looks transparent since the circular boundary of the plate, and even its section behind the cup, is completely visible. This seeming transparency is caused by the reflection in the cup: The mirror of the plate's visible boundary appears as an exact continuation of itself. Similar effects can be seen in Figure 8.1. Is this incidental, or is there a theoretical reason behind it?

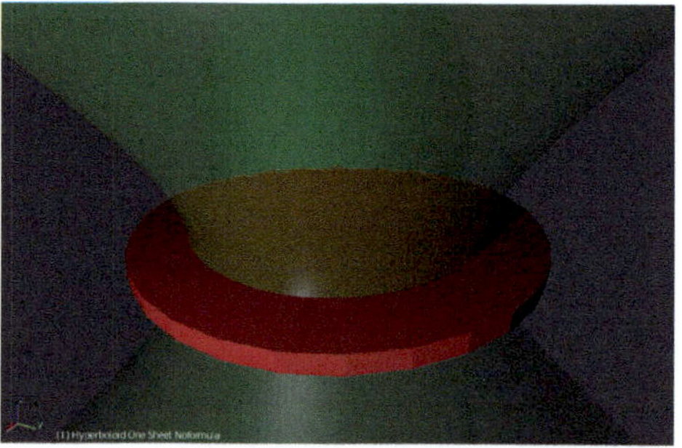

FIGURE 8.1. Why does the bounding circle of the plate continue in the reflection (Picture by courtesy of Marko Knöbl [75]).

Just to clarify the terminology, we would like to emphasize that by *reflection* in a quadric we mean the physical reflection and not the inversion in a quadric.[2] We study the physical reflection in its geometric idealization, which is defined as a transformation applied, in general, to non-directed lines ℓ in the following way: At each point P of intersection with the mirror $\mathcal{R}$, *i.e.*, the reflecting surface, the line ℓ is reflected in the tangent plane τ_P (or in the normal line n_P) to $\mathcal{R}$ at P. The line ℓ can have more than one point of intersection with $\mathcal{R}$, and, hence, more than one image. Note that each tangent line at P to $\mathcal{R}$ – to be more precise, each real line – remains fixed. Yet, each non-real tangent t of $\mathcal{R}$ is also mapped onto itself, except those in isotropic tangent planes. In the latter case,

[1] See `http://imgur.com/N10ESf1`, retrieved August 2019.

[2] The latter is also known under the name 'projective inversion'; it is a rational transformation where corresponding points are conjugate w.r.t. a given quadric and collinear with a given center.

the image of t is indeterminate since, at all points $Q \in t$, normal lines to the isotropic tangent plane lie in the tangent plane.

Reflection in vertical cylinders

We begin with a fairly popular case of a reflection which is often used for producing anamorphoses, *i.e.*, distorted perspective images, which require the viewer to use special devices or a specific point of view to reconstitute the image. A very early and impressive example is depicted in Figure 8.2.

FIGURE 8.2. Left: JEAN FRANÇOIS NICÉRON: *Ritratto di Luigi XIII,* ~ 1635, 50 × 66.5 cm, Palazzo Barberini, Rome – with kind permission of the Gallerie Nazionali di Arte Antica. Right: Photo of the painting together with a reflecting cylinder.

Let a right cylinder $\mathcal{R}$ in a vertical position be the reflector. As illustrated in Figure 8.3, if observed from the center C, a point Q of the horizontal ground plane is visible at $P \in \mathcal{R}$. We call P an *in* $\mathcal{R}$ *reflected image of* Q *w.r.t. the center* C. The surface normal n_P to the cylinder at P is horizontal. Therefore, the two segments PQ and PC of the reflected ray have the same inclination, and n_P is the interior angle bisector of $\sphericalangle QPC$, also when seen in the top view as $\sphericalangle QP'C'$.

As a consequence, for a given center C and a point Q, a reflected image $P \in \mathcal{R}$ has its top view on a *strophoid*, a curve of degree three [124]. This is the locus of points X in the ground plane such that a bisector of the angle QXC' passes through a given center M' which, in our case, coincides with the top view of the axis of $\mathcal{R}$ (Figure 8.4). Obviously, there is a second

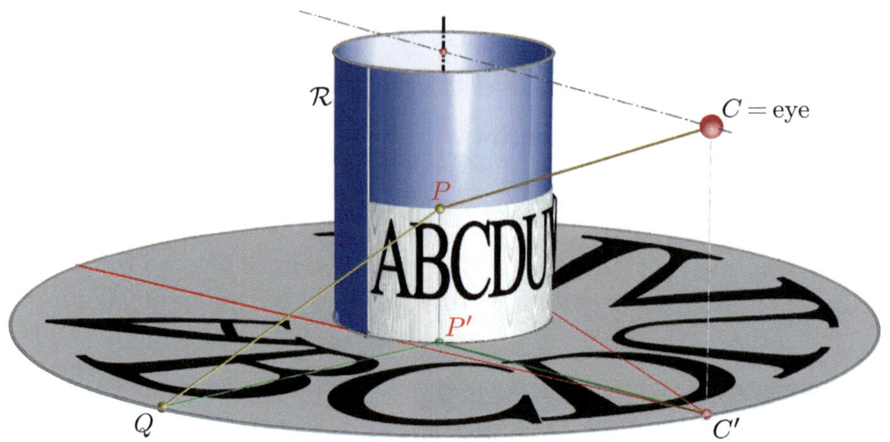

FIGURE 8.3. Reflection in the right cylinder $\mathcal{R}$: The point P is a reflected image of the point Q in the ground plane w.r.t. the center C.

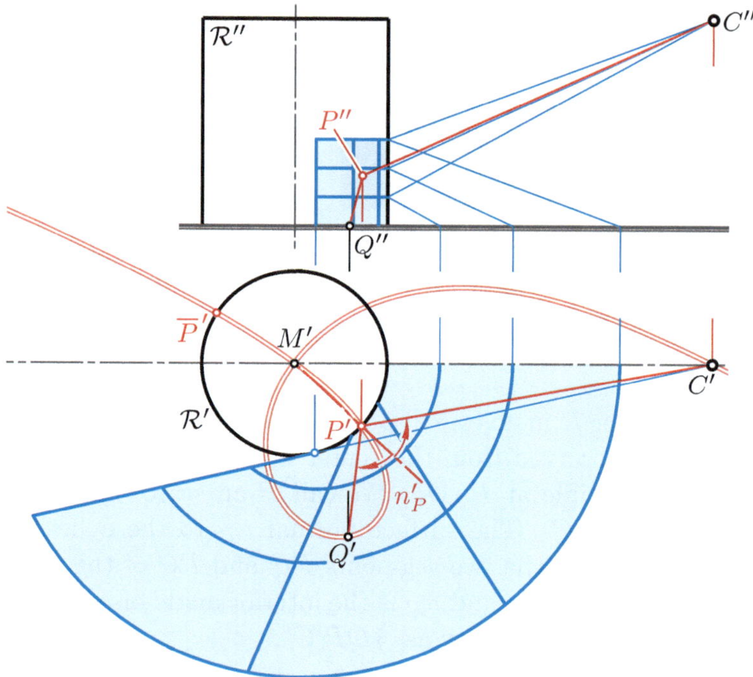

FIGURE 8.4. For a given center C and a point Q in the ground plane, the top views of the reflected images P and $\overline{P}$ lie on a strophoid.

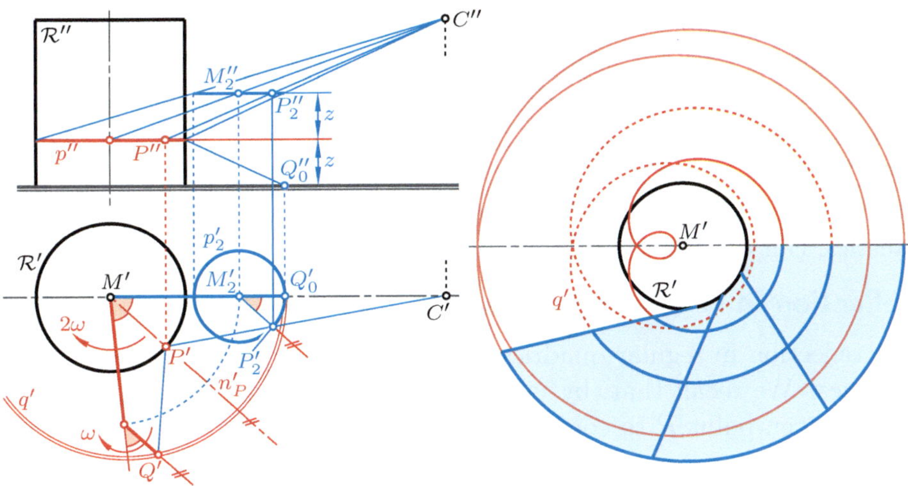

FIGURE 8.5. Parallel circles p on $\mathcal{R}$ are the reflected images of Pascal limaçons q in the ground plane.

point $\overline{P}$ of intersection between the strophoid and the cylinder $\mathcal{R}$ such that the interior angle bisector of $\sphericalangle Q\overline{P}'C'$ passes through M. This shows that the point Q can (theoretically) have two in $\mathcal{R}$ reflected images w.r.t. C. The second one, $\overline{P}$, lies on the back wall.

Figure 8.4 shows the trajectories q of Q when a reflected image P runs on $\mathcal{R}$ along horizontal circles p. These trajectories are circles only in two particular cases: Either $P \in \mathcal{R}$ lies in the ground plane or P has exactly half of the height of C over the ground plane. Otherwise, the trajectories are *Pascal limaçons*.

This can be proved as follows (see Figure 8.5): The reflection at $P \in \mathcal{R}$ acts like the reflection in the surface normal n_P and maps the line $[P, C]$ to the line $[P, Q]$. If P has the height z over the ground plane, then the reflection in n_P maps Q to a point P_2 in the height $2z$ on the line $[P, C]$. Let P run with angular velocity ω along the parallel circle $p \subset \mathcal{R}$. Then, the intersection point P_2 of $[P, C]$ with the plane in the height $2z$ runs with the same angular velocity ω on a horizontal circle p_2 with center M_2 on the cone connecting p with C.

In the top view, we obtain Q' when $P_2' \in p_2'$ is reflected in n_P', which rotates with angular velocity ω about M', the top view of the cylinder's axis. This shows that the trajectory q of Q is traced when a first bar

$M'M_2'$ rotates about M' with angular velocity 2ω while a second bar $M_2'Q_0'$ rotates with the (absolute) velocity ω. The path of the end point of a dyad $M'M_2'Q_0'$ moving this way is a particular trochoid, namely a Pascal limaçon q' [145, p. 155], provided that no moving bar has length zero.

Further properties of the reflection in right cylinders, *e.g.*, concerning caustics, can be found in [45].

Reflection in regular quadrics

The reflection in regular quadrics is closely related to results on confocal quadrics. We recall that, by virtue of Theorem 7.1.2, the tangent cones from a given point P to confocal quadrics are confocal, too.

Let a tangent line ℓ of any quadric $\mathcal{Q}_0$ pass through a point P on the quadric $\mathcal{Q}$ being confocal with $\mathcal{Q}_0$. Then, the reflection of ℓ at P in $\mathcal{Q}$ is again tangent to $\mathcal{Q}_0$, since the tangent plane τ_P to $\mathcal{Q}$ is a plane of symmetry of the cone of tangents drawn from P to $\mathcal{Q}_0$. Thus, we obtain a spatial analogue of a well-known theorem concerning conics (see [46, p. 42, Fig. 9.39]).

Corollary 8.1.1 *Let $\mathcal{Q}$ and $\mathcal{Q}_0$ be two different quadrics in a confocal family. Then, the reflection in $\mathcal{Q}$ maps the line complex of tangents of $\mathcal{Q}_0$ to itself. In particular, the complex of lines meeting any focal conic f of $\mathcal{Q}$ remains fixed.*

We learned in Theorem 7.1.3 that, in general, a given line contacts two surfaces of a confocal family, and the tangent planes at the respective points of contact are orthogonal. The only exceptions are the focal axes ℓ, characterized by the property that the isotropic planes through ℓ are tangent to all confocal quadrics.

Corollary 8.1.1 is the main reason for the optical effects mentioned at the beginning (Figure 8.1): Let a quadric $\mathcal{Q}$ and a central projection with center C be given. If any line ℓ of sight which meets a focal conic f of $\mathcal{Q}$ at a point Q_1 is reflected at the point $P \neq C$ in $\mathcal{Q}$, then the transformed line still meets f at any point Q_2. Therefore, the perspective images of the points Q_1 and P are coinciding where P is the reflected image of Q_2 w.r.t. C. This holds for all $Q_1 \in f$. Therefore, in the perspective view, the focal conic f and its in $\mathcal{Q}$ reflected image w.r.t. C belong to the same conic.

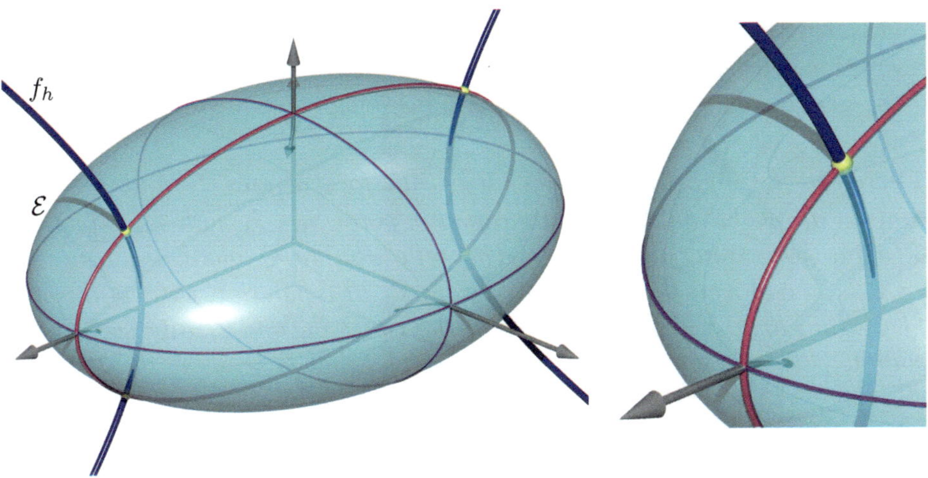

FIGURE 8.6. In the perspective view, the focal hyperbola coincides with its reflection in the ellipsoid $\mathcal{E}$.

The quadric in Figure 8.1 is a one-sheeted hyperboloid of revolution, and f passes through the focal points of the meridians. With Figure 8.6, left, we recall Figure 7.2 showing a triaxial ellipsoid $\mathcal{E}$ and its focal hyperbola f_h. This time, the perspective view is combined with a zoom-in on f_h on the right-hand side of Figure 8.6. In the perspective view, the (in $\mathcal{E}$) reflected image of f_h appears as the exact continuation of the hyperbola f_h.

As a generalization, we can replace the focal conic f with any other quadric in the confocal family and claim:

Corollary 8.1.2 *Let a reflecting quadric $\mathcal{Q}$ be given together with a confocal quadric $\mathcal{Q}_0$. Then, in the perspective image with any center C, the quadric $\mathcal{Q}_0$ and its in $\mathcal{Q}$ reflected image w.r.t. C have coinciding contours. This is also valid when $\mathcal{Q}_0$ degenerates to a focal conic f: The perspective of f coincides with that of its in $\mathcal{Q}$ reflected image.*

Remark 8.1.1 Lines which do not intersect $\mathcal{Q}$ in real points remain fixed under the reflection in $\mathcal{Q}$. Tangent lines of $\mathcal{Q}$ also remain fixed while directed tangent lines change their orientiation. Every other non-directed line has two real points of intersection and, therefore, two images.

Reflecting cones in a quadric

By virtue of Corollary 8.1.1, a line intersecting a pair of focal conics f_1 and f_2 preserves this property after reflection in any quadric being confocal with f_1 and f_2. The set of such lines is the union of cones of revolution with apices on the focal conics. Now we check what happens if the generators of one of these cones are reflected.

Theorem 8.1.1 *Let $\mathcal{Q}$ be a quadric with focal conics f_1 and f_2. The cone $\mathcal{C}_0$ of revolution, which connects any point $S_0 \in f_1$ with f_2, intersects $\mathcal{Q}$ along two conics c_1 and c_2. The reflection in $\mathcal{Q}$ along the conic c_i, $i = 1, 2$, transforms $\mathcal{C}_0$ again into a cone $\mathcal{C}_i$ of revolution passing through f_2 with an apex $S_i \in f_1$. The axes of the cones $\mathcal{C}_0$ and $\mathcal{C}_i$ intersect at the point T_i. This is the apex of the cone which touches $\mathcal{Q}$ along the conic c_i.*

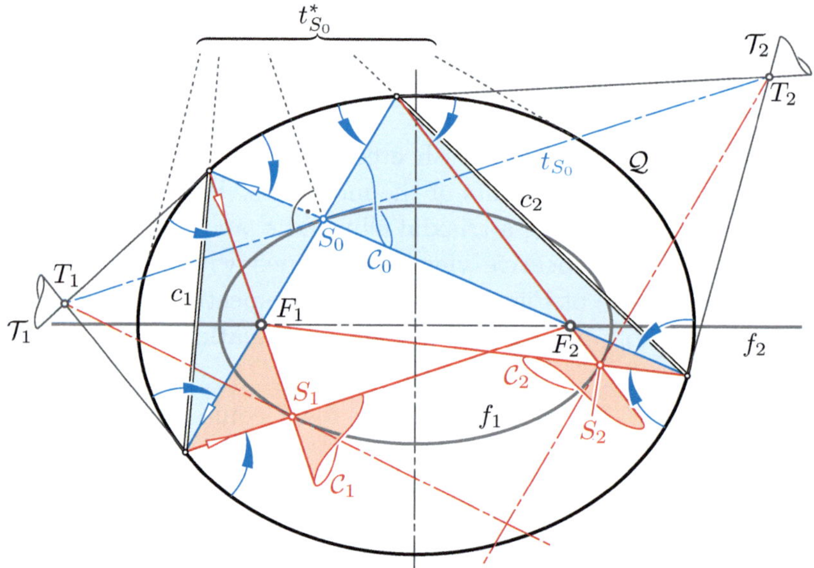

FIGURE 8.7. The reflection in the quadric $\mathcal{Q}$ transforms the right cone with apex $S_0 \in f_1$ into two right cones with apices $S_1, S_2 \in f_1$.

Proof: The tangent cones drawn from the point $S_0 \in f_1$ to the quadrics of the given confocal family are confocal with the cone $\mathcal{C}_0$ connecting S_0 with f_2. Since the latter one is a cone of revolution, they are all cones of revolutions with the proper tangents t_{S_0} to f_1 at S_0 as common axes. These cones are tangent to the isotropic planes through t_{S_0}; the respective lines of contact are isotropic lines in the plane orthogonal to t_{S_0} through S_0.

On the other hand, the poles of a fixed plane w.r.t. the quadrics of a range are collinear. For both isotropic planes through t_{S_0}, which touch all quadrics confocal with $\mathcal{Q}$, the respective points of contact are collinear with S_0, with the point of contact with f_1, and with the respective absolute point as the point of contact with the absolute conic.

Therefore, the quadric $\mathcal{Q}$, like any other confocal quadric, contacts the cone $\mathcal{C}_0$ at two points. Consequently, the curve of intersection $\mathcal{Q} \cap \mathcal{C}_0$ splits into two conics c_1 and c_2, both passing through the points of contact on the line $t^*_{S_0}$, polar to t_{S_0} w.r.t. $\mathcal{Q}$. Figure 8.7 shows the scene after being orthogonally projected into the plane of f_1.

For $i = 1, 2$, the tangent cone $\mathcal{T}_i$ of $\mathcal{Q}$ along c_i has the apex T_i. In accordance with Lemma 7.1.4, the two proper tangents drawn from T_i to f_1 are the focal axes of $\mathcal{T}_i$. One of them is t_{S_0}; the other contacts f_1 at S_i (Figure 8.7). As already noted, the reflection in $\mathcal{T}_i$ transforms planes through t_{S_0} into planes through $[T_i, S_i]$. On the other hand, the reflection in $\mathcal{Q}$ at any point $X \in c_i$ transforms the generator $[S_0, X]$ into a line meeting f_1 (and f_2). Thus, the reflection of $[S_0, X]$ must coincide with $[S_i, X]$, as stated in Theorem 8.1.1. For all $X \in c_i$, the planes spanned by the incoming and outgoing ray, which also contain the surface normal n_X to $\mathcal{Q}$, have the common trace $[S_0, S_i]$ in the plane of f_1. ■

The given proof reveals that Theorem 8.1.1 can be generalized by replacing the focal conic f_2 with any confocal quadric $\mathcal{Q}_0$.

Theorem 8.1.2 *The reflection in any non-revolute quadric $\mathcal{Q}$ transforms each cone of revolution with its apex on a focal curve f and axis tangent to f into two cones of the same type.*

Conversely, this is the only case where two smooth cones with different apices correspond under the reflection in $\mathcal{Q}$.

Remark 8.1.2 (a) Tangent cones of $\mathcal{Q}$ would be trivial examples of cones which remain cones under the reflection in $\mathcal{Q}$. However, they remain fixed, and thus, they are excluded by the condition that the apices of a cone and its reflection should be different.

(b) Figure 8.7 shows the case of an ellipsoid $\mathcal{E}$ with S_0 on the focal ellipse f_1. Theorem 8.1.2 also holds for the focal hyperbola f_2, but then, for S_0 specified in the interior of $\mathcal{E}$, the apices S_1 or S_2 can be outside. In this case, only the extensions of the reflected rays meet at a single point.

Proof: Let $c \subset \mathcal{Q}$ be the smooth curve where rays emanating from the point S_0 are reflected to rays passing through another point S_1.

(i) The normal line of $\mathcal{Q}$ at each point $P \in c$ is an interior or exterior bisector of the angle $\sphericalangle S_0PS_1$. Similar arguments, as used to prove Graves's theorem on quadrics (Figure 7.14), yield that either the sum or the difference of distances $\overline{S_0P}$ and $\overline{S_1P}$ is constant along c. This determines a quadric of revolution $\mathcal{Q}_R$ with focal points S_0 and S_1 which shares with $\mathcal{Q}$ the surface normal at each point $P \in c$ (compare with [46, Fig. 4.17]).

(ii) Since the degree of a curve of contact between the two quadrics $\mathcal{Q}$ and $\mathcal{Q}_R$ cannot be greater than two, the curve c of contact between $\mathcal{Q}_R$ and $\mathcal{Q}$ is part of a conic c.

(iii) By virtue of Theorem 2.4.5, the cones connecting c with S_0 and S_1 are cones of revolution $\mathcal{C}_0$ and $\mathcal{C}_1$, respectively. Therefore, S_0 and S_1 are located on the focal conic of c, which lies in the plane σ of symmetry of c passing through its principal axis ([46, p. 139]).

(iv) The plane σ contains the pole T of the plane of c w.r.t. $\mathcal{Q}$ and $\mathcal{Q}_R$ because of their contact along c. The line connecting T with the center of c is a diameter of $\mathcal{Q}$ and $\mathcal{Q}_R$, and it is polar w.r.t. both quadrics to the line at infinity of the plane of c, which is orthogonal to the diameter through T. Therefore, the plane σ is also a plane of symmetry of $\mathcal{Q}$.

(v) Finally, we prove that the given apices S_0 and S_1 lie on the focal conic $f \subset \sigma$ of $\mathcal{Q}$, and the respective axes s_0 and s_1 of the cones $\mathcal{C}_0$ and $\mathcal{C}_1$ are proper tangents to f. For this purpose, we use the complex extension of the Euclidean 3-space.

The cone of revolution $\mathcal{C}_0$ contacts the isotropic planes through the axis s_0 along the isotropic generators in the plane orthogonal to s_0 through S_0. They intersect the conic c at two complex conjugate points P and $\overline{P}$. All isotropic lines through S_0 form the tangent cone of $\mathcal{Q}_R$ (see [46, p. 457]). Thus, P and $\overline{P}$ are points of c with isotropic tangent planes to $\mathcal{Q}_R$ and also to $\mathcal{Q}$, due to their contact along c. Therefore, the axis s_0 is a focal axis of $\mathcal{Q}$ (Definition 7.1.2) within the plane of symmetry σ, and, consequently, by Lemma 7.1.4, a proper tangent to the focal conic $f \subset \sigma$.

Let T_0 denote its point of contact with f. The contact points of the isotropic planes through T_0 with all quadrics confocal with $\mathcal{Q}$ are located in the plane orthogonal to s_0 through T_0, since they are generators of tangent cones of revolution with apex T_0. On the other hand, we have noted before that the contact points P and $\overline{P}$ with $\mathcal{Q}$ are located in the plane orthogonal to s_0 through S_0. Therefore, the points S_0 and T_0 must be identical on s_0.

The same holds for the second cone $\mathcal{C}_1$. ■

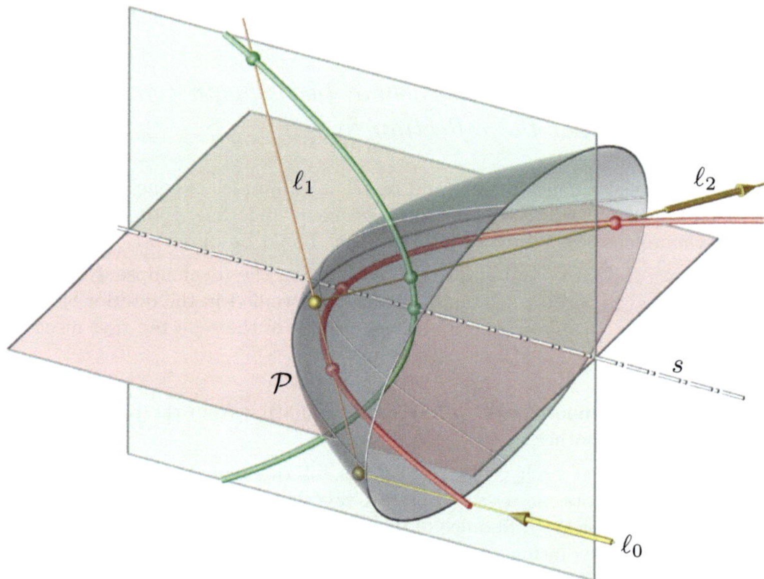

FIGURE 8.8. The reflection in an elliptic (or hyperbolic) paraboloid $\mathcal{P}$ transforms the line ℓ_0 parallel to the axis into a line ℓ_1 meeting both focal parabolas.

From a limiting case of Theorem 8.1.1, we learn how the well-known reflection property of a satellite-TV receiving dish changes when the paraboloid of revolution is replaced with a general elliptic paraboloid $\mathcal{P}$ (Figures 8.8 and 8.9). In the projective setting, the ideal point of the axis s is common to both focal curves of $\mathcal{P}$. Hence, lines parallel to s are reflected onto lines meeting both focal parabolas.

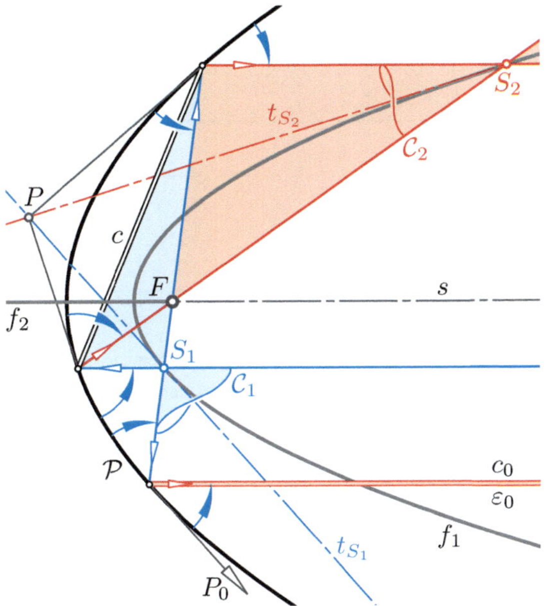

FIGURE 8.9. The reflection in the elliptic paraboloid $\mathcal{P}$ transforms the right cone with apex $S_1 \in f_1$ into the right cone with apex $S_2 \in f_1$ and a pencil of lines parallel to the axis s into the plane ε_0.

Theorem 8.1.3 *Let $\mathcal{P}$ be any paraboloid other than a paraboloid of revolution. Then, the reflection in $\mathcal{P}$ maps all lines ℓ, which are parallel to the axis s of $\mathcal{P}$ to lines meeting both focal parabolas f_1 and f_2 of $\mathcal{P}$. The pencil of those parallels ℓ to s which lie in a plane ε_0 orthogonal to the plane of f_1, is mapped to a cone of revolution with apex $S_1 \in f_1$.*

The latter can also be concluded as follows (see Figure 8.9): Let c_0 denote the parabola $\mathcal{P} \cap \varepsilon_0$. The tangent cone of $\mathcal{P}$ along c_0 is a parabolic cylinder $\mathcal{C}_0$ with apex P_0 at infinity. After an orthogonal projection with center P_0, the cylinder $\mathcal{C}_0$ appears as a parabola $\mathcal{C}_0^n$. In this view, the reflection in $\mathcal{P}$ along c_0 appears as a planar reflection in $\mathcal{C}_0^n$, which transforms lines parallel to the parabola's axis in lines through the focus of $\mathcal{C}_0^n$. This focus coincides with the projection of S_1 which is the point of f_1 with the proper tangent t_{S_1} passing through P_0.

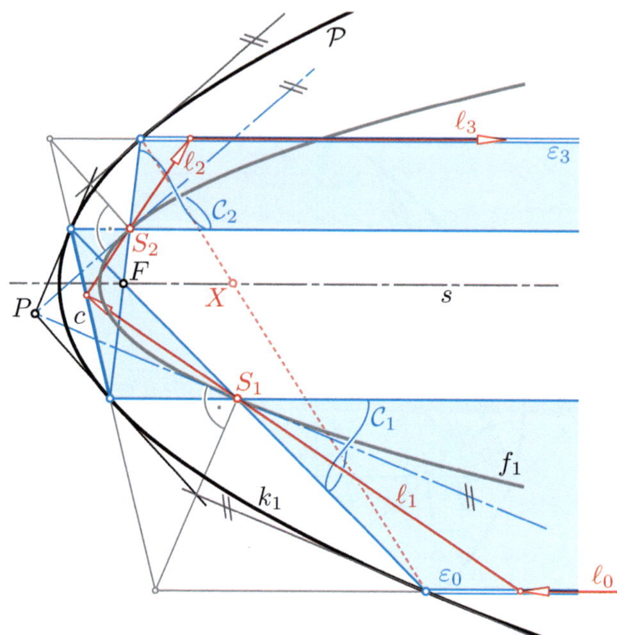

FIGURE 8.10. After at most three consecutive reflections in the elliptic paraboloid $\mathcal{P}$, all lines ℓ parallel to the axis s are sent back to such lines.

Let us once more focus on the orthogonal view in the plane of the focal parabola f_1, as displayed in Figure 8.9. We see that both the cone $\mathcal{C}_1$ with apex S_1 and the cone $\mathcal{C}_2$ with apex S_2 have one contour line parallel to the axis s and the other passing through the focal point F. In the same way as $\mathcal{C}_1$ on the one side is reflected into the pencil of parallels to s in the plane ε_0, the cone $\mathcal{C}_2$ is on the other side reflected into a pencil of parallel lines in a plane ε_3 (note Figure 8.10). The relation between ε_0 and ε_3 is involutive; the connecting line of the vertices of the parabolas $\varepsilon_0 \cap \mathcal{P}$ and $\varepsilon_3 \cap \mathcal{P}$ passes through a fixed point X on the axis s (dotted line in Figure 8.10). The proof is left to the reader as an Exercise (cf. 8.1.1). Thus, we can state a result which goes back to WUNDERLICH [140, p. 18].

Theorem 8.1.4 *After three consecutive reflections in an elliptic paraboloid $\mathcal{P}$, all lines ℓ_0 parallel to the axis s of $\mathcal{P}$ are sent back as lines ℓ_3 parallel to s, provided ℓ_0 is not located in a plane of symmetry of $\mathcal{P}$. Otherwise, line ℓ_0 is already returned after two reflections as a line ℓ_2 parallel to s.*

A similar result holds for hyperbolic paraboloids.

Remark 8.1.3 The two-parameter family of parallels to the axis s of the paraboloid $\mathcal{P}$ consists of all lines orthogonal to a plane. By virtue of the Theorem of Malus and Dupin [106, p. 446], the property of being a *congruence of surface normals* is preserved under reflection in any surface. The surfaces orthogonal to the lines meeting the pair of focal parabolas of $\mathcal{P}$ are parabolic Dupin cyclides [46, p. 147ff]. We recall that the surfaces whose normals intersect an ellipse and its focal hyperbola are general Dupin cyclides.

• **Exercise 8.1.1** Reflection in an elliptic paraboloid $\mathcal{P}$.

With regard to Theorem 8.1.4, prove the following statement (note Figure 8.10): Using Cartesian coordinates in the plane of the focal parabola f_1 with the axis s as x-axis, the product of the y-coordinates of the planes ε_0 and ε_3 equals $-p^3/p_f$, where p_f and p are the respective parameters of the focal parabola f_1 and of the intersection k_1 of the paraboloid $\mathcal{P}$ with the $[x, y]$-plane. The points of intersection of ε_0 and ε_3 with the parabola k_1 define an elliptic involution on k_1 with the center X on s (note [46, p. 251]).

Hint: Note the dilation $f_1 \to k_1$ with center F.

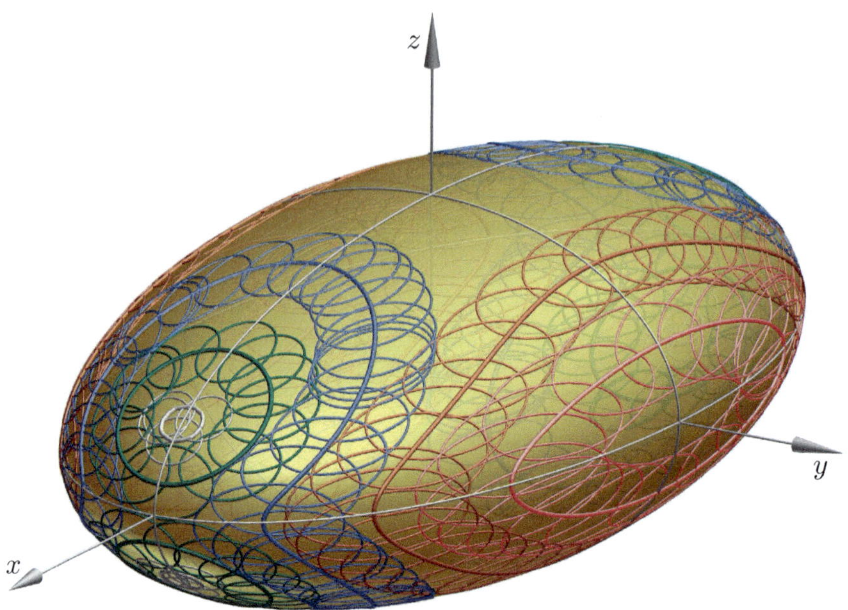

FIGURE 8.11. Moving ellipses on an ellipsoid (courtesy F. GRUBER).

8.2 Moving conics on quadrics

For any given quadric $\mathcal{Q}$, there is a three-parameter family of planes which intersect $\mathcal{Q}$ along conics. However, the size of an ellipse or hyperbola depends only on its two semiaxes. Consequently, on each quadric $\mathcal{Q}$ there exist ellipses or hyperbolas with a one-parameter family of congruent copies on $\mathcal{Q}$. One can expect that some conics can even be moved on the quadric.

The same holds for parabolas: Their size depends on one single length, its parameter, while on hyperboloids and paraboloids, there exists a two-parameter family of planes which intersect along a parabola.

There are well-known examples of movable conics on quadrics. Apart from the trivial case of circles on a sphere, paraboloids are surfaces of translation, even with a continuum of translational nets of parabolas (Theorem 2.3.2). Each parabola on a paraboloid can be embedded in a net of translation. On quadrics of revolution, each planar section remains on the quadric under rotations about the axis. Conversely, each conic symmetric w.r.t. a plane through an axis s in space sweeps a ringlike portion of a quadric when rotated about s (see [119]).

What about general quadrics $\mathcal{Q}$? F. Gruber created a numerical algorithm to visualize the motion of ellipses on an ellipsoid, as depicted on page 91 or in Figure 8.11.[3]

Moving ellipses on an ellipsoid

If we define an ellipsoid $\mathcal{E}$ as the image of a sphere under an affine transformation α, we recognize immediately that on $\mathcal{E}$ any two ellipses e_1, e_2 in parallel planes must be homothetic. The centers lie on the diameter through a point $P \in \mathcal{E}$ with a tangent plane τ_P parallel to the cutting planes. The axes of the conics are parallel to the principal curvature directions at P (see Figure 8.12).

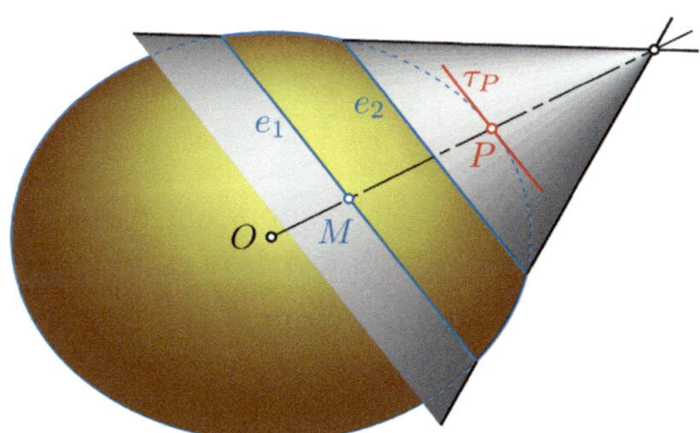

FIGURE 8.12. The sections of an ellipsoid with parallel planes are homothetic.

The same is valid for all other central quadrics $\mathcal{Q}$, due to properties of the polarity in $\mathcal{Q}$. It is well-known that even the Dupin indicatrix at P (see, *e.g.*, [46, p. 120] or [82, p. 75]) is homothetic to the conics in planes parallel to τ_P. This can be confirmed either by straightforward computation using the Taylor expansion of the quadratic polynomial at P or by applying a particular projective transformation which sends $\mathcal{Q}$ to a paraboloid with vertex P, while all points of τ_P remain fixed.

---

[3]The authors have included these pictures in memory of Franz Gruber, University of Applied Arts Vienna, who passed away in 2019 at the age of 47.

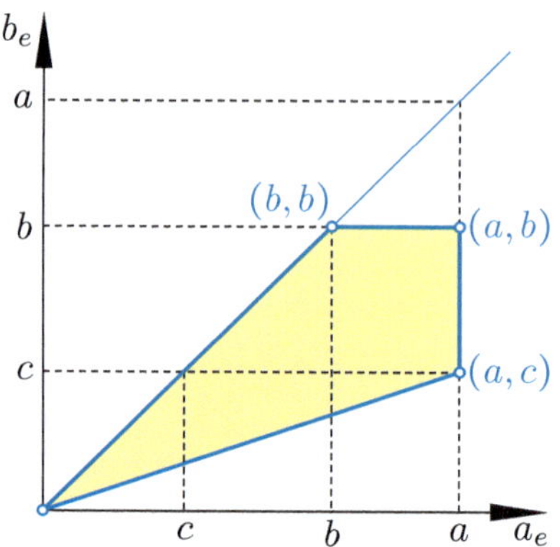

FIGURE 8.13. The semiaxes (a_e, b_e) of ellipses on an ellipsoid $\mathcal{E}$ with semiaxes a, b, and c.

Let $\mathcal{E}$ be a triaxial ellipsoid with semiaxes $a > b > c$. Figure 8.13 shows in a diagram the semiaxes (a_e, b_e) of all ellipses e on $\mathcal{E}$. We notice that the ratio of semiaxes for ellipses on $\mathcal{E}$ is restricted by

$$1 \le \frac{a_e}{b_e} \le \frac{a}{c}. \tag{8.1}$$

It will be shown that all ellipses corresponding to interior points of the marked area in the diagram can be moved on $\mathcal{E}$.

According to the definition of the Dupin indicatrix, the ratio of the principal curvatures κ_1, κ_2 at P is reciprocal to the ratio of the squared semiaxes of the ellipses on $\mathcal{E}$ in planes parallel to τ_P, *i.e.*,

$$a_e : b_e = \sqrt{\kappa_1} : \sqrt{\kappa_2} \quad \text{if} \quad \kappa_1 > \kappa_2. \tag{8.2}$$

The lines of curvature on quadrics and even the principal curvatures are related to confocal quadrics. For that purpose, we recall a few properties of confocal quadrics from Chapter 7.

The one-parameter family of quadrics being confocal with the ellipsoid $\mathcal{E}$ is given by

$$\frac{x^2}{a^2+k} + \frac{y^2}{b^2+k} + \frac{z^2}{c^2+k} = 1, \quad \text{where} \quad k \in \mathbb{R} \setminus \{-a^2, -b^2, -c^2\} \tag{8.3}$$

serves as parameter within the family. In the case $a > b > c > 0$, this family contains triaxial ellipsoids for $-c^2 < k < \infty$, one-sheeted hyperboloids for

$-b^2 < k < -c^2$, and two-sheeted hyperboloids for $-a^2 < k < -b^2$ (note Figure 8.14). As limits for $k \to -c^2$ and $k \to -b^2$, we obtain 'flat' quadrics, the focal ellipse f_e in the plane $z = 0$, and the focal hyperbola f_h in $y = 0$.

The confocal family (8.3) sends through each point $P = (\xi, \eta, \zeta)$ in 'general position', *i.e.*, with Cartesian coordinates $\xi, \eta, \zeta \neq 0$, exactly one ellipsoid, one one-sheeted hyperboloid, and one two-sheeted hyperboloid. The corresponding parameters k define the three *elliptic coordinates* of P. We concentrate on points P on the ellipsoid $\mathcal{E}$ with $k = 0$, and denote the parameters of the two hyperboloids $\mathcal{H}_1$ and $\mathcal{H}_2$ by k_1 and k_2 (Figure 8.14). Therefore, the three Cartesian coordinates (ξ, η, ζ) satisfy

$$\mathcal{E}: \ \frac{\xi^2}{a^2} + \frac{\eta^2}{b^2} + \frac{\zeta^2}{c^2} = 1, \tag{8.4}$$

and for $i = 1, 2$,

$$\mathcal{H}_i: \ \frac{\xi^2}{a^2 + k_i} + \frac{\eta^2}{b^2 + k_i} + \frac{\zeta^2}{c^2 + k_i} = 1 \ \text{for} \ -a^2 < k_2 < -b^2 < k_1 < -c^2. \tag{8.5}$$

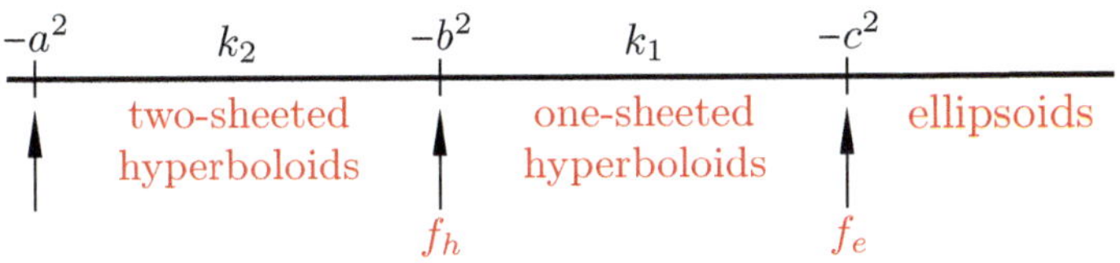

FIGURE 8.14. The boundaries of elliptic coordinates (k_1, k_2) for points $P \in \mathcal{E}$.

From a given triple $(0, k_1, k_2)$ of elliptic coordinates, we can compute the Cartesian coordinates (ξ, η, ζ) of the corresponding points, according to (7.11), via

$$\xi^2 = \frac{a^2(a^2 + k_1)(a^2 + k_2)}{(a^2 - b^2)(a^2 - c^2)}, \quad \eta^2 = \frac{b^2(b^2 + k_1)(b^2 + k_2)}{(b^2 - c^2)(b^2 - a^2)}, \quad \zeta^2 = \frac{c^2(c^2 + k_1)(c^2 + k_2)}{(c^2 - a^2)(c^2 - b^2)}. \tag{8.6}$$

There exist eight such points, symmetric w.r.t. the coordinate planes.

The tangent plane to $\mathcal{E}$ at P is given by

$$\tau_P: \ \frac{\xi}{a^2}x + \frac{\eta}{b^2}y + \frac{\zeta}{c^2}z = 1. \tag{8.7}$$

For its distance h to the center O of $\mathcal{E}$, we obtain

$$h^2 = \overline{O\tau_P}^2 = \frac{1}{\frac{\xi^2}{a^4} + \frac{\eta^2}{b^4} + \frac{\zeta^2}{c^4}}. \tag{8.8}$$

The surface normal n_P to $\mathcal{E}$ at P has the direction vector

$$\mathbf{n}_P := \left(\frac{\xi}{a^2}, \frac{\eta}{b^2}, \frac{\zeta}{c^2}\right) \quad \text{with} \quad \|\mathbf{n}_P\| = \frac{1}{h^2}. \tag{8.9}$$

The surface normals of the hyperboloids $\mathcal{H}_1$ and $\mathcal{H}_2$ at the point P have the direction vectors

$$\mathbf{v}_i := \left(\frac{\xi}{a^2+k_i}, \frac{\eta}{b^2+k_i}, \frac{\zeta}{c^2+k_i}\right), \quad i = 1,2. \tag{8.10}$$

For given $P = (\xi, \eta, \zeta) \in \mathcal{E}$, the parameters k_1 and k_2 of the hyperboloids through P are the two roots of the quadratic equation

$$k^2 + Lk + M = 0 \tag{8.11}$$

with coefficients

$$\begin{aligned} L &= a^2+b^2+c^2-\xi^2-\eta^2-\zeta^2 = \frac{(b^2+c^2)\xi^2}{a^2} + \frac{(c^2+a^2)\eta^2}{b^2} + \frac{(a^2+b^2)\zeta^2}{c^2}, \\ M &= a^2b^2+a^2c^2+b^2c^2-\xi^2(b^2+c^2)-\eta^2(a^2+c^2)-\zeta^2(a^2+b^2) = \frac{a^2b^2c^2}{h^2}, \end{aligned} \tag{8.12}$$

where $h = \overline{O\tau_P}$ satisfies (8.8). Concerning L, the second expression in (8.12) follows from the first one by multiplication with the left-hand side of (8.4). On the other hand, we obtain the last expression of M after multiplying $a^2b^2 + a^2c^2 + b^2c^2$ with the left-hand side of (8.4).

The differences of any two of the equations in (8.4) and (8.5) yield

$$\begin{aligned} &\frac{\xi^2}{a^2(a^2+k_i)} + \frac{\eta^2}{b^2(b^2+k_i)} + \frac{\zeta^2}{c^2(c^2+k_i)} = \langle \mathbf{n}_P, \mathbf{v}_i \rangle = 0, \quad i = 1,2, \quad \text{and} \\ &\frac{\xi^2}{(a^2+k_1)(a^2+k_2)} + \frac{\eta^2}{(b^2+k_1)(b^2+k_2)} + \frac{\zeta^2}{(c^2+k_1)(c^2+k_2)} = \langle \mathbf{v}_1, \mathbf{v}_2 \rangle = 0. \end{aligned} \tag{8.13}$$

This confirms again what has been stated in Theorem 7.1.2: Confocal quadrics form a triply orthogonal system of surfaces. Therefore, according

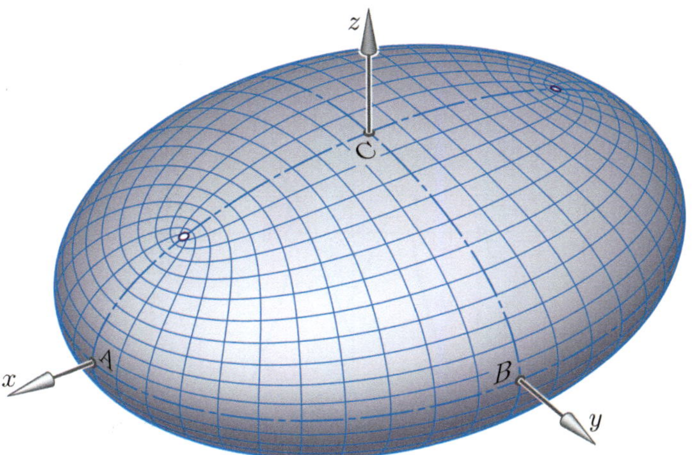

FIGURE 8.15. Lines of curvature on an ellipsoid and the umbilic points.

to DUPIN's theorem, confocal surfaces of different types intersect each other along lines of curvature (Figure 8.15).

At each point $P \in \mathcal{E}$ in general position, the two principal curvature tangents are the surface normals of the two hyperboloids $\mathcal{H}_1$ und $\mathcal{H}_2$ through P, and hence in direction of the vectors $\mathbf{v}_1$ and $\mathbf{v}_2$, as given in (8.10).

The following result is the 3-dimensional generalization of a well-known property of confocal conics, which can be proved by using the quadratic transformation of conjugate orthogonal lines (see, [46, p. 340–343]): The center of curvature at any point P of a given conic is the pole of the tangent t_P w.r.t. the confocal conic through P (Figure 8.16).

Lemma 8.2.1 *Given an ellipsoid $\mathcal{E}$, let $P \in \mathcal{E}$ be a point in general position with the tangent plane τ_P to $\mathcal{E}$. If $\mathcal{H}_1$ and $\mathcal{H}_2$ denote the two confocal hyperboloids passing through P, the poles of τ_P w.r.t. $\mathcal{H}_1$ and $\mathcal{H}_2$ are the centers of curvature of the orthogonal sections of $\mathcal{E}$ through the principal curvature tangents at P.*

Proof: The tangential equation (7.10) of confocal quadrics yields the pole T_i of τ_P w.r.t. $\mathcal{H}_i$:

$$T_i = \left(\frac{\xi(a^2 + k_i)}{a^2}, \ \frac{\eta(b^2 + k_i)}{b^2}, \ \frac{\zeta(c^2 + k_i)}{c^2} \right).$$

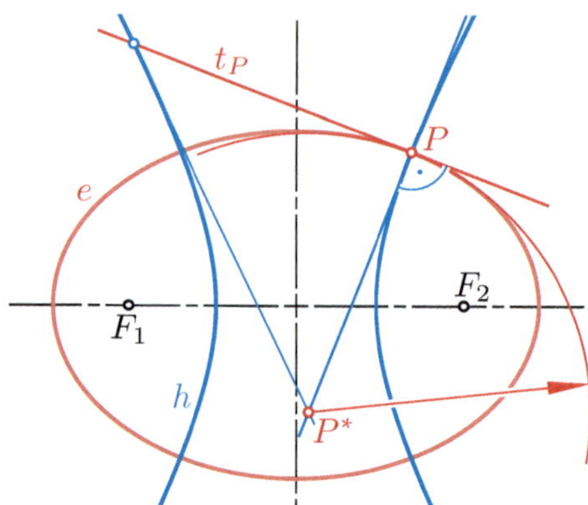

FIGURE 8.16. The center of curvature P^* of the point P w.r.t. the ellipse e is the pole of the tangent t_P w.r.t. the confocal hyperbola h.

The two poles T_1 and T_2 are located on the surface normal n_P at P to $\mathcal{E}$, and can be written as

$$T_i = \begin{pmatrix} \xi \\ \eta \\ \zeta \end{pmatrix} + k_i \begin{pmatrix} \xi/a^2 \\ \eta/b^2 \\ \zeta/c^2 \end{pmatrix}.$$

Thus, their distance to P equals, by (8.8),

$$\overline{PT_i} = |k_i|\sqrt{\frac{\xi^2}{a^4} + \frac{\eta^2}{b^4} + \frac{\zeta^2}{c^4}} = \frac{|k_i|}{h} = -\frac{k_i}{h}. \tag{8.14}$$

The statement of Lemma 8.2.1 means that the principal curvatures of $\mathcal{E}$ at P are

$$\kappa_i = 1/\overline{PT_i} = -\frac{h}{k_i}, \quad i = 1, 2. \tag{8.15}$$

We verify this by comparing it with well-known formulas for the Gaussian curvature K and the mean curvature H ([7, p. 517, footnote] or [141]):

$$K = \kappa_1\kappa_2 = \frac{h^2}{k_1k_2} = \frac{h^2}{M} = \frac{h^4}{a^2b^2c^2},$$

according to (8.11) and (8.12). Furthermore,

$$H = \frac{1}{2}(\kappa_1 + \kappa_2) = -\frac{h}{2}\left(\frac{1}{k_1} + \frac{1}{k_2}\right) = \frac{hL}{2M} = \frac{h^3}{2a^2b^2c^2}L.$$

The points T_1 and T_2 on the surface normal n_P are the *Meusnier points*, *i.e.*, the centers of the *Meusnier spheres*, associated with the principal curvature tangents. Since $k_2 \le k_1 < 0$, the pole T_2 of τ_P w.r.t. the two-sheeted hyperboloid $\mathcal{H}_2$ corresponds to the major axis of the Dupin indicatrix. This major axis has the direction vector $\mathbf{v}_2$ of the surface normal to the two-sheeted $\mathcal{H}_2$ at P, as given in (8.10) (Figure 8.17). ■

We formulated Lemma 8.2.1 for the case of an ellipsoid, but it holds similarly for hyperboloids and even paraboloids.

Lemma 8.2.2 *For each point P with elliptic coordinates $(0, k_1, k_2)$ on the ellipsoid $\mathcal{E}$, the ratio of semiaxes of ellipses $e \subset \mathcal{E}$ in planes parallel to τ_P is*

$$a_e : b_e = \sqrt{\kappa_1} : \sqrt{\kappa_2} = \sqrt{-k_2} : \sqrt{-k_1}.$$

The major axis of e has the direction of the normal vector $\mathbf{v}_2$ to the two-sheeted hyperboloid through point P.

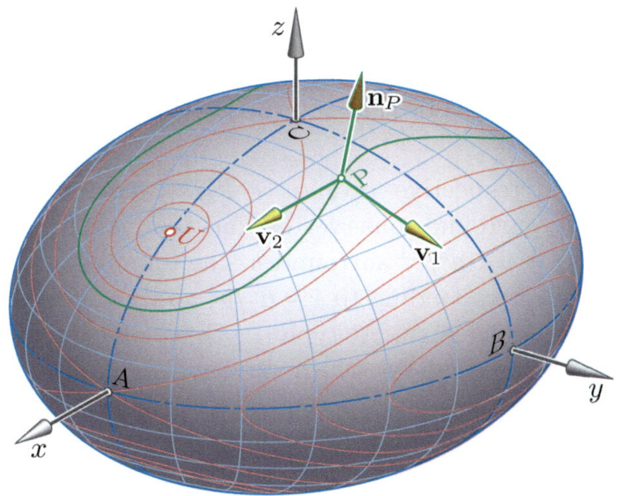

FIGURE 8.17. Curves of constant ratio of principal curvatures $\kappa_1 : \kappa_2$ in the case (i) $ac < b^2$ and direction vectors $\mathbf{v}_1, \mathbf{v}_2$ of the principal curvature tangents at P.

A direct consequence of Lemma 8.2.2 is:

Theorem 8.2.3 *On a triaxial ellipsoid $\mathcal{E}$, the curves of points with constant ratio of principal curvatures are identical with the loci of points with proportional elliptic coordinates k_1 and k_2. These curves are algebraic of degree 8.*

Proof: We can use $t := -k_2$ as a parameter for the curves of constant ratio $\kappa_1 : \kappa_2$. The definition

$$v := \frac{\kappa_1}{\kappa_2} = \frac{k_2}{k_1} = \frac{a_e^2}{b_e^2} = \text{const.} \tag{8.16}$$

yields $k_1 = t/v$. From (8.6) follows for the subarcs in the octant $x, y, z > 0$ the parametrization

$$\mathbf{p}(t) = \left(\sqrt{\frac{a^2(a^2 - t/v)(a^2 - t)}{(a^2 - b^2)(a^2 - c^2)}},\ \sqrt{\frac{b^2(b^2 - t/v)(b^2 - t)}{(b^2 - c^2)(b^2 - a^2)}},\ \sqrt{\frac{c^2(c^2 - t/v)(c^2 - t)}{(c^2 - a^2)(c^2 - b^2)}} \right), \tag{8.17}$$

where, due to (8.5),

$$\max\{b^2, vc^2\} \le t \le \min\{a^2, vb^2\}. \tag{8.18}$$

Reflections in coordinate planes generate closed curves on $\mathcal{E}$. The degree of these curves, as stated in Theorem 8.2.3, is obtained after the elimination of the parameter t in (8.17). ■

The curves of constant ratio $v = \kappa_1/\kappa_2 = k_2/k_1$ were already studied by W. WUNDERLICH [141, p. 680], however without considering the relations (8.15) between the principal curvatures κ_1, κ_2 and the elliptic coordinates k_1, k_2. In this paper [141], even the motion of an 'infinitesimally small ellipse' on the ellipsoid is mentioned.

Figure 8.18 shows the curves in question on a triaxial ellipsoid. Three types of ellipsoids have to be distinguished: (i) $ac < b^2$, (ii) $ac = b^2$, and (iii) $ac > b^2$.

Depending on the respective ratio $v = k_2/k_1$, curves with $1 < v < \min\{a^2/b^2,\, b^2/c^2\}$ surround one vertex on the y-axis. Curves with $\max\{a^2/b^2,\, b^2/c^2\} < v < a^2/c^2$ surround a single umbilic point. The distinguished curves corresponding to $v = a^2/b^2$ and $v = b^2/c^2$ are displayed as dashed lines in Figure 8.18. At type (ii) they coincide, otherwise they bound a family of closed curves which surround pairs of umbilic points.

Moving an ellipse on an ellipsoid

In order to move a given ellipse e with semiaxes (a_e, b_e) on the ellipsoid $\mathcal{E}$ with semiaxes $a > b > c$, we first check the inequality (8.1). If it is satisfied, we determine the locus of points P on $\mathcal{E}$ with the ratio of principal curvatures κ_1, κ_2 according to (8.16). A parametrization $\mathbf{p}(t)$ of this trajectory is provided in (8.17).

Then, for each point P, the center M of e has to be specified between P and the center O of $\mathcal{E}$ such that the major semiaxis equals the given length a_e. This is possible only if e is not bigger than the intersection of $\mathcal{E}$ with the diameter plane parallel to τ_P, which is clarified below.

Lemma 8.2.4 *The semiaxes of the ellipse in the diameter plane parallel to the tangent plane τ_P at the point $P \in \mathcal{E}$ with the elliptic coordinates $(0, k_1, k_2)$ are*

$$a_P = \sqrt{-k_2}, \quad b_P = \sqrt{-k_1}. \tag{8.19}$$

Proof: The diameter plane is spanned by the direction vectors $\mathbf{v}_1$ and $\mathbf{v}_2$ given in (8.10). We look for $\lambda \in \mathbb{R}$ with $\lambda \mathbf{v}_i \in \mathcal{E}$, hence

$$\lambda^2 \left[\frac{\xi^2}{(a^2 + k_i)^2 a^2} + \frac{\eta^2}{(b^2 + k_i)^2 b^2} + \frac{\zeta^2}{(c^2 + k_i)^2 c^2} \right] = 1.$$

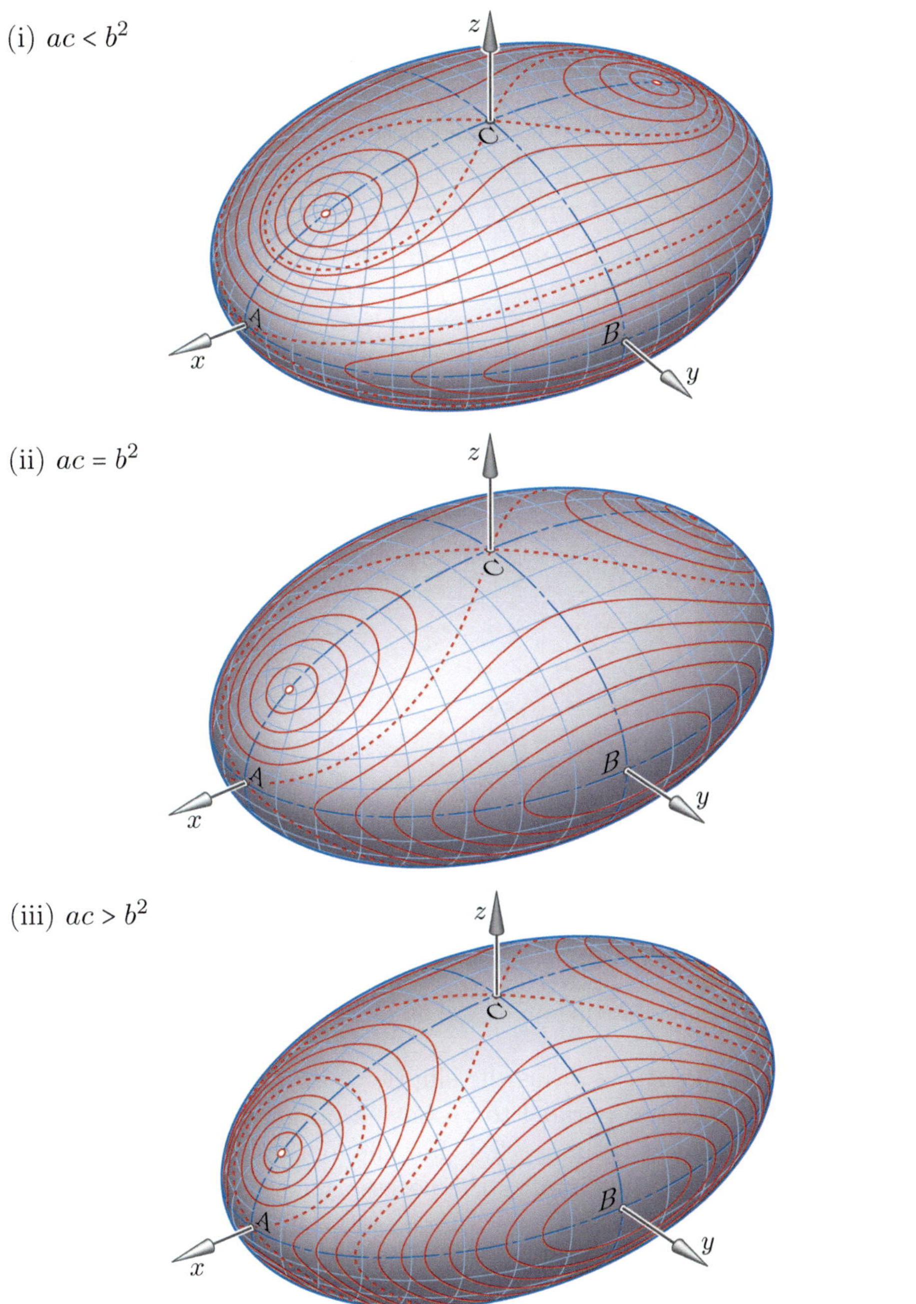

FIGURE 8.18. Curves of constant ratio of principal curvatures $\kappa_1 : \kappa_2$ in the cases (i), (ii), and (iii).

This condition does not change if we subtract from the term in square brackets the left-hand side of the first equation in (8.13), divided by k_i. Thus, we obtain

$$\lambda^2\left[\frac{\xi^2}{(a^2+k_i)^2a^2}-\frac{\xi^2}{k_i(a^2+k_i)a^2}+\dots\right]=1,$$

and, finally,

$$-\frac{\lambda^2}{k_i}\left[\frac{\xi^2}{(a^2+k_i)^2}+\frac{\eta^2}{(b^2+k_i)^2}+\frac{\zeta^2}{(c^2+k_i)^2}\right]=-\frac{\lambda^2}{k_i}\,\|\mathbf{v}_i\|^2=1,$$

hence, $a_P=|\lambda|\,\|\mathbf{v}_2\|=\sqrt{-k_2}$ and $b_P=|\lambda|\,\|\mathbf{v}_1\|=\sqrt{-k_1}$. These equations can already be found in [7, p. 517]. ■

The equations in (8.19) confirm again that the ratio $a_P^2 : b_P^2$ equals that of the elliptic coordinates $k_2 : k_1$ of P, as stated in Lemma 8.2.2.

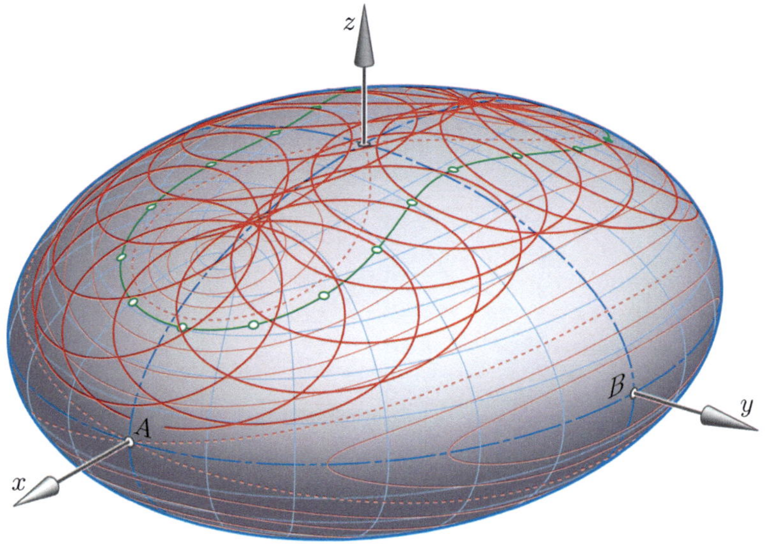

FIGURE 8.19. Moving an ellipse on an ellipsoid of type (i).

Conversely, if the point $P\in\mathcal{E}$ has the elliptic coordinates $(0,k_1,k_2)$, then all ellipses of $\mathcal{E}$ in planes parallel to τ_P have a major semiaxis smaller or equal to a_P, as given in Lemma 8.2.4. For the motion of a given ellipse e with semiaxes (a_e,b_e), this implies the necessary condition

$$a_e\le a_P=\sqrt{-k_2},\ \text{ where } b<\sqrt{-k_2}<a\,. \tag{8.20}$$

All ellipses in planes parallel to τ_P have their principal vertices on an ellipse with the conjugate diameters OP and the major axis of the diametral section. If $\mathbf{p}$ denotes the position vector of P and $\mathbf{m}=\mu\,\mathbf{p}$ with

$0 \le \mu = \sin x < 1$ that of the center M of any ellipse in this family, then its major semiaxis a_e equals $a_P \cos x = a_P\sqrt{1-\mu^2}$, which results in

$$\mu^2 = 1 - \frac{a_e^2}{a_P^2} = 1 - \frac{a_e^2}{t}. \tag{8.21}$$

Suppose that, during the motion of the ellipse e, the scalar μ vanishes, *i.e.*, its center M coincides with the center O of $\mathcal{E}$. Then, the corresponding point P reaches its final pose on the trajectory at $t = -k_2 = a_e^2$. By (8.20), such positions are only possible when $a_e > b$. In order to continue the motion, either the point P jumps to its antipode or the scalar μ becomes negative. Consequently, for the parametrization $\mathbf{p}(t)$ in (8.17), there is an additional restriction for the parameter $t = -k_2$, namely

$$\max\{b^2, vc^2, a_e^2\} \le t \le \min\{a^2, vb^2\}. \tag{8.22}$$

Figure 8.20 shows a motion of an ellipse where only a part of the trajectory of P is used. The marked principal vertex of the moving ellipse e traces a closed curve, as depicted on the left-hand side.

Now we are able to give an explicit parametrization of the motion of the ellipse e with semiaxes a_e and b_e on the ellipsoid $\mathcal{E}$. Based on (8.17) with the parameter t according to (8.22), we obtain for the trajectory of the center M of e

$$\mathbf{m}(t) = \mu(t)\mathbf{p}(t) \quad \text{with} \quad \mu(t) = \sqrt{1 - \frac{a_e^2}{t}}. \tag{8.23}$$

We can express the motion of e in matrix form, in terms of position vectors $\mathbf{x}_m$ w.r.t. the moving space (attached to e) and $\mathbf{x}_f$ w.r.t. the fixed space (attached to $\mathcal{E}$), as

$$\mathbf{x}_f = \mathbf{m}(t) + \mathbf{M}(t)\,\mathbf{x}_m, \quad \text{where } \mathbf{M}(t) = \left[\frac{\mathbf{v}_2}{\|\mathbf{v}_2\|}, \frac{\mathbf{v}_1}{\|\mathbf{v}_1\|}, \frac{\mathbf{n}_P}{\|\mathbf{n}_P\|}\right]. \tag{8.24}$$

The square brackets embrace the column vectors according to (8.10) and (8.9) (note Figure 8.17) in the orthogonal matrix $\mathbf{M}(t)$.

The interval in (8.22) restricts the parameter t for a motion, which keeps the point P within the octant $x, y, z \ge 0$. In order to obtain a closed motion, we have to continue by appropriate reflections in the planes of symmetry of the ellipsoid $\mathcal{E}$.

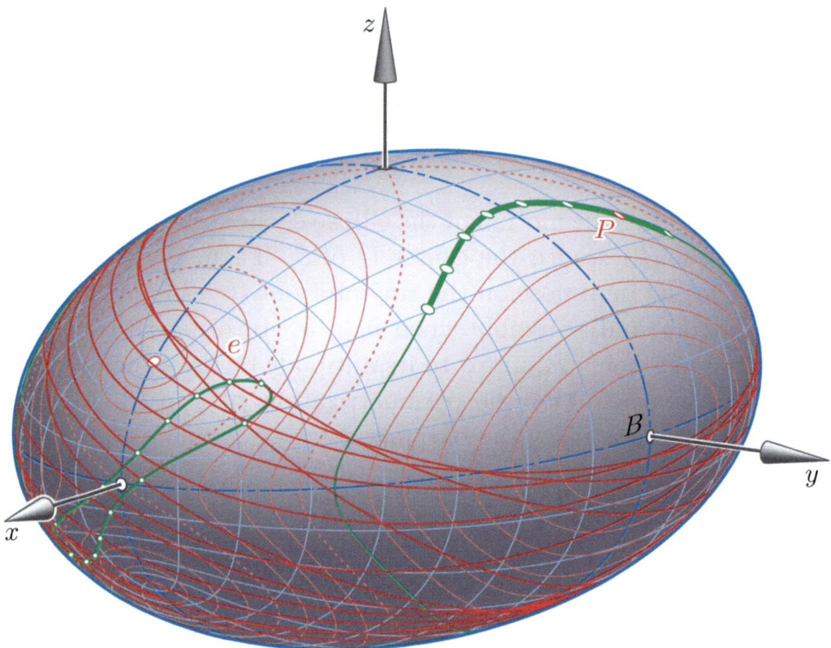

FIGURE 8.20. Motion of the ellipse e on an ellipsoid of type (iii) with $ac > b^2$, together with the trajectories of a principal vertex of e and of the corresponding point P.

Algebraic properties of this motion and some references to this problem are provided in H. BRAUNER's paper [15], such as: The carrier planes of conics moving on a central quadric form a developable of class twelve.

Moving conics on hyperboloids

On hyperboloids and paraboloids as well, the curves of intersection with parallel planes are homothetic. However, the method that was used before for ellipsoids, cannot be applied in all cases, since point the P either does not exist or lies at infinity. Moreover, paraboloids have no center O. Below, we analyse such cases, but we confine ourselves to the motions of ellipses and parabolas on one-sheeted hyperboloids. The remaining cases can be solved similarly.

For ellipses e on a one-sheeted hyperboloid $\mathcal{H}_1$, there is no point $P \in \mathcal{H}_1$ available with the tangent plane τ_P parallel to e. However, we find an appropriate point $\widetilde{P}$ on the *conjugate* two-sheeted hyperboloid $\mathcal{H}_2$ with a

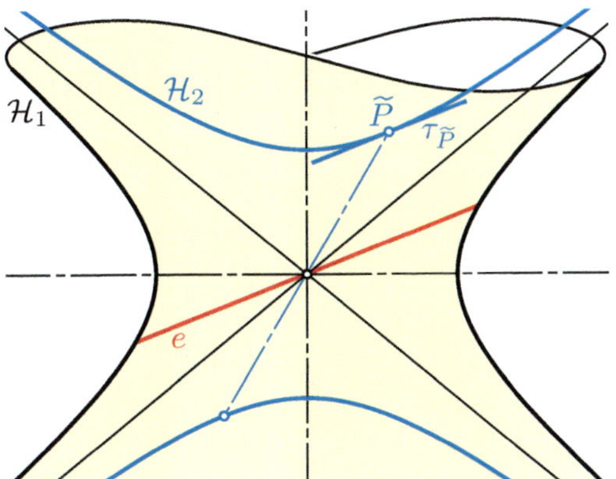

FIGURE 8.21. For ellipses e on a one-sheeted hyperboloid $\mathcal{H}_1$, there does not exist a point $P \in \mathcal{H}_1$ with the tangent plane τ_P parallel to the plane of e.

tangent plane $\tau_{\widetilde{P}}$ parallel to the plane of e (Figure 8.21). The hyperboloid $\mathcal{H}_2$ shares the asymptotic cone with $\mathcal{H}_1$, and therefore, the axes of the ellipse e are parallel to the principal curvature directions of $\mathcal{H}_2$ at $\widetilde{P}$. The two hyperboloids satisfy the respective equations

$$\mathcal{H}_1\colon \frac{x^2}{a^2} + \frac{y^2}{b^2} - \frac{z^2}{c^2} = 1 \quad \text{and} \quad \mathcal{H}_2\colon -\frac{x^2}{a^2} - \frac{y^2}{b^2} + \frac{z^2}{c^2} = 1$$

with $a > b$. The quadrics confocal with $\mathcal{H}_2$ can be written as

$$-\frac{x^2}{a^2 - k} - \frac{y^2}{b^2 - k} + \frac{z^2}{c^2 + k} = 1.$$

Again, this family sends three mutually orthogonal quadrics through each point $\widetilde{P}$ in space outside of the planes of symmetry, one of each type. On the two-sheeted hyperboloid $\mathcal{H}_2$ with $k = 0$, we use the respective parameters k_0 and k_1 of the ellipsoid and the one-sheeted hyperboloid through $\widetilde{P} \in \mathcal{H}_2$ as the elliptic coordinates of $\widetilde{P}$ with

$$k_0 > a^2 \quad \text{and} \quad a^2 > k_1 > b^2.$$

Then, similarly to Lemma 8.2.4, the ellipse $e \in \mathcal{H}_1$ in the diameter plane parallel to $\tau_{\widetilde{P}}$ has the semiaxes

$$a_{\widetilde{P}} = \sqrt{k_0} \quad \text{and} \quad b_{\widetilde{P}} = \sqrt{k_1}.$$

Hence, if any given ellipse with semiaxes a_e and b_e should be moved on $\mathcal{H}_1$, the corresponding point $\widetilde{P} \in \mathcal{H}_2$ has to trace a curve with proportional elliptic coordinates

$$k_0 : k_1 = a_{\widetilde{P}}^2 : b_{\widetilde{P}}^2 = a_e^2 : b_e^2$$

on $\mathcal{H}_2$. Similarly to (8.17), we can parametrize the trajectory of $\widetilde{P}$ by $t := k_0 > a^2$, where

$$v := \frac{k_0}{k_1} = \frac{a_e^2}{b_e^2} = \text{const.},$$

hence $k_1 = t/v$ with $b^2 \leq k_1 \leq a^2$.

For each $\widetilde{P}$, the principal vertices of the ellipses in planes parallel to $\tau_{\widetilde{P}}$ are placed on a hyperbola, for which one principal vertex in the diameter plane and the point $\widetilde{P}$ define conjugate diameters. If $a_e = a_{\widetilde{P}} \cosh x$, then the position vectors $\mathbf{m}$ of the center of the ellipse e and $\widetilde{\mathbf{p}}$ of the point $\widetilde{P}$ are related by $\mathbf{m} = \sinh x \, \widetilde{\mathbf{p}}$. Thus, we obtain

$$\mathbf{m} = \mu \widetilde{\mathbf{p}} \quad \text{with} \quad \mu = \sqrt{\frac{a_e^2}{a_{\widetilde{P}}^2} - 1}\,. \tag{8.25}$$

Finally, this yields a parametrization, similar to (8.24), for the motion of the ellipse e on $\mathcal{H}_1$. As a consequence of (8.25), on the trajectory of $\widetilde{P}$, only points with $a_{\widetilde{P}}^2 = k_0 \leq a_e^2$ are admitted. Therefore, the parameter $t = k_0$ runs the interval

$$\max\{a^2, vb^2\} \leq t \leq \min\{a_e^2, va^2\}.$$

In the case $a_e^2 < va^2$, the same phenomenon appears as depicted in Figure 8.20. When the parameter t reaches a_e^2, then, in order to continue the motion of the ellipse, either the point $\widetilde{P}$ has to jump to its antipode, or the scalar μ in (8.25) must get a negative sign.

Figure 8.22 shows the motion of an ellipse on a one-sheeted hyperboloid. In the depicted case, we have $vb^2 < a^2$ and $va^2 < a_e^2$. Therefore, the center M of the ellipse e never reaches the center O of the ellipsoid. In order to obtain a closed motion, the point $\widetilde{P}$ has to trace twice its closed curve of constant ratio v of elliptic coordinates. After one turn, the moving ellipse e returns into its initial position, but the principal vertices have switched their positions.

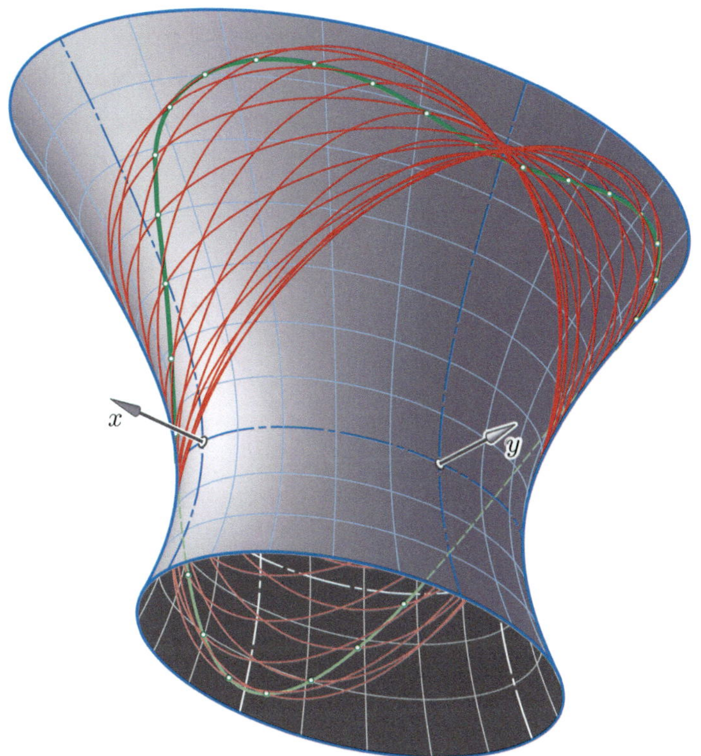

FIGURE 8.22. Motion of an ellipse on a one-sheeted hyperboloid.

Finally, we analyse the motion of parabolas on quadrics. Parabolas exist on hyperboloids and paraboloids. Congruent parabolas on paraboloids are located in families of parallel planes and form a net of translation. Thus, what is left to be studied are parabolas on hyperboloids, and we can confine ourselves to non-revolute hyperboloids.

As shown in Figure 2.23, if the parabola p is the intersection of any plane ε with a hyperboloid $\mathcal{H}$, then p is congruent to the intersection $p_{\mathcal{C}}$ between ε and the asymptotic cone $\mathcal{C}$ of $\mathcal{H}$. The parabola p arises from $p_{\mathcal{C}}$ by a translation along the common axis s. As a consequence, we first have to focus on congruent parabolas $p_{\mathcal{C}} = \varepsilon \cap \mathcal{C}$ on a quadratic cone $\mathcal{C}$.

The plane $\overline{\varepsilon}$ parallel to ε through the apex O of $\mathcal{C}$ contacts $\mathcal{C}$ along a generator $\overline{s}$ which is parallel to the axis s of $p_{\mathcal{C}}$ (see Figure 8.23). The vertex A of $p_{\mathcal{C}}$ is characterized by a tangent t_A orthogonal to s. The tangent plane τ_A to $\mathcal{C}$ at A intersects $\overline{\varepsilon}$ along a line $\overline{t}_A$ through O orthogonal to $\overline{s}$.

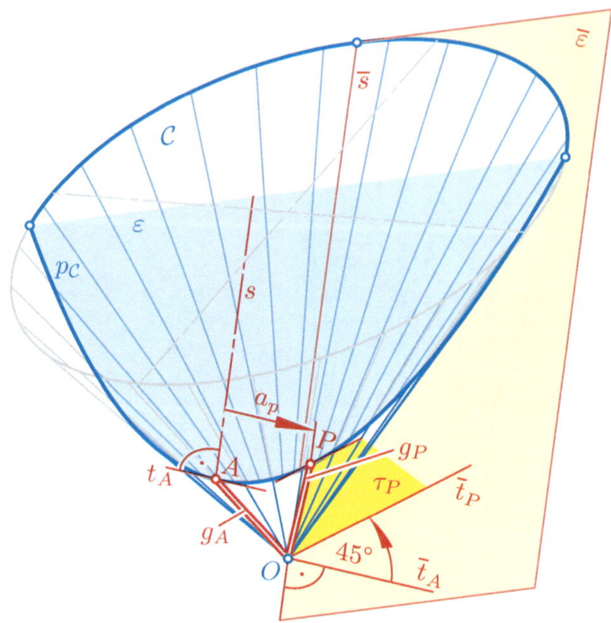

FIGURE 8.23. Determination of the parameter of a parabola p_C as the intersection on the quadratic cone $\mathcal{C}$ with the plane ε.

The second tangent plane through $\overline{t}_A$ to $\mathcal{C}$ contacts $\mathcal{C}$ along a generator g_A, which carries all vertices of parabolas on $\mathcal{C}$ in planes parallel to ε.

The parameter a_p of any parabola equals half of the chord through the focal point perpendicular to the axis (see [46, Fig. 2.2]). Let P denote one of the endpoints, called *parameter point* of the parabola. The tangent t_P at P encloses 45° with the parabola's axis. Therefore, the tangent plane τ_P at a parameter point P of p_C intersects $\overline{\varepsilon}$ along a line $\overline{t}_P$, which meets $\overline{s}$ in O at an angle of 45°. Consequently, P is located on a generator g_P along which the second tangent plane τ_P through $\overline{t}_P$ contacts the cone.

Let the parameter a_p of any parabola on $\mathcal{C}$ be given. Then, for a specified $\overline{s}$ on $\mathcal{C}$ with the tangent plane $\overline{\varepsilon}$, the apex A of the corresponding parabola $p_C \subset \mathcal{C}$ lies on the generator g_A, and all possible tangents t_A are parallel to $\overline{t}_A \subset \overline{\varepsilon}$ and orthogonal to $\overline{s}$. A parameter point P of p_C lies on the generator g_P such that the distance to the axis, measured parallel to $\overline{t}_A$, equals the given parameter a_p. We find P as the point of intersection between g_P and a plane parallel to $[g_A, \overline{s}]$, the polar plane of $\overline{t}_A$ w.r.t. $\mathcal{C}$.

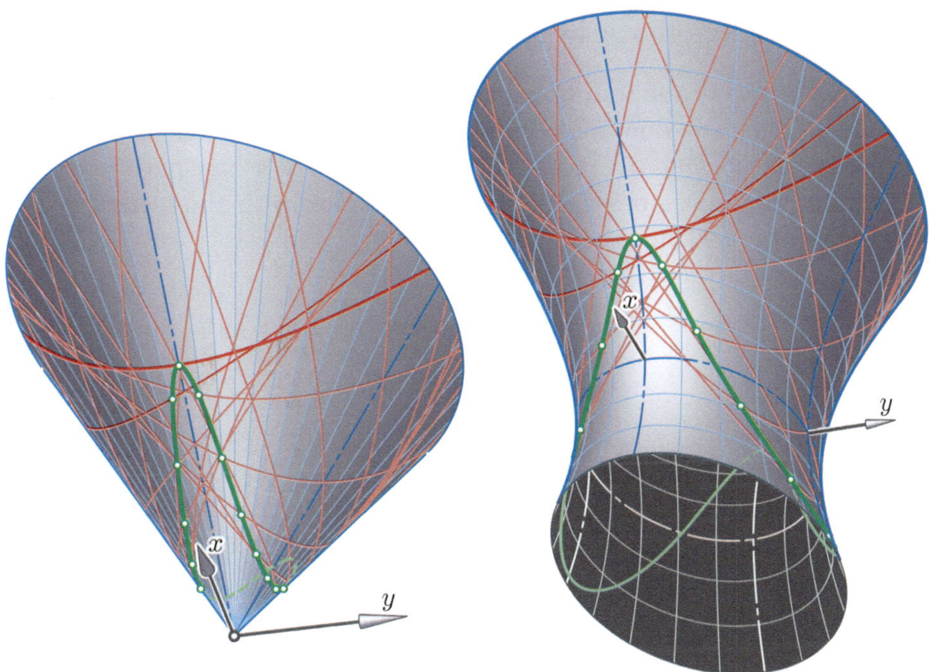

FIGURE 8.24. Motion of a parabola on a quadratic cone (left) and on a one-sheeted hyperboloid (right). The green curves are the trajectories of the vertex.

Finally, the axis s of $p_{\mathcal{C}}$ has one remaining finite point of intersection $\widetilde{A}$ with the given hyperboloid $\mathcal{H}$. The point $\widetilde{A}$ is the vertex of the parabola on $\mathcal{H}$ with the given parameter a_p and with an axis s parallel to $\overline{s} \subset \mathcal{C}$. The translation $A \mapsto \widetilde{A}$ takes $p_{\mathcal{C}}$ to the requested parabola $p \subset \mathcal{H}$.

Variations of $\overline{s}$ on the asymptotic cone $\mathcal{C}$ of $\mathcal{H}$ gives rise to a motion of a parabola on $\mathcal{H}$. For each $a_p > 0$, this motion has two connected components, symmetric w.r.t. to the apex of the cone $\mathcal{C}$. Note that the constructions of g_A, g_P, P, A, and $\widetilde{A}$ are linear. Nevertheless, the motion is not rational since the direction vector of $\overline{t}_A$ needs to be normalized.

In Figure 8.24, poses of a parabola during motions on a quadratic cone and, after respective translations, on a one-sheeted hyperboloid are depicted. The green curves are the respective trajectories of the parabola's vertex.

Infinitesimal kinematics

What can be said about the instantaneous kinematics of the motion of a conic e along any quadric $\mathcal{Q}$? Since e remains on $\mathcal{Q}$, for each pose of e the surface normals n_X to $\mathcal{Q}$ at the points $X \in e$ are path normals. Therefore, they must belong to a linear complex with the instant screw axis as axis of the complex.

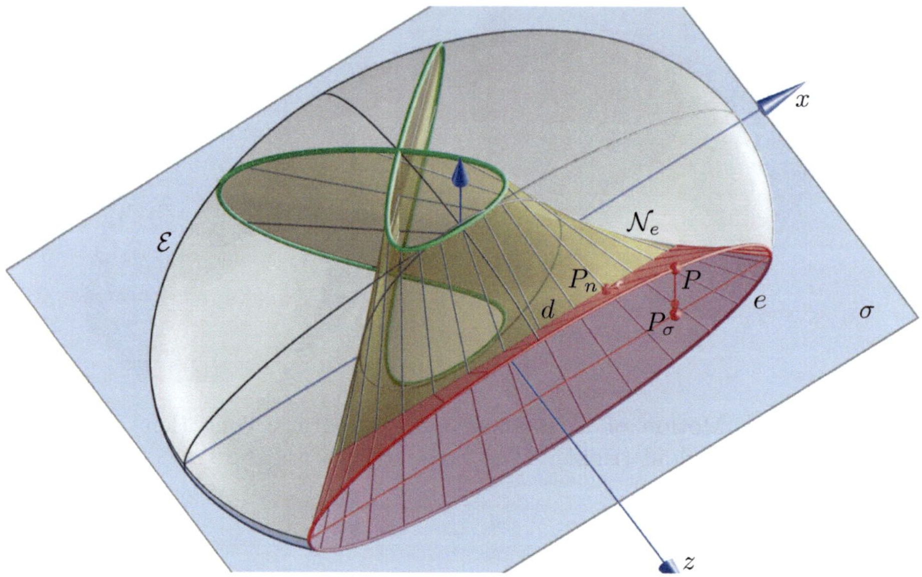

FIGURE 8.25. Surface $\mathcal{N}_e$ of normals to the ellipsoid $\mathcal{E}$ along the ellipse e, together with the remaining curve of intersection $\mathcal{N}_e \cap \mathcal{E}$. Since the plane of e is orthogonal to two planes of symmetry of $\mathcal{E}$, the surface $\mathcal{N}_e$ has two double lines; $d \subset \sigma$ is one of them.

It is well-known (see, *e.g.*, [90, pp. 273–281]) that the normals of a quadric $\mathcal{Q}$ along regular planar sections e are ruled surfaces $\mathcal{N}_e$ of degree four and generally of type III according to Sturm's classification (see Figure 8.25). This holds since there is an affine transformation between the intersection points P_n of the surface normals n_P with any plane σ of symmetry and the orthogonal projections P_σ of the points $P \in e$ into this plane σ (Theorem 2.4.1). Ruled surfaces of this type belong to a linear complex of lines [90, p. 251], and at the same time, they are surface normals along e to the quadratic tangent cone of $\mathcal{Q}$.

There are three particular cases where the linear line complex spanned by the surface normals along e becomes singular or is not uniquely defined:

(a) If the plane ε spanned by the conic e is orthogonal to a plane of symmetry σ of the quadric $\mathcal{Q}$, as depicted in Figure 8.25. Then the ruled surface $\mathcal{N}_e$ of degree four has a double line d in σ. If this is the only double line of $\mathcal{N}_e$, then the linear complex is singular and the instant screw motion of e is a rotation about d. In the particular case, shown in Figure 8.25, the plane of e is even orthogonal to a second plane of symmetry, the plane $x = 0$. Therefore, $\mathcal{N}_e$ has a second double line and is of Sturm type I. In this case the linear complex of path normals is not uniquely defined. This correlates with the fact that, locally, two symmetric motions of the moving ellipse pass through this pose.

(b) Let E denote the pole w.r.t. $\mathcal{Q}$ of the plane ε of e. If E is placed on a focal conic of $\mathcal{Q}$, then the tangent cone along e is a cone of revolution. In this case, all surface normals n_P along e intersect the axis s of this cone. Therefore, the instant screw motion of e is a rotation about s.

(c) If E is a point at infinity, then the tangent cone along e becomes a cylinder and ε is a diameter plane of $\mathcal{Q}$. Since the surface normals along e are orthogonal to the generators of the cylinder, the linear complex is again singular. The instant screw motion is a translation parallel to the generators of the cylinder, and the surface $\mathcal{N}_e$ of normals n_P is of Sturm type V. We have encountered such poses when the scalar μ in (8.21) vanished and the center M of the moving ellipse e coincides with the center O of the ellipsoid.

8.3 Quadrics on skew quadrilaterals

In this section, we shall discuss the quadrics that contain a skew quadrilateral $\mathcal{V} = P_0P_1P_2P_3$ in Euclidean 3-space $\mathbb{E}^3$, *i.e.*, four non-coplanar points $P_1, \ldots, P_4$ together with the lines $[P_i, P_{i+1}]$ where the indices are taken modulo 4.

From the very beginning, the tangent planes τ_i at P_i of all quadrics through $\mathcal{V}$ are known, since $\tau_i = [P_{i-1}, P_i, P_{i+1}]$. Again, the indices are taken modulo 4. Therefore, there are eight elements that determine the quadrics, which makes clear that there exists a one-parameter family of quadrics through $\mathcal{V}$. The one-parameter family is, indeed, a pencil of quadrics: Together with any pair of vertices P_i and P_{i+1} of $\mathcal{V}$, the line $[P_i, P_{i+1}]$ is contained in a quadric on $\mathcal{V}$. Algebraically, this leads to a system of eight linear equations in the ten (but homogeneous) coefficients of the quadric's equation, which generally has a one-dimensional linear space for its solution. Among these quadrics we find two singular ones, *i.e.*, two pairs of planes, $\mathcal{S}_1 = [P_0, P_1, P_2] \cup [P_2, P_3, P_0]$ and $\mathcal{S}_2 = [P_1, P_2, P_3] \cup [P_3, P_0, P_1]$.

Hyperbolic paraboloids on skew quadrilaterals

A hyperbolic paraboloid $\mathcal{P}$ is a (ruled) quadric that touches the ideal plane ω (or plane at infinity). Therefore, we perform the projective closure of $\mathbb{E}^3$ and we study the quadrilateral $\mathcal{V}$ and the quadrics through it in projective 3-space $\mathbb{P}^3(\mathbb{R})$ (see Figure 8.26).

The pencil of quadrics on $\mathcal{V}$ intersects the ideal plane ω along a pencil of conics, the ideal conics of all quadrics in the pencil. The singular quadrics $\mathcal{S}_1$ and $\mathcal{S}_2$ defined by $\mathcal{V}$ intersect ω in pairs (t_1, t_3) and (t_0, t_2) of lines being the ideal lines of the tangent planes T_1, T_3 and T_0, T_2 which form the singular quadrics. These two pairs of straight lines comprise the base of the pencil of ideal conics. The two diagonal lines $e = [t_0 \cap t_1, t_2 \cap t_3]$ and $f = [t_0 \cap t_3, t_1 \cap t_2]$ form the third singular conic in the pencil. The diagonal point $P = e \cap f$ together with $\mathcal{V}$ determines precisely one quadric $\mathcal{P}$ that touches ω at P. The lines e and f are the ideal lines of $\mathcal{P}$ and meet opposite side lines of $\mathcal{V}$, like any pair of concurrent generators of $\mathcal{V}$. In summary, we can say (cf. Corollary 2.3.1):

Theorem 8.3.1 *There exists exactly one hyperbolic paraboloid on a given skew quadrilateral in Euclidean 3-space.*

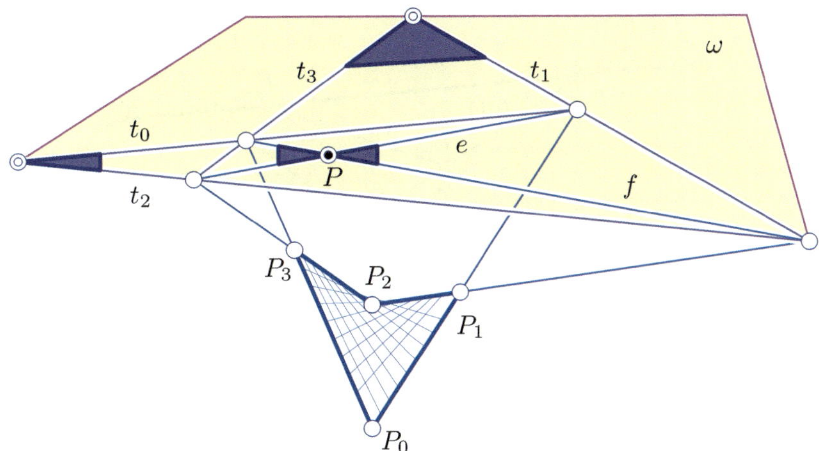

FIGURE 8.26. A skew quadrilateral $\mathcal{V} = P_0P_1P_2P_3$ in the projective closure of $\mathbb{R}^3$ and the determination of a hyperbolic paraboloid through $\mathcal{V}$.

• **Exercise 8.3.1** The ideal point of a hyperbolic paraboloid on a skew quadrilateral.

Assume that the vertices $P_0, \ldots, P_3$ of a skew quadrilateral $\mathcal{V}$ are given by their inhomogeneous coordinate vectors $\mathbf{p}_0, \ldots, \mathbf{p}_3$. Show that

$$(0, \mathbf{p}_0 - \mathbf{p}_1 + \mathbf{p}_2 - \mathbf{p}_3) \in \mathbb{R}^4$$

is a homogeneous coordinate vector of the ideal point of the uniquely determined hyperbolic paraboloid on $\mathcal{V}$. The latter homogeneous coordinate vector is a homogeneous version of (2.42).

The homogeneous coordinates of the ideal points of $\mathcal{V}$'s side lines are $(0, \mathbf{p}_i - \mathbf{p}_{i+1})$ (indices modulo 4). The diagonals e and f can be given in terms of parametrizations with homogeneous coordinates in ω as $\mathbf{e} = \lambda(\mathbf{p}_2 - \mathbf{p}_1) + \mu(\mathbf{p}_0 - \mathbf{p}_3)$ and $\mathbf{f} = \lambda'(\mathbf{p}_0 - \mathbf{p}_1) + \mu'(\mathbf{p}_2 - \mathbf{p}_3)$ with $(\lambda, \mu), (\lambda', \mu') \in \mathbb{R}^2 \setminus \{\mathbf{o}\}$. For $(\lambda, \mu) = (1, 1)$ and $(\lambda', \mu') = (1, 1)$, we obtain the only point $\mathbf{p} = \mathbf{p}_0 - \mathbf{p}_1 + \mathbf{p}_2 - \mathbf{p}_3$ that is incident with both e and f. Thus, $e \cap f = (0, \mathbf{p})$.

In Exercise 8.3.1, we have learned how to find the ideal point $P = (0, \mathbf{p})\mathbb{R}$ of the uniquely determined hyperbolic paraboloid $\mathcal{P}$ that contains the vertices P_i and side lines of a skew quadrilateral $\mathcal{V} = P_0P_1P_2P_3$. There is an affine construction that yields the direction of $\mathcal{P}$'s axis as indicated in Section 2.3.

We still assume that the three-dimensional space is projectively closed. Pairs of opposite sides of the given finite skew quadrilateral define the ideal line of the director planes. With $(ij) = [P_i, P_j] \cap \omega$ where $(i, j) \in \{(0,1), (1,2), (2,3), (3,0)\}$, we define the ideal points of the sides of the skew quadrilaterals. The ideal lines of the director planes are then $e_\infty = [(12), (30)]$ and $f_\infty = [(01), (23)]$ (cf. Figure 8.27). This construction is

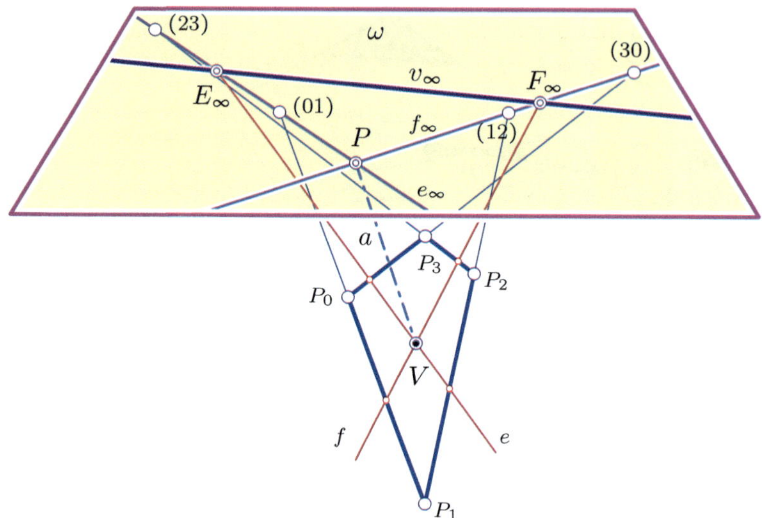

FIGURE 8.27. At the only vertex V of a hyperbolic paraboloid $\mathcal{P}$, the tangent plane $T_V = [e, f]$ is perpendicular to the diameter a (axis). Therefore, the ideal line v_∞ of T_V is the absolute polar of $\mathcal{P}$ with respect to the quadric $\mathcal{P}$.

the three-dimensional analogue of the construction of a parabola's vertex in $\mathbb{P}^2(\mathbb{R})$ as shown in [46, p. 270 ff.].

From Exercise 8.3.1, we can deduce the following

Theorem 8.3.2 *Let $\mathcal{P}$ be a hyperbolic paraboloid and let $\mathcal{Q} = P_0P_1P_2P_3$ be a finite skew quadrilateral on $\mathcal{P}$, i.e., $P_i \in \mathcal{P}$ and $[P_i, P_{i+1}] \subset \mathcal{P}$ for i modulo 4. Then, the line joining the midpoints of the segments P_0P_2 and P_1P_3 is parallel to the axis.*

Conversely, for any skew quadrilateral, the line joining the midpoints of the diagonals of a skew quadrilateral are parallel to the axis of the uniquely determined hyperbolic paraboloid on $\mathcal{Q}$.

Proof: Without loss of generality, we can assume that $\mathcal{P}$ is given by $2z = \frac{x^2}{a^2} - \frac{y^2}{b^2}$. Then, we choose four points P_i on $\mathcal{P}$ such that they form a skew quadrilateral on $\mathcal{P}$ by letting

$$\mathbf{p}_0 = (a(u_0+v_0), b(v_0-u_0), 2u_0v_0), \quad \mathbf{p}_1 = (a(u_1+v_0), b(v_0-u_1), 2u_1v_0),$$
$$\mathbf{p}_2 = (a(u_1+v_1), b(v_1-u_1), 2u_1v_1), \quad \mathbf{p}_3 = (a(u_0+v_1), b(v_1-u_0), 2u_0v_1),$$

where $u_0 \neq u_1$ and $v_0 \neq v_1$. Now,

$$\frac{1}{2}(\mathbf{p}_0 - \mathbf{p}_1 + \mathbf{p}_2 - \mathbf{p}_3) = (0, 0, u_0v_0 - u_0v_1 - u_1v_0 + u_1v_1),$$

which is obviously parallel to the paraboloid's axis pointing to $(0, 0, 1)$.

For the converse statement, we observe the following: Let $(1, \mathbf{p}_i)$ (with $i \in \{0,1,2,3\}$) be the homogeneous coordinates of the vertices of the skew quadrilateral. The midpoints of the diagonals have the coordinate vectors

$$\mathbf{m}_{02} = \frac{1}{2}(\mathbf{p}_0 + \mathbf{p}_2), \quad \mathbf{m}_{13} = \frac{1}{2}(\mathbf{p}_1 + \mathbf{p}_3),$$

and the difference

$$\mathbf{a} = \mathbf{m}_{02} - \mathbf{m}_{13} = \frac{1}{2}(\mathbf{p}_0 - \mathbf{p}_1 + \mathbf{p}_2 - \mathbf{p}_3).$$

According to Theorem 8.3.1, $(0, \mathbf{a})$ are the homogeneous coordinates of the paraboloid's ideal point. ■

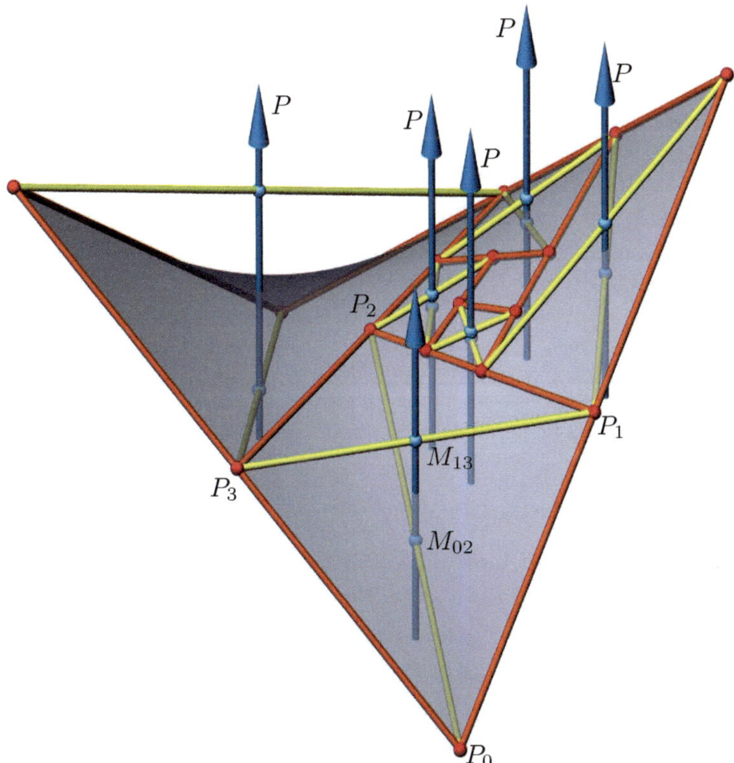

FIGURE 8.28. Each skew quadrilateral $P_0P_1P_2P_3$ is suitable for the determination of the axis of a hyperbolic paraboloid. The midpoints M_{02} and M_{13} of the diagonals lie on a line parallel to the axis.

Figure 8.28 shows that the choice of the skew quadrilateral on a hyperbolic paraboloid has no influence on the direction of the line joining the midpoints of the diagonals. In other words: Each skew quadrilateral on $\mathcal{P}$ determines the same axis direction.

In Exercise 8.3.1, we have learned that $P = e_\infty \cap f_\infty$ is the ideal point of the hyperbolic paraboloid, and therefore, P is the common ideal point of all diameters of $\mathcal{P}$. The vertex V of $\mathcal{P}$ is the surface point whose diameter a is perpendicular to its tangent plane T_V. Thus, the ideal line v_∞ of T_V is the absolute polar (line) of P. Now, $E_\infty = v_\infty \cap e_\infty$ and $F_\infty = v_\infty \cap f_\infty$ are the ideal points for those two generators e and f of $\mathcal{P}$ that pass through V (cf. Figure 8.27). The diameter a of $\mathcal{P}$ is called the *axis* of the hyperbolic paraboloid $\mathcal{P}$. Moreover, by definition and due to the construction of V as the intersection of e and f, the axis a is the polar line of v_∞ with respect to $\mathcal{P}$.

• **Exercise 8.3.2** A hyperbolic paraboloid as a $(1,1)$-Bézier surface.

The uniquely determined hyperbolic paraboloid on the skew quadrilateral $P_0P_1P_2P_3$ equals the $(1,1)$-Bézier surface with the control points $B_{0,0} = P_0$, $B_{0,1} = P_1$, $B_{1,0} = P_2$, and $B_{1,1} = P_3$. Thus, the tensor product representation

$$\mathbf{s}(u,v) = \mathbf{p}_0(1-u)(1-v) + \mathbf{p}_1 u(1-v) + \mathbf{p}_2 uv + \mathbf{p}_3(1-u)v \quad \text{with} \quad (u,v) \in \mathbb{R}^2 \tag{8.26}$$

is a parametrization of the unique hyperbolic paraboloid on the given skew quadrilateral. We define $w_{ij} := \langle \mathbf{p}_i, \mathbf{p}_j \rangle$ for $i,j \in \{0,1,2,3\}$. Now show that the surface point with parameters

$$u = \frac{w_{00} - w_{01} + w_{02} - 2w_{03} + w_{13} - w_{23} + w_{33}}{w_{00} - 2w_{01} + 2w_{02} - 2w_{03} + w_{11} - 2w_{12} + 2w_{13} - 2w_{23} + w_{33}},$$
$$v = \frac{w_{00} - 2w_{01} + w_{02} - w_{03} + w_{11} - w_{12} + w_{13}}{w_{00} - 2w_{01} + 2w_{02} - 2w_{03} + w_{11} - 2w_{12} + 2w_{13} - 2w_{23} + w_{33}}$$

is the vertex of the hyperbolic paraboloid (8.26). Compare this result with (2.43).

Quadrics of revolution on skew quadrilaterals

From Theorem 8.3.1 and Corollary 2.3.1, we know that there exists only one hyperbolic paraboloid on a skew quadrilateral $\mathcal{V}$ in Euclidean 3-space. This raises the question whether there exist quadrics of revolution containing a given skew quadrilateral or not.

A quadric of revolution can be characterized in the complex extension of the projective closure of Euclidean 3-space as a quadric whose ideal curve touches the absolute conic twice. According to what has been said above, a generic skew quadrilateral defines a pencil of conics (pencil of the first kind, see [46, p. 268]). In order to look for quadrics of revolution in the given pencil of quadrics through $\mathcal{V}$, we have to find conics in the pencil in ω that touch the absolute conic twice. This means imposing two conditions on a one-parameter family of conics. Therefore, we cannot find a solution in the generic case (cf. Theorem 2.2.2).

Thus, we assume that the skew quadrilateral displays some symmetries.

Onefold symmetry

The four vertices P_0, P_1, P_2, and P_3 of a skew quadrilateral $\mathcal{V}$ symmetric w.r.t. a plane can be given by their Cartesian coordinate vectors

$$\begin{aligned} \mathbf{p}_0 &= (a,0,0), \quad \mathbf{p}_1 = (0,b,0), \\ \mathbf{p}_2 &= (c,0,d), \quad \mathbf{p}_3 = (0,-b,0), \end{aligned} \tag{8.27}$$

since in this case, the plane of symmetry has to pass through two vertices. In order to determine the quadrics $\mathcal{Q}$ through $\mathcal{V}$, we solve the system of nine linear equations arising from inserting the coordinates of P_i and four further points on the lines $[P_i, P_{i+1}]$ (indices modulo 4) into the most general form a quadric's equation. The ninth point $P \notin [P_i, P_{i+1}]$ (indices modulo 4) shall have the coordinates $\mathbf{p} = (\xi, \eta, \zeta)$.

The equations of the quadrics are thus

$$\begin{aligned} \mathcal{Q}:\ & d\zeta(c\zeta+d\xi)(b^2x^2-a^2y^2-2ab^2x-2ab^2(a+c)z+a^2b^2)- \\ -&(cd(b^2\xi^2-a^2\eta^2-2ab^2\xi+a^2b^2)-2ab^2c(a+c)\zeta-b^2(a^2-c^2)\xi\zeta)z^2- \\ -&(d^2(b^2\xi^2-a^2\eta^2-2ab^2\xi+a^2b^2)-2ab^2d(a+c)\zeta+b^2(a^2-c^2)\zeta^2)xz=0. \end{aligned} \tag{8.28}$$

Now, the Hessian matrix of (8.28) can be diagonalized and shows three mutually distinct entries in its principal diagonal:

$$\mathrm{H}(\mathcal{Q}) = \begin{pmatrix} p+R & 0 & 0 \\ 0 & p-R & 0 \\ 0 & 0 & q \end{pmatrix}, \tag{8.29}$$

where p is a quadratic polynomial and R is the root of a quartic polynomial, both in the variables ξ, η, ζ. We obtain a quadric of revolution if two entries of the diagonal matrix are equal, which yields three cases to be distinguished:

(1) Assuming that the first two eigenvalues of (8.29) are equal leads to $p+R = p-R$ and, thus, $R = 0$. Therefore, $R^2 = 0$ is also an equation of the set of all points P with coordinates (ξ, η, ζ) that lie on quadrics of revolution passing through $\mathcal{V}$.

(2a,b) From $p - R = q$, we get the polynomial expression $R^2 = (p-q)^2$. Since $p + R = q$ yields

$$(p-q)^2 = (-R)^2 = R^2,$$

these two cases coincide.

■ **Example 8.3.1** The unique hyperbolic paraboloid through $\mathcal{Q}$.
Show that

$$(b^2x^2 - a^2y^2 - 2ab^2x + a^2b^2)d^2 + b^2(a-c)^2z^2 - 2b^2d(a-c)xz - 2ab^2d(a+c)z = 0$$

is the equation of the uniquely determined hyperbolic paraboloid on $\mathcal{Q}$ with the vertices (8.27). Its ideal point has the homogeneous coordinates $(0 : a-c : 0 : d)$.

Twofold symmetry

A skew quadrilateral $\mathcal{Q} = P_0P_1P_2P_3$ symmetric w.r.t. two planes can be coordinatized by

$$P_0 = (a,0,0); \quad P_1 = (0,b,-c), \quad P_2 = (-a,0,0), \quad P_3 = (0,-b,-c). \quad (8.30)$$

Forcing the points P_i and also the lines $[P_i, P_{i+1}]$ (indices modulo 4) to lie on a quadric results in a pencil of quadrics, since a ninth independent point P can be chosen freely.

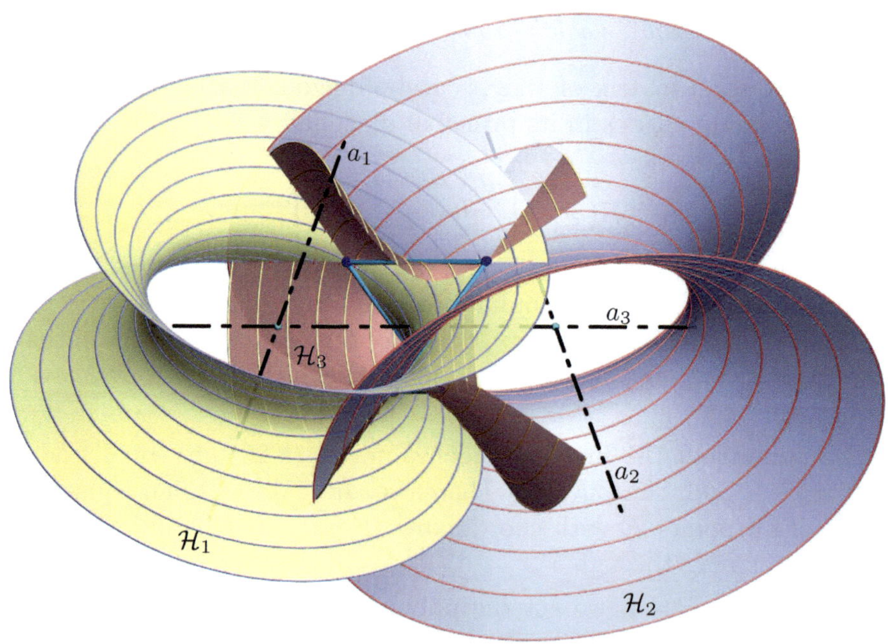

FIGURE 8.29. There exist three (one-sheeted) hyperboloids $\mathcal{H}_1$, $\mathcal{H}_2$, $\mathcal{H}_3$ of revolution with axes a_1, a_2, a_3 through a skew quadrilateral with three planes of symmetry.

With the same methods as in the previous section, we can show that there are three (one-sheeted) hyperboloids $\mathcal{H}_i$ of revolution on the given skew quadrilateral (8.30). Their centers are M_1, M_2, and M_3 with the following Cartesian coordinates

$$\mathbf{m}_1 = \left(0, 0, \frac{-a^2c}{a^2+b^2}\right), \quad \mathbf{m}_2 = \left(0, 0, \frac{a^2}{c}\right), \quad \mathbf{m}_3 = \left(0, 0, -\frac{b^2+c^2}{c}\right).$$

The axes of $\mathcal{H}_1$, $\mathcal{H}_2$, and $\mathcal{H}_3$ are parallel to $(0,0,1)$, $(0,1,0)$, and $(1,0,0)$. The semiaxis lenghts are

$$\begin{array}{llll}
\mathcal{H}_1: & \dfrac{ab}{\sqrt{a^2+b^2}}, & \dfrac{ab}{\sqrt{a^2+b^2}}, & \dfrac{iabc}{a^2+b^2}, \\
\mathcal{H}_2: & \dfrac{a}{c}\sqrt{a^2+c^2}, & \dfrac{iab}{c}, & \dfrac{a}{c}\sqrt{a^2+c^2}, \\
\mathcal{H}_3: & \dfrac{iab}{c}, & \dfrac{b}{c}\sqrt{b^2+c^2}, & \dfrac{b}{c}\sqrt{b^2+c^2}.
\end{array}$$

Figure 8.29 shows these three hyperboloids of revolution on the skew quadrilateral (8.29).

8.4 Rational parametrizations, quadrics as Bézier surfaces

Stereographic projection

The stereographic projection $\sigma: c \to l$ from a conic $c \subset \mathbb{P}^2$ to a line $l \subset \mathbb{P}^2$ can be made a bijective mapping, since we can define the σ-image of the center $C \in c$ of the projection as the intersection of the tangent t_C (of c at C) with l: $\sigma(C) = t_C \cap l$ (cf. [46, p. 231]).

We have used a stereographic projection in [46, p. 231] in order to show that a conic is projectively equivalent to a projective line. As a byproduct, we found a rational parametrization of a conic. In terms of homogeneous coordinates, the parametrization of a conic can be written in terms of polynomials.

Stereographic projections from a quadric $\mathcal{Q}$ in a projective space $\mathbb{P}^n$ with $n > 2$ aim to establish a bijective mapping between $\mathcal{Q}$ and a hyperplane. But they fail to establish bijective mappings, since the exceptional space is *larger*: It is the intersection of $\mathcal{Q}$ with the tangent hyperplane T_C of $\mathcal{Q}$ at C, and as such, it is a quadratic cone.

In the case of a quadric $\mathcal{Q}$ in $\mathbb{P}^n$, we define a *stereographic projection* σ in nearly the same way as in the case of a conic (cf. Figure 2.44). Let $C \in \mathcal{Q}$ be a point in the quadric and let $\mathcal{H}$ be a hyperplane that is different from the tangential hyperplane T_C of $\mathcal{Q}$ at C. The stereographic projection

$$\sigma: \mathcal{Q} \setminus \{T_C\} \to \mathcal{H} \setminus T_C$$

assigns to each point $P \in \mathcal{Q} \setminus T_C$ the uniquely defined point $\sigma(P) = [C, P] \cap \mathcal{H}$ in $\mathcal{H} \setminus T_C$. The mapping σ reaches each point in $\mathbb{P}^n \setminus \{\mathcal{H} \cap T_C\}$. Only if $\dim \mathcal{H} \cap T_C = 0$, can the mapping σ be turned into a bijective mapping by defining $\sigma(C) = \mathcal{H} \cap T_C$. Unfortunately, this is only the case if $n = 2$, and then, $\mathcal{Q}$ is a conic. In the case $n \geq 2$, *i.e.*, $\dim \mathcal{H} \cap T_C \geq 1$, the stereographic projection is bijective only as the mapping $\mathcal{Q} \setminus \{C\} \to \mathcal{H} \setminus T_C$.

Sometimes the stereographic projection is considered to act in the opposite direction, *i.e.*, $\sigma: \mathcal{H} \setminus T_C \to \mathcal{Q} \setminus T_C$. We shall use this notion of a stereographic projection when we derive rational parametrizations of quadrics.

Central quadrics

Usually, stereographic projections are considered only in connection with spheres. Then, the $(n-1)$-dimensional sphere $S^{n-1} \subset \mathbb{F}^n$ is given by its equation

$$\mathcal{S}: \; x_1^2 + \ldots + x_n^2 = 1. \tag{8.31}$$

In most cases, $\mathbb{F} = \mathbb{C}$ or $\mathbb{F} = \mathbb{C}$. Nevertheless, $\mathbb{F}$ can be any commutative field and the characteristic of $\mathbb{F}$ is not restricted to 2.

The center C of the projection shall be given by its coordinate vector $\mathbf{c} = (0, 0, \ldots, 0, -1)$. Now, the image space $\mathcal{H}$ of σ shall be the hyperplane $x_n = 0$. The points H in the hyperplane $\mathcal{H}$ allow the coordinatization $\mathbf{h} = (u_1, u_2, \ldots, u_{n-1}, 0)$ with $u_i \in \mathbb{R}$. The intersection of $\mathcal{S}$ with the line $[C, H]$ with the parametrization $\mathbf{l}(w) = (1-w)\mathbf{c} + w\mathbf{h}$ is obtained by inserting the line's parametrization into (8.31). This yields

$$w^2 \left(1 + \sum_{i=1}^{n-1} u_i^2\right) = 2w.$$

From the latter equation, we can cancel w, since $w = 0$ corresponds to C. So,

$$w = \frac{2}{1+U} \quad \text{with} \quad U = 1 + \sum_{i=1}^{n-1} u_i^2,$$

and therefore,

$$\begin{aligned} \mathbf{s}(u_1, \ldots, u_{n-1}) &= \tfrac{1}{1+U} \left(2u_1, \ldots, 2u_{n-1}, 1-U\right) \\ \text{with} \quad &(u_1, \ldots, u_{n-1}) \in \mathbb{F}^{n-1} \end{aligned} \tag{8.32}$$

is a rational parametrization of the $(n-1)$-dimensional Euclidean unit sphere and, at the same time, a coordinate representation of the stereographic projection from an affine space $\mathbb{F}^{n-1}$ to the $(n-1)$-dimensional Euclidean unit sphere.

Setting $n = 2$ in (8.32) yields the standard rational parametrization of the Euclidean unit circle (cf. [46, Eq. (6.6) on p. 231]). With $n = 3$ in (8.32), we obtain a rational parametrization of the Euclidean unit sphere. This parametrization is by no means the only rational parametrization of the Euclidean unit sphere. Indeed, any rational substitution of the form

$$u_i \to \frac{\nu_i(u_1, \ldots, u_{n-1})}{\delta(u_1, \ldots, u_{n-1})} \quad \text{with} \quad i \in \{1, \ldots, n-1\}$$

with the polynomial functions ν_i and δ_i turns (8.32) into a rational parametrization of the Euclidean $(n-1)$-sphere. However, (8.32) is the rational parametrization of lowest possible degree.

From the parametrization (8.32) of the $(n-1)$-dimensional sphere, we can easily derive rational parametrizations of the $(n-1)$-dimensional ellipsoid $\mathcal{E}$ with the equation

$$\mathcal{E}: \sum_{i=1}^{n} \frac{x_i^2}{a_i^2} = 1 \tag{8.33}$$

with all semiaxes lengths $a_i \neq 0$ by multiplying (8.32) with the diagonal matrix $D = \operatorname{diag}(a_1, \dots, a_n)$.

• **Exercise 8.4.1** Rational parametrizations for hyperboloids.

Compute a rational parametrization of the quadric

$$\mathcal{Q}: \sum_{i=1}^{n-1} \varepsilon_i \frac{x_i^2}{a_i^2} = 1. \tag{8.34}$$

Assume that no semiaxis length a_i is equal to zero. Further, $\varepsilon_i \in \{-1, +1\}$, and it means no restriction to assume that $\varepsilon_n = +1$. Then, we can choose the point C with coordinates $\mathbf{c} = (0, \dots, 0, -a_n)$ as the center of the stereographic projection $\sigma: \mathcal{H} \to \mathcal{Q}$. The hyperplane $\mathcal{H}$ shall now be $x_n = 0$, which admits the coordinatization $\mathbf{h} = (u_1, \dots, u_{n-1}, 0)$ with $u_i \in \mathbb{F}$.

Now, the intersection of $\mathbf{q}(u_1, \dots, u_{n-1}) = (1-w)\mathbf{c} + w\mathbf{h}$ with $\mathcal{Q}$ given by (8.34) is determined by

$$w = \frac{2}{1 + \sum_{i=1}^{n-1} \varepsilon_i \frac{u_i^2}{a_i^2}}$$

and yields a parametrization of (8.34):

$$\mathbf{q}(u_1, \dots, u_n) = \frac{1}{1 + \sum_{i=1}^{n-1} \varepsilon_i \frac{u_i^2}{a_i^2}} \left(2u_1, \dots, 2u_{n-1}, 1 - \sum_{i=1}^{n-1} \varepsilon_i \frac{u_i^2}{a_i^2} \right). \tag{8.35}$$

The thus obtained parametrization can be simplified by the regular reparametrization

$$(u_1, \dots, u_{n-1}) \to (a_i u_1, \dots, a_{n-1} u_{n-1}).$$

If $\mathbb{F}$ is algebraically closed, *i.e.*, for all $i \in \{1, \dots, n-1\}$ there exist the roots $\sqrt{\varepsilon_i}$, then, the affine mapping $\alpha: \mathbb{F}^n \to \mathbb{F}^n$ with the diagonal transformation matrix

$$\operatorname{diag}(\sqrt{\varepsilon_1}\, a_i, \dots, \sqrt{\varepsilon_n}\, a_n)$$

transforms (8.35) into (8.32).

Gaussian plane

The real part u and the imaginary part v of a complex number $w = u + iv$ can be interpreted as Cartesian coordinates in the Euclidean plane, while w is sometimes referred to as the *complex coordinate* of the point (u, v).

The stereographic projections from the south pole $\mathbf{c}_s = (0,0,-1)$ and the north pole $\mathbf{c}_n = (0,0,1)$ onto the Euclidean unit sphere $\mathcal{S}: x^2+y^2+z^2=1$ yield the rational standard parametrizations

$$\begin{aligned} \mathbf{s}_s(u,v) &= \left(\frac{2u}{1+u^2+v^2}, \frac{2v}{1+u^2+v^2}, \frac{1-u^2-v^2}{1+u^2+v^2}\right), \\ \mathbf{s}_n(u,v) &= \left(\frac{2u}{1+u^2+v^2}, \frac{2v}{1+u^2+v^2}, \frac{u^2+v^2-1}{1+u^2+v^2}\right), \end{aligned} \tag{8.36}$$

which are well-defined over $\mathbb{R}^2 \cong \mathbb{C}$. With the help of the conjugate complex number $\overline{w} = u - \mathrm{i}v$ of w, we are able to replace u and v in (8.36). We also join the first two coordinate functions x and y, build the complex number $x + \mathrm{i}y$, and find

$$\mathbf{s}_s = \left(\frac{2w}{1+w\overline{w}}, \frac{1-w\overline{w}}{1+w\overline{w}}\right) \quad \text{and} \quad \mathbf{s}_n = \left(\frac{2w}{1+w\overline{w}}, \frac{w\overline{w}-1}{1+w\overline{w}}\right). \tag{8.37}$$

Now, we read (8.37) as homogeneous coordinates on the complex projective line $\mathbb{P}^1(\mathbb{C})$. Therefore, we can immediately see that the mapping $\iota: \mathbb{C}^2 \to \mathbb{C}^2$ with the coordinate matrix $\mathbf{T} = \mathrm{diag}(1,-1)$ joins the two stereographic images, since $\mathbf{s}_n = \mathbf{T}\mathbf{s}_s$. Moreover, the mapping ι is an involutive projective mapping, because $\mathbf{T}^2 = \mathbf{I}_2$, or because its characteristic cross ratio equals -1 (cf. [46, p. 214]). Thus, the two points $\mathbf{s}_s$ and $\mathbf{s}_n$ are each other's inverses w.r.t. the inversion in the equator $e: x^2+y^2=1$. We can summarize this as follows:

Theorem 8.4.1 *The two stereographic projections from a pair of antipodal centers S and N of a Euclidean sphere $\mathcal{S}$ onto the bisector plane β of SN map a point $P \in \mathcal{S}$ to a pair of points P_S and P_N, which are inverse w.r.t. the great circle $\beta \cap \mathcal{S}$.*

Figure 8.30 illustrates the contents of Theorem 8.4.1 for a point P and the circle of lattitude l.

We can also show the following:

Theorem 8.4.2 *The stereographic projection from a sphere to a plane and back is conformal and preserves circles.*

There exist various proofs of the fact that the stereographic projection is conformal and circle-preserving. We shall try the following:

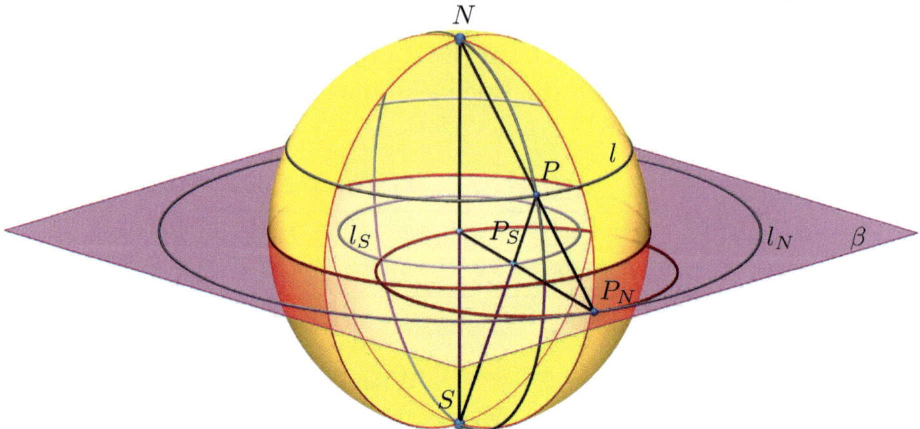

FIGURE 8.30. The pair of stereographic projections from two antipodal centers N and S produce inverse images. The images P_N and P_S are inverse with respect to the equator circle. The same holds true for the two images l_N and l_S of P's circle l of latitude.

Proof: We use the symbols introduced on page 368. Then, let C be any point on the Euclidean unit sphere $\mathcal{S} \subset \mathbb{R}^n$. The tangent hyperplane at C shall be denoted by T_C. Further, the image space $\mathcal{H}$ of the stereographic projection $\sigma : \mathcal{S} \to \mathcal{H}$ shall be precisely that hyperplane through the sphere's center which is orthogonal to $\mathcal{S}$'s diameter through C.

In the complex extension of the projectively closed Euclidean space $\mathbb{R}^n$, the intersection of T_C and $\mathcal{S}$ is a quadratic cone Γ_C with vertex C. The cone Γ_C meets $\mathcal{H}$ along a quadric $\mathcal{I}$, which is empty over the real number field. The quadric $\mathcal{C}$ is fixed under σ, since its entire superspace $\mathcal{H} \cap T_C$ is left fixed under σ. The quadric $\mathcal{I}$ can be viewed as the absolute quadric of the Euclidean geometry in $\mathcal{H}$.

Now let $\mathcal{J} \neq T_c, \mathcal{H}$ with $C \notin \mathcal{J}$ be an arbitrary hyperplane and let $\mathcal{S}_\mathcal{J}$ denote the quadric $\mathcal{J} \cap \mathcal{S}$, which shall not be empty. Clearly, $\mathcal{S}_\mathcal{J}$ is a Euclidean sphere of dimension $n-2$, and the $n-1$-dimensional cone $\Gamma_\mathcal{J}$ projecting $\mathcal{S}_\mathcal{J}$ from C to $\mathcal{H}$ is a quadratic cone. Therefore, the cone's intersection $\mathcal{S}'_\mathcal{J}$ with $\mathcal{H}$ is a quadric in $\mathcal{H}$. Furthermore, $\mathcal{S}'_\mathcal{J}$ and $\Gamma_\mathcal{J}$ share the same subquadric of $\mathcal{I}$, and thus, $\mathcal{S}_\mathcal{J}$ is a Euclidean sphere.

The mapping σ is conformal since the absolute figure of the Euclidean geometry in any transversal hyperplane $\mathcal{J}$ is mapped to the absolute figure of the Euclidean geometry in the image space $\mathcal{H}$. Thus, the measure of angles based on Laguerre's formula (see (2.4) and cf. [44]) remains unchanged. ∎

The above proof is independent of the dimension of the space. In the case of the well-known stereographic projection $\sigma : \mathcal{S}^2 \to \mathbb{R}^2$ from the south pole C to the plane $\mathcal{H} : z = 0$, the cone Γ_C is the pair (i, j) of complex conjugate isotropic generators of $\mathcal{S}^2$ at C, *i.e.*, $j = \bar{i}$. The absolute quadric $\mathcal{I}$ of Euclidean geometry is the pair (I, J) of absolute points $I = (0 : 0 : 1 : \mathrm{i})$ and $J = \overline{I}$ on the ideal line of the projectively closed Euclidean plane $\mathbb{R}^2$. Any planar intersection of $\mathcal{S}^2$ is either a circle or empty.

In the case of the stereographic projection $\mathcal{S}^2 \to \mathbb{R}^2$, we could have also argued as follows:

Proof: (Proof of Theorem 8.4.2.) Assume that $q \in \mathcal{S}^2$ is a planar intersection of $\mathcal{S}^2$, *i.e.*, it is a circle on $\mathcal{S}^2$. The generators of $\mathcal{S}^2$ through the center C of the projection are a pair of isotropic lines i and j in the (Euclidean) tangent plane T_C of $\mathcal{S}^2$ at C. The respective ideal points $I \in i$ and $j \in J$ are mapped to $I' = i \cap \mathcal{H} = I$ and $J' = j \cap \mathcal{H} = J$, which are the absolute points of the Euclidean geometry in the image plane $\mathcal{H}$. The plane $[q]$ spanned by the circle $q \in \mathcal{S}^2$ meets i and j in points which are mapped to I and J in $\mathcal{H}$. The cone $\Gamma_q = C \vee q$ that projects q from C is quadratic and intersects $\mathcal{H}$ in a conic q'. Further, since q's intersections with i and j are mapped to I and J, the conic q' contains I and J. Therefore, q' is a Euclidean circle. ■

When we use the conformality of σ, we could argue in a slightly different way:

Proof: (Proof of Theorem 8.4.2.) The rulings of the tangent cone Ψ_q of $\mathcal{S}^2$ along q intersect q at right angles. These rulings are mapped to a pencil of lines, and the image q' of q intersects the lines of the pencil at right angles. Thus, q' is a circle. (In Example 8.4.1, we show that the orthogonal trajectory of a pencil of lines is indeed a circle.) ■

■ Example 8.4.1 The orthogonal trajectories of a pencil of lines.

In the second proof of Theorem 8.4.2, we used the fact that the orthogonal trajectories of pencils of lines are circles. We shall verify this with the help of a short computation. Assume

$$F(x, y, c) = x - cy = 0$$

is the implicit equation of a pencil of lines. It means no restriction to assume that the pencil is the pencil about the origin of the coordinate system. Implicit differentiation yields $\mathrm{d}F = \mathrm{d}x - c\,\mathrm{d}y = 0$, and thus, $c = \frac{\mathrm{d}x}{\mathrm{d}y}$. This allows us to eliminate c from $F(x, y, c)$. This results in the differential equation whose integral curves are the lines of the pencil:

$$x\,\mathrm{d}y - \mathrm{d}x = 0.$$

The differential equation of the orthogonal trajectories is obtained by replacing $\frac{\mathrm{d}x}{\mathrm{d}y}$ with its negative inverse, *i.e.*, $-\frac{\mathrm{d}y}{\mathrm{d}x}$. Hence, we have

$$x\,\mathrm{d}x + y\,\mathrm{d}y = 0,$$

which integrates to $x^2 + y^2 = c'$ with a constant $c' \in \mathbb{R}$ of integration. The latter equation describes the pencil of concentric circles about $(0, 0)$ (provided that $c' > 0$) as the orthogonal trajectories of the pencil.

• Exercise 8.4.2 The center of an image circle.

We use the south pole $C = (0, 0, -1)$ of the Euclidean unit sphere $\mathcal{S}^2$ as the center of the stereographic projection $\sigma: \mathcal{S}^2 \to \mathcal{H}$ onto the image plane $\mathcal{H}: z = 0$.

Assume that E is the pole of the plane ε w.r.t. the Euclidean unit sphere $\mathcal{S}^2$. Let e denote the circle $\varepsilon \cap \mathcal{S}^2$.

Show that the center of the image circle $\sigma(e)$ of e is the image $\sigma(E)$ of E (cf. Fig. 8.31). The stereographic projection of E is found in the same way as for points on $\mathcal{S}$.

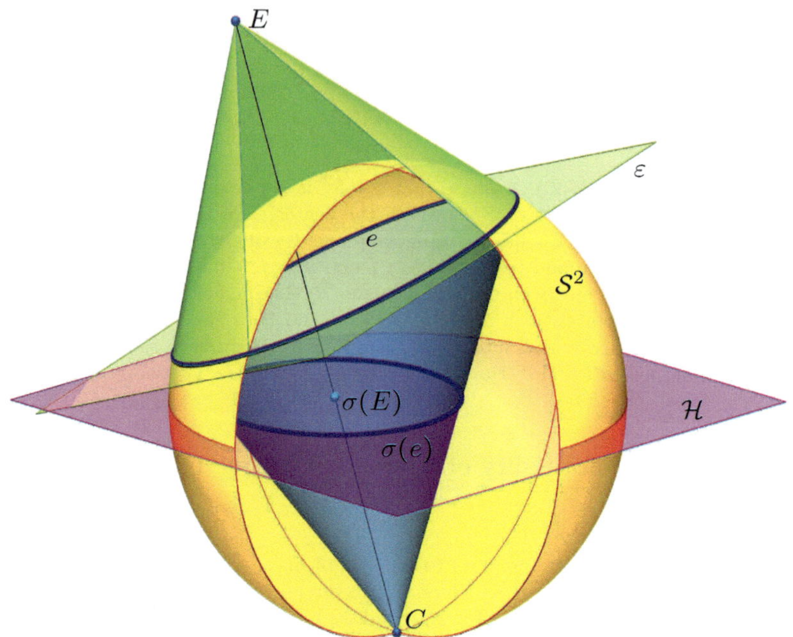

FIGURE 8.31. The stereographic projection maps circles to circles. The pole E of the plane $\varepsilon = [e]$ is mapped to the center of the image circle $\sigma(e)$.

Paraboloid

There is a significant difference between central quadrics and paraboloids, besides the fact that the latter do not have a center. According to Theorem 3.1.1, the equations of parabolic quadrics have the form

$$\mathcal{P}: \sum_{i=1}^{n-1} \varepsilon_i \frac{x_i^2}{a_i^2} = 2x_n \tag{8.38}$$

where $\varepsilon_i \in \{-1, +1\}$ and $a_i \in \mathbb{F} \setminus \{0\}$. Therefore, x_n is always a quadratic polynomial in the variables $x_1, \ldots, x_n$. Hence, parabolic quadrics are the graph surfaces of polynomial functions of degree two.

In the following, we will show that the parametrizations of parabolic quadrics as graphs of a polynomial functions are also the results of stereographic projections. For that purpose, we perform the projective closure of $\mathbb{F}^n$. Again, the transition from affine coordinates to homogeneous ones is given by

$$x_1 \to y_1 y_0^{-1}, \ldots, x_n \to y_n y_0^{-1}.$$

Consequently, (8.38) changes to

$$\sum_{i=1}^{n-1} \varepsilon_i \frac{y_i^2}{a_i^2} = 2y_n y_0,$$

and we can immediately infer that the ideal point X_n of the x_n-axis with the homogeneous coordinates $\mathbf{c} = (0 : \ldots : 0 : 1)$ lies on $\mathcal{P}$. The projection from $\mathcal{H}:\ x_n = 0$ to $\mathcal{Q}$ with center C maps the points in $\mathcal{H}$ to the points $\mathcal{P}$ via

$$(1 : u_1 : u_2 : \ldots : u_{n-1} : 0) \longrightarrow \left(1 : u_1 : u_2 : \ldots : u_{n-1} : \frac{1}{2} \sum_{i=1}^{n-1} \varepsilon_i \frac{y_i^2}{a_i^2}\right).$$

Stereographic projections of singular quadrics

In this section, we will discuss the stereographic projection from a singular quadric only in a special case. We start with the cylinder

$$\Delta:\ x^2 + y^2 = 1$$

and choose the center $C \in \Delta$ given by $\mathbf{c} = (1, 0, 0)$. The image plane $\mathcal{H}$ of σ is the plane $x = 0$. The plane $\mathcal{E}:\ x = 1$ passes through the center C and is parallel to the image plane $\mathcal{H}$, and therefore, its points have no image under σ. According to the theory of linear mappings, the exceptional plane $\mathcal{E}$ can be called the *vanishing plane* of σ.

First, we are interested in the interplay between conics on Δ (ellipses) and their stereographic images in $\mathcal{H}$ (see Figure 8.32). We assume that a plane ε is given by the equation

$$\varepsilon:\ e_0 + e_1 x + e_2 y + e_3 z = 0. \tag{8.39}$$

As long as the normal $\mathbf{e} = (e_1, e_2, e_3)$ of ε is not orthogonal to the z-axis (and thus, ε is not parallel to the generators of Δ), the curve $e = \varepsilon \cap \Delta$ is an ellipse that can be parametrized by substituting $x = \cos t$, $y = \sin t$ into ε's equation and solving if for z.

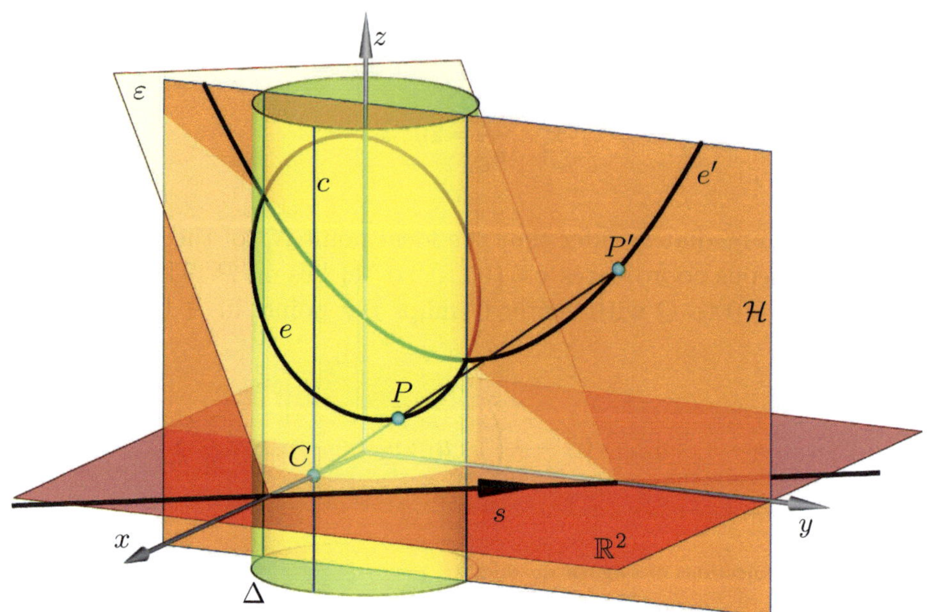

FIGURE 8.32. Each spear s in the Euclidean plane defines a unique oriented plane $\varepsilon \supset s$ with $\sphericalangle(s,\varepsilon) = \frac{\pi}{4}$. The intersection $e = \varepsilon \cap \Delta$ of BLASCHKE's cylinder and ε is an ellipse and is also uniquely assigned to s. The subsequent stereographic projection $\sigma : \Delta \backslash \{\mathcal{E}\} \to \mathcal{H}$ transforms e into a parabola e', *i.e.*, into an isotropic circle.

The stereographic projection $\sigma : \Delta \to \mathcal{H}$ maps each point $X \in \mathbb{R}^3 \setminus \{\mathcal{E}\}$ to a point $X \in \mathcal{H}$ according to

$$\mathbf{x} = (\xi, \eta, \zeta) \mapsto \mathbf{x}' = \left(0, \frac{\eta}{1-\xi}, \frac{\zeta}{1-\xi}\right).$$

Therefore, the ellipse e is mapped to the parabola e' with the parametrization

$$\mathbf{e}' = \left(0, \frac{\sin t}{1-\cos t}, -\frac{e_1 \cos t + e_2 \sin t + e_0}{e_3(1-\cos t)}\right),$$

which can be given in an implicit form as

$$(e_0 + e_1)y^2 + 2e_2 y + 2e_3 z + (e_0 - e_1) = 0. \tag{8.40}$$

It is clear that the ellipses on Δ (not through C) are mapped to parabolas in $\mathcal{H}$: Each ellipse e intersects Δ's generator c through C in exactly one point, which is mapped to the ideal point of the z-axis. It is obvious that the ellipses through C are mapped to straight lines.

In the second step, we impose the following condition on the coefficients e_i of (8.39):

$$e_1^2 + e_2^2 = e_3^2,$$

i.e., the planes described by (8.39) enclose the angle $\alpha = \left|\frac{\pi}{4}\right|$ with the plane $z = 0$. The intersection of such a plane ε with the plane $z = 0$ is a straight line s with the equation

$$s:\ e_0 + e_1 x + e_2 y = 0,$$

that can be oriented with the help of the oriented normal $\mathbf{e}$. Thus, s is an oriented straight line, frequently referred to as a *spear*. Ellipses through C correspond to straight lines in $\mathcal{H}$. Consequently, we have the chain of objects

$$s \mapsto \varepsilon \mapsto e \mapsto e',$$

which shows that the spears of the Euclidean plane $z = 0$ correspond to the parabolas with vertical axes and straight lines in $\mathcal{H}$. The parabolas in $\mathcal{H}$ with common ideal point can be considered as circles of the *isotropic geometry* in $\mathcal{H}$. We summarize the thus obtained relations:

Theorem 8.4.3 *There is a one-to-one and onto correspondence between the oriented lines (spears) in the Euclidean plane and the circles in the isotropic plane.*

The Euclidean Laguerre[4] *geometry (i.e., the geometry of spears in the Euclidean plane) corresponds to the geometry of isotropic circles (in the isotropic plane).*

Laguerre geometry is not restricted to planes. The geometry of oriented hyperplanes in arbitrary Euclidean spaces can also be termed Laguerre geometry (cf. [10, 24]).

In any case, a hyper cylinder can be used as another model of Laguerre geometry. This cylinder is usually called BLASCHKE[5] *cylinder*. A stere-

[4] EDMOND NICOLAS LAGUERRE (1834–1886) was a French mathematician who studied contact transformation (Laguerre transformations), algebraic equations, approximation theory, orthogonal polynomials, elliptic functions, and closure theorems.

[5] WILHELM JOHANN EUGEN BLASCHKE (1885–1962) was an Austrian mathematician. He worked in a huge variety of fields, such as Differential Geometry, non-Euclidean Geometry, kinematics, and integral geometry.

ographic projection from a Blaschke cylinder to a hyperplane $\mathcal{H}$ relating the geometry of oriented hyperplanes with the geometry of isotropic hyperspheres in the isotropic space $\mathcal{H}$ is straightforward. Blaschke's cylinder for the geometry of oriented planes in Euclidean 3-space is a subquadric of Lie's quadric L_2^4, as we shall see in Section 10.2. This is clear since the oriented planes in Euclidean 3-space can also be viewed as oriented spheres with infinitely large radius.

Bézier representation of quadrics

As we have seen, quadrics admit rational parametrizations. If we use homogeneous coordinates, then we are able to multiply the parametrization of a quadric by the common denominator of all coordinate functions. Therefore, the homogeneous representation can be written with polynomials only.

Polynomials are usually given in the monomial basis, but can be rewritten in any other basis. Among the huge amount of bases in the ring $\mathbb{F}[u]$ or $\mathbb{F}[v]$, the Bernstein basis with basis functions

$$\varphi_k(u) = \binom{n}{k}(1-u)^{n-k}u^k, \quad \text{with} \quad k = 0, 1, \ldots, n$$

turned out to be useful. The above polynomials span the space of polynomials of degree n. Moreover, independent of the choice of n, the polynomials φ_k form a partition of unity, *i.e.*, $\sum_{k=0}^{n} \varphi_k = 1$. Choosing $n+1$ control points B_i with coordinate vectors $\mathbf{b}_i \in \mathbb{F}^r$, we can give a parametrization of a Bézier curve as

$$\sum_{k=0}^{n} \mathbf{b}_k \varphi_k(u) \quad \text{with} \quad u \in [0,1],$$

which parametrizes a curve segment starting at B_0 ending at B_n with tangent lines $[B_0, B_1]$ and $[B_{n-1}, B_n]$ at the end points.

A Bézier surface patch of bi-degree (m,n) is described by a parametrization of the form

$$\mathbf{f}(u,v) = \sum_{i=0}^{m} \sum_{j=0}^{n} \varphi_i(u)\varphi_j(v)\mathbf{b}_{i,j}, \quad (u,v) \in \mathbb{F}^2, \tag{8.41}$$

where $\mathbf{b}_{i,j}$ are the coordinate vectors of the control points. The functions $\varphi_i(u)$ and $\varphi_j(v)$ are the Bernstein polynomials of degrees m and n, respectively. In the following, we set $m = n$ even though it is possible to construct Bézier patches of arbitrary bi-degrees (m, n) with $m \neq n$.

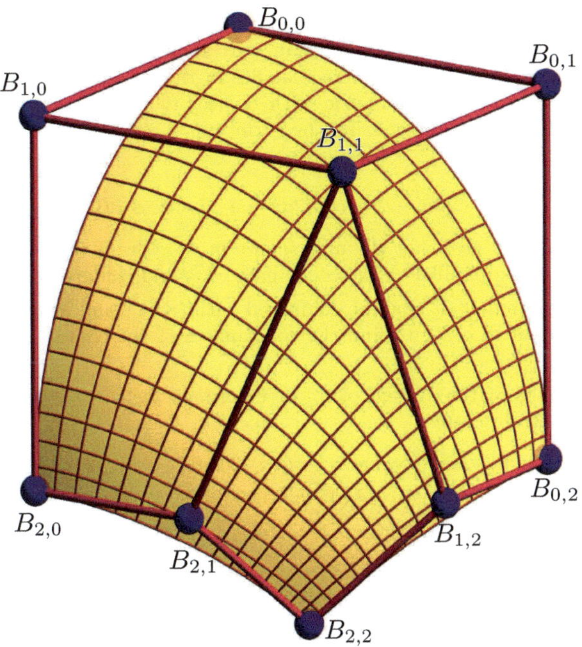

FIGURE 8.33. Rewriting the rational standard parametrization of the Euclidean unit sphere in the Bernstein basis yields the depicted control structure.

In the case $n = 2$, (8.36) equals the rational standard parametrization of the Euclidean unit sphere. If we use homogeneous coordinates, the Euclidean unit sphere can be parametrized by

$$\mathbf{s}(u, v) = (1 + u^2 + v^2,\ 2u,\ 2v,\ 1 - u^2 - v^2), \quad (u, v) \in \mathbb{R}^2.$$

Now, we can compare the latter equation with (8.41), which results in a system of linear equations for all polynomial coordinate functions. The variables in the equations are the homogeneous coordinates of the control points of the desired surface patch. In the present case, we find the homogeneous coordinates of the nine control points as

$$\begin{aligned}
\mathbf{b}_{0,0} &= (1:0:0:1), & \mathbf{b}_{0,1} &= (1:0:1:1), & \mathbf{b}_{0,2} &= (2:0:2:0),\\
\mathbf{b}_{1,0} &= (1:1:0:1), & \mathbf{b}_{1,1} &= (1:1:1:1), & \mathbf{b}_{1,2} &= (2:1:2:0),\\
\mathbf{b}_{2,0} &= (2:2:0:0), & \mathbf{b}_{2,1} &= (2:2:1:0), & \mathbf{b}_{2,2} &= (3:2:2:-1).
\end{aligned}$$

Figure 8.33 shows the surface patch in the sphere that is defined over the parameter domain $[0,1]\times[0,1]$. There, the control points are also shown.

• **Exercise 8.4.3** Representing a symmetric part of the sphere as a Bézier patch

The standard parametrization of the Euclidean unit sphere over the domain $[0,1] \times [0,1]$ describes a rectangular patch on the sphere that lies in the octant described by $x \geq 0$, $y \geq 0$, and $z \geq 0$. Obviously, it is not symmetric with respect to the coordinate planes. We can overcome this difficulty by reparametrizing the surface.

As a first step, show that

$$u \to 2du' - d, \quad v \to 2dv' - d \quad (\text{with } d \neq 0)$$

is a reparametrization that stretches the unit square in the parameter domain to a square of edge length $2d$, with edges parallel to the coordinate axes and centered at $(0,0)$.

Now, the parametrization $\mathbf{s}(u',v')$ can be compared with (8.41), and again, a system of linear equations is to be solved in order to find the coordinates of the control points of the Bézier surface.

Show that the solutions of the system of linear equations yield the following homogeneous coordinates of the control points:

$$\begin{aligned}
&\mathbf{b}_{0,0} = (1{+}2d^2 : -2d : -2d : 1{-}2d^2), && \mathbf{b}_{0,1} = (1 : -2d : 0 : 1), && \mathbf{b}_{0,2} = (1+2d^2, -2d, 2d, 1-2d^2),\\
&\mathbf{b}_{1,0} = (1, 0, -2d, 1), && \mathbf{b}_{1,1} = (1{-}2d^2, 0, 0, 1{+}2d^2), && \mathbf{b}_{1,2} = (1, 0, 2d, 1),\\
&\mathbf{b}_{2,0} = (1{+}2d^2, 2d, -2d, 1{-}2d^2), && \mathbf{b}_{2,1} = (1, 2d, 0, 1), && \mathbf{b}_{2,2} = (1{+}2d^2, 2d, 2d, 1{-}2d^2).
\end{aligned}$$

Figure 8.34 shows this particular control structure together with the thus defined surface patch for $d = \sqrt{2}$ (left) and $d = 5$ (right). Choosing a very large d extends the parameter domain. However, the hole around the opposite pole can never be closed.

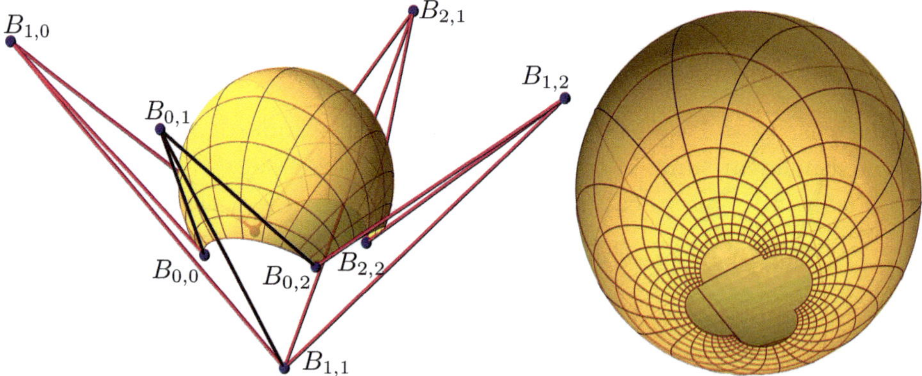

FIGURE 8.34. The sphere as a Bézier surface: Left: The control structure is chosen to be symmetric around the north pole. Right: The quadratic parameter domain can be extended arbitrarily large. A hole still remains as long as $d \in \mathbb{R}$.

• **Exercise 8.4.4** The ellipsoid and the one-sheeted hyperboloid as Bézier surfaces.

Find rational parametrizations of the one-sheeted hyperboloid $\mathcal{Q}_h$ and the triaxial ellipsoid $\mathcal{Q}_e$

$$\mathcal{Q}_h: \frac{x^2}{a^2} + \frac{y^2}{b^2} - \frac{z^2}{c^2} = 1 \quad \text{and} \quad \mathcal{Q}: \frac{x^2}{a^2} + \frac{y^2}{b^2} + \frac{z^2}{c^2} = 1.$$

Use the stereographic projections with the centers C_h and C_e with coordinates $\mathbf{c}_h = (-a, 0, 0)$ and $\mathbf{c}_e = (0, 0, -c)$ onto the planes $\mathcal{H}_h : x = 0$ and $\mathcal{H}_e : z = 0$ in the case of the hyperboloid and the ellipsoid, respectivley.

Then, derive the control points of a quadratic rational Bézier surface patch which is symmetric with respect to the plane $y = 0$ in both cases. An example of either symmetric Bézier patch on $\mathcal{Q}_h$ and on $\mathcal{Q}_e$ is shown in Figure 8.35.

Paraboloids are translation surfaces, as we have learned in Theorem 2.3.2. Therefore, the parametrization of a paraboloid can be written as the sum of parametrizations of two Bézier curves $\mathbf{p}(u, v) = \mathbf{b}(u) + \mathbf{c}(v)$, where $\mathbf{b}(u)$ and $\mathbf{c}(v)$ parametrize two profile curves.

For elliptic paraboloids the two Bézier profile curves are quadratic curves, *i.e.*, these curves are parabolas. The case of a hyperbolic paraboloid has been studied in Section 8.3. There, we have seen that a hyperbolic paraboloid can be parametrized as a Bézier surface of bidegree (1,1).

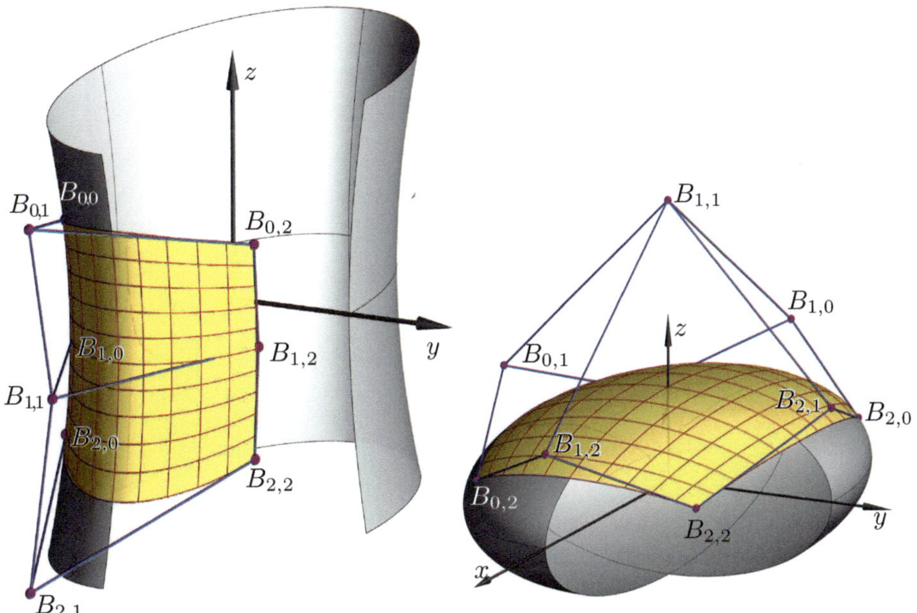

FIGURE 8.35. A symmetric Bézier patch on a one-sheeted hyperboloid (left) and on a triaxial ellipsoid (right).

9 Quadrics and Differential Geometry

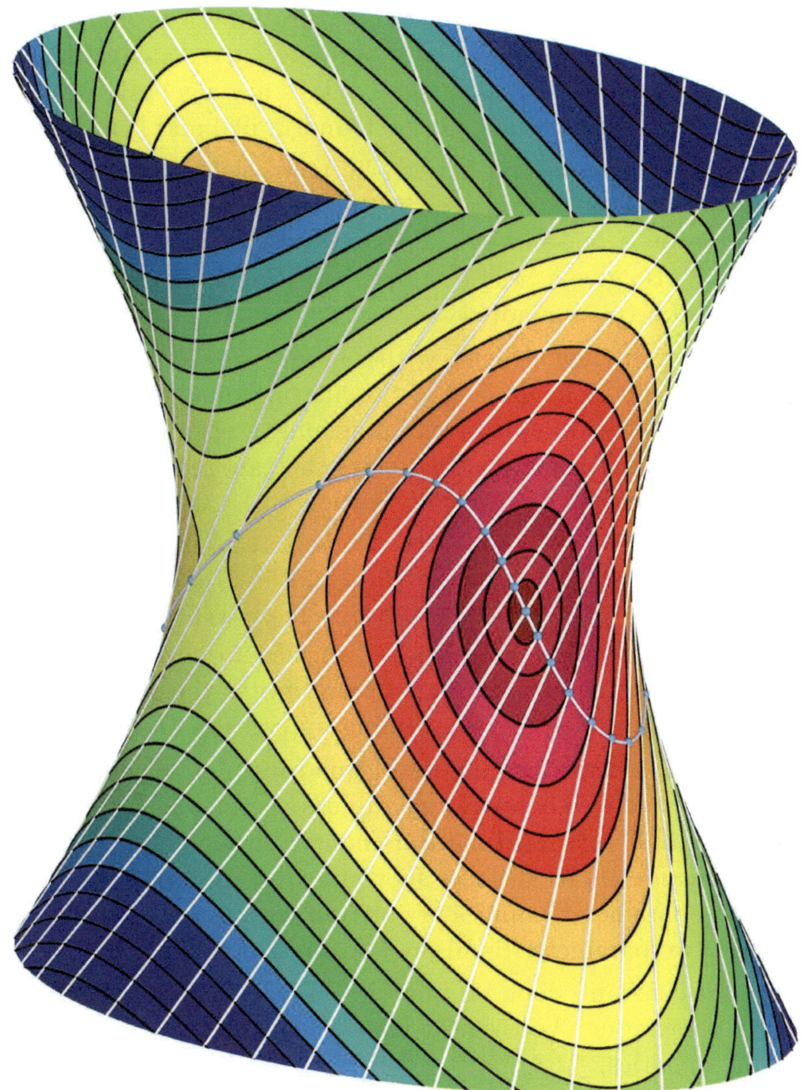

The Gaussian curvature has a minimum at the striction points of the ruled hyperboloid, and therefore, the curves of constant Gaussian curvature touch the hyperboloid's rulings precisely at the striction points.

B. Odehnal et al., *The Universe of Quadrics*,
https://doi.org/10.1007/978-3-662-61053-4_9

Quadrics appear in differential geometry as approximations of surfaces: The second order Taylor approximation of a surface at a regular point P (that is not a flat point) is either an elliptic or a hyperbolic paraboloid, or a parabolic cylinder (cf. [46, p. 124]).

In this chapter we shall study some curvature functions and the distribution of curvatures on quadrics by describing curves along which some curvatures are constant. Clearly, these curves are algebraic since so are quadrics. Some of these results can be found in [95] and [141].

One-sheeted hyperboloids and hyperbolic paraboloids are ruled surfaces in two ways, and as such, they define their own striction curves. The striction curve on a one-sheeted hyperboloid is a quartic of the second kind. We shall also give a brief description of the orthogonal trajectories of the rulings of a ruled quadric.

As we shall see, in line geometry there exist intrinsic approximating quadrics. We want to focus on Lie's osculating quadric and show how to compute it. Higher order differential geometry of ruled surfaces is not particularly useful for ruled quadrics, since they are represented by conics in Klein's model of line geometry (cf. Section 10.1), which is not the case with generic ruled surfaces (cf. [94]).

From a generic point, we can draw up to six real normals to a central quadric and five normals to a paraboloid. The totality of normals to a quadric form the quadric's congruence of surface normals whose envelope is one algebraic surface consisting of two sheets, the so-called central surfaces. We give some results concerning a quadric's offsets.

Envelopes of one-parameter families of quadrics play an important role in many applications. Therefore, we discuss channel surfaces, pipe surfaces, rolling ball blends, and natural channel surfaces. Special types of pseudo-Euclidean channel surfaces show up when we study the envelopes of certain one-parameter families of cones of revolution. We add a short discussion of two-parameter families of spheres.

The curves of constant slope on quadrics of revolution show interesting relations to kinematics. Apparently, these curves can be found as cycloids lifted to the quadrics.

Finally, we devote a small section to algebraic geodesics on quadrics. Among them, we can find quartics of the first and second kind on central quadrics and even geodesic cubics on paraboloids.

9.1 Curvature functions on quadrics

Gaussian and mean curvature

In the following, we shall mainly treat regular quadrics. Singular quadrics in Euclidean 3-space are either cylinders or cones, and thus of constant Gaussian curvature $K_0 = 0$. Their mean curvature equals the non-vanishing principal curvature. Hence, the curves of constant mean curvature agree with the curves of constant principal curvature.

Central quadrics

Without loss of generality, we may assume that central quadrics in Euclidean 3-space may be centered at the origin $O = (0,0,0)$ of a Cartesian coordinate system. Then, they can always be given by their respective equations

$$\begin{aligned} \mathcal{E}: \frac{x^2}{a^2} + \frac{y^2}{b^2} + \frac{z^2}{c^2} &= 1, \\ \mathcal{H}_1: \frac{x^2}{a^2} + \frac{y^2}{b^2} - \frac{z^2}{c^2} &= 1, \\ \mathcal{H}_2: \frac{x^2}{a^2} + \frac{y^2}{b^2} - \frac{z^2}{c^2} &= -1 \end{aligned} \tag{9.1}$$

with $a, b, c \in \mathbb{R}^\star$. In the case of the ellipsoid $\mathcal{E}$, it is admissible to assume that $a > b > c$ holds. We exclude the case of rotationally symmetric quadrics for the moment and focus on triaxial ones. For the cases of one-sheeted hyperboloids $\mathcal{H}_1$ and two-sheeted hyperboloids $\mathcal{H}_2$, it is allowed to assume that $a > b$ holds.

Let us now focus on the triaxial ellipsoid $\mathcal{E}$. The Hesse form of the equation of the polar planes τ of all points $\mathbf{x} = (\xi, \eta, \zeta)$ with respect to $\mathcal{E}$ reads

$$\tau: \frac{\frac{\xi}{a^2}x + \frac{\eta}{b^2}y + \frac{\zeta}{c^2}z - 1}{\sqrt{\frac{\xi^2}{a^4} + \frac{\eta^2}{b^4} + \frac{\zeta^2}{c^4}}} = 0 \tag{9.2}$$

and yields the tangent planes of $\mathcal{Q}$ at $\mathbf{x}$ if, and only if, ξ, η, and ζ satisfy (9.1). Moreover, for any triplet $\mathbf{x} = (\xi, \eta, \zeta)$, (9.2) yields the (oriented) distance h between $\mathcal{Q}$'s tangent planes and the origin of the coordinate system, *i.e.*, the support function $h: \mathrm{S}^2 \to \mathbb{R}$ of $\mathcal{Q}$ by letting $x = y = z = 0$

$$h = -\frac{1}{\sqrt{\frac{\xi^2}{a^4} + \frac{\eta^2}{b^4} + \frac{\zeta^2}{c^4}}}. \tag{9.3}$$

In order to find expressions for the Gaussian and mean curvature of the quadrics (9.1), we derive algebraic parametrizations by solving the quadrics' equations for z. In the case of the ellipsoid $\mathcal{E}$, we find

$$\mathbf{e}(x,y) = (x, y, z(x,y))^{\mathrm{T}} \tag{9.4}$$

with $z(x,y) = c\sqrt{1-\frac{x^2}{a^2}-\frac{y^2}{b^2}}$. The expressions for the one- and two-sheeted hyperboloid are similar. With the restriction $x, y \geq 0$, (9.4) parametrizes only one eighth of the ellipsoid. This is sufficient, since the remaining seven parts are symmetric copies and can be obtained by reflecting (9.4) in the three planes of symmetry of $\mathcal{E}$. With the partial derivatives of $z(x,y)$ w.r.t. the parameters x and y, we have

$$\partial_x z = -\tfrac{c^2 x}{a^2 z}, \quad \partial_y z = -\tfrac{c^2 y}{b^2 z},$$
$$\partial_{xx} z = \tfrac{c^4(y^2-b^2)}{a^2b^2z^3}, \quad \partial_{xy} z = \tfrac{c^4 xy}{a^2b^2z^3}, \quad \partial_{yy} z = \tfrac{c^4(x^2-a^2)}{a^2b^2z^3}.$$

This yields the coordinate functions of the first and second fundamental form on $\mathcal{E}$ at (ξ, η, ζ)

$$g_{11} = 1 + \frac{c^4\xi^2}{a^4\zeta^2}, \quad g_{12} = \frac{c^4\xi\eta}{a^2b^2\zeta^2}, \quad g_{22} = 1 + \frac{c^4\eta^2}{b^2\zeta^2},$$
$$h_{11} = -\frac{c^2(b^2-\eta^2)h}{a^2b^2\zeta^2}, \quad h_{12} = -\frac{c^2\xi\eta\, h}{a^2b^2\zeta^2}, \quad h_{22} = -\frac{c^2(a^2-\xi^2)h}{a^2b^2\zeta^2}$$

where we have used (9.3). The well-known expressions for the Gaussian and mean curvature

$$K = \frac{h_{11}h_{22} - h_{12}^2}{g_{11}g_{22} - g_{12}^2} \quad \text{and} \quad H = \frac{g_{11}h_{22} - 2g_{12}h_{12} + g_{22}h_{11}}{2(g_{11}g_{22} - g_{12}^2)}$$

lead to

$$K = \frac{h^4}{a^2b^2c^2} \quad \text{and} \quad H = \frac{h^3}{2a^2b^2c^2}L \tag{9.5}$$

with $L := (b^2+c^2)\frac{\xi^2}{a^2} + (c^2+a^2)\frac{\eta^2}{b^2} + (a^2+b^2)\frac{\zeta^2}{c^2}$. The latter formula relating the Gaussian curvature K with the support function h of a quadric can be found in [109] and in even older literature. However, the formula (9.5) covers not only the case of the ellipsoid. It also returns the Gaussian curvature of a two-sheeted hyperboloid $\mathcal{H}_2$ with the equation given in

(9.1). A small modification turns (9.5) into the equivalent expression for the Gaussian curvature of the one-sheeted hyperboloid (cf. [90, p. 66]).

The formulas for the two most important curvature functions on $\mathcal{E}$ allow us to formulate

Theorem 9.1.1 *1. The curves of constant Gaussian curvature on a regular central quadric (9.1) are quartic space curves of the first kind.*

2. Along a curve of constant Gaussian curvature on a regular central quadric, the support function h is constant. The developable ruled surface $\mathcal{D}$ enveloped by the center quadric's tangent planes along a curve of constant Gaussian curvature are tangent to a sphere concentric with the quadric.

3. The curves of constant mean curvature on a regular central quadric are algebraic space curves of degree twelve.

Proof:

1. From (9.3) and (9.5), we infer
$$abc\sqrt{K}\left(\frac{\xi^2}{a^4}+\frac{\eta^2}{b^4}+\frac{\zeta}{c^4}\right)=1. \tag{9.6}$$
On a regular quadric, the Gaussian curvature K is always non-zero. Therefore, the latter equation is that of an ellipsoid, even in the case of a one- or two-sheeted hyperboloid. Consequently, the curves of constant Gaussian curvature are the curves of intersection of the given quadrics and the one-parameter family of quadrics (9.6), and thus, they are quartic space curves of the second kind (see Section 6.9).
2. By virtue of (9.5), we see that K is constant if h is constant, and *vice versa*. At each point of a curve of constant Gaussian curvature K_0 on a regular central quadric, the tangent plane has the constant distance
$$h_0=\sqrt{abc\sqrt{K_0}}$$
to the center of the quadric. Hence, these tangent planes are also tangent to a concentric sphere with radius h_0 (cf. Figure 9.1).
3. This follows from counting degrees in the respective equations. ■

Figure 9.1 (right) shows a specific curve k of constant Gaussian curvature on a one-sheeted hyperboloid $\mathcal{H}_2$ together with the tangent developable Δ of $\mathcal{H}_2$ along k. The developable ruled surface Δ is in line contact with the sphere Σ. Consequently, the Dupin indicatrices (cf. [46, p. 120]) i_Δ and i_Σ of Δ and Σ constitute a pencil of conics of the third kind (cf. [46, p. 287]) and the only singular conic in the pencil is a double line. This double line is conjugate to the ruling of Δ. Lines which are conjugate w.r.t. a sphere are orthogonal. Thus, the tangent to s is orthogonal to the ruling of Δ, and therefore, the curve s of contact between Σ and Δ is an orthogonal trajectory of the rulings of Δ.

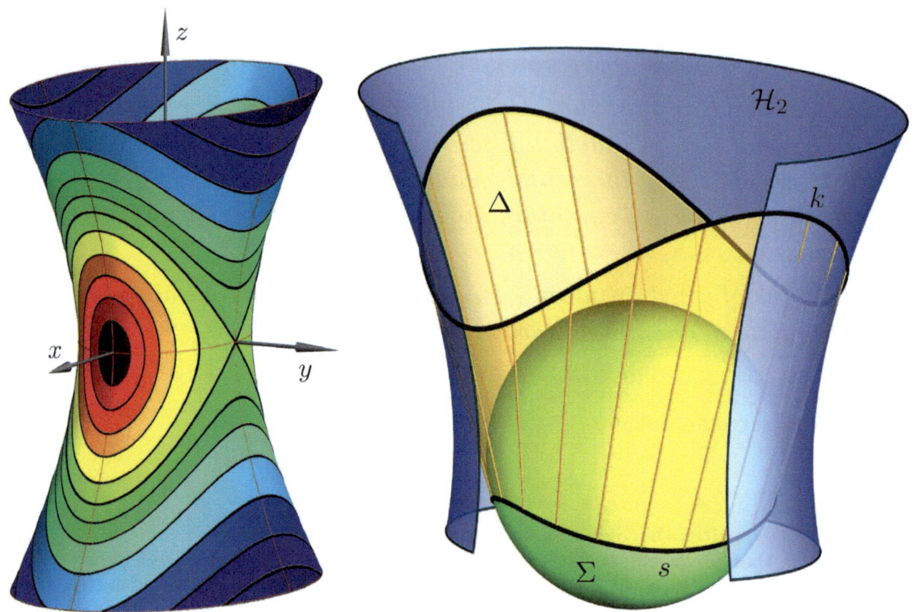

FIGURE 9.1. Left: Curves of constant Gaussian curvature on a one-sheeted hyperboloid. Right: The developable ruled surface Δ tangent to $\mathcal{H}_2$ and the concentric sphere Σ touches $\mathcal{H}_2$ along a curve k of constant Gaussian curvature.

The curve k is an algebraic curve of degree four (cf. Theorem 9.1.1). The same holds true for the curve s. The developable surface Δ is of class four, since it is a developable that is tangent to two surfaces of class two, *i.e.*, to Σ and $\mathcal{H}_2$. The curve s of contact is the polar image of the one-parameter family of tangent planes of Δ w.r.t. Σ. The same holds true for k on $\mathcal{H}_2$. Moreover, this works on any quadric.

As an immediate consequence of Theorem 9.1.1 and (9.6), the curves of constant Gaussian curvature on a regular central quadric appear as conics in the three orthogonal projections onto the quadrics' planes of symmetry. Figure 9.2 illustrates this fact by means of a triaxial ellipsoid.

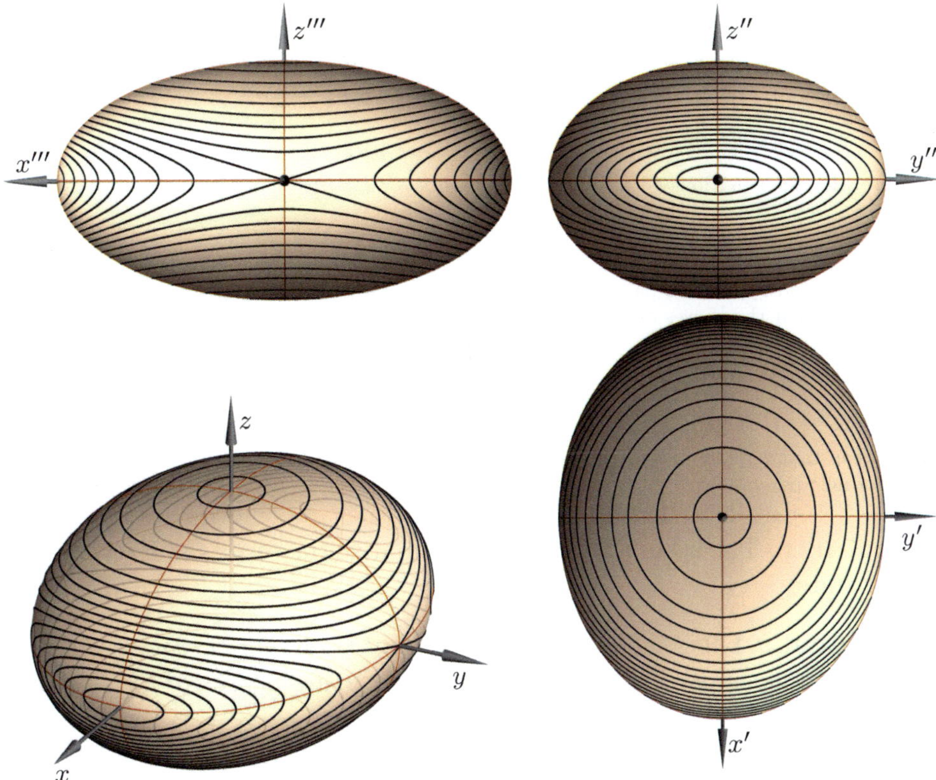

FIGURE 9.2. Top row and bottom right: The curves of constant Gaussian curvature on a triaxial ellipsoid $\mathcal{E}$ are mapped to ellipses and hyperbolas in the orthogonal projections onto $\mathcal{E}$'s planes of symmetry. Bottom left: axonometric view of the curves of constant Gaussian curvature on $\mathcal{E}$.

• **Exercise 9.1.1** One- and two-sheeted hyperboloids.

Show that the formula for the Gaussian curvature of a two-sheeted hyperboloid $\mathcal{H}_2$ agrees with (9.5). For that purpose, use the equation of $\mathcal{H}_2$ given in (9.1) and derive the support function h. Then, compute the first and second fundamental form. Now, use the well-known formula for the Gaussian curvature. Finally, eliminate the parameters and global coordinates x, y, z with help of the expression for the support function.

With the same methods show that the Gaussian curvature function on a one-sheeted hyperboloid equals

$$K = -\frac{h^4}{a^2b^2c^2}. \tag{9.7}$$

Paraboloids

The Gaussian curvature on a paraboloid cannot be expressed in terms of the paraboloid's support function only. However, the curves of constant

Gaussian curvature on paraboloids exhibit a behaviour that is similar to that of curves of constant Gaussian and mean curvature on regular central quadrics.

- **Exercise 9.1.2** Gaussian curvature on elliptic and hyperbolic paraboloids.

Start with the equations

$$\mathcal{P}_e: \frac{x^2}{a^2} + \frac{y^2}{b^2} = 2z \quad \text{and} \quad \mathcal{P}_h: \frac{x^2}{a^2} - \frac{y^2}{b^2} = 2z \tag{9.8}$$

of elliptic and hyperbolic paraboloids, derive an algebraic (rational) parametrization of each, and show that the curves of constant Gaussian curvature and mean curvature on paraboloids are the intersection of either paraboloid with the surfaces

$$K = \pm\frac{a^6b^6}{(a^4b^4 + b^4x^2 + a^4y^2)^2} \quad \text{and} \quad H = \frac{a^2b^2(a^4b^2 \pm a^2b^4 + a^2y^2 \pm b^2x^2)}{2(a^4b^4 + b^4x^2 + a^4y^2)^{\frac{3}{2}}} \tag{9.9}$$

with constant K or H and the upper sign for the elliptic paraboloid and the lower sign for the hyperbolic paraboloid.

Similarly to Theorem 9.1.1, we can summarize the results of Exercise 9.1.2 and state

Theorem 9.1.2 *1. The curves of constant Gaussian curvature on (elliptic or hyperbolic) paraboloids are quartic space curves of the first kind.*

2. The curves of constant mean curvature on (elliptic or hyperbolic) paraboloids are algebraic space curves of degree twelve.

- **Exercise 9.1.3** Curves of constant Gaussian curvature - quartic space curves of the first kind.

Show that the three principal views (top view k', front view k'', right-side view k''') of the curves of constant Gaussian curvature on the triaxial ellipsoid $\mathcal{E}$ from (9.1) have the equations

$$\begin{aligned} k': &\quad x^2(a^2-c^2)a^{-4} + y^2(b^2-c^2)b^{-4} = 1 - c^2h^{-2}, \\ k'': &\quad x^2(a^2-b^2)a^{-4} - z^2(b^2-c^2)c^{-4} = 1 - b^2h^{-2}, \\ k''': &\quad y^2(a^2-b^2)b^{-4} + z^2(a^2-c^2)c^{-4} = a^2h^{-2} - 1, \end{aligned}$$

where the support function h is subject to $a \geq h \geq c$. How do these equations change in the case of a one- or two-sheeted hyperboloid, or even in the case of an elliptic or a hyperbolic paraboloid. The equations of k', k'', or k''' can be considered the equations of projecting cylinders. Figure 9.2 shows the curves of constant Gaussian curvature on a triaxial ellipsoid. Show that the curve with $K = \frac{b^2}{a^2c^2}$ (through the vertices $(0, \pm b, 0)^{\mathrm{T}}$ splits into two ellipses.

Figure 9.3 shows color maps of the mean curvature on a triaxial ellipsoid, on a one-sheeted hyperboloid, on a hyperbolic paraboloid, and on an elliptic paraboloid.

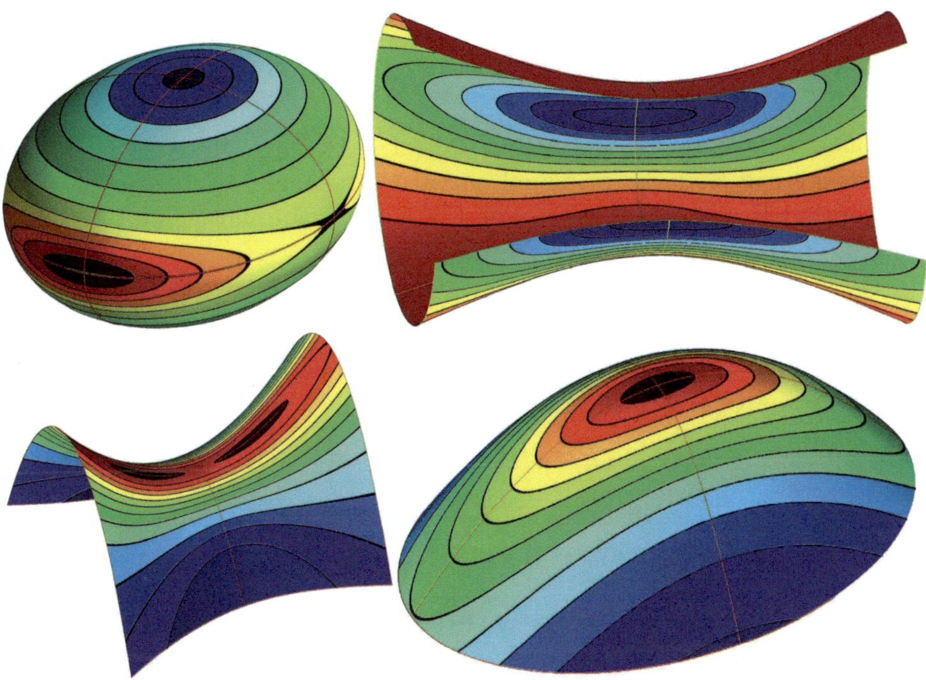

FIGURE 9.3. Color maps of the mean curvature: on a triaxial ellipsoid and a one-sheeted hyperboloid (top row); on an elliptic and a hyperbolic paraboloid (bottom row).

The principal curvatures

Once we have found the Gaussian curvature K and the mean curvature H, we can use the well-known formulas

$$2H = \kappa_1 + \kappa_2 \quad \text{and} \quad K = \kappa_1 \kappa_2,$$

relating them with the principal curvatures κ_1 and κ_2. Thus, each principal curvature is, according to Vieta's theorem, a solution of the quadratic equation

$$\kappa^2 - 2H\kappa + K = 0.$$

Now, we are going to describe the curves of constant principal curvature. These are not to be confused with the curvature lines which are the integral curves of the field of principal curvature tangents and are quartic curves on each regular quadric. This is so, because they are the intersections of the given quadrics with the family of confocal quadrics (see Chapter 7).

We follow an idea by W. WUNDERLICH (cf. [141]) applied to the ellipsoid. First, we recall that the principal curvature tangents at each point P of an ellipsoid $\mathcal{E}$ are orthogonal. For the sake of simplicity, we write the equation of $\mathcal{E}$ from (9.1) in matrix form

$$\mathcal{E}: \ \mathbf{x}^{\mathrm{T}}\mathbf{A}\mathbf{x} = 1 \tag{9.10}$$

with $\mathbf{A} = \operatorname{diag}(a^{-2}, b^{-2}, c^{-2})$ and $\mathbf{x} = (x, y, z)$. The tangent plane T_P at P with coordinates $\mathbf{p} = (\xi, \eta, \zeta)$ has the equation

$$T_P: \ \mathbf{p}^{\mathrm{T}}\mathbf{A}\mathbf{x} = \langle \mathbf{n}, \mathbf{x} \rangle = 1, \tag{9.11}$$

where $\mathbf{n} = \mathbf{A}\mathbf{p}$. The distance $|h|$ between T_P and the center of $\mathcal{E}$ (the origin of the coordinate system) equals

$$-h^{-1} = \|\mathbf{n}\| = \sqrt{\langle \mathbf{n}, \mathbf{n} \rangle} = \sqrt{\mathbf{p}^{\mathrm{T}}\mathbf{A}^2\,\mathbf{p}}. \tag{9.12}$$

We can write down the equation of a sphere $\mathcal{S}$ tangent to $\mathcal{E}$ at P with radius $\varrho > 0$ as

$$\mathcal{S}: \ \langle \mathbf{x} - \mathbf{m}, \mathbf{x} - \mathbf{m} \rangle = \varrho^2, \tag{9.13}$$

where the coordinate vector $\mathbf{m}$ of $\mathcal{S}$'s center satisfies $\mathbf{m} = \mathbf{p} - \varrho h\,\mathbf{n}$.

The intersection q of $\mathcal{E}$ and $\mathcal{S}$ is a quartic of the first kind with a double point at P. The quartic's tangents at the double point P determine the normal curvature $\kappa = \varrho^{-1}$. The two tangents at the double point P coincide, exactly for two specific choices of ϱ, *i.e.*, the two principal curvature radii ϱ_1 and ϱ_2 corresponding to the two principal curvatures κ_1 and κ_2.

It proves useful to change the coordinate frame by applying the translation $\mathbf{x} = \mathbf{y} + \mathbf{p}$. Consequently, the equations of $\mathcal{E}$ and $\mathcal{S}$ from (9.10) and (9.13) turn into

$$\mathcal{E}: \mathbf{y}^{\mathrm{T}}\mathbf{A}\mathbf{y} + 2\mathbf{p}^{\mathrm{T}}\mathbf{A}\mathbf{y} = 0, \tag{9.14}$$

$$\mathcal{S}: \mathbf{y}^{\mathrm{T}}\mathbf{y} - 2\varrho h\,\mathbf{p}^{\mathrm{T}}\mathbf{A}\mathbf{y} = 0. \tag{9.15}$$

We multiply (9.14) by $-\varrho h$, add (9.15), and obtain

$$\Gamma: \ \mathbf{y}^{\mathrm{T}}\mathbf{B}\mathbf{y} = 0, \quad \mathbf{B} := \varrho h\mathbf{A} + \mathbf{I}_3,$$

which is the equation of a quadratic cone Γ centered at P carrying the quartic intersection curve $q = \mathcal{E} \cap \mathcal{S}$. The cone Γ also carries q's tangents at the double point P. The two tangents at the double point coincide, if Γ and $\mathcal{E}$ share the tangent plane T_P at P, *i.e.*, T_P contains its polar line t w.r.t. Γ. Let us assume that t is described by $\mathbf{t}$. Then, we have $\mathbf{t}^{\mathrm{T}}\mathbf{B}\mathbf{y} = 0$ for T_P's polar line t w.r.t. Γ. According to (9.11), $t \subset T_P$ if, and only if,

$$\mathbf{t}^{\mathrm{T}}\mathbf{B} = \mathbf{p}^{\mathrm{T}}\mathbf{A} \quad \Longleftrightarrow \quad \mathbf{t} = \mathbf{B}^{-1}\mathbf{A}\mathbf{p}.$$

Hence, t is contained in Γ (or equivalently, in T_P) if $\mathbf{t}^{\mathrm{T}}\mathbf{B}\mathbf{t} = 0$, *i.e.*,

$$\mathbf{p}^{\mathrm{T}}\mathbf{A}\mathbf{B}^{-1}\mathbf{A}\mathbf{p} = 0.$$

In terms of coordinates, this reads

$$\frac{\alpha^2 x^2}{\alpha + \kappa h} + \frac{\beta^2 y^2}{\beta + \kappa h} + \frac{\gamma^2 z^2}{\gamma + \kappa h} = 0, \tag{9.16}$$

where h satisfies (9.3) and α, β, and γ are shorthand for a^{-2}, b^{-2}, and c^{-2}. We see that (9.16) is a quadratic equation in κ. Its solutions are the principal curvatures of $\mathcal{E}$ at P. The quadratic equation for κ reads

$$\kappa^2 h^4 + M h \kappa + \alpha\beta\gamma = 0 \tag{9.17}$$

with

$$M = \alpha^2(\beta + \gamma)x^2 + \beta^2(\gamma + \alpha)y^2 + \gamma^2(\alpha + \beta)z^2 \tag{9.18}$$

and is of degree eight in the coordinates x, y, z after the middle term is isolated and squared. Hence, we can say

Theorem 9.1.3 *The curves of constant principal curvature on a regular central quadric are algebraic space curves of degree sixteen.*

Proof: The determination of curves of constant principal curvature on one- or two-sheeted hyperboloids is most likely the same as of ellipsoids. ∎

Another way of computing the principal curvatures is shown in Lemma 8.2.1.

Figure 9.4 shows some curves of constant principal curvature on a triaxial ellipsoid $\mathcal{E}$. At first glance, there appear to be two kinds of such curves, since at each point on the ellipsoid, there are two different principal curvatures. There are only four exceptions: the umbilics of $\mathcal{E}$. There, the two principal curvatures have the same value and the two branches of the curves of constant principal curvature coincide. However, the algebraic description of the curves makes it hard to distinguish between these two curves.

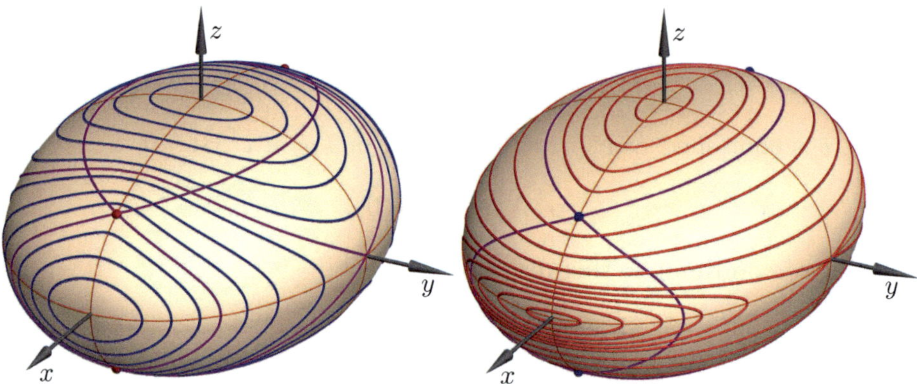

FIGURE 9.4. The curves of constant principal curvature on a triaxial ellipsoid appear to form two separate families. This net of curves has four singularities at the umbilics, and the curves of constant principal curvature through the umbilics have double points there.

• **Exercise 9.1.4** Parametrization of the curves of constant principal curvature.

Show that an eighth of the curves of constant principal curvature can be parametrized by means of algebraic functions. For that purpose, it is useful to interpret h^{-1} as a parameter. Use (9.18) with M = const. and (9.12) with h = const. and determine the eight points of intersection of the thus defined three quadrics. Further, (9.17) has to be solved for M. In fact, this means solving a system of three equations which are linear in x^2, y^2, and z^2 and yields

$$x^2 = \frac{M - \alpha h^2 - \beta\gamma}{\alpha(\beta - \alpha)(\alpha - \gamma)}, \quad y^2 = \frac{M - \beta h^2 - \gamma\alpha}{\beta(\gamma - \beta)(\beta - \alpha)}, \quad z^2 = \frac{M - \gamma h^2 - \alpha\beta}{\gamma(\alpha - \gamma)(\gamma - \beta)}. \tag{9.19}$$

Similar equations can be given for the curves of constant principal curvature on a one- or two-sheeted hyperboloid.

Curves of constant principal curvature on paraboloids

It is easy to verify that the curves of constant principal curvatures on elliptic or hyperbolic paraboloids are algebraic curves of degree sixteen. The elliptic and the hyperbolic paraboloid

$$\mathcal{P}_e : \frac{x^2}{a^2} + \frac{y^2}{b^2} = 2z \quad \text{and} \quad \mathcal{P}_h : \frac{x^2}{a^2} - \frac{y^2}{b^2} = 2z$$

can be written as graphs over the $[x, y]$-plane. The curves of constant principal curvature κ lie on the surface of degree eight

$$\begin{gathered}\kappa^4(b^4x^2+a^4y^2)^4+\\+a^4b^4\kappa^2(a^4y^2+b^4x^2)(2(a^4y^2+b^4x^2)\kappa-2z)(2(a^4y^2+b^4x^2)\kappa+2z)+\\+a^8b^8\kappa^2(6\kappa^2(a^4y^2+b^4x^2)^2-2(a^4\pm a^2b^2+b^4)x^2y^2-3(b^4x^4+a^4y^4))+\\+a^8b^8\kappa^2(4a^4b^4(a^4y^2+b^4x^2)\kappa^2-3a^4b^4(x^2+y^2)-b^8x^2-a^8y^2)+\\+a^{12}b^{12}(a^4\kappa^2-1)(b^4\kappa^2-1)=0\end{gathered}$$

and the respective paraboloid. The upper sign (third row, center term) is to be chosen in the case of an elliptic paraboloid. Figure 9.5 shows curves of constant principal curvature on an elliptic paraboloid (left) and on a hyperbolic paraboloid (right).

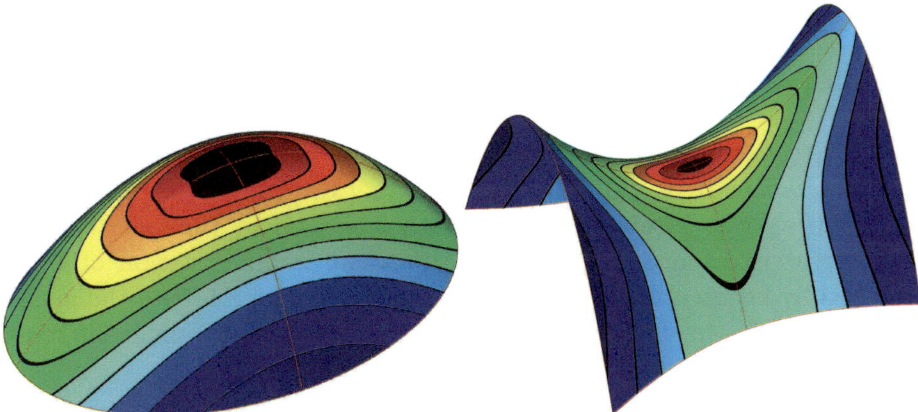

FIGURE 9.5. The curves of constant principal curvature on paraboloids are also of degree sixteen. Left: on an elliptic paraboloid; right: on a hyperbolic paraboloid.

Principal and mean curvature on quadratic cones

We do not have to investigate the curves of constant Gaussian curvature on quadratic cones, since the Gaussian curvature is constantly equal to zero all over the cone. This is due to the fact that one principal curvature that is measured along the rulings of the cone, say κ_2, vanishes.

Therefore, only $\kappa_1 \neq 0$ and the mean curvature equals $H = \frac{1}{2}(\kappa_1 + 0) = \frac{\kappa_1}{2} \neq 0$, which is not constant. Assume that the quadratic cone's equation is given in its normal form

$$\Gamma: \ \frac{x^2}{a^2} + \frac{y^2}{b^2} - \frac{z^2}{c^2} = 0 \tag{9.20}$$

with $a > b > 0$ and $c \neq 0$. Now it is elementary to compute the mean curvature H and to show that the points of constant mean curvature H on Γ also lie on the sextic surface

$$\mathcal{H}: \ 4a^4b^4c^4\left(\frac{x^2}{a^4} + \frac{y^2}{b^4} + \frac{z^2}{c^4}\right)^3 H^2 - (x^2 + y^2 + z^2)^2 = 0. \tag{9.21}$$

The elimination of z from (9.21) with (9.20) allows us to recognize another surprising property of the curves of constant curvature on a quadratic cone. In other words, the top view of the curves $\mathcal{H} \cap \Gamma$ of constant mean curvature on Γ reads

$$\mathcal{H}': \ 4(a^4(b^2+c^2)y^2+b^4(a^2+c^2)x^2)^3H^2 = a^4b^4c^2(b^2(a^2+c^2)x^2+a^2(b^2+c^2)y^2)^2.$$

$\mathcal{H}'$ can be considered as the vertical cylinder through the curve of constant mean curvature H as well as the top view of the curve. Surprisingly, it admits a rational parametrization, which is caused by the fact that the point $(0,0)$ is a four fold point. The rational parametrization of $\mathcal{H}'$ is obtained by intersecting $\mathcal{H}$ with the lines (t, kt) (with $t \in \mathbb{R}$) of the pencil through $(0,0)$. In a first step, this results in a non-rational algebraic parametrization

$$\mathbf{h}(k) = \frac{a^2b^2c(a^2(b^2+c^2)k^2+(a^2+c^2)b^2)}{2H\,(a^4(b^2+c^2)k^2+(a^2+c^2)b^4)^{\frac{3}{2}}}\begin{pmatrix}1\\k\end{pmatrix}, \tag{9.22}$$

which can be transformed into a hyperbolic-trigonometric parametrization by setting $k = \frac{b^2\sqrt{a^2+c^2}}{a^2\sqrt{b^2+c^2}}\sinh l$ that reads

$$\mathbf{h}(l) = \frac{c(a^2 + b^2\sinh^2 l)}{2H(a^2\sqrt{b^2+c^2}\cosh^3 l)}\begin{pmatrix}1\\ \sinh l\end{pmatrix}.$$

Finally, the substitution $\cosh l = \frac{1+t^2}{1-t^2}$ and $\sinh l = \frac{2t}{1-t^2}$ turns the latter into a rational parametrization. We can summarize these results as follows:

Theorem 9.1.4 *The curves of constant mean curvature and the curves of constant principal curvature on a quadratic cone* (9.20) *are algebraic curves of degree twelve. Their top views (orthogonal projection in the direction of the z-axis) are rational curves.*

Proof: So far, we have described the curves of constant mean curvature H = const.. However, for the case of constant principal curvature, we recall that the second principal curvature κ_2 vanishes and $\kappa_1 = 2M$. The sextic surface analogous to (9.21) has a similar equation (just remove the factor 4 on the left-hand side in (9.21)). The rational parametrization is obtained in the same way as for the top view of the curves of constant mean curvature. ∎

Figure 9.6 shows some curves of constant principal curvature on a quadratic cone.

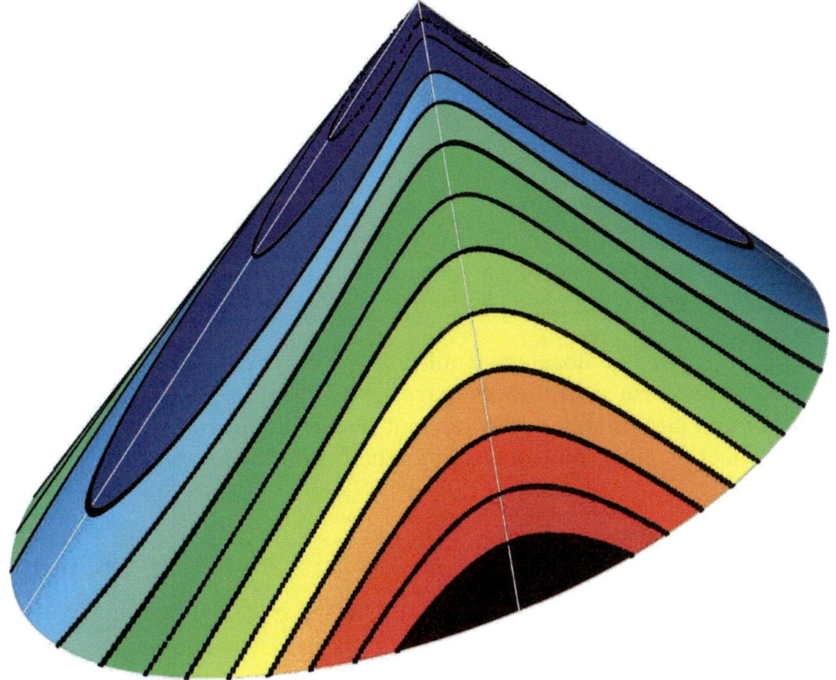

FIGURE 9.6. The curves of constant principal curvature $\kappa_1 \neq 0$ on a quadratic cone Γ have rationally parametrizable top views and are algebraic curves of degree twelve. This is also the case for the curves of constant mean curvature on Γ.

9.2 Quadrics as ruled surfaces

From Section 2.2, we know that, among regular quadrics, the one-sheeted hyperboloid and the hyperbolic paraboloid carry two independent one-parameter families of lines, *i.e.*, the two reguli. So, it seems obvious to start with a mixture of a line geometric and a differential geometric methods, which leads to a treatment of quadrics as ruled surfaces.

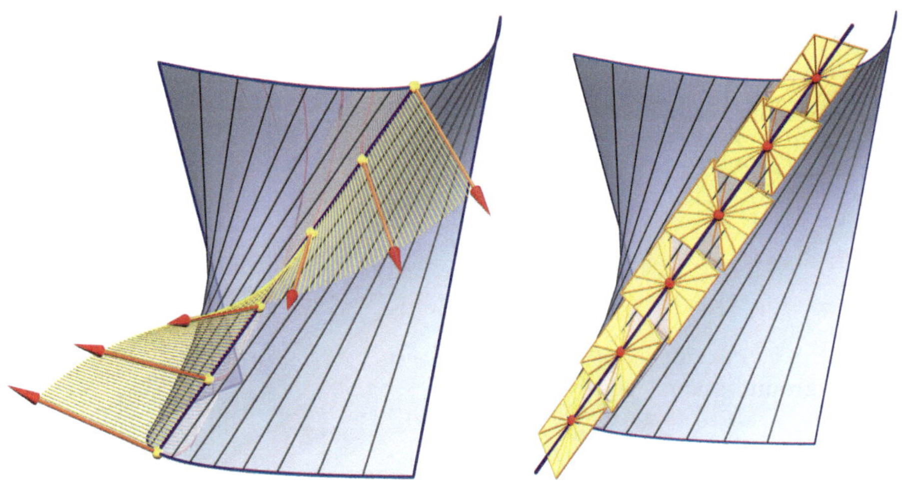

FIGURE 9.7. Left: The normals at all points of a regular non-torsal ruling on any ruled surface form a hyperbolic paraboloid. Right: The mapping between points of contact and the respective tangent planes on any regular and non-torsal ruling is projective, and thus, one-to-one and onto.

Let a ruled surface $\mathcal{R}$ be parametrized by

$$\mathcal{R}(u,v) = \mathbf{l}(u) + v \cdot \mathbf{r}(u), \quad u \in I \subset \mathbb{R},\ v \in \mathbb{R}, \tag{9.23}$$

where $\mathbf{l}: I \to \mathbb{R}^3$ and $\mathbf{r}: I \to \mathbb{R}^3$ are parametrizations of an arbitrary *directrix* and the *direction vector field* of the *rulings*. It means no restriction to assume that the vector function $\mathbf{r}$ is normalized, and thus, $\mathbf{r}: I \to \mathrm{S}^2$ is the *spherical image* of $\mathcal{R}$'s rulings and $v \cdot \mathcal{R}: I \times \mathbb{R} \to \mathbb{R}^3$ is the *director cone* of $\mathcal{R}$.

Here, and in the following, we shall denote derivatives with respect to the parameter u with a dot. The normals of $\mathcal{R}$ along a fixed ruling r, *i.e.*,

$u = u_0 =$ const., depend on v linearly and exclusively. They are parallel to

$$\mathbf{n}(v) = \dot{\mathbf{l}} \times \mathbf{r} + v \cdot \dot{\mathbf{r}} \times \mathbf{r}. \tag{9.24}$$

The latter equation makes clear that there is a projective mapping that assigns to each point on r the tangent plane of $\mathcal{R}$ as long as $\dot{\mathbf{l}} \times \mathbf{r}$ and $\dot{\mathbf{r}} \times \mathbf{r}$ are linearly independent, or equivalently,

$$\det(\dot{\mathbf{l}}, \mathbf{r}, \dot{\mathbf{r}}) \neq 0. \tag{9.25}$$

The mapping that assigns to each point on r the unique tangent plane is usually referred to as the *contact projectivity*. It is degenerate at torsal and singular rulings, *i.e.*, where $\dot{\mathbf{l}} \times \mathbf{r}$ and $\dot{\mathbf{r}} \times \mathbf{r}$ are linearly dependent, and thus, $\det(\dot{\mathbf{l}}, \mathbf{r}, \dot{\mathbf{r}}) = 0$. We shall not discuss this in detail here.

A ruled surface that carries only finitely many or countably many torsal or singular rulings is called a *skew ruled surface*. We shall keep in mind that all regular ruled quadrics (hyperbolic paraboloids and one-sheeted hyperboloids) are skew ruled surfaces.

A ruled surface that carries only torsal rulings is called a *torsal ruled surface*. The most general form of a torsal ruled surface may be composed of at least one of the following basic types: plane, cylinder, cone, and tangent surface of a (true) space curve (see Figure 9.8). A torsal ruled surface is developable, *i.e.*, it can locally be mapped isometrically in a Euclidean plane. Among quadrics, only quadratic cones and cylinders are torsal ruled surfaces. Torsal ruled surfaces are frequently called *developables*.

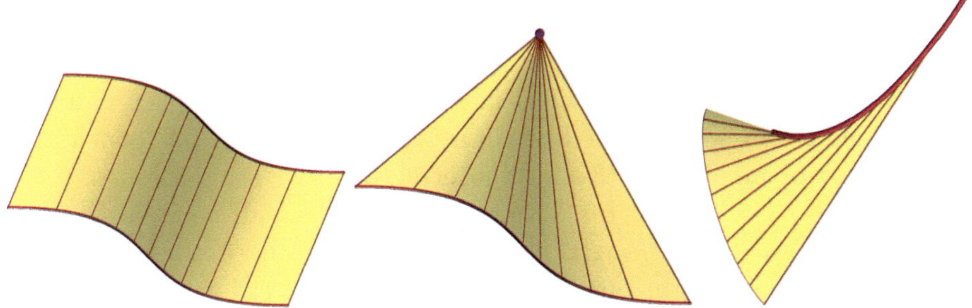

FIGURE 9.8. The basic torsal ruled surfaces (except the plane and from left to right): cylinder, cone, tangent surface of a space curve.

From (9.24), we learn that the tangent plane of any ruled surface along a regular and non-torsal ruling behaves like the tangent planes of a (regular) ruled quadric. The surface tangents of $\mathcal{R}$ along r comprise the set of lines in a *parabolic linear line congruence.* Figure 9.9 shows some of the tangents of a ruled surface $\mathcal{R}$ along a non-torsal ruling r. We shall have a further look at parabolic linear line congruences in Section 10.1.

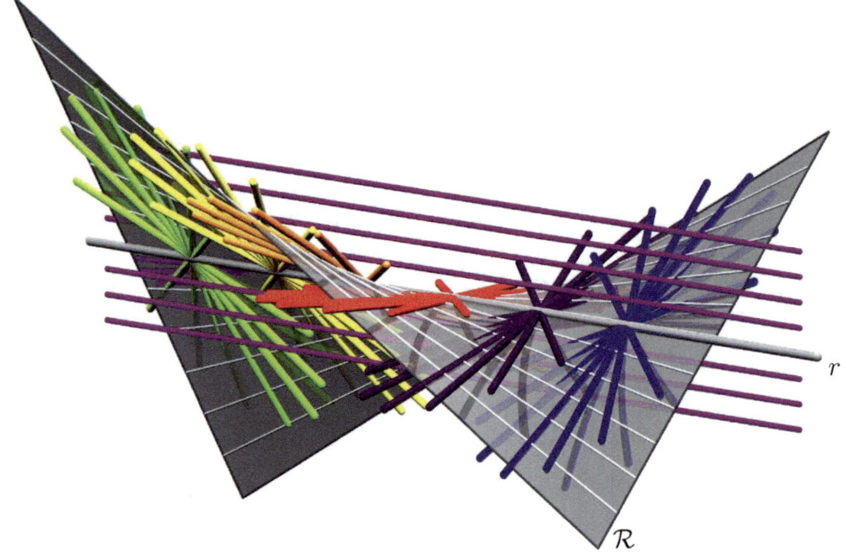

FIGURE 9.9. The surface tangents of a ruled surface $\mathcal{R}$ that touch along a regular non-torsal ruling r form a parabolic linear line congruence. The magenta lines are tangent to $\mathcal{R}$ at the ideal point R_∞ of r and form a pencil in the asymptotic plane.

The two-parameter family of lines in the parabolic congruence can also be obtained as the limit of the lines in a *hyperbolic linear line congruence, i.e.*, the set of all lines in 3-space intersecting two *skew straight lines.* Assume that r_ε is a ruling on $\mathcal{R}$ in a sufficiently[1] small neighborhood of r. For a *skew ruled surface*, it is always guaranteed that r and r_ε are skew. Hence, the straight lines that meet both r and r_ε form a hyperbolic linear line congruence. Once we perform a limit procedure $r_\varepsilon \to r$, *i.e.*,

[1]Here and in the following, *sufficiently* means that functions are still defined at $t \pm \varepsilon$ if they are at t, and that Taylor expansions still exist and converge there if they do at t.

$\varepsilon \to 0$, the hyperbolic congruence of lines that meets $\mathcal{R}$ in two different points converges towards the parabolic congruence of surface tangents that touch $\mathcal{R}$ in points of r.

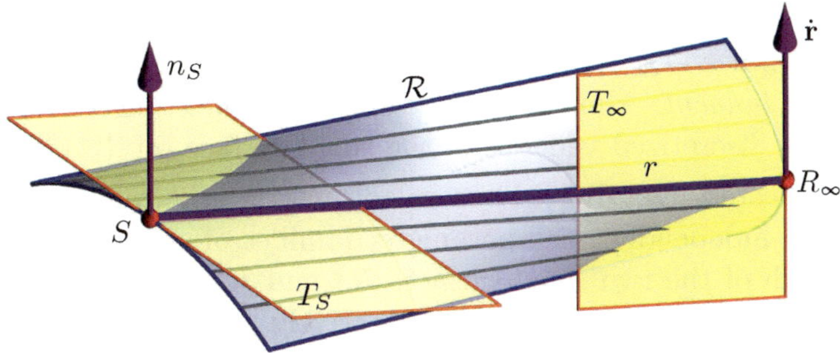

FIGURE 9.10. The central (tangent) plane T_S touches the ruled surface $\mathcal{R}$ at the central point S (on a non-torsal ruling r) and is orthogonal to the asymptotic plane T_∞.

On each (regular) ruling r of a ruled surface $\mathcal{R}$ in Euclidean 3-space $\mathbb{R}^3$, we find a distinguished point called the *central point* S. It is defined with the help of the contact projectivity and the orthogonality in $\mathbb{R}^3$: In the projective closure of $\mathbb{R}^3$, the ruling r has an ideal point R_∞ (point at infinity). The tangent plane T_∞ is spanned by the ruling r and the derivative point $(0, \mathbf{r})$. T_∞ is usually referred to as the *asymptotic plane* of $\mathcal{R}$ at r. The *central (tangent) plane* T_S is said to be orthogonal to T_∞ and is, therefore, uniquely defined in the pencil about r. The contact projectivity along r assigns to T_S a unique point S of contact (cf. Figure 9.10). Consequently, there is a well-defined central point on each ruling, provided that $\mathcal{R}$ is regular at R_∞.

From the parametrization (9.23) of a ruled surface, we can compute the central point via (9.24). While $\mathbf{n}$ depends only linearly on v along a fixed ruling r, it is linearly dependent on $\dot{\mathbf{r}}$ at the central point S. This yields $(\dot{\mathbf{l}} \times \mathbf{r} + v_s \dot{\mathbf{r}} \times \mathbf{r}) \times \dot{\mathbf{r}} = \mathbf{o}$, and therefore,

$$v_S = -\frac{\langle \dot{\mathbf{r}}, \dot{\mathbf{l}} \rangle}{\langle \dot{\mathbf{r}}, \dot{\mathbf{r}} \rangle}. \tag{9.26}$$

The *central curve* or *striction curve* is the union of all *central points* on a ruled surface.

The central point of a regular non-torsal ruling can also be found by means of a limiting procedure: Assume that $\mathbf{r}: I \times \mathbb{R} \to \mathbb{R}^3$ is a parametrization of a ruled surface $\mathcal{R}$ over some interval $I \subset \mathbb{R}$. Choose two rulings $r = \mathbf{r}(t_0)$ and $r_\varepsilon = \mathbf{r}(t_0 + \varepsilon)$ in a sufficiently small neighborhood and compute their common normal. The pedal point $P \in r$ converges towards the central point S of r if $\varepsilon \to 0$ and $r_\varepsilon \to r$ while the common normal converges to the *central tangent*. Almost all invariants of a ruled surface have analogs in properly discretized versions of the ruled surface (cf. [110]).

The one-sheeted hyperboloid is a ruled quadric in two different ways: It carries two independent one-parameter families of straight lines, the two reguli. Each of the two ruled surfaces $\mathcal{R}_1$ and $\mathcal{R}_2$ contained in $\mathcal{Q}$ defines its own striction curve s_1 and s_2, respectively. The two parametrizations

$$\mathcal{R}_{1,2} = \begin{pmatrix} a\cos u \\ b\sin u \\ 0 \end{pmatrix} + u \begin{pmatrix} -a\sin u \\ b\cos u \\ \pm c \end{pmatrix} \quad u \in [0, 2\pi],\ u \in \mathbb{R} \tag{9.27}$$

of the two reguli may be a starting point for the analytic treatment of the differential geometry of $\mathcal{Q}$ as a ruled surface. Both parametrizations given in (9.27) satisfy the equation

$$\mathcal{Q}: \ \frac{x^2}{a^2} + \frac{y^2}{b^2} - \frac{z^2}{c^2} = 1. \tag{9.28}$$

In the first step, we normalize the direction vector field(s)

$$(-a\sin u, b\cos u, \pm c),$$

which yields

$$\mathbf{r}_{1,2} = (-a\sin u, b\cos u, \pm c)/\sqrt{a^2\sin^2 u + b^2\cos^2 u + c^2}.$$

The actual directrix $\mathbf{l} = (a\cos t, b\sin t, 0)$, which is the ellipse in the plane $z = 0$, will be replaced by the striction curve(s) s_1 (or s_2 in the case of $\mathcal{R}_2$). Using (9.26), we find the parametrizations of the two striction curves

$$\mathbf{s}_{1,2} = \frac{1}{(a^2 - b^2)c^2\cos^2 u + b^2(a^2 + c^2)} \begin{pmatrix} a^3(b^2 + c^2)\cos u \\ b^3(a^2 + c^2)\sin u \\ \mp c^3(a^2 - b^2)\sin u\cos u \end{pmatrix}. \tag{9.29}$$

The reparametrization $u \to 2\arctan t$ of (9.29) makes clear that the striction curves on $\mathcal{Q}$ are algebraic, of degree four, and admit even rational parametrizations. In the case of a hyperboloid of revolution, we have $a = b$ and the two striction curves coincide, namely in the gorge circle.

The rational parametrizations of the striction curves on a one-sheeted hyperboloid give rise to another result:

Theorem 9.2.1 *The striction curves of the two reguli on a one-sheeted hyperboloid are quartic space curves of the second kind without singularities.*

Proof: In order to show that the striction curves on a one-sheeted hyperboloid $\mathcal{H}$ are quartic space curves of the second kind (cf. Section 6.9), we have to show that these curves can be found as the intersection of $\mathcal{H}$ with a cubic surface $\mathcal{K}$ sharing two straight lines with $\mathcal{H}$ (9.1) (second equation). Indeed, the pencil of cubic surfaces spanned by

$$\begin{aligned}\mathcal{K}_1 &: b^5c(a^2+c^2)xz + ac^4(b^2-a^2)y^3 - ab^4(a^2+c^2)yz^2 + ab^2c^4(a^2-b^2)y = 0,\\ \mathcal{K}_2 &: bc^5(a^2-b^2)xy + ac^4(a^2-b^2)y^2z + ab^4(a^2+c^2)z^3 + ab^4c^2(a^2+c^2)z = 0\end{aligned}$$

contains a one-parameter family of such surfaces that intersect $\mathcal{H}$ along the striction curve s_1. Two cubic surfaces with similar equations can be given for the second striction curve.

Both surfaces $\mathcal{K}_1$ and $\mathcal{K}_2$ (displayed in Figure 9.11) carry the striction curve s_1 of $\mathcal{H}$. While $\mathcal{K}_1$ carries the rulings $(\pm at, b, \pm ct)$ (with $t \in \mathbb{R}$), $\mathcal{K}_2$ carries the complex conjugate pair in the planes $a^2x^2 + b^2y^2 = 0$.

The striction curve on a (regular) ruled quadric will never show a singularity, since a (regular) ruled quadric carries only regular, non-torsal, and non-inflection rulings. ■

Figure 9.11 shows the two cubic surfaces mentioned in the proof of Theorem 9.2.1.

• **Exercise 9.2.1** Striction curves on hyperbolic paraboloids.

Show that the two striction curves s_1 and s_2 of the two reguli on the hyperbolic paraboloid $\mathcal{P}$

$$\mathcal{P}: \frac{x^2}{a^2} - \frac{y^2}{b^2} = 2z$$

are two parabolas admitting the parametrizations

$$\mathbf{s}_{1,2} = \frac{2}{a^2+b^2}\left(ab^2t, \pm a^2bt, (b^2-a^2)t^2\right), \quad t \in \mathbb{R}.$$

The striction curves s_1 and s_2 lie in the planes

$$ax \pm by = 0.$$

In this case, it is useful to parametrize $\mathcal{P}$ by $\mathbf{p} = (a(u+v), b(u-v), 2uv)$ with $(u,v) \in \mathbb{R}^2$ and extract the two reguli on $\mathcal{P}$.

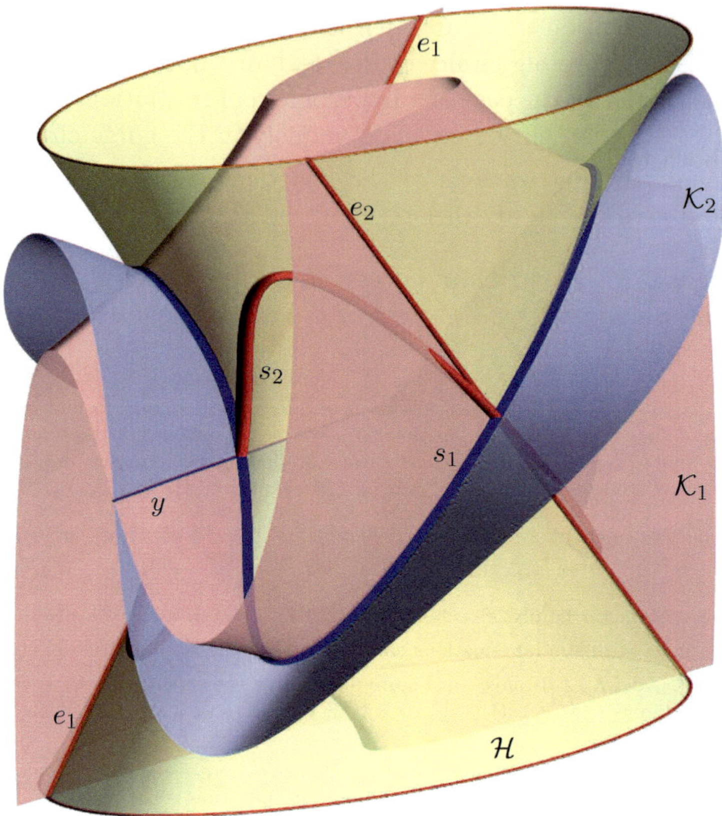

FIGURE 9.11. The two cubic ruled surfaces $\mathcal{K}_1$ (red) and $\mathcal{K}_2$ (blue) through the striction curve s_1 (blue) of the one-sheeted hyperboloid $\mathcal{H}$ also share the y-axis of the underlying Cartesian coordinate system. The two rulings e_1 and e_2 are part of the intersection $\mathcal{K}_1 \cap \mathcal{H}$.

Now, let us return to the one-sheeted hyperboloid. We can show the following result relating the central points on the reguli of a one-sheeted hyperboloid $\mathcal{H}$ with two particular planar intersections of $\mathcal{H}$:

Theorem 9.2.2 *The central points of both reguli on a one-sheeted hyperboloid $\mathcal{H}$ are the midpoints of the ruling's segments bounded by two ellipses $e_1, e_2 \subset \mathcal{H}$, which are the curves of contact of two cylinders of revolution Δ_1, Δ_2 touching $\mathcal{H}$ along theses ellipses.*

Figure 9.12 shows the two striction curves for the two different reguli on a one-sheeted hyperboloid.

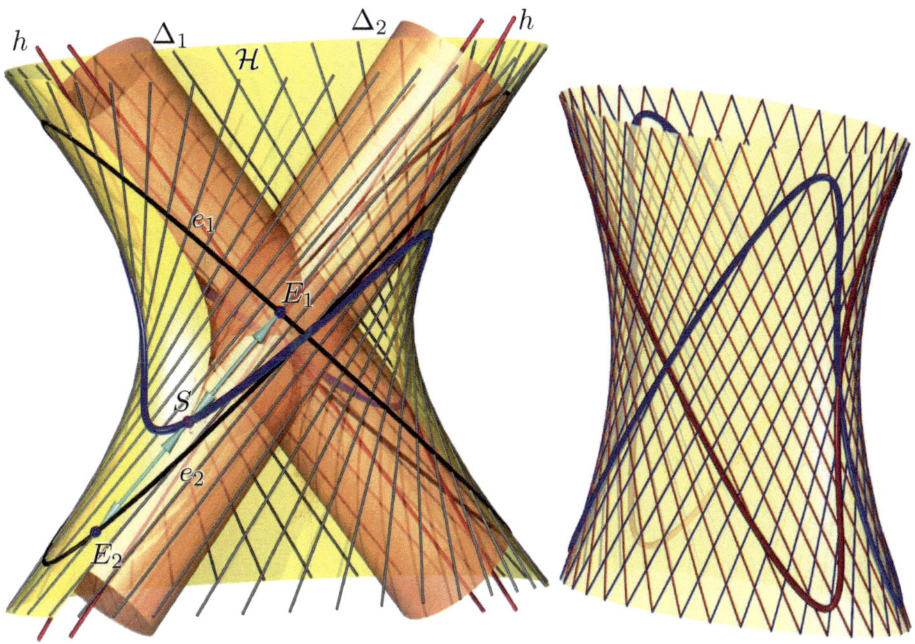

FIGURE 9.12. Left: The central points S (points of the striction curve) on each ruling are the midpoints of the ruling's segments E_1E_2 between the two ellipses e_1, e_2 that are the curves of contact of two cylinders Δ_1, Δ_2 of revolution touching the one-sheeted hyperboloid $\mathcal{H}$. Right: The two striction curves for the two different ruled surfaces on a one-sheeted hyperboloid.

Proof: We may assume that $\mathcal{H}$ is given by (9.28) and $a > b$. A cylinder of revolution that touches $\mathcal{H}$ along an ellipse is one of those quadratic cones whose vertex is an ideal point of one of the focal conics. Consequently, there are two such cylinders, say Δ_1 and Δ_2. Further, Δ_i must have a plane of symmetry with $\mathcal{H}$ in common, and moreover, they both have to touch $\mathcal{H}$ at its vertices $(0, \pm b, 0)$. Therefore, the axes of Δ_i coincide with an asymptote a_i of the focal hyperbola h, and thus, the cylinders have the equations

$$\Delta_{1,2}: \ \frac{y^2}{b^2} + \frac{1}{b^2}\left(y\sqrt{\frac{b^2+c^2}{a^2+c^2}} \pm z\sqrt{\frac{a^2-b^2}{a^2+c^2}}\right)^2 = 1. \tag{9.30}$$

Hence, the intersection curves (ellipses) $e_1 = \Delta_1 \cap \mathcal{Q}$ and $e_2 = \Delta_2 \cap \mathcal{Q}$ are located in the planes

$$c^2\sqrt{a^2+c^2}\,x \pm a^2\sqrt{b^2+c^2}\,z = 0, \tag{9.31}$$

which meet $\mathcal{Q}$ along ellipses because $\mathcal{Q}$'s asymptotic cone $\frac{x^2}{a^2} + \frac{y^2}{b^2} - \frac{z^2}{c^2} = 0$ is steeper.

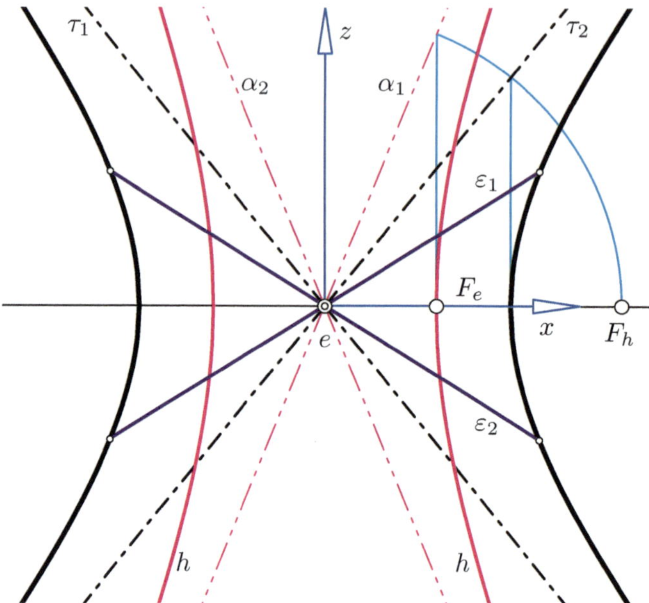

FIGURE 9.13. A front view (orthogonal projection onto $y = 0$) of the planes mentioned in the proof of Theorem 9.2.2 shows two harmonic quadruples of planes: $\mathrm{H}(\tau_1, \tau_2, \alpha_1, \varepsilon_1)$ and $\mathrm{H}(\tau_1, \tau_2, \alpha_2, \varepsilon_2)$.

Now, we intersect each line $\mathcal{R}_i(t)$ (for any $t \in [0, 2\pi[$ and $i = 1, 2$) with each plane ε_j $(j = 1, 2)$ and obtain four parametrizations $\mathbf{e}_{i,j}(t)$ of the ellipses $\varepsilon_j \cap \mathcal{Q}$. Then, it is easy to see that $\mathbf{s}_i = \frac{1}{2}(\mathbf{e}_{i,1} + \mathbf{e}_{i,2})$ with $\mathbf{s}_i$ being the parametrization of the striction curve given in (9.29). ■

Figure 9.12 (left) illustrates the contents of Theorem 9.2.2 for one particular regulus. Note that each plane containing an ellipse (either e_1 or e_2) can be found without explicitly determining the cylinders Δ_i: Let $e = \varepsilon_1 \cap \varepsilon_2$, *i.e.*, the second principal axis of $\mathcal{Q}$ carrying the two real vertices $(0, \pm b, 0)$ of $\mathcal{Q}$. The pencil (of planes) about e contains two tangent planes τ_1, τ_2 of $\mathcal{Q}$'s asymptotic cone. If α_i are now those planes through e that contain the two asymptotes of the focal hyperbola h, then the planes ε_i are the harmonic conjugates of α_i with respect to ε_1 and ε_2. In Figure 9.13, e appears as a point, and so do all planes in the pencil about e.

• **Exercise 9.2.2** The shape of the striction curve.

Show that the front view (orthogonal projection onto $y = 0$) of the striction curve (9.29) has two flat points exactly for $2a^2c^2 = b^2(c^2 - a^2)$. Figure 9.14 shows various appearances of the striction curves of one particular regulus on a one-sheeted hyperboloid depending on the ratio $a : b : c$.

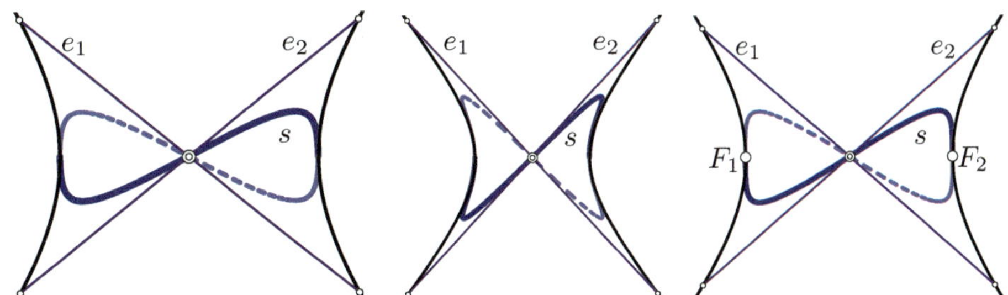

FIGURE 9.14. The front view of the striction curve of a regulus on a one-sheeted hyperboloid has a double point in any case, but besides, it may have four points of inflection (middle) or two flat points (right).

Orthogonal trajectories of the rulings of a quadric

On a ruled quadric $\mathcal{Q}$, we can find two independent one-parameter families of curves that traverse the two families of rulings at a right angle. Assume that

$$\mathbf{f}(u, v) = \mathbf{s}(u) + v\mathbf{r}(u) : \ I \times \mathbb{R} \to \mathbb{R}^3,$$

is a parametrization of one particular regulus on $\mathcal{Q}$ where $\mathbf{s} : \ I \to \mathbb{R}^3$ and $\mathbf{r} : \ I \to \mathrm{S}^2$ are parametrization of the striction curve and the spherical image of the rulings (over some real interval I), then the *striction* (function) $\sigma : \ I \to \mathbb{R}$ of the ruled surface (even if it is not a quadric) is given by

$$\cos\sigma = \langle \dot{\mathbf{s}}, \mathbf{r} \rangle, \tag{9.32}$$

provided that $\|\dot{\mathbf{r}}\| = 1$. It measures the angle between the striction curve and the rulings.

While the striction curve of a closed ruled surface is a closed curve, the orthogonal trajectories of a closed ruled surface are, in general, not closed. According to [62, p. 83], the orthogonal trajectories of a closed ruled surface starting at one point, say P_1, on a particular ruling r_0 meet the ruling r_0 in the well-defined point P_2, which is, in most cases, different from P_1. The distance $l = \overline{P_1 P_2}$ is called the *aperture* and is independent of the trajectory (or independent of the starting point P_1).

Following [62, p. 84], the aperture of the ruled surface can be computed as

$$l = \left| \int \cos\sigma \mathrm{d}u \right|. \tag{9.33}$$

Hence, the orthogonal trajectories of the rulings of (9.23) can be parametrized by

$$\mathbf{o}(u) = \mathbf{s}(u) + v_o\mathbf{r}(u) \tag{9.34}$$

where

$$v_o = -\int \sigma \mathrm{d}u.$$

In the case of the triaxial one-sheeted hyperboloid, the anti-derivative of $\cos\sigma$ can only be expressed in terms of elliptic integrals. The one-sheeted hyperboloid of revolution, *i.e.*, $a = b$, yields the constant value

$$\cos\sigma = \frac{b^2}{\sqrt{b^2 + c^2}}.$$

Thus, the orthogonal trajectories of the rulings of a one-sheeted hyperboloid of revolution can be given in a parametric form. Up to rotations about the z-axis, these curves are

$$\mathbf{o}(t) = \frac{b}{b^2 + c^2}\begin{pmatrix} b^2 t \sin t + (b^2 + c^2)\cos t \\ -b^2 t \cos t + (b^2 + c^2)\sin t \\ -bct \end{pmatrix}. \tag{9.35}$$

The same curve can also be written as

$$\mathbf{o}(t) = \begin{pmatrix} b\cos t \\ b\sin t \\ \frac{b^2}{c} \end{pmatrix} - \frac{b^2 t}{b^2 + c^2}\begin{pmatrix} -b\sin t \\ b\cos t \\ -\frac{b^2}{c} \end{pmatrix}.$$

This shows that the orthogonal trajectories are also located on a helical developable $\mathcal{D}$. $\mathcal{D}$'s curve of regression is a circular helix with radius b, pitch $-\frac{b^2}{c}$, and with the z-axis as its axis.

The top views (orthogonal projection onto $z = 0$) of the orthogonal trajectories of the regulus on a one-sheeted hyperboloid of revolution are spirals with radius function $\varrho(t) = \frac{b}{b^2+c^2}\sqrt{b^4 t^2 + (b^2 + c^2)^2}$.

Figure 9.15 shows the orthogonal trajectories of the rulings on a one-sheeted hyperboloid of revolution (left) and on a triaxial hyperboloid (right). The latter are determined numerically.

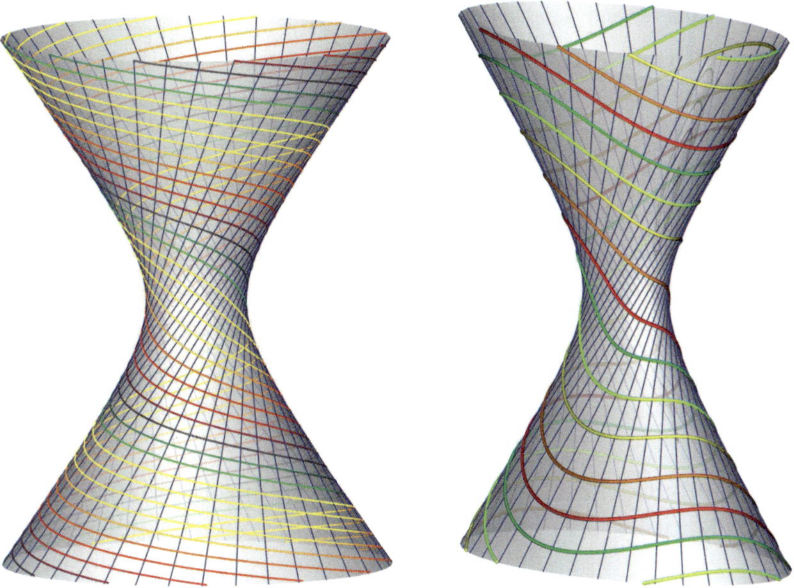

FIGURE 9.15. Some orthogonal trajectories of a regulus on a one-sheeted hyperboloid: a hyperboloid of revolution (left), triaxial (right).

• **Exercise 9.2.3** Orthogonal trajectories on hyperbolic paraboloids.

Use the parametrization of the hyperbolic paraboloid $\mathcal{P}$ given in Exercise 9.2.1 on page 403 and compute the orthogonal trajectories of one of the reguli on $\mathcal{P}$, but admit a parametrization in terms of elementary functions. Figure 9.16 shows some orthogonal trajectories of the hyperbolic paraboloid.

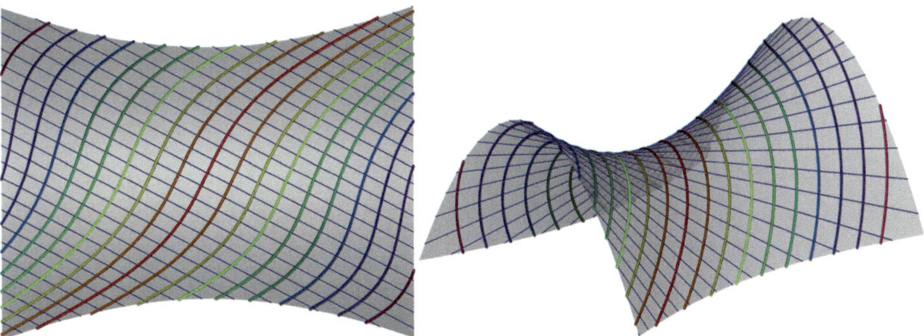

FIGURE 9.16. Some orthogonal trajectories of one family of rulings on a hyperbolic paraboloid shown in a top view (left) and in an axonometric view (right).

What do these curves look like if $\mathcal{P}$ is an orthogonal hyperbolic paraboloid?

Distribution parameter and Gaussian curvature

We have seen that a torsal ruling is characterized by (9.25). Following [62, 151], along each ruling, the *distribution parameter*

$$\delta = \frac{\det(\dot{\mathbf{l}}, \mathbf{r}, \dot{\mathbf{r}})}{\langle \dot{\mathbf{r}}, \dot{\mathbf{r}} \rangle} \tag{9.36}$$

is a measure for the winding of the tangent planes while the point of contact traces the ruling. Herein, the direction vector field $\mathbf{r}$ has to be normalized. Especially for the ruled quadric described by (9.27), we find

$$\delta = abc \frac{(a^2 - b^2)\cos^2 u - (a^2 + c^2)}{c^2(a^2 - b^2)\cos^2 u + b^2(a^2 + c^2)}. \tag{9.37}$$

Now, we compute the Gaussian curvature by inserting (9.27) first into (9.3) and then into (9.7) and find

$$K = -\frac{a^2b^2c^2\left(\frac{\cos^2 u}{b^2} + \frac{\sin^2 u}{a^2} + \frac{1}{c^2}\right)^2}{2a^2b^2c^2\left(\frac{\cos^2 u}{b^2} + \frac{\sin^2 u}{a^2} + \frac{1}{c^2}\right)^2 v^2 + \left(a^2 \sin^2 u + b^2 \cos^2 u + c^2\right)^2}. \tag{9.38}$$

With (9.37) we can eliminate the surface parameter u from (9.38) by solving (9.37) for $\cos^2 u$, which yields

$$\cos^2 u = \frac{b(a^2 + c^2)(ac + b\delta)}{c(a^2 - b^2)(ab - c\delta)} \quad \text{and} \quad \sin^2 u = \frac{a(b^2 + c^2)(bc + a\delta)}{c(a^2 - b^2)(c\delta - ab)}.$$

Substituting the latter expressions for $\cos^2 u$ and $\sin^2 u$ into (9.38), we find that the distribution parameter and the Gaussian curvature of $\mathcal{Q}$ are related via LAMARLE's[2] formula

$$K = -\frac{\delta^2}{(v^2 + \delta^2)^2}, \tag{9.39}$$

where $v \in \mathbb{R}$ is the surface parameter along the ruling with $v = 0$ at the central point and $\delta : I \subset \mathbb{R} \to \mathbb{R}$ is the distribution parameter.

It is worth mentioning that LAMARLE's formula (9.39) is valid for all ruled surfaces (see [62, 90]).

---

[2] ERNEST LAMARLE (1806–1879), Belgian mathematician.

For the example displayed in Figure 9.17, we have chosen $a = \frac{5}{4}$, $b = c = 1$. Thus, (9.37) yields $\delta(0) = -\frac{bc}{a} = -\frac{4}{5}$ and $\delta(\frac{\pi}{2}) = -\frac{ac}{b} = -\frac{5}{4}$. Hence, with (9.39), we find the Gaussian curvature at the vertices $(a,0,0)$ and $(0,b,0)$ to be $K(0,0) = -\frac{25}{16}$ and $K(\frac{\pi}{2},0) = -\frac{16}{25}$.

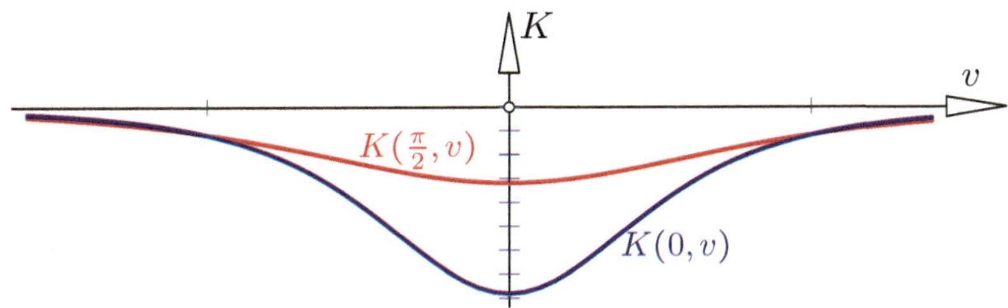

FIGURE 9.17. The Gaussian curvature K at the vertex generators $u = 0$ (vertex $(a,0,0)$, blue) and $u = \frac{\pi}{2}$ (vertex $(0,b,0)$, red) for $a = \frac{5}{4}$, $b = c = 1$.

• **Exercise 9.2.4** Distribution parameter of a hyperbolic paraboloid.

Start with one of the two parametrizations

$$\mathbf{p}(u,v) = (au, bv, \frac{1}{2}(u^2 - v^2)) \quad \text{or} \quad \mathbf{p}(u,v) = (a(u+v), b(u-v), 2uv) \quad \text{with} \quad (u,v) \in \mathbb{R}^2$$

of the hyperbolic paraboloid

$$\mathcal{P}: \ \frac{x^2}{a^2} - \frac{y^2}{b^2} = 2z$$

and compute the distribution parameter δ. For that purpose, the direction vector field has to be normalized. Show further that the Gaussian curvature at the vertex $(0,0,0)$ is

$$K_{\min} = -\frac{1}{a^2b^2}$$

(of course independent of the parametrization).

As can be seen from (9.39), for any ruled surface, and thus, also for ruled quadrics, the Gaussian curvature considered as a function on a particular ruling r is a rational function with $\lim\limits_{v\to\infty} = 0$. It attains its minimum

$$K_{\min} = -\delta^{-2}$$

at the central point S, *i.e.*, at $v = 0$ (cf. Figure 9.17).

Consequently, the curves of constant Gaussian curvature on $\mathcal{Q}$ (and on any ruled surface) touch the rulings at the central points (see Figure 9.18).

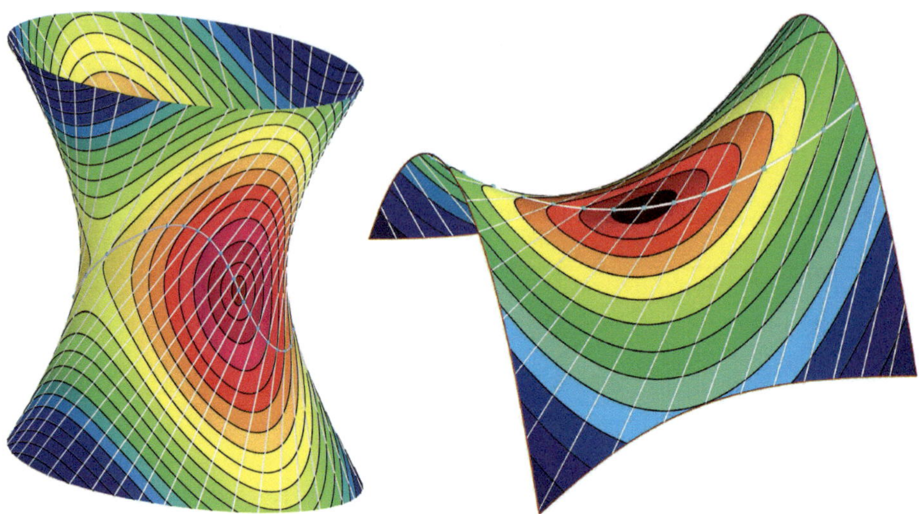

FIGURE 9.18. The curves of constant Gaussian curvature touch the rulings at their central points and all these contact points line up on the striction curve, since at the central point (or striction point), the Gaussian curvature has its minimum along the ruling. Left: one-sheeted hyperboloid, right: hyperbolic paraboloid.

9.3 Lie's osculating quadric

In [46, p. 124], we have discussed various appearances of conics in differential geometry. We have seen that the Dupin indicatrix is a frequently used tool in local differential geometry. Further, the Dupin indicatrix can be seen as the orthogonal projection of a planar section of a local second order Taylor approximation of a surface. The distance of the section to the tangent plane is closely related to the factor of similarity of the indicatrix.

Among the many osculating quadrics of a surface $\mathcal{S}$ in three-dimensional space, there is a special type associated with any regular, non-torsal, and non-inflection ruling r of a ruled surface $\mathcal{R}$ in projective 3-space. It is called *Lie's osculating* quadric $\mathcal{O}$ and osculates $\mathcal{R}$ along the chosen ruling r. $\mathcal{O}$ does not only share the tangent planes with $\mathcal{R}$ along r, it also shares the asymptotic tangents with $\mathcal{R}$ along r. These asymptotic tangents comprise one set of rulings on $\mathcal{O}$.

We may start the discussion of Lie's osculating quadric $\mathcal{O}$ in Euclidean 3-space $\mathbb{R}^3$. However, Lie's osculating quadric, or to be more precise, the osculating regulus of a ruled quadric along a regular non-torsal and non-inflection ruling can also be defined in the projective setting and is, thus, an object that is related to a ruled surface in a projectively invariant way. Klein's model of line geometry which will be sketched roughly in Section 10.1, is a proper framework for that.

As we have seen on page 401, the central point can be found via limit procedure that is applied to the common normals of two sufficiently close rulings. We can go one step further. Similarly to the limit of a circle through three close points on a (planar or space) curve (cf. [46, p. 92]), we can choose two sufficiently close rulings of a smooth ruled surface $\mathcal{R}$ in the vicinity of a (regular, non-torsal, non-inflection) ruling r. In fact, C^2 smoothness of $\mathcal{R}$ would be sufficient. Assume that

$$\mathbf{f}(u,v) = \mathbf{l}(u) + v\mathbf{r}(u) \quad (u,v) \in \mathbb{R}^2$$

with $\mathbf{l}: I \subset \mathbb{R} \to \mathbb{R}^3$ and $\mathbf{r}: I \to \mathrm{S}^2$ is a parametrization of a ruled surface. Let $t_0 \in I$ be some parameter and further $\varepsilon > 0$ such that $t_0 \pm \varepsilon \in I$. Then, $r_{-\varepsilon} = \mathbf{f}(t_0 - \varepsilon, v)$ and $r_\varepsilon = \mathbf{f}(t_0 + \varepsilon, v)$ are the rulings close to r. The symmetry of the ε-vicinity is not necessary, but helpful.

In Chapter 4, we have learned that there exists a unique regular ruled quadric on three mutually skew straight lines. The same holds true for $r_{-\varepsilon}$, r, r_ε and we denote this quadric by $\mathcal{O}_\varepsilon$. Now, with $\varepsilon \to 0$, we see that $\mathcal{O}_\varepsilon$ attains a limit which will, in general, be a quadric $\mathcal{O}$.

From the parametrization (9.3) of a ruled surface $\mathcal{R}$ by means of a directrix $\mathbf{l}$ and a direction vector field $\mathbf{r}$, we can easily derive a parametrization in terms of Plücker coordinates $(\mathbf{r}, \overline{\mathbf{r}}) : I \to M_2^4$. (For details, especially on the Plücker quadric $M_2^4 \subset \mathbb{P}^5$, see Section 10.1.)

For what follows, we need

Lemma 9.3.1 *Assume we are given three mutually skew straight lines a, b, c in Euclidean 3-space. Let $A = (\mathbf{a}, \overline{\mathbf{a}})$, $B = (\mathbf{b}, \overline{\mathbf{b}})$, $C = (\mathbf{c}, \overline{\mathbf{c}})$ be their Plücker coordinates and, furthermore, let $\mathbf{x} = (x, y, z)$ be inhomogeneous Cartesian coordinates. Then, an inhomogeneous equation of the uniquely defined ruled quadric $\mathcal{Q}$ on a, b, c is given by*

$$\begin{aligned} &\det(\mathbf{x}, \mathbf{a}, \mathbf{b})\langle \mathbf{x}, \overline{\mathbf{c}}\rangle + \langle \mathbf{x}, \overline{\mathbf{a}}\rangle(\langle \mathbf{b}, \overline{\mathbf{c}}\rangle - \langle \overline{\mathbf{b}}, \mathbf{c}\rangle) + \\ &+ \det(\mathbf{x}, \mathbf{b}, \mathbf{c})\langle \mathbf{x}, \overline{\mathbf{a}}\rangle + \langle \mathbf{x}, \overline{\mathbf{b}}\rangle(\langle \mathbf{c}, \overline{\mathbf{a}}\rangle - \langle \overline{\mathbf{c}}, \mathbf{a}\rangle) + \\ &+ \det(\mathbf{x}, \mathbf{c}, \mathbf{a})\langle \mathbf{x}, \overline{\mathbf{b}}\rangle + \langle \mathbf{x}, \overline{\mathbf{c}}\rangle(\langle \mathbf{a}, \overline{\mathbf{b}}\rangle - \langle \overline{\mathbf{a}}, \mathbf{b}\rangle) + \\ &+ \det(\overline{\mathbf{a}}, \overline{\mathbf{b}}, \overline{\mathbf{c}}) = 0. \end{aligned} \tag{9.40}$$

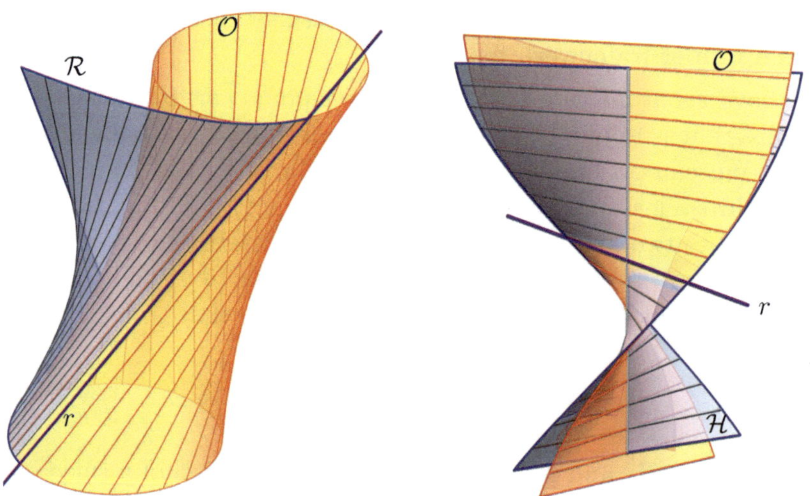

FIGURE 9.19. Lie's osculating quadric $\mathcal{O}$ along a regular non-torsal ruling: at a quartic ruled surface $\mathcal{R}$ (left), at a helicoid $\mathcal{H}$ (right).

Proof: Assume P is a point on $\mathcal{Q}$ and has the Cartesian coordinates $\mathbf{p}$. Then, there exists a line l through P with Plücker coordinates $(\mathbf{l}, \bar{\mathbf{l}})$ from the regulus $\mathcal{R}^\star$ complementary to the regulus $\mathcal{R}$ spanned by a, b, and c. We find $(\mathbf{l}, \bar{\mathbf{l}})$ as the intersection of the two planes $\alpha = [a, P]$ and $\beta = [b, P]$. Using (10.12), we find the homogeneous coordinates of the planes α and β

$$\alpha = (\langle \bar{\mathbf{a}}, \mathbf{p} \rangle, \mathbf{p} \times \mathbf{a} - \bar{\mathbf{a}}), \quad \beta = (\langle \bar{\mathbf{b}}, \mathbf{p} \rangle, \mathbf{p} \times \mathbf{b} - \bar{\mathbf{b}})$$

and therefore, the Plücker coordinates of l computed with (10.6) and (10.8) read

$$(\mathbf{l}, \bar{\mathbf{l}}) = \left((\mathbf{p} \times \mathbf{a} - \bar{\mathbf{a}}) \times (\mathbf{p} \times \mathbf{b} - \bar{\mathbf{b}}); \langle \bar{\mathbf{a}}, \mathbf{p} \rangle (\mathbf{p} \times \mathbf{b} - \bar{\mathbf{b}}) - \langle \bar{\mathbf{b}}, \mathbf{p} \rangle (\mathbf{p} \times \mathbf{a} - \bar{\mathbf{a}}) \right).$$

Since $l \in \mathcal{R}^\star$, it intersects all lines of its complementary regulus $\mathcal{R}$. Therefore, it also intersects c and, according to (10.9), the Plücker coordinates of l and c satisfy $\langle \mathbf{c}, \bar{\mathbf{l}} \rangle + \langle \bar{\mathbf{c}}, \mathbf{l} \rangle = 0$. In this latter intersection condition, we replace $\mathbf{p}$ with $\mathbf{x}$ and obtain (9.40). ■

In terms of the Klein model of the manifold of lines in 3-space (as will be explained in Section 10.1), the points A, B, C corresponding via their Plücker coordinates to the three lines span a plane in $\mathbb{P}^5$.

- **Exercise 9.3.1** Singular quadrics on three lines.

If one, two, or even all three lines have mutual non-empty intersections, then (9.40) returns a singular quadratic equation.

Discuss all configurations of three lines and compute normal forms of the thus arising singular or degenerate quadrics using (9.40).

With Lemma 9.3.1 we can show

Theorem 9.3.1 *Let $\mathcal{R}$ be a C^2 ruled surface in 3-space given by a parametrization in Plücker coordinates $\mathcal{R}(t) = (\mathbf{r}, \bar{\mathbf{r}}) : I \subset \mathbb{R} \to M_2^4$. Define $R := \mathcal{R}(t_0)$, $\dot{R} := \dot{\mathcal{R}}(t_0)$, and $\ddot{R} = \ddot{\mathcal{R}}(t_0)$ (the three derivative points of order 0, 1, and 2). Assume that $R = (\mathbf{r}, \bar{\mathbf{r}})$ is a regular (R, $\dot{R}$ linearly independent), non-torsal ($\Omega(\dot{R}, \dot{R}) \neq 0$, cf. (10.10)), and non-inflection ($[R, \dot{R}, \ddot{R}] \not\subset M_2^4$) ruling. Then, an inhomogeneous equation of Lie's osculating quadric of $\mathcal{R}$ along R is given by (9.40) with $A = R$, $B = \dot{R}$, and $C = \ddot{R}$.*

Proof: The proof becomes more clear once the Klein model of line geometry is established (see Section 10.1). However, the plane $\pi = [R, \dot{R}, \ddot{R}]$ is the osculating plane of a curve r in the Plücker quadric M_2^4. The plane π intersects M_2^4 along a conic o which is, by assumption, a regular conic that is not contained in a generator plane of M_2^4. Therefore, o is in second order contact and represents a ruled surface that is in second order contact with $\mathcal{R}$ along R. In Section 10.1, we will see that regular conics in M_2^4 represent reguli. ■

• **Exercise 9.3.2** Osculating quadrics of helical ruled surfaces.

The composition of a uniform rotation about the z-axis of a Cartesian coordinate system with a uniform translation about the same axis results in a helical motion. In the underlying coordinate system, the helical motion $\mu:\ \mathbb{R}^3 \to \mathbb{R}^3$ shall be given by

$$\mu:\ \mathbf{x} \mapsto \mathbf{x}' = \begin{pmatrix} \cos t & -\sin t & 0 \\ \sin t & \cos t & 0 \\ 0 & 0 & 1 \end{pmatrix} \mathbf{x} + \begin{pmatrix} 0 \\ 0 \\ pt \end{pmatrix}, \tag{9.41}$$

where $t \in \mathbb{R}$ and $p \in \mathbb{R} \setminus \{0\}$ is the pitch of the helical motion.

A straight line L that allows the parametrization

$$\mathbf{l}(u) = (d, u, ku) \quad \text{with} \quad u \in \mathbb{R},$$

with constants $d, k \in \mathbb{R}$ undergoing the helical motion μ traces a *helical ruled surface*

$$\mathcal{R} = \begin{pmatrix} d\cos t - u\sin t \\ d\sin t + u\cos t \\ pt + ku \end{pmatrix} \quad \text{with} \quad (t, u) \in \mathbb{R}^2.$$

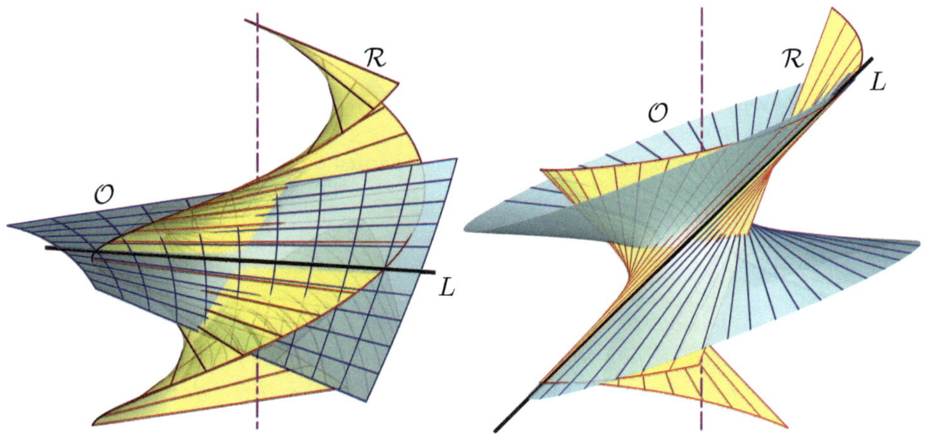

FIGURE 9.20. Left: Lie's osculating quadric $\mathcal{O}$ along L at a right open helical ruled surface $\mathcal{R}$ ($k = 0$, $d \neq 0$) is a hyperbolic paraboloid. Right: Lie's osculating quadric $\mathcal{O}$ along L at a skew open helical ruled surface $\mathcal{R}$ ($k \neq 0$, $d \neq 0$) is a one-sheeted hyperboloid.

Show that Lie's osculating quadric $\mathcal{O}$ along L has an equation of the form

$$\mathcal{O}:\ k(p - dk)x^2 - k(dk - 2p)y^2 - 2pzy + dz^2 + 2p(dk - p)x + d(dk - p)(dk - 2p) = 0. \tag{9.42}$$

For helical ruled surfaces, it is sufficient to write down this particular quadric at $t = 0$. Lie's osculating quadric $\mathcal{O}_t$ along any other ruling can be obtained from (9.42) by applying the helical motion (9.41) to (9.42) and it is, therefore, congruent to $\mathcal{O}$ (cf. Figure 9.20).

$\mathcal{O}$ is a double plane if R is a helical developable, *i.e.*, the torsal ruled surface traced by the tangents of a helix. If the slope k of R's generators equals zero, R is a right helical ruled surface and Lie's osculating quadric along any ruling is a hyperbolic paraboloid.

9.4 Normals to a quadric

In [46, Section 9.3, p. 402], we have seen that a generic point in the plane of a conic sends up to four normals to the conic. The computation of the four pedal points is reduced to the computation of the common points of the conic and an equilateral hyperbola.

Now, we want to discuss the three-dimensional case by means of a triaxial ellipsoid $\mathcal{Q} \subset \mathbb{R}^3$ with the equation

$$\mathcal{Q}: \frac{x^2}{a^2} + \frac{y^2}{b^2} + \frac{z^2}{c^2} - 1 = 0 \tag{9.43}$$

with $a, b, c \in \mathbb{R}^\star$ being its semiaxes lengths.

Let $\mathbf{p} = (\xi, \eta, \zeta) \in \mathbb{R}^3$ be a point P off the quadric and assume further that the normals from P to $\mathcal{Q}$ hit the surface at yet undetermined points F with coordinates $\mathbf{f} = (u, v, w)$. Then, $\mathcal{Q}$'s normals $\mathbf{n} = \mathrm{grad}\mathcal{Q}(\mathbf{f})$ at F and the vectors $\mathbf{p} - \mathbf{f}$ are linearly dependent, which yields

$$\varrho\frac{u}{a^2} = \xi - u, \quad \varrho\frac{v}{b^2} = \eta - v, \quad \varrho\frac{w}{c^2} = \zeta - w, \tag{9.44}$$

where $\varrho \in \mathbb{R}^\star$. The scaling factor ϱ is to be determined. We solve (9.44) for the coordinates u, v, w of the pedal point F, find

$$u = \frac{\xi}{1 + \frac{\varrho}{a^2}}, \quad v = \frac{\eta}{1 + \frac{\varrho}{b^2}}, \quad w = \frac{\zeta}{1 + \frac{\varrho}{c^2}}, \tag{9.45}$$

insert them into (9.43), since $F \in \mathcal{Q}$, and obtain

$$\frac{\xi^2}{a^2\left(1 + \frac{\varrho}{a^2}\right)^2} + \frac{\eta^2}{b^2\left(1 + \frac{\varrho}{b^2}\right)^2} + \frac{\zeta^2}{c^2\left(1 + \frac{\varrho}{c^2}\right)^2} = 1. \tag{9.46}$$

The latter equation is of degree six in the unknown ϱ. Thus, we have:

Theorem 9.4.1 *A generic point $P \in \mathbb{R}^3$ sends up to six (real) normals to an ellipsoid $\mathcal{Q} \subset \mathbb{R}^3$.*

• **Exercise 9.4.1** Normals to a paraboloid.

The determination of normals from a generic point to a paraboloid differs slightly from the analogous problem with ellipsoids and hyperboloids. Assume that we are given an elliptic and a hyperbolic paraboloid:

$$\mathcal{P}_e: \frac{x^2}{a^2} + \frac{y^2}{b^2} = 2z, \quad \mathcal{P}_h: \frac{x^2}{a^2} - \frac{y^2}{b^2} = 2z \tag{9.47}$$

with $a, b \in \mathbb{R}^\star$.

Now show that there are no more than five normals from a generic point P with coordinates $\mathbf{p} = (\xi, \eta, \zeta)$ to any paraboloid. These five normals correspond to the zeros of the quintic equations

$$\frac{\xi^2}{a^2\left(1+\frac{\varrho}{a^2}\right)^2} \pm \frac{\eta^2}{b^2\left(1\pm\frac{\varrho}{b^2}\right)^2} - 2(\zeta+\varrho) = 0 \tag{9.48}$$

with either the plus or the minus, the latter in the case of the hyperbolic paraboloid. Figure 9.21 (right) shows two different points P_3 and P_5 with their normals to an elliptic paraboloid. At P_5, five normals meet, while at P_3 there are only three concurrent normals.

The reality of all six solutions of (9.46) is guaranteed at least in the interior of the ellipsoid $\mathcal{E}$ if we choose P in the quadric's center. Figure 9.21 (left) shows a triaxial ellipsoid and three points P_2, P_4, and P_6 that differ in the number of normals of $\mathcal{E}$ passing through these points.

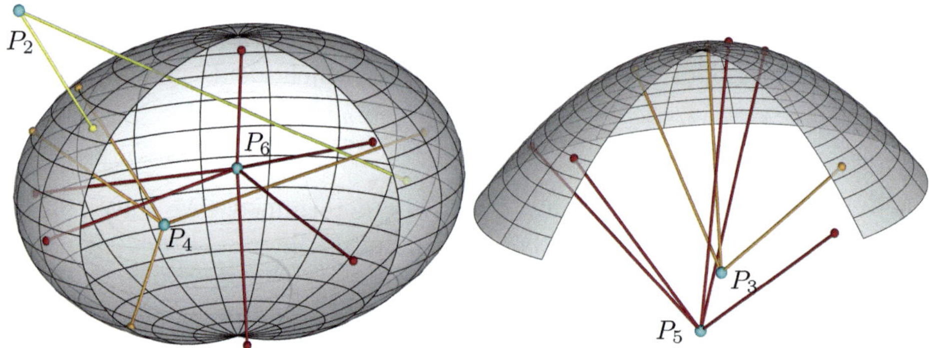

FIGURE 9.21. Left: A generic point sends six normals to a triaxial ellipsoid. The points P_2, P_4, P_6 have two, four, six real pedal points and lie on just as many normals. Right: A generic point sends five normals to an elliptic paraboloid. The points P_3 and P_5 have three and five pedal points.

Focal surfaces, central surfaces

Congruences of lines are the line geometric analoga to surfaces in a 3-space (see, *e.g.*, [83, 106, 151]). Among the congruences of lines, the *normal congruences*, *i.e.*, the congruence of all normals of a surface play an outstanding role. The normals of an ellipsoid $\mathcal{E}$ also form a congruence $\mathcal{N}$ of surface normals. Usually, the congruence of surface normals of any surface envelopes a surface consisting of two sheets. These surfaces are called *focal surfaces of the congruence*, or, especially in connection with the normals of a surface, the *central surfaces*. Figure 9.23 shows the two different sheets of the central surface of a triaxial ellipsoid.

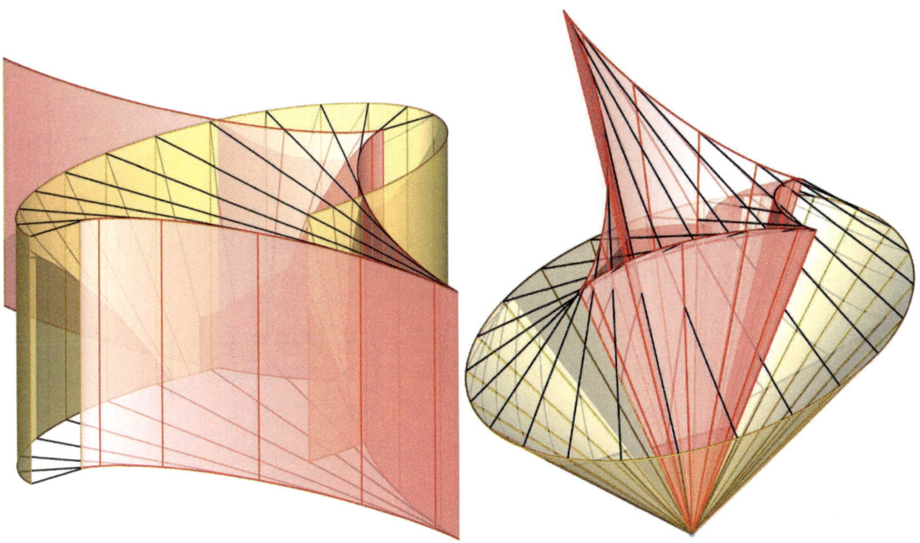

FIGURE 9.22. Left: The central surface of an elliptic cylinder is a cylinder whose orthogonal cross section is the evolute of the cross section of the elliptic cylinder together with the ideal line of all planes orthogonal to the rulings of both cylinders. Right: The central surface of a quadratic cone (not of revolution) is an algebraic cone of class four plus a conic in the ideal plane.

In special cases, the focal surfaces of a congruence coincide. Sometimes, one or even both sheets of the focal surface degenerate to curves or even points. For example, the normals of a Dupin cyclide intersect a pair of focal conics (cf. [46, p. 137]). The normals of a sphere form a star of lines, and therefore, the focal surface of the congruence of normals of a sphere is a single point. The normals of a cylinder of revolution meet the cylinder's axis at a right angle and, at the same time, the common ideal line of all planes orthogonal to the axis. So, the congruence of surface normals of a cylinder of revolution consists of a pair of skew straight lines one of which is even a line at infinity. In the case of an elliptic cylinder, the focal surfaces are a straight line and a further cylinder whose cross section (orthogonal to the cylinder's rulings) is an evolute of the cross section of the given cylinder (cf. Figure 9.22, left). The central surface of a quadratic cone that differs from a cone of revolution consists of a cone of algebraic class four and a conic in the ideal plane (Figure 9.22, right).

We can describe the ellipsoid's normal congruence $\mathcal{N}$ by a parametrization of the lines' point sets: Starting from the trigonometric standard parametrization of $\mathcal{E}$

$$\mathbf{e}(u,v) = \begin{pmatrix} a\sin u\cos v \\ b\sin u\sin v \\ c\cos u \end{pmatrix} \tag{9.49}$$

over the domain $[0,\pi]\times[0,2\pi[$, we find the direction of $\mathcal{E}$'s normals with (9.43) as $\operatorname{grad}\mathcal{Q}(\mathbf{e})$ and arrive at

$$\mathcal{N}(u,v) = \begin{pmatrix} a\sin u\cos v \\ b\sin u\sin v \\ c\cos u \end{pmatrix} + w\begin{pmatrix} a^{-1}\sin u\cos v \\ b^{-1}\sin u\sin v \\ c^{-1}\cos u \end{pmatrix} \tag{9.50}$$

with $w\in\mathbb{R}$, $(u,v)\in[0,\pi]\times[0,2\pi[$.

The parametrization of the congruence $\mathcal{N}: \mathbb{R}^3 \to \mathbb{R}^3$ has a two-dimensional manifold of singularities, *i.e.*, the focal surface(s) of the congruence.

Mathematically, these singularities are best described as the set of zeros of the determinant of Jacobian of $\mathcal{N}$. Therefore, we compute the Jacobian

$$\mathrm{J}_{\mathcal{N}} = (\partial_u\mathcal{N}, \partial_v\mathcal{N}, \partial_w\mathcal{N}) \tag{9.51}$$

and, finally, its determinant $\det \mathrm{J}_{\mathcal{N}} =: f$

$$\begin{aligned} f = {} & w^2 + (a^2(1-\sin^2 u\cos^2 v) + b^2(1-\sin^2 u\sin^2 v) + c^2\sin^2 u)w + \\ & + a^2b^2\cos^2 u + a^2c^2\sin^2 u\sin^2 v + b^2c^2\sin^2 u\cos^2 v. \end{aligned} \tag{9.52}$$

The polynomial f in (9.52) is quadratic in w, *i.e.*, the parameter of points on the surface normals. We can expect up to two real singular points of the mapping $\mathcal{N}: [0,\pi]\times[0,2\pi[\times\mathbb{R} \to \mathbb{R}^3$. Indeed, it can be shown that, on each line of a congruence of surface normals, there are two real points. It turns out that these points are the Meusnier points of the principal curvature tangents at $\mathbf{e}(u,v)$ (note Exercise 9.4.2). In Lemma 8.2.1, these points have been identified as the poles of the tangent planes w.r.t. confocal hyperboloids. (Only in the case of a sphere, all the singular points coincide, the center of the sphere.)

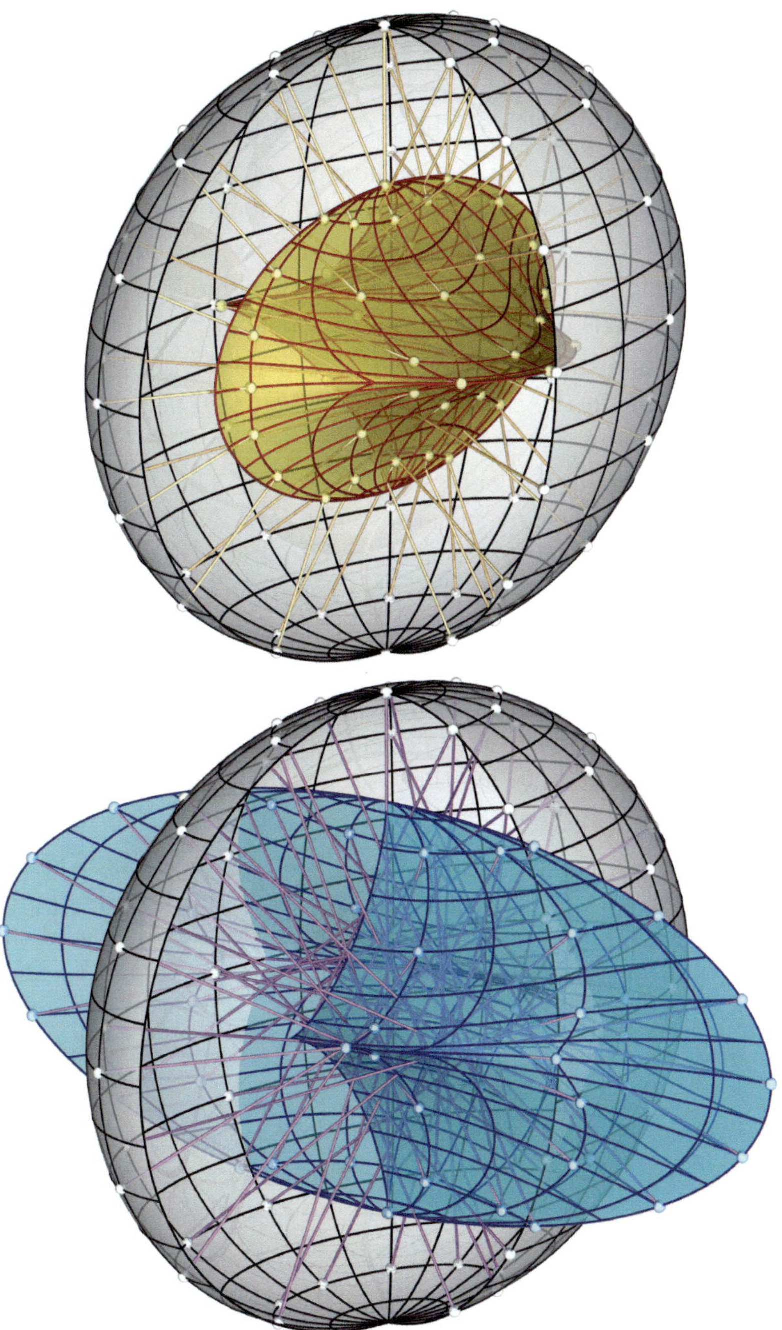

FIGURE 9.23. The central surface of an ellipsoid is the focal surface of the ellipsoid's normal congruence. The central surface has two sheets (yellow and blue). Here, the semiaxes lengths are $a = 9$, $b = 5$, $c = 7$. It is possible to choose the axes lengths such that both sheets of the focal surface are either inside or outside the ellipsoid, or only one sheet is inside.

At a singular point, $\mathcal{N}$ is of rank two, and thus, the singular points form two surfaces $\mathcal{C}_1$ and $\mathcal{C}_2$ which can be seen as two sheets of one algebraic variety, at least for algebraic surfaces, especially for quadrics. (There are exceptions, some of which have been described above):

(1) If $\mathcal{N}$ is the congruence of surface normals of a Dupin cyclide, then the set of singular points consists of a pair of focal conics or a conic and a straight line.

(2) For surfaces of revolution, one sheet of the singular set is a straight line, *i.e.*, the axis of revolution.)

(3) The sphere is a surface of revolution. Since all the sphere's normals are concurrent in the center, the central surface degenerates and is just a single point.

The surfaces $\mathcal{C}_1$ and $\mathcal{C}_2$ are called *focal surfaces.* This is justified by considering the lines of the congruence as light rays. Near the focal surfaces, the density of the light rays increases, and finally, a caustic appears as the envelope of all light rays. On the other hand, the congruence of surface normals of a surface in (Euclidean) 3-space consists of lines tangent to two (focal) surfaces.

The zeros of the polynomial $f(w)$ correspond to the singular points on each line in the congruence $\mathcal{N}$, and thus, to the two *focal points.* Therefore, the polynomial f is called the *focal polynomial* of the congruence $\mathcal{N}$.

• **Exercise 9.4.2** Congruence of surface normals, normal curvatures, and focal points.

Assume $\mathbf{s}(u^1, u^2): D \subset \mathbb{R}^2 \to \mathbb{R}^3$ is a regular parametrization of a surface $\mathcal{S} \subset \mathbb{R}^3$. Then, $\mathbf{n}: D \to S^2$ is the unit normal vector field along $\mathcal{S}$ induced by the Euclidean structure on $\mathbb{R}^3$, *i.e.*,

$$\mathbf{n}(u^1, u^2) = \frac{1}{\sqrt{G}} \partial_1 \mathbf{s} \times \partial_2 \mathbf{s},$$

where $G(u^1, u^2)$ equals the determinant of the first fundamental form of $\mathcal{S}$. The congruence $\mathcal{N}$ of $\mathcal{S}$'s surface normals can now be parametrized by

$$\mathbf{f}(u^1, u^2, u^3) = \mathbf{s} + u^3 \cdot \mathbf{n}$$

with $u^3 \in \mathbb{R}$ and $(u^1, u^2) \in D$. The focal polynomial f of $\mathcal{N}$, as defined in (9.51), then equals

$$f = \sqrt{G}((u^3)^2 K - 2Hu^3 + 1),$$

wherein $H, K: D \to \mathbb{R}$ are the mean and Gaussian curvature (functions). The zeros of f are

$$u^3_{1,2} = \frac{H \pm \sqrt{H^2 - K}}{K},$$

which are the inverses of the principal curvature functions. So, the central surfaces are the focal surfaces of the congruence $\mathcal{N}$ of surface normals and the loci of the principal curvature centers (Meusnier points) of $\mathcal{F}$ at the same time.

• **Exercise 9.4.3** Central surface of a quadratic cone.

Derive a parametrization of the normal congruence of the quadratic cone

$$\Gamma: \frac{x^2}{a^2} + \frac{y^2}{b^2} - \frac{z^2}{c^2} = 0. \tag{9.53}$$

Compute the proper part $\mathcal{F}$ of the focal surfaces of the congruence. For the sake of simplicity, set $\alpha := b^2 + c^2$, $\beta := c^2 + a^2$, and $\gamma := a^2 - b^2$ and show that $\mathcal{F}$ is a cone with the same vertex as Γ, which can be parametrized by

$$\mathbf{f}(u,v) = \frac{1}{abc(\alpha + \gamma\cos^2 u)} \begin{pmatrix} bc\beta\gamma v\cos^3 u \\ -ca\gamma\alpha v\sin^3 u \\ ab\alpha\beta v \end{pmatrix} \tag{9.54}$$

with $(u,v) \in [0, 2\pi[\, \times\mathbb{R}$ and has the equation

$$\mathcal{C}: \ (a^2\alpha^2x^2 + b^2\beta^2y^2 - c^2\gamma^2z^2)^3 + 27a^2b^2c^2\alpha^2\beta^2\gamma^2x^2y^2z^2 = 0. \tag{9.55}$$

An alternative approach to the equation of the cone $\mathcal{C}$ as the central surface of the cone (9.53) uses the dual representation of the cone. We can consider Γ as the set $\Gamma^\star$ of its tangent planes. From the parametrization

$$\mathbf{g}(u,v) = v(a\cos u, b\sin u, c)$$

with $(u,v) \in [0, 2\pi[\, \times\mathbb{R}$ of Γ, we derive the equations of Γ's tangent planes

$$\tau: \ bc\cos u x + ca\sin u y - abz = \langle \mathbf{n}, \mathbf{x}\rangle = 0$$

with $\mathbf{x} = (x,y,z)$, which form a one-parameter family, and therefore, they depend on a single parameter. Through each generator e of Γ there exists a unique plane $\tau^\perp$ orthogonal to τ (cf. Fig. 9.24). These planes are tangent to the central surface $\mathcal{C}$ of Γ. Since e aims in the direction $\mathbf{e} = (bc\cos u, ca\sin u, -ab)$, the planes $\tau^\star$ are perpendicular to

$$\mathbf{e}\times\mathbf{n} = (-a\alpha\sin u, b\beta\cos u, c\gamma\sin u\cos u) = \mathbf{n}^\star.$$

The envelope of all planes $\tau^\star : \langle \mathbf{x}, \mathbf{n}^\star\rangle = 0$ can be computed by differentiating the equation of $\tau^\star$ w.r.t. the parameter u which yields $\dot\tau^\star : \langle \mathbf{x}, \dot{\mathbf{n}}^\star\rangle = 0$. Then, the intersection of $\tau^\star$ and $\dot\tau^\star$ yields a parametrization of the cone $\mathcal{C}$. The subsequent implicitization yields (9.55).

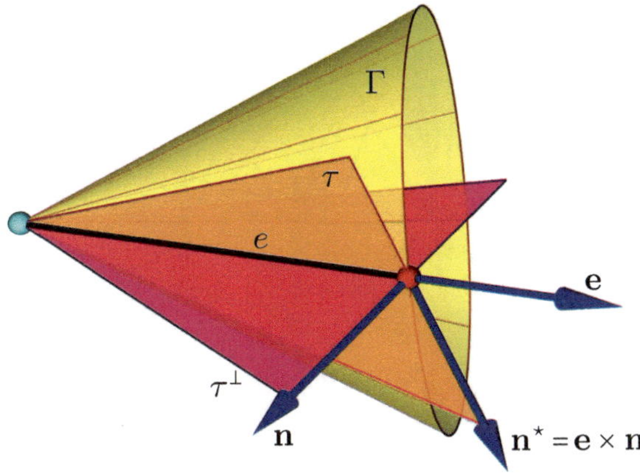

FIGURE 9.24. The enevelope of the one-parameter family of planes $\tau^\perp$ (through the generators and orthogonal to the tangent planes τ of the cone Γ) is the central surface $\mathcal{C}$ of Γ.

• **Exercise 9.4.4** Central surface of paraboloids.

Compute the focal polynomial analogous to (9.52) for the two paraboloids

$$\mathcal{P}_e : \frac{x^2}{a^2} - \frac{y^2}{b^2} = 2z \quad \text{and} \quad \mathcal{P}_h : \frac{x^2}{a^2} - \frac{y^2}{b^2} = 2z.$$

Derive parametrizations of the focal surfaces of the respective normal congruences. A hyperbolic paraboloid with the two sheets of its central surface is shown in Figure 9.25.

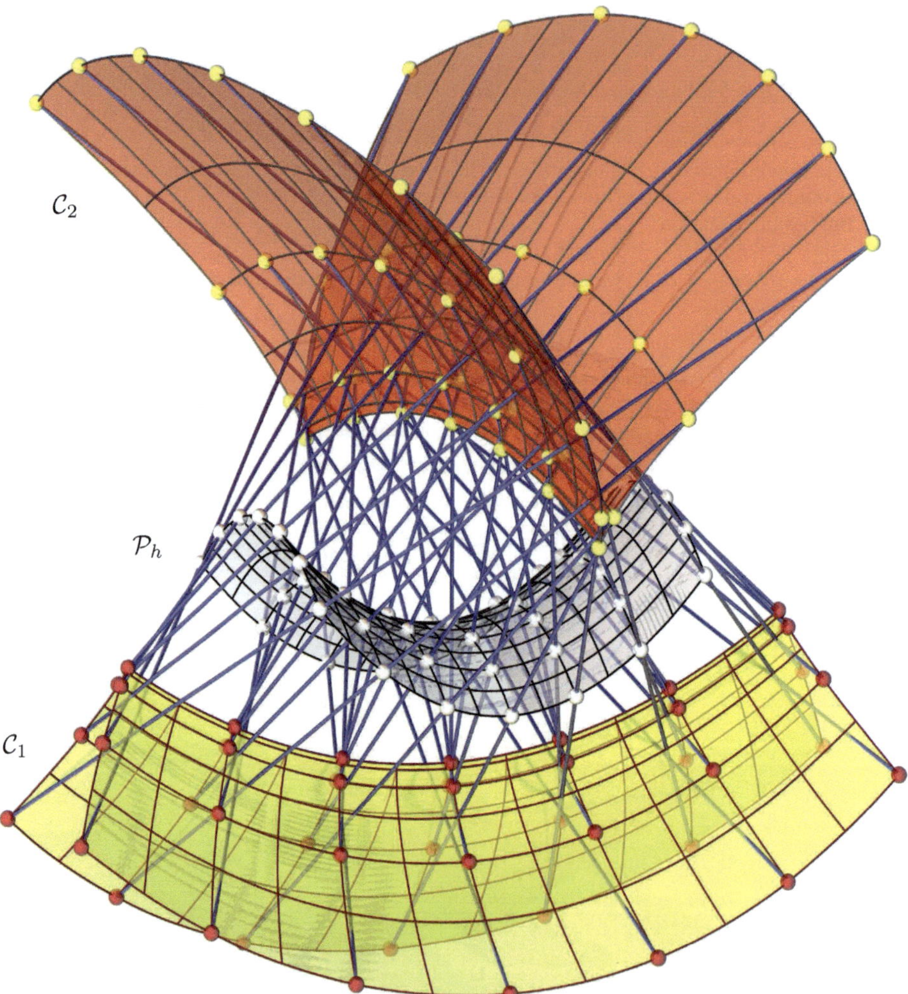

FIGURE 9.25. A hyperbolic paraboloid $\mathcal{P}_h$ with the two sheets of its central surface. Both sheets have a sharp ridge, each of which corresponds to the principal sections of $\mathcal{P}$.

• **Exercise 9.4.5** Central surface of an orthogonal hyperbolic paraboloid.

Show that both sheets of the central surface $\mathcal{C}$ of the hyperbolic paraboloid $\mathcal{P}:\ xy = z$ can be described by the implicit equation of degree nine:

$$\begin{aligned}\mathcal{C}:\ &4xyz(x^2y^2z^2 - 372xyz + 1344) - 27(x^4y^4 + x^4z^4 - y^4z^4)+\\ &+192(xyz+4)(x^4+y^4+z^4) - 6(x^2y^2z^2 + 244xyz - 128)(x^2+y^2-z^2)+\\ &+256(x^6+y^6-z^6) - 48(xyz+58)(x^2y^2 - x^2z^2 - y^2z^2)+\\ &-96(x^4y^2 - x^4z^2 + x^2y^4 + x^2z^4 - y^4z^2 + y^2z^4) + 256 = 0.\end{aligned}$$

Show that this surface has the pair of focal parabolas

$$p_1:\ x = y,\quad y^2 - 4z - 4 = 0 \quad \text{and} \quad p_2: x = -y,\quad y^2 + 4z - 4 = 0$$

as sharp ridges, *i.e.*, as curves of singular points (cf. Figure 9.26).

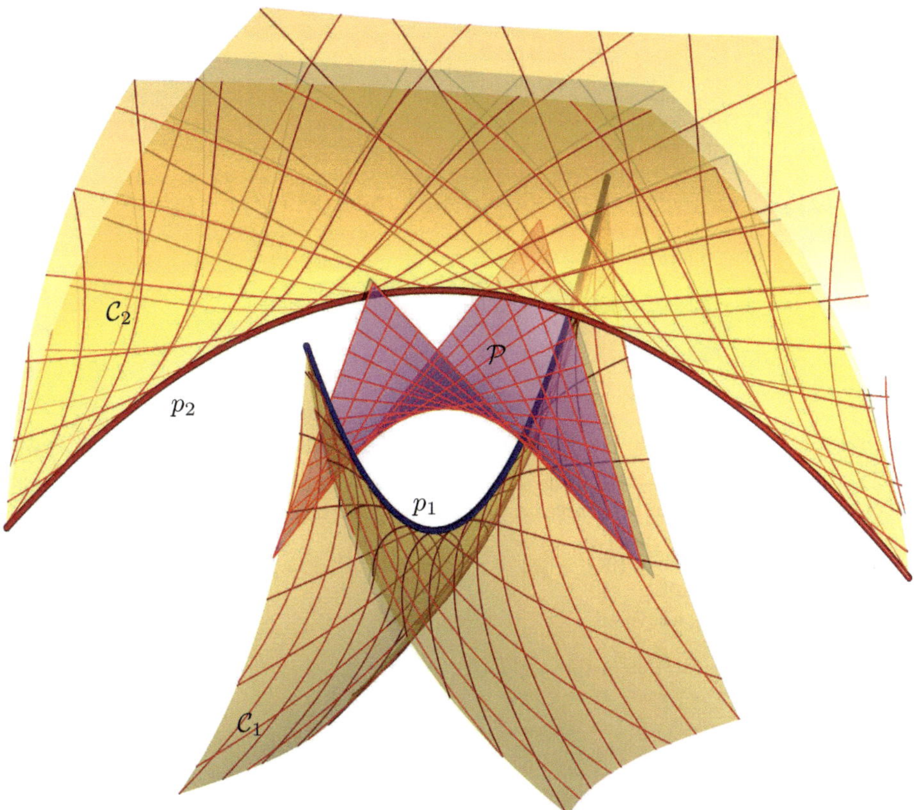

FIGURE 9.26. The two sheets $\mathcal{C}_1$ and $\mathcal{C}_2$ of the central surface $\mathcal{C}$ of the hyperbolic paraboloid $\mathcal{P}$ have parabolas as sets of their singular point. The two parabolas p_1 and p_2 comprise a pair of focal conics.

• **Exercise 9.4.6** Central surfaces of quadrics of revolution.

One sheet of the central surface of a quadric of revolution is a (part of a) straight line.

1. The central surface of the paraboloid of revolution
$$\mathcal{P}:\ x^2+y^2=2pz$$
with $p\in\mathbb{R}^\star$ consists of the half-line $\mathbf{c}_1=\left(0,0,\frac{u^2}{2p}+p\right)$ with $u\in\mathbb{R}$ (or, equivalently $x=y=0$, $z\geq p$). The second part $\mathcal{C}_2$ is a surface of revolution sharing the axis with the paraboloid $\mathcal{P}$. A meridian curve of $\mathcal{C}_2$ is a semi-cubic parabola that is the evolute of the coplanar meridian of $\mathcal{P}$ (see Figure 9.27, left). An equation of $\mathcal{C}_2$ reads
$$8(z-p)^3=27p(x^2+y^2).$$

2. In the case of the one-sheeted hyperboloid of revolution
$$\mathcal{H}_1:\ \frac{x^2+y^2}{a^2}-\frac{z^2}{c^2}=1$$
with $a,c\in\mathbb{R}^\star$, the first branch of the central surface is the z-axis. The doubly curved part $\mathcal{C}_2$ of the central surface can be described by
$$\mathcal{C}_2:\ Q^3+(a^2+c^2)^2(-3Q^2+3(a^2+c^2)^2Q-27a^2c^2(x^2+y^2)z^2-(a^2+c^2)^4)=0,$$
where $Q=-a^2x^2-a^2y^2+c^2z^2$.

3. For the two-sheeted hyperboloid of revolution
$$\mathcal{H}_2:\ \frac{x^2+y^2}{a^2}-\frac{z^2}{c^2}=-1$$
with $a,c\in\mathbb{R}^\star$, the straight part of the central surface equals $\mathbf{c}_1=\left((0,0,\frac{a^2+c^2}{c}\cosh u\right)$ with $u\in\mathbb{R}$ (or $x=y=0$, $z\geq\frac{a^2+c^2}{c}$). The second sheet is a surface of revolution with the evolute of one of $\mathcal{H}_2$'s meridian hyperbolas as its meridian and has an equation of the form
$$\mathcal{C}_2:\ Q^3+(a^2+c^2)^2(3Q^2+3(a^2+c^2)^2Q+27a^2c^2(x^2+y^2)z^2+(a^2+c^2)^4)=0$$
with the same Q as above.

4. The ellipsoid of revolution with $a\neq c$ and $a,c\in\mathbb{R}^\star$
$$\mathcal{E}:\ \frac{x^2+y^2}{a^2}+\frac{z^2}{c^2}=1$$
has the (non-trivial) central surface
$$\mathcal{C}:\ Q^3-(a^2-c^2)^2\left(3Q^2-3(a^2-c^2)Q-27a^2c^2(x^2+y^2)z^2+(a^2-c^2)^4\right)=0,$$
where $Q=a^2x^2+a^2y^2+c^2z^2$.

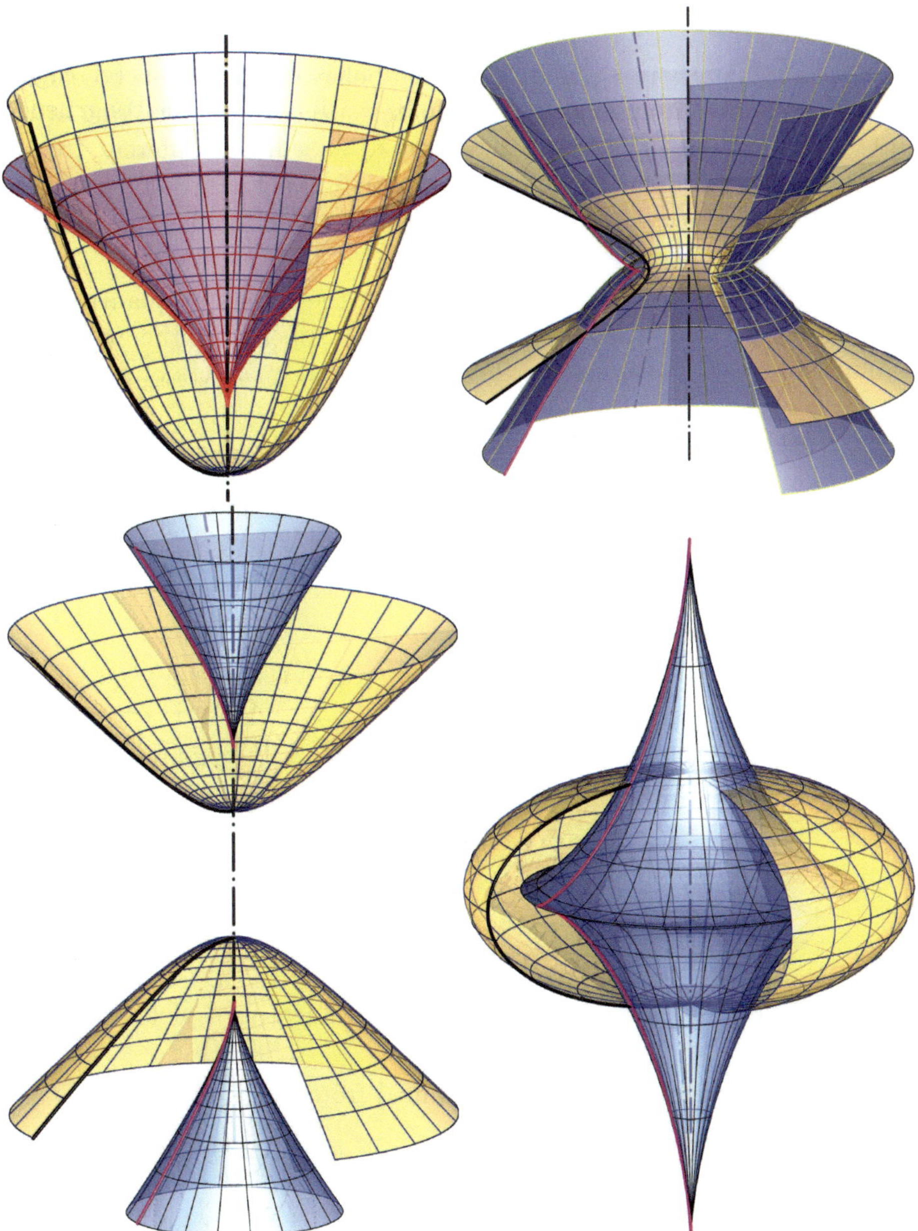

FIGURE 9.27. Quadrics of revolution with their doubly curved central surfaces. Top row: paraboloid (left), one-sheeted hyperboloid (right); bottom row: two-sheeted hyperboloid (left), ellipsoid (right).

Offsets to quadrics

In the case of conics in a plane, it is often believed that offsets to ellipses are ellipses. This impression may originate especially from looking at the outer offset curves. However, the offset of an ellipse (different from a circle) is not an ellipse, as we have seen in [46, p. 117]. The offsets of an ellipse at distances $\pm d$ are two branches of one algebraic curve of degree 8.

It takes a little bit more work to determine the offsets of quadrics than those of conics. Only the offset surfaces of the sphere and the cylinder of revolution are of the same type, *i.e.*, spheres and cylinders of revolution.

Regular central quadrics

First, we shall consider the case of the ellipsoid. For the sake of simplicity, we assume that the ellipsoid is centered at the origin of the coordinate system and has the equation

$$\mathcal{Q}: \ \frac{x^2}{a^2} + \frac{y^2}{b^2} + \frac{z^2}{c^2} = 1. \tag{9.56}$$

The *offset surface* $\mathcal{E}_d$ of $\mathcal{E}$ at distance $d \in \mathbb{R}$ can easily be parametrized as

$$\mathbf{e}_d(u,v,d) = \begin{pmatrix} a\sin u\cos v \\ b\sin u\sin v \\ c\cos u \end{pmatrix} + \frac{d}{W}\begin{pmatrix} \frac{1}{a}\sin u\cos v \\ \frac{1}{b}\sin u\sin v \\ \frac{1}{c}\cos u \end{pmatrix}, \tag{9.57}$$

where $W := \sqrt{\frac{1}{a^2}\sin^2 u\cos^2 v + \frac{1}{b^2}\sin^2 u\sin^2 v + \frac{1}{c^2}\cos^2 u}$. A positive distance $d \in \mathbb{R}$ leads to the outer offsets, while a negative d yields the inner offsets, and $\mathcal{E}_0 = \mathcal{E}$.

An implicit equation of $\mathcal{E}_d$ can be obtained from

$$\begin{aligned} a^2(x - a\sin u\cos v)^2W^2 &= d^2\sin^2 u\cos^2 v, \\ b^2(y - b\sin u\sin v)^2W^2 &= d^2\sin^2 u\sin^2 v, \\ c^2(z - c\cos u)^2W^2 &= d^2\cos^2 u \end{aligned}$$

by substituting $\cos u = \frac{1-U^2}{1+U^2}$, $\sin u = \frac{2U}{1+U^2}$, $\cos v = \frac{1-V^2}{1+V^2}$, $\sin v = \frac{2V}{1+V^2}$, and the subsequent elimination of U, V. It turns out that (9.57) parametrizes a one-parameter family of surfaces of algebraic degree twelve. There is no ellipsoid among them, irrespective of the choice of $d \neq 0$.

Note that the thus obtained equation describes the algebraic completion $\mathcal{E}_d^\star$ of $\mathcal{E}_d$. It also contains $\mathcal{E}_{-d}$, since the distance d appears only squared in the equation.

We will not provide the equation $\mathcal{E}_d^\star$ of the offset surfaces in full length. However, the terms of degree twelve are

$$T_{12} = (x^2+y^2+z^2)^2(a^2b^2z^2+a^2c^2y^2+b^2c^2x^2)^2 \cdot \mathbf{X}^{\mathrm{T}} \underbrace{\begin{pmatrix} \alpha^2 & -\alpha\beta & -\alpha\gamma \\ -\alpha\beta & \beta^2 & -\beta\gamma \\ -\alpha\gamma & -\beta\gamma & \gamma^2 \end{pmatrix}}_{=:\mathbf{Q}} \mathbf{X}, \tag{9.58}$$

where $\mathbf{X} = (x^2, y^2, z^2)$ and $\alpha := b^2 - c^2$, $\beta := c^2 - a^2$, and $\gamma := a^2 - b^2$.

The coordinates x, y, and z can be viewed as homogeneous coordinates in the ideal plane. Then, $T_{12} = 0$ is the equation of the intersection of the offset surface(s) $\mathcal{E}_d^\star$ with the ideal plane. We can see that the offsets of an ellipsoid pass twice through the absolute conic of Euclidean geometry, which makes it a part of the offsets' double curve independent of the offset distance $d \in \mathbb{R}^\star$, since the factor $x^2+y^2+z^2$ appears with multiplicity two and d does not appear in any factor of T_{12}.

Furthermore, the ideal conic of the ellipsoid which is given by the second factor $a^2b^2z^2+a^2c^2y^2+b^2c^2x^2$ in (9.58) (also with multiplicity two) turns out to be a further part of the curve at infinity and the double curve.

The quartic factor $\mathbf{X}^{\mathrm{T}}\mathbf{Q}\mathbf{X}$ can be rewritten as

$$\begin{aligned} &(\alpha x^2 - \beta y^2 - \gamma z^2)^2 - 4\beta\gamma y^2 z^2 = \\ &= (\alpha x^2 - \beta y^2 - \gamma z^2 - 2\sqrt{\beta\gamma}yz)(\alpha x^2 - \beta y^2 - \gamma z^2 + 2\sqrt{\beta\gamma}yz) \end{aligned}$$

and yields the equation of four lines when set equal to zero. In case that either $a = b$, or $b = c$, or $c = a$, $\mathbf{X}^{\mathrm{T}}\mathbf{Q}\mathbf{X} = 0$ is the equation of a pair of lines with multiplicity two.

We can summarize these results in

Corollary 9.4.1 *The ideal curve (intersection with the plane at infinity) of the offset surface of a central quadric $\mathcal{Q}$ is a cycle of algebraic degree* *twelve consisting of the absolute circle of Euclidean geometry (multiplicity two), the ideal conic of $\mathcal{Q}$ (multiplicity two) and four straight lines.*

If $\mathcal{Q}$ is a quadric of revolution, i.e., either $a = b$, or $b = c$, or $c = a$, then the four lines in the cycle form a pair of lines with multiplicity two.

Proof: The result remains to be verified for two- and one-sheeted hyperboloids. This can be done in the same way as for ellipsoids, with some signs of coefficients changing. ■

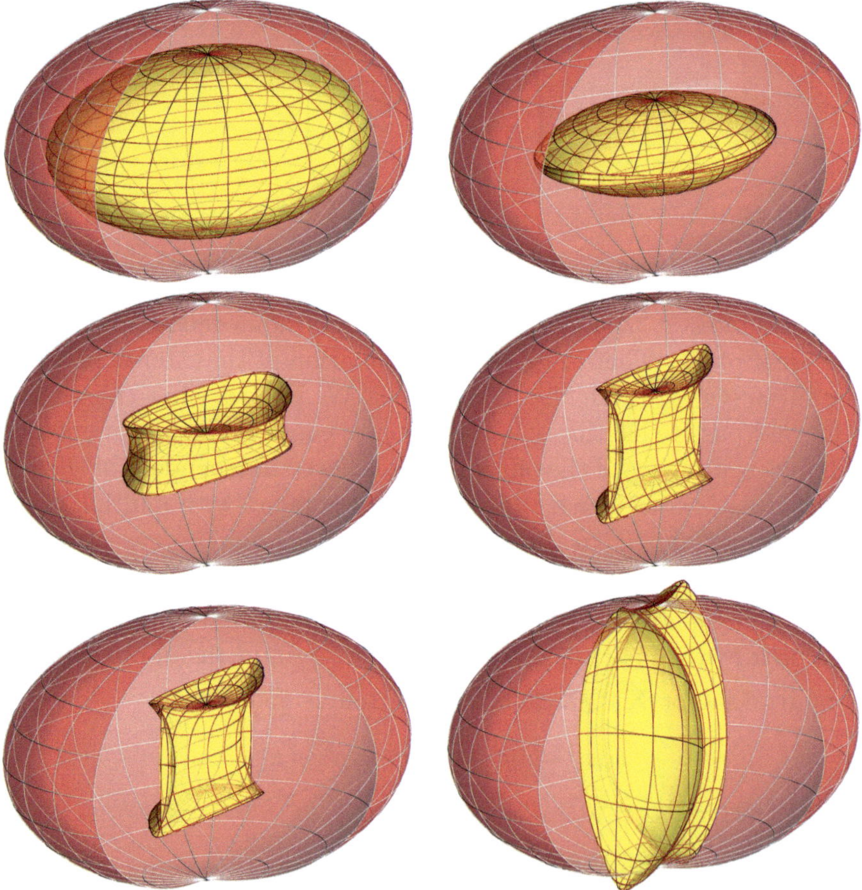

FIGURE 9.28. The red ellipsoid with $a = 2$, $b = 1.5$, and $c = 1$ with offsets at distances $d = -\frac{1}{3}, -\frac{2}{3}, -1, -\frac{4}{3}, -\frac{5}{3}, -2$ (from top to bottom and from left to right).

Figure 9.28 shows some interior offsets of an ellipsoid. (Here, interior means that the surface normal is oriented outwards and the offset distance

d is negative.) As can be seen, the offsets may have singular curves (sharp ridges) or even isolated singularities. The term interior offset may lead to the somewhat wrong impression that interior offsets of an ellipsoid are totally in the interior of the ellipsoid. Choosing a sufficiently large d, the interior offset can be completely outside the quadric $\mathcal{E}$.

The outer offsets (surface normal of points outwards and d is positive) look like ellipsoids, but they are not. If d is growing larger, $\mathcal{E}_d$ increasingly assumes the shape of a sphere, for the difference between the *major axes length* is always $2d = \text{const.}$, but the ratios

$$a + d : b + d : c + d$$

converge to 1 as d approaches infinity.

Once the equation(s) of the ellipsoids offset surfaces is known, it is an elementary task to show

Corollary 9.4.2 *The self-intersections of the offset surface $\mathcal{E}_d^\star$ of an ellipsoid $\mathcal{Q}$ consist of the absolute conic of Euclidean geometry, the ideal conic of the ellipsoid, and three conics in the planes of symmetry of $\mathcal{Q}$:*

$$\begin{aligned} s_{xy}: &\ c^2\alpha x^2 - c^2\beta y^2 + \alpha\beta(c^2 - d^2) = 0, \\ s_{xz}: &\ b^2\alpha x^2 - b^2\gamma z^2 - \alpha\gamma(b^2 - d^2) = 0, \\ s_{yz}: &\ a^2\beta y^2 - a^2\gamma z^2 + \beta\gamma(a^2 - d^2) = 0. \end{aligned}$$

Proof: From Corollary 9.4.1, we know that the absolute conic together with the ideal conic of $\mathcal{Q}$ are part of the double curve.

The three conics mentioned above come along as the intersections of the offset surface $\mathcal{E}_d^\star$ with the planes of symmetry of $\mathcal{Q}$ $z = 0$, $y = 0$, and $x = 0$, which are also the planes of symmetry of $\mathcal{Q}$'s offsets. The conics' equations appear as factors with multiplicity two.

The remaining factors have multiplicity one and are the offset curves of the principal conics on $\mathcal{Q}$ joining $\mathcal{Q}$'s vertices. ■

The intersections of the offset surfaces $\mathcal{E}_d^\star$ with the three planes of symmetry do not only contain the conics mentioned in Corollary 9.4.2. They also contain the offset curves of the principal sections of the ellipsoid. This also holds true for any other quadric, no matter if it is regular or singular.

The three conics given in Corollary 9.4.2 are not to be confused with the sharp ridges on the offset surfaces. These sharp ridges consist of singular points of the offset surfaces and are the analogues to cusps on the evolutes of planar curves.

Offsets of paraboloids

The offset surfaces $\mathcal{P}_d$ and $\mathcal{P}_{-d}$ of the paraboloids

$$\mathcal{P}: \frac{x^2}{a^2} + \frac{y^2}{b^2} = 2z$$

at distance $\pm d \in \mathbb{R} \setminus \{0\}$ belong to an algebraic surface $\mathcal{P}_d^\star$ of degree ten. We do not provide their equations in full length, because only little can be read off from them.

The leading terms are independent of d and yield an equation of the intersection $\mathcal{P}_\omega = \mathcal{P}_d \cap \omega$ of the offset surfaces with the plane ω at infinity

$$\mathcal{P}_\omega: (x^2 + y^2 + z^2)(x^2 + y^2)^2(a^2y^2 + b^2x^2)^2 = 0,$$

which tells us that the offset of paraboloids always pass through the absolute conic with the equation $x^2 + y^2 + z^2 = 0$. There are two factors with multiplicity two in $\mathcal{P}$'s equation:

(1) $x^2 + y^2 = (x + \mathrm{i}y)(x - \mathrm{i}y) = 0$ is the equation of a complex conjugate pair of ideal lines passing through the ideal point of the z-axis.

(2) The factor $a^2y^2 + b^2x^2 = (ay + \mathrm{i}bx)(ay - \mathrm{i}bx) = 0$ describes a further pair of complex conjugate lines. The latter lines are the paraboloid's rulings in the ideal plane. Clearly, these rulings form a complex conjugate pair if $\mathcal{P}$ is elliptic, *i.e.*, $b^2 > 0$. In the case of a hyperbolic paraboloid, these lines form a real pair.

It is worth noting that paraboloids are *LN surfaces*, which means that the coordinate functions of the normal vector field derived from a certain parametrization are linear functions in the surface parameters. The elliptic as well as the hyperbolic paraboloid share this property with all quadratically parametrizable triangular Bézier surfaces (see Section 8.4).

The standard parametrizations

$$\mathbf{p}(u, v) = (2au, 2bv, 2(u^2 \pm v^2)) \quad \text{with} \quad (u, v) \in \mathbb{R}$$

and $ab \neq 0$ for the elliptic and hyperbolic paraboloid $\frac{x^2}{a^2} \pm \frac{y^2}{b^2} = 2z$ yield the normal vector field

$$\mathbf{n}(u, v) = (-2bu, \mp 2av, ab).$$

A complete classification of quadratically parametrizable triangular Bézier surfaces can be found in [104].

Sometimes it is necessary to reparametrize a surface such that the normal vector field becomes linear (see [102]). It turned out that these reparametrizations are, in principal, planar quadratic Cremona transformations (cf. [46, p. 329 ff.]) applied in the parameter domain of the quadratically parametrized surface. All three types of planar quadratic Cremona transformations occur when reparametrizing all eight types of quadratically parametrizable triangular Bézier surfaces in 3-space. This includes parabolic cylinders, quadratic cones, paraboloids, the Whitney umbrella (an affine version of Plücker's conoid), and Cayley's cubic ruled surface.

LN surfaces could be seen as generalizations of paraboloids. If the dual surface $\widehat{\mathcal{F}}$ of an LN surface $\mathcal{F}$ is of algebraic degree n, *i.e.*, $\mathcal{F}$ is of class n, then the ideal plane ω is an $(n-1)$-*fold point* of $\widehat{\mathcal{F}}$. Paraboloids are of class two and the ideal plane is a tangent plane.

Singular central quadrics

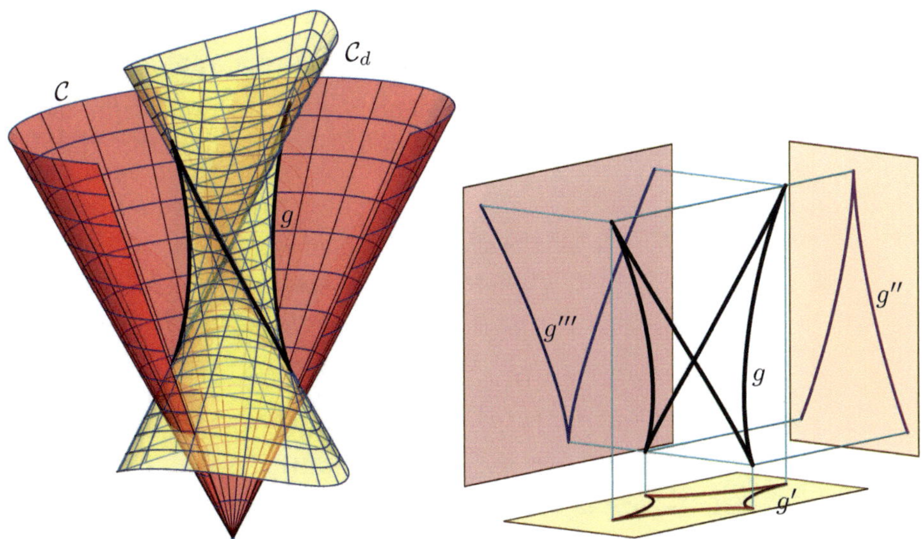

FIGURE 9.29. Left: Offset $\mathcal{C}_d$ of an elliptic (not right circular) cone $\mathcal{C}$ with $a = 2$, $b = 1.5$, $c = 1$ at distance $d = -\frac{2}{3}$. The offset is a developable (neither a cone nor a cylinder) with a non-trivial curve g of regression and has self-intersections. Right: The curve g of regression and its three principal views.

Each cone is a developable (ruled) surface and so are quadratic cones. Thus, a quadratic cone $\mathcal{C}$ is also the envelope of a one-parameter family of planes. The construction of the offset surface $\mathcal{C}_d$ of $\mathcal{C}$ at d can be viewed in two ways:

(1) It is the envelope of all spheres of constant (and fixed) radius $d \in \mathbb{R}$ centered in the points of $\mathcal{C}$

(2) It is the enevelope of the one-parameter family of planes at distance d from the tangent planes of $\mathcal{C}$ (and parallel to the tangent planes of $\mathcal{C}$).

Therefore, offset surfaces of (quadratic) cones are developable.

The offset $\mathcal{C}_d$ of a cone $\mathcal{C}$ will, in general, no longer be a cone. Only if $\mathcal{C}$ is a cone of revolution, then its offset is also a cone of revolution. Figure 9.29 shows a quadratic cone with one of its offsets.

The offsets of quadratic cones can be parametrized in the same manner as shown for the ellipsoid in (9.57). Subsequently, an implicit equation can be derived, as shown for the ellipsoid below (9.57). It turns out that the algebraic completion $\mathcal{C}_d^\star$ of the offset $\mathcal{C}_d$ and $\mathcal{C}_{-d}$ of the quadratic cone

$$\mathcal{C}: \ \frac{x^2}{a^2} + \frac{y^2}{b^2} - \frac{z^2}{c^2} = 0 \tag{9.59}$$

is an algebraic surface of degree eight. Again, we have make use of the abbreviations $\alpha := b^2 + c^2$, $\beta := c^2 + a^2$, and $\gamma := a^2 - b^2$ and define additionally $C := a^2\alpha + b^2\beta$, $B := a^2\alpha + c^2\gamma$, $A := b^2\beta - c^2\gamma$. Then, the equation of $\mathcal{C}_d^\star$ reads

$$\begin{gathered}
\mathcal{C}_d^\star: \ b^4c^4\alpha^2x^8 + a^4c^4\beta^2y^8 + a^4b^4\gamma^2z^8 + \\
+(c^4x^4y^4 + d^4\gamma^2z^4)(c^4\gamma^2 + 6a^2b^2\alpha\beta) + \\
+(b^4x^4z^4 + d^4\beta^2y^4)(b^4\beta^2 + 6a^2c^2\alpha\gamma) + \\
+(a^4y^4z^4 + d^4\alpha^2x^4)(a^4\alpha^2 - 6b^2c^2\beta\gamma) + \\
+2c^4C(b^2\alpha x^4 + a^2\beta y^4)x^2y^2 - 2b^4B(c^2\alpha x^4 + a^2\gamma z^4)x^2z^2 + \\
-2a^4A(c^2\beta y^4 - b^2\gamma z^4)y^2z^2 + \\
-2b^2c^2d^2\alpha^2Ax^6 - 2a^2c^2d^2\beta^2By^6 - 2a^2b^2d^2\gamma^2Cz^6 - 2d^2CABx^2y^2z^2 + \\
-2(b^2c^2x^4y^2z^2 - b^2d^2\gamma z^4x^2 + c^2d^2\beta y^4x^2 - d^4\beta\gamma z^2y^2)(3a^4\alpha^2 - b^2c^2\beta\gamma) + \\
-2(a^2c^2y^4x^2z^2 + c^2d^2\alpha x^4y^2 + a^2d^2\gamma z^4y^2 + d^4\alpha\gamma x^2z^2)(3b^4\beta^2 + a^2c^2\alpha\gamma) + \\
+2(a^2b^2z^4x^2y^2 - a^2d^2\beta y^4z^2 - b^2d^2\alpha x^4z^2 + d^4\alpha\beta x^2y^2)(3c^4\gamma^2 + a^2b^2\alpha\beta) + \\
-2d^6\alpha\beta\gamma^2Cx^2 - 2d^6\alpha\beta^2\gamma By^2 + 2d^6\alpha^2\beta\gamma Az^2 + d^8\alpha^2\beta^2\gamma^2 = 0.
\end{gathered}$$

The equation of the ideal curve $\mathcal{C}_\infty$, *i.e.*, the intersection of $\mathcal{C}_d^\star$ with the plane ω at infinity

$$\mathcal{C}_\infty: \ \left((\alpha x^2+\beta y^2+\gamma z^2)^2-4\alpha\gamma x^2z^2\right)(a^2b^2z^2-a^2c^2y^2-b^2c^2x^2)^2$$

shows that the ideal conic of $\mathcal{C}$ and $\mathcal{C}_d^\star$ is a curve of multiplicity two on all offset surfaces.

Since $\mathcal{C}_d$ is a developable, it has a curve g_d of regression. The curve g determines the surface $\mathcal{C}_d$, for it is the envelope of g_d's osculating planes.

A parametrization $\mathbf{g}_d: I\subset\mathbb{R}\to\mathbb{R}^3$ of the curve g_d of regression can be computed from the dual representation $\widehat{\mathcal{C}}_d$ of $\mathcal{C}$ or from a parametrization of $\mathcal{C}_d$. The dual representation can be given in the form

$$\widehat{\mathcal{C}}_d: \ \langle\mathbf{x}-(\mathbf{c}+d\,\mathbf{n}),\mathbf{n}\rangle=0,$$

where $d\in\mathbb{R}$ is the offset distance, $\mathbf{x}=(x,y,z)$, $\mathbf{c}$ is a parametrization of the cone $\mathcal{C}$ (*e.g.*, $\mathbf{c}=(av\cos u, bv\sin u, cv)$), and $\mathbf{n}$ is the unit normal vector field of $\mathcal{C}$ derived from the parametrization $\mathbf{c}$. $\widehat{\mathcal{C}}_d$ depends on one parameter and for each parameter, say $u_0\in I$, it delivers the equation of a tangent plane of the offset surface $\mathcal{C}_d$, provided that $d\in\mathbb{R}$ is fixed.

The curve g_d of regression (for fixed d) consists of the common points of $\widehat{\mathcal{C}}_d$ and its first and second derivative $\dot{\widehat{\mathcal{C}}}_d$ and $\ddot{\widehat{\mathcal{C}}}_d$. We skip the details of this lengthy, but rather simple computation. So, we arrive at

$$\mathbf{g}_d(u)=\frac{d}{\sqrt{(a^2\alpha-c^2\gamma\cos^2u)^3}}\begin{pmatrix}-cb\beta\gamma\cos^3u\\ ac\gamma\alpha\sin^3u\\ -ab\alpha\beta\end{pmatrix}. \tag{9.60}$$

We summarize our results in

Theorem 9.4.2 *The offset surface $\mathcal{C}_d$ at distance $d\in\mathbb{R}\setminus\{0\}$ of the quadratic cone $\mathcal{C}$ given by* (9.59) *is a developable ruled surface. Its algebraic completion $\mathcal{C}_d^\star$ is of degree eight and passes twice through the ideal conic of $\mathcal{C}$. The curve of regression of $\widehat{\mathcal{C}}_d$ is an algebraic curve of degree 6 and consists of two disjoint parts parametrizable by* (9.60) *as $\mathbf{g}_d$ and $\mathbf{g}_{-d}$.*

Finally, we shall note that offsets of quadratic cylinders are cylinders. Offsets of quadratic cylinders are quadratic only if the cylinders are cylinders of revolution. Offsets of elliptic, parabolic, hyperbolic cylinders are cylinders whose profile curves (cross sections in planes orthogonal to the generators) are offsets to conics. From [46, p. 118 & 119], we know that offsets of ellipses and hyperbolas are algebraic curves of degree eight, while those of parabolas are only of degree six. The same holds true for offset surfaces of elliptic, hyperbolic, and parabolic cylinders.

One-parameter families of quadrics

A one-parameter family of quadrics can simply be defined by the equations of all quadrics in the family: Assume that all coefficients in the equation of a quadric are univariate functions. Smoothness and differentiability of the family of quadrics is then defined by the smoothness and differentiability of the coefficient functions.

In what follows, we shall not deal with the most general case. We will select some simple examples that have applications in various fields such as the production of pipings and milling. Some inner geometric applications, *i.e.*, applications to geometric problems and explanations of geometric facts will appear in a natural way.

Let $\mathbf{A}(t): I \subset \mathbb{R} \to \mathbb{R}^{4\times 4}$ be a one-parameter family of (not necessarily regular) symmetric 4×4 matrices. If now $\mathbf{x} = (x_0, x_1, x_2, x_3)$ are homogeneous coordinates in $\mathbb{P}^3(\mathbb{R})$, then

$$\mathcal{Q}(t): \ \mathbf{x}^{\mathrm{T}}\mathbf{A}(t)\mathbf{x} = 0 \tag{9.61}$$

are the equations of the quadrics in the one-parameter family. Usually, such a family of quadrics has an envelope.

Although (9.61) depends on one parameter, an implicit equation of the envelope can be derived. For that purpose, we compute the derivative of (9.61) with respect to the parameter t and denote it by

$$\dot{\mathcal{Q}}(t): \ \mathbf{x}^{\mathrm{T}}\dot{\mathbf{A}}(t)\mathbf{x} = 0. \tag{9.62}$$

The points of the envelope satisfy both (9.61) and (9.62). Thus, we *only* have to eliminate the parameter t from both equations.

The intersection curve $c = \mathcal{Q} \cap \dot{\mathcal{Q}}$ is called the *characteristic curve* at t. The quadric $\mathcal{Q}$ touches the envelope along c. Since $\mathcal{Q}$ is a quadric, its derivative will, in general, also be a quadric. (Later, when we deal with special examples, we will mostly confine ourselves to cases where $\dot{\mathcal{Q}}$ is a plane.) Consequently, the characteristics of envelopes are algebraic curves which are of degree four at most.

If the coordinate functions of $\mathbf{A}(t)$ in (9.61) are of degree

$$n := \max_{i,j}(\deg(a_{ij}(t))),$$

then the degree of the envelope $\mathcal{E}$ is bounded by

$$\deg(\mathcal{E}) \leq 6n - 2.$$

In the vast majority of examples that we will see, the actual degree of the envelope is much lower than this upper bound.

An implicit equation of the envelope may be interesting from the theoretical point of view. In practical applications, an implicit equation of relatively high degree may not be satisfactory. Moreover, most CAD systems cannot deal with such implicit equations. In order to overcome such difficulties, a lot of research has been done on this topic (see [101]).

Envelopes of spheres – one-parameter families

The setting of the previous section will now be simplified. We deal with Euclidean spheres in three-dimensional space $\mathbb{R}^3$. It is sufficient to use Cartesian coordinates $\mathbf{x} = (x, y, z)$ and to assume that $\mathbf{m}(t) = (m_x(t), m_y(t), m_z(t)) : I \subset \mathbb{R} \to \mathbb{R}^3$ is a parametrization of a curve m, and further that $r(t) : I \to \mathbb{R}$ is a function defined over the same interval. The curve m is usually called the *spine curve* of the envelope $\mathcal{E}$ of the spheres

$$\mathcal{S} : \langle \mathbf{x} - \mathbf{m}, \mathbf{x} - \mathbf{m} \rangle - r^2 = 0. \tag{9.63}$$

According to the previous section, the envelope of (9.63) is found by intersecting $\mathcal{S}(t)$ with $\dot{\mathcal{S}}$ for all $t \in I$. We find that the derivatives

$$\dot{\mathcal{S}} : \langle \dot{\mathbf{m}}, \mathbf{x} - \mathbf{m} \rangle + \dot{r}\, r = 0 \tag{9.64}$$

are planes. Therefore, the characteristic curves of the envelope $\mathcal{E}$ of a one-parameter family of spheres are circles. Here, we have to distinguish between two cases:

(1) The radius function is constant. Then, the plane (9.64) contains the center M of the sphere for all $t \in I$. Thus, the envelope $\mathcal{E}$ can be viewed as the *offset* of the spine curve. These envelopes are called *pipe surfaces.* Since the carrier plane $\dot{\mathcal{S}}$ of the characteristic circle c contains the center of the sphere (at each $t \in I$), the characteristic circle is a great circle of the sphere and the envelope $\mathcal{E}$ is covered by a one-parameter family of

congruent circles. So, there is a one-parameter family of distinguished conics (circles) on each pipe surface.

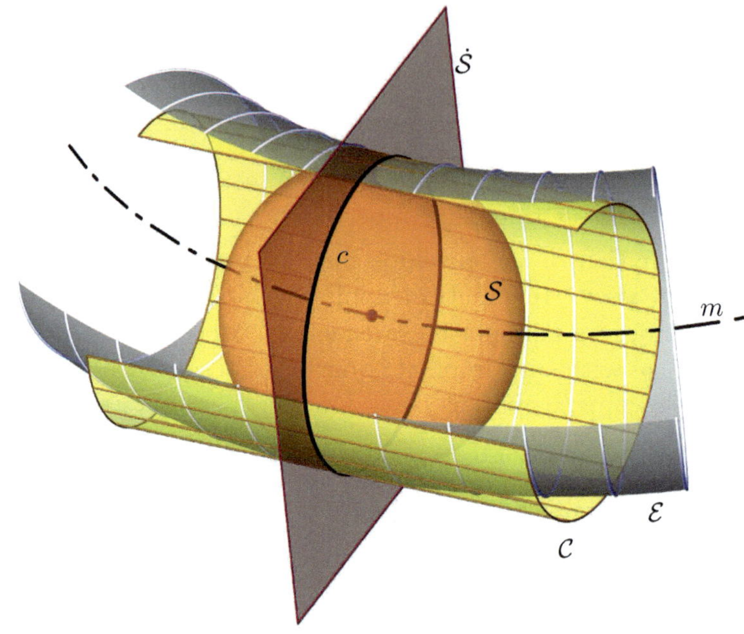

FIGURE 9.30. A pipe surface $\mathcal{E}$ can be seen as an envelope in two different ways: as the envelope of a one-parameter family $\mathcal{S}(t)$ of spheres or as the envelope of a one-parameter family of cylinders $\mathcal{C}(t)$ of revolution. The characteristic circles c are curvature lines on $\mathcal{E}$.

At each $t \in I$, the envelope $\mathcal{E}$ and the sphere $\mathcal{S}(t)$ touch along a great circle c of $\mathcal{S}(t)$. Therefore, the tangent planes of $\mathcal{E}$ along c envelope a cylinder $\mathcal{C}$ of revolution (cf. Figure 9.30). Consequently, a pipe surface is also the envelope of a certain one-parameter family of congruent cylinders of revolution.

The surface normals of $\mathcal{E}$ along c constitute a pencil of lines about the center of $\mathcal{S}$. Therefore, the ruled surface of normals along c is developable, and thus, the characteristic curves on $\mathcal{E}$ are lines of curvature.

(2) If the radius function r is not constant, then $\dot{r} \neq 0$ and the plane (9.64) does not pass through the sphere's center. $\dot{\mathcal{S}}$ meets $\mathcal{S}$ along a small circle c, the common tangent planes of $\mathcal{S}$ and the envelope $\mathcal{E}$ along c envelope a cone Γ_c of revolution (cf. Figure 9.31). In this case, the envelope $\mathcal{E}$

belongs to the class of *channel surfaces.* Channel surfaces are also the envelopes of certain one-parameter families of cones of revolution.

It is worth pointing out that envelopes of one-parameter families of cones of revolution are not necessarily channel surfaces. We shall have a look at surfaces of constant slope later in Section 9.4.

The center N of c can be found as the intersection of $\dot{\mathcal{S}}$'s normal $\mathbf{n}(w) = \mathbf{m} + w\dot{\mathbf{m}}$ through M and is, therefore,

$$\mathbf{n} = \mathbf{m} - \frac{\dot{r}\,r}{\langle \dot{\mathbf{m}}, \dot{\mathbf{m}} \rangle}\dot{\mathbf{m}}.$$

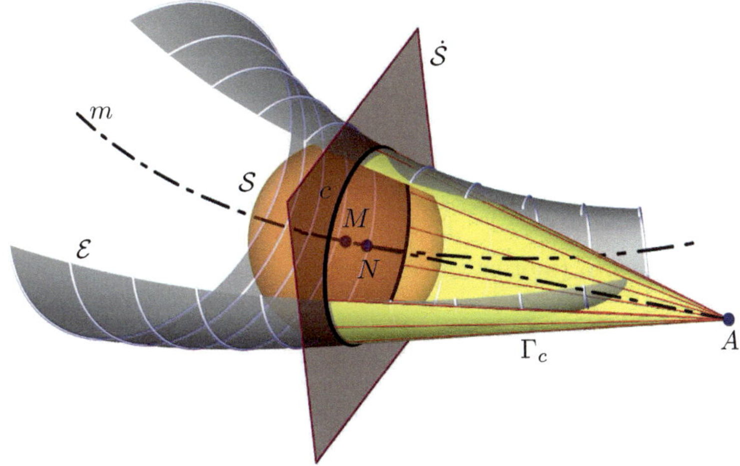

FIGURE 9.31. The envelope of a one-parameter family of spheres with variable radius is a channel surface. Each sphere $\mathcal{S}$ of the family touches the envelope $\mathcal{E}$ along a characteristic circle c whose carrier plane $\dot{\mathcal{S}}$ does not contain the spheres center. Along the characteristic circle c, both $\mathcal{E}$ and $\mathcal{S}$ are tangent to a cone Γ_c of revolution. The characteristics of $\mathcal{E}$ comprise one family of lines of curvature on $\mathcal{E}$.

The characteristic circle is of radius

$$r_c = \sqrt{r^2 - \overline{MN}^2} = r\sqrt{1 - \frac{\dot{r}^2}{\langle \dot{\mathbf{m}}, \dot{\mathbf{m}} \rangle}}.$$

Note that in the case $\overline{MN} > r$ the sphere $\mathcal{S}$ has no real contact with the envelope. The apex A of the tangent cone Γ_c is the pole of $\dot{\mathcal{S}}$ with respect to $\mathcal{S}$.

We insert $\mathbf{n} = \mathbf{m} + w\,\dot{\mathbf{m}}$ into the polar form $\langle \mathbf{x} - \mathbf{m}, \mathbf{x}' - \mathbf{m}\rangle = r^2$ of $\mathcal{S}$ and find the polar plane π_N of N w.r.t. $\mathcal{S}$ as $\dot{r}\langle \mathbf{x} - \mathbf{m}, \dot{\mathbf{m}}\rangle + r\langle \dot{\mathbf{m}}, \dot{\mathbf{m}}\rangle = 0$. The intersection of $\mathbf{n} = \mathbf{m} + w\,\dot{\mathbf{m}}$ and π_N yields Γ_c's apex A with the coordinate vector

$$\mathbf{a} = \mathbf{m} - \frac{r}{\dot{r}}\dot{\mathbf{m}}. \tag{9.65}$$

Clearly, the surface normals of $\mathcal{E}$ along c also form a cone of revolution, which makes c a line of curvature on $\mathcal{E}$.

Among channel surfaces, we find a further class of special surfaces: Assume that the spine curve is a straight line parametrized by $\mathbf{m}(t) = (0, 0, t)$ (which means no loss of generality, it is only an appropriate choice of the coordinate system). Then, (9.63) simplifies to

$$\mathcal{S}:\ x^2 + y^2 + (z - t)^2 = r^2(t) \tag{9.66}$$

and (9.64) becomes

$$\dot{\mathcal{S}}:\ -(z - t) = \dot{r}(t)\,r(t). \tag{9.67}$$

Inserting (9.67) into (9.66) yields

$$x^2 + y^2 = r^2(t)(1 - \dot{r}^2(t))\,t^{-2},$$

which gives rise to a representation of the envelope $\mathcal{E}$ in cylinder coordinates

$$\mathcal{E}:\ \varrho(t) = r(t)\sqrt{1 - \dot{r}^2(t)}, \quad \text{and} \quad z(t) = t - r(t)\dot{r}(t). \tag{9.68}$$

The surfaces described by (9.68) are invariant w.r.t. all rotations about the z-axis, since all intersections with planes $x =$ const. are circles. These surfaces are usually referred to as *surfaces of revolution*. From the explanations in (2), it is clear that surfaces of revolution are also envelopes of one-parameter families of coaxial cones of revolution. Only the vertex and the angle of aperture change, see Figure (9.32).

Further, the characteristic circles – here usually referred to as parallel circles – comprise one family of lines of curvature. Clearly, the intersections of $\mathcal{E}$ with planes through the axis a – called meridians – are orthogonal trajectories of the parallel circles, and thus, they form the second family of lines of curvature on $\mathcal{E}$.

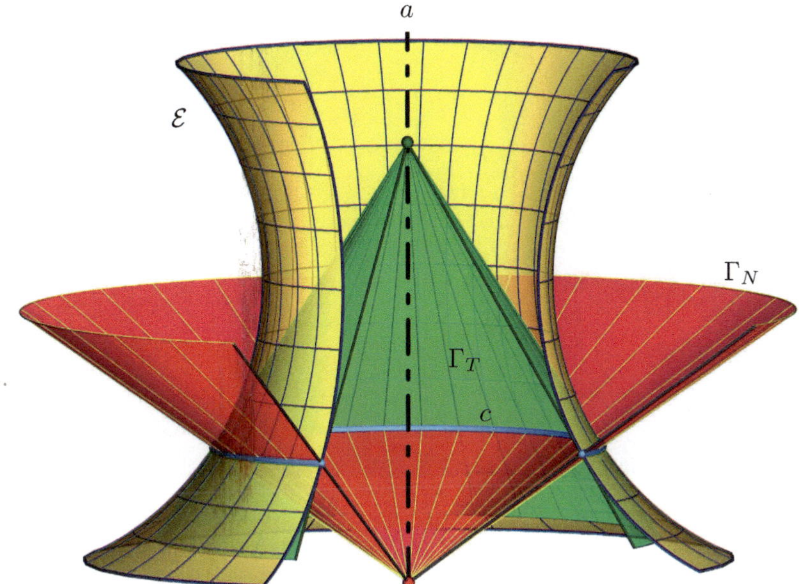

FIGURE 9.32. A surface $\mathcal{E}$ of revolution is a channel surface, for it is the envelope of a one-parameter family of coaxial spheres. $\mathcal{E}$ is also the envelope of a one-parameter family of coaxial cones Γ_T of revolution. Along each parallel circle c, the surface normals of $\mathcal{E}$ form a cone Γ_N of revolution. The axis a of $\mathcal{E}$ is the common axis of all tangent cones and normal cones of $\mathcal{E}$ along parallel circles. The parallel circles and the meridians are the lines of curvature on $\mathcal{E}$.

Rolling ball blends

Pipe surfaces appear frequently in pipe manufacturing. There, a smooth join between two (not necessarily) cylindrical pipes $\mathcal{P}_1$ and $\mathcal{P}_2$ is to be modeled in order to make the transition smooth, without creases, and waterproof or even steam-tight. However, rolling ball blends are dense, which is important for piping under pressure.

The technique uses the following idea: While a sphere $\mathcal{S}$ with constant radius d is assumed to roll on both pipes simultaneously, it leaves a trace c_i on either pipe. Precisely that part of the envelope $\mathcal{E}$ of all spheres in this one-parameter family that lies between the two traces c_1 and c_2 is the *blending surface*. In the case of a constant radius, the pipe surface (part) $\mathcal{E}$ is called *constant radius rolling ball blend*; otherwise it is called a *variable radius rolling ball blend*. The latter are studied especially for those cases where planes and natural quadrics are to be joined (cf. [80]).

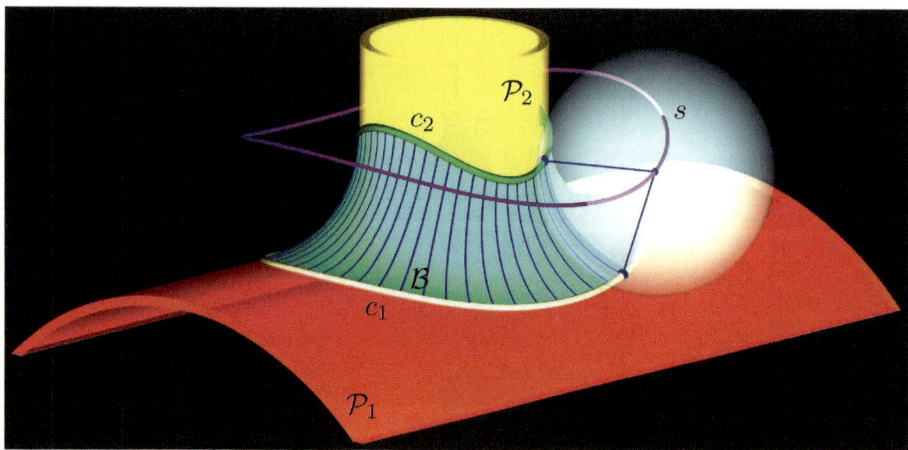

FIGURE 9.33. Rolling ball blends with constant radius are pipe surfaces. The spine curve of the pipe is the intersection of the offsets of the surfaces to blend. Each characteristic curve joins two points of contact of the blending surface $\mathcal{B}$ and the surfaces $\mathcal{P}_1$ and $\mathcal{P}_2$ to blend.

We shall discuss the case of a constant radius in more detail. The spine c of the blending pipe surface $\mathcal{E}$ is the intersection of the offset surfaces $\mathcal{P}_1(d)$ and $\mathcal{P}_2(d)$ of the two pipes at distance d (see Figure 9.34). In the case of two cylinders of revolution $\mathcal{P}_1$ and $\mathcal{P}_2$, the offsets are again cylinders of revolution and the intersection is a quartic curve. (Degeneracies will not occur with ordinary pipe connections.) The curves c_1 and c_2 of contact between the rolling ball blend $\mathcal{E}$ and the pipes are the set of pedal points from s to $\mathcal{P}_1$ and $\mathcal{P}_2$. The blend $\mathcal{B} \subset \mathcal{E}$ is that part of the surface $\mathcal{E}$ enveloped by a one-parameter family of congruent spheres of radius d that lies between the two contact curves c_1 and c_2.

The construction (shown in Figure 9.34) can be enhanced by adding the tangents to the contact curves. Assuming a fixed speed of the rolling ball at a certain position, the tangent vector (velocity vector) can be decomposed into its two components in the directions of the tangents to the respective contact curves. Here, in the case of cylinders, an axial similarity transforms the two components into the tangent vectors at the corresponding points of contact on c_1 and c_2.

An alternative approach to the tangents of c_1 and c_2 uses the fact that the Dupin indicatrices (see [46, p. 120]) of pairs of surfaces that are in line contact form a pencil of conics of the third kind (cf. [46, p. 287]).

The indicatrix $i_{\mathcal{P}_i}$ of each cylinder $\mathcal{P}_i$ is a pair of parallel lines, while the indicatrices $i_{\mathcal{E}}$ of $\mathcal{E}$ at each point of c_i is a hyperbola (at least in that part showing up in Figure 9.34). The common chord of $i_{\mathcal{P}_i}$ and $i_{\mathcal{E}}$ is the tangent to c_i.

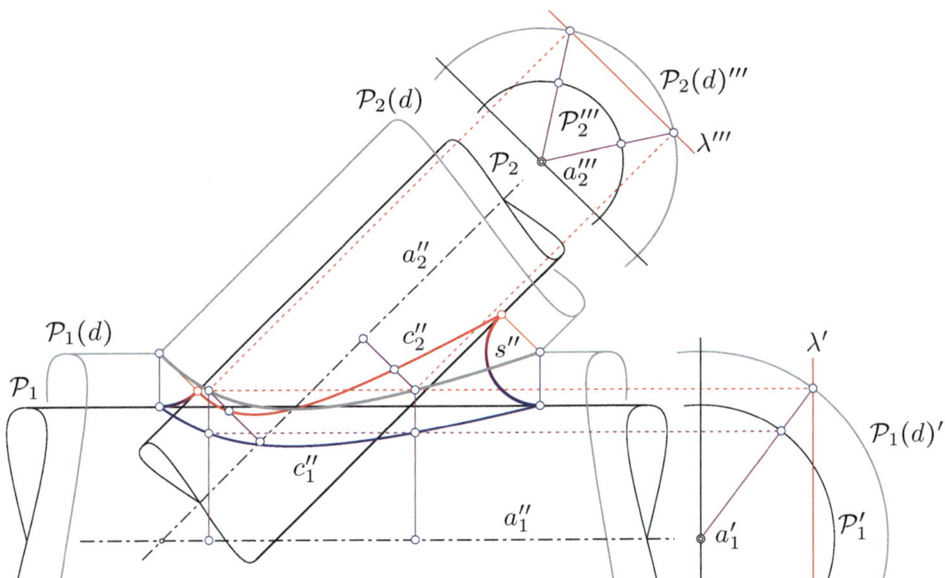

FIGURE 9.34. Construction of the rolling ball blend between two cylinders $\mathcal{P}_1$, $\mathcal{P}_2$ of revolution with intersecting axes a_1 and a_2 and radii r_1 and r_2: The radius d of the rolling sphere equals the offset distance of either cylinder. The offset $\mathcal{P}_i(d)$, $i = 1, 2$ (cylinders of revolution with radii $r_i + d$) intersect in the spine curve s of the blending pipe surface $\mathcal{E}$. The displayed construction uses the orthogonal projection onto the plane $[a_1, a_2]$ spanned by a_1 and a_2 (front view) with two different edge views that show either axis as a point, *i.e.*, a_1' and a_2'''. Planes $\lambda \parallel [a_1, a_2]$ intersect the offsets $\mathcal{P}_i(d)$ along generators which appear as points in the edge views, but as lines in the front view. The points of the spine curve s are constructed as the intersections of generators of $\mathcal{P}_i(d)$. The points of the contact curves c_i between the rolling balls and either $\mathcal{P}_i$ lie on the cylinder's normals through the points on the spine curve.

Figure 9.33 shows a rolling ball blend between two cylinders of revolution. This example is sufficient in order to explain the construction of a rolling ball blend.

Natural channel surface

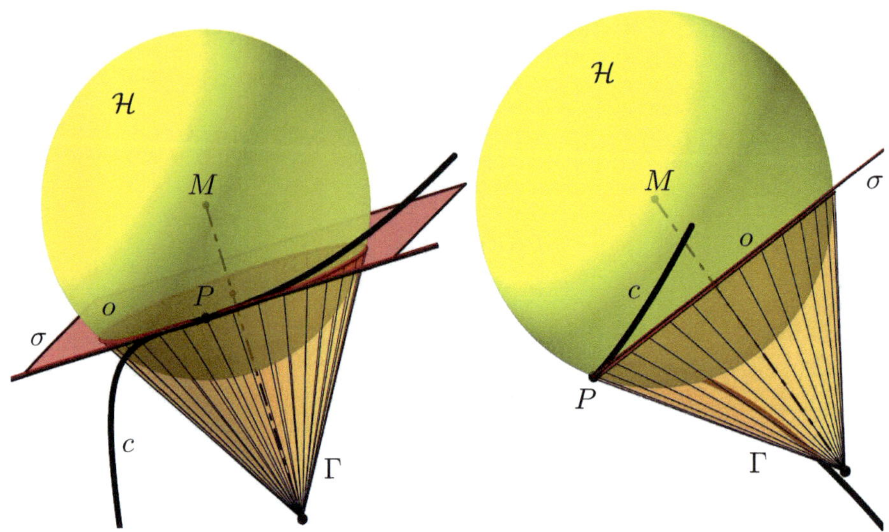

FIGURE 9.35. Left: The hyperosculating sphere $\mathcal{H}$ of c at a regular point P (that is neither a point of inflection nor a handle point). The osculating circle o lies on $\mathcal{H}$ and in the osculating plane σ of c at P. The tangent of c is tangent to o and $\mathcal{H}$ as well. Right: $\mathcal{H}$ is tangent to a cone of revolution Γ along o. The cone Γ is also tangent to the natural channel surface of c (cf. Theorem 9.4.4).

Assume that a C^3 space curve c is given by an arc length parametrization $\mathbf{c} : I \subset \mathbb{R} \to \mathbb{R}^3$. With $\{\mathbf{c}_1, \mathbf{c}_2, \mathbf{c}_3\}$ (satisfying $\langle \mathbf{c}_i, \mathbf{c}_j \rangle = \delta_{ij}$), we denote the Frenet frame of c and the derivatives of these vectors are linear combinations in the Frenet frame, since they satisfy the Frenet equations:

$$\begin{aligned} \dot{\mathbf{c}}_1 &= \kappa \mathbf{c}_2, \\ \dot{\mathbf{c}}_2 &= -\kappa \mathbf{c}_1 + \tau \mathbf{c}_3, \\ \dot{\mathbf{c}}_3 &= -\tau \mathbf{c}_2. \end{aligned} \tag{9.69}$$

Here and in the following, $\dot{}$ indicates derivation with respect to the arc length parameter of c. Further, $\kappa, \tau : I \subset \mathbb{R}$ are the *curvature* and *torsion* (functions) of c. At any regular point P of c which is neither a point of inflection ($\kappa \neq 0$) nor a handle point ($\tau \neq 0$), we can ask for a sphere $\mathcal{H}$ through P that has the highest order contact there. Clearly, the order of the contact will be three at most. Higher order contact can only be expected at special points. Therefore, the sphere $\mathcal{H}$ is called *hyperosculating sphere.*

First, we determine the center M with coordinate vector $\mathbf{m}$ and the radius r and set

$$\mathcal{H}: \ \langle \mathbf{c} - \mathbf{m}, \mathbf{c} - \mathbf{m} \rangle = r^2. \tag{9.70}$$

We differentiate once, twice, thrice, insert (9.69), and obtain

$$\begin{gathered} \dot{\mathcal{H}} : \langle \mathbf{c}_1, \mathbf{c} - \mathbf{m} \rangle = 0, \quad \ddot{\mathcal{H}} : \kappa \langle \mathbf{c}_2, \mathbf{c} - \mathbf{m} \rangle + 1 = 0, \\ \dddot{\mathcal{H}} : \kappa\tau \langle \mathbf{c}_3, \mathbf{c} - \mathbf{m} \rangle - \frac{\dot{\kappa}}{\kappa^2} = 0. \end{gathered} \tag{9.71}$$

We recall that the *curvature radius* (function) ϱ of c and the curvature κ are related via $\kappa\varrho = 1$ (at each $t \in I$). Hence, $\dot{\varrho} = -\dot{\kappa}\,\kappa^{-2}$.

The derivatives of $\mathcal{H}$ (9.71) allow us to write the vector $\mathbf{c} - \mathbf{m}$ as a linear combination of the Frenet vectors: From each single equation, we obtain the component of the vector $\mathbf{c} - \mathbf{m}$ in the direction of a vector $\mathbf{c}_i$. Thus, we have

$$\mathbf{m} = \mathbf{c} + \varrho\,\mathbf{c}_2 + \frac{\dot{\varrho}}{\tau}\mathbf{c}_3 \tag{9.72}$$

and the radius r of $\mathcal{H}$ equals

$$r = \sqrt{\varrho^2 + \frac{\dot{\varrho}^2}{\tau^2}}. \tag{9.73}$$

Finally, the hyperosculating sphere $\mathcal{H}$ has the equation

$$\mathcal{H}: \ \left\| \mathbf{x} - \mathbf{c} - \varrho\,\mathbf{c}_2 - \frac{\dot{\varrho}}{\tau}\mathbf{c}_3 \right\|^2 = \varrho^2 + \frac{\dot{\varrho}^2}{\tau^2} \tag{9.74}$$

(cf. [81]). As an intermediate result, we can formulate

Theorem 9.4.3 *At each regular point P (that is neither a point of inflection nor a handle point) on a C^3 space curve c, there exists a uniquely defined sphere $\mathcal{H}$ which is in third order contact with c at P. The equation of $\mathcal{H}$ is given by* (9.74).

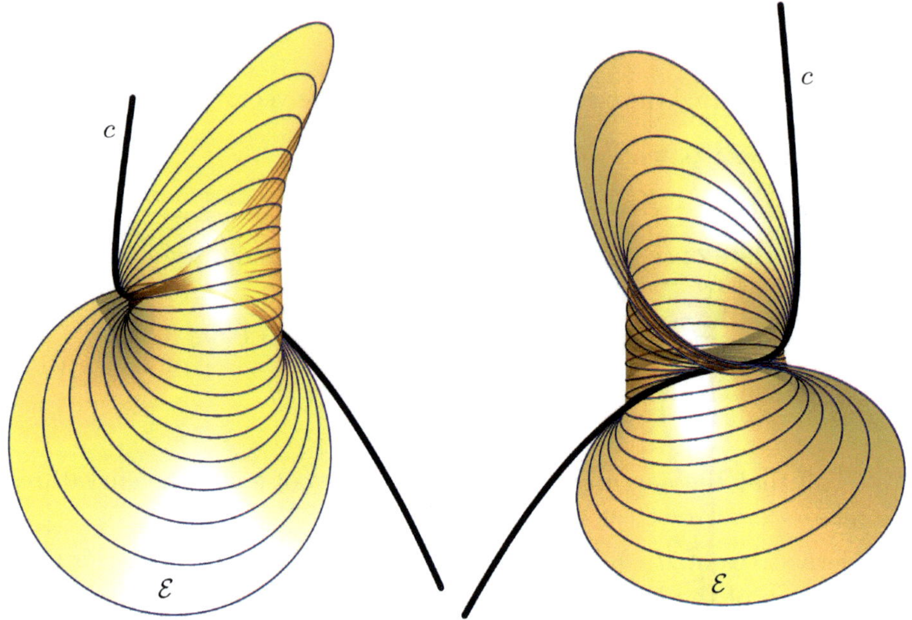

FIGURE 9.36. Two views of the natural channel surface $\mathcal{E}$ of a cubic curve c. The cubic c is the locus of all singular points of $\mathcal{E}$. The osculating circles of c are the characteristic circles of $\mathcal{E}$.

The hyperosculating spheres of a space curve form a one-parameter family as long as the curve is not spherical, *i.e.*, the curve is not entirely contained in its hyperosculating sphere. So, it is obvious to expect an envelope, which is called the *natural channel surface* of the given space curve. We can show

Theorem 9.4.4 *The hyperosculating spheres of a regular non-spherical space curve c which is free of inflection points and handle points in the desired interval envelop a channel surface $\mathcal{E}$. The characteristic circles are osculating circles of c. The points of c are singular points on $\mathcal{E}$.*

Proof: Clearly, there exists an envelope $\mathcal{E}$ of all hyperosculating spheres $\mathcal{H}$ of c. The following will show that this envelope indeed consists of a real branch.

We only have to show that the osculating circles of c are the characteristic circle of $\mathcal{E}$. For that purpose, we insert (9.74) into (9.64) and make use of (9.69). Finally, some minor manipulations yield

$$\dot{\mathcal{H}}: \ \langle \mathbf{x} - \mathbf{c}, \mathbf{c}_3 \rangle = 0, \tag{9.75}$$

which is the equation of the osculating planes of c for all t. Inserting the latter equation into the equations of the hyperosculating sphere (9.74) results in the equation of the intersection $\mathcal{H} \cap \ddot{\mathcal{H}}$:

$$o: \langle \mathbf{x} - \mathbf{c} - \varrho \mathbf{c}_2, \mathbf{x} - \mathbf{c} - \varrho \mathbf{c}_2 \rangle = \varrho^2.$$

Taking the linear equation (9.75) of $\dot{\mathcal{H}}$ into account, we recognize o as the osculating circle of c.

We differentiate (9.75) once again, use (9.69), and find

$$\ddot{\mathcal{H}}: \langle \mathbf{x} - \mathbf{c}, \mathbf{c}_2 \rangle = 0,$$

which is the equation of the rectifying plane of c. The osculating plane σ and the rectifying plane ϱ of a space curve c at a certain point $P = \mathbf{c}(t)$ intersect in the tangent T of c at P. Since T is tangent to $\mathcal{H}$, P is a double point on $\mathcal{E}$, and thus, a singular point on $\mathcal{E}$. ■

Figure 9.36 illustrates the contents of Theorem 9.4.4. A cubic curve c with its natural channel surface swept by c's osculating circles is shown. Figure 9.35 shows that, along the characteristic circle of the natural channel surface of a space curve c, the tangent planes envelop a cone of revolution.

■ **Example 9.4.1** Natural pipe of a helix.

A helix h (right circular helix) is a curve of constant slope, curvature, and torsion, winding up a cylinder of revolution. Since κ and τ of h are constant, the radius of $\mathcal{H}$ is, too. Thus, the natural channel surface $\mathcal{E}$ is a helical pipe surface that is invariant under the helical motion which generates h. An example is shown in Figure 9.37.

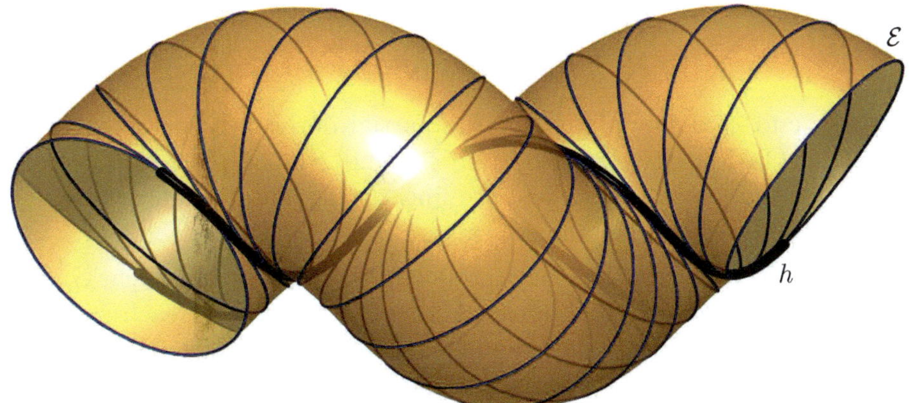

FIGURE 9.37. The natural channel surface $\mathcal{E}$ of a helix h is a pipe surface, since h's curvature and torsion are constant. The helix is a sharp ridge on $\mathcal{E}$.

Envelopes of spheres – two-parameter families

The geometry of two-parameter families of spheres compares to the geometry of two-parameter families of lines (congruences of lines). Both the two-parameter families of spheres and the two-parameter families of lines can be considered as *two dimensional surfaces*. This will become clearer, when we describe the geometry of spheres and lines as geometries that take place on quadrics in Sections 10.1 and 10.2.

In the following, we shall try a more elementary differential geometric approach to two-parameter families of spheres. Sometimes, algebraic aspects are of more importance. For example, rational envelopes of one- and even two-parameter families of spheres are of interest and can be constructed using deep knowledge from pure geometry and classical geometries (see, for example, [103]). However, the case of two-parameter families with variable radius function still bears some difficulties.

Let us assume that

$$\mathcal{S}(u,v): \langle \mathbf{x} - \mathbf{m}(u,v), \mathbf{x} - \mathbf{m}(u,v)\rangle = r^2(u,v) \tag{9.76}$$

are the equations of the spheres in a two-parameter family. This means that $\mathbf{m}(u,v): D \subset \mathbb{R}^2 \to \mathbb{R}^3$ is a parametrization of a surface in $\mathbb{R}^3$ and $r(u,v)$; $I \subset \mathbb{R} \to \mathbb{R}$ is a real-valued function.

If the spheres $\mathcal{S}$ in (9.76) produce an envelope, we shall find the points of contact as the intersection of $\mathcal{S}$ with its two partial derivatives

$$\begin{aligned} \mathcal{S}_{,u}&: \langle \mathbf{m}_{,u}, \mathbf{x} - \mathbf{m}\rangle + r_{,u}\, r = 0,\\ \mathcal{S}_{,v}&: \langle \mathbf{m}_{,v}, \mathbf{x} - \mathbf{m}\rangle + r_{,v}\, r = 0, \end{aligned} \tag{9.77}$$

where $_{,u}$ and $_{,v}$ is short-hand for the partial derivatives with respect to u and v. Obviously, the derivatives $\mathcal{S}_{,u}$ and $\mathcal{S}_{,v}$ are planes. These are not parallel unless $\mathbf{m}_{,u}$ and $\mathbf{m}_{,v}$ are linearly dependent, *i.e.*, the surface parametrized by $\mathbf{m}$ is singular.

We assume that the surface of centers is regular in D and then, $\mathcal{S}_{,u}$ and $\mathcal{S}_{,v}$ intersect in a straight line a. The contact points of each sphere $\mathcal{S}$ and the envelope $\mathcal{E}$ are the common points of $\mathcal{S}$ and a (see Figure 9.38). Consequently, there are up to two such points that are real, and therefore, the envelope consists of two sheets at most.

We can give a condition on $\mathbf{m}$ and r so that the two contact points are either a real pair, or fall in one (real) point, or are a pair of complex

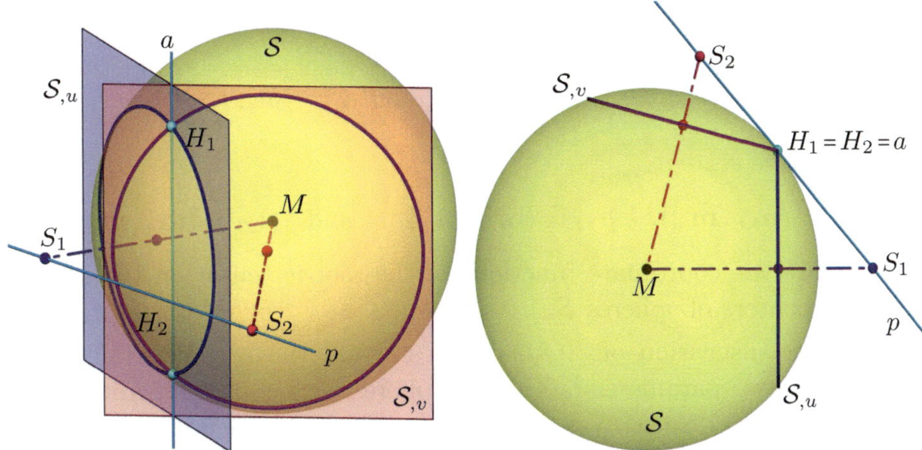

FIGURE 9.38. Left: The points of contact H_1 and H_2 of the envelope $\mathcal{E}$ of the two-parameter family of spheres are the intersection points of $a = \mathcal{S}_{,u} \cap \mathcal{S}_{,v}$. Right: a configuration with precisely one point of contact, seen in the direction of a.

conjugate points. The poles of the two planes $\mathcal{S}_{,u}$ and $\mathcal{S}_{,v}$ shall be denoted by S_1 and S_2. The line $p = [S_1, S_2]$ is the polar of $a = \mathcal{S}_{,u} \cap \mathcal{S}_{,v}$ with regard to $\mathcal{S}$. Intersecting pairs of polars with regard to a quadric are conjugate surface tangents. According to (9.65), the coordinate vectors of the poles S_1 and S_2 are

$$\mathbf{s}_1 = \mathbf{m} - \frac{r}{r_{,u}}\mathbf{m}_{,u} \quad \text{and} \quad \mathbf{s}_2 = \mathbf{m} - \frac{r}{r_{,v}}\mathbf{m}_{,v}, \tag{9.78}$$

provided that neither $r_{,u} = 0$ nor $r_{,v} = 0$. Then, we derive the Plücker coordinates $(\mathbf{p}, \overline{\mathbf{p}})$ of p (see p. 478) using (9.78) and the Plücker coordinates $(\mathbf{a}, \overline{\mathbf{a}})$ of a (see p. 482) with (9.77) and find

$$\begin{aligned}(\mathbf{p}, \overline{\mathbf{p}}) &= (\mathbf{s}_2 - \mathbf{s}_1, \mathbf{s}_1 \times \mathbf{s}_2) = \\ &= \left(\tfrac{r}{r_{,u}}\mathbf{m}_{,u} - \tfrac{r}{r_{,v}}\mathbf{m}_{,v},\ \tfrac{r^2}{r_{,u}r_{,v}}\mathbf{m}_{,u} \times \mathbf{m}_{,v} - \tfrac{r}{r_{,u}}\mathbf{m}_{,u} \times \mathbf{m} - \tfrac{r}{r_{,v}}\mathbf{m} \times \mathbf{m}_{,v}\right),\end{aligned} \tag{9.79}$$

and subsequently,

$$(\mathbf{a}, \overline{\mathbf{a}}) = (\mathbf{m}_{,u} \times \mathbf{m}_{,v}, (rr_{,u} - \langle \mathbf{m}, \mathbf{m}_{,u}\rangle)\mathbf{m}_{,v} - (rr_{,v} - \langle \mathbf{m}, \mathbf{m}_{,v}\rangle)\mathbf{m}_{,u}). \tag{9.80}$$

Following (10.9), the two lines a and p intersect if, and only if, their Plücker coordinates $(\mathbf{a}, \overline{\mathbf{a}})$ and $(\mathbf{p}, \overline{\mathbf{p}})$ satisfy $\langle \mathbf{a}, \overline{\mathbf{p}}\rangle + \langle \overline{\mathbf{a}}, \mathbf{p}\rangle = 0$, and thus, with (9.80), we find

$$\frac{r^2}{r_{,u}r_{,v}}(\mu_{11}\mu_{22} - \mu_{12}^2 - r_{,u}^2\mu_{11} + 2r_{,u}r_{,v}\mu_{12} - r_{,v}^2\mu_{22}) = 0. \tag{9.81}$$

Here, we have set

$$\mu_{11} := \langle \mathbf{m}_{,u}, \mathbf{m}_{,u} \rangle, \quad \mu_{12} := \langle \mathbf{m}_{,u}, \mathbf{m}_{,v} \rangle, \quad \text{and} \quad \mu_{22} := \langle \mathbf{m}_{,v}, \mathbf{m}_{,v} \rangle,$$

i.e., μ_{ij} are the coordinate functions of the metric tensor on the surface $\mathcal{M}$ of all centers of spheres $\mathcal{S}$.

A complete discussion of (9.81) is far beyond the scope of this book. The case of two-parameter families of spheres with constant radius is not covered by (9.81), since by assumption $r_{,u} \neq 0$ and $r_{,v} \neq 0$.

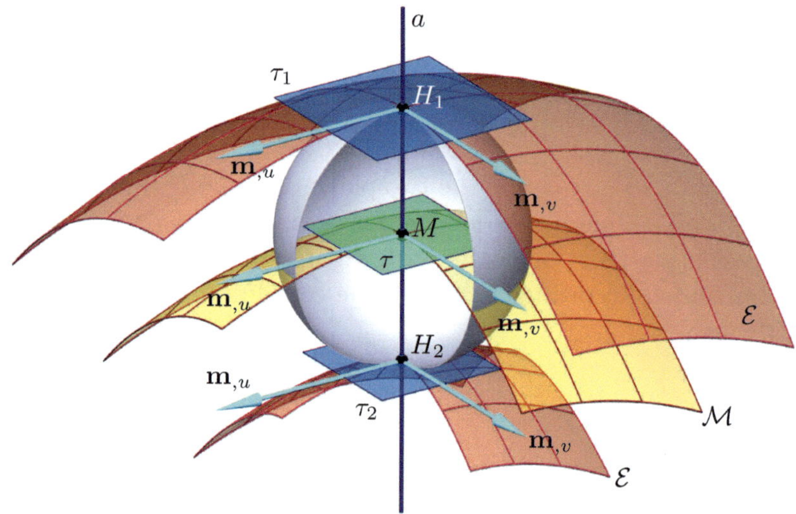

FIGURE 9.39. The offset $\mathcal{E}$ at a distance r of a surface $\mathcal{M}$ consists of two sheets. The points H_1 and H_2 on $\mathcal{E}$ lie on the surface normal a of $\mathcal{M}$ at M and $\overline{MH_1} = \overline{MH_2} = d$.

In the case of r = const., the points (9.78) become the ideal points $(0, \mathbf{m}_{,u})\mathbb{R}$ and $(0, \mathbf{m}_{,v})\mathbb{R}$ of two (linearly independent) surface tangents of $\mathcal{M}$ and the tangent cones Γ_u and Γ_v of $\mathcal{S}$ along $\mathcal{S} \cap \mathcal{S}_{,u}$ and $\mathcal{S} \cap \mathcal{S}_{,v}$ become cylinders of revolution. Finally, the condition (9.81) on the reality of contact points of $\mathcal{S}$ and $\mathcal{E}$ can be simplified to

$$\mu_{11}\mu_{22} - \mu_{12}^2 \geq 0. \tag{9.82}$$

This is equivalent to the regularity of $\mathcal{M}$ at M, the center of that particular sphere $\mathcal{S}$. In this case, the planes $\mathcal{S}_{,u}$ and $\mathcal{S}_{,v}$ contain the center of $\mathcal{S}$ (cf. (9.77)). Therefore, $a = \mathcal{S}_{,u} \cap \mathcal{S}_{,v}$ also passes through $\mathcal{S}$'s center and the two points H_1 and H_2 of the envelope $\mathcal{E}$ form an antipodal pair. Moreover, since $\mathbf{a} = \mathbf{m}_{,u} \times \mathbf{m}_{,v}$ (according to (9.80)), the line a is the surface normal of $\mathcal{M}$ at M (see Figure 9.39).

Therefore, the tangent planes of the envelope $\mathcal{E}$ at H_1 and H_2 are parallel to the tangent plane of $\mathcal{M}$ at M. Thus, the envelope consists of all points at distance $\pm r$ from $\mathcal{M}$ and is, therefore, the *offset* of $\mathcal{M}$ at distance r. (The fact that the tangent planes of $\mathcal{E}$ at H_1 and H_2 and of $\mathcal{M}$ at M are parallel is the reason why offsets are sometimes called *parallel surfaces*; "Parallelflächen" in German.)

Surfaces of constant slope

Surfaces of constant slope are developable (ruled) surfaces whose tangent planes subtend a fixed and constant angle β with a (horizontal) base plane π. These surfaces can be obtained as envelopes in two different ways:

(1) as the envelopes of planes that enclose a fixed angle with the base plane π or

(2) as the envelopes of cones of revolution with a fixed angle of aperture and vertical axes.

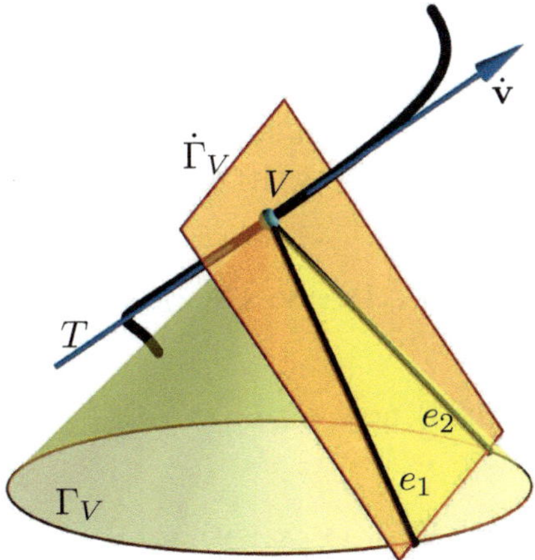

FIGURE 9.40. A cone Γ_V of constant slope in a one-parameter family together with its derivative $\dot{\Gamma}_V$: In this case, the derivative plane $\dot{\Gamma}_V$ intersects Γ_V in two real generators e_1 and e_2, along which the envelope $\mathcal{E}$ touches the cone.

We will deal with the second way, since this is more quadrics-related. For the sake of simplicity, we let $\pi: \ z = 0$ and the lead is parallel to the z-axis of the underlying Cartesian coordinate system. The slope angle shall be

$$0 < \beta < \frac{\pi}{2}.$$

Then, the angle of aperture α of a generic cone of constant slope equals $\pi - 2\beta$ and the equation of such a cone reads

$$\Gamma_V: \ (x - v_x)^2 + (y - v_y)^2 - k^2(z - v_z)^2 = 0, \tag{9.83}$$

provided that $\mathbf{v} = (v_x, v_y, v_z)$ is the cone's vertex and $k = \cot \beta$.
Let us assume that

$$\langle \mathbf{u}, \mathbf{v} \rangle_{\mathrm{pe}} := u_1 v_1 + u_2 v_2 - k^2 u_3 v_3 \tag{9.84}$$

is a family of pseudo-Euclidean scalar products. The value $k \in \mathbb{R} \setminus \{0\}$ is just a parameter and has the geometric meaning we have just described. With $\mathbf{x} = (x, y, z)$, we will denote the vector of indetermined coordinates. Let now $\mathbf{v}: \ I \subset \mathbb{R} \to \mathbb{R}^3$ be the parametrization of a curve in $\mathbb{R}^3$, *i.e.*, v_x, v_y, and v_z shall be univariate functions depending on some parameter t. All involved functions shall be of a sufficiently high differentiablity class. Then, the equation of Γ_V can be written in the slightly simpler form

$$\Gamma_V: \ \langle \mathbf{x} - \mathbf{v}, \mathbf{x} - \mathbf{v} \rangle_{\mathrm{pe}} = 0 \tag{9.85}$$

and the derivative of (9.83) now reads

$$\dot{\Gamma}_V: \ \langle \dot{\mathbf{v}}, \mathbf{x} - \mathbf{v} \rangle_{\mathrm{pe}} = 0. \tag{9.86}$$

Figure 9.41 shows the appearance of developables in a real world problem, namely, in road layouts.

FIGURE 9.41. Ruled surfaces of constant slope play an important role in many real world problems. The layout of roads show developables of constant slope through the curbsides. At each point on the curbside, a cone of constant slope is centered and the envelope comes in a natural way, since the angle of repose (for gravel, macadam, etc.) is constant.

Obviously, $\dot{\Gamma}_V$ is a plane through V for any $t \in I$ (cf. Figure 9.40). Therefore, the derivative of Γ_V intersects Γ_V along two of its generators and the envelope $\mathcal{E}$ of Γ_V touches each of the cones along these generators.

The contact of $\mathcal{E}$ and Γ_V takes place along two real rulings if $\langle \dot{\mathbf{v}}, \dot{\mathbf{v}} \rangle > 0$, along exactly one ruling if $\langle \dot{\mathbf{v}}, \dot{\mathbf{v}} \rangle = 0$, and along a pair of complex conjugate rulings otherwise.

Since the cones are developables, *i.e.*, their tangent planes along the rulings are fixed, the tangent planes of $\mathcal{E}$ along the contact rulings are also fixed. Thus, $\mathcal{E}$ is also a developable.

We summarize our results as follows:

Theorem 9.4.5 *The envelopes $\mathcal{E}$ of one-parameter families of cones of constant slope w.r.t. a plane π are developables of constant slope w.r.t. π. The curve v traced by the cones' vertices is a double curve on $\mathcal{E}$.*

Remark 9.4.1 The cone Γ_V given by (9.83) (or (9.85)) is a pseudo-Euclidean null sphere. Thus, the envelope of the cones Γ_V is a very special case of a pseudo-Euclidean channel surface. The plane (9.86) is the pseudo-Euclidean normal plane of the vector $\dot{\mathbf{v}}$ at $\mathbf{v}(t)$.

The most general forms of pseudo-Euclidean channel surfaces are the envelopes of one-parameter families of pseudo-Euclidean spheres. From the Euclidean point of view, these spheres are one- or two-sheeted hyperboloids of revolution with congruent asymptotic cones.

Pseudo-Euclidean space, cylcographic mapping

We have centered a cone Γ_V of revolution with a vertical axis and fixed angle of aperture at each point V of a curve. Indeed, we can do this at each point in the three-dimensional space $\mathbb{R}^3$.

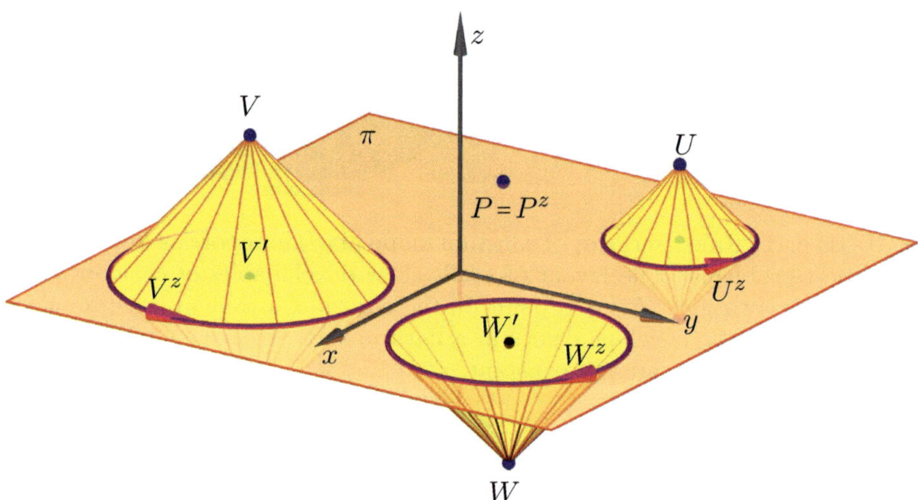

FIGURE 9.42. The cyclographic mapping ζ maps a point V in 3-space to oriented circles V^z in the Euclidean plane π. Points $P \in \pi$ agree with their image $P = P^z$ that carries both orientations.

If we intersect the cones Γ_V with the plane $\pi: \ z = 0$, we get circles V^z centered at $V' = (v_x, v_y, 0)$, *i.e.*, the top view of V with radius $r_V = |k \cdot v_z|$ (see Figure 9.42). The radius r_V can be equipped with a sign by $\operatorname{sgn} r_V := \operatorname{sgn} v_Z$. Loosely speaking, this means that the sign is positive if V is above π ($v_z > 0$). The circle V^z shall be oriented in the mathematical sense according to the sign of its radius. In this way, we find a mapping $\zeta: \mathbb{R}^3 \to \pi$ of points in three-dimensional affine space to oriented circles in the Euclidean plane π. The ambient space of π is referred to the cyclographic model of Euclidean Laguerre geometry.

The mapping ζ is usually called the *cyclographic mapping*. For the sake of simplicity, we let $k = 1$, which turns the cones Γ_V into cones with a right angle of aperture. From the computational as well as from the constructive point of view, this turned out to be beneficial (cf. [44, 89]).

The cyclographic mapping ζ does not only relate oriented circles in the Euclidean plane with points in a three-dimensional space. Oriented straight lines (spears) in π correspond to planes with slope ± 1.

The pseudo-Euclidean metric (9.84) returns the square of the pseudo-Euclidean length of a vector $\mathbf{v}$. Depending on whether $\langle \mathbf{v}, \mathbf{v} \rangle$ is greater, equal, or less than zero, the vector $\mathbf{v}$ is called *spacelike*, *lightlike* or *isotropic*, or *timelike*. The generators of each cone Γ_V of constant slope are isotropic or light-like in terms of the pseudo-Euclidean geometry in the cyclographic model space. Since the metric (9.84) is of signature $(++-)$, it is common to denote the cyclographic model space by $\mathbb{R}^{2,1}$.

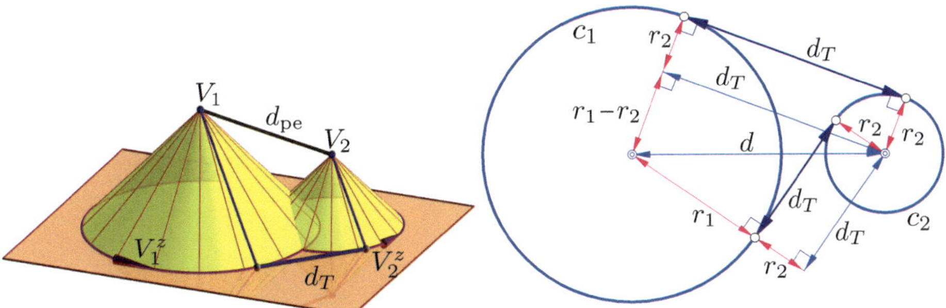

FIGURE 9.43. Left: The pseudo-Euclidean distance d_{pe} between two points P and Q equals the tangential distance d_T between the two oriented cyclographic image circles. Right: Two non-oriented circles c_1 and c_2 have up to two real tangential distances d_T (on the inner and outer tangents). The square is real in any case.

The pseudo-Eculidean distance between points in $\mathbb{R}^{2,1}$ has a geometric relative in the Euclidean plane π. Let c_1 and c_2 be two Euclidean circles and assume that they have at least one real common tangent T. The respective radii shall be r_1 and r_2 and the centers may be at distance d. Then, the square of the *tangential distance* d_T of c_1 and c_2 equals

$$d_T^2 = d^2 - (r_1 \pm r_2)^2, \tag{9.87}$$

depending on whether T is an inner or outer tangent of c_1 and c_2 as illustrated in Figure 9.43. Now, with (9.84) and $k = 1$, it is easily verified that the pseudo-Euclidean distance of the corresponding points $C_1 = (0, 0, r_1)$ and $C_2 = (d, 0, r_2)$ in the pseudo-Euclidean model space $\mathbb{R}^{2,1}$ fulfills (9.87).

Finally, we remark that setting $k = c$ (the speed of light) would bring the pseudo-Euclidean metric (9.84) closer to special relativity.Indeed, the pseudo-Euclidean space is a model for special relativity. Then, the space $\mathbb{R}^{2,1}$ is called *Minkowski*[3] *space*.

Projection with a quadratic complex

Later, in Section 10.1, we shall see that the cyclographic mapping $\zeta : \mathbb{R}^{2,1} \to \pi$ can be viewed as a projection with a quadratic complex of lines. The manifold of lines that enclose a fixed angle β with a plane in Euclidean 3-space is three-dimensional. All these lines meet a well-defined conic c_β in the plane ω at infinity. The cones Γ_V that map points V to oriented circles V^z in π are complex cones (cones of revolution), *i.e.*, exactly those lines of the quadratic complex (of constant slope) that pass through the point V. The complex cones share the conic $c_\beta \subset \omega$.

[3] Hermann Minkowski (1864–1909) was a German mathematician. He started working in number theory, diophantine approximation, and convex geometry. Minkowski was one of the first to discover that non-Euclidean geometries fit well with Einstein's special relativity.

9.5 Curves of constant slope on quadrics

General remarks on curves of constant slope

A curve $c \subset \mathbb{R}^3$ is called a *curve of constant slope* if its tangent (unit vector) $\mathbf{c}_1$ encloses a fixed angle, say β, with a fixed direction $\mathbf{l}$. The direction $\mathbf{l}$ is called the *lead* and will, henceforth be assumed to be $\mathbf{l} = (0,0,1)$. The angle is then called the *slope* or *inclination*, which is ambiguous, since the term slope also denotes the value $k = \tan\beta$.

It is admissible to assume that the parametrization $\mathbf{c}$ depends on the arc length s on c. Further, the Frenet frame $(\mathbf{c}_1, \mathbf{c}_2, \mathbf{c}_3)$ satisfies the Frenet equations (9.69), and c is free from handle points and points of inflection.

By assumption, we have $\langle \mathbf{c}_1, \mathbf{l}\rangle = \cos\beta = \text{const.}$ and we can differentiate once obtaining $\kappa\langle \mathbf{c}_2, \mathbf{l}\rangle = 0$. This tells us that the principal normal of c is always orthogonal to the lead. Differentiating once more, we find

$$\dot{\kappa}\langle \mathbf{c}_2, \mathbf{l}\rangle + \kappa\langle -\kappa\mathbf{c}_1 + \tau\mathbf{c}_3, \mathbf{l}\rangle = 0,$$

which simplifies to $-\kappa\cos\beta + \tau\langle \mathbf{c}_3, \mathbf{l}\rangle = 0$, because of $\mathbf{l} = \cos\beta\,\mathbf{c}_1 + \sin\beta\,\mathbf{c}_3$. Finally,

$$\frac{\mathrm{d}}{\mathrm{d}s}\langle \mathbf{c}_3, \mathbf{l}\rangle = -\tau\langle \mathbf{c}_2, \mathbf{l}\rangle = 0,$$

from which we can infer that the scalar product $\langle \mathbf{c}_3, \mathbf{l}\rangle$ is constant. Thus, we can state

Theorem 9.5.1 *For curves of constant slope, the ratio of the curvature and the torsion is constant.*

It is easy to see that the assumption of a constant ratio $\kappa : \tau$ implies a curve of constant slope.

■ **Example 9.5.1** Another characterization of curves of constant slope.

Show that a curve c of constant slope with an at least four times differentiable arc length parametrization $\mathbf{c} : I \subset \mathbb{R} \to \mathbb{R}^3$ has the property $\det(\ddot{\mathbf{c}}, \dddot{\mathbf{c}}, \ddddot{\mathbf{c}}) = 0$.

By assumption, $\mathbf{c}$ is an arc length parametrization. So, $\dot{\mathbf{c}} = \mathbf{c}_1$, and according to (9.69), $\ddot{\mathbf{c}} = \kappa\mathbf{c}_2$, $\dddot{\mathbf{c}} = -\kappa^2\mathbf{c}_1 + \dot{\kappa}\mathbf{c}_2 + \kappa\tau\mathbf{c}_3$, and $\ddddot{\mathbf{c}} = -(3\dot{\kappa}\kappa + \kappa^3)\mathbf{c}_1 + (\ddot{\kappa} - \kappa\tau)\mathbf{c}_2 + (2\dot{\kappa}\tau + \kappa\dot{\tau})\mathbf{c}_3$. This yields

$$\det(\ddot{\mathbf{c}}, \dddot{\mathbf{c}}, \ddddot{\mathbf{c}}) = -\kappa^3(\dot{\kappa}\tau - \kappa\dot{\tau}) = \kappa^3\tau^2\left(\frac{\kappa}{\tau}\right)^{\cdot} = 0,$$

since $\kappa\tau^{-1} = \text{const.}$

Curves of constant slope on cylinders and cones

In the following, we shall extract the special subclasses of the curves of constant slope that can be found on quadrics. The most simple cases of curves of constant slope on quadrics of revolution are those where the quadrics are singular. On the cylinder of revolution

$$\mathbf{z}(u,v) = (R\cos v, R\sin v, u), \quad (u,v) \in \mathbb{R}^2, \ R \in \mathbb{R}^\star,$$

the curves of constant slope are helices

$$\mathbf{h}(t) = (R\cos t, R\sin t, pt + c), \quad t \in \mathbb{R}, \quad c, p \in \mathbb{R},$$

including the case of the parallel circles for $p = 0$ with slope $k = 0$ and the generators of the cylinder with slope $k = \infty$ if $p = \infty$. The helices on the cylinder of revolution are invariant with respect to the helical motion with pitch p. Further, the helices on the cylinder of revolution allow a discrete group of automorphic collineations, *i.e.*, the translations

$$\tau : \ (x,y,z) \mapsto (x,y,z + 2j\pi) \quad \text{with} \quad j \in \mathbb{Z}.$$

The curves of constant slope on a cone of revolution with vertical axes include the parallel circles (slope $k = 0$) and the generators of the cone. The latter have maximal slope that equals the slope of the cone. Much more interesting are the remaining curves of constant slope on the cone of revolution. These curves are called *cylindro-conical spirals* and are invariant with respect to one-parameter subgroups of the group of equiform transformations. The cone

$$\mathbf{c}(u,v) = (u\cos v, u\sin v, ku) \quad (u,v) \in \mathbb{R}^2, \ k \in \mathbb{R} \setminus \{0\}$$

with $\omega = 2\text{arccot}\, k$ for its angle of aperture carries the cylindro-conical spirals

$$\mathbf{s}(t) = \mathrm{e}^{pt}(\cos t, \sin t, k)$$

and all congruent copies obtained by rotating one of the curves about the axis of the cone (z-axis). The spiral parameter p and the slope angle β are related via

$$p = \pm \frac{1}{\sqrt{k^2 \tan^2 \beta - 1}}.$$

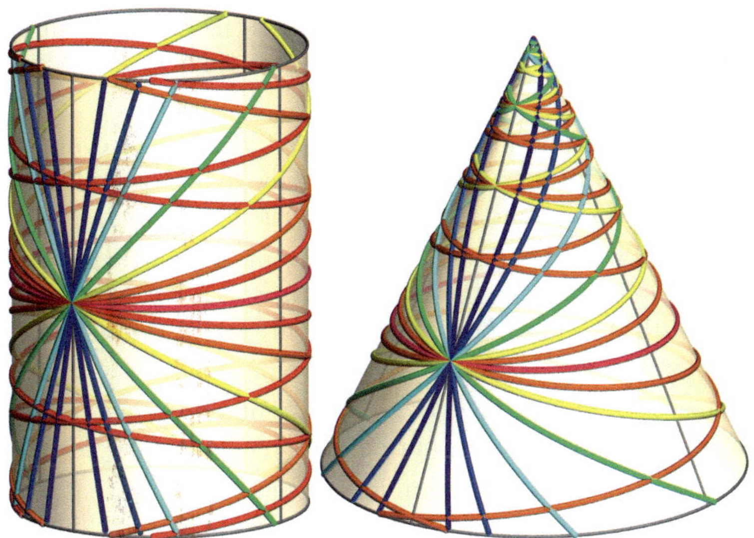

FIGURE 9.44. Curves of constant slope on a cylinder of revolution (left) are helices (including circles and rulings). Curves of constant slope on a cone of revolution (right) are cylindro-conical spirals (including parallel circles and rulings).

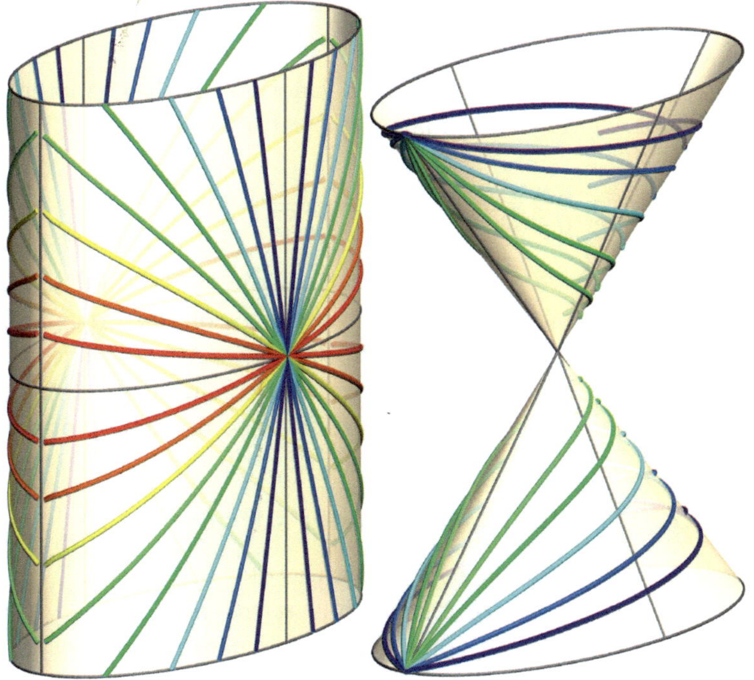

FIGURE 9.45. Curves of constant slope on an elliptic cylinder (left) and on an elliptic cone (right). The majority of these curves is not even algebraic.

These curves can also be generated from the one above by applying all central similarities (with the cone's vertex for their center) to that particular curve.

Figure 9.44 (left) shows some helices (on a cylinder of revolution) with various slopes. Figure 9.44 (right) shows some cylindro-conical spirals as curves of constant slope on a cone of revolution.

The spirals shown in this figure emanate from one particular point. Except the parallel circles on the cone, all curves of constant slope would eventually reach the apex of the cone, but only after infinitely many windings about the cone's axis.

The visual appearance of the curves of constant slope on an elliptic cylinder does not differ much from that of ordinary helices (cf. Figure 9.45).

However, their analytic description is not comparable to that of ordinary helices. We determine the curves of constant slope on the elliptic cylinder

$$\mathcal{C}: \ \frac{x^2}{a^2} + \frac{y^2}{b^2} = 1, \quad a, b \neq 0.$$

In order to do so, we parametrize the curve b of constant slope by

$$\mathbf{b}(t) = (a\cos t, b\sin t, u(t)), \quad t \in \mathbb{R}$$

where $u(t): \ I \subset \mathbb{R} \to \mathbb{R}$ is a real-valued function which is differentiable at least once. This yields the ordinary differential equation

$$\frac{\dot{u}}{\sqrt{a^2\sin^2 t + b^2\cos^2 t + \dot{u}^2}} = \cos\beta,$$

whose solutions involve elliptic integrals of the third kind. Figure 9.45 shows a numerical approximation of the curves of constant slope on an elliptic cylinder (left) and on an elliptic cone (right). The analogous ordinary differential equation for the curves of constant slope on an elliptic cone behaves equally bad, *i.e.*, its solutions also involve elliptic integrals of third order.

Quadrics of revolution and Enneper's theorem

The curves of constant slope on the quadrics of revolution are well-known. They can rather easily be parametrized in terms of elementary functions, and are in close relation to conics and its generalizations as have seen in [46, p. 105–108]. The case of a cylinder and a cone of revolution has already been discussed. So we will study paraboloids, one- and two-sheeted hyperboloids, and ellipsoids of revolution. We can give their equations in cylinder coordinates (r, z) as

$$\mathcal{P}: \; r^2 = 2pz, \quad \mathcal{H}_{1,2}: \; \frac{r^2}{a^2} - \frac{z^2}{c^2} = \pm 1, \quad \mathcal{E}: \; \frac{r^2}{a^2} + \frac{z^2}{c^2} = 1, \tag{9.88}$$

provided that the z-axis of the underlying coordinate system coincides with the axis of rotation.

Let us call the orthogonal projection in the direction of the z-axis a top view. The image plane may be any plane orthogonal to the z-axis. We want to describe the top views c' of the curves c of constant slope on the quadrics (9.88) by means of a *pseudo natural equation* that relates the distance function $r: \; I \subset \mathbb{R} \to \mathbb{R}$ of the points on c' to the origin (= orthogonal projection of the axis) with the curvature radius (function) $\varrho(s): \; I \subset \mathbb{R} \to \mathbb{R}$ of c' and the arc length s on c'.

Therefore, we assume c' to be the path of the planar motion Σ/Σ_0 performed by the Frenet frame of c' which is considered to be part of the moving system Σ. Let M denote the instantaneous pole of Σ/Σ_0 and let $\omega > 0$ be the angular velocity. By assumption, $\mathbf{c}$ is an arc length parametrization of the top view c', and thus, $\|\dot{\mathbf{c}}\| = 1$ and for each point P on c', we have $\omega \cdot \overline{MP} = \|\dot{\mathbf{c}}\| = 1$. Clearly, $\overline{MP} = \varrho$ equals the curvature radius of c' at P, and therefore, $\omega\varrho = 1$.

The path normal vector of P equals $\mathbf{v} = \frac{1}{\varrho}(x, y - \varrho)$, with (x, y) being the coordinates of P, and thus, the tangent vector equals

$$\tfrac{1}{\varrho}(y - \varrho, -x) = (\dot{x}, \dot{y}). \tag{9.89}$$

The distance function r and the coordinates (x, y) of P depend on the arc length s and are related via

$$r^2 = x^2 + y^2. \tag{9.90}$$

Implicit differentiation with respect to s yields

$$\dot{r}\, r = \dot{x}\, x + \dot{y}\, y. \tag{9.91}$$

We insert (9.89) into (9.91), obtain

$$\dot{r}\, r = \tfrac{1}{\varrho}(x(y - \varrho) + y(-x)) = -x, \tag{9.92}$$

differentiate once again, and using (9.89), we find

$$\dot{r}^2 + \ddot{r}\,r = \dot{x} = -\tfrac{1}{\varrho}(y - \varrho).$$

Finally, the latter equation is solved for y and yields

$$y = \varrho\,(1 - \dot{r}^2 - \ddot{r}\,r). \tag{9.93}$$

Now, we eliminate x and y with (9.92) and (9.93) from (9.90) and arrive at the pseudo natural equation of c (a kind of the natural equation of c'):

$$\varrho = r\sqrt{1 - \dot{r}^2}(1 - \dot{r}^2 - \ddot{r}\,r)^{-1}. \tag{9.94}$$

Equation (9.94) allows us to find the top views of the curves of constant slope on the quadrics of revolution from (9.88). Once the top views are known, the entire curves can easily be determined by lifting them onto the quadrics. We can summarize this in

Theorem 9.5.2 (Enneper*'s*[4] *theorem:*) *Let* $\mathbf{l} = (0,0,1)$ *be the lead which is parallel to the axis of a quadric* $\mathcal{Q}$ *of revolution. Then, the top view of a curve of constant slope on* $\mathcal{Q}$ *is*

(1) a circle if $\mathcal{Q}$ *is a cylinder,*

(2) a logarithmic spiral if $\mathcal{Q}$ *is a cone,*

(3) the involute of a circle if $\mathcal{Q}$ *is a paraboloid,*

(4) an epicycloid if $\mathcal{Q}$ *is an ellipsoid,*

(5a) a hypocycloid if $\mathcal{Q}$ *is a one-sheeted hyperboloid and the slope is larger than that of the asymptotic cone,*

(5b) a straight line if $\mathcal{Q}$ *is a one-sheeted hyperboloid and the slope is equal to that of the asymptotic cone,*

(5c) a hypercycloid if $\mathcal{Q}$ *is a one-sheeted hyperboloid and the slope is smaller than that of the asymptotic cone,*

(6) a paracycloid if $\mathcal{Q}$ *is a two-sheeted hyperboloid.*

Proof: (1) The case of the cylinder is clear, since the top view of a cylinder (orthogonal projection in the direction of its axis) is a circle.

---

[4] Alfred Enneper (1830–1895) was a German mathematician who worked on Differential Geometry, especially on minimal surfaces. The famous Enneper (minimal) surface is named after him.

(2) The generators of the cone are curves of constant slope $\cot\frac{\alpha}{2}$ (maximal slope; α is the angle of aperture) and appear as a pencil of lines in the top view. The parallel circles correspond to $k = 0$. Both the generators and the parallel circles can be seen as limiting cases of logarithmic spirals. In terms of cylinder coordinates, the cone's equation equals

$$r^2 - z^2 \tan^2 \frac{\alpha}{2} = 0.$$

For a generic curve c of constant slope $0 < k < \cot\frac{\alpha}{2}$, we have $z = ks$, and thus, $r^2 - k^2 s^2 = 0$, *i.e.*, either $r = ks$ or $r = -ks$. (It is sufficient to consider only one of the latter two equations.) With (9.94), we find the natural equation or the Mannheim curve (cf. [46, p. 105–107])

$$\varrho(s) = k \cdot s, \tag{9.95}$$

where $k = \tan\beta$. From the curvature radius function $\kappa(s) = \frac{1}{\varrho(s)}$, we can derive a parametrization of the curve c' which is uniquely determined up to Euclidean motions. For that purpose, we use the techniques and results from [46, p. 107]. Consequently, the curve determined by (9.95) is a logarithmic spiral (as long as $k \neq 0, \pm 1$), see also [46, Table 3.1, p. 108].

(3–6) From (9.88) and with $z = ks$, we find the natural equations of the top views of the curves of constant slope

$$\mathcal{P}' : \varrho(s) = \sqrt{kp(2s - kp)}, \quad \mathcal{H}'_{1,2} : \pm \frac{a^2k^2 - c^2)^2}{a^2c^4}\varrho^2 \pm \frac{k^2(a^2k^2 - c^2)}{c^4}s^2 = 1,$$
$$\mathcal{E}' : \frac{(a^2k^2 + c^2)^2}{a^2c^4}\varrho^2 + \frac{k^2(a^2k^2 + c^2)}{c^4}s^2 = 1.$$

Now, the obvious restrictions on k have to be discussed in order to distinguish the cases, especially for (5). Then, a comparison with [46, Table 3.1, p. 108] confirms the result. ■

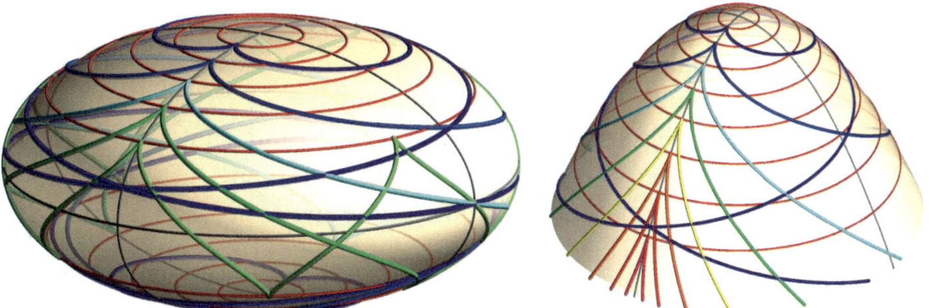

FIGURE 9.46. Left: curves of constant slope on an ellipsoid of revolution. Right: curves of constant slope on a paraboloid of revolution.

Figure 9.46 shows curves of constant slope on an ellipsoid and on a paraboloid of revolution. On the ellipsoid, the curves of constant slope for a fixed k fill a region that is bounded by those two parallel circles whose tangent cone have the same slope. On the paraboloid of revolution, the curves of fixed constant slope have their cusps on the parallel circle where the tangent cone has the prescribed slope and the orbit of all curves with the prescribed slope is bounded by just one parallel circle.

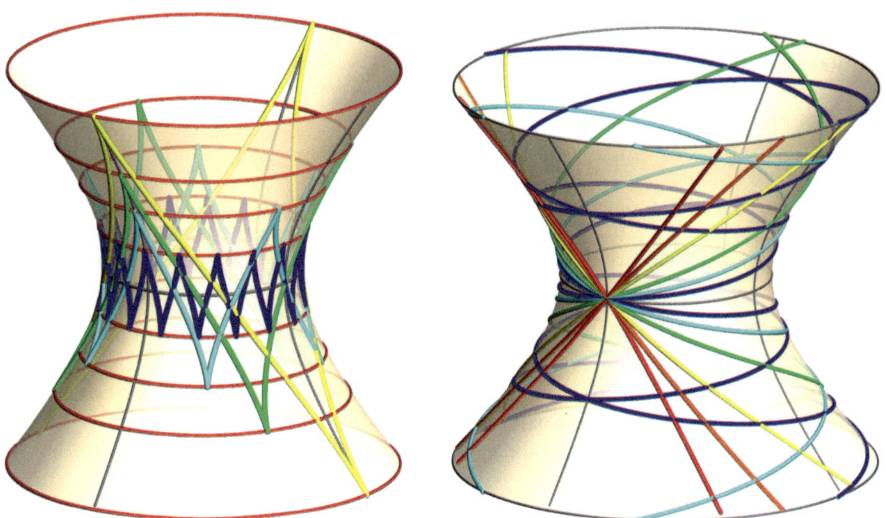

FIGURE 9.47. Curves of constant slope on a one-sheeted hyperboloid: steeper than the asymptotic cone (left), flatter than the asymptotic cone (right).

The curves of constant slope on a one-sheeted hyperboloid belong to three different classes:

(1) If the slope is larger than that of the asymptotic cone, the curves of constant slope remain within a strip bounded by two parallel circles. Along these parallel circles, the tangent planes, and thus, the tangent cones agree in the slope with the curve of constant slope (cf. Figure 9.47, left). These curves of constant slope are algebraic, rational, and closed if the ratio of the radii of the polhodes of the trochoid motion generating the top view c' is rational.

(2) The curves with a slope smaller than that of the asymptotic cone wind around the hyperboloid infinitely often (Figure 9.47, right). These curves are transcendental in any case and their top views are hypercycloids. (These curves shall not be confused with hypocycloids, cf. [46, p. 107]).

(3) Curves with a slope equal to that of the asymptotic cone are straight lines, and thus, the generators on a one-sheeted hyperboloid of revolution are also curves of constant slope.

■ **Example 9.5.2** Curves of constant slope on horizontal parabolic cylinders.

In the following, the lead shall point in the direction $\mathbf{l} = (0, 0, 1)$ and the curves of constant slope are assumed to have the slope angle $\beta \in]0, \frac{\pi}{2}[$. We want to determine explicit parametrizations of the curves of constant slope on the two types of parabolic cylinders (with $p \in \mathbb{R}^+$)

$$\mathcal{C}_1 : z^2 - 2py = 0 \quad \text{and} \quad \mathcal{C}_2 : y^2 + 2pz = 0$$

with rulings parallel to the x-axis. For that purpose, we assume that the real-valued function $u_i(t) : I \subset \mathbb{R} \to \mathbb{R}$ is the x-coordinate function of the desired curve of constant slope b_i on either cylinder. A parametrization of b_i is then given by

$$\mathbf{b}_1 = (u_1(t), 2pt^2, 2pt) \quad \text{and} \quad \mathbf{b}_2 = (u_2(t), 2pt, 2pt^2).$$

Now, the ordinary differential equation for $u_i(t)$ in order to make b_i curves of constant slope equals $\langle \dot{\mathbf{b}}_i, \mathbf{l} \rangle = \|\dot{\mathbf{b}}_i\| K$ where $K := \cos\beta$, which, in the present cases, reads explicitly

$$\frac{2p}{\sqrt{\dot{u}_1(t)^2 + 4p^2 + 16p^2t^2}} = K \quad \text{and} \quad \frac{4pt}{\sqrt{\dot{u}_2(t)^2 + 4p^2 + 16p^2t^2}} = K.$$

These differential equations are solved by

$$\begin{aligned} u_1(t) &= \pm \tfrac{K^2-1}{2K^2} p \arctan \tfrac{2Kt}{\sqrt{1-K^2-4K^2t^2}} \mp \tfrac{pt}{K}\sqrt{1-K^2-4K^2t^2} + C, \\ u_2(t) &= \pm \tfrac{pK}{2\sqrt{K^2-1}} \arctan \tfrac{2\sqrt{K^2-1}t}{\sqrt{4(1-K^2)t^2-K^2}} \mp \tfrac{pt}{K}\sqrt{4(1-K^2)t^2-K^2} + C, \end{aligned}$$

where either the upper or the lower sign is to be chosen and C is an arbitrary constant. In both cases, we use substitutions in order to simplify the representation of the curves:

$$\sigma_1 : t \to \frac{\sqrt{1-K^2}}{2K} \sin\tau \quad \text{and} \quad \sigma_2 : t \to \frac{K}{2\sqrt{1-K^2}} \cosh\tau.$$

Finally, we obtain

$$\begin{aligned} \mathbf{b}_1 &= \tfrac{p(1-K^2)}{2K^2} \left(\pm(\tau + \sin\tau\cos\tau), \sin^2\tau, \tfrac{2K}{\sqrt{1-K^2}} \sin\tau \right), \\ \mathbf{b}_2 &= \tfrac{pK}{2\sqrt{1-K^2}} \left(\pm(\tau + \sinh\tau\cosh\tau), 2\cosh\tau, \tfrac{1}{2\sqrt{1-K^2}} \cosh^2\tau \right), \end{aligned}$$

where we have cut out the integration constants, since they only cause a shift of the curves in the direction of the cylinder's generators. This does not change the shape and differential geometric properties of these curves. Figure 9.48 shows some curves of constant slope on the horizontal cylinders $\mathcal{C}_1$ and $\mathcal{C}_2$.

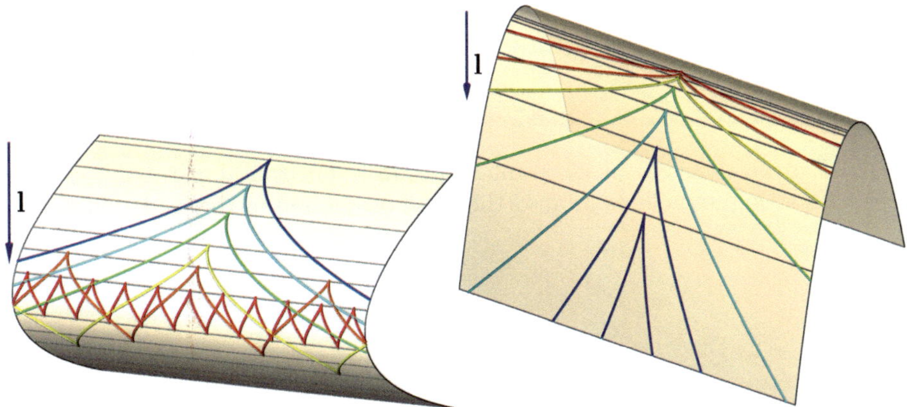

FIGURE 9.48. Curves of constant slope on horizontal parabolic cylinders: The lead $\mathbf{l}$ is orthogonal to the uniquely determined plane of symmetry (left). The lead $\mathbf{l}$ is orthogonal to the *vertex ruling* (right).

The curves b_1 on $\mathcal{C}_1$ somehow fit with those described in Theorem 9.5.2, although a parabolic cylinder can hardly be viewed as a quadric of revolution. Parametrizations of the top views of the curves b_1 can be obtained by setting the third coordinate function in $\mathbf{b}_1$ equal to zero. Ignoring those solutions corresponding to the lower sign yields

$$\mathbf{b}_2' = \frac{p(1-K^2)}{2K^2}\left(\tau + \sin\tau\cos\tau, \sin^2\tau\right) = \frac{p(1-K^2)}{4K^2}\left(2\tau + \sin 2\tau, 1 - \cos 2\tau\right).$$

Then, the reparametrization $\tau \to \frac{1}{2}t$ produces

$$\mathbf{b}_2' = \frac{p(1-K^2)}{2K^2}\left(t + \sin t, 1 - \cos t\right). \tag{9.96}$$

The curve (9.96) is called *(straight) cycloid* and it is the path of a point of a circle that rolls on a straight line. Figure 9.49 shows the top view of the parabolic cylinder $\mathcal{C}_2$ from Figure 9.48 (right). The cycloids that show up in this orthogonal projection touch the contour line (vertex generator) and their cusps gather on rulings of $\mathcal{C}_2$ where the tangent planes have the prescribed slope.

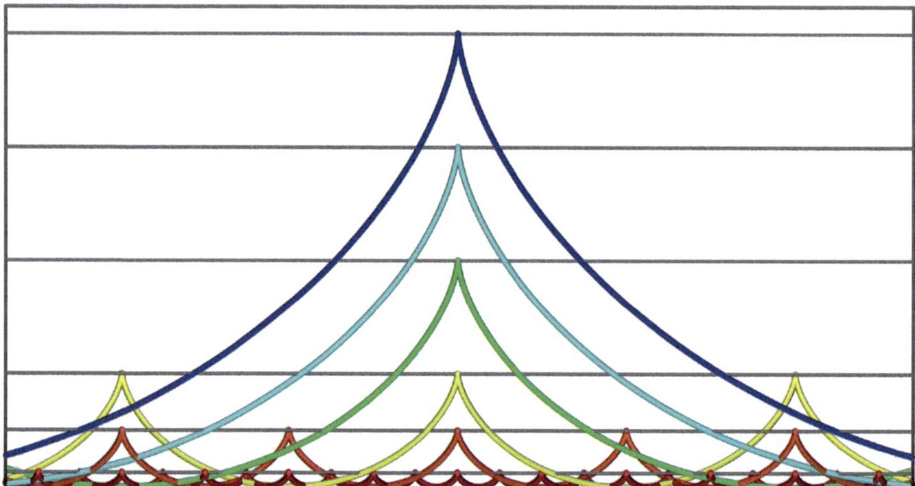

FIGURE 9.49. Curves of constant slope on a horizontal parabolic cylinder appear as straight cycloids in an orthogonal projection in the lead's direction.

9.6 Geodesics on quadrics

In this section, we shall select some special results on geodesics. We will focus on algebraic and rational geodesics whose existence is rather surprising. In the case of geodesics on a surface of revolution (and also on quadrics of revolution), CLAIRAUT's relation $r(t)\cos\varphi(t) = \text{const.}$ relates the radius function $r(t)$ (distance of the points on the geodesic to the axis of rotation) and the course angle $\varphi(t)$ of a geodesic. Unfortunately, even in the apparently simple case of ellipsoids of revolution, the parametrization of geodesics needs, in general, elliptic integrals.

On triaxial ellipsoids, a generic geodesic makes infinitely many circuits around the ellipsoid and touches a pair of opposite lines of curvature. Figure 9.50 shows the typical behavior of a generic geodesic on a triaxial ellipsoid. The tangents of the geodesics on an arbitrary quadric $\mathcal{Q}$ touch a certain quadric $\mathcal{Q}' \neq \mathcal{Q}$ from the geodesic family, as we have seen in Section 7.1 (note Theorem 7.1.4).

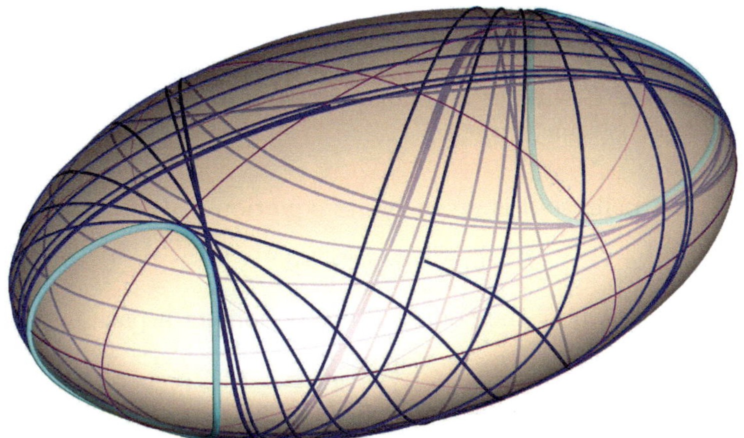

FIGURE 9.50. A geodesic on a triaxial ellipsoid (blue) may wind infinitely many times around the ellipsoid. Thereby, it touches a pair of opposite lines of curvature (cyan).

Much is said about the periodicity and behavior of geodesics (cf. [76]). On ellipsoids of revolution, most of the geodesics traverse a strip bounded by two parallel circles. The geodesics on an ellipsoid that emanate from a point may even have a star-shaped envelope (see [69]).

The geodesics on the triaxial ellipsoid fall into three classes: (1) Circumpolar geodesics encircle the north and south pole without reaching them. (2) The transpolar geodesics accumulate at the north and south pole. (3) The umbilical geodesics join the umbilics of the triaxial ellipsoid and, when returning to the initial umbilic, they come closer to the plane containing the four umbilics.

In the following, we shall give just the results on algebraic geodesics on quadrics. Details can be found in [40, 76, 99, 100]. It is clear that the principal sections of a quadric $\mathcal{Q}$ are geodesic curves on $\mathcal{Q}$, for their osculating plane is everywhere normal to the quadric.

Algebraic geodesics on a triaxial ellipsoid

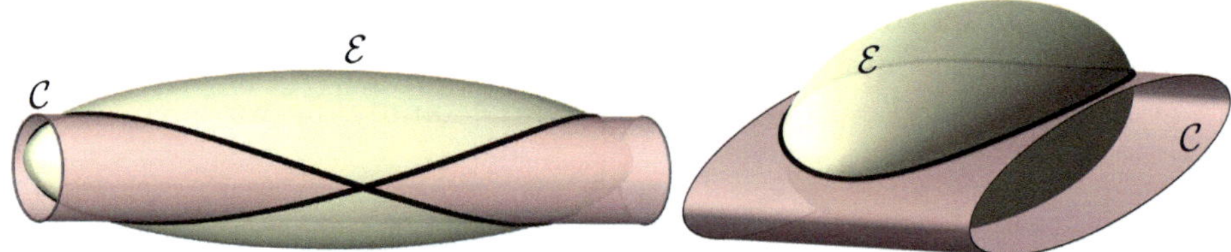

FIGURE 9.51. Closed algebraic geodesics on triaxial ellipsoids: a geodesic with a double point (left), without a double point (right). In both cases, we have chosen the shape parameter $a = 4$.

Assume that an ellipsoid is given by its equation

$$\mathcal{E}:\ b_1x^2 + b_2y^2 + b_3z^2 = 1,$$

where $b_i \in \mathbb{R}^\star$. Let now the coordinate functions of a parametrization of a curve $g \subset \mathcal{E}$ be

$$x(t) = c_x \cdot f_x(t), \quad y(t) = c_y \cdot f_y(t), \quad z(t) = c_z \cdot f_z(t) \tag{9.97}$$

with

$$f_x(t) = \frac{b - \sin^2 t}{a - \sin^2 t}, \quad f_y(t) = \frac{\sin t \cos t}{a - \sin^2 t}, \quad f_z(t) = \frac{\cos t}{a - \sin^2 t}, \tag{9.98}$$

and

$$\begin{gathered} b = \frac{a-2}{2a-1}, \quad b_1 = 4(a^2 - a + 1), \quad b_2 = (2a-1)^2, \quad b_3 = (a-2)^2, \\ c_x = \frac{(a-1)\sqrt{b_2}}{a(a+1)\sqrt{b_1}}, \quad c_y = \frac{\sqrt{a}\sqrt{b_1 b_3}}{(a+1)\sqrt{b_2}}, \quad c_z = \frac{\sqrt{b_1 b_2}}{(a+1)\sqrt{b_3}}. \end{gathered} \tag{9.99}$$

Now, the following result is elementary to verify:

Theorem 9.6.1 *The algebraic space curves on the triaxial ellipsoid $\mathcal{E}$ defined by (9.97), (9.98), and (9.99) are geodesics and quartic space curves of the first kind. They can also be found as the intersection of $\mathcal{E}$ and the quadratic cylinder*

$$\mathcal{C}: \; a_1(x+a_0)^2 + a_2 y^2 = 1, \tag{9.100}$$

where

$$a_0 = \frac{\sqrt{b_2}}{a(a+1)\sqrt{b_1}}, \quad a_1 = \frac{16a^2(a+1)}{b_1}, \quad a_2 = \frac{4(a-1)(a+1)^2 b_2}{\sqrt{b_1 b_3}}.$$

Figure 9.51 (left) shows one example of the geodesic curves described in Theorem 9.6.1.

Proof: In order to show that $\mathbf{g} = (x(t), y(t), z(t))$ (with the coordinate functions from (9.98) and constants from (9.99)) is a parametrization of a geodesic on $\mathcal{E}$, it is sufficient to show that the principal normals of g are scalar multiples of the normals of $\mathcal{E}$ precisely at the points of $g \subset \mathcal{E}$.

We recall that a curve g on the quadric $\mathcal{E}$ is a quartic of the first kind if there exists at least one further quadric $\mathcal{C} \neq \mathcal{E}$ through g (cf. Section 6.9). In the present case, we can implicitize the top view $\mathbf{g}'$ of $\mathbf{q}$ (orthogonal projection onto the plane $z=0$) and find (9.100). ■

• **Exercise 9.6.1** Geodesics without double points on a triaxial ellipsoid.

Analogously to Theorem 9.6.1, show that the ellipsoid $\mathcal{Q}: \; b_1x^2 + b_2y^2 + b_3z^2 = 1$ carries closed algebraic geodesics without double point if the shape parameters and coefficients of $\mathcal{E}$'s equation in (9.98) and (9.99) are changed to

$$b_1 = (a-2)^2, \quad b_2 = (a+1)^2, \quad b_3 = 4(a^2-a+1),$$
$$c_1 = \frac{\sqrt{a^2-1}\sqrt{b_3}}{(2a-1)\sqrt{b_1}}, \quad c_2 = \frac{\sqrt{a(a-2)}\sqrt{b_3}}{(2a-1)\sqrt{b_2}}, \quad c_3 = \tfrac{1}{4}\sqrt{b_3}$$

while b remains unchanged, *i.e.*, $b = (a-2)(2a-1)^{-1}$.

Show further that these geodesics are also quartic curves of the first kind by verifying that they are also located on the cylinders

$$\mathcal{C}: \frac{x^2}{l_x^2} + \frac{(z-a_z)^2}{l_z^2} = 1$$

with

$$l_x^2 = \frac{(a+1)b_3}{4b_1(2a-1)^2}, \quad l_z^2 = \frac{b_3}{16(2a-1)^2(a-1)^2}, \quad a_z = \frac{a\sqrt{b_1}}{(a-1)(2a-1)\sqrt{b_3}}.$$

An example of this type of geodesic on an ellipsoid is shown in Figure 9.51 (right).

The examples of closed algebraic geodesics on a triaxial ellipsoid given above are not the only examples. However, they are the simplest examples. The geodesics on a triaxial ellipsoid can be much more complicated than those described in Theorem 9.6.1 or in Exercise 9.6.1 (cf. Figure 9.52).

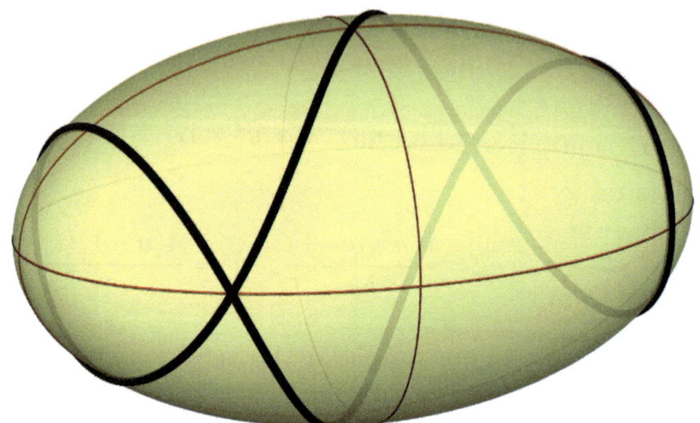

FIGURE 9.52. A closed algebraic geodesic on a triaxial ellipsoid.

Geodesics on a one-sheeted hyperboloid

The simplest examples of geodesics on a ruled quadric are the rulings of the quadric. Since all points on a ruling are inflection points, any normal of a ruling is a principal normal which includes the normals of the surface as well. There exist algebraic geodesics on one-sheeted hyperboloids which belong to a class of curves that we have already described in Section 6.9. We assume that the two-sheeted hyperboloid $\mathcal{H}$ is given by the equation

$$\mathcal{H}:\ b_1x^2 + b_2y^2 + b_3z^2 = 1, \tag{9.101}$$

where the coefficients b_i depend on a parameter $a \in \mathbb{R} \setminus \{-2, -\frac{1}{2}, 1\}$ such that

$$b_1 = (2a+1)^2, \quad b_2 = (2+a)^2, \quad b_3 = -(1-a)^2. \tag{9.102}$$

Then, geodesics on $\mathcal{H}$ can be parametrized by

$$\mathbf{g}(t) = \frac{1}{\cos^2 t - a\sin^2 t}\begin{pmatrix} \dfrac{\cos t}{(1+2a)} \\ \dfrac{a\sin t}{(2+a)} \\ \dfrac{(a+1)\cos t\sin t}{(a-1)} \end{pmatrix}, \quad t \in [0, 2\pi[\,. \tag{9.103}$$

The following property of these geodesics is easily verified:

Theorem 9.6.2 *The algebraic space curves g defined by (9.103) on the one-sheeted hyperboloids (9.101) defined by (9.102) are geodesics and quartic space curves of the second kind.*

Proof: We have to show that the curves parametrized by (9.103) are located on at least one cubic surface that shares two straight lines with $\mathcal{H}$. For that purpose, we reparametrize $\mathbf{g}$ from (9.103) with $t \to 2\arctan u$. Then, we can compute a Groebner basis in the ideal spanned by the polynomials which are simply the numerators of $x - \mathbf{g}_1(u)$, $y - \mathbf{g}_2(u)$, and $z - \mathbf{g}_3(u)$. This yields, among others, the equation of the cubic surface $\mathcal{C}$

$$\mathcal{C}: (a+1)(a+2)^3y^3 - a(a+2)(a-1)^2yz^2 + a(2a+1)(a-1)xz - (a+1)(a+2)y = 0$$

that shares two straight lines with $\mathcal{H}$ in the planes $(a+2)y = \pm 1$. ■

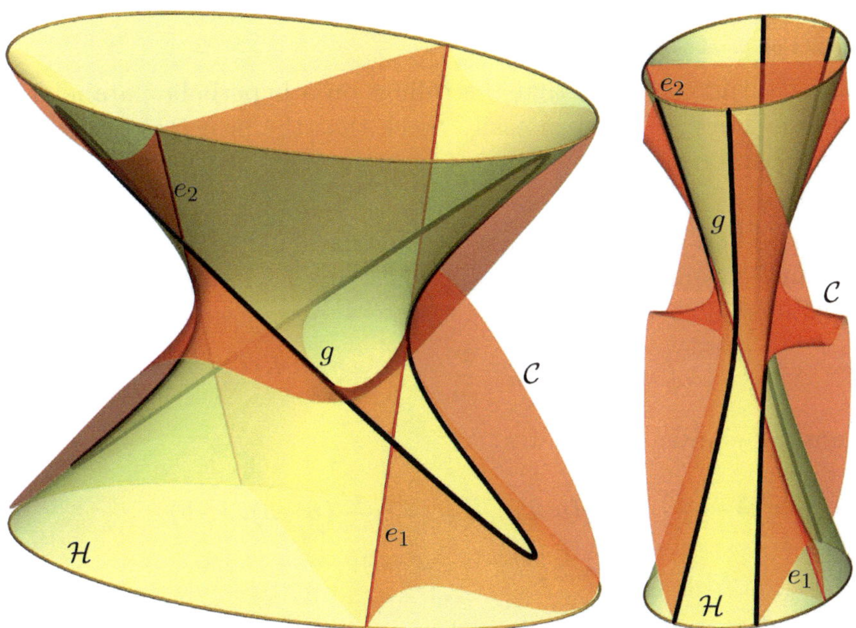

FIGURE 9.53. Left: The rational geodesic g on the one-sheeted hyperboloid $\mathcal{H}$ is a quartic of the second kind. The curve g is a part of the intersection of the hyperboloid $\mathcal{H}$ and a cubic surface $\mathcal{C}$. The two rulings e_1 and e_2 are also common to both $\mathcal{H}$ and $\mathcal{C}$. Right: Another geodesic of the same type shows that these curves can have real points at infinity.

Figure 9.53 (left) shows a rational geodesic on a one-sheeted hyperboloid $\mathcal{H}$. We have chosen $a = -20$ in (9.103) for the curve g on the left-hand side. The example on the right-hand side shows that geodesics (different from lines) on a one-sheeted hyperboloid may have ideal points, which is clear, since the common denominator $\cos^2 t - a\sin^2 t$ of all coordinate functions in (9.103) may have real zeros $t = \operatorname{arccot}(\pm\sqrt{a})$, provided that $a > 0$.

• **Exercise 9.6.2** Quartics of the second kind as geodesics on a one-sheeted hyperboloid.
Show that the cubic surface in the proof of Theorem 9.6.2 can be replaced by

$$\mathcal{K}:\ (a^2-1)(a+2)^2y^2z - a(a-1)^3z^3 + (2a+1)(a+2)(a+1)xy - a(a-1)z = 0.$$

The straight lines in $\mathcal{H}\cap\mathcal{K}$ lie in the planes

$$(2a+1)^2x^2 - (a+2)^2y^2 = 0.$$

Geodesics on paraboloids

It is not surprising at all that the rulings on a hyperbolic paraboloid are geodesics. Further, one can easily check that the principal sections, *i.e.*, the two parabolas in the planes of symmetry of the paraboloid are also geodesics.

In order to find geodesics on a hyperbolic paraboloid $\mathcal{P}_h$ which are different from $\mathcal{P}$'s rulings, we follow [49] and write down a parametrization of the paraboloid

$$\mathcal{P}_h:\ \frac{x^2}{a} - \frac{y^2}{b} = 2z$$

in terms of elliptic coordinates

$$\begin{gathered} x^2 = -\tfrac{a}{a+b}(a+u)(a+v), \quad y^2 = -\tfrac{b}{a+b}(u-b)(v-b), \\ z = \tfrac{1}{2}(b-a-u-v), \end{gathered} \tag{9.104}$$

where $u < -a$ and $v > b$ describes the parameter domain.

Then, the first fundamental form (metric tensor) of $\mathcal{P}_h$ reads

$$\mathrm{d}s^2 = \frac{1}{4}(u-v)\left(\frac{u\mathrm{d}u^2}{(u+a)(u-b)} - \frac{v\mathrm{d}v^2}{(v+a)(v-b)}\right). \tag{9.105}$$

The equation of the geodesics on $\mathcal{P}_h$ is given by

$$\int\sqrt{\frac{u}{(u+a)(v-b)(u+c)}}\,\mathrm{d}u = \int\sqrt{\frac{v}{(v+a)(v-b)(v+c)}}\,\mathrm{d}v, \tag{9.106}$$

with $c \in \mathbb{R}$ being a constant. The integrals in (9.106) are elliptic. In [49] it was shown that (9.106) is the equation of an algebraic curve if

$$\varrho := \sqrt{1 + \frac{b}{a}}$$

is rational. Provided the rationality of ϱ, we can substitute

$$u \to \frac{b}{a(1 - U^2)}, \quad v \to \frac{b}{a(1 - V^2)} \tag{9.107}$$

and integrate (9.106), which, indeed, yields the implicit equation

$$(1 - U)^p (1 - V)^p (\varrho + U)^q (\varrho + V)^q = C(1 + U)^p (1 + V)^p (\varrho - U)^q (\varrho - V)^q, \tag{9.108}$$

where p and q are the numerator and denominator of the rational number $\varrho = \frac{p}{q}$. With the inverse of (9.107) and (9.104) and some rounds of squaring, we can turn (9.108) into an implicit algebraic equation $E(x, y) = 0$ in the coordinates x and y. For fixed p and q, $E(x, y) = 0$ is the equation of an algebraic cylinder $\mathcal{C}$ that intersects the paraboloid $\mathcal{P}_h$ along an algebraic geodesic.

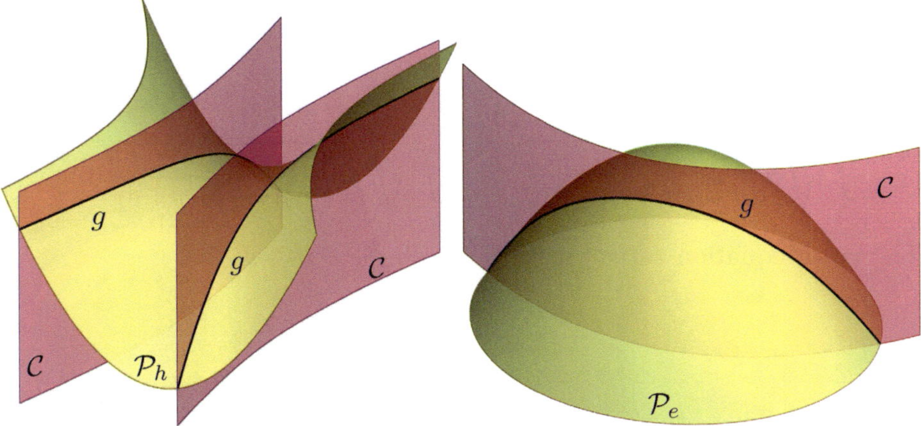

FIGURE 9.54. Left: The geodesic g on the hyperbolic paraboloid $\mathcal{P}_h$ is a cubic parabolic hyperbola, for it has two different ideal points (cf. Example 6.1.1). Right: The geodesic g on the elliptic paraboloid $\mathcal{P}_e$ is the intersection of $\mathcal{P}_e$ with a parabolic cylinder $\mathcal{C}$.

• **Exercise 9.6.3** A cubic geodesic on a hyperbolic paraboloid.

Show that the choice $a = 1$, $b = 3$, $\varrho = 2$, $p = 2$, and $q = 1$ in (9.108) yields the cubic parabolic hyperbola with the parametrization

$$\mathbf{q}(t) = \frac{1}{2t^2}\left(t^3, \sqrt{3}t(t^2-2), t^2-1\right)$$

as a geodesic curve on the hyperbolic paraboloid $\mathcal{P}_h : x^2 - \frac{y^2}{3} = 2z$. Show further that the cylinder $\mathcal{C}: 2\sqrt{3}xy - 6x^2 + 3 = 0$ carries g (cf. Figure 9.54).

We find algebraic geodesics on elliptic paraboloids in nearly the same way as on hyperbolic paraboloids. The parametrization of the elliptic paraboloid $\mathcal{P}_e$ in terms of elliptic coordinates

$$\mathcal{P}_e : \frac{x^2}{a} + \frac{y^2}{b} = 2z$$

reads

$$\begin{aligned} x^2 = -\tfrac{a}{a-b}(u+a)(v+a), \quad y^2 = -\tfrac{b}{b-a}(u+b)(v+b), \\ 2z = -(u+v+a+b), \end{aligned} \tag{9.109}$$

where $u < -a$ and $-a < v < -b$. The metric tensor on the elliptic paraboloid then reads

$$\mathrm{d}s^2 = \frac{u-v}{4}\left(\frac{u\mathrm{d}u^2}{(u+a)(u+b)} - \frac{v\mathrm{d}v^2}{(v+a)(v+b)}\right),$$

and (9.106) changes to

$$\int \sqrt{\frac{u}{(u+a)(u+b)(u+c)}}\,\mathrm{d}u = \int \sqrt{\frac{v}{(v+a)(v+b)(v+c)}}\,\mathrm{d}v.$$

The latter equation turns out to be algebraic if

$$\varrho = \varrho := \sqrt{1 - \frac{b}{a}}$$

is rational. Then, the integration is followed by the change of variables, which yields an algebraic equation of the cylinder that meets $\mathcal{P}_e$ along an algebraic geodesic. An example of an algebraic geodesic on an elliptic paraboloid is shown in Figure 9.54.

10 Line Geometry, Sphere Geometry, Kinematics

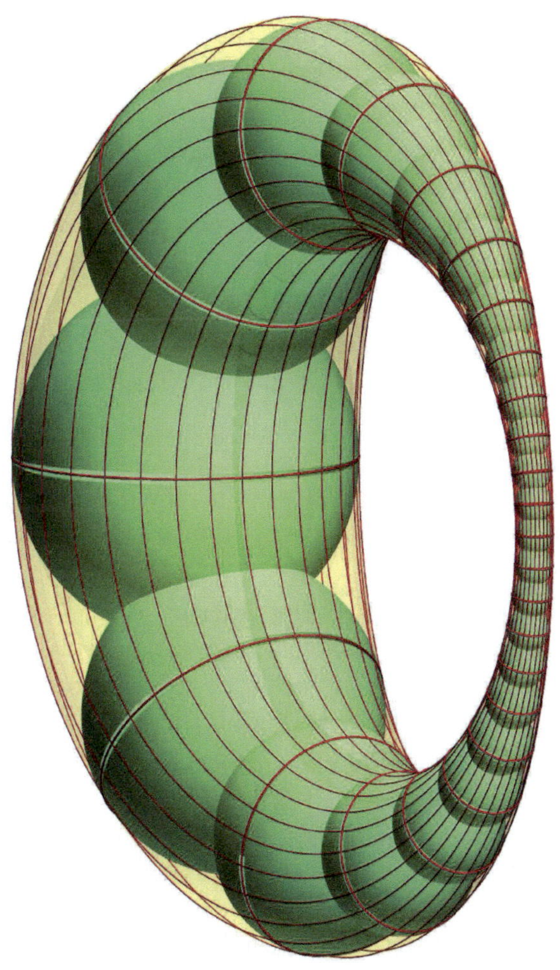

A Dupin cyclide is a quartic and cyclic surface. It is the envelope of a one-parameter family of spheres. In Lie's model of sphere geometry, it is represented by a conic. Lie's line-sphere-mapping maps a conic in Lie's quadric to a conic on Plücker's quadric which corresponds to a regulus in the manifold of lines. Each regulus defines a ruled quadric, for example, a one-sheeted hyperboloid. Consequently, up to Lie's line-sphere-mapping, there is no difference between a Dupin cyclide and a ruled quadric.

B. Odehnal et al., *The Universe of Quadrics*,
https://doi.org/10.1007/978-3-662-61053-4_10

In this section, we show that quadrics in higher dimensional spaces may serve as models for certain geometries. We shall do this by means of three prominent examples:

(1) Line geometry deals with lines in three-dimensional spaces (Euclidean, non-Euclidean, projective, ...) and can be seen as a geometry on a four-dimensional quadric M_2^4 (called *Plücker's quadric*) in projective five-space.

Especially the intersections of M_2^4 with linear subspaces correspond to linear manifolds of lines in 3-space, such as pencil and stars of lines, ruled planes, reguli, linear congruences, and linear complexes.

We shall also study quadratic complexes each of which defines a certain pencil of quadrics with Plücker's quadric.

(2) The geometry of spheres in Euclidean spaces can be studied on a four-dimensional quadric L_2^4 (called Lie's quadric) which also spans a projective space of five dimensions.

However, the model quadric in the case of sphere geometry differs slightly from that of line geometry. We shall see that the models of line and sphere geometry can be merged by a complex collinear transformation, *i.e.*, Lie's line-sphere-mapping.

We shall close this section with a brief description of the sphere geometric analogues to the basic line geometric objects. This will lead to pencils of spheres, Dupin cyclides, a classification of linear complexes of spheres, and the contact transformations acting on the set of Lie spheres.

(3) Finally, the geometry of Euclidean motions can fruitfully be studied on a six-dimensional quadric called *Study's quadric* S_2^6. The geometry on this particular quadric will also be described in brief. The rather curious phenomenon of triality that to a certain degree generalizes the notion of duality in projective spaces can be observed on S_2^6. Usually, triality is considered as a description of some special features of the Dynkin diagram D_4 and the associated Lie group Spin(8) (cf. [1]).

All the models of these geometries turned out to be very useful not only for theoretical studies, but also for many applications in surface modeling and kinematics (cf. [20, 24, 50, 65, 98, 106, 134]).

10.1 Line Geometry

The most important difference between line geometry and projective geometry in the classical sense lies in the change of the point of view. While in projective geometry, the point is the base element of geometry and each object in this geometry is considered as a set of points, this is no longer the case in line geometry. Here, the base element is the straight line in three-dimensional space. A point can then be considered as the carrier of a *star of lines* (two-dimensional manifold of lines through that particular point). A plane is no longer the hyperplane in the ambient space. It is a *ruled plane*, *i.e.*, the two-dimensional manifold of lines in a plane (cf. Figure 10.1).

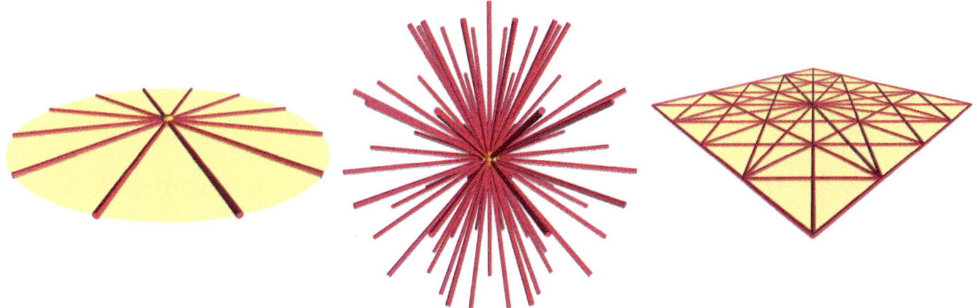

FIGURE 10.1. The three basic building blocks of line geometry (from left to right): a pencil of lines (one-dimensional), a star of lines, and a ruled plane (both two-dimensional).

We shall first coordinatize the manifold of lines in three-dimensional spaces. This gives rise to a point model embedded in a five-dimensional projective space $\mathbb{P}^5$, usually referred to as the *Klein model*[1] of the geometry of lines. There, a four-dimensional quadric $M_2^4 \subset \mathbb{P}^5$ serves as a point model of the set of lines in projective 3-space. M_2^4 is called *Plücker's quadric*,[2] and as such, it contains a huge variety of subquadrics of various dimensions, each of which corresponds to a certain configuration or manifold of lines in (projective) 3-space.

[1] Felix Klein (1849–1925), German mathematician, presented the Erlangen program, which considers a geometry as the theory of invariants of groups of transformations.

[2] Julius Plücker (1801–1868), German mathematician and physicist, worked in the field of line geometry and algebraic curves.

Coordinatization of lines - Plücker coordinates

Let us start in Euclidean 3-space $\mathbb{E}^3$ and assume that points P are given by Cartesian coordinates $\mathbf{p}$. A straight line $l = [P, Q] \subset \mathbb{R}^3$ spanned by two (different) points P and Q can be described by a parametrization $\mathbf{x}(t) = \mathbf{p} + t(\mathbf{q} - \mathbf{p})$ with $t \in \mathbb{R}$. This representation of l depends on the choice of P and Q. We want to end this arbitrariness and define $\mathbf{l} = \mathbf{q} - \mathbf{p}$ as a *direction vector* of l and introduce the *momentum vector*[3] $\bar{\mathbf{l}} := \mathbf{p} \times \mathbf{q}$.

Now, we arrange the two vectors $\mathbf{l}$ and $\bar{\mathbf{l}}$ in one vector

$$(\mathbf{l}, \bar{\mathbf{l}}) = (l_1, l_2, l_3; l_4, l_5, l_6) \in \mathbb{R}^6, \tag{10.1}$$

which is called the *Plücker coordinate* vector of the line l. The Plücker coordinates of a line do not depend on the choice of P and Q on the line l. Replacing each by $\mathbf{p}' = \mathbf{p} + \sigma(\mathbf{q} - \mathbf{p})$ and $\mathbf{q}' = \mathbf{p} - \tau(\mathbf{q} - \mathbf{p})$ with arbitrary $\sigma, \tau \in \mathbb{R}$ with $\sigma \neq \tau$, we find

$$\mathbf{l}' = \mathbf{q}' - \mathbf{p}' = (\tau - \sigma)\mathbf{l} \quad \text{and} \quad \bar{\mathbf{l}}' = \mathbf{p}' \times \mathbf{q}' = (\tau - \sigma)\bar{\mathbf{l}},$$

which results in a scaling of the original Plücker coordinates.

The two vectors $\mathbf{l}$ and $\bar{\mathbf{l}}$ satisfy a bilinear condition, which is clear from the very definition:

$$M_2^4 : \langle \mathbf{l}, \bar{\mathbf{l}} \rangle = l_1 l_4 + l_2 l_5 + l_3 l_6 = 0. \tag{10.2}$$

On the other hand, each sextuple $(l_1, \ldots, l_6) \in \mathbb{R}^6 \backslash \{\mathbf{0}\}$ that satisfies (10.2) corresponds to an oriented straight line in Euclidean 3-space. Equation (10.2) is sometimes referred to as the *Plücker condition.*

- **Exercise 10.1.1** Pedal point of a line.

Assume that $(\mathbf{l}, \bar{\mathbf{l}}) \in \mathbb{R}^6$ are the Plücker coordinates of a line l in Euclidean 3-space. Show that the point P with Cartesian coordinates

$$\mathbf{p} = \mathbf{l} \times \bar{\mathbf{l}} / \langle \mathbf{l}, \mathbf{l} \rangle \tag{10.3}$$

is a point on l and, further, that the line l is orthogonal to the line $[P, O]$ (O being the origin of the coordinate system). The point P is called the *pedal point* of l.

[3]The momentum vector has a physical meaning. It is, indeed, the angular momentum at the origin of the coordinate system induced by a line-bound force acting along l.

The Plücker coordinates of a line can be multiplied with an arbitrary non-vanishing factor $\lambda \in \mathbb{R}^\star$, which leaves the geometric object unchanged. Therefore, the Plücker coordinates can be made homogeneous and allow for an interpretation as homogeneous coordinates in projective five-space $\mathbb{P}^5(\mathbb{R})$. Then, (10.2) is the equation of a ruled quadric with signature $(+++---)$ or $(3,3)$ (cf. Section 4). Thus, M_2^4 described by equation (10.2) carries two independent three-parameter families of two-dimensional subspaces (planes).

Lines that pass through the origin of the underlying coordinate system have the momentum vector $\bar{\mathbf{l}} = \mathbf{0}$. From the definition of Plücker coordinates of lines in Euclidean 3-space, it is clear that the direction vector $\mathbf{l}$ cannot vanish. Nevertheless,

$$\omega: \ l_1 = l_2 = l_3 = 0$$

is a plane contained in M_2^4, but its points do not correspond to lines in Euclidean 3-space. We can summarize in

Theorem 10.1.1 *Each straight line l in Euclidean 3-space $\mathbb{R}^3$ can be represented by a sextuple $(\mathbf{l}, \bar{\mathbf{l}}) = (l_1, \ldots, l_6) \in \mathbb{R}^6 \setminus \{\mathbf{0}\}$ that satisfies* (10.2). *Each point on $M_2^4 \setminus \omega$ corresponds to a straight line in Euclidean 3-space.*

The advantage of the Klein model for the study of the geometry of lines in (Euclidean) 3-space lies in the fact that lines in 3-space are now simply represented by points. Thus, manifolds of lines are submanifolds of M_2^4.

- **Exercise 10.1.2** Pencils of lines are lines in M_2^4.

Show that a pencil of lines (all lines in a plane through a fixed point) are represented by a line in M_2^4. Assume that P with coordinates $\mathbf{p}$ is the vertex of the pencil and assume that $P \notin [Q, R]$ where Q and R have the Cartesian coordinate vectors $\mathbf{q}$ and $\mathbf{r}$. Then, an affine parametrization of $[P, Q]$ is given by $\mathbf{x}(t) = (1-t)\mathbf{q} + t\mathbf{r}$ (with $t \in \mathbb{R}$). The Plücker coordinates $(\mathbf{l}, \bar{\mathbf{l}})$ of the lines through P are now $\mathbf{l} = \mathbf{q}(1-t) + \mathbf{r} - \mathbf{p}$ and $\bar{\mathbf{l}} = \mathbf{p} \times \mathbf{q}(1-t) + \mathbf{p} \times \mathbf{r}\, t$. Obviously, the parametrization of the set of points in M_2^4 is linear.

- **Exercise 10.1.3** Stars of lines and ruled planes.

Use Exercise 10.1.2 in order to show that the lines of a star of lines, *i.e.*, all lines in 3-space passing through a fixed point, correspond to points in a plane that is entirely contained in $M_2^4 \setminus \omega$.

Show that the lines of a ruled plane, *i.e.*, the two-parameter family of lines in an arbitrary plane in 3-space, also correspond to the points in a plane in $M_2^4 \setminus \omega$.

In the Exercises 10.1.2 and 10.1.3, we have seen that there are linear submanifolds of M_2^4 that correspond to linear manifolds of lines in 3-

space. It can also be shown that each line in M_2^4 corresponds to a pencil of lines in 3-space.

It is worth noting that there are two kinds of planes in M_2^4: *Planes of the first kind* correspond to *stars of lines*, while *planes of the second kind* correspond to *ruled planes*. All pairs of planes of the same kind (whether first or second) intersect in a point (corresponding to the common line of two stars or two ruled planes). Two planes of different kinds are either skew or share a line (cf. Theorem 4.3.2): If the vertex of the star is located in the ruled plane, then the star and the ruled plane share a pencil of lines. If the vertex of the star is not located in the ruled plane, then the star and the ruled plane have no line in common, and thus, the corresponding planes in M_2^4 are skew (cf. Figure 10.2).

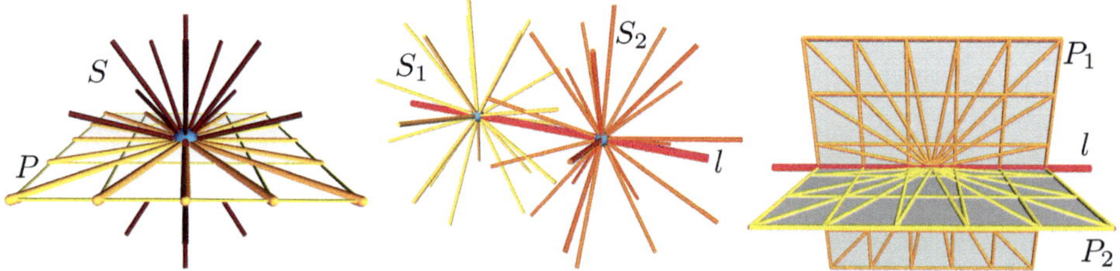

FIGURE 10.2. Pairs of planes of equal and different kind in M_2^4. Left: Two planes σ and π in M_2^4 of different kinds that intersect in a line correspond to a star S of lines and a ruled plane P sharing a pencil of lines. Middle: Two planes σ_1 and σ_2 (of M_2^4) of the first kind always share a point L and correspond to two stars S_1 and S_2 of lines which always share a line l. Right: Two planes π_1 and π_2 (of M_2^4) of the second kind always share a point L and correspond to two ruled planes P_1 and P_2 that have a straight line l in common.

In the next section, we shall recreate the Klein model of the geometry of lines starting in projective 3-space. This allows us to treat manifolds of lines more uniformly, since it is no longer necessary to differ between lines at infinity and proper lines. The exceptional set ω will then be removed.

Projective Line Geometry

Let us now shift to a projective space $\mathbb{P}^3(\mathbb{F})$ over some commutative field with char$\mathbb{F} \neq 2$. Assume that $\mathbf{x} = (x_0, x_1, x_2, x_3)$ and $\mathbf{y} = (y_0, y_1, y_2, y_3)$ (with $\mathbf{x}, \mathbf{y} \in \mathbb{F}^4 \setminus \{\mathbf{0}\}$) being linearly independent). Then, $\mathbf{x}\mathbb{F}$ and $\mathbf{y}\mathbb{F}$ can be considered the homogeneous coordinates of points in $\mathbb{P}^3(\mathbb{F})$. The line $l = [X, Y]$ spanned by X and Y can be parametrized by $\mathbf{p}(\lambda, \mu) = \lambda\mathbf{x} + \mu\mathbf{y}$

with $\lambda : \mu \neq 0 : 0$, which is again a representation that depends on arbitrarily chosen points on the line. In order to free the line's representation from the coordinate of the arbitrarily chosen points X and Y, we look at the matrix

$$\begin{pmatrix} x_0 & x_1 & x_2 & x_3 \\ y_0 & y_1 & y_2 & y_3 \\ x_0 & x_1 & x_2 & x_3 \\ y_0 & y_1 & y_2 & y_3 \end{pmatrix}, \tag{10.4}$$

which is clearly of rank two. Consequently, its determinant and the determinants of all its 3×3-submatrices vanish. This is not the case for its 2×2-submatrices: At least one of the 2×2-submatrices of (10.4) is regular, and therefore, its determinant does not vanish.

Among all 2×2-submatrices, there are only six relevant ones,

$$S_{ij} := \begin{pmatrix} x_i & x_j \\ y_i & y_j \end{pmatrix} \text{ with } (i,j) \in \{(0,1),(0,2),(0,3),(2,3),(3,1),(1,2)\}, \tag{10.5}$$

and it is admissible to assume that at least one of them is regular. Now we define the coordinates of the line l via

$$l = (l_1, l_2, l_3; l_4, l_5, l_6) = (l_{01}, l_{02}, l_{03}; l_{23}, l_{31}, l_{12}) \text{ with } l_{ij} := \det S_{ij}. \tag{10.6}$$

An elementary computation shows that the thus defined coordinates of a line also satisfy (10.2).

This gives rise to

Theorem 10.1.2 *Each line l in projective 3-space can be mapped to a point on a regular ruled quadric M_2^4 in projective five-space. Each point on M_2^4 corresponds to precisely one line in $\mathbb{P}^3(\mathbb{F})$.*

Proof: The only thing that remains to be shown is that each point $(l_1, \dots, l_6) \in M_2^4$ corresponds to a line in $\mathbb{P}^3(\mathbb{F})$. For that purpose, we write down the matrix

$$\begin{pmatrix} 0 & l_1 & l_2 & l_3 \\ -l_1 & 0 & l_6 & -l_5 \\ -l_2 & -l_6 & 0 & l_4 \\ -l_3 & l_5 & -l_4 & 0 \end{pmatrix} \tag{10.7}$$

and observe that each row contains the homogeneous coordinates of a point in one of the coordinate planes. The above matrix is of rank two as can be shown by means of elementary matrix operations together with (10.2), which is valid for all $l \in M_2^4$.

Consequently, any pair of points out of these four can be chosen in order to recompute the Plücker coordinates of l. By virtue of (10.2), this yields a scalar multiple of $(l_1, \dots, l_6)$. ■

The mapping γ that maps a line $l \subset \mathbb{P}^3$ to the corresponding point $L \in M_2^4$ is called *Klein mapping*. The quadric M_2^4 is the first non-trivial example of a *Grassmann manifold*,[4] which in this context would be denoted by $\mathcal{G}_{3,1}$. Moreover, M_2^4 is the only hypersurface among the non-trivial Grassmannians.

• **Exercise 10.1.4** Independence of Plücker coordinates.

Show that the homogeneous Plücker coordinates of a line $\mathbf{x}(\lambda,\mu) = \lambda\mathbf{p}+\mu\mathbf{q}$ with $(\lambda,\mu) \in \mathbb{F}^2\backslash\{\mathbf{0}\}$ are independent of the choice of P and Q on l by replacing $\mathbf{p}$ and $\mathbf{q}$ by two points $\mathbf{x}(\lambda_1,\mu_1)$ and (λ_2,μ_2) with $\lambda_1\mu_2 - \lambda_2\mu_1 \neq 0$.

• **Exercise 10.1.5** Switching between affine and projective 3-space.

Assume that P and Q are points in $\mathbb{F}^3$ with coordinate vectors $\mathbf{p}$ and $\mathbf{q}$. Perform the projective closure by assigning the homogeneous coordinates $(1,\mathbf{p})$ and $(1,\mathbf{q})$ to the points P and Q. Show that the Plücker coordinates defined in (10.1) and (10.6) agree. One point, say Q, can be replaced by the line's ideal point, and the Plücker coordinates remain unchanged.

• **Exercise 10.1.6** Points and lines at infinity.

Assume that P and Q are points in the plane at infinity given by $x_0 = 0$. Show that the lines at infinity are characterized by Plücker coordinates with the first three entries equal to zero (while at least one of the remaining coordinates does not vanish). This shows that the plane ω at infinity considered as a ruled plane is given by the equation $l_1 = l_2 = l_3 = 0$.

• **Exercise 10.1.7** Lines as intersections of planes, axis coordinates.

Each line l can be considered as the intersection of two different planes ε and φ. Assume that the planes are given by their homogeneous coordinates $(e_0,\ldots,e_3)\mathbb{F}$ and $(f_0,\ldots,f_3)\mathbb{F}$. Then, the *axis coordinates* $(\mathbf{a},\overline{\mathbf{a}})$ of the line $l = \varepsilon \cap \varphi$ can be computed in the same way as the (Plücker) line coordinates as defined in (10.5) and (10.6).

Show that the axis coordinates and the line coordinates agree up to a permutation:

$$\mathbf{a} = \overline{\mathbf{l}} \quad \text{and} \quad \overline{\mathbf{a}} = \mathbf{l}. \tag{10.8}$$

The polar system of M_2^4

Assume that l and m are two different lines that intersect in one point, say P, with inhomogeneous coordinate vector $\mathbf{p} \in \mathbb{F}^3$, and the respective (non-zero) direction vectors are $\mathbf{l}$ and $\mathbf{m}$. Then, the momentum vectors are $\overline{\mathbf{l}} = \mathbf{p} \times \mathbf{l}$ and $\overline{\mathbf{m}} = \mathbf{p} \times \mathbf{m}$. Now, the Plücker coordinates $(\mathbf{l},\overline{\mathbf{l}})$ and $(\mathbf{m},\overline{\mathbf{m}})$ satisfy

$$\langle \mathbf{l},\overline{\mathbf{m}}\rangle + \langle \mathbf{m},\overline{\mathbf{l}}\rangle = 0, \tag{10.9}$$

[4] Hermann Grassmann (1809–1877), German mathematician, physicist, teacher, and linguist. He founded the theory of linear spaces and tensors in his famous book *Die Ausdehnungslehre* (Engl.: The theory of Linear Extensions.) first published in 1844, extended and improved in 1862.

since $\langle \mathbf{l}, \overline{\mathbf{m}} \rangle + \langle \mathbf{m}, \overline{\mathbf{l}} \rangle = \det(\mathbf{l}, \mathbf{p}, \mathbf{m}) + \det(\mathbf{m}, \mathbf{p}, \mathbf{l}) = 0$. On the other hand, if (10.9) is fulfilled for two vectors $(\mathbf{l}, \overline{\mathbf{l}})$ and $(\mathbf{m}, \overline{\mathbf{m}})$ that additionally satisfy (10.2)), then these two sextuples are the Plücker coordinates of two intersecting lines.

Let $L = (l_1, \dots, l_6)\mathbb{F}$ and $M = (m_1, \dots, m_6)\mathbb{F}$ be the points of M_2^4 which correspond to the lines $(\mathbf{l}, \overline{\mathbf{l}})$ and $(\mathbf{m}, \overline{\mathbf{m}})$. Then, (10.9) reads

$$\begin{aligned} \Omega(L, M) &:= \langle \mathbf{l}, \overline{\mathbf{m}} \rangle + \langle \overline{\mathbf{l}}, \mathbf{m} \rangle = \\ &= l_1 m_4 + l_2 m_5 + l_3 m_6 + l_4 m_1 + l_5 m_2 + l_3 m_6 = 0. \end{aligned} \tag{10.10}$$

The latter equation gives the polar form of the quadric M_2^4. Thus, two intersecting lines l and m correspond to a pair of points L and M conjugate w.r.t. M_2^4. For any point $L \in M_2^4$ and $X = (x_1, \dots, x_6)\mathbb{F}$, $\Omega(L, X) = 0$ is the equation of the tangential hyperplane of M_2^4 at L. If $C \notin M_2^4$, then $\Omega(C, X) = 0$ is the equation of the polar hyperplane w.r.t. M_2^4.

Linear complexes of lines

Now, we can prescribe an arbitrary point $C = (c_1, \dots, c_6)\mathbb{F}$ in the Klein model and ask for all points $X = (x_1, \dots, x_6)\mathbb{F}$ conjugate w.r.t. M_2^4. Obviously, there are two cases to be distinguished:

(1) If $\Omega(C, C) = 0$, then C corresponds to a straight line in 3-space. The set of all X with $\Omega(C, X) = 0$ is a three-dimensional manifold $\mathcal{C}$ of straight lines intersecting the line $(\mathbf{c}, \overline{\mathbf{c}})$. $\mathcal{C}$ is called a *singular linear line complex.* The straight line $(\mathbf{c}, \overline{\mathbf{c}})$ is called the *axis of the complex.* The Klein image of the set of lines in singular linear line complex is a three-dimensional quadratic cone with vertex $C = (c_1, \dots, c_6)\mathbb{F}$, which is the intersection of M_2^4 with its tangential hyperplane at C. Figure 10.3 (left) shows some lines of a singular linear line complex with a proper axis a.

(2) If $\Omega(C, C) \neq 0$, then $(\mathbf{c}, \overline{\mathbf{c}})$ is not the Plücker coordinate vector of a straight line in three-dimensional space. The manifold described by $\Omega(C, X) = 0$ is the polar hyperplane of C w.r.t. M_2^4, and its points X that additionally satisfy $\Omega(X, X) = 0$ are the straight lines of a so-called *regular linear line complex.* The Klein image of the set of lines in a regular line complex is a regular three-dimensional quadric in M_2^4. Figure 10.3 (right) shows some lines of a regular linear line complex. Later, we shall see that these line manifolds are in a close relation to kinematics (see Theorem 10.1.3 and Exercise 10.1.1).

FIGURE 10.3. Left: The lines of a singular linear line complex meet a fixed line a and can be arranged in stars with their vertices on a or in pencils with their planes through a. The lines parallel to a form the star around a's ideal point A_∞. Right: The lines in a regular linear line complex are the null lines in a null polarity, which can also be viewed as the instantaneous path normals of a uniform Euclidean motion.

- **Exercise 10.1.8** Generators of the Klein image of linear line complexes.

Assume that we are given a three-dimensional quadratic cone in M_2^4 with vertex $C = (1, 0, \ldots, 0)\mathbb{F}$. Show that the points of the cone represent the lines in a singular linear complex. The represented singular linear line complex contains pencils and stars of lines as well as ruled planes.

Show that the regular linear line complex $C = (1, 0, 0, 1, 0, 0)\mathbb{F}$ contains pencils of lines, but no stars and no ruled planes. Use the fact that the Klein image of a regular linear line complex is a three-dimensional regular ruled quadric. Therefore, its index (*i.e.*, the maximal dimension of subspaces in the quadric) is one at most.

A regular linear line complex $C = (c_1, \dots, c_6)\mathbb{F}$ defines a regular correlation κ, *i.e.*, a collineation from $\mathbb{P}^3(\mathbb{F})$ onto its dual space $\widehat{\mathbb{P}}^3(\mathbb{F})$ (cf. Section 6.3) given by the regular skew symmetric matrix

$$\mathbf{C} = \begin{pmatrix} 0 & c_4 & c_5 & c_6 \\ -c_4 & 0 & c_3 & -c_2 \\ -c_5 & -c_3 & 0 & c_1 \\ -c_6 & c_2 & -c_1 & 0 \end{pmatrix}. \tag{10.11}$$

The matrix $\mathbf{C}$ is regular, for C is regular by assumption and $\det \mathbf{C} = 4\langle \mathbf{c}, \overline{\mathbf{c}}\rangle^2$.

According to Theorem 4.2.2, the correlation κ is a null polarity, since each point P is incident with its image plane. The null polarity κ assigns to each point a plane, called the *null plane*. The adjoint correlation $\widehat{\kappa}: \widehat{\mathbb{P}}^3 \to \mathbb{P}^3$ assigns to each plane in $\mathbb{P}^3$ a unique point called the *null point*.

We have seen another null polarity associated with the cubic space curves in Chapter 6 (cf. Theorem 6.3.1 on page 227).

• **Exercise 10.1.9** Singular null polarity.

Study the linear mapping $\mathbf{x}\mathbb{F} \mapsto \mathbf{u}\mathbb{F} = \mathbf{C}\mathbf{x}$ with a singular skew symmetric matrix $\mathbf{C}$. What is the meaning of $\ker \mathbf{C}$?

• **Exercise 10.1.10** Incidence conditions of points and lines.

Let a point P be given by homogeneous coordinates $(p_0, \mathbf{p})\mathbb{F}$ and let further $(\mathbf{l}, \overline{\mathbf{l}})\mathbb{F}$ be the homogeneous Plücker coordinates of a straight line in $\mathbb{P}^3$. Now, show that

$$\langle \mathbf{p}, \overline{\mathbf{l}}\rangle = 0 \quad \text{and} \quad -p_0\overline{\mathbf{l}} + \mathbf{p} \times \mathbf{l} = \mathbf{0} \tag{10.12}$$

are necessary and sufficient for $P \in l$. The matrix form of these conditions is dual to that given in (10.7). We made use of these conditions in the proof of Theorem 6.5.1.

The left-hand sides of the four linear equations (10.12) evaluate to the homogeneous coordinates of the plane spanned by P and l if $P \notin l$, while the *dual equations* (10.7) yield the homogeneous coordinates of the intersection point of a line and a plane.

• **Exercise 10.1.11** Axis of a regular linear line complex.

Let $C = (c_1, \dots, c_6)$ be the coordinate vector of a regular linear line complex, *i.e.*, $\Omega(C, C) = 2\langle \mathbf{c}, \overline{\mathbf{c}}\rangle \neq 0$. Use (10.11) and show that the null point of the ideal plane has the homogeneous coordinate vector $U = (0, \mathbf{c})$.

Then, compute the polar line a of the absolute polar u of U and show that the Plücker coordinates of a are

$$(\mathbf{a}, \overline{\mathbf{a}}) = (\mathbf{c}, \overline{\mathbf{c}} - p\mathbf{c}) \tag{10.13}$$

with the *pitch of the complex* defined by

$$p = \frac{\langle \mathbf{c}, \overline{\mathbf{c}}\rangle}{\langle \mathbf{c}, \mathbf{c}\rangle}. \tag{10.14}$$

Linear complexes of lines play an important role in Euclidean kinematics:

Theorem 10.1.3 *At each regular instant of a one-parameter Euclidean motion, the path normals form a regular or singular linear line complex called the* instantaneous complex.

Proof: A smooth one-parameter Euclidean motion can be described by

$$\mathbf{x}'(t) = \mathbf{A}(t)\mathbf{x} + \mathbf{a}(t), \tag{10.15}$$

where $t \in I \subset \mathbb{R}$ with an orthonormal matrix $\mathbf{A}(t) \in \mathrm{SO}(3)$, all of whose entries are smooth functions depending on the parameter t and the path of the origin $\mathbf{a}(t) : I \to \mathbb{R}^3$.

We differentiate once w.r.t. t (we mark derivatives with a dot and suppress the instant t), which yields the velocity vector $\mathbf{v}(\mathbf{x})$ of $\mathbf{x}$:

$$\mathbf{v}(\mathbf{x}) = \dot{\mathbf{x}}' = \dot{\mathbf{A}}\mathbf{x} + \dot{\mathbf{a}}. \tag{10.16}$$

In the latter equation, $\mathbf{v}$ is expressed in terms of $\mathbf{x}$ in the moving system. It is more convenient if we express $\mathbf{v}$ in the fixed system. For that purpose, we solve (10.15) for $\mathbf{x}$ and find $\mathbf{x} = \mathbf{A}^{\mathrm{T}}(\mathbf{x}' - \mathbf{a})$. This is true, for $\mathbf{A} \in \mathrm{SO}(3)$, and thus, $\mathbf{A}^{-1} = \mathbf{A}^{\mathrm{T}}$. Moreover, since $\mathbf{A}\mathbf{A}^{\mathrm{T}} = \mathbf{E}_3 =$ const., we obtain by differentiation w.r.t. t:

$$\dot{\mathbf{A}}\mathbf{A}^{\mathrm{T}} + \mathbf{A}\dot{\mathbf{A}}^{\mathrm{T}} = \dot{\mathbf{A}}\mathbf{A}^{\mathrm{T}} + \left(\dot{\mathbf{A}}\mathbf{A}^{\mathrm{T}}\right)^{\mathrm{T}},$$

and thus,

$$\dot{\mathbf{A}}\mathbf{A}^{\mathrm{T}} = -\left(\dot{\mathbf{A}}\mathbf{A}^{\mathrm{T}}\right)^{\mathrm{T}} = \mathbf{0}.$$

So, $\mathbf{C} := \dot{\mathbf{A}}\mathbf{A}^{\mathrm{T}}$ is skew symmetric and the linear mapping $\mathbf{x} \mapsto \mathbf{C}\mathbf{x}$ can be written as a cross product $\mathbf{C}\mathbf{x} = \mathbf{c} \times \mathbf{x}$. Inserting this into (10.16) and defining $\overline{\mathbf{c}} := \dot{\mathbf{a}} - \dot{\mathbf{A}}\mathbf{A}^{\mathrm{T}}\mathbf{a}$, we arrive at

$$\mathbf{v} = \mathbf{c} \times \mathbf{x} + \overline{\mathbf{c}}. \tag{10.17}$$

Finally, the path normals may have Plücker coordinates $(\mathbf{n}, \overline{\mathbf{n}})$ that satisfy

$$\begin{aligned}\langle \mathbf{v}, \mathbf{n}\rangle &= \langle \mathbf{c} \times \mathbf{x}, \mathbf{n}\rangle + \langle \overline{\mathbf{c}}, \mathbf{n}\rangle = \\ &= \det(\mathbf{c}, \mathbf{x}, \mathbf{n}) + \langle \overline{\mathbf{c}}, \mathbf{n}\rangle = \langle \mathbf{c}, \overline{\mathbf{n}}\rangle + \langle \overline{\mathbf{c}}, \mathbf{n}\rangle = \Omega(C, N) = 0,\end{aligned}$$

which confirms that their Plücker coordinates satisfy a homogeneous linear equation. ■

▪ **Example 10.1.1** Helical motion, rotation, translation – the one-parameter subgroups of the group of Euclidean motions.

1. Use (10.15) and assume that

$$\mathbf{A}(t) = \begin{pmatrix} \cos t & -\sin t & 0 \\ \sin t & \cos t & 0 \\ 0 & 0 & 1 \end{pmatrix} \quad \text{and} \quad \mathbf{a}(t) = \begin{pmatrix} 0 \\ 0 \\ qt \end{pmatrix}$$

   with $t \in \mathbb{R}$ and $q \in \mathbb{R}^\star$. Compute the coordinates of the instantaneous complex of the thus defined *helical motion* and show that it does not depend on t: $\mathbf{c} = (0, 0, 1)$ and $\overline{\mathbf{c}} = (0, 0, p)$. Determine the axis with (10.13) and the pitch with (10.14).

   We use extended Cartesian coordinates and write

$$\begin{pmatrix} 1 \\ \mathbf{x} \end{pmatrix} \mapsto \begin{pmatrix} 1 \\ \mathbf{x}' \end{pmatrix} = \begin{pmatrix} 1 \\ \mathbf{A}\mathbf{x} + \mathbf{a} \end{pmatrix} = \underbrace{\begin{pmatrix} 1 & \mathbf{o}^{\mathrm{T}} \\ \mathbf{a} & \mathbf{A} \end{pmatrix}}_{=:\underline{\mathbf{A}}} \begin{pmatrix} 1 \\ \mathbf{x} \end{pmatrix}.$$

Now, the corresponding tangent vector of the Lie group equals

$$\mathbf{T} = \begin{pmatrix} 0 & 0 & 0 & 0 \\ 0 & 0 & -1 & 0 \\ 0 & 1 & 0 & 0 \\ p & 0 & 0 & 0 \end{pmatrix},$$

and the exponential mapping yields $\underline{\mathbf{A}} = \exp(t\mathbf{T})$.

2. We can apply (10.15) to a simpler kind of uniform motion by assuming

$$\mathbf{A} = \begin{pmatrix} \cos t & -\sin t & 0 \\ \sin t & \cos t & 0 \\ 0 & 0 & 1 \end{pmatrix} \quad \text{and} \quad \mathbf{a} = \mathbf{o}.$$

Compute the coordinates of the instantaneous complex of the thus defined *rotation* and show that it does not depend on t: $\mathbf{c} = (0,0,1)$ and $\overline{\mathbf{c}} = \mathbf{o}$. Determine the axis with (10.13) and the pitch with (10.14).

In the present case, it is not necessary to switch to a representation in terms of extended Cartesian coordinates. The matrix $\mathbf{C}$ of the linear mapping $\mathbb{R}^3 \to \mathbb{R}^3$ with $\mathbf{x} \mapsto \mathbf{c} \times \mathbf{x}$ equals

$$\mathbf{C} = \begin{pmatrix} 0 & -1 & 0 \\ 1 & 0 & 0 \\ 0 & 0 & 0 \end{pmatrix}.$$

$\mathbf{C}$ is a tangent vector of the Lie group of Euclidean motions and $\mathbf{A} = \exp(t\mathbf{C})$.

3. Finally, let us assume that $\mathbf{A} = \mathbf{E}_3$ and $\mathbf{a} = (0,0,pt)$ with constant $p \in \mathbb{R}^\star$. Compute the coordinates of the instantaneous complex of the thus defined *translation* and show that it does not depend on t: $\mathbf{c} = (0,0,0)$ and $\overline{\mathbf{c}} = (0,0,p)$. Determine the axis with (10.13) and the pitch with (10.14).

Again, it is favorable to use the extended coordinate representation. Now the tangent vector of the Lie group reads

$$\mathbf{T} = \begin{pmatrix} 0 & 0 & 0 & 0 \\ 0 & 0 & 0 & 0 \\ 0 & 0 & 0 & 0 \\ p & 0 & 0 & 0 \end{pmatrix}$$

with $p \neq 0$, and the exponential map applied to $\mathbf{T}$ yields a parametrization of the one-parameter subgroup, *i.e.*, the translation.

Beyond line geometry

The definition of Plücker coordinates as done in (10.6) is a special case of a more general concept. In fact, here we have built the *alternating tensor product* $\mathbb{F}^4 \wedge \mathbb{F}^4$ of the vector space $\mathbb{F}^4$ with itself. According to the universal factorization property, $\mathbb{F}^4 \wedge \mathbb{F}^4 \cong \mathbb{F}^6$. For any two elements $\mathbf{p}, \mathbf{q} \in \mathbb{F}^4$, the element $\mathbf{p} \wedge \mathbf{q} \in \mathbb{F}^6$ is a two-tensor, and as such, an element of the *exterior algebra* $\bigwedge \mathbb{F}^4 = \bigoplus_{i=0}^{4} \bigwedge^i \mathbb{F}^4$ of $\mathbb{F}^4$.

The presence of an analytical model is not a necessary ingredient for the construction of Grassmannians. Grassmannians can be generated in a purely synthetical way, and recursively, starting from trivial Grassmannians which are identical with projective space (see [21]).

Most of the time in this and in previous sections, we restrict ourselves to analytical models based on the real or complex number field. However, finite fields of arbitrary characteristic can be chosen. Grassmannians over finite fields show many special features which can never be observed with those over the real or complex number field. For example, if $\mathrm{char}\mathbb{F} = 2$, the polarities are null polarities, and vice versa. Each quadric and especially $\mathcal{G}_{3,1}(\mathbb{F})$ has a *nucleus*, *i.e.*, a projective subspace which is incident with all tangent hyperplanes of the quadric (see for example [57]).

Collinear automorphisms and quadratic submanifolds

Collinear transformations in projective 3-space $\mathbb{P}^3(\mathbb{F})$ are induced by linear automorphisms of $\mathrm{GL}(\mathbb{F},4)$. Assume that $\kappa: \mathbb{P}^3(\mathbb{F}) \to \mathbb{P}^3(\mathbb{F})$ is given by one of its coordinate matrices $\mathbf{A} \in \mathbb{F}^{4\times 4}$ and acts on the set of points in the following way:

$$\kappa: \ \mathbf{x}\mathbb{F} \mapsto \mathbf{x}'\mathbb{F} = (\mathbf{A}\mathbf{x})\mathbb{F}. \tag{10.18}$$

For any collineation, the transformation matrix $\mathbf{A}$ is unique only up to a non-vanishing factor. Collineations map lines in 3-space to lines in 3-space. The Plücker coordinates $(l_1, \ldots, l_6)$ of a line $l = [X, Y]$ change according to

$$l'_{ij} = \sum_k a_{ik}x_k \sum_l a_{jl}y_l - \sum_k a_{jk}x_k \sum_l a_{il}y_l.$$

Thus, $\kappa: \mathbb{P}^3(\mathbb{F}) \to \mathbb{P}^3(\mathbb{F})$ induces a collineation κ^γ in the Klein model. In fact, κ^γ is an automorphic collineation, since the Plücker coordinates l'_{ij} of $l' = \kappa(l)$ are linear in the Plücker coordinates l_i of l and can also be computed from the homogeneous coordinates of $X' = \kappa(X)$ and $\kappa(Y)$ according to (10.6).

The coordinate matrix of the induced collineation κ^γ is simply denoted by

$$\mathbf{A} \owedge \mathbf{A} \in \mathrm{GL}(\mathbb{F}, 6) \tag{10.19}$$

and usually called the *alternating Kronecker product* of $\mathbf{A}$ with itself.

Besides collineations, there are correlations acting on projective 3-space that leave lines invariant. To be more precise: A correlation maps a line considered as a range of points to a line considered as the carrier of a pencil of planes. Since correlations in $\mathbb{P}^3(\mathbb{F})$ are also induced by linear mappings $\mathbb{F}^4 \to \mathbb{F}^4$, the line preserving transformations induced in the Klein model are also described by matrices in $\mathbb{F}^{6\times 6}$. For example, null polarities induce automorphic harmonic homologies of M_2^4. The matrices of the induced mappings are also computed according to (10.19). The action of the mapping induced in the Klein model is also a left-multiplication, but this time with a subsequent switch of coordinates according to Exercise 10.1.7.

Theorem 10.1.4 *Projective collineations $\mathbb{P}^3(\mathbb{F}) \to \mathbb{P}^3(\mathbb{F})$ induce automorphic collineations of M_2^4 that preserve the types of generators of M_2^4. Projective correlations $\mathbb{P}^3(\mathbb{F}) \to \mathbb{P}^3(\mathbb{F})$ induce automorphic collineations of M_2^4 that interchange generators of the first and second kind.*

Each automorphic collineation of M_2^4 induces either a projective collineation or a projective correlation in $\mathbb{P}^3(\mathbb{F})$.

Normal forms of linear complexes of lines

With the notion of the induced mappings, we are able to show that linear line complexes can be given in normal forms. We will not do this in detail here. It is more or less a subject of linear algebra.

In Example 10.1.1, we have seen that the three basic one-parameter subgroups of the group of Euclidean motions correspond to three standard types of linear complexes:

(1) The complex $C = (0,0,1;0,0,p)$ attached to the helical motion (rotation about the z-axis with additional proportional translation in the same direction) is a regular linear line complex. Its equation in terms of homogeneous coordinates in the Klein model of line geometry reads $x_6 + px_3 = 0$. This hyperplane of $\mathbb{P}^5(\mathbb{R})$ intersects M_2^4 along a regular ruled quadric whose equation can be given in the form

$$C^\gamma:\ x_6 + px_3 = 0, \quad x_1x_4 + x_2x_5 - px_3^2 = 0.$$

The second equation is the equation of a regular ruled quadric and can be transformed into its normal form by the linear substitution

$$\tau:\ x_1 = y_1 + y_4,\ x_4 = y_1 - y_4,\ x_2 = y_2 + y_5,\ x_5 = y_2 - y_5,\ x_3 = y_3 = 0$$

and changes to

$$Q:\ y_1^4 + y_2^2 - py_3^2 - y_4^2 - y_5^2 = 0.$$

The index of Q (the maximal dimension of subspaces on Q) equals one, and indeed, there is a three-parameter family of straight lines on Q corresponding to the pencils of null lines in the complex C.

(2) The singular complex $C = (0,0,1;0,0,0)$ corresponds to the rotation about the z-axis. Its homogeneous equation reads $T: \; x_6 = 0$, which is the equation of the tangential hyperplane of M_2^4 at $C = (0,0,1;0,0,0)$. According to the results in Chapter 4, the intersection of a quadric with its tangential hyperplane T at point C is a quadratic cone Γ_C whose vertex contains the point C of contact. The equation of $\Gamma_C \subset M_2^4$ equals

$$\Gamma_C: \; x_6 = 0, \quad x_1x_4 + x_2x_5 = 0.$$

The normal form is again found by linear substitution

$$\tau: \; x_1 = y_1 + y_4, \; x_4 = y_1 - y_5, \; x_2 = y_2 + y_5, \; x_5 = y_2 - y_5$$

and reads

$$\Gamma_C: \; y_1^2 + y_2^2 - y_4^2 - y_5^2 = 0.$$

Obviously, Γ_C is a quadratic cone erected on a two-dimensional ruled quadric with a point-shaped vertex. Therefore, this cone has two independent one-parameter families of two-dimensional generators corresponding to the stars of lines and ruled planes contained in the complex C. Clearly, the two families of generators on Γ_C stem from the two different families of generators on M_2^4.

(3) Finally, the complex $C = (0,0,0;0,0,1)$ corresponding to the translation is also singular. From the projective point of view, it does not differ from the complex related to the rotation in (2). There is only a difference from the viewpoint of Euclidean geometry. The complex consists of all straight lines intersecting a particular ideal line.

Pencils of linear complexes of lines

The study of submanifolds of the manifold of lines usually starts with the discussion of pencils of linear line complexes. For that purpose, we assume that C_1 and C_2 are two linear line complexes, and thus, they can also be considered as points in the Klein model. Then, the linear one-parameter family of linear complexes

$$P(t_1, t_2) = t_1C_1 + t_2C_2 \quad \text{with } (t_1, t_2) \in \mathbb{R}^2 \setminus \{0, 0\} \tag{10.20}$$

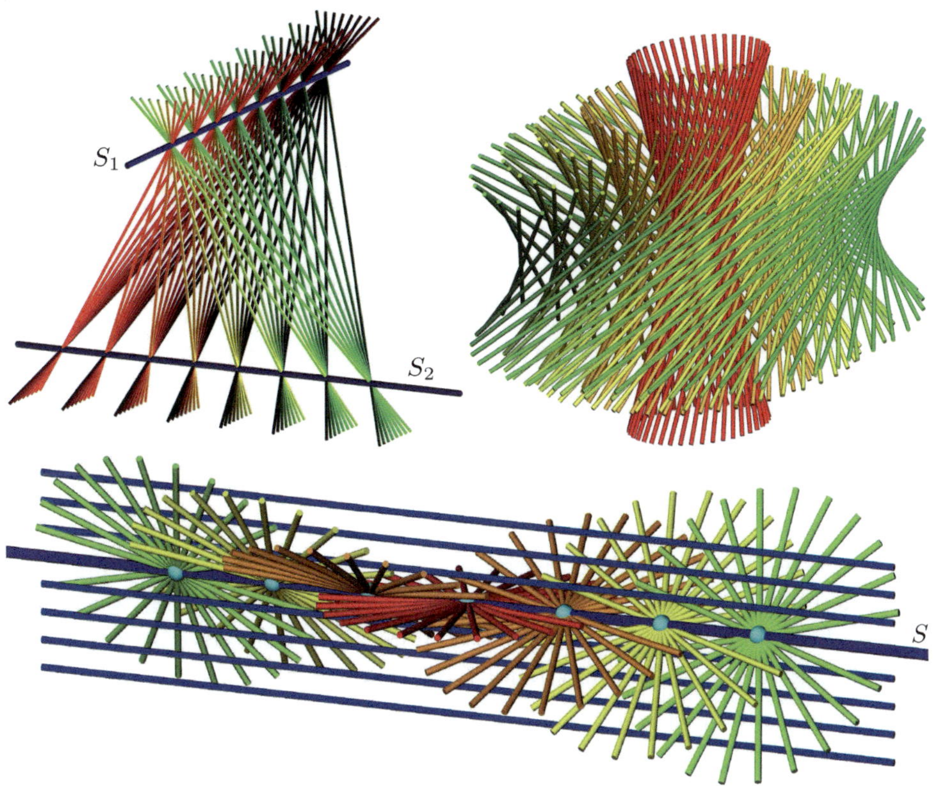

FIGURE 10.4. Top, left: A hyperbolic linear line congruence consists of all straight lines that meet a pair of skew lines (axes) S_1 and S_2 and can be decomposed into two independent one-parameter families of pencils of lines. Top, right: An elliptic linear line congruence consists of all lines that meet a pair of skew and complex conjugate lines (axes) S_1 and S_2. The lines in this congruence (spread) can be arranged into two one-parameter families of reguli. Bottom: A parabolic linear line congruence consists of a one-parameter family of pencils of lines whose carrier planes wind about the only axis while the pencils' vertices are located on the axis S. The pencil about S's ideal point fills the asymptotic plane. In the affine part, we can see some of these lines as parallels to S.

is a straight line in $\mathbb{P}^5(\mathbb{R})$, and the distinction of its relative position w.r.t. M_2^4 gives a classification of pencils of linear line complexes.

In a first step, we study the zeros of the quadratic equation

$$\Omega(P,P)(t_1,t_2) = 0, \tag{10.21}$$

which yield the singular linear line complexes in the pencil. The straight lines shared by all complexes of the pencil P of complexes intersect the singular linear line complexes, since the bilinear form Ω characterizes intersecting lines. Naturally, (10.21) has either two real solutions (1), a pair of complex conjugate solutions (2), one real double solution (3), or is identically zero (4).

The hyperbolic linear line congruence

Without loss of generality, we may assume that

$$S_1 = (1,0,0;0,0,0) \quad \text{and} \quad S_2 = (0,0,0;1,0,0)$$

are the singular linear line complexes in the pencil (10.20). Since $\Omega(S_1, S_2) = 1 \neq 0$, S_1 and S_2 are skew straight lines in $\mathbb{P}^3(\mathbb{R})$. In fact, any pair of skew lines can be mapped to that particular pair via a collineation and then serve as carriers of the pencil P.

The totality of straight lines in the pencil is the two parameter family of straight lines intersecting S_1 and S_2. Such line configurations are called *hyperbolic linear line congruences*. The lines S_1 and S_2 are called the *axes* of the congruence and do not belong to the congruence.

The Klein image of the lines in the hyperbolic linear line congruence are found as the intersection of the 3-space

$$D: \ \Omega(S_1, X) = \Omega(S_2, X) = 0 \quad \Longleftrightarrow \quad x_4 = x_1 = 0$$

with M_2^4. Note that D is polar to the line P with regard to M_2^4.[5]

Therefore, the Klein image of the hyperbolic linear line congruence is the two-dimensional ruled quadric in M_2^4 with the equation

$$H: \ x_1 = x_4 = 0, \quad x_2x_5 + x_3x_6 = 0.$$

The two one-parameter families of rulings (reguli) on H correspond to the two different and independent one-parameter families of pencils of lines in the hyperbolic linear line congruence. These pencils arise from projecting each point on either axis with all points on the other axis (see Figure 10.4). The axes of the hyperbolic linear line congruence do not belong to the congruence.

[5] The pencil P of linear complexes of lines is represented by a straight line in $\mathbb{P}^5$, which we shall also denote by P in order not to overload the text with symbols.

■ **Example 10.1.2** Axes of a pencil of linear complexes of lines.

Assume that a pencil P of linear complexes of lines is given by the two regular linear complexes C_1 and C_2 of lines with their homogeneous coordinates

$$\mathbf{c}_1 = (1,0,0;1,0,0)\mathbb{R} \quad \text{and} \quad \mathbf{c}_2 = (0,1,0;0,-1,0)\mathbb{R}.$$

The pencil P of complexes spanned by C_1 and C_2 can be parametrized by

$$\mathbf{p}(\lambda,\mu) = \lambda\mathbf{c}_1 + \mu\mathbf{c}_2 = (\lambda,\mu,0;\lambda,-\mu,0)\mathbb{R} \quad \lambda:\mu \neq 0:0.$$

To each linear complex, there exists a uniquely determined axis. With (10.13), we find the homogeneous Plücker coordinates of the axes of all complexes in the pencil P as

$$(\mathbf{a},\bar{\mathbf{a}}) = (\lambda(\lambda^2+\mu^2), \mu(\lambda^2+\mu^2), 0; 2\lambda\mu^2, -2\lambda^2\mu, 0)\mathbb{R}.$$

The homogeneous parameter $\lambda:\mu$ can be replaced by an inhomogeneous parameter if we set $\lambda = \cos t$ and $\mu = \sin t$. This yields

$$(\mathbf{a},\bar{\mathbf{a}}) = (\cos t, \sin t, 0; 2\cos t\sin^2 t, -2\cos^2 t\sin t, 0)\mathbb{R}.$$

The Plücker representation of the thus described ruled surface can be turned into a standard parametrization $\mathbf{r}(t,w): D \subset \mathbb{R}^2 \to \mathbb{R}^3$. For that purpose, we compute the set of pedal points with (10.3) in order to find a directrix. This yields

$$\mathbf{r}(t,w) = \begin{pmatrix} 0 \\ 0 \\ -\sin 2t \end{pmatrix} + w \begin{pmatrix} \cos t \\ \sin t \\ 0 \end{pmatrix}. \tag{10.22}$$

The cubic ruled surface parametrized by (10.22) is called *Plücker's conoid.* In Section 11.3, we shall see this surface as the inverse of the surface of all osculating circles of all normal sections at a regular surface point. An example of Plücker's conoid is depicted in Figure 11.12. In Section 2.2, the relations of Plücker's conoid to one-sheeted hyperboloids of revolution have been disclosed.

The elliptic linear line congruence

If (10.21) has a pair of complex conjugate solutions, the singular linear line complexes also form a complex conjugate pair. The same is true for their axes, and the *elliptic linear line congruence* consists of all real lines meeting a skew and complex conjugate pair of straight lines.

A special case shows up if we choose

$$S_1 = (1,\mathrm{i},0;1,\mathrm{i},0) \quad \text{and} \quad S_2 = \overline{S_1},$$

and the equation of the polar space of $[S_1,S_2]$ with regard to M_2^4 equals

$$D:\ x_1 + x_4 = x_2 + x_5 = 0,$$

which intersects M_2^4 in the two-dimensional oval quadric

$$O:\ x_1^2 + x_2^2 - x_3x_6 = 0.$$

Since O is oval and regular, it does not carry a single straight line. Consequently, there are no pencils of lines in the elliptic linear line congruence. The elliptic linear line congruence can be decomposed into two independent one-parameter families of reguli (see Example 10.1.3). The special case presented here is invariant w.r.t. rotations about the z-axis (cf. Figure 10.4). As in the hyperbolic case, the axes are not part of the congruence.

■ **Example 10.1.3** Elliptic linear line congruence with rotational symmetry.

We can find a parametrization of the elliptic linear line congruence defined by the two linear homogeneous equations

$$x_1 + x_4 = x_2 + x_5 = 0$$

by setting $x_1(u,v) = -u\sin v$, $x_2 = u\cos v$ which yields $u^2 - x_3x_6 = 0$. We can assume that $x_3(u,v) = 1$ and find $x_6 = u^2$, and therefore, $\mathbf{E}(u,v) = (-u\sin v, u\cos v, 1, u\sin v, -u\cos v, u^2)$ with $(u,v) \in \mathbb{R} \times [0, 2\pi[$ is an inhomogeneous parametrization of the elliptic linear line congruence in terms of Plücker coordinates. The set of points on the lines of the congruences can be found using (10.3). This results in

$$\mathbf{e}(u,v) = \begin{pmatrix} u\cos v \\ u\sin v \\ 0 \end{pmatrix} + w \begin{pmatrix} -u\sin v \\ u\cos v \\ 1 \end{pmatrix}, \quad (u,v,w) \in \mathbb{R} \times [0, 2\pi[\times \mathbb{R}.$$

From the latter parametrization, we can eliminate either the pair of parameters (u,w) or the pair (v,w) and obtain the implicit equations

$$\mathcal{P}_h(v): \ (xz - y)\cos v + (yz + x)\sin v = 0,$$

$$\mathcal{H}(u): \ x^2 + y^2 - u^2(z^2 + 1) = 0$$

of a one-parameter family of hyperbolic paraboloids $\mathcal{P}_h(v)$ and one-sheeted hyperboloids $\mathcal{H}(u)$. This shows that the lines of this particular elliptic linear line congruence can be arranged in two independent one-parameter families of reguli. In fact, this is a projective property. Figure 10.5 shows reguli from both families in the elliptic linear line congruence.

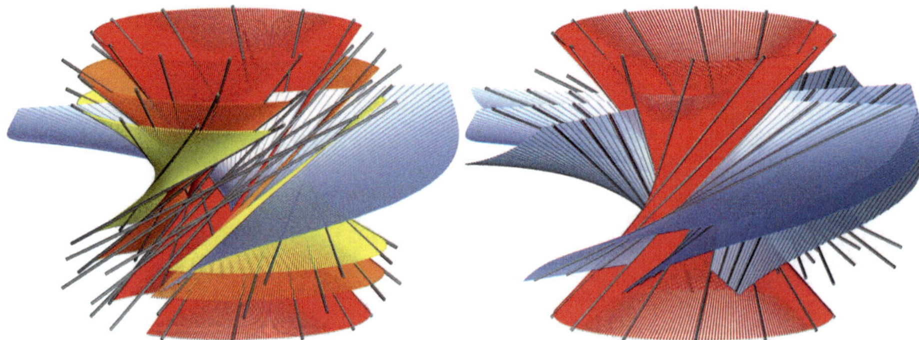

FIGURE 10.5. An elliptic linear line congruence can be viewed as the union of two independent one-parameter families of reguli.

The parabolic linear line congruence

A double solution of (10.21) is characteristic for a tangent P. The linear line congruence defined by a tangent of M_2^4 is called *parabolic*. In this case, there is only one axis whose Klein image is the point of contact of P with M_2^4. The axis of the parabolic congruence is a part of the congruence.

In Section 9.2 (p. 400), we have learned that the tangents of a ruled surface along a regular non-torsal ruling form a parabolic linear line congruence. The projective generation is also explained there. Figure 10.4 shows an example of a parabolic linear line congruence. Note that the ideal point of the axis is mapped to the asymptotic plane under the fundamental projectivity acting along the axis. The pencil of lines in the asymptotic plane consists of lines parallel to the congruence's axis.

The singular linear line congruence

The *singular linear line congruence* consists of a star of lines and a ruled plane sharing a pencil of lines (cf. Figure 10.2). The line P (10.20) is completely contained in M_2^4, and its polar 3-space with regard to M_2^4 intersects M_2^4 in a pair of planes in M_2^4. The pair of planes consists of a plane of the first and second kind, and the union of both can be seen as a singular quadric in M_2^4.

Conics in M_2^4 and ruled quadrics

One-dimensional quadrics, *i.e.*, conics in the sense of planar geometry, can be found as the intersections of planes (two-dimensional subspaces) with M_2^4. Unfortunately, the number of different conics to be distinguished does not match the number of different relative positions of planes w.r.t. M_2^4. We try to give a complete list of cases. However, we shall discuss only the relevant cases in more detail.

A plane π in any projective space is the span of three independent points C_1, C_2, and C_2. In the context of line geometry, the three points C_i are the Klein images of linear complexes and the plane π parametrized by

$$\mathbf{P}(\lambda_0, \lambda_1, \lambda_2) = \lambda_1 C_1 + \lambda_2 C_2 + \lambda_3 C_3 \quad \text{with} \quad (\lambda_1, \lambda_2, \lambda_3) \in \mathbb{R}^3 \setminus \{\mathbf{0}\} \tag{10.23}$$

can be viewed as the Klein image of a linear two-parameter family of complexes of lines.

The lines belonging to this linear complex manifold are the common points of the plane (10.23) and M_2^4. Therefore, we have to discuss the ternary quadratic form

$$\Omega(\mathbf{P},\mathbf{P}) = \boldsymbol{\lambda}^{\mathrm{T}} \begin{pmatrix} \omega_{11} & \omega_{12} & \omega_{13} \\ \omega_{12} & \omega_{22} & \omega_{23} \\ \omega_{13} & \omega_{23} & \omega_{33} \end{pmatrix} \boldsymbol{\lambda} = 0, \tag{10.24}$$

where $\boldsymbol{\lambda} = (\lambda_1, \lambda_2, \lambda_3)^{\mathrm{T}}$ and $\omega_{ij} := \Omega(C_i, C_j)$ (with $i, j \in \{1,2,3\}$). Now we can distinguish the following cases:

1. Ruled quadrics

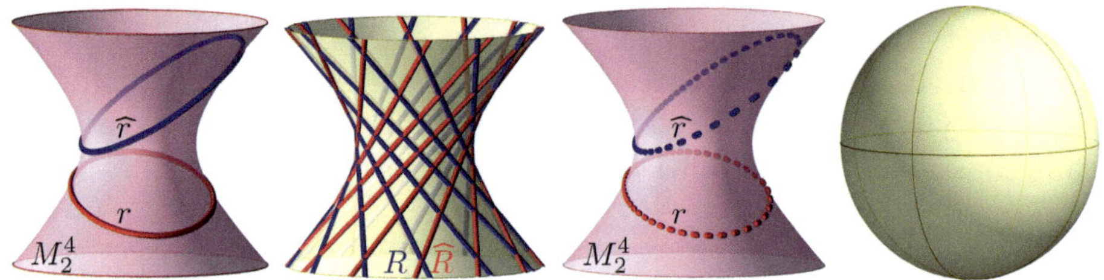

FIGURE 10.6. Pairs $(r,\widehat{r})$ of polar conics in M_2^4 correspond to pairs $(R,\widehat{R})$ of complementary reguli. These are real in ruled quadrics (left) and complex in oval quadrics (right).

The quadratic form (10.24) is regular and has real zeros. Then, (10.24) describes a conic r in M_2^4. The conic r is not contained in a plane of M_2^4, otherwise (10.24) would be the zero form.

The points of r correspond to the one-parameter family R of straight lines, *i.e.*, a regulus, in a regular (real) ruled quadric $Q \subset \mathbb{P}^3(\mathbb{R})$. The polar plane $\widehat{\pi}$ of $\pi = [r]$ is skew to π and intersects M_2^4 along a second conic $\widehat{r}$.

The points of $\widehat{r}$ correspond to the one-parameter family $\widehat{R}$ of lines of the regulus complementary to R (cf. Figure 10.6, left).

Due to the polar reciprocity of $[r]$ and $[\widehat{r}]$, each line of R intersects each line of $\widehat{R}$, and vice versa.

2. Oval quadrics

The quadratic form (10.24) is regular, but has no real zero. Still, (10.24) describes a conic r in M_2^4 which is not contained in a plane of M_2^4 due to the same reasons as in (1).

Again, the plane has a polar plane, and the intersections of the pair of planes with M_2^4 results in a pair of complex conics r and $\widehat{r}$. These are the

Klein images of a pair of complementary but complex conjugate reguli R and $\widehat{R}$ that cover an oval quadric (cf. Figure 10.6, right).

3. Entangled pencils of lines

If (10.24) is degenerate (without being completely zero), it can split into two linear factors both describing straight lines l and m in M_2^4. Since (10.24) does not completely vanish, the plane π is not contained in M_2^4 and the two straight lines l and m are the Klein images of a pair of *entangled pencils* of lines. The pencils have different vertices (cf. Figure 10.7), different carrier planes, and a common line s, for the coplanar lines l and m must have a point S in common.

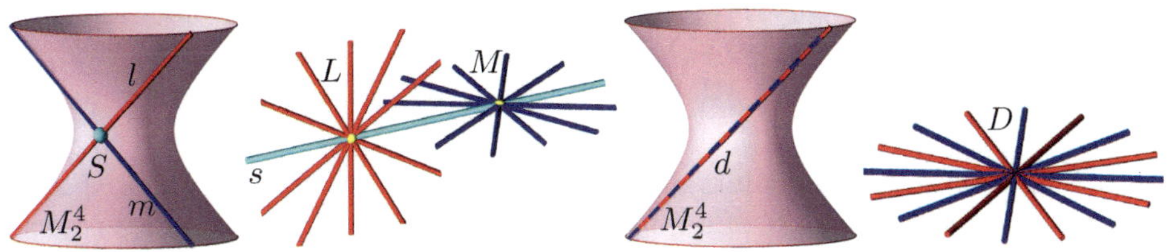

FIGURE 10.7. Left: A pair $(l,m) \subset M_2^4$ of lines with $[l,n] \not\subset M_2^4$ corresponds to a pair (L,M) of entangled pencils of lines. The line s joining the pencils' vertices is common to both pencils. Right: A double line $d \subset M_2^4$ corresponds to a pencil D of lines with multiplicity two.

4. Double pencil

A singular form (10.24) can be a full square of a linear factor. If this is the case, (10.24) describes a *double line* d corresponding to a *double pencil* D as a highly degenerate quadric (see Figure 10.7, right).

5. The most degenerate cases

There are some more appearances of regular and singular conics in M_2^4 which cannot be found by discussing (10.24) although their carrier planes can be given in the form (10.23).

(1) If the conic $r \subset M_2^4$ is contained in a plane π_1 of the first kind, then its points are the Klein images of the rulings of a regular quadratic cone if r is regular. Since the plane $\pi_1 = [r]$ is self-polar, there is no further family of straight lines on the quadratic cone (cf. Figure 10.8, left).

(2) In the case that $\pi_2 = [r]$ is a plane of the second kind, the points on r correspond to the tangents of a conic in a ruled plane. Actually, this

is a dual conic, *i.e.*, the set of tangents of a regular conic in a plane (cf. Figure 10.8, right).

(3) In either case (1) or (2), the conic r can split into a pair of straight lines. These are the Klein images of pencils of lines sharing either the vertices or the carrier planes.

(4) If the singular conic r from the cases (1) and (2) is a double line, then it is the Klein image of double pencil.

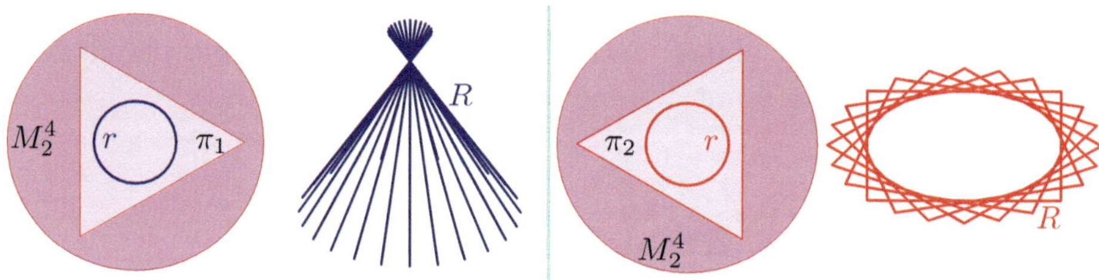

FIGURE 10.8. Left: The points of a conic r in a plane $\pi_1 \subset M_2^4$ of the first kind correspond to the rulings R of a quadratic cone. Right: The points of a conic r in a plane $\pi_2 \subset M_2^4$ of the second kind correspond to the tangents R of a conic.

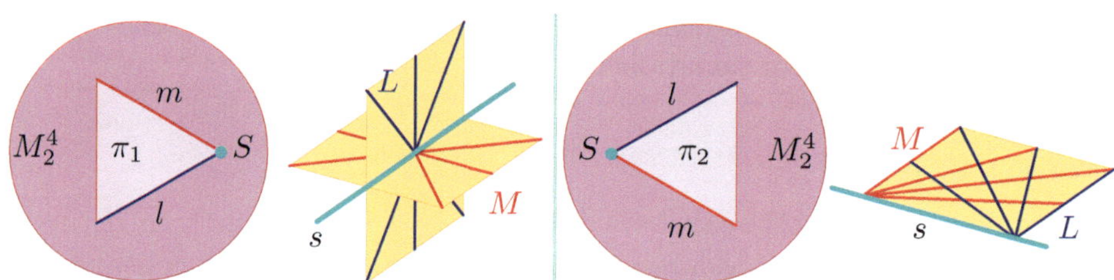

FIGURE 10.9. Left: Two lines l and m that span a plane $\pi_1 \subset M_2^4$ of the first kind correspond to a pair of pencils L and M in a common star of lines and share a line s which corresponds to $S = l \cap m \in M_2^4$. Right: Two lines l and m that span a plane $\pi_2 \subset M_2^4$ of the second kind correspond to a pair of pencils L and M in a common ruled plane. The pencils have different vertices and share the straight line s.

■ **Example 10.1.4** Cones without any real point – except the vertex.

The quadratic cone $\Gamma: x^2 + y^2 + z^2 = 0$ carries only one real point. Its vertex is the point $V = (0,0,0) \in \Gamma$. The generators of Γ are the isotropic lines emanating from V. The Klein image of these lines gather on the conic $(\cos t, \sin t, \mathrm{i}; 0, 0, 0)$ (with $t \in [0, 2\pi[$), which lies completely in a plane of the first kind. Γ is called *isotropic cone*.

Its generators (rulings) are isotropic lines, *i.e.*, they meet the plane at infinity in points of the absolute conic of Euclidean geometry. For any pair of different points $P \neq Q$ on an isotropic line, the Euclidean distance function evaluates to zero: $\overline{PQ} = 0$ (cf. [46, p. 253]).

Isotropic cones play a much more important role in pseudo-Euclidean geometries, where they are referred to as *light cones*, which separate timelike and spacelike points (see Example 10.1.5).

■ **Example 10.1.5** Real cones and pseudo-Euclidean geometry.

Making a small change in the equation of the cone that we have looked at in Example 10.1.4, we arrive at $\Gamma : x^2 + y^2 - z^2 = 0$, which describes a cone of revolution (from the Euclidean point of view). It is an isotropic cone in the sense of pseudo-Euclidean geometry (see Section 9.4). The rulings of Γ meet the plane at infinity in real points on a conic that is common to all cones of revolution with the same angle of aperture, provided that their axes are parallel.

The cone Γ separates the lines through its vertex. In the sense of pseudo-Euclidean geometry, the lines in the interior of Γ are timelike, while the lines outside are spacelike, and the lines on Γ are timelike. In physics, the pseudo-Euclidean space is usually referred to as Minkowski space.

In Section 9.4, we use the isotropic cones of a pseudo-Euclidean space $\mathbb{R}^{2,1}$ in order to project points to oriented circles in a Euclidean plane.

■ **Example 10.1.6** Conics without real points.

The conic $n : x^2 + y^2 + 1 = 0$ is empty over the real number field. In the complex extension of $\mathbb{P}^2(\mathbb{R})$, we can parametrize it by $\mathbf{n} = \mathrm{i}(\cos t, \sin t)$ (with $t \in [0, 2\pi[$). The tangents of n form the dual conic $n^\star$ and can be given by $\mathrm{i}x \cos t + \mathrm{i}y \sin t + 1 = 0$ or in Plücker coordinates $(\mathrm{i} \sin t, -\mathrm{i} \cos t, 0; 0, 0, 1)$. The latter describes a conic without any real point in a plane of the second kind in M_2^4.

Quadratic complexes

A *quadratic complex of lines* is the set of lines whose (homogeneous) Plücker coordinates fulfill a homogeneous quadratic equation (besides (10.2)). Therefore, the study of quadratic complexes can be reduced to the study of pencils of quadrics in $\mathbb{P}^5$, with M_2^4 being one of the quadrics.

A generic quadratic complex depends on 19 constants: The symmetric matrix of the complex's equation has $\frac{1}{2}(6+1)6 = 21$ relevant entries, and only their ratio matters. Because of the homogeneity, one degree of freedom has to be subtracted. Further, the Plücker condition (10.2) is also to be imposed on the Plücker coordinates of the complex lines, which again reduces the number of free constants about one.

The classification of quadratic complexes is based on the theory of *characteristics* and *elementary divisors* (cf. [87]). We shall give a sketch of this classification. It is similar to the classification of pencils of quadrics as done in Section 5.1, since the quadrics in $\mathbb{P}^5$ defining the complex and M_2^4 span a pencil of quadrics.

Assume that $\mathbf{A} \in \mathbb{F}^{6\times 6}$ with $\mathbf{A} = \mathbf{A}^{\mathrm{T}}$ is the coefficient matrix of a quadratic form in six homogeneous variables. Then, the equation of a quadratic complex $\mathcal{K}$ of lines reads

$$\mathcal{K}: \ \mathbf{x}^{\mathrm{T}}\mathbf{A}\mathbf{x} = 0. \tag{10.25}$$

The quadratic form describing M_2^4 has the coefficient matrix

$$\mathbf{M} = \begin{pmatrix} \mathbf{0} & \mathbf{I}_3 \\ \mathbf{I}_3 & \mathbf{0} \end{pmatrix},$$

and the pencil of quadrics spanned by M_2^4 and $\mathcal{K}$ is parametrized by

$$\mathcal{P}: \ \mathbf{x}^{\mathrm{T}}(\mathbf{A} + \lambda\mathbf{M})\mathbf{x} = 0. \tag{10.26}$$

The pencils of quadrics (10.26) in projective five-space can now be classified by means of the number and type of singular quadrics in the pencil. Therefore, we compute the characteristic polynomial $\Delta(\lambda)$ given by

$$\Delta(\lambda) = \det(\mathbf{A} + \lambda\mathbf{M}). \tag{10.27}$$

For any solution λ_i of (10.27) with multiplicity k, the rank of $\mathbf{A} + \lambda_i\mathbf{M}$ can be five or less. If, for any solution λ_i, the rank of $\mathbf{A} + \lambda_i\mathbf{M}$ is exactly five, we collect the multiplicities of the zeros of (10.27) and enclose them in square brackets: [...]. It is obvious that there are eleven such cases, for six can be written as the sum of integers less than or equal to six in eleven different ways.

This rather rough classification can be refined as follows: Assume that Δ' and Δ'' are the minors of fourth and fifth order of Δ. If now λ_i is a root of (10.27) with multiplicity $k > 1$ and $\mathrm{rk}(A + \lambda_i M) = 4$, then λ_i is a common root of all Δ' with multiplicity k'. Then, $k > k'$ and $k - k' \geq k'$. So, $k - k'$ roots are lost during the transition from Δ' to Δ''. Now we add the pair $(k - k', k')$ to the above list [...] of multiplicities. This process can be iterated if the rank of Δ drops further. The number of types to be distinguished equals 49, in general, and it equals 57 if we add the most singular cases to our list (see [152]).

Special examples of quadratic complexes

We shall complete this section with some general remarks on quadratic complexes of lines and some examples of quadratic complexes of lines that appear frequently in geometry. Some of the complexes in this section are discussed in [42] with a special focus on the singular surfaces.

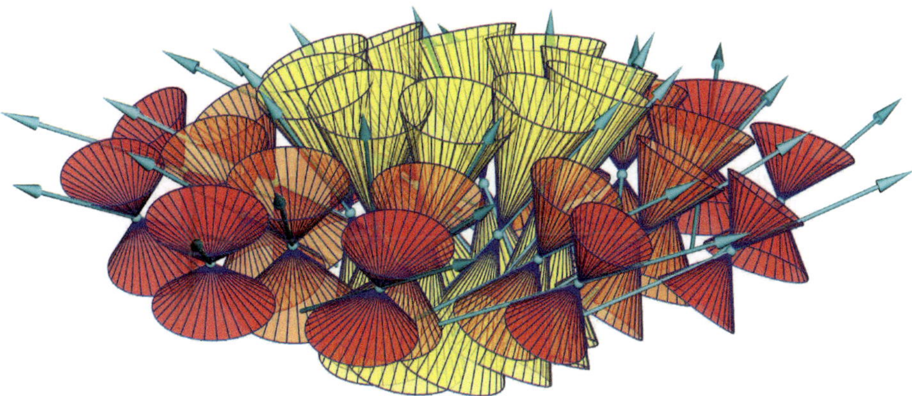

FIGURE 10.10. The lines of a quadratic complex that are incident with a particular point form a quadratic cone, namely the complex cone. Here, the underlying quadratic complex is the path tangent complex of an equiform motion. For some points, the tangent cones are displayed. Each cone contains the path tangent of its apex, which is indicated by an arrow.

A *complex cone* is the set of lines of a complex (whether quadratic or not) that pass through a given point (cf. Figure 10.10). In other words, the complex cone is the intersection of the quadratic complex with a star of lines. Dual to the complex cone, we have the notion of a *complex curve*, which is the set of lines of the complex that lie in a chosen plane. We can also say that the complex curve is the intersection of the complex with a ruled plane. The complex curve is a curve in the dual projective plane, *i.e.*, it is the set of tangents to a planar curve.

In the case of linear complexes of lines, both the complex cones and the complex curves are always pencils of lines. In the case of quadratic complexes, the complex cones are quadratic ones and the complex curves are the set of tangents to a conic including degenerate cases.

Tangents of a quadric

The tangents of any surface form a complex of lines, *i.e.*, the two-parameter manifold of pencils of surface tangents which is, in total, a three-parameter manifold of lines. It is obvious that the tangents of an algebraic surface are themselves an algebraic variety.

Let $\mathbf{Q} \in \mathbb{F}^{4\times 4}$ be a symmetric matrix. Then, $\mathcal{Q}: \ \mathbf{x}^{\mathrm{T}}\mathbf{Q}\mathbf{x} = 0$ is a quadric in $\mathbb{P}^3(\mathbb{F})$. A straight line $l = [P, Q]$ parametrized by $\mathbf{l}(\lambda, \mu) = \lambda\mathbf{p} + \mu\mathbf{q}$ is tangent to $\mathcal{Q}$ if, and only if,

$$(\mathbf{p}^{\mathrm{T}}\mathbf{Q}\mathbf{q})^2 - (\mathbf{p}^{\mathrm{T}}\mathbf{Q}\mathbf{p})(\mathbf{q}^{\mathrm{T}}\mathbf{Q}\mathbf{q}) = 0.$$

Now, we replace $\mathbf{p}$ and $\mathbf{q}$ by an arbitrary pair of independent rows from (10.7) and arrive at a homogeneous equation (in terms of Plücker coordinates) of the quadratic *complex of tangents of a quadric*. At first glance, the equation of the complex seems to be quartic, but (10.2) can be cancelled. The coefficient matrix $\mathbf{A}$ of the quadratic equation of the complex can be computed from $\mathbf{Q}$ as $\mathbf{Q} \oslash \mathbf{Q}$ as explained above in (10.19).

For any point P off the quadric Q, the complex cone Γ_P is the union of all tangents from P to the quadric, *i.e.*, Γ_P is the tangent cone from P to Q. The complex curves in an arbitrary non-tangential plane π are the tangents of Q in π, and therefore, the tangents to the curve $Q \cap \pi$.

▪ **Example 10.1.7** Tangent complexes of singular and regular quadrics.

The projective classification of quadrics in $\mathbb{P}^3(\mathbb{R})$ yields the following normal forms:

$$x_0^2 + x_1^2 + x_2^2 + x_3^2 = 0 \ \text{ empty}, \quad x_0^2 + x_1^2 + x_2^2 - x_3^2 = 0 \ \text{ oval}, \quad x_0^2 + x_1^2 - x_2^2 - x_3^2 = 0 \ \text{ ruled},$$
$$x_0^2 + x_1^2 + x_2^2 = 0 \ \text{ empty cone}, \quad x_0^2 + x_1^2 - x_2^2 = 0 \ \text{ regular cone},$$
$$x_0^2 + x_1^2 = 0 \ \text{ complex conjugate pair of planes}, \quad x_0^2 - x_1^2 = 0 \ \text{ pair of real planes}.$$

The case of a double plane is not worth being mentioned here.

The coefficient matrices of the equations of the respective tangent complexes are (in the same ordering)

$$\operatorname{diag}(1,1,1,1,1,1), \quad \operatorname{diag}(1,1,-1,-1,-1,1), \quad \operatorname{diag}(-1,1,1,1,1,1),$$
$$\operatorname{diag}(1,1,0,0,0,1), \quad \operatorname{diag}(1,-1,0,0,0,-1),$$
$$\operatorname{diag}(1,0,0,0,0,0), \quad \operatorname{diag}(1,0,0,0,0,0).$$

The respective characteristics are [(111)(111)].

▪ **Example 10.1.8** Tangent complexes for quadrics in principal position.

The characteristics of the quadratic complexes of tangents to quadrics will not be altered if the underlying coordinate system is changed. For the basic types of (regular) quadrics

$$\frac{x^2}{a^2} + \frac{y^2}{b^2} + \frac{z^2}{c^2} = 1, \quad \frac{x^2}{a^2} + \frac{y^2}{b^2} - \frac{z^2}{c^2} = 1, \quad \frac{x^2}{a^2} + \frac{y^2}{b^2} - \frac{z^2}{c^2} = -1,$$
$$\frac{x^2}{a^2} + \frac{y^2}{b^2} = 2z, \quad \frac{x^2}{a^2} - \frac{y^2}{b^2} = 2z,$$

the coefficient matrices of the tangent complexes are

$$\operatorname{diag}(a^{-1},b^{-2},c^{-2};-(bc)^{-1},-(ca)^{-2},-(ab)^{-2}),\operatorname{diag}(a^{-1},b^{-2},-c^{-2};(bc)^{-1},(ca)^{-2},-(ab)^{-2}),$$
$$\operatorname{diag}(a^{-1},b^{-2},-c^{-2};-(bc)^{-1},-(ca)^{-2},-(ab)^{-2}),$$
$$\begin{pmatrix} 0 & 0 & 0 & 0 & a^{-2} & 0 \\ 0 & 0 & 0 & -b^{-2} & 0 & 0 \\ 0 & 0 & 1 & 0 & 0 & 0 \\ 0 & -b^{-2} & 0 & 0 & 0 & 0 \\ a^{-2} & 0 & 0 & 0 & 0 & 0 \\ 0 & 0 & 0 & 0 & 0 & -(ab)^{-2} \end{pmatrix}, \begin{pmatrix} 0 & 0 & 0 & 0 & a^{-2} & 0 \\ 0 & 0 & 0 & b^{-2} & 0 & 0 \\ 0 & 0 & 1 & 0 & 0 & 0 \\ 0 & b^{-2} & 0 & 0 & 0 & 0 \\ a^{-2} & 0 & 0 & 0 & 0 & 0 \\ 0 & 0 & 0 & 0 & 0 & (ab)^{-2} \end{pmatrix}.$$

The complex of tangents of a Euclidean sphere with radius R (oval quadric with $a = b = c = R$) is called *complex of constant width.* Its equation reads $R^2\langle \mathbf{l}, \mathbf{l}\rangle - \langle \bar{\mathbf{l}}, \bar{\mathbf{l}}\rangle = 0$.

Complexes of constant slope

Lines that enclose a certain fixed angle φ with a fixed plane π form a *complex of constant slope.* It is easily verified that the Plücker coordinates $(l_1, \dots, l_6)$ of such a complex fulfill

$$l_1^2 + l_2^2 - l_3^2 \tan^2\varphi = 0, \tag{10.28}$$

provided that the plane π equals $z = 0$ in Euclidean 3-space and the slope equals $\tan\varphi$.

Especially the case $\varphi = \frac{\pi}{4}$ is of theoretical interest. The thus defined complex consists of all lines that meet the plane π under an angle of $\frac{\pi}{4}$. In the projective closure of Euclidean 3-space, we can say that this complex consists of all straight lines that meet a real conic in the ideal plane.

The *cyclographic mapping* uses this complex in order to project points P in $\mathbb{R}^3$ to oriented Euclidean circles P^z in the plane π. The complex cones in the complex of constant slope are cones of revolution with vertical axes, *i.e.*, their axes are orthogonal to the plane π and the complex cone Γ_P (centered at P) intersects π in the (oriented) circle P^z with its center being the orthogonal projection P' of P onto π. The signed radius of P^z equals the z-coordinate of P^z (oriented Euclidean distance from P to the image plane). The circle P^z is called the *cyclographic image* of P (cf. Section 9.4).

The inverse of the cyclographic mapping relates Euclidean circle geometry (in the plane π) with the pseudo-Euclidean geometry in the space $\mathbb{R}^3$, which is often referred to as Minkowski space and then denoted by $\mathbb{R}^{2,1}$.

■ **Example 10.1.9** The complex of isotropic lines in Euclidean 3-space.

The complex of isotropic lines in Euclidean 3-space is best described in the projective closure and complex extension of Euclidean 3-space. Therefore, we use homogeneous coordinates $(x_0 : x_1 : x_2 : x_3) \neq (0:0:0:0)$ in order to describe points. The coordinates x_i can be even complex numbers if necessary.

The complex of isotropic lines consists of all lines that meet the absolute conic $x_1^2 + x_2^2 + x_3^2 = 0$ in the ideal plane $x_0 = 0$. Therefore, the Plücker coordinates $(l_1, \ldots, l_6)$ of these lines fulfill $l_1^2 + l_2^2 + l_3^2 = 0$. A projection with the help of this complex to the plane $\pi : x_3 = 0$ maps points $P = (1 : x_P : y_P : z_P)$ to *empty circles* $(x - x_P)^2 + (y - y_P)^2 + z_P^2 = 0$, *i.e.*, circles with a real center $(x_P, x_P, 0)$ and a purely imaginary radius $\mathrm{i}z_P$ (cf. [46, Exc. 7.1.1, p. 264]). The projection with the complex of isotropic lines is usually called *isotropic projection.*

At first glance, such a projection seems to be useless. However, in Descriptive Geometry, the isotropic projection finds its application when constructing *directly in* perspective images.

A perspective image is the result of a central projection from a point O (eye) onto a plane $\pi \not\ni O$ (image plane). The isotropic projection of the eye O to π clearly yields an empty circle. The polarity w.r.t. this circle can be realized as the anti-polarity w.r.t. to its *real representative* o (see [46, Exc. 7.1.1, p. 264, 265] or Theorem 4.2.11). In this context, the circle o is called the *distance circle* of the perspective image.

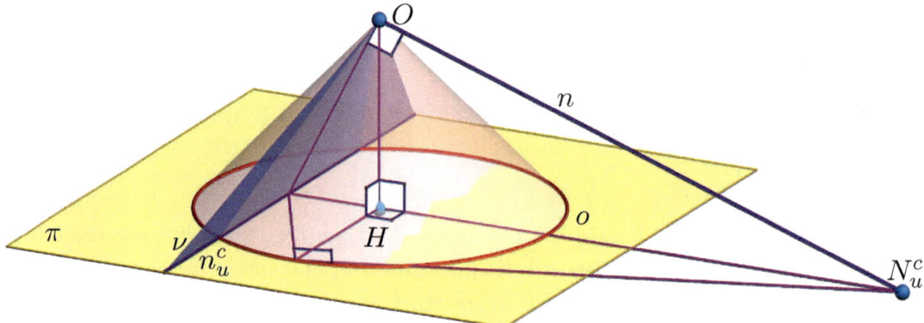

FIGURE 10.11. The Euclidean orthogonality in the star about O is projected to the anti-polarity w.r.t. the distance circle o in the image plane π from the eye point O. The plane ν and the line n are orthogonal, and thus, the respective vanishing line n_u^c and vanishing point N_u^c are antipolar w.r.t. the circle o.

Lifting the anti-polarity w.r.t. o to the star of lines about O, yields the Euclidean orthogonality in the star about O, which allows us to find the vanishing points of directions orthogonal to given planes, or the other way around (cf. Figure 10.11).

Figure 10.12 illustrates how the anti-polarity w.r.t. the distance circle o is used in constructions within perspective images.

Path tangents of a Euclidean or an equiform motion

According to Theorem 10.1.3, the path normals of a Euclidean motion at any regular instant form a linear line complex. This is not the case for

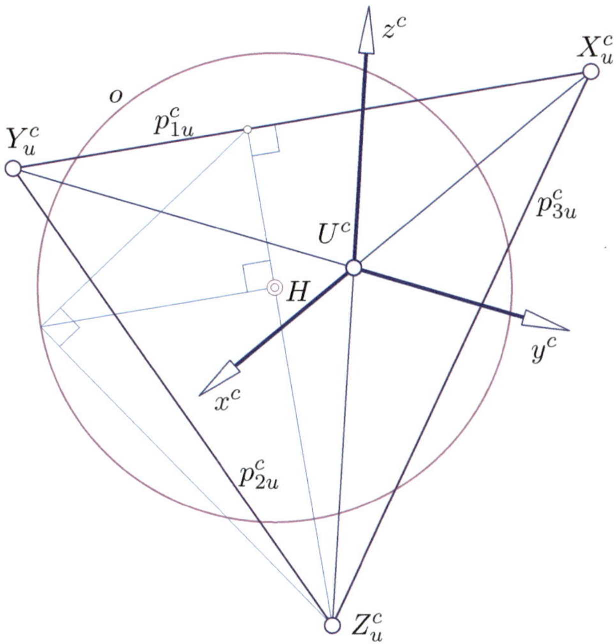

FIGURE 10.12. The vanishing points of X_u^c, Y_u^c, Z_u^c of the axes of a Cartesian frame are the vertices of an antipolar triangle of the distance circle o. The principal (vanishing) point H (orthogonal projection of the eye point O onto the image plane π) is the orthocenter of the triangle of the three vanishing points.

the path tangents although the velocity vector field at any regular instant is linear, which is clear by virtue of (10.17).

At this point we will extend the study of path tangents from the case of purely Euclidean motions to the case of equiform motions that covers the Euclidean case. An equiform motion $\mu: \mathbb{R}^3 \to \mathbb{R}^3$ is the composition of a Euclidean motion with a dilatation. With $\mathbf{A}: I \subset \mathbb{R} \to \mathrm{SO}_3$, $\mathbf{a}: I \to \mathbb{R}^3$, and $\alpha: I \to \mathbb{R}^\star$, we can describe a one-parameter equiform motion transforming points X to points X' by

$$\mathbf{x}' = \alpha \mathbf{A}\mathbf{x} + \mathbf{a}. \tag{10.29}$$

Differentiating once w.r.t. t and rewriting $\dot{\mathbf{x}}' = \mathbf{v}(\mathbf{x})$ in terms of $\mathbf{x}' = \frac{1}{\alpha}\mathbf{A}^{\mathrm{T}}(\mathbf{x} - \mathbf{a})$, *i.e.*, in the fixed system, we arrive at a linear representation of the velocity vector field as

$$\dot{\mathbf{x}}' = \mathbf{v}(\mathbf{x}) = \gamma \mathbf{x}' + \mathbf{c} \times \mathbf{x}' + \overline{\mathbf{c}}, \tag{10.30}$$

where we have set

$$\gamma = \dot{\alpha}\alpha^{-1}, \quad \mathbf{c} \times \mathbf{x}' = \mathbf{A}\dot{\mathbf{A}}^{\mathrm{T}}\mathbf{x}', \quad \text{and} \quad \overline{\mathbf{c}} = \dot{\mathbf{a}} - \dot{\alpha}\alpha^{-1}\mathbf{a} - \mathbf{c} \times \mathbf{a}.$$

(10.30) also covers the special case of the vector field of Euclidean motions as given in (10.17).

This time we are not interested in the study of path normals. In the case of equiform motions, it has been revealed that the study of *line elements* (lines plus incident points) is much more fruitful (see [58, 93]).

We return to the linear velocity vector field (10.30) and let $\mathbf{x} = (x, y, z)$. In [93], it has been shown that the suitable choice of a coordinate system transforms the triple $(\mathbf{c}, \overline{\mathbf{c}}, \gamma)$ into one of four normal forms.

Without loss of generality, it can always be assumed that

$$\mathbf{c} = (0, 0, 1), \quad \overline{\mathbf{c}} = (0, 0, p), \quad \text{with} \quad p, \gamma \in \mathbb{R}.$$

Then, the velocity of the point X (with coordinate vector $\mathbf{x}$) can be written as

$$\mathbf{v}(\mathbf{x}) = (\gamma x - y, \gamma y + x, \gamma z + p),$$

and the path tangents (considered as straight lines) have Plücker coordinates

$$T = (\gamma x - y, \gamma y + x, \gamma z + p; py - xz, -px - yz, x^2 + y^2). \tag{10.31}$$

An implicit equation of (10.31) can be found by homogenizing the Cartesian point coordinates with

$$x \to X_1 X_0^{-1}, \quad y \to X_2 X_0^{-1}, \quad z \to X_3 X_0^{-1}$$

and taking into account that the Plücker coordinates $(t_1, \ldots, t_6)$ of T have to fulfill (10.2). This yields

$$\gamma(t_1 t_5 - t_2 t_4) + p(t_1^2 + t_2^2) - t_3 t_6 = 0, \tag{10.32}$$

which is obviously a quadratic complex. For $\gamma = 0$ and $p \neq 0$, (10.32) is the equation of the instantaneous tangent complex of a Euclidean motion. The equation of the quadratic complex of path tangents of a pure spiral motion is obtained if we insert $\gamma \neq 0$ and $p = 0$ into (10.32). The complex (10.32) is of characteristic [222].

Collineation complexes

Assume that

$$\kappa\colon \mathbb{P}^3(\mathbb{F}) \to \mathbb{P}^3(\mathbb{F}) \quad \text{with} \quad \mathbf{x} \mapsto \mathbf{y} = \mathbf{A}\mathbf{x}$$

is a projective collineation. Then, the lines joining each point $X = \mathbf{x}\mathbb{F}$ with its κ-image $Y = \mathbf{y}\mathbb{F}$ form a quadratic complex of lines. This is clear, since $\mathbf{x} \wedge \mathbf{y}$ is quadratic in the homogeneous coordinates $\mathbf{x} = (x_0, x_1, x_2, x_3)$ of X, and so are the Plücker coordinates $(l_1, \dots, l_6)$ of the lines $l = [X, Y]$. The elimination of x_0, x_1, x_2, and x_3 from the parametrization of the complex yields a quadratic form that is the complex's equation.

Tetrahedral complexes

A tetrahedron $\mathcal{T}$ is a polyhedron with four vertices and four faces. Naturally, it has six edges. A straight line l that intersects no edge of $\mathcal{T}$ will meet the four faces F_i in four mutually distinct points L_i ($i \in \{0,1,2,3\}$). Then, the cross ratio $\operatorname{cr}(L_0, L_1, L_2, L_3) = \delta$ of these four points will be any value in $\mathbb{F} \setminus \{0,1\}$. The set of all lines intersecting the faces of $\mathcal{T}$ at a fixed cross ratio $\delta \in \mathbb{F} \setminus \{0,1\}$ is called a *tetrahedral complex.*

We will give an equation of a tetrahedral complex based on a suitable coordinatization. For that purpose, we assume that the vertices of $\mathcal{T}$ coincide with the base points of the projective frame. Provided that the lines in question have the Plücker coordinates $(l_1, \dots, l_6)$, the homogeneous coordinates of its intersections L_0, ..., L_3 with the faces of $\mathcal{T}$ given by $x_0 = 0$, ..., $x_3 = 0$ can be read off from the rows (or columns) of (10.7). Using L_0 and L_3 as the base points $\mathbf{b}_0 = (1,0)\mathbb{F}$ and $\mathbf{b}_1 = (0,1)\mathbb{F}$ of the projective frame on l, the points L_1 and L_2 obtain their homogeneous coordinates on l:

$$\mathbf{l}_1 = (-l_5, l_1)\mathbb{F} \quad \text{and} \quad \mathbf{l}_2 = (l_4, l_2)\mathbb{F}.$$

With the cross ratio formula (5.6) from [46, p. 202], we compute $\operatorname{cr}(L_0, L_1, L_2, L_3)$, equate this with the chosen value $\delta \in \mathbb{F} \setminus \{0,1\}$, and find the equation of the complex

$$l_2 l_5 + \delta l_3 l_6 = 0.$$

Axes of a quadric

Each point P in a three-dimensional metric space sends a unique normal n_P to its polar plane π_P w.r.t. a regular quadric $\mathcal{Q}$ (cf. Figure 10.13).

Thus, the totality $\mathcal{A}$ of all such lines is a three-dimensional submanifold of the manifold of lines, *i.e.*, a complex of lines. It is called the *complex of axes of* $\mathcal{Q}$. This complex contains the congruence of surface normals of $\mathcal{Q}$, since the points of the quadric are incident with their polar planes, the tangent planes of $\mathcal{Q}$. It is clear that the principal axes are among the axes of a quadric.

In order to give an analytic description of this complex, we may assume that a quadric $\mathcal{Q}$ with center is given by

$$\mathcal{Q}: \quad \frac{x^2}{a} + \frac{y^2}{b} + \frac{z^2}{c} = 1 \tag{10.33}$$

with non-vanishing and non-proportional constants a, b, c. Note that this includes ellipsoids and hyperboloids as well. The case of a quadric without center (elliptic or hyperbolic paraboloid) is postponed to Exercise 10.1.12.

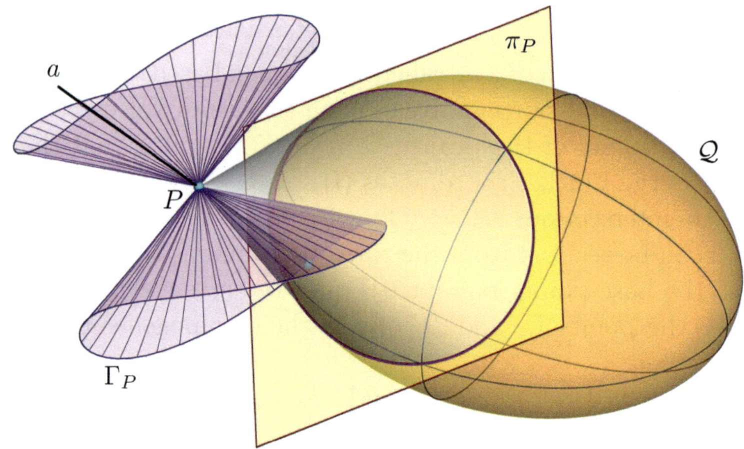

FIGURE 10.13. The axes a of a quadric $\mathcal{Q}$ through a point P are orthogonal to the polar plane π_P of P with regard to the quadric. The complex cone Γ_P at P of the complex of axes is a quadratic cone centered at P.

For a generic point $P = (\xi, \eta, \zeta)$, its polar plane π_P with regard to $\mathcal{Q}$ has the equation

$$\pi_P: \quad \frac{x\xi}{a} + \frac{y\eta}{b} + \frac{z\zeta}{c} = 1,$$

and the normal n_P from P to π_P can be parametrized by

$$\mathbf{n}_P(\lambda) = \begin{pmatrix} \xi \\ \eta \\ \zeta \end{pmatrix} + \lambda \begin{pmatrix} \frac{\xi}{a} \\ \frac{\eta}{b} \\ \frac{\zeta}{c} \end{pmatrix},$$

which in terms of Plücker coordinates reads

$$\begin{aligned} &(l_1, l_2, l_3; l_4, l_5, l_6) = \\ &= \left(\tfrac{\xi}{a}, \tfrac{\eta}{b}, \tfrac{\zeta}{c}; \left(\tfrac{1}{c} - \tfrac{1}{b}\right)\eta\zeta, \left(\tfrac{1}{a} - \tfrac{1}{c}\right)\zeta\xi, \left(\tfrac{1}{b} - \tfrac{1}{a}\right)\xi\eta\right). \end{aligned} \tag{10.34}$$

Eliminating the parameters ξ, η, ζ, we find the implicit equation of the complex

$$\mathcal{A}:\ al_1l_4 + bl_2l_5 + cl_3l_6 = 0. \tag{10.35}$$

Obviously, $\mathcal{A}$ is a quadratic complex. Moreover, with (10.34) and the cross ratio formula [46, Eq. (5.6) on p. 202]), we find that

$$\operatorname{cr}(L_0, L_1, L_2, L_3) = \frac{a - c}{a - b} = \text{const.}, \tag{10.36}$$

which makes the complex $\mathcal{A}$ a tetrahedral complex.

According to Exercise 7.1.3, the poles of a fixed plane w.r.t. the quadrics of a confocal family are located on a line orthogonal to π. Therefore, the complex does not change if $\mathcal{Q}$ is replaced by a confocal quadric. Consequently, it comprises all surface normals of the confocal family. By Theorem 7.1.2, the complex cones contain triples of mutually orthogonal lines and, are therefore, equilateral (cf. [46, p. 459]).

• **Exercise 10.1.12** Complex of axes of paraboloids.

Show that the equation of the complex of axes of a paraboloid

$$\mathcal{P}:\ \frac{x^2}{a} + \frac{y^2}{b} = 2z$$

with $a, b \in \mathbb{R}^*$ is proportional to

$$\mathcal{A}:\ (a - b)l_1l_2 + l_3l_6 = 0.$$

The case of an elliptic paraboloid is covered if both constants a and b have the same sign; otherwise the hyperbolic paraboloid is given.

• **Exercise 10.1.13** Complex cone of the complex of axes.

Let P be a generic point with Cartesian coordinates $\mathbf{p} = (\xi, \eta, \zeta)$, and let $\mathcal{Q}$ be the quadric (10.33). Further, with $\mathbf{x} = (x, y, z)$, we denote Cartesian coordinates.

Show that the complex cone $\Gamma_P \subset \mathcal{A}$ through P (in the axis complex (10.35)) has the equation

$$\Gamma_P: \ (\mathbf{x}-\mathbf{p})^{\mathrm{T}} \begin{pmatrix} 0 & (a-b)\zeta & (c-a)\eta \\ (a-b)\zeta & 0 & (b-c)\xi \\ (c-a)\eta & (b-c)\xi & 0 \end{pmatrix} (\mathbf{x}-\mathbf{p}) = 0$$

and is an equilateral cone. Γ_P degenerates if, and only if, P lies in a plane of symmetry of $\mathcal{Q}$ or if, and only if, $\mathcal{Q}$ is a quadric of revolution.

Orthogonal polars w.r.t. a quadric

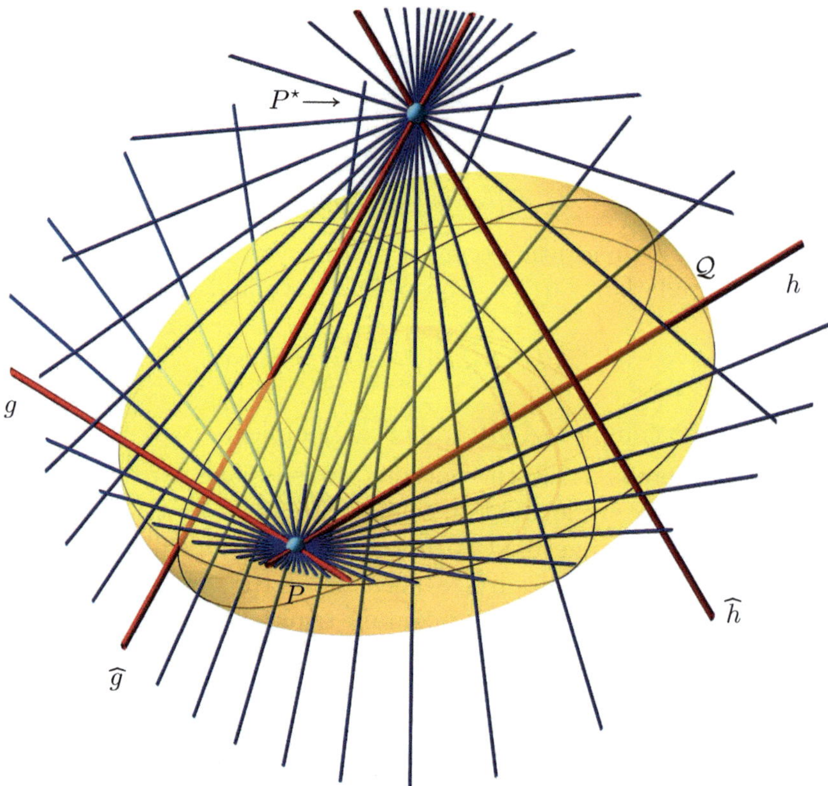

FIGURE 10.14. In a pencil of lines and its polar pencil, we find exactly two pairs $(g, \widehat{g})$ and $(h, \widehat{h})$ of orthogonal polars with regard to the quadric Q.

If the polar line $l^\star$ of a line l w.r.t. a quadric $\mathcal{Q}$ is orthogonal to l, then the plane through $l^\star$ orthogonal to l has its pole on l. Hence, l belongs to the complex of axes. Conversely, if l belongs to this complex, then its polar line $l^\star$ lies in a plane orthogonal to l and is, therefore, orthogonal to l. Thus, we obtain no new line complex.

■ **Example 10.1.10** Orthogonal polars w.r.t. an ellipsoid.

Assume that an ellipsoid $\mathcal{Q}$ is given by (10.33), which is different from a sphere. Let now $(l_1,\ldots,l_6)$ be the Plücker coordinates of a straight line l in the (preferably projectively closed) ambient space of $\mathcal{Q}$. The Plücker coordinates of the polar line $l^\star$ by applying the polarity with the coordinate matrix

$$\mathbf{T} = \operatorname{diag}(-1, a^{-1}, b^{-1}, c^{-1})$$

to two points L_O and L_3 on l. The homogeneous coordinates of L_O and L_3 are taken from (10.7), and then we find the homogeneous coordinates of the respective polar planes as $\pi(L_O) = (0, l_1 a^{-1}, l_2 b^{-1}, l_3 c^{-1})$ and $\pi(L_3) = (l_3, l_5 a^{-1}, -l_4 b^{-1}, 0)$. The polar line $l^\star$ of l equals

$$l^\star = \pi(L_O) \cap \pi(L_3) = (al_4, bl_5, cl_6; -bcl_1, -cal_2, -abl_3).$$

The condition that polars are orthogonal is simply the orthogonality of the two direction vectors, which yields the complex's equation (10.35). Some lines of the complex are shown in Figure 10.14. Thus, the complex of pairs of orthogonal polars with regard to a quadric is a tetrahedral complex. The lines of the complex intersect the faces of the fundamental tetrahedron in quadruples of points with the cross ratio given in (10.36).

Harmonic complex, Battaglini's complex

Let $\mathcal{Q}_1$ and $\mathcal{Q}_2$ be two quadrics which have a common polar tetrahedron. The common polar tetrahedron shall be the base of a projective frame. Within this frame, both quadrics have the simplest equations, *i.e.*, the corresponding quadratic forms are diagonalized simultaneously. In the case of two ellipsoids, we have

$$\mathcal{Q}_1: \frac{x^2}{a_1^2} + \frac{y^2}{b_1^2} + \frac{z^2}{c_1^2} = 1 \quad \text{and} \quad \mathcal{Q}_2: \frac{x^2}{a_2^2} + \frac{y^2}{b_2^2} + \frac{z^2}{c_2^2} = 1. \tag{10.37}$$

We assume that neither a pair out of (a_1, b_1, c_1) nor a pair out of (a_2, b_2, c_2) consists of equal numbers, *i.e.*, quadrics of revolution are excluded.

Let now $(l_1,\ldots,l_6)$ be the Plücker coordinates of an arbitrary straight line l. This line will intersect the quadric $\mathcal{Q}_i$ in two points $S_{i,1}$ and $S_{i,2}$ with $i \in \{1,2\}$. We aim at a description of all lines such that the four points $S_{1,1}$, $S_{1,2}$, $S_{2,1}$, and $S_{2,2}$ are harmonic, *i.e.*, that they form the cross ratio -1.

An affine parametrization of l can be given by intersecting the third and fourth plane through l given in (10.11) (just replace c_i with l_i and multiply from the left with $(x_0,\ldots,x_3)^{\mathrm{T}}$). We obtain

$$\mathbf{l}(t) = \begin{pmatrix} 0 \\ -l_6 l_1^{-1} \\ l_5 l_1^{-1} \end{pmatrix} + t \begin{pmatrix} 1 \\ l_2 l_1^{-1} \\ l_3 l_1^{-1} \end{pmatrix}, \tag{10.38}$$

where $t \in \mathbb{R}$ is an affine parameter on l. Inserting (10.38) into (10.37), we find two quadratic equations whose solutions shall be denoted by $t_{i,j}$ (with $i, j \in \{1, 2\}$) and correspond to the common points $S_{i,j}$ of l and $\mathcal{Q}_i$. According to Vieta's Lemma, the solutions $t_{i,j}$ satisfy

$$\begin{aligned} t_{i,1} + t_{i,2} &= -\frac{2a_i^2(b_i^2 l_3 l_5 - c_i^2 l_2 l_6)}{a_i^2 b_i^2 l_3^2 + a_i^2 c_i^2 l_2^2 + b_i^2 c_i^2 l_1^2}, \\ t_{i,1} \cdot t_{i,2} &= \frac{a_1^2(b_1^2 l_5^2 + c_1^2 l_6^2 - b_1^2 c_1^2 l_1^2)}{a_1^2 b_1^2 l_3^2 + a_1^2 c_1^2 l_2^2 + b_1^2 c_1^2 l_1^2} \end{aligned} \tag{10.39}$$

with $i = 1, 2$. The parameter values $t_{i,j}$ can be transformed into homogeneous coordinates of the points $S_{i,j}$ on the line l by simply setting $(1, t_{i,j})$. Then, we use formula (5.6) from [46, p. 202] and find that the four values $t_{i,j}$ have to satisfy

$$2(t_{1,1}t_{1,2} + t_{2,1}t_{2,2}) - (t_{1,1} + t_{1,2})(t_{2,1} + t_{2,2}) = 0 \tag{10.40}$$

in order to make $(S_{1,1}, S_{1,2}, S_{2,1}, S_{2,2})$ a harmonic quadruple (in that particular ordering).

Now, we insert (10.39) into (10.40), use (10.2), and arrive at the equation

$$\begin{aligned} \mathcal{H}: &\, b_1^2 b_2^2 c_1^2 c_2^2 (a_1^2 + a_2^2) l_1^2 + a_1^2 a_2^2 c_1^2 c_2^2 (b_1^2 + b_2^2) l_2^2 + a_1^2 a_2^2 b_1^2 b_2^2 (c_1^2 + c_2^2) l_3^2 = \\ &= a_1^2 a_2^2 (b_1^2 c_2^2 + b_2^2 c_1^2) l_4^2 + b_1^2 b_2^2 (a_1^2 c_2^2 + a_2^2 c_1^2) l_5^2 + c_1^2 c_2^2 (a_1^2 b_2^2 + a_2^2 b_1^2) \end{aligned} \tag{10.41}$$

of the complex of lines that meet a pair of quadrics in four harmonic points.

Obviously, the complex $\mathcal{H}$ (10.41) is a quadratic complex. In the literature, it is frequently referred to as Battaglini's complex.[6]

Let us have a look at the complex cones of $\mathcal{H}$. For that purpose, we simplify the equation of the complex by using abbreviations for the coefficients of l_i^2, *i.e.*, we write $\mathcal{H}$'s equation in the form

$$\mathcal{H}: \ \alpha_1 l_1^2 + \ldots + \alpha_6 l_6^2 = 0.$$

For a point P with Cartesian coordinates $\mathbf{p} = (\xi, \eta, \zeta) \in \mathbb{R}^3$, we assume that the star of lines through P is parametrized by $\mathbf{s} = \mathbf{p} + (u, v, w)$ with

[6] Giuseppe Battaglini (1826–1894), Italian mathematician.

(u,v,w) varying freely in $\mathbb{R}^3$. The Plücker coordinates of the lines in the star are $(u,v,w;\eta w-\zeta v,\zeta u-\xi w,\xi v-\eta u)$. Inserting the latter into (10.41) yields a condition on u, v, and w such that the parametrization of the star becomes the parametrization of the complex cone through P. After the subsequent elimination of the remaining parameters u and v, we obtain the equation of the complex cone Γ_P. With $\mathbf{x}=(x,y,z)$, it reads

$$\Gamma_P:(\mathbf{x}-\mathbf{p})^{\mathrm{T}}\begin{pmatrix}\alpha_6\eta^2+\alpha_5\zeta^2+\alpha_1 & -\alpha_6\eta\xi & -\alpha_5\xi\zeta\\ -\alpha_6\eta\xi & \alpha_6\xi^2+\alpha_4\zeta^2+\alpha_2 & -\alpha_4\eta\zeta\\ -\alpha_5\xi\zeta & -\alpha_4\eta\zeta & \alpha_4\eta^2+\alpha_5\xi^2+\alpha_3\end{pmatrix}(\mathbf{x}-\mathbf{p})=0$$

and degenerates if, and only if,

$$\begin{aligned}\mathcal{K}:\alpha_1\alpha_5\alpha_6\xi^4+\alpha_2\alpha_4\alpha_6\eta^4+\alpha_3\alpha_4\alpha_5\zeta^4+\\ +\alpha_6(\alpha_1\alpha_4+\alpha_2\alpha_5)\xi^2\eta^2+\alpha_5(\alpha_1\alpha_4+\alpha_3\alpha_6)\xi^2\zeta^2+\\ +\alpha_4(\alpha_2\alpha_3+\alpha_3\alpha_6)\eta^2\zeta^2+\alpha_1\alpha_2\alpha_3=0.\end{aligned} \tag{10.42}$$

The quartic surface $\mathcal{K}$ is the locus of all points $P\in\mathbb{R}^3$ where the complex cone Γ_P degenerates and is, therefore, called the *singular surface of the complex.* The surface $\mathcal{K}$ is called *Kummer's surface*[7] an example of which is displayed in Figure 10.15. It is a quartic surface and has up to 16 real nodes, which are the vertices of four tetrahedra. The sixteen nodes form a *Kummer configuration*: This is an arrangement of sixteen points and sixteen planes such that each plane carries six points and each of the points belongs to six planes.

Usually, the equation of Kummer's surface is given in the form

$$(x^2+y^2+z^2-\alpha^2)^2-\beta\lambda_1\lambda_2\lambda_3\lambda_4=0, \tag{10.43}$$

where λ_i are the equations of the four planes building the fundamental tetrahedron. The constant $\alpha\in\mathbb{R}$ can be chosen freely and $\beta=(2\alpha^2-1)(2-\alpha^2)^{-1}$. The surface depicted in Figure 10.15 is determined by $\alpha=3/\sqrt{8}$. The surfaces given in (10.42) and (10.43) differ only by a collinear transformation.

It is worth mentioning that there exists a certain case of degeneracy: If one of the quadrics (10.37) becomes the ideal conic of Euclidean geometry,

[7] Ernst Eduard Kummer (1810–1893), German mathematician who worked in various fields of mathematics such as number theory, function theory, and algebraic geometry. From the line-geometric point of view, his most interesting work deals with line congruences (see [83]).

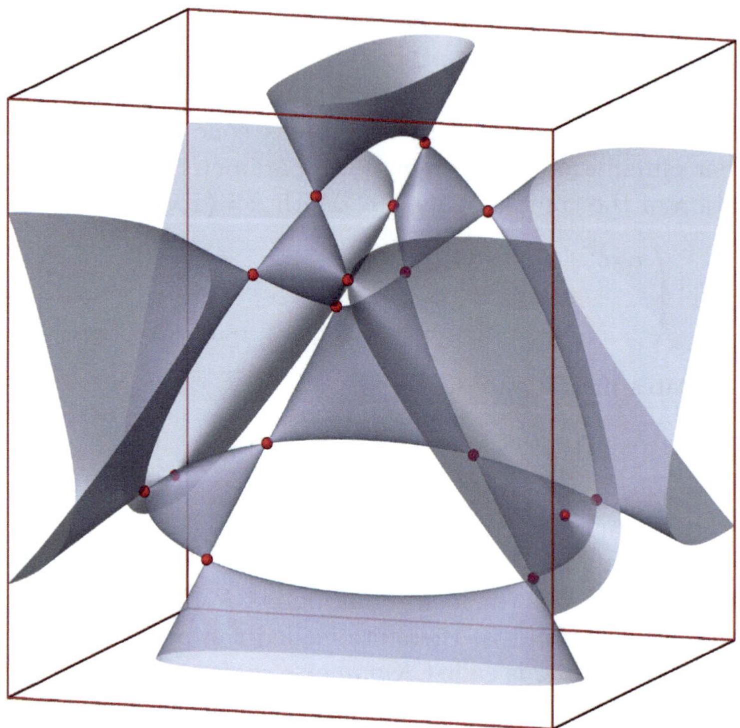

FIGURE 10.15. Kummer's surface with $\alpha = 3/\sqrt{8}$ with 16 real nodes (displayed as small red spheres).

then (10.42) is the equation of *Fresnel's wave surface*[8] (see Figure 10.16). Assume that we are given an ellipsoid $\mathcal{Q}$ with center $(0,0,0)$, semi axes lengths $a, b, c \in \mathbb{R}^+$, and the equation as given in (10.33). The planes through the center of $\mathcal{Q}$ have the equations

$$\varepsilon : \; e_1 x + e_2 y + e_3 z = 0, \tag{10.44}$$

where we may assume that the planes' normal vectors $\mathbf{e} = (e_1, e_2, e_3)$ are unit vectors, *i.e.*, $e_1^2 + e_2^2 + e_3^2 = 1$. Each plane ε intersects the quadric in a conic q_ε with major semiaxis a_ε and minor semiaxis b_ε. Applying $\pm a_\varepsilon$ and $\pm b_\varepsilon$ on the normal $\mathbf{e}$, we obtain four points $W_{1,2} = \pm a_\varepsilon \mathbf{e}$ and $W_{3,4} = \pm b_\varepsilon \mathbf{e}$, which belong to one algebraic variety.

[8] AUGUSTIN JEAN FRESNEL (1788–1827) was a French physicist and engineer who founded the wave theory of light.

The principal axes lengths are the extrema of the function $f := x^2 + y^2 + z^2$ under the two side conditions (10.44) and (10.33). Geometrically speaking, we are looking for a sphere $\mathcal{S}$ of radius r concentric with the quadric $\mathcal{Q}$ such that the two surfaces touch. Hence, we build the Lagrange function

$$F = x^2 + y^2 + z^2 - \lambda\left(\frac{x^2}{a} + \frac{y^2}{b} + \frac{z^2}{c} - 1\right) - 2\mu(e_1 x + e_2 y + e_3 z) \qquad (10.45)$$

with the two Lagrangian multipliers λ and μ. A necessary, but by no means sufficient, criterion for F to be minimal at some point (x, y, z) is that the partial derivatives of F w.r.t. x, y, and z vanish, *i.e.*,

$$\begin{gathered} \tfrac{\partial F}{\partial x} = x\left(1 - \tfrac{\lambda}{a}\right) - \mu e_1 = 0, \quad \tfrac{\partial F}{\partial y} = y\left(1 - \tfrac{\lambda}{b}\right) - \mu e_2 = 0, \\ \tfrac{\partial F}{\partial z} = z\left(1 - \tfrac{\lambda}{c}\right) - \mu e_3 = 0. \end{gathered} \qquad (10.46)$$

■ **Example 10.1.11** Constructive approach to Fresnel's wave surface.

Multiplying the first equation with x, the second with y, the third with z, and summing them, we find

$$\lambda = x^2 + y^2 + z^2 = r^2, \qquad (10.47)$$

since (10.33) and (10.44) hold. Thus, each of the points W_i on the wave surface can be written as $\mathbf{w} = r\mathbf{e}$. We solve (10.46) w.r.t. x, y, z, and arrive at

$$x = \frac{\mu e_1 a}{a^2 - r^2}, \quad y = \frac{\mu e_2 b}{b^2 - r^2}, \quad z = \frac{\mu e_3 c}{c^2 - r^2}, \qquad (10.48)$$

where λ is replaced by $r^2 = x^2 + y^2 + z^2$ according to (10.47). From the components of $\mathbf{x} = r\mathbf{e}$, we can infer

$$e_1 = \frac{x}{r}, \quad e_2 = \frac{y}{r}, \quad e_3 = \frac{z}{r}$$

and after multiplication of the first, second, and third equation in (10.48), we compute their sum, which yields a first form of the equation of Fresnel's wave surface as

$$\mathcal{W}: \quad \frac{ax^2}{a - x^2 - y^2 - z^2} + \frac{by^2}{b - x^2 - y^2 - z^2} + \frac{cz^2}{c - x^2 - y^2 - z^2} = 0, \qquad (10.49)$$

or, by subtracting from $x^2 + y^2 + z^2 = r^2$, we find a second version

$$\mathcal{W}: \quad \frac{x^2}{x^2 + y^2 + z^2 - a} + \frac{y^2}{x^2 + y^2 + z^2 - b} + \frac{z^2}{x^2 + y^2 + z^2 - c} = 0. \qquad (10.50)$$

It is also possible to give a parametrization of $\mathcal{W}$:

$$x = \pm\sqrt{\frac{(a-u)(bc-v)}{(c-a)(a-b)}}, y = \pm\sqrt{\frac{(b-u)(ca-v)}{(a-b)(b-c)}}, z = \pm\sqrt{\frac{(c-u)(ab-v)}{(b-c)(c-a)}}.$$

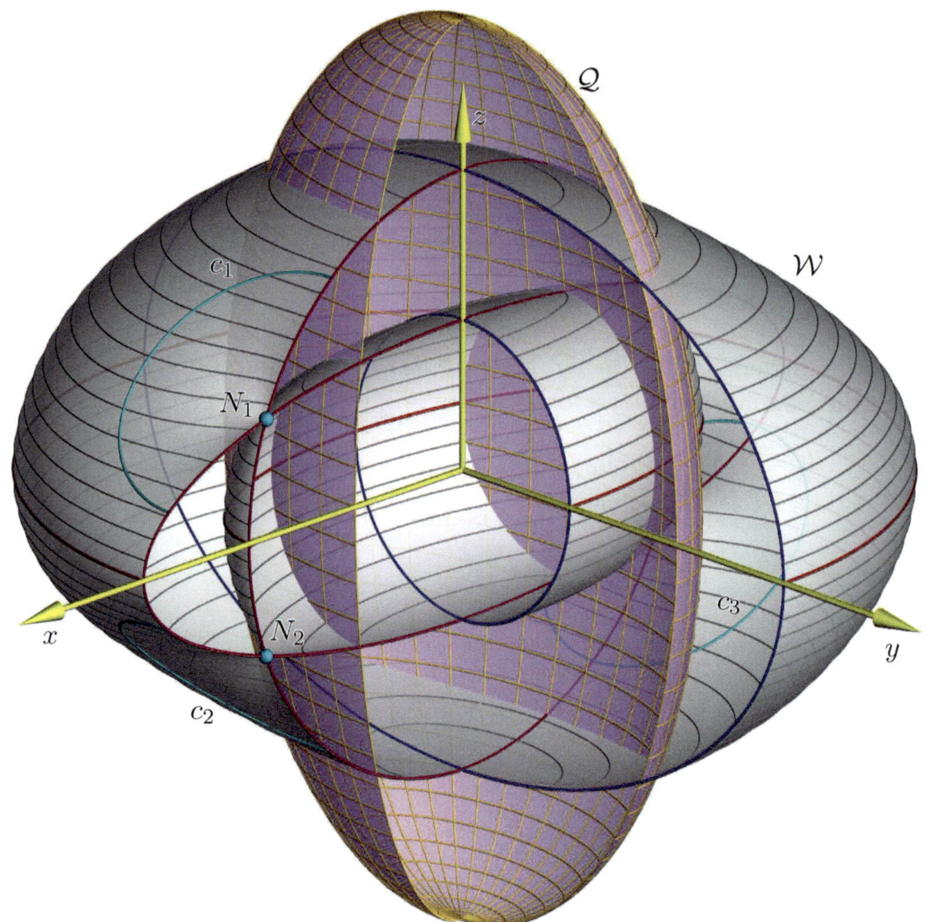

FIGURE 10.16. The quartic Fresnel wave surface $\mathcal{W}$ has four conical nodes N_1, $\ldots$, N_4 corresponding to the umbilics of an ellipsoid $\mathcal{Q}$. $\mathcal{W}$ intersects its planes of symmetry along circles and ellipses (red and blue). There are four circles c_1, $\ldots$, c_4 on Fresnel's surface that are the intersection of the four real double tangent planes with $\mathcal{W}$.

Painvin's complex

The tangent planes of a regular quadric $\mathcal{Q}$ form the dual quadric $\mathcal{Q}^\star$. A quadric $\mathcal{Q}$ sends two tangent planes through each line that is not tangent to $\mathcal{Q}$, and thus, a quadric is said to be of *rank* two. This notion of rank is not to be confused with the rank of a matrix, especially with the rank of the coefficient matrix of a quadric. By assumption, $\mathcal{Q}$ is regular, and therefore, the coefficient matrix of its equation is of full rank, *i.e.*, it is four.

In a metric space, we may ask for all lines a that send a pair of orthogonal tangent planes to the quadric $\mathcal{Q}$.[9] This yields one additional condition on the Plücker coordinates of the lines a (besides (10.2)). Consequently, the lines a form a complex of lines.

Let the homogeneous Plücker coordinates of the lines a be $(a_1, \ldots, a_6)$. According to (10.7) and (10.8), the homogeneous coordinates of planes through a are linear combinations of the rows (or columns) of

$$\begin{pmatrix} 0 & a_4 & a_5 & a_6 \\ -a_4 & 0 & a_3 & -a_2 \\ -a_5 & -a_3 & a_5 & a_1 \\ -a_6 & a_2 & -a_1 & 0 \end{pmatrix}.$$

We let $\mathbf{a}_0 = (0, a_4, a_5, a_6)$ and $\mathbf{a}_3 = (a_6, -a_2, a_1, 0)$ be the homogeneous coordinate vectors of two planes α_0 and α_3 from the pencil about a. Then, a homogeneous parametrization of a (in terms of homogeneous coordinates) with the homogeneous parameter $\lambda : \mu \neq 0 : 0$ equals

$$\mathbf{a}(\lambda, \mu) = \lambda \mathbf{a}_0 + \mu \mathbf{a}_3. \tag{10.51}$$

The latter equation can also be viewed as a coordinatization of the pencil of planes about a. The tangent planes of $\mathcal{Q}$ through a are the *intersection points* of the line (10.51) with the dual quadric $\mathcal{Q}^\star$. If we assume that $\mathcal{Q}$ is given by (10.33), then the coefficient matrix of $\mathcal{Q}^\star$, when written in homogeneous plane coordinates, equals $\mathbf{M} = \text{diag}(-1, a, b, c,)$. Let further

$$\omega_{ij} := \mathbf{a}_i^{\mathrm{T}} \mathbf{M} \mathbf{a}_j \tag{10.52}$$

for $i, j \in \{0, 3\}$. Note that $\omega_{ij} = \omega_{ji}$, for $\mathbf{A} = (\omega_{ij})$ is symmetric. Then, the coordinates (λ, μ) in the pencil about a have to satisfy

$$\omega_{00}\lambda^2 + 2\omega_{03}\lambda\mu + \omega_{33}\mu^2 = 0$$

and are, thus,

$$\lambda_1 : \mu_1 = (-\omega_{03} + R) : \omega_{00} \quad \text{and} \quad \lambda_2 : \mu_2 = (-\omega_{03} - R) : \omega_{00} \tag{10.53}$$

[9] Note that the edges of boxes as displayed in Figure 2.9 belong to a complex of lines related to an ellipsoid.

with $R = \sqrt{\omega_{03}^2 - \omega_{00}\omega_{33}}$. The condition that the two tangent planes τ_1 and τ_2 with coordinate vectors $\mathbf{t}_1 = \mathbf{a}(\lambda_1, \mu_1)$ and $\mathbf{t}_2 = \mathbf{a}(\lambda_2 : \mu_2)$ are orthogonal is simply

$$\langle \lambda_1 \overline{\mathbf{a}}_0 + \mu_1 \overline{\mathbf{a}}_3, \lambda_2 \overline{\mathbf{a}}_0 + \mu_2 \overline{\mathbf{a}}_3 \rangle = 0$$

with $\overline{\mathbf{a}}_i$ being the vector $\mathbf{a}_i$ with the first row removed, since this gives the normal vector of the plane α_i. Inserting (10.53) into the latter equation, we find

$$\omega_{00}(a_1^2 + a_2^2) - 2\omega_{03}(a_1 a_5 - a_2 a_4) + \omega_{33}(a_4^2 + a_5^2 + a_6^2) = 0,$$

where the coefficients ω_{ij} still depend on the Plücker coordinates a_i of a. Substituting (10.52) yields the equation of the quadratic complex of lines that send pairs of orthogonal tangent planes to the quadric (10.33)

$$\mathcal{P}:\ (b+c)a_1^2 + (c+a)a_2^2 + (a+b)a_3^2 = a_4^2 + a_5^2 + a_6^2. \tag{10.54}$$

The complex $\mathcal{P}$ is called *Painvin's complex.*[10] For details and more results on quadratic complexes, see [70, 152].

• **Exercise 10.1.14** Frenel's surface as the singular surface of Painvin's complex.

Show that the singular surface, *i.e.*, the locus of points with degenerate complex cones of the complex (10.54), is the wave surface by Fresnel.

[10]Louis Felix Painvin (1826–1875?) was a French mathematician, teacher at the lycée de Douai, and assistant professor for mechanics and experimental physics at the Sorbonne.

10.2 Sphere Geometry

Sphere geometry deals with a geometry where Euclidean spheres are the basic objects. Each object in Euclidean 3-space (or even in n-dimensional space) is considered as a manifold of Euclidean spheres. This includes points as spheres of radius zero, planes as spheres with infinite radius, and, of course, ordinary spheres. A Euclidean circle in 3-space can be viewed as the intersection of two Euclidean spheres, and thus, the circle can be seen as the carrier of an elliptic or hyperbolic pencil of spheres, depending on whether the circle is real or not. Depending on the approach to sphere geometry, spheres may be oriented or not.

We shall describe Lie's model of the geometry of oriented spheres in Euclidean 3-space. Generalizations to quadric models of sphere geometries in Euclidean spaces of arbitrary dimensions are straightforward (cf. [24]). Elementary geometric aspects as well as the construction of a Grassmannian over Lie's quadric can be found in [30] and shall not be discussed here.

Sphere coordinates, Lie's quadric

There are various ways to a proper coordinatization of the manifold of spheres. We shall start with the equation

$$S:\ (s_6 - s_4)(x^2 + y^2 + z^2) - 2s_1x - 2s_2y - 2s_3z + (s_4 + s_6) = 0 \quad (10.55)$$

of a Euclidean sphere S. By completing to full squares for each variable, we find the center M and the radius r of S as

$$\mathbf{m} = \frac{1}{s_6 - s_4}(s_1, s_2, s_3) \quad \text{and} \quad r = \frac{s_5}{s_6 - s_4}, \quad (10.56)$$

provided that $s_6 \neq s_4$. Here, we have implicitly defined s_5 such that

$$L_2^4:\ \Omega_L(S,S) := s_1^2 + s_2^2 + s_3^2 + s_4^2 - s_5^2 - s_6^2 = 0. \quad (10.57)$$

The six values $(s_1, \ldots, s_6)$ are the *Lie*[11] *coordinates of a Euclidean sphere* if they satisfy (10.57). On the other hand, we can show that any sextuple satisfying (10.57) can be interpreted as the coordinates of a Euclidean

[11] Sophus Lie (1842–1899), Norwegian mathematician, famous for his works on partial differential equations, continuous groups, and minimal surfaces.

sphere in 3-space. The fifth coordinate s_5 can be equipped with a sign which can be used to express the sphere's orientation.

In (10.56), we assumed $s_6 \neq s_4$. Dropping this assumption enables us to describe oriented planes in Euclidean 3-space. With $s_4 = s_6$, neither the center nor the radius of the sphere are determined. If now s_1, s_2, s_3, and s_5 still satisfy (10.57) without being simultaneously equal to zero, then

$$(s_1, s_2, s_3, s_4, s_5, s_4)$$

is the *Lie coordinate vector of an oriented plane.* The orientation of the plane equals the orientation of its normal (s_1, s_2, s_3).

Points can also be described by Lie coordinates: If

$$s_5 = 0$$

and the remaining s_i satisfy (10.57), then the point M from (10.56) is considered as a sphere of radius zero, and thus, it is often called a *null sphere.* Sometimes, it is much better to view these kinds of spheres as the isotropic cone emanating from M.

The coordinate vector

$$\mathbf{u} = (0, 0, 0, 1, 0, 1)$$

describes a point U on Lie's quadric. Unfortunately, $s_4 - s_6 = 0$ and $s_5 = 0$. Therefore, $\mathbf{u}$ does neither describe a sphere nor a point. Although it satisfies $s_4 - s_6$, it does not represent a plane, since the first three entries are equal to zero, and thus, the normal vector is undefined. Usually, U is called the *ideal element* or the *improper point*, sometimes denoted by ∞. Adding the point U to the set of spheres, planes, and points in Euclidean 3-space creates the set of *Lie spheres.*

The six coordinates of a sphere are subject to the homogeneous quadratic form (10.57). Moreover, $(s_1, \ldots, s_6)$ themselves can be made homogeneous, since the multiplication of $\mathbf{s} = (s_1, \ldots, s_6)$ with a non-vanishing factor $\lambda \in \mathbb{R} \setminus \{\mathbf{0}\}$ does not change the center and radius as given in (10.56). This allows us to interpret $(s_1, \ldots, s_6) \in \mathbb{R}^6 \setminus \{\mathbf{0}\}$ as homogeneous coordinates in $\mathbb{P}^5(\mathbb{R})$. Then, (10.57) is the equation of a quadric $L_2^4 \subset \mathbb{P}^5(\mathbb{R})$ which is called *Lie's quadric.*

Lie's quadric is of index one, *i.e.*, the maximal dimension of subspaces entirely contained in L_2^4 equals one (cf. Theorem 4.3.1). In fact, there are real lines in L_2^4. Later, we shall study the linear and quadratic submanifolds of L_2^4.

We collect all the obtained results in the following theorem:

Theorem 10.2.1 *Each oriented sphere (including oriented planes and points) in Euclidean 3-space can be mapped to a unique point on the four-dimensional quadric L_2^4 (called Lie's quadric), which is of index one, and thus, it carries straight lines as maximal subspaces. Each point on L_2^4 corresponds to an oriented sphere (including oriented planes and points) in Euclidean 3-space.*

Proof: It remains to be shown that each point on L_2^4 corresponds to an oriented sphere (oriented plane or point). From the sextuple $(s_1, \ldots, s_6)$, we can try to recover the center and radius of a sphere with (10.56). If now $s_6 - s_4 \neq 0$ and $s_5 = 0$, the radius of the sphere equals 0 and $(s_1, \ldots, s_6)$ describes a point. If $s_6 - s_4 \neq 0$ and $s_5 \neq 0$, (10.56) yields the center and the radius of the sphere. In the case $s_6 = s_4$, the quadratic term of the sphere's equation (10.55) vanishes and the remainder is the equation of a plane, which is oriented in the sense of (s_1, s_2, s_3). ■

■ **Example 10.2.1** Sphere coordinates from center and radius – a parametrization of L_2^4.

We want to express the sphere coordinates $(s_1, \ldots, s_6)$ in terms of the center M with coordinate vector $\mathbf{m} = (m_1, m_2, m_3)$ and the radius r.

From (10.56) we can deduce the following system of linear equations:

$$m_1(s_6 - s_4) = s_1, \quad m_2(s_6 - s_4) = s_2, \quad m_3(s_6 - s_4) = s_3, \quad r(s_6 - s_4) = s_5.$$

We insert s_1, s_2, s_3, and s_5 into (10.57), and with $m_1^2 + m_2^2 + m_3^2 = \langle \mathbf{m}, \mathbf{m} \rangle$, we find

$$\langle \mathbf{m}, \mathbf{m} \rangle (s_6 - s_4)^2 - r^2 (s_6 - s_4)^2 + s_4^2 - s_6^2 =$$
$$(s_6 - s_4)(\langle \mathbf{m}, \mathbf{m} \rangle (s_6 - s_4) - r^2 (s_6 - s_4) - s_4 - s_6) = 0.$$

By assumption, $(s_1, \ldots, s_6)$ are the coordinates of a sphere, *i.e.*, $s_6 - s_4 \neq 0$. Thus, we can cancel this factor from the latter equation and arrive at a single homogeneous linear equation in s_4 and s_6, which yields

$$(s_1, \ldots, s_6) = \lambda \left(2\mathbf{m}, \langle \mathbf{m}, \mathbf{m} \rangle - r^2 - 1, 2r, \langle \mathbf{m}, \mathbf{m} \rangle - r^2 + 1 \right), \quad \lambda \in \mathbb{R}. \tag{10.58}$$

The polar system of Lie's quadric

Lie's quadric is a regular quadric. Therefore, it has a regular polar system. We shall disclose the sphere geometric meaning of this polar system. Let S and T be two Euclidean spheres with Lie coordinates $(s_1, \ldots, s_6)$ and $(t_1, \ldots t_6)$. According to (10.56), the respective centers and radii are

$$\begin{gathered} \mathbf{m}_S = \tfrac{1}{s_6 - s_4}(s_1, s_2, s_3), \quad \mathbf{m}_T = \tfrac{1}{t_6 - t_4}(t_1, t_2, t_3), \\ r_S = \tfrac{s_5}{s_6 - s_4}, \quad r_T = \tfrac{t_5}{t_6 - t_4}. \end{gathered} \tag{10.59}$$

For a common point X of S and T, the Law of Cosines yields

$$r_S^2 + r_T^2 - 2r_S r_T \cos\varphi = \|\mathbf{m}_S - \mathbf{m}_T\|^2, \tag{10.60}$$

where φ is the angle enclosed by the oriented normals of S and T at the point X. Inserting (10.59) into (10.60), and by taking into account that (10.57) holds true for both sextuples $(s_1,\dots,s_6)$ and $(t_1,\dots,t_6)$, we find

$$s_1t_1 + s_2t_2 + s_3t_3 + s_4t_4 - s_5t_5\cos\varphi - s_6t_6 = 0 \tag{10.61}$$

after cutting out the factor $(s_6 - s_4)(t_6 - t_4)$, which is not equal to zero, since S and T are spheres.

If the two spheres S and T are in oriented contact, then $\varphi = 0$ and $\cos\varphi = 1$. Consequently, (10.61) changes to

$$\Omega_L(S,T) = s_1t_1 + s_2t_2 + s_3t_3 + s_4t_4 - s_5t_5 - s_6t_6 = 0, \tag{10.62}$$

and the points S and T in the Lie model of sphere geometry turn out to be conjugate with respect to L_2^4. It is also possible to show that any pair of points conjugate with respect to L_2^4 corresponds to a pair of spheres in Euclidean 3-space which are in oriented contact. We can summarize:

Theorem 10.2.2 *Each pair of Euclidean spheres in oriented contact corresponds to a pair of points conjugate with regard to L_2^4, and vice versa.*

For each pair of oriented Euclidean spheres and for each pair of planes, (10.61) yields their oriented intersection angle.

Proof: The only thing that remains to be shown is the second part of Theorem 10.2.2 for the case of a pair of planes.

As explained on p. 520, oriented planes in Euclidean 3-space can be described by sextuples $(s_1, s_2, s_3, s_4, s_5, s_4)$. Since these coordinates are homogeneous, they can be scaled by any non-vanishing factor, for example, by $1/\sqrt{s_1^2 + s_2^2 + s_3^2}$. Then, (10.61) agrees with the elementary formula for the angle subtended by two unit vectors. ■

Lie's line-sphere-mapping

The two quadrics M_2^4 (10.2) and L_2^4 (10.57) are hyperquadrics in a projective space of five dimensions. Over the real number field, they differ in their signatures and indices.

The linear substitution

$$\begin{aligned} &l_1 \to s_1 + \mathrm{i}s_4,\quad l_2 \to s_2 + s_5,\quad l_3 \to s_3 + s_6,\\ &l_4 \to s_1 - \mathrm{i}s_4,\quad l_5 \to s_2 - s_5,\quad l_6 \to s_3 - s_6 \end{aligned} \tag{10.63}$$

transforms the quadratic form (10.2) into (10.57) and describes a collineation in the complex extension of $\mathbb{P}^5(\mathbb{R})$. Each point $(l_1, \ldots, l_6) \in M_2^4$ corresponds to precisely one point $(s_1, \ldots, s_6) \in L_2^4$. In other words, each line in the real projective 3-space is mapped to an oriented sphere in Euclidean 3-space.

The mapping (10.63) is called *Lie's line-sphere-mapping* and shows the close relation between line geometry and sphere geometry. Since the conjugate position with respect to quadrics is invariant under collineations, pairs of intersecting lines are mapped to pairs of spheres in oriented contact.

Automorphisms of Lie's quadric

The group of automorphisms of Lie's quadric is simply the group of projective collineations $\mathbb{P}^5 \to \mathbb{P}^5$ that leave L_2^4 invariant. Speaking in terms of Linear Algebra, these transformations are induced by all regular linear mappings $\mathbb{R}^6 \to \mathbb{R}^6$ that leave the quadratic form (10.57) invariant. It is clear that such automorphisms interchange the points on L_4^2 while they preserve the conjugacy of points, and thus, they preserve the oriented contact of the corresponding objects in Euclidean 3-space. Thus, the automorphic collineations of L_2^4 induce *contact transformations* in the set of Euclidean spheres. We shall have a look at some of them.

Central similarity

Without loss of generality, we may assume that the center of a central similarity ζ equals the origin of the Cartesian coordinate system. The scaling factor of all lengths shall be $\alpha \in \mathbb{R}^\star$. Then, the coordinates $\mathbf{x}$ of a point X are transformed according to

$$\mathbf{x} \mapsto \mathbf{x}' = \alpha \mathbf{x}.$$

The center M (with coordinates $\mathbf{m}$) and the radius r of a sphere S change to $\mathbf{m}' = \alpha \mathbf{m}$ and $r' = \alpha r$. Now, we use (10.56) and (10.58), derive the homogeneous sphere coordinates $\mathbf{s}'$ of the image sphere S', and obtain

$$\mathbf{s}' = \left(2\alpha s_1, 2\alpha s_2, 2\alpha s_3, (\alpha^2+1)s_4+(\alpha^2-1)s_6, 2\alpha s_5, (\alpha^2-1)s_4+(\alpha^2+1)s_6\right),$$

where we have used (10.57). It is clear and easily verified that $S' = \mathbf{s}'\mathbb{R}$ is also point on L_2^4. Note that the sphere coordinates of S' are linear in the sphere coordintes of S. Thus, the central similarity induces an automorphic collineation of L_2^4.

- **Exercise 10.2.1** Fixed spaces of the induced collineation.

Show that the characteristic polynomial of the linear automorphism of L_2^4 induced by a central similarity with scaling factor $\alpha \in \mathbb{R}^\star$ equals

$$\chi = (\lambda - 2\alpha)^2(\lambda - 2)(\lambda - 2\alpha^2).$$

Let $\{\mathbf{b}_i\}$ with $i \in \{1,2,3,4,5,6\}$ denote the canonical basis of $\mathbb{R}^6$. Then, compute the invariant eigenspaces (fixed subspaces of the induced collineation) and show that these are

$$E(2\alpha) = [\mathbf{b}_1, \mathbf{b}_2, \mathbf{b}_4, \mathbf{b}_6], \quad E(2\alpha^2) = [\mathbf{b}_4 + \mathbf{b}_6], \quad E(2) = [\mathbf{b}_4 - \mathbf{b}_6].$$

Note that $E(2)$ describes the ideal element ∞, which is common to all Lie spheres and is not altered by contact transformations.

Moreover, we observe that $E(2\alpha^2)$ is the coordinate vector of the isotropic cone Γ_0 emanating from the center $(0,0,0)$ of the similarity.

It is clear that the Lie images of ∞ and Γ_0 are among the fixed elements of the induced collineation. Both are left invariant under the central similarity.

Inversion in a Euclidean sphere

Again, we start with a sphere S centered at M (with coordinates $\mathbf{m}$) and radius $r \in \mathbb{R}^\star$. An equation of S in terms of Cartesian coordinates can be given in the form

$$S:\ \langle \mathbf{x} - \mathbf{m}, \mathbf{x} - \mathbf{m}\rangle - r^2 = \langle \mathbf{x}, \mathbf{x}\rangle - 2\langle \mathbf{x}, \mathbf{m}\rangle + \langle \mathbf{m}, \mathbf{m}\rangle - r^2 = 0. \qquad (10.64)$$

It means no restriction to assume that we invert in the Euclidean unit sphere, since the results are invariant under scalings and the coordinate system can be chosen properly. Hence, the inversion ι in the Euclidean unit sphere transforms the coordinate vector $\mathbf{x}$ of a point X according to

$$\iota:\ \mathbf{x} \mapsto \mathbf{x}' = \frac{1}{\langle \mathbf{x}, \mathbf{x}\rangle}\mathbf{x}. \qquad (10.65)$$

Since ι is involutive (in fact, it is birational, cf. [46, Ch. 7]), we can find the equation of $S' = \iota(S)$ by simply substituting (10.65) into (10.64) and find

$$\langle \mathbf{x}, \mathbf{x}\rangle \left(\langle \mathbf{m}, \mathbf{m}\rangle - r^2\right) - 2\langle \mathbf{x}, \mathbf{m}\rangle + 1 = 0. \qquad (10.66)$$

By completing to full squares, we find the center and the radius of S' as

$$\mathbf{m}' = \frac{1}{\langle \mathbf{m}, \mathbf{m}\rangle - r^2}\mathbf{m}, \quad r' = \frac{r}{\langle \mathbf{m}, \mathbf{m}\rangle - r^2}.$$

With (10.56) and (10.57), we have $\langle \mathbf{m}, \mathbf{m}\rangle - r^2 = (s_6 + s_4)/(s_6 - s_4)$, and the center and radius of S' take the form

$$\mathbf{m}' = \frac{1}{s_6 + s_4}\begin{pmatrix} s_1 \\ s_2 \\ s_3 \end{pmatrix}, \quad r' = \frac{s_5}{s_6 + s_4}.$$

As in the case of the central similarity, we use (10.56) and (10.58) and compute the homogeneous sphere coordinates of the image sphere S'. This yields

$$S' = (s_1, s_2, s_3, -s_4, s_5, s_6). \tag{10.67}$$

Again, we can see that the homogeneous coordinates of S' are linear in the homogeneous coordinates of S. Therefore, the inversion in the Euclidean unit sphere induces an automorphic collineation of L_2^4.

In the case of the inversion, the induced automorphism of L_2^4 is apparently simpler than in the case of the central similarity. Now, the transformation matrix of the collineation equals $\mathrm{diag}(1,1,1,-1,1,1)$, and therefore, the collineation is involutive. There are two different eigenvalues: $\lambda_1 = 1$ and $\lambda_2 = -1$, the first of which corresponds to the fixed space $\mathcal{A}: x_4 = 0$, the second corresponds to the eigenvector $\mathbf{c} = (0,0,0,1,0,0)$ which describes the center C of the automorphic perspective collineation. It can easily be checked that the characteristic cross ratio (cf. [46, p. 214]) of this collineation equals -1, which is clear for involutive mappings. Clearly, $C = \mathbf{c}$ is the pole of $\mathcal{A}$ w.r.t. L_2^4.

We have:

Lemma 10.2.1 *The inversion in the Euclidean unit sphere induces a projective reflection (in a hyperplane of $\mathbb{P}^5$), i.e., a harmonic homology that leaves L_2^4 invariant.*

Translations and rotations

The previous two subsections were dedicated to the automorphic collineations of L_2^4 that were induced by central similarities and inversions that were centered at the origin of the coordinate frame. In a very similar manner, we can show that a translation

$$\tau: \mathbf{x} \mapsto \mathbf{x}' = \mathbf{x} + \mathbf{t}$$

with a constant vector $\mathbf{t} \in \mathbb{R}^3$ and a rotation

$$\varrho: \mathbf{x} \mapsto \mathbf{x}' = \mathbf{R}\mathbf{x}$$

with $\mathbf{R} \in \mathrm{SO}(3)$ induce linear automorphisms of L_2^4.

• **Exercise 10.2.2** Automorphisms of L_2^4 induced by translations and rotations.

Assume that $\mathbf{t} \in \mathbb{R}^3$ is a constant vector and $\mathbf{R} \in SO(3)$ is a constant orthogonal matrix. Compute the automorphic collineations induced by the translation $\tau: \mathbf{x} \mapsto \mathbf{x}' = \mathbf{x} + \mathbf{t}$ and the rotation $\varrho: \mathbf{x} \mapsto \mathbf{x}' = \mathbf{R}\mathbf{x}$ and show that the respective transformation matrices $\mathbf{T}_\tau, \mathbf{T}_\varrho \in \mathbb{R}^{6\times 6}$ are

$$\mathbf{T}_\tau = \begin{pmatrix} 2\mathbf{I}_3 & -\mathbf{t} & \mathbf{0} & \mathbf{t} \\ 2\mathbf{t}^{\mathrm{T}} & 2 - \langle \mathbf{t}, \mathbf{t} \rangle & 0 & \langle \mathbf{t}, \mathbf{t} \rangle \\ \mathbf{0}^{\mathrm{T}} & 0 & 2 & 0 \\ 2\mathbf{t}^{\mathrm{T}} & -\langle \mathbf{t}, \mathbf{t} \rangle & 0 & 2 + \langle \mathbf{t}, \mathbf{t} \rangle \end{pmatrix}, \quad \mathbf{T}_\varrho = \begin{pmatrix} \mathbf{R} & \mathbf{0} \\ \mathbf{0} & \mathbf{I}_3 \end{pmatrix}.$$

Show that the invariant subspaces are straight lines and their polar spaces w.r.t. L_2^4 in both cases.

• **Exercise 10.2.3** Reflections in planes.

It is sufficient to study the action of the reflection σ in a plane ε through the origin of the coordinate system. We assume that ε has the equation $\langle \mathbf{e}, \mathbf{x} \rangle = 0$ with the unit vector $\mathbf{e}$ as its normal. Show that the induced collineation has the transformation matrix

$$\mathbf{T}_\sigma = \left(\begin{array}{c|ccc} \mathbf{I}_3 - 2\mathbf{e}^{\mathrm{T}}\mathbf{e} & & \mathbf{0} & \\ \hline & 1 & 0 & 0 \\ \mathbf{0} & 0 & -1 & 0 \\ & 0 & 0 & 1 \end{array}\right).$$

Show that $\mathbf{T}_\sigma$ is similar to $\mathrm{diag}(1,1,1,1,-1,-1)$. Again, the invariant subspaces consist of a straight line and its polar space w.r.t. L_2^4.

We collect the results of the latter subsections:

Theorem 10.2.3 *The group of equiform transformations and the group of conformal transformations in Euclidean 3-space induces a group of automorphic collineations of Lie's quadric L_2^4.*

Proof: The case of the equiform transformations is clear: An equiform transformation is the product of a central similarity and four reflections in planes at most.

According to Liouville's Theorem, the group of conformal transformations in Euclidean 3-space is generated by inversions in spheres (including reflections in planes, *i.e.*, inversions in spheres with infinite radius and central similarities). ■

Parabolic pencils of spheres

According to (5.12) in Section 5.5, the equations of spheres in a parabolic pencil can be given by

$$\mathcal{P}: \ x^2 + y^2 + z^2 - 2mx = 0, \quad m \in \mathbb{R}.$$

The centers of the spheres trace the straight line $\mathbf{m} = (m, 0, 0)$ (with $m \in \mathbb{R}$) and the radii are $R = m$. We derive the Lie coordinate vector of the spheres in the pencil $\mathcal{P}$ with (10.58) and find

$$\mathbf{p}(m) = (2m, 0, 0, -1, 2m, 1), \quad m \in \mathbb{R}$$

or, after introducing the homogeneous parameters by setting $m = m_1 m_0^{-1}$

$$\mathbf{p}(m_0, m_1) = (2m_1 : 0 : 0 : -m_0 : 2m_1 : m_0), \quad m_0 : m_1 \neq 0 : 0. \quad (10.68)$$

Obviously, this is the parametrization of a straight line in L_2^4 and the point $\mathbf{n} = \mathbf{p}(0,1) = (0,0,0,1,0,-1)$ corresponds to the only null sphere in $\mathcal{P}$, while $\mathbf{r} = \mathbf{p}(1,0) = (1,0,0,0,1,0)$ corresponds to the only oriented plane ϱ, *i.e.*, to the common tangent plane (and radical plane) of all spheres in $\mathcal{P}$.

So, we see that a parabolic pencil of spheres corresponds to a straight line l_2^4. Conversely, we can state:

Theorem 10.2.4 *Each parabolic pencil of spheres corresponds to a straight line in L_2^4 and each straight line L_2^4 corresponds to a parabolic pencil of spheres.*

Proof: The first statement is explained above.

The converse is true, since L_2^4 is a homogeneous space, *i.e.*, for each pair $p \neq q$ of straight lines in L_2^4, there exists at least one automorphic projective collineation of L_2^4 that interchanges these lines. Thus, for each line $q \subset L_2^4$, we can find at least one collineation $\kappa : L_2^4 \to L_2^4$ that maps q to the line (10.68). ■

• **Exercise 10.2.4** The generic parabolic pencil of spheres.

Write down the Lie coordinates of a generic parabolic pencil of spheres. Assume that $(\mathbf{a}, \overline{\mathbf{a}})$ are normalized Plücker coordinates of a straight line a in Euclidean 3-space, *i.e.*, besides (10.2) they satisfy $\langle \mathbf{a}, \mathbf{a} \rangle = 1$. Let $\mathbf{n} = \mathbf{a} \times \overline{\mathbf{a}} + \alpha \mathbf{a}$ with $\alpha \in \mathbb{R}$ be the null sphere $\mathcal{N}$ on the axis a of the parabolic pencil. The pair $(\mathcal{N}, a)$ determines the parabolic pencil, and therefore, the coordinates $(\mathbf{a}, \overline{\mathbf{a}}, \alpha) \in \mathbb{R}^7$ of the line element $(\mathcal{N}, a)$ (cf. [93]) are a suitable coordinate representation of the parabolic pencil of spheres.

Show that the Lie coordinates of the parabolic pencil $\mathcal{P}$ on $(\mathcal{N}, a)$ are

$$\mathbf{p}(t) = (2(\mathbf{a} \times \overline{\mathbf{a}} + \alpha \mathbf{a} + t\mathbf{a}), \|\overline{\mathbf{a}}\|^2 + \alpha^2 - 1 + 2\alpha t, 2t, \|\overline{\mathbf{a}}\|^2 + \alpha^2 + 1 + 2\alpha t) \quad \text{with } t \in \mathbb{R}.$$

Elliptic, hyperbolic pencils and concentric spheres

In Section 5.5, we have given the normal forms of elliptic and hyperbolic pencils of spheres and pencils of concentric spheres. With $\varrho = 1$ in (5.12), these normal forms read

$$\mathcal{E}, \mathcal{H}: \; x^2 + y^2 + z^2 - 2mx \mp 1 = 0, \quad m \in \mathbb{R},$$
$$\mathcal{C}: \; x^2 + y^2 + z^2 = m^2, \quad m \in \mathbb{R},$$

or in terms of Lie coordinates, we have

$$\begin{aligned} \mathcal{E}: \mathbf{e}(m) &= (m,0,0,-1,\sqrt{1+m^2},0), \\ \mathcal{H}: \mathbf{h}(m) &= (m,0,0,0,\sqrt{1-m^2},1), \\ \mathcal{C}: \mathbf{c}(m) &= (0,0,0,1+m^2,-2m,m^2-1), \end{aligned}$$

where $m \in \mathbb{R}$. As can be seen from the parametrizations, the elliptic and hyperbolic pencil of spheres as well as the pencil of concentric spheres correspond to one-dimensional quadratic submanifolds of L_2^4.

Indeed, $\mathbf{e}$, $\mathbf{h}$, and $\mathbf{c}$ parametrize regular conics in L_2^4, since $[\mathcal{E}]$, $[\mathcal{H}]$, and $[\mathcal{C}]$ are planes in $\mathbb{P}^5$ that intersect L_2^4 along regular conics: The equations of the carrier planes are

$$\begin{aligned} [\mathcal{E}]&: x_2 = x_3 = x_6 = 0, \\ [\mathcal{H}]&: x_2 = x_3 = x_4 = 0, \\ [\mathcal{C}]&: x_1 = x_2 = x_3 = 0. \end{aligned}$$

None of them is tangent to L_2^4, since each is skew to its polar plane

$$\begin{aligned} [\mathcal{E}]^\star &: x_1 = x_4 = x_5 = 0, \\ [\mathcal{H}]^\star &: x_1 = x_5 = x_6 = 0, \\ [\mathcal{C}]^\star &: x_4 = x_5 = x_6 = 0. \end{aligned}$$

Linear complexes of spheres

In Section 10.1, we have defined a linear complex of lines as the three-dimensional manifold of lines whose homogeneous Plücker coordinates satisfy a homogeneous linear equation.

In analogy to that, we shall define linear complexes of spheres: Let $\mathbf{c} \in \mathbb{R}^6 \setminus \{\mathbf{0}\}$, then a *linear complex of spheres* C is the set of all spheres X with Lie coordinates $\mathbf{x} = (x_1, \dots, x_6)$ that satisfy

$$\mathcal{C}: \ \Omega_L(C, X) = 0.$$

The classification of linear complexes of spheres is different from that of linear complexes of lines. While the linear complexes of lines can be arranged in two different classes, *i.e.*, the singular and the regular linear complexes of lines, there are three different types of linear complexes of spheres.

The symmetric bilinear form (10.62) allows us to distinguish between three different types of linear complexes of spheres: The complex C is called *elliptic, hyperbolic, or parabolic* if $\Omega_L(C, C)$ is greater, smaller, or equal to zero. This is equivalent to classifying vectors $\mathbf{c} \in \mathbb{F}^6$ as spacelike, timelike, or lightlike w.r.t. the metric (10.57). We shall describe the three types of linear complexes of spheres based on simple examples. The general version of either type can be obtained by applying a Lie transformation to the representatives.

(1) A typical example of an *elliptic linear complex of spheres* is the set of all oriented spheres that intersect a given plane π at right angles. This includes the points in π as null spheres, all oriented planes which are orthogonal to π, and the oriented spheres that are centered at π. The group $\mathrm{O}(3,2)$ fixes the complex as a whole and interchanges the Lie spheres in the complex.

(2) A simple example of a *hyperbolic linear complex of spheres* is the three-dimensional manifold of null spheres, *i.e.*, the points in 3-space. However, the 3-dimensional manifold of spheres with equal and equally signed radius is a hyperbolic linear complex of spheres. It is the group of Möbius transformations that fixes this manifold as a whole and interchanges oriented spheres.

(3) The three-dimensional manifold of spheres which are in oriented contact with a given sphere S constitute a *parabolic linear complex of spheres.* The given sphere S may either be a null sphere, a proper sphere, or the improper point ∞. In the latter case, the complex consists of all oriented planes in Euclidean 3-space. It is the group of Laguerre transformations that leaves a parabolic linear complex of spheres fixed (as a whole).

The general form of a parabolic complex of spheres is the manifold of spheres that intersect a given sphere at a constant (and fixed) angle, which can be seen from (10.61).

• **Exercise 10.2.5** Normal forms of linear complexes of spheres.

Show that the equations of an elliptic, a hyperbolic, and a parabolic linear complex of spheres can be given in the following normal forms

$$\begin{aligned} x_3 &= 0 && \text{elliptic complex,} \\ r(x_4 - x_6) + x_5 &= 0 && \text{hyperbolic complex, } r \in \mathbb{R} \\ x_4 - r x_5 &= 0 && \text{parabolic complex, } r \in \mathbb{R}. \end{aligned}$$

Hint: In the case of an elliptic complex assume that the plane π which is intersected by all spheres of the complex at right angles equals $z = 0$. For the hyperbolic complex, we can take either the complex of null spheres or the complex of spheres with radius $r \in \mathbb{R}^\star$. In the case of the parabolic complex, it means no restriction to assume that the basic sphere S equals the Euclidean unit sphere.

Dupin cyclides

In Section 10.1 (p. 495 ff.), we have seen that reguli are represented by conics in Plücker's quadric M_2^4. These conics were not allowed to lie entirely within a plane of the first or second kind of M_2^4. Since Lie's quadric L_2^4 is of index one, regular conics of L_2^4 can only be planar intersections of L_2^4. Thus, the sphere geometric analogues to reguli are uniquely determined by three points in L_2^4 that span a plane which does not intersect L_2^4 along two lines.

However, it would be too simple to carry over the results from line geometry to sphere geometry by just applying Lie's line-sphere-mapping. This approach would ignore that Lie's line sphere mapping is not a real collineation, *i.e.*, it takes a by-pass over the complex number field. Furthermore, on page 528, we have seen that elliptic and hyperbolic pencils and families of concentric spheres correspond to conics in L_2^4.

Therefore, we stick back to a more elementary approach to the sphere geometric analogues of reguli and study one-parameter families of oriented spheres that touch three given oriented spheres. As explained above, we assume that the three given spheres $\mathcal{R}$, $\mathcal{S}$, and $\mathcal{T}$ correspond to three points on L_2^4 that span a plane ε which intersects L_2^4 along a regular conic e. The polar plane φ of ε w.r.t. L_2^4 intersects L_2^4 along a regular conic f, since ε does so too. The points on f correspond to oriented spheres in $\mathbb{R}^3$ that are in oriented contact with the given spheres $\mathcal{R}$, $\mathcal{S}$, $\mathcal{T}$ simultaneously. The spheres corresponding to the points of f play the same role as the rulings of a complementary regulus. On the other hand, we can choose three independent oriented spheres $\mathcal{R}^\star$, $\mathcal{S}^\star$, and $\mathcal{T}^\star$ which correspond to three points on f. All oriented spheres which are in simultaneous oriented contact with $\mathcal{R}^\star$, $\mathcal{S}^\star$, and $\mathcal{T}^\star$ correspond to the points of e (cf. Figure 10.17 and also [46, Fig. 4.21, p. 137]).

The envelope $\mathcal{D}$ of the spheres corresponding to the points of e equals the envelope of the spheres corresponding to the points of f. The surface $\mathcal{D}$ is called *Dupin cyclide* (cf. [42], and in connection with pairs of focal conics in [46, p. 137]).

The different types of Dupin cyclides with zero, one, or two, singularities can be obtained as envelopes of one-parameter families of spheres if we

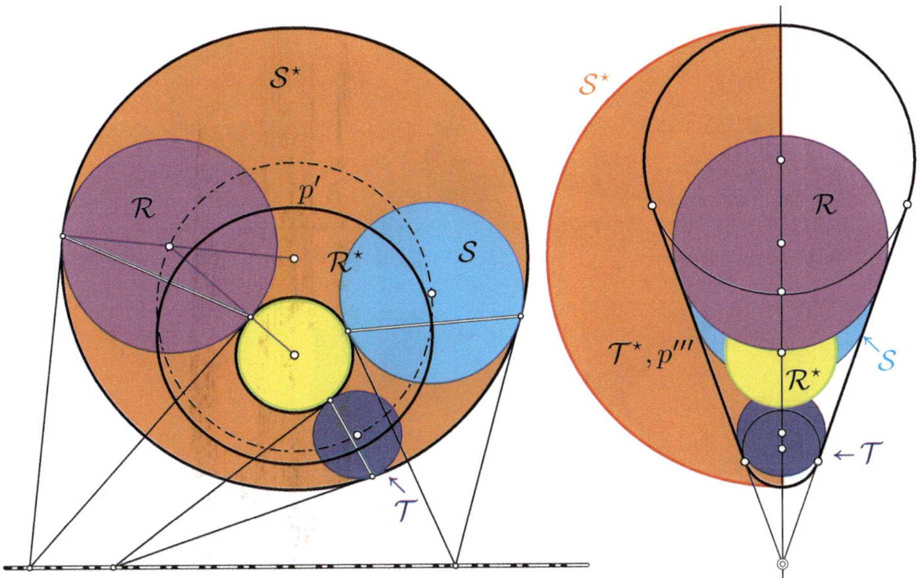

FIGURE 10.17. A ring-shaped Dupin cyclide $\mathcal{D}$ as the envelope of a one-parameter family of oriented spheres which are simultaneously tangent to three oriented spheres $\mathcal{R}$, $\mathcal{S}$, and $\mathcal{T}$. The complementary family of spheres that also envelopes $\mathcal{D}$ contains the spheres $\mathcal{R}^\star$, $\mathcal{S}^\star$, and $\mathcal{T}^\star$. The latter is an oriented plane that touches $\mathcal{D}$ along a circle p of parabolic surface points (shown in a front and side view).

choose zero, one, or two null spheres in the triple $(\mathcal{R}, \mathcal{S}, \mathcal{T})$. (Dupin cyclides with singular points can occur even if none of the given spheres is a null sphere.) Figure 10.18 shows the parabolic and non-parabolic versions: ring-, horn-, and needle-shaped cyclides. These are obtained from ring-shaped tori, cones, and cylinders of revolution by inversions in a sphere.

• **Exercise 10.2.6** Dupin cyclides as inverses of tori, cylinders, and cones of revolution.

Let d, r, R, m, n, k be real constants. Show that the inversion of a torus $\mathcal{T}: \ ((x-d)^2 + y^2 + z^2 - R^2 - r^2)^2 - 4R^2(r^2 - z^2) = 0$, a cylinder of revolution $\mathcal{C}: \ (x-d)^2 + y^2 = 1$, or a cone of revolution $\mathcal{D}: \ (x-m)^2 + (y-m)^2 = k^2 z^2$ in the Euclidean unit sphere yields all different types of parabolic and non-parabolic Dupin cyclides. The parabolic Dupin cyclides occur if the center of the inversion lies on the surface to be inverted. Assume that r, R, and k are fixed and derive conditions on d or (m, n) in order to create parabolic cyclides.

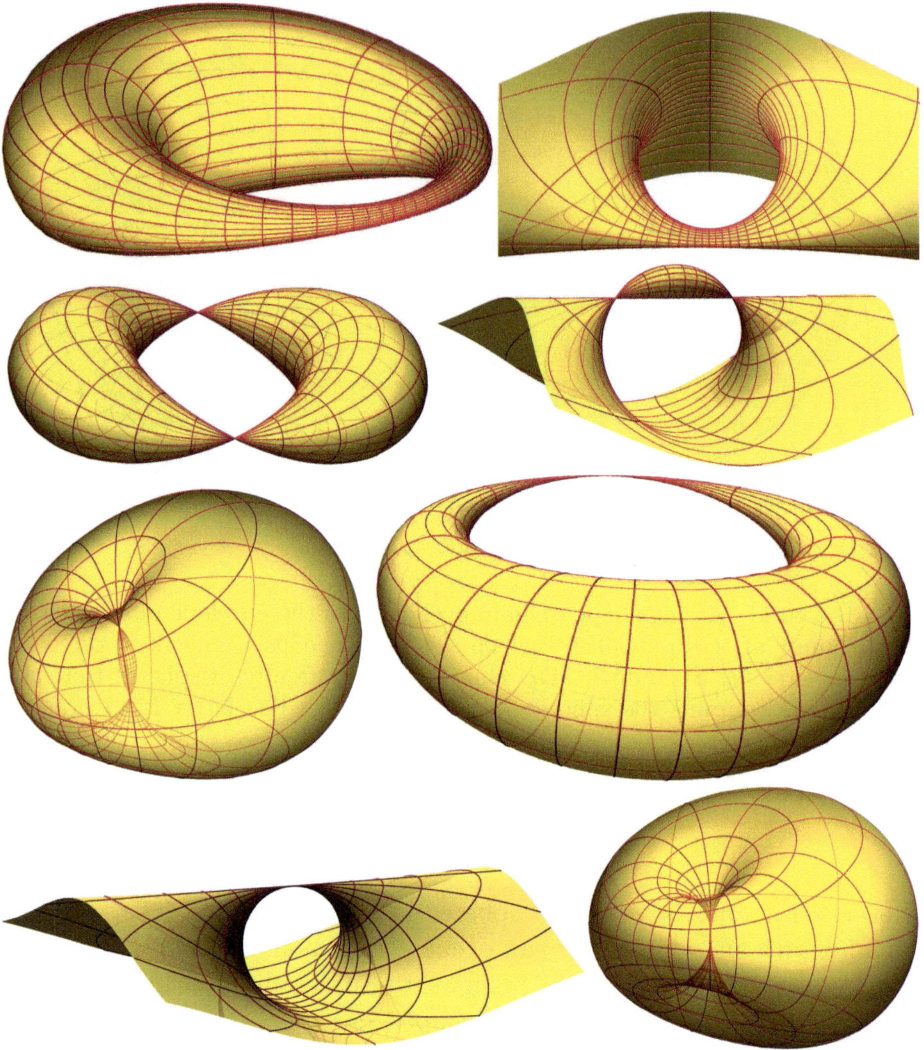

FIGURE 10.18. Dupin cyclides are the sphere geometric analogues of ruled quadrics. They are the envelopes of one-parameter families of spheres in two ways and can be obtained from tori, cylinders, and cones of revolution by inverting them in a Euclidean sphere. From top-left to bottom-right: ring cyclide, parabolic ring cyclide, horn cyclide, parabolic horn cyclide, spindle cyclide, needle cyclide, parabolic needle cyclide, thorn cyclide.

10.3 Kinematics

This section deals with Kinematics and specific models of the group of Euclidean motions in 3-space. We start with Euclidean motions having one fixed point, *i.e.*, rotations about an axis. These can either be described with the help of orthogonal matrices or unit quaternions. The restriction to position vectors of unit length leads to a description of spherical motions and spherical kinematics, which are closely related to elliptic geometry. Then, we present a further model of the geometry of lines in 3-space, which is more suited for the handling of oriented straight lines, frequently referred to as spears. Euclidean motions turn out to be in a close relation to pairs of spears, rather than points. Dual quaternions prove to be an elegant tool for the analytic treatment of spatial kinematics. Moreover, the dual quaternions deliver a coordinatization of the group of Euclidean motions in 3-space. This coordinatization leads to a representation of the group of Euclidean motions. The quadric model of the group of Euclidean motions that arises here is called Study's quadric. This six-dimensional quadric is also interesting from the viewpoint of quadrics. In this ruled quadric, we can observe a geometric phenomenon that generalizes the principle of duality known from projective geometry to a principle of triality.

Motions with one fixed point – rotations

A regular 3×3 matrix $\mathbf{A}$ is called *orthogonal* if it satisfies $\mathbf{A}^{\mathrm{T}}\mathbf{A} = \mathbf{I}_3$. It describes an orientation-preserving transformation $\beta : \mathbb{R}^3 \to \mathbb{R}^3$ if $\det \mathbf{A} = +1$, and it is orientation-reversing if $\det \mathbf{A} = -1$. The set of all orientation-preserving transformations constitute a group, usually denoted by $\mathrm{SO}(3)$. First, we show:

Lemma 10.3.1 *For each* $\mathbf{A} \in \mathrm{SO}(3)$*,* $\mathbf{A} - \mathbf{I}_3$ *is singular, i.e.,* $+1$ *is an eigenvalue of* $\mathbf{A}$*. Further,* $\mathrm{tr}\mathbf{A} = 1 + 2\cos\varphi$ *with* $\varphi \in [0, \pi]$.

Proof: If 1 is an eigenvalue of A, then $\mathbf{A} - \mathbf{I}_3$ is singular and $\det(\mathbf{A} - \mathbf{I}_3) = 0$. Since $\mathbf{A} \in \mathrm{SO}(3)$, we have $\mathbf{A}^{\mathrm{T}}\mathbf{A} = \mathbf{I}_3$, and thus, $\mathbf{A}^{\mathrm{T}}(\mathbf{A} - \mathbf{I}_3) = \mathbf{I}_3 - \mathbf{A}^{\mathrm{T}} = (\mathbf{I}_3 - \mathbf{A})^{\mathrm{T}}$.

Further, the determinant is multipliciative, and so, $\det(\mathbf{A}^{\mathrm{T}}(\mathbf{A} - \mathbf{I}_3)) = \det \mathbf{A}^{\mathrm{T}} \det(\mathbf{A} - \mathbf{I}_3) = \det(\mathbf{A} - \mathbf{I}_3)$ hold. On the other hand, $\det(\mathbf{I}_3 - \mathbf{A}) = (-1)^3 \det(\mathbf{A} - \mathbf{I}_3)$. Hence, $\det(-\mathbf{I}_3) = -\det(\mathbf{A} - \mathbf{I}_3)$. Consequently, $\det(\mathbf{A} - \mathbf{I}_3) = 0$ and $\mathbf{A} - \mathbf{I}_3$ is singular.

In terms of an eigenvector basis, the linear mapping described by $\mathbf{A}$ can be given in normal form. Clearly, the matrix $\mathbf{A}$ is then diagonal dominant, and the right lower 2×2-submatrix is also orthogonal. The 2×2-submatrix can be given in the form

$$\begin{pmatrix} \cos\varphi & -\sin\varphi \\ \sin\varphi & \cos\varphi \end{pmatrix}.$$

Therefore, we can write $\mathbf{A}$ as

$$\mathbf{A} = \begin{pmatrix} 1 & 0 & 0 \\ 0 & \cos\varphi & -\sin\varphi \\ 0 & \sin\varphi & \cos\varphi \end{pmatrix}. \tag{10.69}$$

Now, we can easily verify that $\mathrm{tr}\mathbf{A} = 1 + 2\cos\varphi$. ■

From Lemma 10.3.1 we can infer that each orientation-preserving orthogonal transformation in Euclidean 3-space does not only fix the origin of the coordinate system. Each point of the one-dimensional subspace $a = [\mathbf{a}]$ spanned by the eigenvector $\mathbf{a}$ that corresponds to the eigenvalue 1 is fixed. Moreover, we can show that each point X with coordinates $\mathbf{x}$ and its image $\mathbf{X}'$ with $\mathbf{x}' = \mathbf{A}\mathbf{x}$ satisfies

$$\sphericalangle([X,a],[X',a]) = \varphi.$$

Consequently, each $\mathbf{A} \in \mathrm{SO}(3)$ describes a Euclidean *rotation* with *axis* a and rotation angle φ.

In the following Lemma, we give a more geometric description of a rotation independent of a special coordinate system.

Lemma 10.3.2 *Let* $\mathbf{a} \in \mathbb{R}^3$ *be a unit vector in the direction of the axis* a *of a rotation with angle* φ. *For each point* $\mathbf{x}$, *its image* $\mathbf{x}'$ *under the rotation about* a *through the angle* φ *can be given in vector form as*

$$\mathbf{x}' = \mathbf{x}\cos\varphi + (1 - \cos\varphi)\langle\mathbf{a},\mathbf{x}\rangle\mathbf{a} + \sin\varphi\,\mathbf{a} \times \mathbf{x}. \tag{10.70}$$

Proof: The two vectors $\mathbf{a}$ and $\mathbf{x}$ define a right-handed basis with mutually orthogonal vectors. We choose $\mathbf{a}$ as the third vector, $\mathbf{a} \times \mathbf{x}$ as the second vector, and $(\mathbf{a} \times \mathbf{x}) \times \mathbf{a}$ completes the right-handed frame, as shown in Figure 10.19.

Now, we can write $\mathbf{x}'$ in this particular basis as

$$\mathbf{x}' = \alpha(\mathbf{a} \times \mathbf{x}) \times \mathbf{a}\cos\varphi + \beta\mathbf{a} \times \mathbf{x}\sin\varphi + \gamma\mathbf{a}.$$

It is clear that γ equals the component of $\mathbf{x}$ in the direction of $\mathbf{a}$. Since $\|\mathbf{a}\| = 1$, $\gamma = \langle\mathbf{a},\mathbf{x}\rangle$.

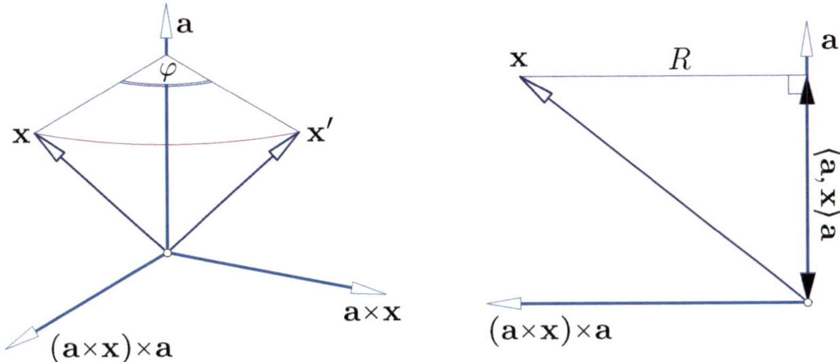

FIGURE 10.19. The vector representation of a rotation uses the Cartesian coordinate system that is defined by $\mathbf{a}$ and $\mathbf{x}$ (left); the radius of the path circle of a point X (right).

We observe that $(\mathbf{a} \times \mathbf{x}) \times \mathbf{a} = \mathbf{x} - \langle \mathbf{a}, \mathbf{x} \rangle \mathbf{a}$. Therefore, $\|(\mathbf{a} \times \mathbf{x}) \times \mathbf{a}\| = \|\mathbf{x} - \langle \mathbf{a}, \mathbf{x} \rangle \mathbf{a}\| = \sqrt{\langle \mathbf{x}, \mathbf{x} \rangle - \langle \mathbf{a}, \mathbf{x} \rangle^2}$. Further, the radius R of the orbit of $\mathbf{x}$ equals $R = \sqrt{\langle \mathbf{x}, \mathbf{x} \rangle - \langle \mathbf{a}, \mathbf{x} \rangle^2}$, as can be seen from Figure 10.19. Consequently, $\alpha = R \cos \varphi / \|\mathbf{x} - \langle \mathbf{a}, \mathbf{x} \rangle \mathbf{a}\| = \cos \varphi$ and $\beta = R \sin \Phi / \|\mathbf{x} - \langle \mathbf{a}, \mathbf{x} \rangle \mathbf{a}\| = \sin \varphi$. So, we have

$$\mathbf{x}' = (\mathbf{x} - \langle \mathbf{a}, \mathbf{x} \rangle \mathbf{a}) \cos \varphi + \mathbf{a} \times \mathbf{x} \sin \varphi + \langle \mathbf{a}, \mathbf{x} \rangle \mathbf{a},$$

which proves Lemma 10.3.2 after regrouping some terms. ■

A brief introduction to quaternions

In this short section, we shall define quaternions, a skew field of hypercomplex numbers that were originally introduced by W.R. HAMILTON[12] in 1843 and are, therefore, often called Hamilton quaternions. Nevertheless, the quaternions were independently discovered in 1840 by B.O. RODRIGUES[13] when he studied rotations in three-dimensional space.

A *quaternion* $\mathbf{Q}$ is a hypercomplex number that can always be written in the form

$$\mathbf{Q} = q_0 \cdot 1 + q_1 \mathrm{i} + q_2 \mathrm{j} + q_3 \mathrm{k},$$

---

[12] WILLIAM ROWAN HAMILTON (1805–1865) was an Irish mathematician and physicist. He found the multiplication rules for quaternions during a walk along the Royal Canal and immediately engraved them into Brougham Bridge in Dublin. Hamilton worked on geometric optics and mechanics.

[13] BENJAMIN OLINDE RODRIGUES (1795–1851) was a French mathematician, banker, and social reformer. He discovered quaternions while he was studying representations of the group of rotations. Thereby, he rediscovered the Euler angles.

where $\{1, \mathrm{i}, \mathrm{j}, \mathrm{k}\}$ are four independent basis elements (vectors) and q_i are real numbers called the *components* or *coordinates* of the quaternion. Usually, the basis element 1 is not written explicitly.

The addition of two quaternions is defined component-wise, *i.e.*, the sum $\mathbf{P}+\mathbf{Q}$ of two quaternions $\mathbf{P}$ and $\mathbf{Q}$ is defined as

$$\mathbf{P}+\mathbf{Q} = (p_0+q_0) + (p_1+q_1)\mathrm{i} + (p_2+q_2)\mathrm{j} + (p_3+q_3)\mathrm{k}.$$

The multiplication of a quaternion $\mathbf{Q}$ with a real scalar λ is simply

$$\lambda\mathbf{Q} = \lambda q_0 + \lambda q_1\mathrm{i} + \lambda q_2\mathrm{j} + \lambda q_3\mathrm{k},$$

which makes

$$\mathbb{H} = \{q_0 + q_1\mathrm{i} + q_2\mathrm{j} + q_3\mathrm{k} \mid q_i \in \mathbb{R}\}$$

a four-dimensional vector space of the real numbers. Obviously, $\mathbb{R}$ is a one-dimensional and $\mathbb{C}$ is a two-dimensional subspace of $\mathbb{H}$ (both considered as real vector spaces). Quaternions with $q_0 = 0$ are called *vector quaternion*, while quaternions with $q_1 = q_2 = q_3 = 0$ are called *scalar quaternions*.

However, $\mathbb{H}$ is also a four-dimensional $\mathbb{R}$-algebra. The multiplication of quaternions is well-defined if we prescribe the multiplication rules obeyed by the basis elements 1, i, j, k:

$$\mathrm{i}^2 = \mathrm{j}^2 = \mathrm{k}^2 = -1, \quad \mathrm{ij} = \mathrm{k} \quad \text{(cyclic)}.$$

Now, it is easily verified that $\mathbb{H}$ is a skew field.

Each quaternion $\mathbf{Q} = q_0 + q_1\mathrm{i} + q_2\mathrm{j} + q_3\mathrm{k}$ has a *conjugate quaternion*

$$\widetilde{\mathbf{Q}} = q_0 - q_1\mathrm{i} - q_2\mathrm{j} - q_3\mathrm{k}.$$

This allows us to define the *norm of a quaternion* as

$$N(\mathbf{Q}) = \widetilde{\mathbf{Q}}\mathbf{Q} = \mathbf{Q}\widetilde{\mathbf{Q}} = q_0^2 + q_1^2 + q_2^2 + q_3^2 \in \mathbb{R}.$$

The *length of a quaternion* equals

$$\|\mathbf{Q}\| = \sqrt{N(\mathbf{Q})} \in \mathbb{R}.$$

The norm as well as the length of a quaternion $\mathbf{Q}$ equals 0 if, and only if, $\mathbf{Q} = \mathbf{0}$, *i.e.*, the quaternion equals the zero quaternion.

In kinematics, especially *unit quaternions* play an important role. We find a unit quaternion $\mathbf{Q}^\star$ associated with a given quaternion $\mathbf{Q} \neq \mathbf{0}$ as

$$\mathbf{Q}^\star = \frac{1}{\|\mathbf{Q}\|}\widetilde{\mathbf{Q}}.$$

In many cases, it proved useful to write quaternions in the form

$$\mathbf{Q} = q_0 + \mathbf{q},$$

where $\mathbf{q} = (q_1, q_2, q_3)$ is the vector part of the quaternion and the plus sign indicates just a formal sum. The sum $\mathbf{P} + \mathbf{Q}$ of two quaternions $\mathbf{P}$ and $\mathbf{Q}$ then reads $\mathbf{P} + \mathbf{Q} = (p_0 + q_0) + (\mathbf{p} + \mathbf{q})$. The multiplication of a quaternion $\mathbf{Q}$ with a scalar $\lambda \in \mathbb{R}$ yields $\lambda\mathbf{Q} = \lambda q_0 + \lambda\mathbf{q}$.

• **Exercise 10.3.1** Multiplication of quaternions.

Assume that $\mathbf{P} = p_0 + p_1\mathrm{i} + p_2\mathrm{j} + p_3\mathrm{k}$ and $\mathbf{Q} = q_0 + q_1\mathrm{i} + q_2\mathrm{j} + q_3\mathrm{k}$ are two quaternions. Show that the product $\mathbf{PQ}$ of $\mathbf{P} = p_0 + \mathbf{p}$ and $\mathbf{Q} = q_0 + \mathbf{q}$ can be written in the form

$$\mathbf{PQ} = p_0 q_0 - \langle \mathbf{p}, \mathbf{q} \rangle + p_0\mathbf{q} + q_0\mathbf{p} + \mathbf{p} \times \mathbf{q}, \tag{10.71}$$

where $\langle \cdot, \cdot \rangle$ denotes the canonical scalarproduct in $\mathbb{R}^3$ and $\times$ indicates the thus induced exterior product of vectors in $\mathbb{R}^3$.

• **Exercise 10.3.2** Norm and length of a quaternion, unit quaternions.

1. Show that the norm $N(\mathbf{Q})$ of a quaternion $\mathbf{Q} = q_0 + \mathbf{q}$ can be computed by means of the scalar q_0 and the vector part $\mathbf{q}$ as
$$N(\mathbf{Q}) = q_0^2 + \langle \mathbf{q}, \mathbf{q} \rangle.$$
2. Show that the norm for quaternions is *multiplicative*, *i.e.*,
$$N(\mathbf{PQ}) = N(\mathbf{P})N(\mathbf{Q})$$
by first showing that
$$\widetilde{\mathbf{PQ}} = \widetilde{\mathbf{Q}}\widetilde{\mathbf{P}}.$$
3. Let $\mathbf{a} \in \mathbb{R}^3$ be a unit vector and let $\alpha \in \mathbb{R}$. Show that
$$\mathbf{A} = \cos\alpha + \sin\alpha\,\mathbf{a} \tag{10.72}$$
is a unit quaternion.

• **Exercise 10.3.3** Scalar product and cross product in terms of quaternions.

The expression (10.71) of the quaternion product in terms of vectors and scalars allows us to extract the cross product of the vectorial parts: Show that the crossproduct of the vector parts $\mathbf{p}$ and $\mathbf{q}$ of two quaternions $\mathbf{P}$ and $\mathbf{Q}$ can be found as

$$\frac{1}{2}(\mathbf{PQ} - \mathbf{QP}) = \mathbf{p} \times \mathbf{q}. \tag{10.73}$$

If the two quaternions are pure vector quaternions, *i.e.*, $\mathbf{P} = \mathbf{p}$ and $\mathbf{Q} = \mathbf{q}$, then we can derive the canonical scalar product of the vector parts. Show that

$$\frac{1}{2}(\mathbf{PQ} + \mathbf{QP}) = -\langle \mathbf{p}, \mathbf{q} \rangle \tag{10.74}$$

holds for pure vector quaternions.

Quaternions and spherical motions

In this section we can unite the results from Lemma 10.3.2 and Exercise 10.3.2. For that purpose, we describe points X in $\mathbb{R}^3$ by pure vector quaternions $\mathbf{X} = \mathbf{x}$.

We can show the following result:

Theorem 10.3.1 *Let $\mathbf{a} \in \mathbb{R}^3$ be a unit vector and let $\alpha \in [0, \pi]$ from which we can build the unit quaternion $\mathbf{A} = \cos\frac{\alpha}{2} + \sin\frac{\alpha}{2}\mathbf{a}$. The image X' of a point X under the rotation about the axis $[\mathbf{a}]$ through the angle α is given by the pure vector quaternion*

$$\mathbf{x}' = \mathbf{A}\mathbf{x}\widetilde{\mathbf{A}}. \tag{10.75}$$

Proof: We expand the product given in (10.75) in the vector representation and find

$$\begin{aligned} \mathbf{x}' &= (\cos\tfrac{\alpha}{2} + \sin\tfrac{\alpha}{2}\mathbf{a})\mathbf{x}(\cos\tfrac{\alpha}{2} - \sin\tfrac{\alpha}{2}\mathbf{a}) = \\ &= (-\sin\tfrac{\alpha}{2}\langle\mathbf{a},\mathbf{x}\rangle + \cos\tfrac{\alpha}{2}\mathbf{x} + \sin\tfrac{\alpha}{2}\mathbf{a}\times\mathbf{x})(\cos\tfrac{\alpha}{2} - \sin\tfrac{\alpha}{2}\mathbf{a}) = \\ &\cos^2\tfrac{\alpha}{2}\mathbf{x} + \sin\tfrac{\alpha}{2}\cos\tfrac{\alpha}{2}\mathbf{a}\times\mathbf{x} + \sin^2\tfrac{\alpha}{2}\langle\mathbf{a},\mathbf{x}\rangle\mathbf{a} - \cos\tfrac{\alpha}{2}\sin\tfrac{\alpha}{2}\mathbf{x}\times\mathbf{a} - \sin^2\tfrac{\alpha}{2}(\mathbf{a}\times\mathbf{x})\times\mathbf{a}. \end{aligned} \tag{10.76}$$

Now, we use the fact that $\mathbf{a}\times\mathbf{x} = -\mathbf{x}\times\mathbf{a}$ and $(\mathbf{a}\times\mathbf{x})\times\mathbf{a} = \mathbf{x} - \langle\mathbf{a},\mathbf{x}\rangle\mathbf{a}$. Further, $2\sin\frac{\alpha}{2}\cos\frac{\alpha}{2} = \sin\alpha$ and $2\sin^2\frac{\alpha}{2} = 1 - \cos\alpha$. Thus, (10.76) simplifies to

$$\mathbf{x}' = \cos\alpha\mathbf{x} + (1 - \cos\alpha)\langle\mathbf{a},\mathbf{x}\rangle\mathbf{a} + \sin\alpha\,\mathbf{a}\times\mathbf{x}.$$

Obviously, the latter equation agrees with the vector representation of the same rotation as given in (10.70). ■

Theorem 10.3.1 renders a representation of SO(3) in the group-theoretical sense. We can find a further representation which is interesting from the theoretical point of view and relates the group of spherical motions with quadrics. Therefore, we can state and prove:

Theorem 10.3.2 *The three-parameter family of orthogonal matrices describing rotations in Euclidean 3-space can be parametrized by*

$$\mathbf{Q} = \begin{pmatrix} q_0^2 + q_1^2 - q_2^2 - q_3^2 & 2(q_1q_2 - q_0q_3) & 2(q_1q_3 + q_0q_2) \\ 2(q_1q_2 + q_0q_3) & q_0^2 - q_1^2 + q_2^2 - q_3^2 & 2(q_2q_3 - q_0q_1) \\ 2(q_1q_3 - q_0q_2) & 2(q_2q_3 + q_0q_1) & q_0^2 - q_1^2 - q_2^2 + q_3^2 \end{pmatrix} \tag{10.77}$$

with $q_i \in \mathbb{R}$ and $q_0^2 + q_1^2 + q_2^2 + q_3^2 = 1$.

Proof: Let $\mathbf{Q} = q_0 + q_1\mathrm{i} + q_2\mathrm{j} + q_3\mathrm{k}$ be a unit quaternion, *i.e.*, $q_0^2 + q_1^2 + q_3^2 + q_3^2 = 1$. Then, we use (10.75) and compute $\mathbf{x}' = \mathbf{Q}\mathbf{x}\widetilde{\mathbf{Q}}$ with $= x_1\mathrm{i} + x_2\mathrm{j} + x_3\mathrm{k}$. This yields a linear mapping $\mathbb{R}^3 \to \mathbb{R}^3$ with the coordinate matrix $\mathbf{A} \in \mathbb{R}^{3\times3}$, and the coefficients of i, j, k are the rows of the $\mathbf{A}$. ■

(10.77) can also be seen as a rational parametrization, and thus, as a rational representation of the group SO(3). Note that the quaternions $\mathbf{A}$ and $-\mathbf{A}$ determine the same group element.

■ **Example 10.3.1** Orthogonal Cartesian frames with rational coordinate representation.

The matrix representation (10.77) delivers more than just rationally parametrized rotation matrices. Since all these matrices are orthogonal, we can interpret the columns (as well as the rows) as the coordinate vectors of the three vectors of a Cartesian frame with equal rational coordinates and rational lengths $\sqrt{q_0^2+q_1^2+q_2^2+q_3^2}$. Moreover, each row (or column) gives rise to a Pythagorean quadruple. For example, the following choices for $q_0:q_1:q_2:q_3$

$$
\begin{array}{lll}
q_0=2,q_1=1,q_2=1,q_3=1, & q_0=2,q_1=2,q_2=2,q_3=1, & q_0=3,q_1=1,q_2=1,q_3=1,\\
q_0=3,q_1=2,q_2=1,q_3=1, & q_0=3,q_1=2,q_2=2,q_3=1, & q_0=3,q_1=2,q_2=2,q_3=2,\\
q_0=3,q_1=3,q_2=2,q_3=1, & q_0=3,q_1=3,q_2=3,q_3=2, & q_0=4,q_1=1,q_2=1,q_3=1,\\
q_0=4,q_1=2,q_2=1,q_3=1, & q_0=4,q_1=2,q_2=2,q_3=1, & q_0=4,q_1=3,q_2=1,q_3=1,\\
q_0=4,q_1=3,q_2=3,q_3=1, & q_0=4,q_1=3,q_2=3,q_3=2, & q_0=4,q_1=3,q_2=3,q_3=3,
\end{array}
$$

lead to the Cartesian frames below:

$$
\begin{pmatrix} 3 & -2 & 6\\ 6 & 3 & -2\\ -2 & 6 & 3 \end{pmatrix},\
\begin{pmatrix} 3 & 4 & 12\\ 12 & 3 & -4\\ -4 & 12 & -3 \end{pmatrix},\
\begin{pmatrix} 2 & -1 & 2\\ 2 & 2 & -1\\ -1 & 2 & 2 \end{pmatrix},
$$

$$
\begin{pmatrix} 11 & -2 & 10\\ 10 & 5 & 10\\ -2 & 14 & 5 \end{pmatrix},\
\begin{pmatrix} 4 & 1 & 8\\ 7 & 4 & 4\\ -4 & 8 & 1 \end{pmatrix},\
\begin{pmatrix} 5 & -4 & 20\\ 20 & 5 & -4\\ -4 & 20 & 5 \end{pmatrix},
$$

$$
\begin{pmatrix} 13 & 6 & 18\\ 18 & 3 & -14\\ -6 & 22 & -3 \end{pmatrix},\
\begin{pmatrix} 5 & 6 & 30\\ 30 & 5 & -6\\ -6 & 30 & -5 \end{pmatrix},\
\begin{pmatrix} 15 & -6 & 10\\ 10 & 15 & -6\\ -6 & 10 & 15 \end{pmatrix},
$$

$$
\begin{pmatrix} 9 & -2 & 6\\ 6 & 6 & -7\\ -2 & 9 & 6 \end{pmatrix},\
\begin{pmatrix} 15 & 0 & 20\\ 16 & 15 & -12\\ -12 & 20 & 9 \end{pmatrix},\
\begin{pmatrix} 23 & -2 & 14\\ 14 & 7 & -22\\ -2 & 26 & 7 \end{pmatrix},
$$

$$
\begin{pmatrix} 15 & 10 & 30\\ 26 & 15 & -18\\ -18 & 30 & -1 \end{pmatrix},\
\begin{pmatrix} 6 & 1 & 18\\ 17 & 6 & -6\\ -6 & 18 & 1 \end{pmatrix},\
\begin{pmatrix} 7 & -6 & 42\\ 42 & 7 & -6\\ -6 & 42 & 7 \end{pmatrix}.
$$

According to Theorem 10.3.2, each rotation can be represented by a point (q_0,q_1,q_2,q_3) on the three-dimensional unit sphere $\mathrm{S}^3 \subset \mathbb{E}^4$ (see Figure 10.20).

A three-dimensional sphere is sometimes called *glome*. Frequently shown pictures of glomes are two-dimensional images of stereographic projections of the three-dimensional sphere S^3 into a hyperplane of $\mathbb{R}^4$. These images are conformal images of S^3, but circles and two-dimensional spheres through the center of the projection are straightened.

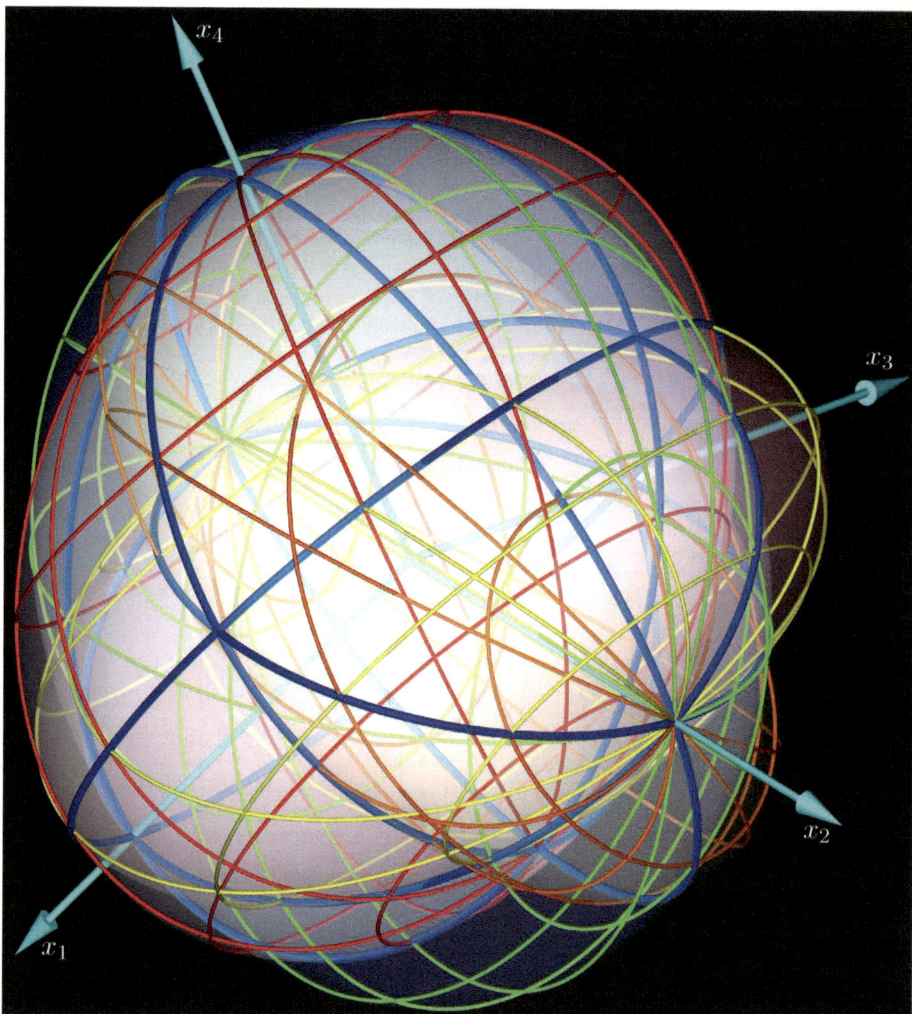

FIGURE 10.20. Circles and two-dimensional spheres in the three-dimensional sphere: This is a perspective image of S^3 into $\mathbb{R}^3$. The contours of the three-dimensional subspheres contained in the coordinate hyperplanes appear as hollow ellipsoids, while the three-dimensional contour of S^3 is not shown.

■ **Example 10.3.2** Euler angles.

To each pair of Cartesian coordinate frames in Euclidean 3-space with coinciding origins, there exists a unique Euclidean (spherical) motion moving one frame into the other. The coordinate matrix $\mathbf{T} \in SO(3)$ of the linear mapping transforming one frame into the other can be obtained as the product of three orthogonal matrices $\mathbf{R}_\alpha$, $\mathbf{R}_\beta$, $\mathbf{R}_\gamma$ describing three independent rotations about three (in general) linearly independent axes through three different rotation angles α, β, and γ, as shown in Figure 10.21.

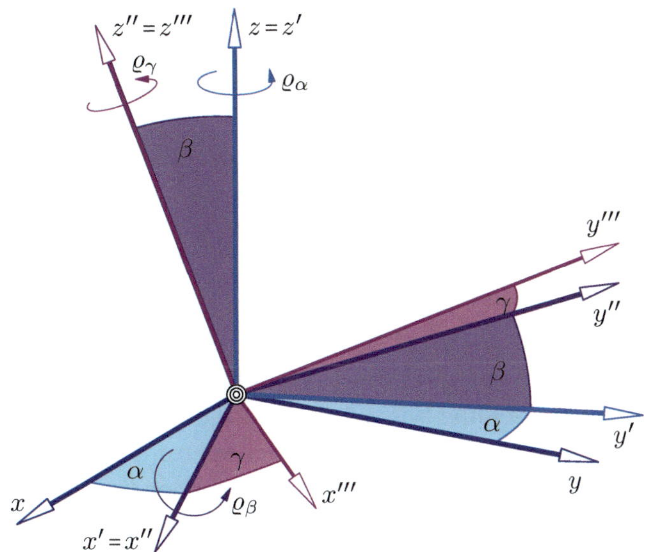

FIGURE 10.21. The composition of three rotations equals a generic spherical motion.

Usually, the first axis of rotation equals the z-axis of the underlying coordinate system and the first angle of rotation equals α. Then, the new frame (x', y', z') is rotated about the x'-axis through the angle β. Finally, the second frame (x'', y'', z'') is rotated about the z''-axis through the angle γ.

We write down the unit quaternions **A**, **B**, **C** describing the three rotations:

$$\mathbf{A} = \cos\frac{\alpha}{2} + \sin\frac{\alpha}{2}\mathrm{k}, \quad \mathbf{B} = \cos\frac{\beta}{2} + \sin\frac{\beta}{2}\mathrm{i}, \quad \mathbf{C} = \cos\frac{\gamma}{2} + \sin\frac{\gamma}{2}\mathrm{k}.$$

Then, we compute the quaternion $\mathbf{Q} = \mathbf{ABC}$ as

$$\mathbf{Q} = \cos\frac{\beta}{2}\cos\frac{\alpha+\gamma}{2} + \sin\frac{\beta}{2}\cos\frac{\alpha-\gamma}{2}\mathrm{i} + \sin\frac{\beta}{2}\sin\frac{\alpha-\gamma}{2}\mathrm{j} + \cos\frac{\beta}{2}\sin\frac{\alpha+\gamma}{2}\mathrm{k}.$$

With the pure vector quaternion $\mathbf{X} = x\mathrm{i} + y\mathrm{j} + z\mathrm{k}$ representing a point X and (10.75), we find the image X' of X under the product of the three given rotations as

$$\begin{aligned}\mathbf{X}' = ((-\cos\beta\sin\gamma\sin\alpha + \cos\gamma\cos\alpha)x + (-\cos\beta\cos\gamma\sin\alpha - \sin\gamma\cos\alpha)y + \sin\alpha\sin\beta z)\mathrm{i} + \\ ((\cos\beta\sin\gamma\cos\alpha + \cos\gamma\sin\alpha)x + (\cos\beta\cos\gamma\cos\alpha - \sin\gamma\sin\alpha)y + -\sin\beta\cos\alpha z)\mathrm{j} + \\ (\sin\gamma\sin\beta x + \sin\beta\cos\gamma y + \cos\beta z)\mathrm{k}.\end{aligned}$$

The entries of the first, second, and third row of the transformation matrix of the rotation are the coefficients of x, y, z. Therefore, the matrix reads

$$\mathbf{T} = \begin{pmatrix} -\cos\beta\sin\gamma\sin\alpha + \cos\gamma\cos\alpha & -\cos\beta\cos\gamma\sin\alpha - \sin\gamma\cos\alpha & \sin\alpha\sin\beta \\ \cos\beta\sin\gamma\cos\alpha + \cos\gamma\sin\alpha & \cos\beta\cos\gamma\cos\alpha - \sin\gamma\sin\alpha & -\sin\beta\cos\alpha \\ \sin\gamma\sin\beta & \sin\beta\cos\gamma & \cos\beta \end{pmatrix}. \tag{10.78}$$

The three elementary rotations ϱ_α, ϱ_β, and ϱ_β can also be described by the three rotation matrices

$$\mathbf{R}_\alpha = \begin{pmatrix} \cos\alpha & -\sin\alpha & 0 \\ \sin\alpha & \cos\alpha & 0 \\ 0 & 0 & 1 \end{pmatrix}, \quad \mathbf{R}_\beta = \begin{pmatrix} 1 & 0 & 0 \\ 0 & \cos\beta & -\sin\beta \\ 0 & \sin\beta & \cos\beta \end{pmatrix}, \quad \mathbf{R}_\gamma = \begin{pmatrix} \cos\gamma & -\sin\gamma & 0 \\ \sin\gamma & \cos\gamma & 0 \\ 0 & 0 & 1 \end{pmatrix}.$$

Their product equals $\mathbf{T} = \mathbf{R}_\alpha \mathbf{R}_\beta \mathbf{R}_\gamma$.

We note that (10.78) gives a trigonometric representation of the group SO(3).

Oriented lines in Euclidean 3-space – spears

In Section 10.1 we have seen one possible way to coordinatize the manifold of lines in a three-dimensional space. We can impose a *normalization condition* on the Plücker coordinates $(\mathbf{l}, \bar{\mathbf{l}})$ of a line L in order to specialize the geometry we are dealing with. In the case of Euclidean line geometry, we may always assume that the direction vector $\mathbf{l}$ of a straight line is a unit vector, which means that the Plücker coordinates of a line L in Euclidean space satisfy two quadratic equations:

$$\langle \mathbf{l}, \mathbf{l} \rangle = 1 \quad \text{and} \quad \langle \mathbf{l}, \bar{\mathbf{l}} \rangle = 0. \tag{10.79}$$

Note that the second equation of (10.79) is the so-called Plücker condition (10.2) that is fulfilled by Plücker coordinates of a line regardless of the geometry in the ambient space. Oriented straight lines in Euclidean 3-space are sometimes referred to as *spears*.

Replacing $\langle \mathbf{l}, \mathbf{l} \rangle = 1$ by $\langle \mathbf{l}, \mathbf{l} \rangle + \langle \bar{\mathbf{l}}, \bar{\mathbf{l}} \rangle = 1$ is a suitable normalization for line geometry in elliptic 3-space (see [11, 91]). With the definition of the *left* and *right spherical kinematic image* of a line $L = (\mathbf{l}, \bar{\mathbf{l}})$ as $\mathbf{l}^l = \mathbf{l} - \bar{\mathbf{l}}$ and $\mathbf{l}^r = \mathbf{l} + \bar{\mathbf{l}}$, one obtains a mapping from the manifold of lines in elliptic 3-space to $S^2 \times S^2$. It can be shown that the group of isometries in elliptic 3-space is isomorphic to $SO(3) \times SO(3)$ (cf. [11, 91]).

Now, we follow a concept by E. Study[14] (cf. [131]): We combine the direction vector $\mathbf{l}$ and the momentum vector $\bar{\mathbf{l}}$ to a *dual vector* by

$$\mathbf{L} = \mathbf{l} + \varepsilon \bar{\mathbf{l}}. \tag{10.80}$$

[14] Christian Hugo Eduard Study (1862–1930) was a German mathematician and geometer. His research was dedicated to ternary forms, hypercomplex numbers, and line and sphere geometry. He also contributed to quantum chemistry.

Therein, the symbol ε marks the second part of $\mathbf{L}$ as the dual vector part. Whatever we may call it, ε obeys the rule

$$\varepsilon^2 = 0, \quad \varepsilon \neq 0. \tag{10.81}$$

Thus, $\mathbf{L} = \mathbf{l} + \varepsilon\bar{\mathbf{l}}$ is an element of the three-dimensional module $\mathbb{D}^3$, and each component is an element of the ring of *dual numbers*. By assumption, the Plücker coordinates $(\mathbf{l}, \bar{\mathbf{l}})$ of the line L fulfill (10.79), and therefore,

$$\langle \mathbf{L}, \mathbf{L} \rangle = \langle \mathbf{l} + \varepsilon\bar{\mathbf{l}}, \mathbf{l} + \varepsilon\bar{\mathbf{l}} \rangle = \underbrace{\langle \mathbf{l}, \mathbf{l} \rangle}_{=1} + 2\varepsilon \underbrace{\langle \mathbf{l}, \bar{\mathbf{l}} \rangle}_{=0} + \underbrace{\varepsilon^2}_{=0} \langle \bar{\mathbf{l}}, \bar{\mathbf{l}} \rangle = 1.$$

This shows that $\mathbf{L}$ is a dual unit vector. Moreover, we can establish a one-to-one and onto mapping between the manifold of oriented lines in Euclidean 3-space and the points on the *dual unit sphere.*

Without explicitly stating so, we have extended the canonical scalar product to dual values. In fact, we can do so with all analytic functions $f: I \subset \mathbb{R} \to \mathbb{R}$. Then, the dual extension $\underline{f}$ of f can be written as

$$\underline{f}(t) = f(t) + \varepsilon f'(t), \quad \forall t \in I,$$

which is the Taylor series of f cut off right after the linear term, since

$$\varepsilon^2 = \varepsilon^3 = \ldots = 0.$$

The dual extension of the canonical scalar product and the thus induced cross product allow a simple computation of distances, angles, and the common oriented normal of a pair of spears:

Theorem 10.3.3 *Let* $\mathbf{L}$ *and* $\mathbf{M}$ *be two dual unit vectors describing two straight lines* L *and* M *in Euclidean 3-space enclosing the oriented angle* φ *and having the oriented distance* $d \in \mathbb{R}$.

We define the dual number $\Phi := \varphi + \varepsilon d$ *as the dual angle between* $\mathbf{L}$ *and* $\mathbf{M}$. *Then, the cosine of the dual angle* Φ *equals*

$$\langle \mathbf{L}, \mathbf{M} \rangle = \cos\Phi, \tag{10.82}$$

and the oriented common normal spear N *has the dual coordinates*

$$\mathbf{L} \times \mathbf{M} = \sin\Phi\, \mathbf{N}. \tag{10.83}$$

Proof:

1. The spear coordinates $\mathbf{L} = \mathbf{l} + \varepsilon\bar{\mathbf{l}}$ and $\mathbf{M} = \mathbf{m} + \varepsilon\overline{\mathbf{m}}$ satisfy (10.79) and we compute

$$\langle \mathbf{L}, \mathbf{M}\rangle = \langle \mathbf{l}, \mathbf{m}\rangle + \varepsilon(\langle \mathbf{l}, \overline{\mathbf{m}}\rangle + \langle \mathbf{m}, \bar{\mathbf{l}}\rangle). \tag{10.84}$$

The two unit direction vectors $\mathbf{l}$ and $\mathbf{m}$ yield $\langle \mathbf{l}, \mathbf{m}\rangle = \cos\varphi$. In order to find the geometric meaning of the second summand, we recall that the momentum vectors $\bar{\mathbf{l}}$ and $\overline{\mathbf{m}}$ of the lines L and M can be computed as $\bar{\mathbf{l}} = \mathbf{p} \times \mathbf{l}$ and $\overline{\mathbf{m}} = \mathbf{q} \times \mathbf{m}$ with arbitrary points $P \in L$ and $Q \in M$. Now, the dual part of (10.84) reads

$$\langle \mathbf{l}, \overline{\mathbf{m}}\rangle + \langle \mathbf{m}, \bar{\mathbf{l}}\rangle = \det(\mathbf{l}, \mathbf{q}, \mathbf{m}) + \det(\mathbf{m}, \mathbf{p}, \mathbf{l}) = \det(\mathbf{l}, \mathbf{m}, \mathbf{p} - \mathbf{q}) = \langle \mathbf{l} \times \mathbf{m}, \mathbf{p} - \mathbf{q}\rangle.$$

The cross product and the angle φ subtended by a pair of unit vectors are related via

$$\|\mathbf{l} \times \mathbf{m}\| = \sin\varphi.$$

Since $\langle \mathbf{l} \times \mathbf{m}, \mathbf{p} - \mathbf{q}\rangle$ yields the orthogonal projection of $\mathbf{p} - \mathbf{q}$ onto $\mathbf{l} \times \mathbf{m}$, we have

$$\langle \mathbf{l} \times \mathbf{m}, \mathbf{p} - \mathbf{q}\rangle = -d\sin\varphi.$$

Note that $(\mathbf{l}, \mathbf{m}, \mathbf{q}-\mathbf{p})$ form a right-handed tripod, which causes the sign of the determinant. Finally, the dual extension of the cosine equals

$$\cos\Phi = \cos(\varphi + \varepsilon d) = \cos\varphi - \varepsilon d\sin\varphi.$$

2. We compute the unnormalized cross product of the two dual unit vectors representing the lines L and M and find

$$\mathbf{L} \times \mathbf{M} = (\mathbf{l} + \varepsilon\bar{\mathbf{l}}) \times (\mathbf{m} + \varepsilon\overline{\mathbf{m}}) = \mathbf{l} \times \mathbf{m} + \varepsilon(\mathbf{l} \times \overline{\mathbf{m}} + \bar{\mathbf{l}} \times \mathbf{m}).$$

The direction vector $\mathbf{n}$ of the common normal N of L and M is a scalar multiple of $\mathbf{l} \times \mathbf{m}$, *i.e.*, $\mathbf{n} = \sin\varphi\,\mathbf{l} \times \mathbf{m}$ with $\varphi = \measuredangle(\mathbf{l}, \mathbf{m})$. It is admissible to assume that $P \in L$ and $Q \in M$ are the pedal points of the common normal of L and M. Now, we expand the dual part of $\mathbf{L} \times \mathbf{M}$ and find

$$\mathbf{l} \times \overline{\mathbf{m}} + \bar{\mathbf{l}} \times \mathbf{m} = \langle \mathbf{p}, \mathbf{m}\rangle\mathbf{l} \underbrace{-\langle \mathbf{l}, \mathbf{m}\rangle\mathbf{p} + \langle \mathbf{l}, \mathbf{m}\rangle\mathbf{q}}_{=d\sin\varphi\,\mathbf{n}} - \langle \mathbf{q}, \mathbf{l}\rangle\mathbf{m}.$$

We can add $\langle \mathbf{q} - \mathbf{p}, \mathbf{l}\rangle\mathbf{m} = \langle \mathbf{q}, \mathbf{l}\rangle\mathbf{m} - \langle \mathbf{p}, \mathbf{l}\rangle\mathbf{m} = 0$ to the right-hand side of the latter equation without altering the result. This yields

$$\mathbf{l} \times \overline{\mathbf{m}} + \bar{\mathbf{l}} \times \mathbf{m} = d\sin\varphi\,\mathbf{n} + \langle \mathbf{p}, \mathbf{m}\rangle\mathbf{l} \underbrace{-\langle \mathbf{q}, \mathbf{l}\rangle\mathbf{m} + \langle \mathbf{q}, \mathbf{l}\rangle\mathbf{m}}_{=0} - \langle \mathbf{p}, \mathbf{l}\rangle\mathbf{m}.$$

The momentum vector $\overline{\mathbf{n}}$ of the common normal N (of L and M) can be computed either from P or Q. Therefore, $\overline{\mathbf{n}} = \mathbf{p} \times \mathbf{n} = \mathbf{q} \times \mathbf{n}$. Consequently, $\langle \mathbf{p}, \mathbf{m}\rangle\mathbf{l} - \langle \mathbf{p}, \mathbf{l}\rangle\mathbf{m} = \mathbf{p} \times (\mathbf{l} \times \mathbf{m}) = \sin\varphi\,\overline{\mathbf{n}}$, which results in $\mathbf{l} \times \overline{\mathbf{m}} + \bar{\mathbf{l}} \times \mathbf{m} = d\sin\varphi\,\mathbf{n} + \sin\varphi\,\overline{\mathbf{n}}$, and thus,

$$\mathbf{L} \times \mathbf{M} = d\sin\varphi\,\mathbf{n} + \varepsilon(d\cos\varphi\,\mathbf{n} + \sin\varphi\,\overline{\mathbf{n}}). \tag{10.85}$$

Now, we can easily verify that

$$\sin\Phi\,\mathbf{N} = (\sin\varphi + \varepsilon d\cos\varphi)(\mathbf{n} + \varepsilon\overline{\mathbf{n}}) = \sin\varphi\mathbf{n} + \varepsilon(d\cos\varphi\mathbf{n} + \sin\varphi\overline{\mathbf{n}}).$$ ■

■ **Example 10.3.3** Normalization of a dual vector.

A generic dual vector is not a dual unit vector. Moreover, for an arbitrarily given dual vector, we cannot expect that the real part and the dual part form a pair of orthogonal vectors so that $\langle \mathbf{v}, \overline{\mathbf{v}} \rangle \neq 0$ holds.

Sometimes it is useful to normalize a dual vector. Therefore, we compute the norm of a dual vector $\mathbf{V} = \mathbf{v} + \varepsilon \overline{\mathbf{v}}$ as the scalar product of $\mathbf{V}$ with itself and find

$$N(\mathbf{V}) = \langle \mathbf{V}, \mathbf{V} \rangle = \langle \mathbf{v}, \mathbf{v} \rangle + 2\varepsilon \langle \mathbf{v}, \overline{\mathbf{v}} \rangle. \tag{10.86}$$

As long as $\mathbf{v} \neq 0$, we can define the length of the dual vector $\mathbf{V}$ as

$$\|\mathbf{V}\| = \sqrt{N(\mathbf{V})}, \tag{10.87}$$

and the dual unit vector then equals

$$\mathbf{V}^\star = \frac{1}{\|\mathbf{V}\|}\mathbf{V}.$$

We compute $\|\mathbf{V}\|$ with the extension of the function $f: \mathbb{R}^+ \to \mathbb{R}^+$, $f: x \mapsto \sqrt{x}$ to dual values. The extension of the square root to dual numbers reads

$$\sqrt{x + \varepsilon y} = \sqrt{x} + \varepsilon \frac{y}{2\sqrt{x}}.$$

Applying this to (10.86), we find

$$\|\mathbf{V}\| = \|v\| + \varepsilon \frac{\langle \mathbf{v}, \overline{\mathbf{v}} \rangle}{\|v\|},$$

where $\|v\|$ is the Euclidean length of $\mathbf{v}$ and $\langle \cdot, \cdot \rangle$ denotes the canonical scalar product as well as its dual counterpart. The normalization of a vector needs the multiplication of the vector with the reciprocal of the length. For that purpose, we need the dual extension of the real-valued function $f: \mathbb{R}^\star \to \mathbb{R}$, $f: x \mapsto \frac{1}{x}$. This is simply

$$\frac{1}{x + \varepsilon y} = \frac{1}{x} - \varepsilon \frac{y}{x^2},$$

and therefore,

$$\frac{1}{\|\mathbf{V}\|} = \frac{1}{\|v\|} - \varepsilon \frac{\langle \mathbf{v}, \overline{\mathbf{v}} \rangle}{\|\mathbf{v}\|^3}.$$

Finally, the dual unit vector $\mathbf{V}^\star$ to $\mathbf{V}$ equals

$$\mathbf{V}^\star = \left(\frac{1}{\|v\|} - \varepsilon \frac{\langle \mathbf{v}, \overline{\mathbf{v}} \rangle}{\|\mathbf{v}\|^3} \right) (\mathbf{v} + \varepsilon \overline{\mathbf{v}}) = \frac{1}{\|\mathbf{v}\|}\mathbf{v} + \frac{\varepsilon}{\|\mathbf{v}\|} \left(\overline{\mathbf{v}} - \frac{\langle \mathbf{v}, \overline{\mathbf{v}} \rangle}{\|\mathbf{v}\|^2}\mathbf{v} \right).$$

• **Exercise 10.3.4** Oriented common normal.

Use the computations from Example 10.3.3 in order to verify (10.83) in Theorem 10.3.3.

• **Exercise 10.3.5** A circle of spears.

Assume that $\mathbf{L} = \mathbf{l} + \varepsilon \overline{\mathbf{l}}$ and $\mathbf{M} = \mathbf{m} + \varepsilon \overline{\mathbf{m}}$ are two dual unit vectors. Let further $\Phi = \varphi + \varepsilon \overline{\varphi}$. Show that, for any dual angle $T = t + \varepsilon \overline{t}$, the vector

$$\mathbf{V}(T) = \cos T\, \mathbf{L} + \sin T\, \mathbf{M}$$

is a dual unit vector.

The manifold of all dual unit vectors is called the dual unit sphere. It can be viewed as the dual extension of the Euclidean unit sphere and serves as another point model of the manifold of oriented straight lines in Euclidean 3-space.

Euclidean displacements and spears

In this section we shall see how Euclidean displacements act on dual unit vectors describing spears. For that purpose, we assume that $\mathbf{A} \in SO(3)$ is an orthogonal matrix with $\det \mathbf{A} = +1$ and $\mathbf{a} \in \mathbb{R}^3$ as a vector.

The Cartesian coordinate vector $\mathbf{x}$ of a point X is transformed under a Euclidean displacement (motion) $\beta : \mathbb{R}^3 \to \mathbb{R}^3$ according to

$$\mathbf{x}' = \mathbf{A}\mathbf{x} + \mathbf{a}.$$

Now, we assume that $\mathbf{L} = \mathbf{l}+\varepsilon\bar{\mathbf{l}}$ is a dual unit vector representing a straight line L in Euclidean 3-space. We are looking for the dual unit vector $\mathbf{L}' = \mathbf{l}' + \varepsilon\bar{\mathbf{l}}'$ of the line $L' = \beta(L)$. It is clear that the direction vector $\mathbf{l}$ changes to $\mathbf{l}' = \mathbf{A}\mathbf{l}$, since the translation has no effect on the direction. What about the momentum vector? We apply β to the pedal point P of L and use (10.3) from Exercise 10.1.1 in order to find the Cartesian coordinates of P. This yields $\mathbf{p} = \mathbf{l}\times\bar{\mathbf{l}}$, and further, $\mathbf{p}' = \mathbf{A}\mathbf{p}+\mathbf{a} = \mathbf{A}\mathbf{l}\times\bar{\mathbf{l}}+\mathbf{a}$.

This enables us to compute the momentum vector $\bar{\mathbf{l}}'$ of L' as

$$\bar{\mathbf{l}}' = \mathbf{p}' \times \mathbf{l}' = (\mathbf{A}\mathbf{l} \times \bar{\mathbf{l}} + \mathbf{a}) \times \mathbf{A}\mathbf{l} = \mathbf{A}(\mathbf{l} \times \bar{\mathbf{l}}) \times \mathbf{A}\mathbf{l} + \mathbf{a} \times \mathbf{A}\mathbf{l}.$$

Since $\mathbf{A} \in SO(3)$, we have $\mathbf{A}(\mathbf{l} \times \bar{\mathbf{l}}) \times \mathbf{A}\mathbf{l} = \mathbf{A}((\mathbf{l} \times \bar{\mathbf{l}}) \times \mathbf{l})$. Further, with (10.79), we find $(\mathbf{l} \times \bar{\mathbf{l}}) \times \mathbf{l} = \bar{\mathbf{l}}$.

The cross product $\mathbf{a} \times \mathbf{A}\mathbf{l}$ induces a linear mapping $\mathbb{R}^3 \to \mathbb{R}^3$ that can be written as $\mathbf{A}^{\times}\mathbf{l}$ with $\mathbf{A}^{\times} \in SO(3)$. Therefore, the momentum vector of L' reads

$$\bar{l}' = \mathbf{A}\bar{\mathbf{l}} + \mathbf{A}^{\times}\mathbf{l}.$$

From the pair $(\mathbf{A}, \mathbf{A}^{\times})$ of matrices, we can build the dual matrix $\underline{\mathbf{A}} = \mathbf{A} + \varepsilon\mathbf{A}^{\times}$. Now, we can write the transformation of the dual unit vector as the dual matrix product

$$\mathbf{L}' = \underline{\mathbf{A}}\mathbf{L},$$

since $\underline{\mathbf{A}}\mathbf{L} = (\mathbf{A} + \varepsilon\mathbf{A}^{\times})(\mathbf{l} + \varepsilon\bar{\mathbf{l}}) = \mathbf{A}\mathbf{l} + \varepsilon(\mathbf{A}\bar{\mathbf{l}} + \mathbf{A}^{\times}\mathbf{l}) = \mathbf{L}'$ holds.

We can summarize our results as follows:

Theorem 10.3.4 *A Euclidean displacement that maps Cartesian coordinates* $\mathbf{x}$ *of points* X *to points* X' *in Euclidean 3-space according to* $\mathbf{x} \mapsto \mathbf{x}' = \mathbf{A}\mathbf{x} + \mathbf{a}$ *induces an isometric mapping of the dual unit sphere.*

Dual quaternions and Study's quadric

Euclidean motions and dual quaternions

Assume we are given a Euclidean motion $\beta: \mathbb{R}^3 \to \mathbb{R}^3$, which is the product of a rotation about an axis through the origin of the underlying coordinate system and a translation in the direction of the vector $\mathbf{t} \in \mathbb{R}^3$.

According to (10.75), the coordinates $\mathbf{x}$ of a point X are transformed to the coordinates $\mathbf{x}'$ of a point X' via

$$\mathbf{x}' = \mathbf{R}\mathbf{X}\widetilde{\mathbf{R}} + \mathbf{t}, \tag{10.88}$$

where $\mathbf{R}$ is the unit quaternion (*i.e.*, $\mathbf{R}\widetilde{\mathbf{R}} = 1$) describing the rotation and the coordinate vectors $\mathbf{x}$ and $\mathbf{x}'$ are identified with pure vector quaternions $\mathbf{X}$ and $\mathbf{X}'$.

Now, we are going to look for the action of β on the coordinates $(\mathbf{l}, \bar{\mathbf{l}})$ of a spear L. For that purpose, we choose two arbitrary points X_1 and X_2 (with respective coordinate vectors $\mathbf{x}_1$ and $\mathbf{x}_2$) on L. Without loss of generality, the Euclidean distance of X_1 and X_2 equals 1. Then, the spear coordinates of L are

$$\mathbf{l} = \mathbf{x}_2 - \mathbf{x}_1 \quad \text{and} \quad \bar{\mathbf{l}} = \mathbf{x}_1 \times \mathbf{l} + \mathbf{x}_1 \times \mathbf{x}_2.$$

From this, we can immediately infer

$$\mathbf{l}' = \mathbf{x}_2' - \mathbf{x}_1' = \mathbf{R}(\mathbf{x}_2 - \mathbf{x}_1)\widetilde{\mathbf{R}}, \tag{10.89}$$

and moreover, with (10.73), we find

$$\bar{l}' = \mathbf{x}_1' \times \mathbf{x}_2' = \frac{1}{2}(\mathbf{X}_1\mathbf{X}_2 - \mathbf{X}_2\mathbf{X}_1),$$

where $\mathbf{X}_1$ and $\mathbf{X}_2$ are the pure vector quaternions describing X_1 and X_2. Considering $\mathbf{t}$ as a pure vector quaternion $\mathbf{T}$ yields

$$\begin{aligned} 2\bar{\mathbf{l}}' &= (\mathbf{R}\mathbf{X}_1\widetilde{\mathbf{R}} + \mathbf{t})(\mathbf{R}\mathbf{X}_2\widetilde{\mathbf{R}} + \mathbf{t}) - (\mathbf{R}\mathbf{X}_2\widetilde{\mathbf{R}} + \mathbf{t})(\mathbf{R}\mathbf{X}_1\widetilde{\mathbf{R}} + \mathbf{t}) = \\ &= \mathbf{R}\underbrace{(\mathbf{X}_1\mathbf{X}_2 - \mathbf{X}_2\mathbf{X}_1)}_{=\bar{\mathbf{l}}}\widetilde{\mathbf{R}} + \mathbf{R}\underbrace{(\mathbf{X}_1 - \mathbf{X}_2)}_{=-\mathbf{l}}\widetilde{\mathbf{R}}\mathbf{T} + \mathbf{T}\mathbf{R}\underbrace{(\mathbf{X}_2 - \mathbf{X}_1)}_{=\mathbf{l}}\widetilde{\mathbf{R}}. \end{aligned}$$

Now, we define

$$\overline{\mathbf{R}} := \frac{1}{2}\mathbf{T}\mathbf{R},$$

which causes $\widetilde{\overline{\mathbf{R}}} = -\frac{1}{2}\widetilde{\mathbf{R}}\mathbf{T}$, and therefore,

$$\mathbf{T} = 2\overline{\mathbf{R}}\widetilde{\mathbf{R}} = -2\mathbf{R}\widetilde{\overline{\mathbf{R}}} \tag{10.90}$$

is a pure vector quaternion and can be identified with $\mathbf{t}s$. So, we have

$$\bar{\mathbf{l}}' = \mathbf{R}\overline{\mathbf{L}}\widetilde{\mathbf{R}} + \mathbf{R}\mathbf{L}\widetilde{\overline{\mathbf{R}}} + \overline{\mathbf{R}}\mathbf{L}\widetilde{\mathbf{R}}. \tag{10.91}$$

In order to simplify the notation, the rules for the transformation of the direction vector (10.89) and the momentum vector (10.91) can be combined to a *dual quaternion*

$$\underline{\mathbf{R}} := \mathbf{R} + \varepsilon\overline{\mathbf{R}},$$

where $\varepsilon^2 = 0$ still holds. The conjugate dual quaternion $\underline{\widetilde{\mathbf{R}}}$ of $\underline{\mathbf{R}}$ is defined as $\underline{\widetilde{\mathbf{R}}} := \widetilde{\mathbf{R}} + \varepsilon\widetilde{\overline{\mathbf{R}}}$. This simplifies the equation of transformation of spear coordinates to

$$\mathbf{L}' = \underline{\mathbf{R}}\,\mathbf{L}\,\underline{\widetilde{\mathbf{R}}}, \tag{10.92}$$

which really evaluates to

$$\mathbf{l}' + \varepsilon\bar{\mathbf{l}}' = (\mathbf{R} + \varepsilon\overline{\mathbf{R}})(\mathbf{l} + \varepsilon\bar{\mathbf{l}})(\widetilde{\mathbf{R}} + \varepsilon\widetilde{\overline{\mathbf{R}}}) = \mathbf{R}\mathbf{l}\widetilde{\mathbf{R}} + \varepsilon(\overline{\mathbf{R}}\mathbf{l}\widetilde{\mathbf{R}} + \mathbf{R}\bar{\mathbf{l}}\widetilde{\mathbf{R}} + \mathbf{R}\mathbf{l}\widetilde{\overline{\mathbf{R}}}).$$

We can easily verify that $\underline{\mathbf{R}}$ is a *dual unit quaternion*:

$$\underline{\mathbf{R}}\,\underline{\widetilde{\mathbf{R}}} = (\mathbf{R} + \varepsilon\overline{\mathbf{R}})(\widetilde{\mathbf{R}} + \varepsilon\widetilde{\overline{\mathbf{R}}}) = \underbrace{\mathbf{R}\widetilde{\mathbf{R}}}_{=1} + (\varepsilon\underbrace{\mathbf{R}\widetilde{\overline{\mathbf{R}}} + \overline{\mathbf{R}}\widetilde{\mathbf{R}}}_{=0}) = 1,$$

by definition and in accordance with (10.90).

Let the Euclidean motion (10.88) be a rotation about the axis a with spear coordinates $(\mathbf{a}, \overline{\mathbf{a}})$. For an arbitrary point S on the axis of rotation a, we have $\overline{\mathbf{a}} = \mathbf{s} \times \mathbf{a}$. According to Theorem 10.3.1, we have

$$\mathbf{x}' - \mathbf{s} = \mathbf{R}(\mathbf{x} - \mathbf{s})\widetilde{\mathbf{R}},$$

where $\mathbf{R} = \cos\frac{\varphi}{2} + \sin\frac{\varphi}{2}\mathbf{a}$. We expand the latter equation and arrive at

$$\begin{aligned} \mathbf{x}' &= \mathbf{R}\mathbf{x}\widetilde{\mathbf{R}} + \mathbf{s} - \mathbf{R}\mathbf{s}\widetilde{\mathbf{R}} = \mathbf{R}\mathbf{x}\widetilde{\mathbf{R}} + (\mathbf{s}\mathbf{R} - \mathbf{R}\mathbf{s})\widetilde{\mathbf{R}} = \\ &= \mathbf{R}\mathbf{x}\widetilde{\mathbf{R}} + 2\sin\tfrac{\varphi}{2}(\mathbf{s} \times \mathbf{a})\,\widetilde{\mathbf{R}} = \mathbf{R}\mathbf{x}\widetilde{\mathbf{R}} + 2\sin\tfrac{\varphi}{2}\overline{\mathbf{a}}\widetilde{\mathbf{R}}, \end{aligned}$$

and finally, at

$$\mathbf{x}' = \mathbf{R}\,\mathbf{x}\,\widetilde{\mathbf{R}} + 2\sin\frac{\varphi}{2}\overline{\mathbf{a}}\,\widetilde{\mathbf{R}}. \qquad (10.93)$$

Now, we superimpose the rotation with a translation with the distance d in the direction $\mathbf{a}$. This means that we have to add $d\mathbf{a}$ to $\mathbf{x}'$. With (10.90), the displacement vector equals $\mathbf{t} = 2\sin\frac{\varphi}{2}\overline{\mathbf{a}}$, and with (10.93), we find

$$\begin{aligned} \overline{\mathbf{R}} &= \tfrac{1}{2}\mathbf{t}\,\mathbf{R} = \sin\tfrac{\varphi}{2}\overline{\mathbf{a}} + \tfrac{d}{2}\mathbf{a}\,\mathbf{R} = \\ &= \sin\tfrac{\varphi}{2}\overline{\mathbf{a}} + \tfrac{d}{2}\mathbf{a}\left(\cos\tfrac{\varphi}{2} + \sin\tfrac{\varphi}{2}\mathbf{a}\right) = -\tfrac{d}{2}\sin\tfrac{\varphi}{2} + \tfrac{d}{2}\cos\tfrac{\varphi}{2}\mathbf{a} + \sin\tfrac{\varphi}{2}\overline{\mathbf{a}}. \end{aligned}$$

We combine the partial results to dual quaternions and dual vectors, which yields

$$\begin{aligned} \underline{\mathbf{R}} &= \mathbf{R} + \varepsilon\overline{\mathbf{R}} = \\ &= \left(\cos\tfrac{\varphi}{2} + \sin\tfrac{\varphi}{2}\mathbf{a}\right) + \varepsilon\left(-\tfrac{d}{2}\sin\tfrac{\varphi}{2} + \tfrac{d}{2}\cos\tfrac{\varphi}{2}\mathbf{a} + \sin\tfrac{\varphi}{2}\overline{\mathbf{a}}\right) = \\ &= \left(\cos\tfrac{\varphi}{2} - \varepsilon\tfrac{d}{2}\sin\tfrac{\varphi}{2}\right) + \left(\sin\tfrac{\varphi}{2} + \varepsilon\tfrac{d}{2}\cos\tfrac{\varphi}{2}\right)(\mathbf{a} + \varepsilon\overline{\mathbf{a}}). \end{aligned} \qquad (10.94)$$

The dual extension of the trigonometric functions

$$\cos\left(\frac{\varphi}{2} + \varepsilon\frac{d}{2}\right) = \cos\frac{\varphi}{2} - \varepsilon d\sin\frac{\varphi}{2}, \quad \sin\left(\frac{\varphi}{2} + \varepsilon\frac{d}{2}\right) = \sin\frac{\varphi}{2} + \varepsilon d\cos\frac{\varphi}{2},$$

the dual vector $\mathbf{A} = \mathbf{a} + \varepsilon\overline{\mathbf{a}}$ representing the axis, and the dual angle $\mathbf{\Phi} = \varphi + \varepsilon d$ allow us to simplify (10.94) to

$$\underline{\mathbf{R}} = \cos\frac{\mathbf{\Phi}}{2} + \sin\frac{\mathbf{\Phi}}{2}\mathbf{A}. \qquad (10.95)$$

We can summarize the results of *Study's representation of Euclidean motions* in:

Theorem 10.3.5 *The uniquely defined helical motion with the axis* $\mathbf{A} = \mathbf{a} + \varepsilon\overline{\mathbf{a}}$ *and the dual angle* $\Phi = \varphi + \varepsilon d$ *maps the oriented straight line (spear)* $\mathbf{G} = \mathbf{g} + \varepsilon\overline{\mathbf{g}}$ *to the oriented straight line*

$$\mathbf{G}' = \underline{\mathbf{R}}\,\mathbf{G}\,\underline{\widetilde{\mathbf{R}}}, \qquad (10.96)$$

where $\underline{\mathbf{R}}$ *equals the dual quaternion given in* (10.95).

The comparison between (10.75) and (10.96) shows that results on spherical motions have counterparts in the group of Euclidean motions. In any case, it is sufficient to replace the ordinary Hamiltonian quaternions in formulas of spherical kinematics with dual quaternions in order to obtain the analogous results in the group of motions (in general). This is called *Study's principle* (see [65, 131]).

It is also possible to show the following result:

Theorem 10.3.6 *1. Each dual unit quaternion $\underline{\mathbf{R}}$ determines a helical motion (including the case of a rotation or a translation) and can be written as*

$$\underline{\mathbf{R}} = \cos\frac{\mathbf{\Phi}}{2} + \sin\frac{\mathbf{\Phi}}{2}\mathbf{A}.$$

2. With $\underline{\mathbf{R}} = (r_0+\varepsilon\overline{r_0})+(r_1+\varepsilon\overline{r_1})\mathrm{i}+(r_2+\varepsilon\overline{r_2})\mathrm{j}+(r_3+\varepsilon\overline{r_3})\mathrm{k}$ and the normalization condition $\underline{\mathbf{R}}\,\widetilde{\underline{\mathbf{R}}} = 1$, the components of the dual unit quaternion are subject to two quadratic conditions:

$$r_0^2 + r_1^2 + r_2^2 + r_3^2 = 1 \quad \text{and} \quad r_0\overline{r_0} + r_1\overline{r_1} + r_2\overline{r_2} + r_3\overline{r_3} = 0. \tag{10.97}$$

The matrix representation $\mathbf{x}' = \mathbf{x} \mapsto \mathbf{R}\mathbf{x} + \mathbf{t}$ of the corresponding Euclidean motion is given by the rotation matrix (10.77) *and the translation vector*

$$\mathbf{t} = 2\begin{pmatrix} r_0\overline{r_1} - r_1\overline{r_0} + r_2\overline{r_3} - r_3\overline{r_2} \\ r_0\overline{r_2} - r_1\overline{r_3} + r_2\overline{r_0} + r_3\overline{r_1} \\ r_0\overline{r_3} - r_1\overline{r_2} + r_2\overline{r_1} - r_3\overline{r_0} \end{pmatrix}.$$

We skip the proof of Theorem 10.3.6, since it can be found in textbooks on kinematics (cf. [65]).

• **Exercise 10.3.6** Conversion formulas for Study parameters and matrix entries.

Assume that we are given a Euclidean motion $\beta : \mathbb{R}^3 \to \mathbb{R}^3$, $\beta(\mathbf{x}) = \mathbf{x}' = \mathbf{A}\mathbf{x}+\mathbf{a}$ (with $\mathbf{A} \in SO(3)$ and $\mathbf{a} \in \mathbb{R}^3$). We can write the transformation of the Cartesian coordinates in the extended matrix notation as

$$\begin{pmatrix} 1 \\ \mathbf{x}' \end{pmatrix} = \begin{pmatrix} 1 & \mathbf{0}^{\mathrm{T}} \\ \mathbf{a} & \mathbf{A} \end{pmatrix}\begin{pmatrix} 1 \\ \mathbf{x} \end{pmatrix}. \tag{10.98}$$

Let a_{ij} denote the entries of $\mathbf{A}$, and let further $\mathbf{a} = (a_{10}, a_{20}, a_{30})$.

Show that the entries of the extended transformation matrix in (10.98) and the components r_i and $\overline{r_i}$ (Study parameters) of the dual unit quaternion mentioned in Theorem 10.3.6 are related via

$$\begin{aligned} r_0 : r_1 : r_2 : r_3 = 1 + a_{11} + a_{22} + a_{33} : a_{23} - a_{32} : a_{31} - a_{13} : a_{12} - a_{21} = \\ a_{23} - a_{32} : 1 + a_{11} - a_{22} - a_{33} : a_{12} + a_{21} : a_{31} + a_{13} = \\ a_{31} - a_{13} : a_{12} + a_{21} : 1 - a_{11} + a_{22} - a_{33} : a_{23} + a_{32} = \\ a_{12} - a_{21} : a_{31} + a_{13} : a_{23} + a_{32} : 1 - a_{11} - a_{22} + a_{33}, \\ 2\overline{r_0} = \ r_1 a_{10} + r_2 a_{20} + r_3 a_{30}, \\ 2\overline{r_1} = -r_0 a_{10} + r_3 a_{20} - r_2 a_{30}, \\ 2\overline{r_2} = -r_3 a_{10} - r_0 a_{20} + r_1 a_{30}, \\ 2\overline{r_3} = \ r_2 a_{10} - r_1 a_{20} - r_0 a_{30}. \end{aligned} \tag{10.99}$$

Study's quadric

The first equation in (10.97) is simply the equation of a three-dimensional sphere S^3. From the viewpoint of quadrics, the second equation in (10.97) is more interesting. It allows us to interpret the coordinates of a dual unit quaternion as homogeneous coordinates of points in a seven-dimensional projective space $\mathbb{P}^7(\mathbb{R})$. For what follows, we relabel the coordinates as follows:

$$r_i = x_{i+1}, \quad \overline{r_i} = x_{4+i+1} \quad \text{with} \quad i \in \{0,1,2,3\}.$$

Then,

$$\mathcal{S}_2^6 : \ x_1 x_5 + x_2 x_6 + x_3 x_7 + x_4 x_8 = 0 \tag{10.100}$$

can be seen as the equation of a six-dimensional quadric $\mathcal{S}_2^6 \subset \mathbb{P}^7$. The quadric $\mathcal{S}_2^6$ is called *Study's quadric*.

The mapping that assigns a unique point in $\mathcal{S}_2^6$ to each Euclidean motion is called *Study's kinematic mapping*. Conversely, each point on Study's quadric (except those in an exceptional 3-space) corresponds to a certain Euclidean motion, and thus, to two dual unit quaternions $\pm\underline{\mathbf{R}}$. The components r_i and $\overline{r_i}$ of the dual unit quaternion $\underline{\mathbf{R}}$ are called the *Study parameters* of the thus determined Euclidean motion SE(3).

It is worth noting that k-parameter Euclidean motions correspond to k-dimensional submanifolds of $\mathcal{S}_2^6$. Especially the kinematic image of a one-parameter motion is a curve on Study's quadric. Straight lines correspond either to continuous rotations about a fixed axis or to translations in a fixed direction. Conics on $\mathcal{S}_2^6$ are the kinematic images of the coupler motion of Bennett mechanisms. We will have a closer look at Bennett mechanisms in a later section.

The signature of the quadratic form (10.100) equals $(++++----)$ or $(4,4)$ (cf. Section 4) and tells us that this quadric is of index three. Thus, there are three-dimensional (projective) subspaces contained in $\mathcal{S}_2^6$.

In connection with the geometry of motions, the generator space $U: x_1 = x_2 = x_3 = x_4 = 0$ plays a special role. The points in U do not correspond to Euclidean motions. Therefore, U is frequently called *exceptional space.*

We shall study the geometry on this quadric in more detail in order to get an understanding of phenomena on higher dimensional quadrics that cannot be observed within lower dimensional ones. For that purpose, we apply a stereographic projection

$$\sigma: \mathcal{S}_2^6 \setminus T_C\mathcal{S}_2^6 \to \mathbb{P}^6$$

from a point $C \notin U$ to a hyperplane $\mathbb{P}^6 \neq T_C\mathcal{S}_2^6$.

We let $C = (1,0,0,0,0,0,0,0)\mathbb{R}$, and the image space $\mathbb{P}^6$ shall be given by $x_1 = 0$. The tangent hyperplane $T_C\mathcal{S}_2^6$ of $\mathcal{S}_2^6$ at C has the equation $x_5 = 0$. The intersection $\Gamma_C := T_C\mathcal{S}_2^6 \cap \mathcal{S}_2^6$ is a quadratic cone with vertex C. A director quadric of the cone Γ_C is the quadric $Q_2^4 := T_C\mathcal{S}_2^6 \cap \mathbb{P}^6 \cap \mathcal{S}_2^6$. Since the tangent hyperplane $T_C\mathcal{S}_2^6$ and the image hyperplane $\mathbb{P}^6$ of σ intersect in a projective five-space $\mathbb{P}^5$ (with the equation $x_1 = x_5 = 0$), the quadric Q_2^4 is given by

$$Q_2^4: \ x_2x_6 + x_3x_7 + x_4x_8 = 0.$$

It is projectively equivalent to Plücker's quadric given by (10.2) in Section 10.1.

This fact helps us to study the families of generator spaces contained in $\mathcal{S}_2^6$. The inverse σ^{-1} of the chosen stereographic projection maps the two-dimensional generators (planes) of Q_2^4 to the three-dimensional generators (*solids*) of $\mathcal{S}_2^6$.

For the sake of simplicity, we use some terminology from Section 10.1. There are two kinds of planes in Q_2^4, called planes of the first and second kind.

Two planes E_1 and E_2 of the same kind (corresponding to stars of lines or ruled planes) intersect in one point, say P. The (inverse) stereographic projection σ^{-1} sends them to two solids $S_1, S_2 \in \mathcal{S}_2^6$ of the same kind. Depending on whether $P \in \mathbb{P}^5$ or $P \notin \mathbb{P}^5$, the two solids S_1 and S_2 are skew or intersect in a straight line.

Two planes E and F in Q_2^4 of different kinds are either skew or share a line (since stars of lines and ruled planes either have no line or a pencil of lines in common). Therefore, σ^{-1} maps E and F to two solids, say S and T, of different kinds that either share a point or, if they intersect, a plane.

We collect our results in the following:

Theorem 10.3.7 *The quadric $\mathcal{S}_2^6$ carries two six-parameter families of three-dimensional generator spaces.*

Two generators of the same kind are either skew or intersect along a straight line.

Two generators of different kinds meet either in a single point or they share a plane.

The results of Theorem 10.3.7 hold true for any projectively equivalent six-dimensional quadric. The properties of the generators of $\mathcal{S}_2^6$ were deduced with the help of a stereographic projection. This technique applies to quadrics in any projective space, which gives rise to similar results for ruled quadrics of maximal index in projective spaces of odd dimensions (note Theorem 4.3.2).

Within a quadric of the projective type of $\mathcal{S}_2^6$, we can find a geometric phenomenon that somehow generalizes the principle of duality. It is described in detail in [23, 132] and also in [135], where an analytic description with the help of Cayley octaves is given. We shall use a synthetic approach like the one found in [21, 135].

According to Theorem 10.3.7, $\mathcal{S}_2^6$ is a six-dimensional manifold in three ways: considered as a manifold of its (1) points, (2) 3-spaces of the first kind, and (3) 3-spaces of the second kind. We call two different points *incident*, if they are conjugate w.r.t. $\mathcal{S}_2^6$. Two solids of the same kind are called *incident* if they intersect in a straight line. Two solids of different kinds are called *incident* if they share a plane. A pair consisting of a point and a 3-space is called *incident* if the point lies in the 3-space. Elements which are not incident are called *skew*. Consequently, there are two different types of incidence: the incidence relation between elements of different kinds or the same kind.

Now we have the following:

Theorem 10.3.8 *Principle of triality: A theorem formulated only with the terms points, 3-spaces of the first kind, and 3-spaces of the second kind (in $\mathcal{S}_2^6$) remains valid if the terms points, 3-spaces of the first kind, and 3-space of the second kind are arbitrarily permuted.*

The principle of triality generalizes the principle of duality. Loosely speaking: While a duality has to be applied twice to a geometric object in order to get back the initial object, a triality has to be applied a third time.

We shall investigate some simple geometric figures related in the triality:

Theorem 10.3.9 *A straight line that is entirely contained in $\mathcal{S}_2^6$ is* self-trial, *i.e., it does not change (up to projective transformations) if the principle of triality (Theorem 10.3.8) is applied to it. A straight line in $\mathcal{S}_2^6$ carries a one-parameter family of points, 3-spaces of the first kind, and 3-spaces of the second kind, and each element is incident with any other element.*

Proof: The polar space of a straight line l in $\mathcal{S}_2^6$ w.r.t. $\mathcal{S}_2^6$ is a projective five-space $\mathbb{P}^5$ that intersects $\mathcal{S}_2^6$ along a quadratic cone P_2^4 which carries l as the set of its singular points. The cone P_2^4 emerges as the connection of l with a two-dimensional ruled quadric Q_2^2. Since Q_2^2 carries two independent one-parameter families of straight lines (the reguli), the cone carries two one-parameter families of three-dimensional generator spaces all of which contain l. (Q_2^2's rulings of the first/second kind correspond to $\mathcal{S}_2^6$'s 3-spaces of the first/second kind through l.) Thus, through each point of l, there are exactly two solids, one of each kind. Therefore, l can be considered as a one-parameter family of points or as incident 3-spaces of the first or second kind. Since $l \subset \mathcal{S}_2^6$, any pair of points on l are conjugate w.r.t. $\mathcal{S}_2^6$. Each point on l is contained in any of the 3-spaces through l. Any two solids of the same kind through l intersect in l. Any two solids of different kinds through l intersect in a plane through l. ■

A two-dimensional subspace in $\mathcal{S}_2^6$ can be viewed as the intersection of two incident 3-spaces (generators of different kinds), since there is one solid of the first and one of the second kind through it. On the other hand, the plane can also be considered as a ruled plane. The object that corresponds to the ruled plane (totality of lines in that projective plane) in the triality is a star of lines (of the first or second kind) in a solid (of first or second kind) that is determined in that solid with the incident point. Each point P in the plane is the carrier of a pencil of lines, and each line is the carrier of a range of points. Therefore, each solid of the second kind that is incident with a point and a solid of the first kind contains a pencil of lines; and each line of the pencil is part of a one-parameter family of such pencils. Consequently, each such line is, together with P and the incident solid of the first kind, contained in the solid of the second kind through P. So, we have:

Theorem 10.3.10 *A ruled plane in $\mathcal{S}_2^6$ corresponds to a star of lines (first or second kind) contained in the solid (of the first or second kind) through the carrier plane of the ruled plane.*

Bennet mechanisms

It was G.T. BENNETT in 1914 [4] who detected a flexible kinematic chain consisting of four cyclically ordered systems $\Sigma_1, \dots, \Sigma_4$ with revolute joints between them. Below, we prove both its flexibility and its particular role with regard to Study's quadric.

We start with a *skew isogram* in $\mathbb{E}^3$, *i.e.*, a non-planar quadrangle $ABCD$ (with position vectors $\mathbf{a}, \dots, \mathbf{d}$), where opposite sides are of equal lengths, $a := \overline{AB} = \overline{CD}$ and $b := \overline{BC} = \overline{DA}$. A skew isogram can be obtained by bending a planar parallelogram about one diagonal through an angle γ with $0 < \gamma < 180°$ (see Figure 10.22).

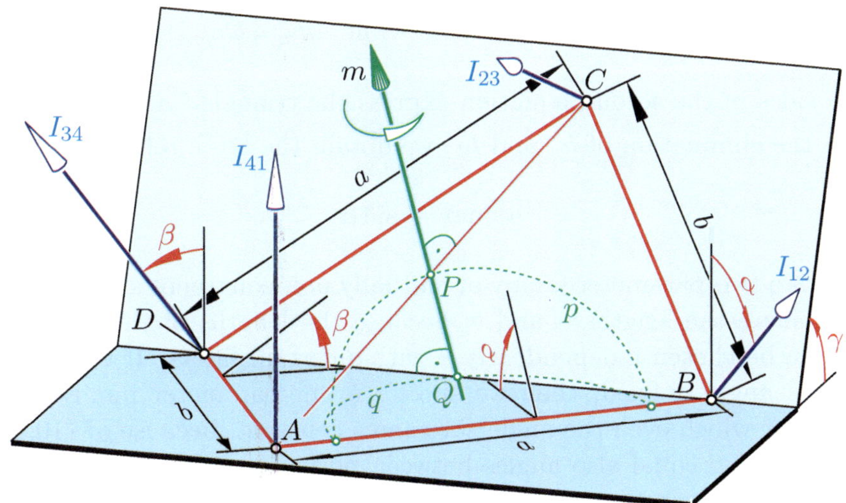

FIGURE 10.22. A skew isogram $ABCD$ with its side lengths and angles.

Each skew isogram has an axis of symmetry. In our case, this means that a 180° rotation about the line m which connects the midpoints P and Q of the diagonals interchanges A with C and B with D. In order to prove this, we denote the position vectors of the involved points $A, \dots, Q$ with

$\mathbf{a}, \ldots, \mathbf{q}$ and factorize the left-hand sides in two equations

$$\langle \mathbf{a}-\mathbf{b}, \mathbf{a}-\mathbf{b}\rangle - \langle \mathbf{c}-\mathbf{d}, \mathbf{c}-\mathbf{d}\rangle \quad \text{and} \quad \langle \mathbf{b}-\mathbf{c}, \mathbf{b}-\mathbf{c}\rangle - \langle \mathbf{d}-\mathbf{a}, \mathbf{d}-\mathbf{a}\rangle = 0$$

according to $x^2 - y^2 = (x+y)(x-y)$. This yields, because of $\mathbf{a}-\mathbf{b}+\mathbf{c}-\mathbf{d} = 2(\mathbf{p}-\mathbf{q})$,

$$\langle \mathbf{a}-\mathbf{b}-\mathbf{c}+\mathbf{d}, \mathbf{p}-\mathbf{q}\rangle = 0, \ \langle \mathbf{b}-\mathbf{c}-\mathbf{d}+\mathbf{a}, \mathbf{p}-\mathbf{q}\rangle = 0,$$

and finally, $\langle \mathbf{a}-\mathbf{c}, \mathbf{p}-\mathbf{q}\rangle = \langle \mathbf{b}-\mathbf{d}, \mathbf{p}-\mathbf{q}\rangle = 0$, which confirms the orthogonality between m and both diagonals AC and BD.

The convex hull of the skew isogram is a tetrahedron with the basis DAB and the apex C. Let α and β denote the dihedral angles along the base edges AB and DA, respectively (Figure 10.22). When h_a and h_b denote the heights on the sides with length a and b, respectively, in the congruent triangles ABC and CDA, then the height of the apex C over the base plane can be expressed in two ways as

$$h_a \sin\alpha = h_b \sin\beta, \quad \text{while} \quad a h_a = b h_b.$$

Both sides in the second equation express the triangles' areas.

After the elimination of h_a and h_b, we obtain the basic relation

$$a \sin\beta = b \sin\alpha. \tag{10.101}$$

There is a two-parameter family of mutually non-congruent skew isograms which share the lengths a and b, because the lengths of the two diagonals can be chosen independently – within certain limits. If we keep the dihedral angle α fixed, then we specify a one-parameter family of isograms, for which even the angle β remains constant, because of (10.101). Since α and β equal also angles between normals to the respective faces of the tetrahedron (note Figure 10.22), our one-parameter family represents flexions of a kinematic revolute chain with four links (Figure 10.23). Each side of our isogram represents one link Σ_i of this chain. At each vertex, the line orthogonal to the adjacent sides is the axis $I_{i\,i+1}$ of rotation between the adjacent links Σ_i and Σ_{i+1} (cf. [146]).

According to [20], the Bennett isogram is the only closed 4R kinematic chain which is flexible.

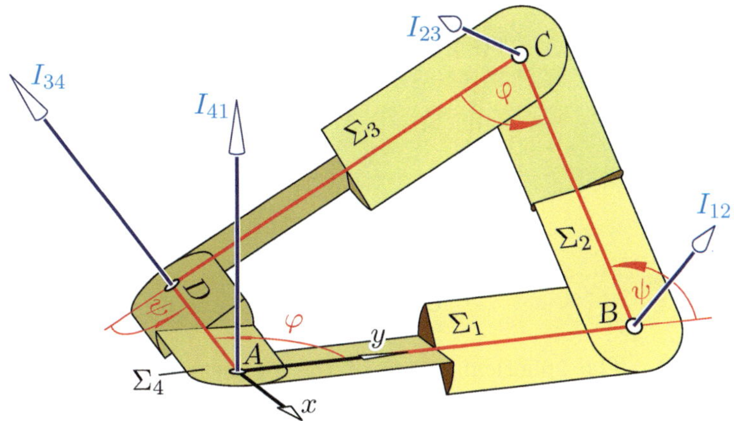

FIGURE 10.23. The Bennett isogram as a flexible 4R kinematic chain. The relative motion Σ_3/Σ_1 with input angle φ and output angle ψ corresponds to two conics on Study's quadric.

In the following, we focus on the relative motion of two opposite systems and prove that the associated kinematic image on Study's quadric consists of two conics (cf. [20, 50, 130]).

Suppose that the system Σ_1 with the isogram's side AB is kept fixed. A rotation of Σ_4 about I_{41} through an 'input angle' φ results in a rotation of Σ_2 about I_{12} through the 'output angle' ψ. Let us choose the aligned pose with $\varphi = \sphericalangle BAD = 0$ and D on $[A, B]$ as the initial pose. In this pose, C lies on the extension of the side AB. Thus, we specify the output angle φ complementary to $\sphericalangle ABC$ (see Figure 10.23).

Let us protract the unit vectors along the axes of rotation $I_{12}, \ldots, I_{41}$ from a fixed point. Then, we obtain the spherical image of the Bennett isogram consisting of four points $I_{12}, \ldots, I_{41}$ on the unit sphere, which is again a *spherical isogram* (Figure 10.24): Opposite sides have the same spherical lengths α or β. We note that the side AB of the Bennett isogram is orthogonal to the diameter plane through the points I_{41} and I_{12} on the unit sphere. Since the same holds for the other sides, the input angle φ and the output angle ψ show also up in the spherical image between two adjacent sides of spherical isogram.

Now, we refer to a well-known result on the motions of spherical isograms, interpreted as particular cases of spherical coupler motions (see [123]):

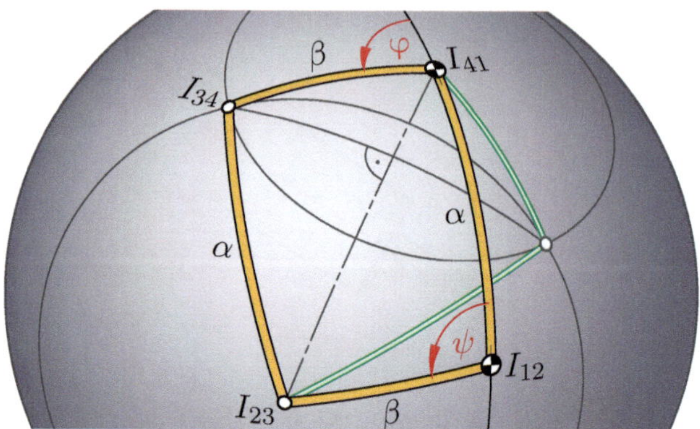

FIGURE 10.24. The spherical image of the Bennett isogram is a spherical isogram.

The transmission of φ to ψ, which, in general, is a 2-2-correspondence, splits into two projectivities $\varphi \mapsto \psi$ with

$$\tan\frac{\psi}{2} = \frac{\sin(\alpha-\beta)}{\sin\alpha \pm \sin\beta}\tan\frac{\varphi}{2}. \tag{10.102}$$

In the aligned poses $\varphi = 0$ and $\varphi = 180°$, we obtain as output angles $\psi = 0$ and $\psi = 180°$, respectively. The two transmission functions in (10.102) result in two different motions Σ_3/Σ with bifurcations in the aligned poses.[15] Only in the case $a = b$ and, consequently, $\alpha = \beta$, one motion splits into two pure rotations, when $D = B$ or $C = A$.

Theorem 10.3.11 *The relative motion between two opposite systems of a Bennett isogram splits into two smooth motions with bifurcations in the aligned poses. Each motion corresponds to a conic on Study's quadric.*

Proof: In the initial pose Σ_3^0 of Σ_3, the side DC is aligned with AB, and the axis I_{34} is in a position I_{34}^0. Now, we can move Σ_3^0 into a general pose of Σ_3 by applying two rotations; firstly, the rotation about I_{34}^0 through $-\psi$, and secondly, the rotation about I_{41} through φ. Then by virtue of (10.95), the two rotations are represented by the dual quaternions

$$\underline{\mathbf{R}}_1 = \cos\frac{\psi}{2} - \sin\frac{\psi}{2}\mathbf{A}_1 \text{ and } \underline{\mathbf{R}}_2 = \cos\frac{\varphi}{2} + \sin\frac{\varphi}{2}\mathbf{A}_2,$$

where the dual vector quaternions $\mathbf{A}_1$ and $\mathbf{A}_2$ represent the two axes I_{34}^0 and I_{41}, respectively. Due to (10.96), the quaternion product $\underline{\mathbf{R}} := \underline{\mathbf{R}}_2\,\underline{\mathbf{R}}_1$ determines the kinematic image.

---

[15]Figure 10.24 illustrates why for a given angle ψ there exist two solutions for the corresponding angle φ.

We use a Cartesian frame with the origin A, the y-axis in direction to B and the z-axis along I_{41}. Consequently, we obtain

$$\mathbf{A}_1 = (-\sin\beta + \varepsilon b\cos\beta)\mathrm{i} + (\cos\beta + \varepsilon b\sin\beta)\mathrm{k} \text{ and } \mathbf{A}_2 = \mathrm{k}.$$

It does not affect the kinematic image when the unit quaternion $\underline{\mathbf{R}}$ is multiplied with the real scalar $\lambda := 1/(\cos\frac{\psi}{2}\cos\frac{\varphi}{2})$. Moreover, by virtue of the transmission function in (10.102), we can introduce a motion parameter

$$t := \tan\frac{\varphi}{2} \text{ with } \tan\frac{\psi}{2} = ct \text{ where } c = \text{const.}$$

This results in

$$\lambda\underline{\mathbf{R}} = [1 + t\mathrm{k}]\,[1 - ct(-\sin\beta + \varepsilon b\cos\beta)\mathrm{i} - ct(\cos\beta + \varepsilon b\sin\beta)\mathrm{k}]\,,$$

and further,

$$\begin{aligned}\lambda\underline{\mathbf{R}} \;=\; & \big(1 + ct^2(\cos\beta + \varepsilon b\sin\beta)\big) + ct(\sin\beta - \varepsilon b\cos\beta)\mathrm{i} + ct^2(\sin\beta - \varepsilon b\cos\beta)\mathrm{j} \\ & + t\,(1 - c\cos\beta - \varepsilon bc\sin\beta)\,\mathrm{k}.\end{aligned}$$

This confirms that the kinematic image can be represented in the form $\mathbf{v}(t) = \mathbf{v}_1 + t\mathbf{v}_2 + t^2\mathbf{v}_3$ with linearly independent vectors

$$\mathbf{v}_1 = (1,0,0,0;\,0,0,0,0),\;\; \mathbf{v}_2 = (0, c\sin\beta, 0, 1 - c\cos\beta;\, 0, -bc\cos\beta, 0, -bc\sin\beta),$$
$$\mathbf{v}_3 = (c\cos\beta, 0, c\sin\beta, 0;\, bc\sin\beta, 0, -bc\cos\beta, 0).$$

Thus, we obtain a conic on Study's quadric. ■

There is another relation between the Bennet isogram and quadrics: If the system Σ_1 with the side AB is kept fixed, then the point $D \in \Sigma_4$ traces a circle with the center A, radius b, and axis I_{41}. On the other hand, $C \in \Sigma_2$ runs on a congruent circle with the center B and axis I_{12}. Consequently, the midpoints P of AC and Q of BD trace two circles p and q with radii $b/2$, which meet on the line $[A, B]$ (Figure 10.22). Since on each circle, the tangent of the half center angle serves as a projective coordinate on the circle, each transmission function in (10.102) defines a projectivity between p and q, which keeps the common points on $[A, B]$ fixed. Therefore, the axes m of symmetry (Figure 10.22) trace a regulus on a one-sheeted hyperboloid passing through the two circles p and q. For each smooth motion Σ_3/Σ_1, the poses of Σ_3 can be obtained by reflecting Σ_1 in the generators of a regulus. Motions of this type are called *line-symmetric* (note also [50]). They were intensively studied by J. Krames; the paper [78] deals with the Bennett isogram.

Remark 10.3.1 1. A complete classification of the motions corresponding to non-degenerate conics on Study's quadric can be found in [50].

2. For H.-P. Schröcker, the Bennett isogram was the starting point for a new approach to the analysis of closed kinematic chains with revolute joints (cf. [113, 114]).

11 Some generalizations of quadrics

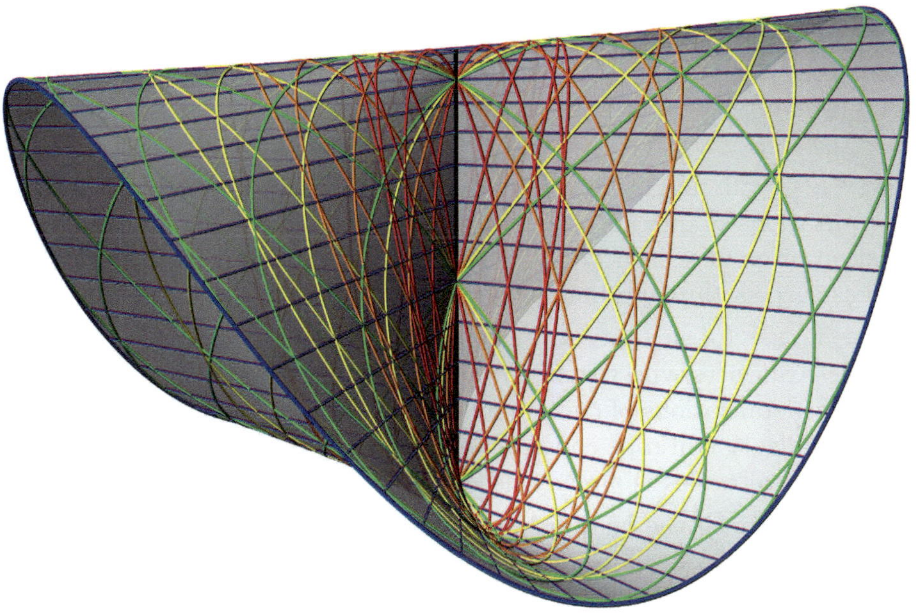

Plücker's conoid appears to be a generalization of quadrics. It carries a two-parameter family of ellipses cut out by cylinders of revolution that contain the conoid's straight directrix. If we consider the rulings as degenerate forms of conics, then there is a further one-parameter family of conics on the surface.

B. Odehnal et al., *The Universe of Quadrics*,
https://doi.org/10.1007/978-3-662-61053-4_11

In this chapter, we try to present ideas for generalizing quadrics. Quadrics in three-dimensional space carry a three-parameter family of conics: Each conic on a quadric is a planar intersection of the quadric, and thus, there are as many conics on a quadric as there are planes in a three-dimensional space.

So, we can ask for the totality of surfaces that carry only one- or two parameter families of conics. Several prominent examples have been studied intensively, due to a lot of applications and interesting properties. Unfortunately, we have to restrict ourselves to only a few classes of surfaces.

However, we can find many other classes of surfaces that can be viewed as generalizations of quadrics. Altering the powers of the monomials in the equation of quadric in normal form yields so-called superquadrics. Only spheres carry a three-parameter family of circles.

So, we can ask for surfaces that carry only one- or two-parameter families of circles. Surfaces whose points fulfill distance relations could be considered as generalizations of conics, and thus, of quadrics, too. Nevertheless, the generalization of rope constructions of conics to that of quadrics using a pair of focal conics is already described in Section 7.2.

We will not touch the wide field of supercyclides (cf. [25, 32, 33]). Only the case of Dupin cyclides was briefly addressed earlier in Section 10.2. The more general type of supercylides has a lot of applications in CAGD, carries several independent families of circles, and could, therefore, be seen as generalizations of quadrics. This is also the case for surfaces that admit quadratic (polynomial or rational) parametrizations in terms of affine coordinates. These surfaces are not necessarily quadrics (cf. [104]).

11.1 Müller's cubic surface

We define a mapping of points in Euclidean 3-space by prescribing a fixed cylinder Δ of revolution

$$\Delta : \ x^2 + y^2 = 1 \tag{11.1}$$

with the axis $a = (0, 0, t)$ (where $t \in \mathbb{R}$). To each point $P = (\xi, \eta, \zeta) \in \mathbb{R}^3 \setminus \{a\}$, we apply the *axial inversion* in Δ by inverting its top view, *i.e.*, the orthogonal projection $P' = (\xi, \eta, 0)$ onto the plane $\pi : \ z = 0$ in the circle $c = \Delta \cap \pi$ while keeping the third coordinate fixed. In terms of Cartesian coordinates, this inversion can be written as

$$P = (\xi, \eta, \zeta) \mapsto \eta(P) = \left(\frac{\xi}{\xi^2 + \eta^2}, \frac{\eta}{\xi^2 + \eta^2}, \zeta \right).$$

We switch to homogeneous coordinates by letting $\xi = x_1 x_0^{-1}$, $\eta = x_2 x_0^{-1}$, $\zeta = x_3 x_0^{-1}$, and arrive at a mapping $\mathbb{P}^3(\mathbb{R}) \setminus \{E\} \to \mathbb{P}^3(\mathbb{R})$, which reads

$$\eta : \ (x_0 : x_1 : x_2 : x_3) \mapsto \left(x_0(x_1^2 + x_2^2) : x_0^2 x_1 : x_0^2 x_2 : x_3(x_1^2 + x_2^2) \right). \tag{11.2}$$

The mapping η is cubic, as can be seen from its coordinate representation. The exceptional set E consist of $a \subset \mathbb{P}^3(\mathbb{R})$, the ideal line of the plane $x_3 = 0$, and the pair of complex conjugate ideal lines $[(0 : 1 : \pm \mathrm{i} : 0), (0 : 0 : 0 : 1)]$.

With the coordinate representation (11.2), we can study the action of η on planes. Let $\alpha : \ a_0 x_0 + a_1 x_1 + a_2 x_2 + a_3 x_3 = 0$ be a plane in $\mathbb{P}^3(\mathbb{R})$ (*i.e.*, $a_0 : a_1 : a_2 : a_3 \neq 0 : 0 : 0 : 0$). Then, the η-image of α can be given by an equation in terms of homogeneous coordinates as

$$\eta(\alpha) : \ a_0 x_0 (x_1^2 + x_2^2) + a_1 x_0^2 x_1 + a_2 x_0^2 x_2 + a_3 x_3 (x_1^2 + x_2^2) = 0, \tag{11.3}$$

which shows that a generic plane is mapped to a cubic surface. An example of such a cubic surface is depicted in Figure 11.1.

In order to simplify the analysis of these cubic surfaces, we let $a_0 : a_1 : a_2 : a_3 = 0 : 1 : 0 : -1$ and turn back to Cartesian coordinates (x, y, z), which results in

$$\mathcal{M} : \ z = \frac{x}{x^2 + y^2}. \tag{11.4}$$

The assumption on the coefficients a_i means no restriction. The various other cubic surfaces that arise for different choices of a_i are equiform

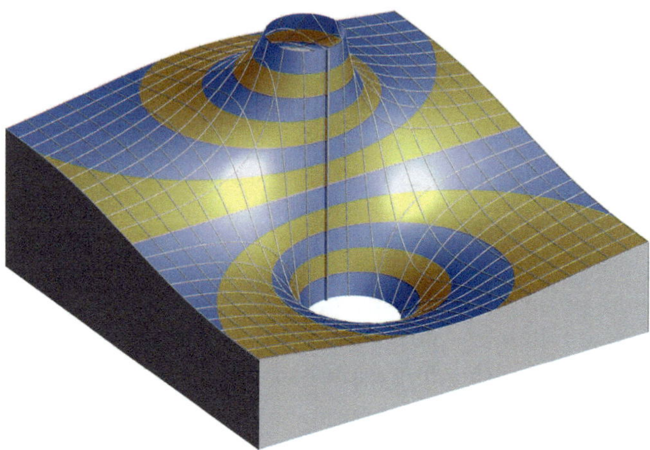

FIGURE 11.1. Müller's cubic surface carries two independent one-parameter families of conics: The horizontal intersections are circles. The intersections with planes from the pencil about a are equilateral hyperbolas with a as a common asymptote.

copies of the surface $\mathcal{M}$ given in (11.4). The surface $\mathcal{M}$ is called *Müller's surface* and was first studied by E. MÜLLER[1] in [88].

It is easy to see that planes $z = h$ (with constant $h \in \mathbb{R}^\star$) intersect $\mathcal{M}$ along circles $h(x^2 + y^2) - x = 0$ whose centers $\left(\frac{1}{2h}, 0, h\right)$ line up on an equilateral hyperbola in the plane $y = 0$. In the special case $h = 0$, the cubic intersection curve degenerates: It is the union of the y-axis and the doubly counted line at infinity of the horizontal planes.

The planes $mx + ny = 0$ in the pencil about the z-axis meet $\mathcal{M}$ along equilateral hyperbolas

$$\mathbf{h}(t) = \left(-nt, mt, -\frac{n}{t(m^2 + n^2)}\right), \quad t \in \mathbb{R} \setminus \{0\}$$

and the z-axis. In this particular pencil of planes, there are two distinguished planes: The plane $x = 0$ meets $\mathcal{M}$ along the y-axis and touches $\mathcal{M}$ along the z-axis, which is, therefore, of multiplicity two. The plane $y = 0$ intersects Müller's surface in the equilateral hyperbola $xz = 1$ and its asymptote $x = 0$.

[1] EMIL ADALBERT MÜLLER (1861–1927) was an Austrian geometer. His research focused on line geometry, sphere geometry, and relative Differentil geometry.

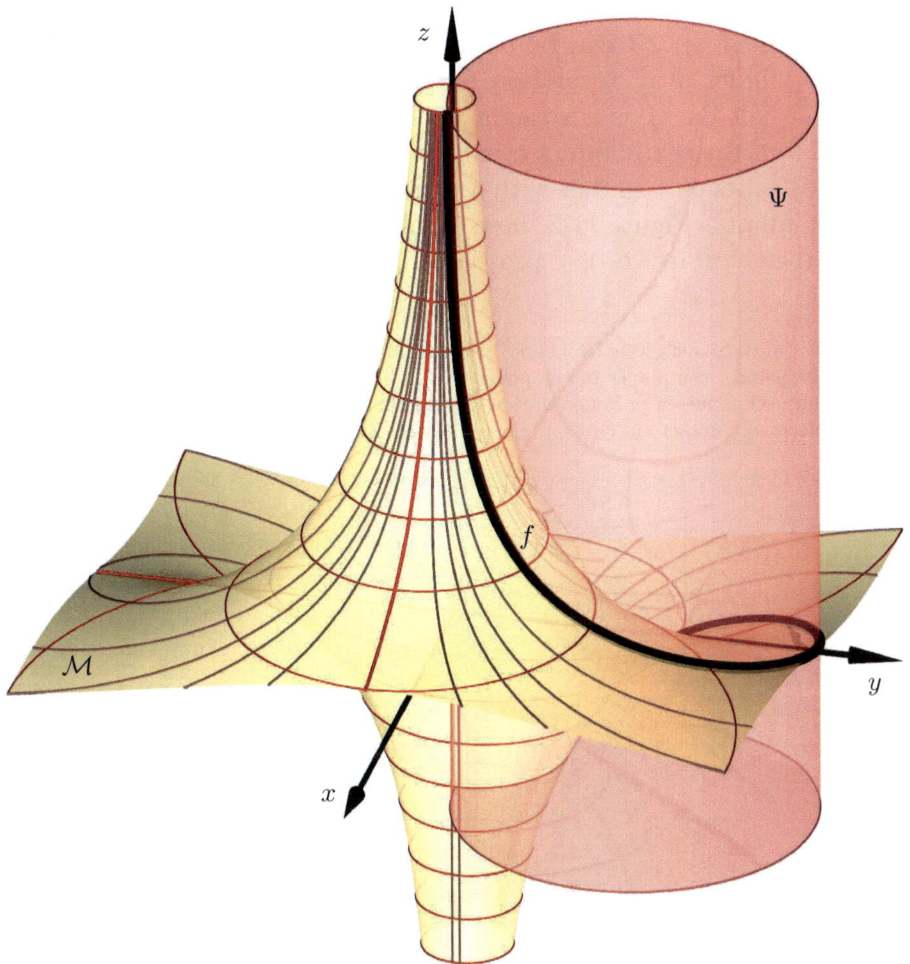

FIGURE 11.2. Each fall line f (except the z-axis) on Müller's cubic surface is a part of the intersection of $\mathcal{M}$ and a cylinder Ψ of revolution that touches the $[x, z]$-plane along the z-axis. The z-axis is also a fall line on $\mathcal{M}$, but it is contained in any of the abovementioned cylinders.

In a top view of Müller's surface, we would see the horizontal curves $h(x^2 + y^2) - x = 0$ as circles. Moreover, the pattern built by these circles is a so-called parabolic pencil of circles (cf. [46, p. 327 & Figure 7.56, left]). Consequently, its orthogonal trajectories would also appear as the circles in a parabolic pencil. The second pencil is then the top view of the *fall lines*, *i.e.*, the path of raindrops running down the surface. Therefore, each fall line f on $\mathcal{M}$ can be obtained as the intersection of $\mathcal{M}$ and a vertical cylinder Ψ of revolution touching the $[x, z]$-plane along the z-axis. The fall line f is a straight cubic circle (cf. Chapter 6, Example 6.2.2). From the sextic intersection of the projection cylinder Ψ and $\mathcal{M}$, the z-axis splits off together with a pair of complex conjugate lines in the plane at infinity. Figure 11.2 shows a fall line f on Müller's cubic surface $\mathcal{M}$ together with its vertical projection cylinder Ψ.

• **Exercise 11.1.1** Ellipses on Müller's surface.

On Müller's surface $\mathcal{M}$ given by (11.4), we can find a further family of conics. Since $\mathcal{M}$ contains the y-axis, the planes in the pencil about that line meet $\mathcal{M}$ along the y-axis and a further conic. Compute a parametrization of $\mathcal{M}$ such that the parameter curves are ellipses (different from the horizontal circles) and equilateral hyperbolas (cf. Figure 11.3).

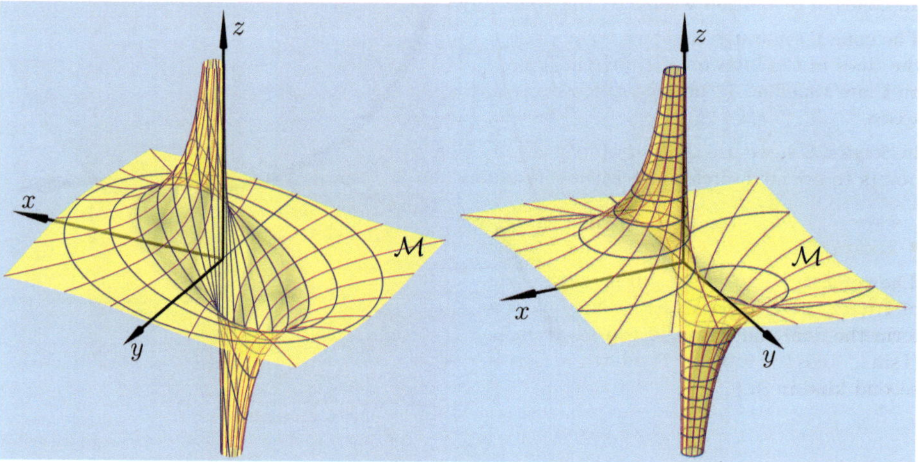

FIGURE 11.3. Left: The ellipses on $\mathcal{M}$ are found as intersections with planes through the y-axis. Right: The horizontal intersections are circles. The intersections with planes from the pencil about a are equilateral hyperbolas with a as a common asymptote.

Müller's surface $\mathcal{M}$ shares a rather curious property with the orthogonal hyperbolic paraboloid. Along the ellipses mentioned in Exercise 11.1.1, $\mathcal{M}$ is touched by rational (and thus, algebraic) minimal surfaces (see [96]). On the other hand, orthogonal hyperbolic paraboloids are touched along their curves of constant Gaussian curvature(s) by algebraic minimal surfaces (cf. [97]).

11.2 Superquadrics

Some small changes in the equation of conics lead to a variety of curves that could be seen as generaliztions of conics (cf. [46, p. 113 ff.]). It is reasonable to try such modifications with quadrics. Raising the powers in the equations of quadrics yields *superquadrics* as analogues to the Lamé curves in the plane. Again, the shape of the surfaces mainly depends on whether the exponents of the monomials are even or odd. Unlike in the case of planar curves, there are much more possible combinations of signs in the equations of superquadrics, comparable to the various possible signatures of quadratic forms.

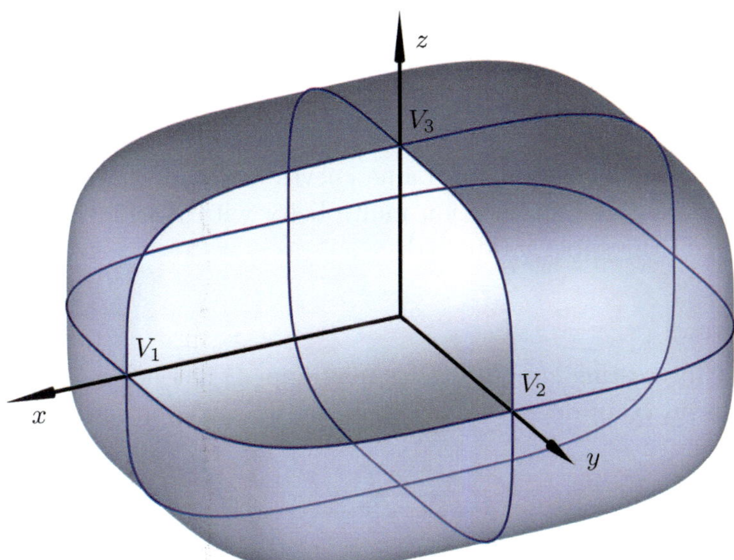

FIGURE 11.4. A superellipsoid of degree four. Its six vertices are joined by Lamé curves which are the intersection curves with the coordinate planes that also serve as planes of symmetry.

We call a surface with an equation of the form

$$\mathcal{S}: \ \sigma_1 \frac{x_1^d}{a_1^d} + \sigma_2 \frac{x_2^d}{a_2^d} + \ldots + \sigma_n \frac{x_n^d}{a_n^d} = \sigma_{n+1} \tag{11.5}$$

a *superquadric* of degree $d \in \mathbb{N}^\star$ in n-dimensional Euclidean space. Here and in the following, we assume $\sigma_i \in \{-1, 1\}$ (where $i \in \{1, \dots, n+1\}$) and $a_j \in \mathbb{R}$ (with $j \in \{1, \dots, n\}$ such that not all a_i vanish simultaneously.

Especially, if $\sigma_i = +1$ and $a_i = r$ for all i, $\mathcal{S}$ is a *supersphere* of radius r. A supersphere is the unit sphere for an L^d norm. Later, we shall extend the range of the exponent d to arbitrary rational numbers. However, in these cases, we can hardly call d the degree of the superquadric.

Obviously, the thus defined superquadrics are symmetric with respect to the coordinate hyperplanes. Therefore, it is reasonable to define the common point of all the symmetry hyperplanes as the *center of the superquadric.*

Setting the left-hand side of (11.5) equal to zero, we obtain the equation of the *asymptotic cone* of the superquadric.

Like quadrics, superellipsoids have no real points in common with the ideal plane. Therefore, the degree of a superellipsoid is always even and the signs σ_i are always positive for $i \in \{1, \dots, n+1\}$. Superhyperboloids intersect the ideal plane along Lamé curves and superparaboloids touch the ideal plane. The intersection multiplicity with the ideal plane equals the degree of the superparaboloid.

Superellipsoids

The most interesting superquadrics may appear in Euclidean 3-space. A superellipsoid $\mathcal{S}_e$ of degree d may be given by the equation

$$\mathcal{S}_e: \ \frac{x^d}{a^d} + \frac{y^d}{b^d} + \frac{z^d}{c^d} = 1. \tag{11.6}$$

As can be seen in Figure 11.4, the superellipsoids tend to follow the faces of the smallest bounding cuboid built by the tangent planes at the vertices

$$V_{1,2} = (\pm a, 0, 0), \quad V_{3,4} = (0, \pm b, 0), \quad V_{5,6} = (0, 0, \pm c).$$

The intersection curves of the superellipsoid (11.6) with the coordinate planes are Lamé curves (cf. [46, p. 113 ff.]). Furthermore, the six vertices V_i of the superellipsoid $\mathcal{S}_e$ are *flat points* in the sense of local differential geometry of surfaces, *i.e.*, the shape operator there equals the zero matrix.

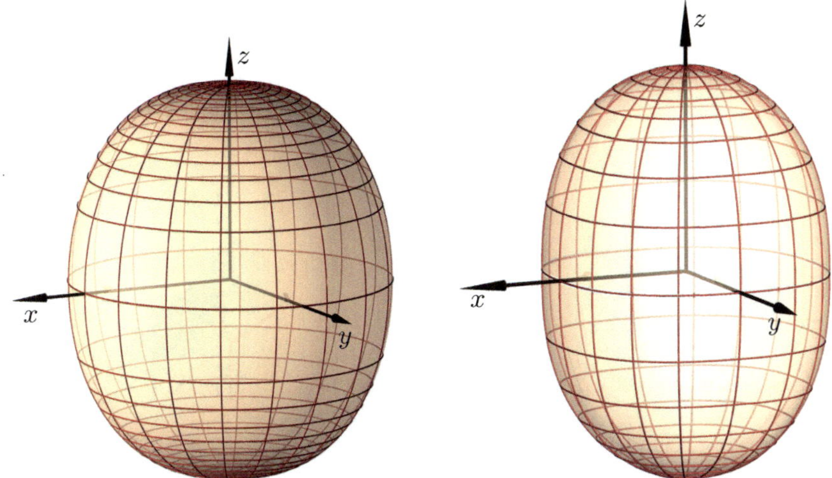

FIGURE 11.5. Left: The Super-Egg by PIET HEIN is a surface of revolution. Right: The "superellipsoid" with $d = \frac{5}{2}$ is an algebraic surface of degree 20 and carries no circles, but a huge variety of Lamé curves.

■ **Example 11.2.1** Piet Hein's *Super-Egg*.

The Danish scientist, inventor, and writer PIET HEIN (1905–1996) suggested the name *Super-Egg* for the surface defined by the equation

$$\left|\frac{\sqrt{x^2+y^2}}{3}\right|^{2.5} + \left|\frac{z}{4}\right|^{2.5} = 1. \tag{11.7}$$

This surface is invariant under rotations about the z-axis. Its meridian curve is the classical *Superellipse* as named by PIET HEIN (cf. [46, p. 114]). A Super-Egg is shown in Figure 11.5 (left). Nevertheless, (11.7) is not a superellipsoid in the strict sense (cf. (11.6)).

Equation of a superellipsoid of odd degrees

Figure 11.6 shows a surface that has an equation very similar to that of a superellipsoid, with $d = 3$:

$$\left(\frac{x}{a}\right)^3 + \left(\frac{y}{b}\right)^3 + \left(\frac{z}{c}\right)^3 = 1. \tag{11.8}$$

Like the Lamé curves of odd degree, this surface of odd degree is not bounded. Removing the constant 1 in (11.8), we obtain the equation of its *asymptotic cone* emanating from the origin of the coordinate system. In this case, the origin of the coordinate system is not a center of the surface, since the surface is not invariant under reflections in $(0,0,0)$. In the one-parameter family of real rulings of the asymptotic cone, we find the asymptotes of the Lamé curves (of odd degree) which show up as the intersection of the surface with the coordinate planes (see Figure 11.6).

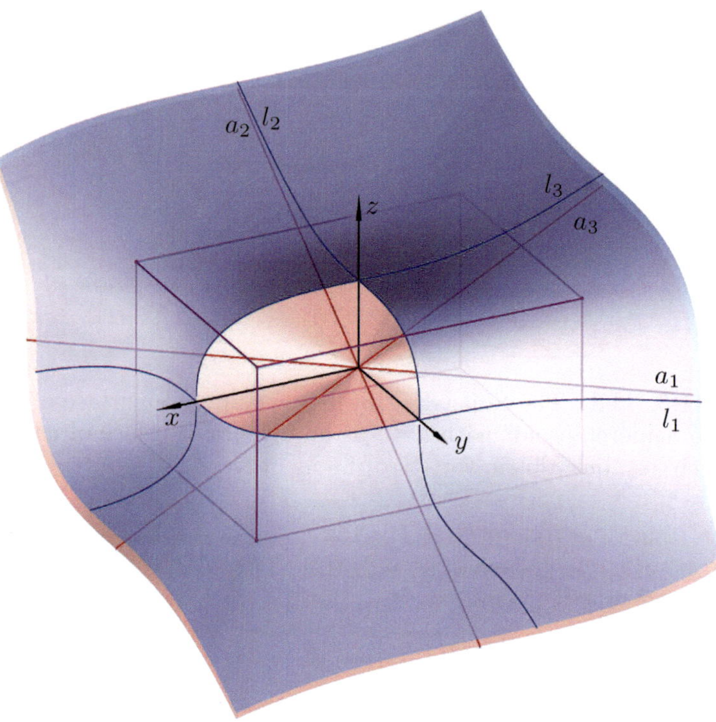

FIGURE 11.6. The superquadric of degree three (blue) is not a superellipsoid. The real asymptotic cone (red) carries the asymptotes a_i of all planar intersections l_i of the cubic superellipsoid, provided that the planes contain the apex of the asymptotic cone.

Superhyperboloids

The equation of a *superhyperboloid* $\mathcal{S}_h$ is obtained from the equation of a hyperboloid by altering the monomials' exponents. Similar to (11.6), we have

$$\mathcal{S}_h : \ \frac{x^d}{a^d} + \frac{y^d}{b^d} - \frac{z^d}{c^d} = \pm 1. \tag{11.9}$$

First, we assume that d is even. This causes the surfaces (11.9) to behave similarly to ordinary hyperboloids. For example, the intersection with the ideal plane is a Lamé curve that carries a real point.

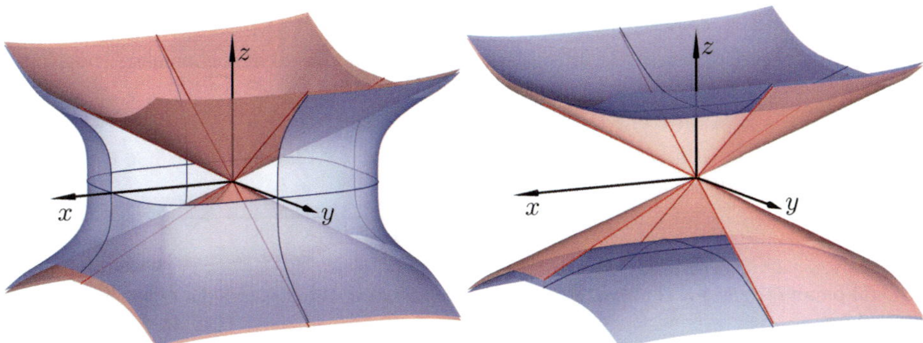

FIGURE 11.7. Superhyperboloids of degree four (blue): one-sheeted superhyperboloid (left), two-sheeted (right). The red surface is the superhyperboloids' asymptotic cone.

Clearly, the superhyperboloids are symmetric with respect to the coordinate planes, and the origin of the coordinate system is the center of the surface. Depending on whether the right-hand side of (11.9) equals +1 or −1, $\mathcal{S}_h$ is either two-sheeted or one-sheeted. The intersections with the planes of symmetry are Lamé curves.

On the one-sheeted surface (cf. Figure 11.7), the four real vertices $(\pm a, 0, 0)$ and $(0, \pm b, 0)$ are flat points in the differential geometric sense. The two-sheeted surface has only two vertices $(0, 0, \pm c)$ (which are also flat points).

Setting the left-hand side equal to zero, we obtain the asymptotic cone of the superhyperboloid. This cone is a projection of the superhyperboloid's Lamé curve in the ideal plane with the surface's center. The cone's generators are the asymptotes of all planar intersection curves of $\mathcal{S}_h$ with planes through the center.

Rational exponents

• **Exercise 11.2.1** The "superellipsoid" algebraically close to the Super-Egg.

Show that the superellipsoid with the rational exponent $d = \frac{5}{2}$ and equation in the normal form

$$\left(\frac{x}{a}\right)^{\frac{5}{2}} + \left(\frac{y}{b}\right)^{\frac{5}{2}} + \left(\frac{z}{c}\right)^{\frac{5}{2}} = 1$$

is actually of degree 20. Show further that an implicit polynomial equation reads

$$64\left(\frac{xyz}{abc}\right)^5 = \left(\left(1 - \left(\frac{x}{a}\right)^5 - \left(\frac{y}{b}\right)^5 - \left(\frac{z}{c}\right)^5\right)^2 - 4\left(\left(\frac{xy}{ab}\right)^5 + \left(\frac{xz}{ac}\right)^5 + \left(\frac{yz}{bc}\right)^5\right)\right)^2.$$

Figure 11.5 (right) shows an example of that particular surface of degree 20.

■ **Example 11.2.2** Various parametrizations of superquadrics.

One eighth of a superellipsoid (11.6) can be described with the aid of trigonometric functions. The surface points with $x > 0$, $y > 0$, and $z > 0$ can be reached by

$$\mathbf{s}(u,v) = (a\sin^q u\cos^q v, b\sin^q u\sin^q v, c\cos^q u) \tag{11.10}$$

with $(u,v) \in]0, \frac{\pi}{2}]^2$, $a, b, c, \in \mathbb{R}^+$, and arbitrary $q \in \mathbb{Q}^\star$. At least in the first octant, the parametrizations satisfy

$$\left(\frac{x}{a}\right)^{\frac{2}{q}} + \left(\frac{y}{b}\right)^{\frac{2}{q}} + \left(\frac{z}{c}\right)^{\frac{2}{q}} = 1.$$

Obviously, $q = 1$ yields ordinary ellipsoids. For integer powers q, the surfaces (11.10) admit even rational parametrizations that can be obtained by inserting the Weierstrass representations of sin and cos.

Some surfaces of that kind for different choices of q are shown in Figure 11.8.

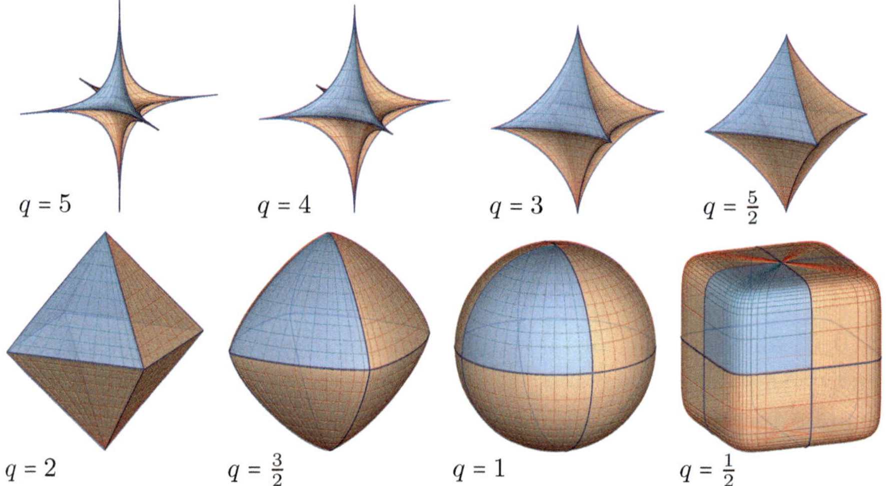

FIGURE 11.8. Superquadrics obtained from (11.10) with different q and $a = b = c$.

• **Exercise 11.2.2** Polynomial parametrization of a star.

Show that the superellipsoid with $d = \frac{1}{2}$, or equivalently, $q = 4$ (cf. Figure 11.8, second from left in the top row), can be parametrized by

$$\mathbf{s}(u,v) = (au^2, bv^2, c(1-u-v)^2) \quad \text{with} \quad (u,v) \in [0,1]^2. \tag{11.11}$$

Therefore, it is a Bézier surface. Show that a control structure, *i.e.*, the nine control points $B_{i,j}$ (with $i, j = 0, 1, 2$), may consist of the following points:

$$\mathbf{b}_{0,0} = \begin{pmatrix} 0 \\ 0 \\ c \end{pmatrix}, \mathbf{b}_{0,1} = \mathbf{b}_{1,0} = \begin{pmatrix} 0 \\ 0 \\ 0 \end{pmatrix}, \mathbf{b}_{0,2} = \mathbf{b}_{1,2} = \begin{pmatrix} 0 \\ b \\ 0 \end{pmatrix}, \mathbf{b}_{1,1} = \begin{pmatrix} 0 \\ 0 \\ -\frac{c}{2} \end{pmatrix}, \mathbf{b}_{2,0} = \mathbf{b}_{2,1} = \begin{pmatrix} a \\ 0 \\ 0 \end{pmatrix}, \mathbf{b}_{2,2} = \begin{pmatrix} a \\ b \\ c \end{pmatrix}.$$

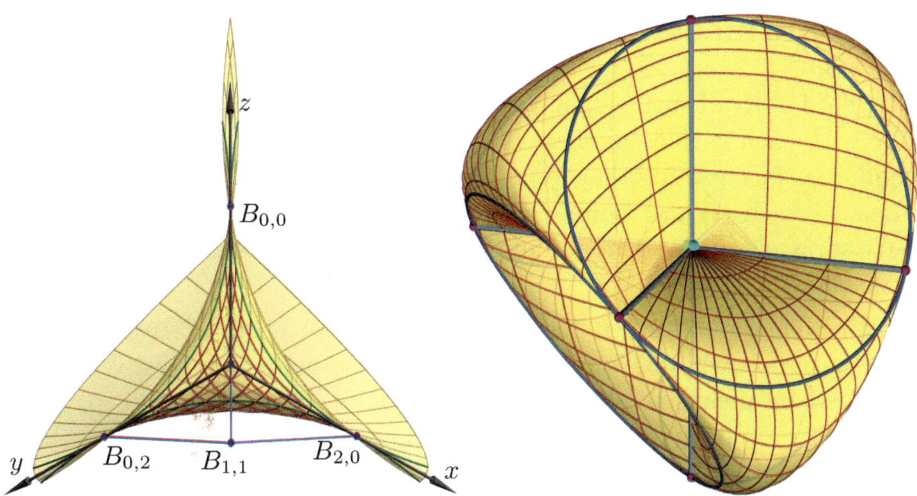

FIGURE 11.9. Left: The surface (11.11) with three parabolas as contact curves of double tangent planes. The part with the red parameter lines is one eighth of the corresponding superquadric. Right: The Roman surface is a collinear image of the surface on the left, and the parabolas are now circles. The three double lines which concur in the triple point are also shown.

Figure 11.9 shows a part of the surface parametrized by (11.11). It is a so-called *Roman surface* or *Steiner surface*, which was intensively studied by the Swiss mathematician JAKOB STEINER (1796–1864). It is one of a huge variety of quadratically parametrizable surfaces that has found a lot of applications in Computer Aided Geometric Design (cf. [29]).

The Roman surface (11.11) is a collinear image of the classical version

$$\mathcal{R}: \; x^2y^2 + y^2z^2 + z^2x^2 - xyz = 0, \tag{11.12}$$

which is the image of the Euclidean unit sphere $x^2 + y^2 + z^2 = 1$ under the quadratic transformation $(x, y, z) \mapsto (yz, zx, xy)$. Removing the six singular points from $\mathcal{R}$, we obtain an immersion of the projective plane into $\mathbb{R}^3$. Figure 11.9 (right) shows an image of $\mathcal{R}$ based on the trigonometric parametrization which is found via that of the Euclidean unit sphere by applying the abovementioned quadratic transformation.

■ **Example 11.2.3** Some more superhyperboloids.

Altering the powers in the equations of one- and two-sheeted hyperboloids, we end up with surfaces whose equations read

$$\mathcal{H}: \; \left(\frac{x}{a}\right)^d + \left(\frac{y}{b}\right)^d - \left(\frac{z}{c}\right)^d = \pm 1. \tag{11.13}$$

Show that these surfaces can, at least in parts, be parametrized by either

$$(a\cosh^q u\cos^q v, b\cosh^q u\sin^q v, c\sinh^q u) \quad \text{or} \quad (a\sinh^q u\cos^q v, b\sinh^q u\sin^q v, c\cosh^q u) \tag{11.14}$$

depending on whether there is a + or a − in the right-hand side of (11.13).

11.3 Surfaces of osculating circles

Another family of surfaces that carry conics comes along in a very natural way with each surface in three-dimensional Euclidean space. Let $\mathcal{S}$ be a regular surface given either by an implicit equation or a parametrization. Then, at each regular point P, the surface has a tangent plane τ_P, and each surface tangent $t \subset \tau_P$ defines its own normal section and normal curvature κ_n.

Usually, one distinguishes between three different types of surface points: elliptic, parabolic, and hyperbolic ones. These points can easily be characterized by means of their Dupin indicatrices $i_P(k)$ (see [46, p. 120]), *i.e.*, the normal curvature diagrams drawn in the tangent plane. The points in these diagrams are found by applying the distance $\pm\sqrt{k\varrho_t}$ on each tangent $t \subset \tau_p$ where $k \in \mathbb{R}^\star$ is an arbitrarily chosen constant, preferably equal to 1, and $\varrho_t = 1/\kappa_t$ is the normal curvature of $\mathcal{S}$ at P in the direction of t. In [46, p. 123], we have seen that $i_P(k)$ are ellipses, pairs of parallel lines, or pairs of conjugate hyperbolas, depending on whether P is elliptic, parabolic, or hyperbolic. In the local coordinate system in τ_P centered at P, the indicatrices have the equations

$$\begin{aligned} i_P(k): \ & \frac{x^2}{\varrho_1} + \frac{y^2}{\varrho_2} = \pm 1 \qquad && P \text{ is elliptic,} \\ i_P(k): \ & \frac{x^2}{\varrho_1} - \frac{y^2}{\varrho_2} = \pm 1 \qquad && P \text{ is hyperbolic,} \\ i_P(k): \ & \frac{x^2}{\varrho_1} = \pm 1 \qquad && P \text{ is parabolic,} \end{aligned}$$

if the axes of the coordinate system coincide with the principal tangents.

For what follows, we may neither consider flat points nor umbilics.

We construct a surface swept by a one-parameter family of circles in the following way: Let n_p be the surface normal of $\mathcal{S}$ at P, and let further $\nu_P(t)$ be that particular plane through the surface normal that contains the surface tangent t. Then, $\mathcal{S} \cap \nu_P(t) = c_n$ is the normal section in the direction of t. According to Euler's formula (cf. [46, p. 122]), the normal curvature for a normal section through a surface tangent t that encloses the angle v with the first principal tangent can be computed by

$$\kappa_t = \kappa_1 \cos^2 v + \kappa_2 \sin^2 v \tag{11.15}$$

with κ_1 and κ_2 being the principal curvatures at P. Now, we draw the osculating circle of each normal section in the normal plane ν. Doing this for all $\nu \in [0, 2\pi]$, we get a smooth one-parameter family of circles that sweep a surface $\mathcal{C}$. Figure 11.10 shows examples of the surface $\mathcal{C}$ at an elliptic point and at a hyperbolic point.[2] The parabolic case is illustrated in Figure 11.11.

A parametrization of $\mathcal{O}$ can easily be given in the local frame built by the principal tangents and the surface normal at P and reads

$$\mathbf{c}(u,v) = \frac{1}{\kappa_1 \cos^2 v + \kappa_2 \sin^2 v} \begin{pmatrix} \cos v \cos u \\ \sin v \cos u \\ 1 + \sin u \end{pmatrix} \tag{11.16}$$

with $(u, v) \in [0, 2\pi[\times[0, \pi]$. At an umbilic, $\varrho_1 = \varrho_2$ and (11.16) is a sphere. The flat point can be seen as a limiting case of an umbilic with $\varrho_1 = \varrho_2 = \infty$, and (11.16) degenerates to a plane, *i.e.*, it coincides with the tangent plane.

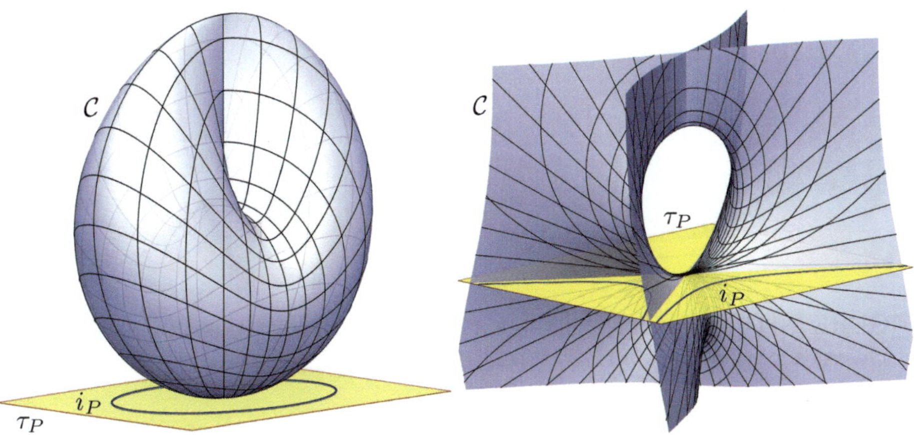

FIGURE 11.10. The surface of osculating circles of normal sections at an elliptic surface point P (left) and at a hyperbolic point P (right).

Implicit equations of the surfaces (11.16) can be obtained by eliminating the parameters u and v. This yields the quartic equation

$$\mathcal{C}: \; (\kappa_1 x^2 + \kappa_2 y^2)(x^2 + y^2 + z^2) - 2(x^2 + y^2)z = 0. \tag{11.17}$$

[2]The German name of the surface $\mathcal{C}$ is *Oskulatorie*, which expresses the fact that it carries the osculating circles of all normal sections at a certain point P on a surface. The English word *osculatory* would be an improper translation, although $\mathcal{C}$ is osculatory.

The intersection of $\mathcal{C}$ with the plane ω at infinity can be found by homogenizing (11.17) by setting

$$x \to x_1 x_0^{-1}, \quad y \to x_2 x_0^{-1}, \quad z \to x_3 x_0^{-1}$$

and subsequently setting $x_0 = 0$. This yields the quartic cycle

$$c\colon \; (\kappa_1 x_1^2 + \kappa_2 x_2^2)(x_1^2 + x_2^2 + x_3^2) = 0.$$

The first factor is the equation of a pair of complex conjugate lines, a real double line, or a pair of real lines, depending on whether P is elliptic, parabolic, or hyperbolic. The second factor is the equation of the absolute conic of Euclidean geometry. Therefore, $\mathcal{C}$ is cyclic.

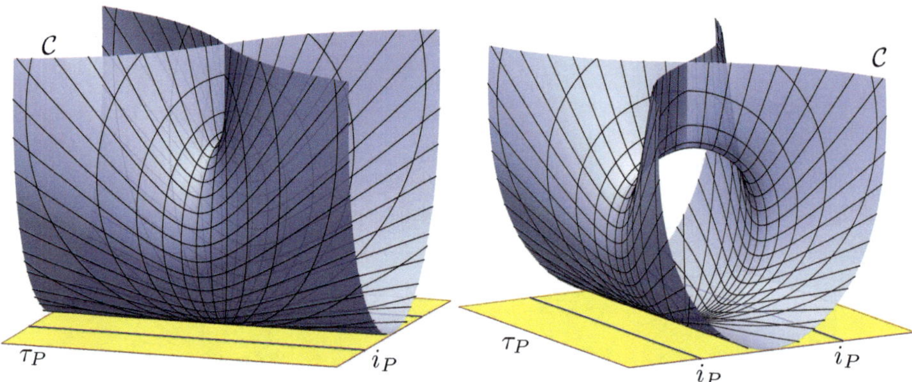

FIGURE 11.11. Two views of the surface $\mathcal{C}$ of the normal sections' osculating circles at a parabolic point.

11.4 Plücker's conoid

The fact that the surface $\mathcal{C}$ of all osculating circles is cyclic has another consequence and leads to another surface that carries, besides a one-parameter family of straight lines, a two-parameter family of conics. Let $\Sigma: x^2 + y^2 + z^2 = 1$ be the Euclidean unit sphere which is centered at the surface point $P \in \mathcal{C}$. The inversion in Σ transforms the local Cartesian coordinates according to

$$x \to \frac{x}{x^2 + y^2 + z^2}, \quad y \to \frac{y}{x^2 + y^2 + z^2}, \quad z \to \frac{z}{x^2 + y^2 + z^2}. \tag{11.18}$$

Since the circles on $\mathcal{C}$ pass through P, *i.e.*, the center of the inversion, the one-parameter family of circles on $\mathcal{C}$ is mapped to a one-parameter family of straight lines, which makes the image surface $\mathcal{P}$ of $\mathcal{C}$ a ruled surface. Substituting (11.18) into (11.16) yields an equation of the cubic ruled surface

$$\mathcal{P}: \ x^2\kappa_1 + y^2\kappa_2 - 2(x^2 + y^2)z = 0. \tag{11.19}$$

We have already met this surface in the previous section, in Section 2.2, and in Section 10.1 as the set of axes of linear complexes of lines when they trace a pencil of linear complexes: It is Plücker's conoid. The fact that the surface of osculating circles and Plücker's conoid are each other's inverses was recognized in [107]. Figure 11.12 shows Plücker's conoid with some of its rulings.

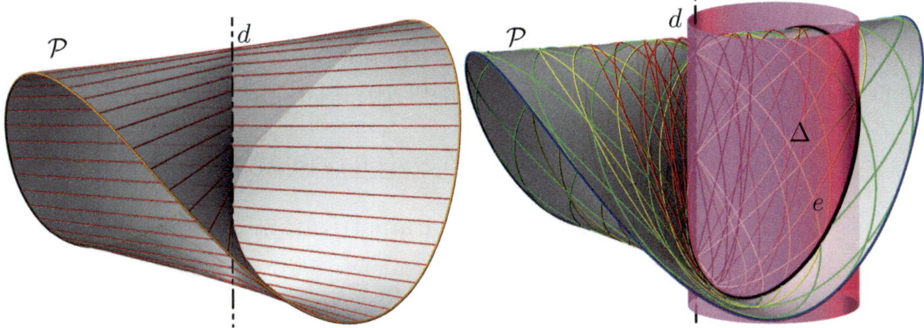

FIGURE 11.12. Left: Plücker's conoid is a cubic ruled surface with a straight line as its double curve. Right: Each ellipse e on Plücker's conoid can be found by intersecting $\mathcal{P}$ with a cylinder Δ of revolution through the directrix d.

The straight lines on $\mathcal{P}$ are the images of the circles on $\mathcal{C}$. For each $v \in [0, 2\pi[$, these lines lie in the planes

$$z = \frac{1}{2}(\kappa_1 \cos^2 v + \kappa_2 \sin^2 v), \tag{11.20}$$

and there, they have the equations

$$(y \cos v - x \sin v)(y \cos v + x \sin v) = 0.$$

Consequently, all generators of $\mathcal{P}$ meet the z-axis of the coordinate system at a right angle. Therefore, the z-axis is one directrix of $\mathcal{P}$, and the ideal line of all planes parallel to $z = 0$ is another directrix. Thus, $\mathcal{P}$ is a *conoidal ruled surface.* Moreover, the z-axis is the *double curve* of $\mathcal{P}$, since there are two lines meeting it in each plane (11.20).

We find the two-parameter family of conics by intersecting $\mathcal{P}$ with cylinders of revolution

$$\Delta: \; (x - r\cos\varphi)^2 + (y - r\sin\varphi)^2 = r^2 \tag{11.21}$$

through the z-axis, which is the directrix and double curve of Plücker's conoid $\mathcal{P}$ at the same time. According to Bézout's theorem, the intersection of Δ and $\mathcal{P}$ is of degree six. Δ and $\mathcal{P}$ share the z-axis which is the double curve of $\mathcal{P}$. Therefore, it is one component of $\Delta \cap \mathcal{P}$ and has multiplicity two. Furthermore, we find that Δ and $\mathcal{P}$ share the pair of complex conjugate lines $x_0 = x_1^2 + x_2^2 = 0$ in the plane at infinity. Consequently, there remains a component of $\Delta \cap \mathcal{P}$ that is of degree two, *i.e.*, it is a conic.

We can parametrize these conics, and thereby also Plücker's conoid, by inserting a parametrization of Δ into (11.19). Setting

$$x(t) = r(\cos t + \cos\varphi) \quad \text{and} \quad y(t) = r(\sin t + \sin\varphi),$$

(11.19) returns $z(t)$, and therefore, a parametrization of the ellipses on $\mathcal{P}$

$$\mathbf{e}(t) = \begin{pmatrix} r(\cos t + \cos\varphi) \\ r(\sin t + \sin\varphi) \\ \frac{1}{4}\left((\kappa_1 - \kappa_2)\cos(t + \varphi) + \kappa_1 + \kappa_2\right) \end{pmatrix}. \tag{11.22}$$

11.5 Sum and product of distances to fixed points

Sum of distances

We have learned in [46, p. 108] that conics can be defined as sets of points satisfying certain distance relations. For example, an ellipse is the set of points X whose distance to the foci F_1 and F_2 sum up to the positive constant value $2a$. In [46, p. 108], we used this definition as a starting point for a more general notion of curves with foci in the plane. However, in three-dimensional space, such distance relations seldomly lead to quadrics, except in the case of Jacobi's focal property (cf. Theorem 2.4.3). Clearly, the set of all points X with fixed distance $r > 0$ to a fixed point F is a sphere with radius r centered at F, and thus, it is a quadric. Further, the set of all points being equidistant to two skew straight lines in Euclidean 3-space $\mathbb{R}^3$ is an orthogonal hyperbolic paraboloid (cf. Theorem 2.3.6).

However, we shall try the following: We assume that we are given a finite set of points F_i with $i \in \{1, \dots, N\}$ with Cartesian coordinate vectors $\mathbf{f}_i$ in Euclidean 3-space $\mathbb{R}^3$. A point X with Cartesian coordinates $\mathbf{x} = (x, y, z)$ has the distances $d_i = \|\mathbf{f}_i - \mathbf{x}\|$. We define the *multifocal surface* $\mathcal{M}$ as the set of all points X that satisfy

$$\sum_{i=1}^{N} d_i = A = \text{const.}, \tag{11.23}$$

where $A \in \mathbb{R} \setminus \{0\}$. The points F_i can be called *foci* of $\mathcal{M}$.

In the case of only two foci, the multifocal surfaces are quadrics: Assume that $F_1 = (-e, 0, 0)$ and $F_2 = (e, 0, 0)$ are the two foci. With (11.23), we have

$$\sqrt{(x+e)^2 + y^2 + z^2} + \sqrt{(x-e)^2 + y^2 + z^2} = A,$$

or as an implicit equation,

$$\mathcal{M}: \ 4(A^2 - 4e^2)x^2 + 4A^2(y^2 + z^2) = A^2(A^2 - 4e^2),$$

which is either an ellipsoid of revolution (if $|A| > 2e$), a double plane (if $|A| = 2e$), or a two-sheeted hyperboloid of revolution (if $|A| < 2e$). Note that the two-sheeted hyperboloid does not satisfy the geometric constraints. It is only part of the algebraic completion. In any case, the

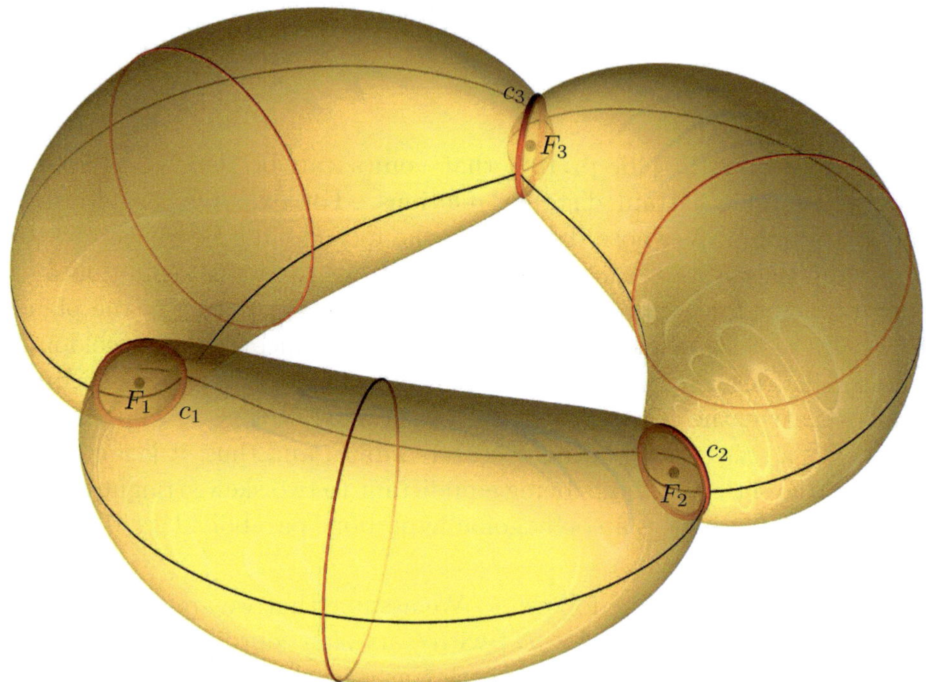

FIGURE 11.13. A trifocal surface on an equilateral triangle of foci F_1, F_2, F_3 with $\overline{F_iF_j} = 1$ and $A = \frac{1}{12}$. The intersections with the bisector planes of the focal triangle consist of circles c_i centered at the foci F_i besides some more complicated algebraic curves.

axis of revolution equals the x-axis of the underlying coordinate system. $\mathcal{M}$ degenerates to the double plane $x = 0$ if $A = 0$. For $A = \pm 2e$, $\mathcal{M}$ is also singular: Then, its equation becomes $y^2+z^2 = (y+\mathrm{i}z)(y-\mathrm{i}z) = 0$, *i.e.*, a pair of complex conjugate planes through the real line $y = z = 0$ (the x-axis). Finally, we remark that squaring is not an equivalence transformation of an equation. Therefore, the implicit equation of $\mathcal{M}$ describes a superset of the multifocal surface. This is especially the case if the number of foci increases.

In the case of three foci, the rotational symmetry is lost (as long as the three foci are not collinear). Figure 11.13 shows an example of a trifocal surface with four planes of symmetry. With the three distances $d_i = \overline{XF_i}$, we have $d_1 + d_2 + d_3 = A$, which gives the implicit equation

$$\left(\left(A^2 - \sum_i d_i^2\right)^2 - 4\sum_{i<j} d_i^2 d_j^2\right)^2 = 64 d_1^2 d_2^2 d_3^2 A^2 \tag{11.24}$$

after two rounds of squaring and inserting the initial linear relation $d_1 + d_2 + d_3 = A$ between the distances. Since each d_i^2 is quadratic in x, y, and z, (11.24) is an algebraic surface of degree eight.

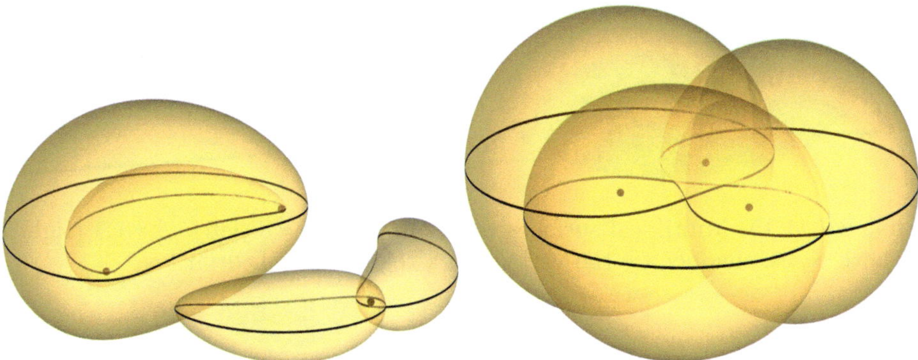

FIGURE 11.14. Trifocal surfaces with non-equilateral focal triangles and different choices of A show different shapes of self-intersections. The banana-shaped surface in the interior of the example on the left-hand side is an artefact that comes from multiple squaring and plotting the implicit equation. Sometimes, the surfaces consist of several disconnected components.

Figure 11.15 shows two surfaces whose points X are defined as those whose distances to four different vertices F_1, F_2, F_3, and F_4 forming a regular tetrahedron sum up to a constant A. In these particular cases, the foci have the Cartesian coordinate vectors

$$\mathbf{f}_1 = (1,1,1), \quad \mathbf{f}_2 = (-1,-1,1), \quad \mathbf{f}_3 = (-1,1,-1), \quad \mathbf{f}_4 = (1,-1,-1).$$

The sum A of distances $d_i = \overline{XF_i}$ equals 8.0 (in Figure 11.15, left) and 7.4 (in Figure 11.15, right).

A result similar to that in Corollary 11.5.1 cannot be given for the surfaces defined by (11.23). The computation of the implicit polynomial equation in the case of sums of square roots needs several rounds of squaring and substituting intermediate phases of the equation, which sometimes involves elimination processes.

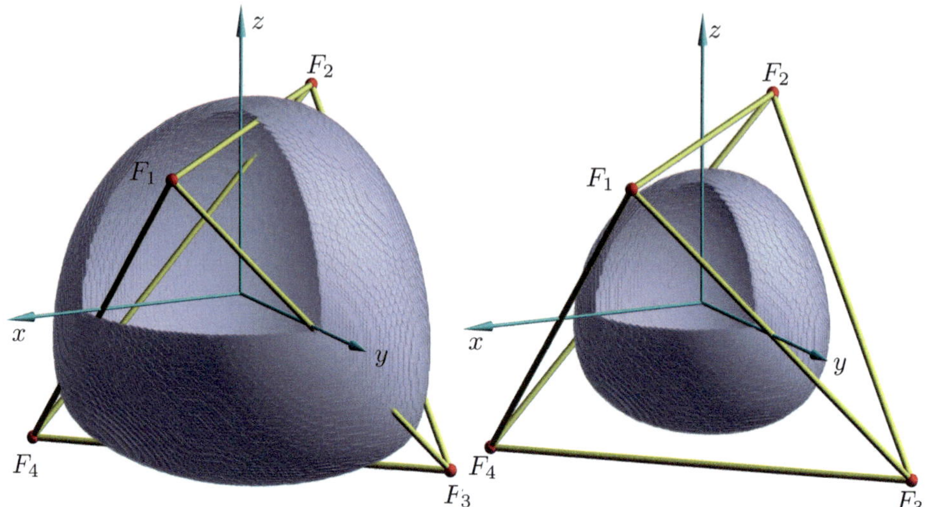

FIGURE 11.15. Multifocal surfaces with four foci: $A = 8$ (left), $A = 7.4$ (right). In both cases, the foci are the vertices of a regular tetrahedron.

Multifocal surfaces tend to become spherical the larger the constant A is (cf. Figure 11.16).

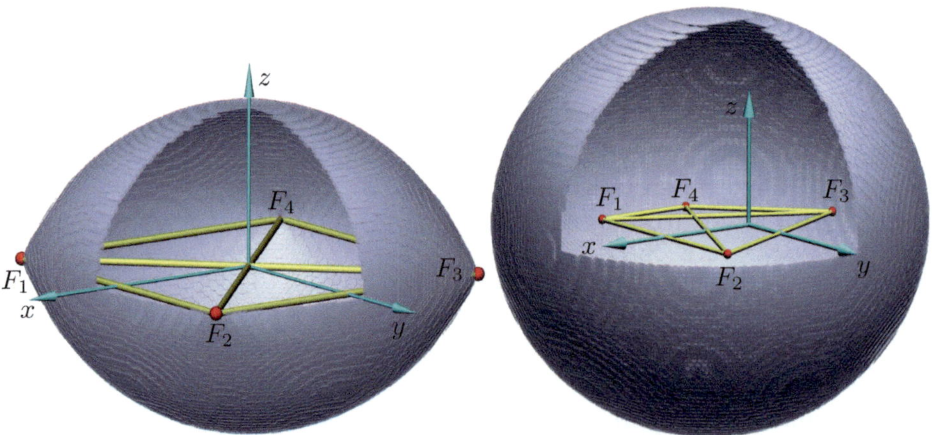

FIGURE 11.16. Multifocal surfaces with four foci: $A = 6.8$ (left), $A = 9$ (right). In both cases the foci form a planar quadrilateral. Left: $F_1F_2F_3F_4$ is a square. Right: The focal quadrilateral is an irregular quadrilateral.

Multifocal surfaces share a property with the conics, and thus, with the quadrics in a plane. It is a property that these surfaces share with the multifocal curves (cf. [46, p. 108]):

Theorem 11.5.1 *Let F_i with Cartesian coordinates $\mathbf{f}_i$ and $i = 1, \ldots, n$ be the focal points of a multifocal surface as defined by (11.23) with an arbitrary constant $A \in \mathbb{R}$. Then, the normal vectors $\mathbf{n_x}$ at points X (with coordinates $\mathbf{x}$) of the multifocal surface defined by (11.23) are independent of the choice of the constant (sum of distances) and aim in the direction*

$$\mathbf{n_x} = \sum_{i=1}^{n} \frac{\mathbf{f}_i - \mathbf{x}}{\|\mathbf{f}_i - \mathbf{x}\|}. \tag{11.25}$$

Proof: First, we note that $\mathbf{n_x} = \mathrm{grad} F|_{\mathbf{x}}$ yields a normal vector at a point X of a surface defined by an implicit equation $F = 0$. For each i,

$$\mathrm{grad}\|\mathbf{f}_i - \mathbf{x}\| = \mathrm{grad}\sqrt{\langle \mathbf{f}_i - \mathbf{x}, \mathbf{f}_i - \mathbf{x}\rangle} = \frac{\partial}{\partial \mathbf{x}}\sqrt{\langle \mathbf{f}_i - \mathbf{x}, \mathbf{f}_i - \mathbf{x}\rangle} = \frac{\mathbf{x} - \mathbf{f}_i}{\|\mathbf{f}_i - \mathbf{x}\|}.$$

The linearity of the gradient confirms (11.25). ■

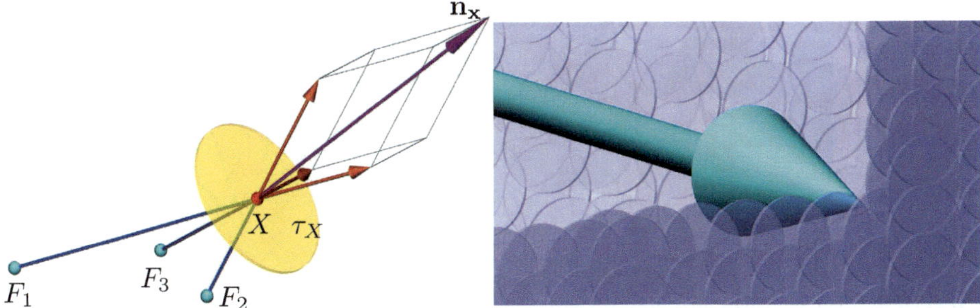

FIGURE 11.17. Left: The normal vector $\mathbf{n_x}$ at a point X to the multifocal surface defined by (11.23) is the sum of the normalized vectors aiming from the foci F_i towards X. Right: A close-up of Figure 11.16 (right) shows that the picture of the multifocal surface is represented by a lot of cylindrical discs with axes parallel to the surface normals.

Figure 11.17 shows how the image of the multifocal surfaces in Figure 11.15 and Figure 11.16 with four foci were produced. It is a C^{-1} representation which is obtained by evaluating the distance function on a cuboid grid and a subsequently collecting those points satisfying (11.23).

Products of distances

We generalize the focal properties by replacing sums by products. In the plane, we obtained spiric curves (cf. [46, p. 111]). Obviously, we can do so in any metric space. We confine ourselves to the three-dimensional Euclidean space. In that way, surfaces whose points X with distances $d_i = \overline{F_i X}$ to a finite number of focal points F_i $(i = 0, \ldots, n)$ satisfy

$$\prod_{i=1}^{N} d_i = A = \text{const.}, \tag{11.26}$$

where $A \in \mathbb{R}$ is a real constant that can also be called *multifocal surfaces* and the points F_i shall be their foci.

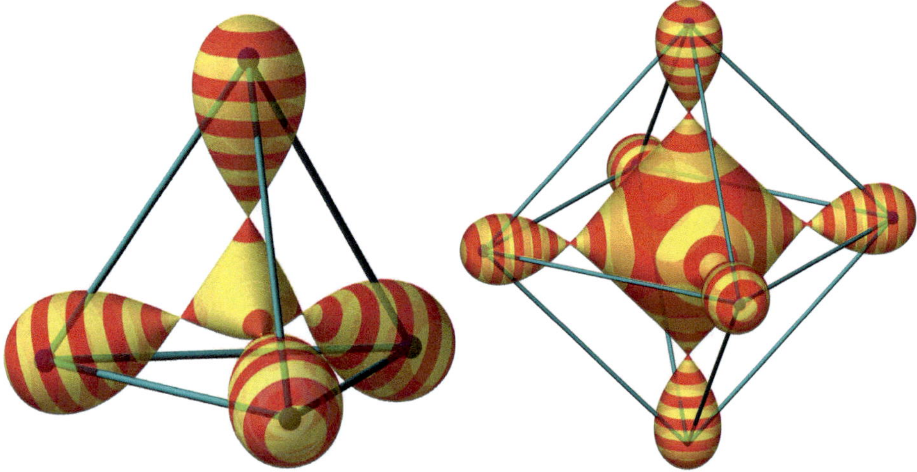

FIGURE 11.18. Foci chosen as the vertices of a tetrahedron (left) and as the vertices of an octahedron (right).

Algebraically, these surfaces differ much from the multifocal surfaces defined by sums of distance in (11.23). For example, an implicit equation is easily obtained by squaring the left- and right-hand side of (11.26). This results in

Corollary 11.5.1 *The algebraic degree of a multifocal surface that satisfies* (11.26) *with n foci is $2n$ at most.*

The normals at regular points of the multifocal surfaces defined by (11.23) can be computed via (11.25). For multifocal surfaces defined by (11.26), the normals can be found in a comparably simple manner. In analogy to Theorem 11.5.1, we have

Theorem 11.5.2 *Let F_i with Cartesian coordinates $\mathbf{f}_i$ and $i = 1, \ldots, n$ be the focal points of a multifocal surface as defined by (11.26) with an arbitrary constant $A \in \mathbb{R}$. Then, the normal vectors $\mathbf{n_x}$ at points X (with coordinates $\mathbf{x}$) of the multifocal surface defined by (11.26) are independent of the choice of the constant A (product of distances) and aim in the direction*

$$\mathbf{n_x} = \sum_{i=1}^{n} \frac{\mathbf{f}_i - \mathbf{x}}{\|\mathbf{f}_i - \mathbf{x}\|^2}.$$

Proof: An implicit equation of the multifocal surfaces in question (defined by the constant product of distances) is given in (11.26), and since $\text{grad}F|_{\mathbf{x}} = \mathbf{n_x}$ and $d_i = \|\mathbf{x} - \mathbf{f}_i\|$, we have

$$\mathbf{n_x} = \text{grad} \prod_{i=1}^{n} d_i = \sum_{i=1}^{n} \text{grad } d_i \prod_{\substack{j=1 \\ j \neq i}}^{n} d_j = \prod_{\substack{j=1 \\ j \neq i}}^{n} \frac{\mathbf{x} - \mathbf{f}_i}{\sqrt{\langle \mathbf{x} - \mathbf{f}_i, \mathbf{x} - \mathbf{f}_i \rangle}} = \prod_{j=1}^{n} d_j \sum_{i=1}^{n} \frac{\mathbf{x} - \mathbf{f}_i}{d_i^2}.$$

The factor $\prod_{j=1}^{n} d_j = A$ is constant and common to all summands. Furthermore, it does not change the direction of the vector $\mathbf{n_x}$, and therefore, it can be ignored. ■

The Figures 11.18 and 11.19 show some multifocal surfaces of the type described by (11.26). The framework built by the focal points is displayed in some cases. As can be seen, such surfaces tend to round the framework of the focal points. However, the constant A to which the distances of points multiply can be chosen such that the surfaces show some singularities or even split into several disconnected components.

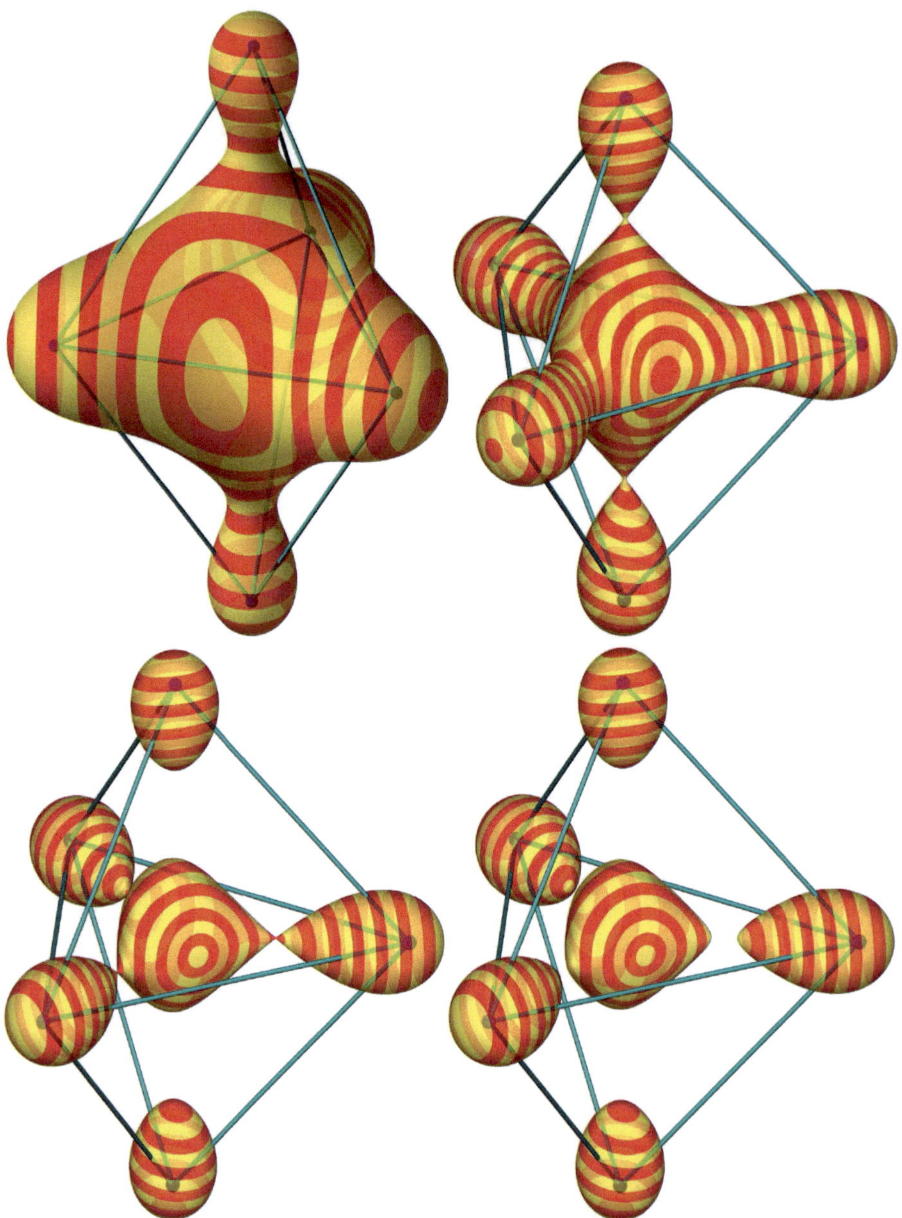

FIGURE 11.19. The nearly regularly distributed five vertices of a multifocal surface (with a constant product of distances) define surfaces with one, three, or six simply connected components.

References

[1] J.F. ADAMS: Spin(8)*, Triality,* F_4 *and all that.* In: S. Hawking and M. Roček (eds.), Superspace and supergravity, Cambridge University Press, Cambridge, 1981, 435–445.

[2] H.F. BAKER: *Principles of Geometry, vol. III, Solid Geometry.* Cambridge University Press, Cambridge 1923.

[3] G.T. BENNETT: *Deformable Octahedra.* Proc. London math. Soc., Sec. Series **10** (1912), 309–343.

[4] G.T. BENNETT: *The skew isogram mechanism.* Proc. Lond. Math. Soc., Sec. Series **13** (1914), 151–173.

[5] M. BERGER: *Geometry I, II.* Translated from the French by M. Cole and S. Levy. Springer, Berlin, Heidelberg, 1987.

[6] M. BERGER: *Geometry Revealed: A Jacob's Ladder to Modern Higher Geometry.* Springer, Berlin, Heidelberg, 2010.

[7] L. BIANCHI: *Vorlesungen über Differentialgeometrie.* B.G. Teubner, Leipzig, 1899.

[8] R. BIX: *Conics and Cubics.* Springer, New York, NY 2006.

[9] W. BLASCHKE: *Über affine Geometrie XXVI: Wackelige Achtflache.* Math. Z. **6** (1920), 85–93.

[10] W. BLASCHKE: *Differentialgeometrie der Kreise und Kugeln.* Vorlesungen über Differentialgeometrie, Grundlehren der mathematischen Wissenschaften 3, Springer, Berlin, Heidelberg, 1929.

[11] W. BLASCHKE: *Kinematik und Quaternionen.* VEB Dt. Verlag der Wissenschaften, Berlin, 1961.

[12] A.I. BOBENKO, Y.B. SURIS: *Discrete Differential Geometry: Integrable Structure.* Graduate Studies in Mathematics 98, American Mathematical Society, Providence, RI, 2009.

[13] W. BÖHM: *Die Fadenkonstruktionen der Flächen zweiter Ordnung.* Math. Nachr. **13** (1955), 151–156.

[14] O. BOTTEMA: *Die Bahnkurven eines merkwürdigen Zwölfstabgetriebes.* Österr. Ingen. Arch. **14** (1960), 218–222.

[15] H. BRAUNER: *Quadriken als Bewegflächen.* Monatsh. Math. **59** (1955), 45–63.

[16] H. BRAUNER: *Über die Projektion mittels der Sehnen einer Raumkurve 3. Ordnung.* Mh. Math. **59** (1955), 258–273.

[17] H. BRAUNER: *Konstruktive Durchführung der durch die Sehnen einer Raumkurve 3. Ordnung vermittelten Abbildung des Raumes auf eine Ebene.* Mh. Math. **60** (1956), 231–248.

[18] H. BRAUNER: *Geometrie Projektiver Räume I, II.* Bibliographisches Institut, Wien, Zürich, 1976.

[19] E. BRIESKORN, H. KNÖRRER. *Plane Algebraic Curves.* Birkhäuser, Basel, 1986.

[20] K. BRUNNTHALER, H.-P. SCHRÖCKER, M. HUSTY: *A New Method for the Synthesis of Bennett Mechanisms.* Proceedings of CK 2005, International Workshop on Computational Kinematics, Cassino, 2005.

[21] W. BURAU: *Mehrdimensionale projektive und höhere Geometrie.* VEB Dt. Verlag der Wissenschaften, Berlin, 1961.

B. Odehnal et al., *The Universe of Quadrics*,
https://doi.org/10.1007/978-3-662-61053-4

[22] W. Burau: *Algebraische Kurven und Flächen I, II.* Sammlung Göschen, Walter de Gruyter & Co, Berlin, 1962.

[23] É. Cartan: *Le principe de dualité et la théorie des groups simples et semisimples.* Bull. des sc. math. **49** (1925), 361–374.

[24] T. E. Cecil: *Lie Sphere Geometry.* Universitext, Springer, New York, NY, 1992.

[25] V. Chandru, D. Dutta, C. M. Hoffmann: *On the geometry of Dupin cyclides.* The Visual Computer **5** (1989), 277–290.

[26] M. Chasles: *Propriétés générales des arcs d'une section conique, dont la difference est rectifiable.* Comptes Rendus hebdomadaires de séances de l'Académie des sciences **17** (1843), 838–844.

[27] M. Chasles: *Les lignes géodésiques et les lignes de courbure des surfaces du second degré.* Journ. de Math. **11** (1846), 5–20.

[28] W.K. Clifford: *Preliminary sketch of biquaternions.* Proc. Lond. Math. Soc. **4** (1871-1873), 381–395.

[29] A. Coffman, A. Schwartz, C. Stanton: *The Algebra and Geometry of Steiner and other Quadratically Parametrizable Surfaces.* Comp. Aided Geom. Design **3**/13 (1996), 257–286.

[30] J.L. Coolidge: *A Treatise on the Circle and the Sphere.* Oxford University Press, Oxford, 1916.

[31] H.S.M. Coxeter: *Non-Euclidean Geometry.* 5th ed., University of Toronto Press, Toronto, 1965.

[32] W. L. F. Degen: *On the origin of supercyclides.* In: R. Cripps (ed.), *The Mathematics of Surfaces VIII.* Information Geometers, Winchester, 1998, 297–312.

[33] W.L.F. Degen: *Cyclides.* In: G. Farin, J. Hoschek, K.-S. Kim (eds.), *Handbook of Computer Aided Geometric Design.* North Holland, Elsevier, Amsterdam, 2002, 575–601.

[34] I.I. Dincă: *Thread configurations for ellipsoids.* arXiv:0902.1421v2 [math.DG], 10/2/2010.

[35] F. Dingeldey: *Kegelschnitte und Kegelschnittsysteme.* Encyklopädie der math. Wiss., B.G. Teubner, Leipzig, 1903, III.2.1, no. C1, 1–160.

[36] A.C. Dixon: *On certain deformable frameworks.* Mess. Math. **29** (1899/1900), 1–21.

[37] W. v. Dyck: *Katalog mathematischer und mathematisch-physikalischer Modelle, Apparate und Instrumente.* C. Wolf & Sohn, München, 1892.

[38] G.E. Farin: *Curves and Surfaces for CAGD:* A Practical Guide (5th ed.). Morgan Kaufmann, Burlington, MA, 2002.

[39] R.T. Farouki: *Pythagorean-Hodograph Curves:* Algebra and Geometry Inseparable. Springer Science & Business Media (2008), 141–178.

[40] Y.N. Fedorov: *Algebraic closed geodesics on a triaxial ellipsoid.* Reg. Chaot. Dyn. **10**/4 (2005), 1–23.

[41] K. Fladt: *Analytische Geometrie spezieller ebener Kurven.* Akademische Verlagsgesellschaft, Frankfurt, 1962.

[42] K. Fladt, A. Baur: *Analytische Geometrie spezieller Flächen und Raumkurven.* Vieweg, Braunschweig, 1975.

[43] D. Genin, B. Khesin, S. Tabachnikov: *Geodesics on an ellipsoid in Minkowski space.* L'Enseignement Mathématique (2) **53**/3-4 (2007), 307–331.

[44] O. Giering: *Vorlesungen über höhere Geometrie.* Vieweg, Wiesbaden, 1982.

[45] G. Glaeser: *Reflections on Spheres and Cylinders of Revolution.* J. Geometry Graphics **3**/2 (1999), 121–139.

[46] G. Glaeser, H. Stachel, B. Odehnal: *The Universe of Conics.* From the ancient Greeks to 21st century developments. Springer Spectrum, Berlin, Heidelberg, 2016.

[47] G.H. Golub, C.F. van Loan: *Matrix Computations.* (4th ed.). J. Hopkins University Press, Baltimore, MD, 2013.

[48] P.M. Gruber, J.M. Wills (eds.): *Handbook of Convex Geometry, Vol. A, B.* North Holland, Amsterdam, 1993.

[49] J. Hadamard: *Sur la forme des linges géodésiques á l' infini et sur les géodésiques des surfaces réglées du second ordre.* Bull. Soc. Math. France **28** (1898), 195–216.

[50] M. Hamann: *Line-symmetric motions with respect to reguli.* Mech. Machine Theory **46**/7 (2011), 960–974.

[51] H. Havlicek: *Zur Theorie linearer Abbildungen I.* J. Geom. **16** (1981), 152–167.

[52] H. Havlicek: *Zur Theorie linearer Abbildungen II.* J. Geom. **16** (1981), 168–180.

[53] H. Havlicek: *Lineare Algebra für Technische Mathematiker.* Heldermann, Lemgo, 2006.

[54] H. Havlicek, S. Pasotti, S. Pianta: *Clifford-like parallelisms.* J. Geom. **110** (2018), article no. 2, 18 pages.

[55] H. Havlicek, G. Weiss: *Altitudes of a tetrahedron and traceless quadratic forms.* Amer. Math. Monthly **110** (2003), 679–693.

[56] D. Hilbert, St. Cohn-Vossen: *Anschauliche Geometrie.* 2nd ed., Springer, Berlin, 1996. English translation: *Geometry and the Imagination.* Chelsea Publishing, reprinted by American Mathematical Society, Providence, RI, 1999.

[57] J.W.P. Hirschfeld: *Projective Geometries Over Finite Fields.* Oxford Mathematical Monographs, Oxford, 1998.

[58] M. Hofer, B. Odehnal, H. Pottmann, T. Steiner, J. Wallner: *3D shape recognition and reconstruction based on line element geometry.* In: 10th IEEE International Conference on Computer Vision, Beijing, PRC (2005). IEEE Computer Society, vol. 2, 1532–1538.

[59] F. Hohenberg: *Drei Paare von Torusflächen, die einen Kugelschnitt gemein haben.* Sitzungsber. Abt. II, Österr. Akad. Wiss. math.-naturw. Kl., **182** (1974), 203–221.

[60] Á. Horváth: *Central quadrics in the Euclidean n-space – metric and focal properties.* Beitr. Algebra Geom., https://doi.org/10.1007/s13366-018-0413-7.

[61] Á. Horváth, I. Prok: *On the constructibility of the axes of an ellipsoid (The construction of Chasles in practice).* J. Geom. **109**/29 (2018), https://doi.org/10.1007/s00022-018-0437-z.

[62] J. Hoschek: *Liniengeometrie.* Bibliographisches Institut, Zürich, 1971.

[63] J. Hosche, D. Lasser: *Fundamentals of Computer Aided Geometric Design.* A.K. Peters, Ltd., Natick, MA, 1993.

[64] M. Husty, H. Sachs: *Abstandsprobleme zu windschiefen Geraden I.* Sitzungsber., Abt. II, österr. Akad. Wiss., Math.-Naturw. Kl. **203** (1994), 31–55.

[65] M. Husty, A. Karger, H. Sachs, W. Steinhilper: *Kinematik und Robotik.* Springer, Berlin, Heidelberg, 1997.

[66] J. Ivory: *On the Attractions of homogeneous Ellipsoids.* Phil. Trans. of the Royal Society of London, 1809, 345–372.

[67] I. Izmestiev, S. Tabachnikov: *Ivory's Theorem revisited.* Journal of Integrable Systems **2**/1 (2017), xyx006 (2017).

[68] D.G.J. Jacobi: *Note von der geodätischen Linie auf dem Ellipsoid und den verschiedenen Anwendungen einer merkwürdigen analytischen Substitution.* Crelle's Journal **19** (1839), 309–313.

[69] C.G.J. Jacobi: *Über die Curve, welche alle von einem Punkte ausgehenden geodätischen Linien eines Rotationsellipsoides berührt.* In: K.T.W. Weierstrass (ed.), Jacobi's Gesammelte Werke, Reimer, Berlin, 1891, 72–87.

[70] C.M. Jessop: *A Treatise on the Line Complex.* Cambridge University Press, Cambr., 1903.

[71] F. Kadeřávanek, J. Klíma, J. Kounovský: *Deskriptivní Geometrie II*, Knihovna spisu matematických a fysikálních, vol. 17, JČMF, Prague, 1932.

[72] B. Khesin, S. Tabachnikov: *Pseudo-Riemannian geodesics and billiards.* Adv. Math. **221** (2009), 1364–1396.

[73] F. Klein: *Vorlesungen über höhere Geometrie.* 3. Auflage (edited by W. Blaschke), Springer, Berlin, 1926, reprint 1968.

[74] F. Klein: *Vorlesungen über nichteuklidische Geometrie.* 3. Aufl., Berlin, 1928.

[75] M. Knöbl: *The Transparent Cup Theorem.* Retrieved from [`http://karuga.eu/transparent-cup.html`], 2016.

[76] H. Knörrer: *Geodesics on the Ellipsoids.* Inventiones math. **59**/2 (1980), 119–143.

[77] V.A. Korotkiy: *Construction of a Nine-Point Quadric Surface.* J. Geometry Graphics **22** (2018), 183–193.

[78] J. Krames: *Zur Geometrie des Bennett'schen Mechanismus, Über symmetrische Schrotungen V.* Sitzungsber., Abt. II, österr. Akad. Wiss., Math.-Naturw. Kl. **146** (1937), 159–173.

[79] J.L. Krames: *Eine bemerkenswerte Eigenschaft der Viviani-Kurve.* Elem. Math. **25** (1970), 1–7.

[80] K. Karčiauskas, R. Krasauskas: *Rational rolling ball blending of natural quadrics.* Mathematical Modelling and Analysis **5**/1 (2000), 97–107.

[81] E. Kruppa: *Analytische und konstruktive Differentialgeometrie.* Springer, Wien, 1957.

[82] W. Kühnel: *Differential Geometry, Curves Surfaces Manifolds.* 3rd ed., American Mathematical Society, 2015.

[83] E.E. Kummer: *Allgemeine Theorie der gradlinigen Strahlensysteme.* J. reine u. angew. Math. **57** (1860), 189–230.

[84] G. Loria: *Spezielle algebraische und transzendente ebene Kurven I & II.* B.G. Teubner, Leipzig, Berlin, 1911.

[85] H. Maehara: *Geometry of frameworks.* Yokohama Math. J. **47** (1999), spec. Issue, 41–65.

[86] H. Maehara, N. Tokushige: *When does a planar bipartite framework admit a continuous deformation?* Theor. Comput. Sci. **263** (2001), no. 1-2, 345–354.

[87] F. Meyer: *Invariantentheorie.* Encyklopädie der math. Wiss., B.G. Teubner, Leipzig 1898, IB2, I 3, 327–334.

[88] E. Müller: *Die achsiale Inversion.* Dt. Math.-Verh. **25** (1916), 209–251.

[89] E. Müller, J.L. Krames: *Vorlesungen über Darstellende Geometrie.* Band II. Die Zyklographie. B.G Teubner, Leipzig, Wien, 1929.

[90] E. Müller, J.L. Krames: *Vorlesungen über Darstellende Geometrie.* Band III: Konstruktive Behandlung der Regelflächen. B.G. Teubner, Leipzig, Wien, 1931.

[91] H.R. Müller: *Sphärische Kinematik.* VEB Dt. Verlag der Wissenschaften, Berlin, 1962.

[92] B. Odehnal: *Zur geometrischen Erzeugung linearer Geradenabbildungen.* Sitzungsber., Abt. II, Österr. Akad. Wiss. math.-naturw. Kl. **213** (2004), 43–69.

[93] B. Odehnal, H. Pottmann, J. Wallner: *Equiform kinematics and the geometry of line elements.* Beitr. Algebra Geom. **47**/2 (2006), 567–582.

[94] B. Odehnal: *Note on flecnodes.* J. Geom. Graphics **13**/1 (2009), 29–40.

[95] B. Odehnal: *Curvature functions on a one-sheeted hyperboloid.* Proc. 16th Internat. Conf. Geometry and Graphics, Aug. 4–8, 2014, Innsbruck/Austria, article No. 076.

[96] B. Odehnal: *Rational minimal surfaces tangent to E. Müller's cubic surface.* G – slovenský Časopis pre Geometriu a Grafiku **14**/28 (2017), 35–46.

[97] B. Odehnal: *On algebraic minimal surfaces.* KoG **20** (2016), 61–78.

[98] B. Odehnal. *Hermite interpolation with ruled and channel surfaces.* G - slovenský Časopis pre Geometriu a Grafiku **14**/28 (2017), 35–58.

[99] A.M. Perelomov: *Some examples of algebraic geodesics on quadrics.* J. Nonlinear Mathem. Physics **16**/1 (2009), 1–5.

[100] A.M. Perelomov: *Some examples of algebraic geodesics on quadrics II.* J. Nonlinear Mathem. Physics **17**/4 (2010), 423–428.

[101] M. Peternell: *Rational Parametrizations for Envelopes of Quadric Families.* PhD thesis, University of Technology, Vienna, Austria, 1997.

[102] M. Peternell, B. Odehnal: *Convolution surfaces of quadratic triangular Bézier surfaces.* Comp. Aided Geom. Design **25** (2008), 116–129.

[103] M. Peternell, B. Odehnal, L.M. Sampoli: *On quadratic two-parameter families of spheres and their envelopes.* Comp. Aided Geom. Design **25** (2008), 342–355.

[104] J. Peters, U. Reif: *The 42 equivalence classes of quadratic surfaces in affine n-space.* Comp. Aided Geom. Design **15**/5 (1998), 459–473.

[105] J. Phillips: *General Spatial Involute Gearing.* Springer, Berlin, Heidelberg, 2003.

[106] H. Pottmann, J. Wallner: *Computational Line Geometry.* Springer, Berlin, 2001.

[107] G. Rösler: *Zur Differentialgeometrie des Flächenelementes dritter Ordnung.* PhD thesis, Universität Stuttgart, 1969.

[108] S. Rowan: *Anvanced Linear Algebra.* 3rd ed., Springer, New York, NY, 2008.

[109] G. Salmon, W. Fiedler: *Die Elemente der analytischen Geometrie des Raumes.* B.G. Teubner, Leipzig, 1863.

[110] R. Sauer: *Differenzengeometrie.* Springer, Berlin, Heidelberg, New York, 1970.

[111] H. Schaal: *Die affin invarianten rechten Winkel.* Sitzungsber., Abt. II, Österr. Akad. Wiss., math.-naturwiss. Kl. **197**/8-10 (1988), 399–416.

[112] M. Schilling: *Catalog mathematischer Modelle.* 7. Auflage, Martin Schilling, Leipzig, 1911.

[113] H.P. Schröcker, B. Jüttler: *Motion interpolation with Bennett-biarcs.* In: A. Kecskeméthy, A. Müller (eds.): *Proceedings of Computational Kinematics (CK 2009)*, Springer, 2009, 101–108.

[114] G. Hegedüs, J. Schicho, H.-P. Schröcker: *Factorization of rational curves in the Study quadric and revolute linkages.* Mech. Mach. Theory **69**/1 (2013), 142–152.

[115] J.G. Semple, G.T. Kneebone: *Algebraic Projective Geometry.* Oxford University Press, Oxford, 1998 (reprint of the 1st ed., 1952).

[116] H. Stachel: *Bemerkungen über zwei räumliche Trilaterationsprobleme.* Z. Angew. Math. Mech. **62** (1982), 329–341.

[117] H. Stachel: *Unendlich viele Kugeln durch vier Tangenten.* Math. Pann., **6** (1995), 55–66.

[118] H. Stachel: *Flexing models of quadrics.* Proc. of Seminars on Computational Geometry, vol. 5 (1996), 103–113.

[119] H. Stachel: *Remarks on A. Hirsch's Paper concerning Villarceau Sections.* J. Geometry Graphics **6** (2002), 133–139.

[120] H. Stachel: *Ivory's Theorem in the Minkowski Plane.* Math. Pannonica **13** (2002), 11–22.

[121] H. Stachel: *Configuration Theorems on Bipartite Frameworks.* Rend. Circ. Mat. Palermo, II. Ser., **70** (2002), 335–351.

[122] H. Stachel, J. Wallner: *Ivory's Theorem in Hyperbolic Spaces.* Sib. Math. J. **45**, no. 4 (2004), 785–794. (russian: Sib. Mat. Zh. **45** (2004), no. 4, 946–959.)

[123] H. Stachel: *A kinematic approach to Kokotsakis meshes.* Comput. Aided Geom. Des. **27** (2010), 428–437.

[124] H. Stachel: *Strophoids are auto-isogonal cubics.* G – Slovak Journal for Geometry and Graphics **12**/24 (2015), 45–59.

[125] O. Staude. *Über Fadenconstructionen des Ellipsoides.* Mathematische Annalen **20** (1882), 147–185.

[126] O. Staude: *Die Focaleigenschaften der Flächen zweiter Ordnung.* B.G. Teubner, Leipzig, 1896.

[127] O.J. Staude: *Flächen 2. Ordnung und ihre Systeme und Durchdringungskurven.* Encyklopädie der math. Wiss., B.G. Teubner, Leipzig, 1915, III.2.1, no. C2, 161–256.

[128] K. Strubecker: *Über die Schraubungen des elliptischen Raumes.* Sitzungsber., Abt. II, österr. Akad. Wiss., Math.-Naturw. Kl. bf 139 (1930), 421–450.

[129] K. Strubecker: *Differentialgeometrie, III. Theorie der Flächenkrümmung.* Sammlung Göschen Bd. 1180/1180a, Walter de Gruyter & Co, Berlin, 1959.

[130] E. Study: *Von den Bewegungen und Umlegungen.* Math. Annal. **39**/4 (1891), 441–565.

[131] E. Study: *Geometrie der Dynamen. Die Zusammensetzung von Kräften und verwandte Gegenstände der Geometrie.* B.G. Teubner, Leipzig, 1903.

[132] E. Study: *Grundlagen und Ziele der analytischen Kinematik.* Sitzungsber. d. Berl. Math. Ges. **12** (1913), 36–60.

[133] S. Tabachnikov: *Geometry and Billiards.* American Mathematical Society, Providence, RI, 2005.

[134] E.A. Weiss: *Einführung in die Liniengeometrie und Kinematik.* B.G. Teubner, Berlin, Leipzig, 1935.

[135] E.A.Weiss: *Oktaven, Engelscher Komplex, Trialitätsprinzip.* Math. Z. **44**/4 (1938), 580–611.

[136] W. Whiteley: *Rigidity and scene analysis.* In: J.E. Goodman, J. O'Rourke (eds.): *Handbook of Discrete and Computational Geometry,* 2nd ed., Chapman & Hall/CRC, Boca Raton 2004, 1327–1354.

[137] H. Wieleitner: *Spezielle ebene Kurven.* Sammlung Schubert, Bd. LVI, Göschen'sche Verlagshandlung, Leipzig, 1908.

[138] *H. Wieners Sammlung mathematischer Modelle.* B.G. Teubner, Leipzig, 1905.

[139] *H. Wieners und P. Teutleins Sammlungen mathematischer Modelle.* 2. Ausg., B.G. Teubner, Leipzig, Berlin, 1912.

[140] W. Wunderlich: *Spiegelung am elliptischen Paraboloid.* Monatsh. Math. **52** (1948), 13–37.

[141] W. Wunderlich: *Überblick über die Krümmgsverhältnisse des Ellipsoid.* Österr. Z. Vermessungsw. (Festschrift E. Doležal) 1952, 673–681.

[142] W. Wunderlich: *Über die Torusloxodromen.* Monatsh. Math. **56** (1952), 313–334.

[143] W. Wunderlich: *Starre, kippende, wackelige und bewegliche Achtflache.* Elemente Math. **20** (1965), 25–32.

[144] W. Wunderlich: *Darstellende Geometrie I, II.* Bibliographisches Institut, Mannheim, 1966, 1967.

[145] W. Wunderlich: *Ebene Kinematik.* Bibliographisches Institut, Mannheim, 1970.

[146] W. Wunderlich: *Starre, kippende, wackelige und bewegliche Gelenkvierecke im Raum.* Elem. Math. **26** (1971), 73–83.

[147] W. Wunderlich: *On deformable nine-bar linkages with six triple joints.* Proc. Nederl. Akad. Wetensch. **79** (1976), 257–262.

[148] W. Wunderlich: *Gefährliche Annahmen der Trilateration und bewegliche Fachwerke I, II.* Z. Angew. Math. Mech. **57** (1977), 297–304 and 363–367.

[149] W. Wunderlich: *Orthogonale Erzeugendenpolygone auf einschaligen Hyperboloiden.* Monatsh. Math. **89** (1980), 163–170.

[150] W. Wunderlich: *Über Ausnahmefachwerke, deren Knoten auf einem Kegelschnitt liegen.* Acta Mechanica **47** (1983), 291–300.

[151] K. Zindler: *Liniengeometrie mit Anwendungen I & II.* G.J. Göschen'sche Verlagshandlung, Leipzig, 1906.

[152] K. Zindler: *Algebraische Liniengeometrie.* Encyklopädie der math. Wiss., B.G. Teubner, Leipzig 1915, III.2.2, no. C8, 973–1228.

Index

B. Odehnal et al., *The Universe of Quadrics*,
https://doi.org/10.1007/978-3-662-61053-4

D

E

F

N

O

P

Q

T

GPSR Compliance

The European Union's (EU) General Product Safety Regulation (GPSR) is a set of rules that requires consumer products to be safe and our obligations to ensure this.

If you have any concerns about our products, you can contact us on ProductSafety@springernature.com

In case Publisher is established outside the EU, the EU authorized representative is:

Springer Nature Customer Service Center GmbH
Europaplatz 3
69115 Heidelberg, Germany

Batch number: 10371042

Printed by Printforce, the Netherlands